Feature Papers in Stem Cells

Feature Papers in Stem Cells

Editor

Mehdi Najar

MDPI

Basel • Beijing • Wuhan • Barcelona • Belgrade • Novi Sad • Cluj • Manchester

Editor
Mehdi Najar
Université Libre de Bruxelles
Brussels
Belgium

Editorial Office
MDPI
St. Alban-Anlage 66
4052 Basel, Switzerland

This is a reprint of articles from the Special Issue published online in the open access journal *Cells* (ISSN 2073-4409) (available at: https://www.mdpi.com/journal/cells/special_issues/feature_stemcell).

For citation purposes, cite each article independently as indicated on the article page online and as indicated below:

Lastname, A.A.; Lastname, B.B. Article Title. *Journal Name* **Year**, *Volume Number*, Page Range.

ISBN 978-3-7258-0363-7 (Hbk)
ISBN 978-3-7258-0364-4 (PDF)
doi.org/10.3390/books978-3-7258-0364-4

Contents

About the Editor

Mehdi Najar

In 2006, Dr. Mehdi Najar set up a new research program at the Laboratory of Experimental Hematology (Institut Jules Bordet) by addressing the biological properties of Stem/Stromal as well as other progenitor cells isolated from the bone marrow. In 2011, he successfully earned a PhD degree in Biomedical and Pharmaceutical Sciences at the Faculty of Medicine of the "Université Libre de Bruxelles" (ULB). Within the Laboratory of Clinical Cell Therapy, he undertook a new challenge by investigating new alternative tissue sources for isolating Stem/progenitor cells with immuno-trophic features. In addition to his fundamental research, he expanded his expertise by following several theoretical and practical training programs related to innovative Stem Cell Technology, Bio-banking, regulations including GDPR, quality manufacturing process (GMP), and clinical research practice (e.g., GCP). In 2017, he joined the Osteoarthritis Research Unit at the "Centre de Recherche du Centre Hospitalier de l'Université de Montréal (Canada)" as a visiting investigator. At the same time, he has been actively involved in different laboratories by developing many projects, training and supervising new people, and establishing and facilitating national/international networks. Dr. Mehdi Najar also participates in the management of several inter-university research programs related to cell therapy and their translational clinical use.

Preface

As a Guest Editor, I am pleased to present this Special Issue "Feature Papers in Stem Cells", addressed to the scientific community who are currently researching Stem Cells. Edited under my direction, this Special Issue belongs to the "Stem Cells" section of the Cells journal. A group of experts working in the field have contributed to this Special Issue, providing a total of 19 papers to be published. I hope that this Special Issue will make a significant contribution to the field through its discussions and opinions about new outcomes linked to the fundamental and applied advances in Stem Cell research.

Mehdi Najar
Editor

cells

Editorial

Fundamental and Applied Advances in Stem Cell Therapeutic Research

Makram Merimi [1,2,†], Saida Rahmani [1,2,†], Ahmed Afailal Tribak [1,†], Fatima Bouhtit [1,2,‡], Hassan Fahmi [3,‡] and Mehdi Najar [3,4,*,‡]

1 Experimental Hematology, Jules Bordet Institute, Unive and nd nd ité Libre de Bruxelles, 1070 Bruxelles, Belgium; makram.merimi.cri@gmail.com (M.M.); saida.ramani@gmail.com (S.R.); tribakafailal.ahmed@gmail.com (A.A.T.); bouhtitfatima@gmail.com (F.B.)
2 Genetics and Immune Cell Therapy Unit, Faculty of Sciences, University Mohammed Premier, Oujda 60000, Morocco
3 Osteoarthritis Research Unit, University of Montreal Hospital Research Center (CRCHUM), Department of Medicine, University of Montreal, Montreal, QC H2X 0A9, Canada; h.fahmi@umontreal.ca
4 Laboratory of Clinical Cell Therapy, Jules Bordet Institute, Université Libre de Bruxelles, 1070 Brussels, Belgium
* Correspondence: mnajar@ulb.ac.bCe
† These authors contributed equally to this work.
‡ Equal senior authors.

Citation: Merimi, M.; Rahmani, S.; Afailal Tribak, A.; Bouhtit, F.; Fahmi, H.; Najar, M. Fundamental and Applied Advances in Stem Cell Therapeutic Research. *Cells* **2022**, *11*, 1976. https://doi.org/10.3390/cells11121976

Received: 13 May 2022
Accepted: 6 June 2022
Published: 20 June 2022

We are pleased to present this Special Issue of *Cells*, entitled 'Feature Papers in Stem Cells'. We hope that this collection of papers may contribute greatly to this field by discussing and presenting new outcomes of basic and translational stem cell-based regenerative medicine research. The rapid progress in the field of stem cell research has laid strong foundations for their use in regenerative medicine applications involving injured or diseased tissues. Cellular therapy aims to replace damaged resident cells by restoring cellular and molecular environments suitable for tissue repair and regeneration. Growing evidence indicates that some of the observed therapeutic outcomes of stem cell-based therapy are due to paracrine effects (including extracellular vesicles), rather than long-term engraftment or the survival of transplanted cells [1]. Embryonic and induced pluripotent stem cells (ESCs and iPSCs), as well as adult stem cells, hold great promise for future cell replacement therapies. Among other candidates, mesenchymal stem/stromal cells (MSCs) represent a critical component of stromal niches known to be involved in tissue homeostasis [2]. Additional evidence suggests that MSCs originate from perivascular cells—principally pericytes that are vascular mural cells—within multiple human organs including lung, adipose tissue and placenta [3]. Accordingly, MSCs play a crucial role during lung development by interacting with the airway epithelium, and also during lung regeneration and remodeling after injury, particularly in chronic obstructive pulmonary disease [4]. During tissue healing, MSCs may exhibit several therapeutic functions to support the repair and regeneration of injured tissue. The process underlying these effects likely involves the migration and homing of MSCs, as well as their immune-tropic functions [5]. Interestingly, tissue-nonspecific alkaline phosphatase (ALP) (TNSALP), a ubiquitous membrane-bound glycoprotein capable of providing inorganic phosphate by catalyzing the hydrolysis of organic phosphate esters, or removing inorganic pyrophosphate that inhibits calcification, is highly expressed in juvenile cells, such as pluripotent stem cells (i.e., ESCs (iPSCs) and somatic stem cells (i.e., MSCs), and is involved in their maintenance and differentiation [6]. Understanding and controlling these cellular products requires in-depth knowledge of their maintenance mechanisms and their exit from undifferentiated states in specific biomaterials mimicking native niches. An interesting approach has been established for differentiating porcine epiblast stem cells (pEpiSCs) into proliferating and functional endothelial cells (ECs). Functional tests revealed that the generated ECs could be used in in vitro assays to examine angiogenesis or cellular responses to various vascular diseases [7]. In another setting, a male mouse model for high running performance recruited myogenic precursor cells/SATCs with lower activation thresholds that responded more rapidly to external stimuli and were more primed for differentiation at the expense of more primitive cells. Satellite cells (SATCs), known as the most abundant skeletal muscle stem cells, play a main role in muscle plasticity, including in the adaptive response following physical activity [8]. In parallel, using pluripotent stem cells (PSCs) to generate hepatocytes is preferable because of their availability and scalability. However, the efficient maturation of PSC-derived hepatocytes towards functional units in liver organoids (LOs) remains a challenging subject. The incorporation of cell-sized microparticles (MPs) derived from the liver extracellular matrix (ECM) provides an enriched tissue-specific microenvironment for the further maturation of hepatocytes inside LOs [9]. This approach has led to the improvement of hepatocyte-like cells in terms of gene expression and function, CYP activities, albumin secretion, and the metabolism of xenobiotics. An experimental basis for the application of stem cells in the treatment of keloids, a pathological scar observed during wound healing, has been developed. Moreover, a co-culture method has been set up to investigate the influence and mechanism of human dental pulp stem cells (HDPSCs) on keloid fibroblast properties [10]. HDPSCs inhibited the migration, the synthesis of the extracellular matrix, and the expression of pro-fibrotic genes within human keloid fibroblasts (HKFs), while promoting the expression of anti-fibrotic genes. Therefore, it can be concluded that HDPSCs can themselves be used as a tool for restraining/hindering the initiation or progression of fibrotic tissue. Mechanistically, new findings have established ten eleven translocation 1 (Tet1) as a regulator of embryonic stem cell (ESC) proliferation by suppressing p21 to ensure a rapid G1-to-S progression [11]. Moreover, Zscan4, which is highly upregulated in telomerase-deficient late-generation mouse ESCs and human alternative lengthening of telomeres (ALT) cancer cells, has been shown to contribute to the telomere maintenance of those cells without telomerase activities [12]. Several features are still to be identified and resolved for improving the safety and efficiency of stem cell-based therapy, in particular for the use of biological delivery systems. Thus, a systematic literature review investigates the potential of therapy with MSCs associated with fibrin glue on the regeneration of the central or peripheral nervous system [13]. Recently, various strategies using a hydrogel-based system, both as encapsulated stem cells and as biocompatible patches loaded with stem cells and applied at the tissue damage site, were developed for regenerating the infarcted myocardium [14]. Joint engineering, representing a potential tool for cartilage regeneration, is an interdisciplinary field that aims to recreate a neo-tissue whose physical and biochemical properties are close to those of the native tissue. It combines cells, biomaterials, and environmental factors. A particular focus on the extrusion bioprinting of cellularized hydrogels for articular cartilage tissue engineering has been discussed [15]. Furthermore, approaches for optimizing standard MSC culture protocols during this essential primary step of in vitro expansion are required. Several studies suggest some improvements in culture media components (amino acids, ascorbic acid, glucose level, growth factors, lipids, platelet lysate, trace elements, serum, and xenogeneic components) as well as culture conditions and processes (hypoxia, cell seeding, and dissociation during passaging) in order to preserve MSC phenotypes and functionality during the primary phase of in vitro culture [16]. Collectively, this Special Issue, managed and supervised by Dr. Mehdi Najar, successfully gathers a great collection of research articles and reviews highlighting recent fundamental and applied advances in different types of stem cells.

Author Contributions: M.M. and M.N. conceived and designed the editorial. M.M., M.N., S.R., A.A.T., F.B. and H.F. have made a substantial, direct and intellectual contribution to the work. All authors listed contributed to manuscript writing, revision, reading, and approved the submitted version. All authors have read and agreed to the published version of the manuscript.

Funding: This work was funded by Generation Life Foundation, 'Fonds Lambeau-Marteaux', 'Fonds National de la Recherche Scientifique (FNRS)', 'Télévie', 'Les Amis de l'Institut Jules Bordet', 'La Chaire en Arthrose de l'Université de Montréal, The Arthritis Society (SOG-20-0000000046) and The Canadian Institutes of Health Research (PJT 175-1110).

Acknowledgments: We would like to thank the Cell Therapy Unit Team for their inspiring dialogs.

Conflicts of Interest: The authors declare that the research was conducted in the absence of any commercial or financial relationships that could be construed as a potential conflict of interest.

References

1. Jarrige, M.; Frank, E.; Herardot, E.; Martineau, S.; Darle, A.; Benabides, M.; Domingues, S.; Chose, O.; Habeler, W.; Lorant, J.; et al. The Future of Regenerative Medicine: Cell Therapy Using Pluripotent Stem Cells and Acellular Therapies Based on Extracellular Vesicles. *Cells* **2021**, *10*, 240. [CrossRef] [PubMed]
2. Najar, M.; Melki, R.; Khalife, F.; Lagneaux, L.; Bouhtit, F.; Moussa Agha, D.; Fahmi, H.; Lewalle, P.; Fayyad-Kazan, M.; Merimi, M. Therapeutic Mesenchymal Stem/Stromal Cells: Value, Challenges and Optimization. *Front. Cell Dev. Biol.* **2021**, *9*, 716853. [CrossRef] [PubMed]
3. Li, C.; Zhao, H.; Wang, B. Mesenchymal stem/stromal cells: Developmental origin, tumorigenesis and translational cancer therapeutics. *Transl. Oncol.* **2021**, *14*, 100948. [CrossRef] [PubMed]
4. Nasri, A.; Foisset, F.; Ahmed, E.; Lahmar, Z.; Vachier, I.; Jorgensen, C.; Assou, S.; Bourdin, A.; De Vos, J. Roles of Mesenchymal Cells in the Lung: From Lung Development to Chronic Obstructive Pulmonary Disease. *Cells* **2021**, *10*, 3467. [CrossRef] [PubMed]
5. Merimi, M.; El-Majzoub, R.; Lagneaux, L.; Moussa Agha, D.; Bouhtit, F.; Meuleman, N.; Fahmi, H.; Lewalle, P.; Fayyad-Kazan, M.; Najar, M. The Therapeutic Potential of Mesenchymal Stromal Cells for Regenerative Medicine: Current Knowledge and Future Understandings. *Front. Cell Dev. Biol.* **2021**, *9*, 661532. [CrossRef] [PubMed]
6. Sato, M.; Saitoh, I.; Kiyokawa, Y.; Iwase, Y.; Kubota, N.; Ibano, N.; Noguchi, H.; Yamasaki, Y.; Inada, E. Tissue-Nonspecific Alkaline Phosphatase, a Possible Mediator of Cell Maturation: Towards a New Paradigm. *Cells* **2021**, *10*, 3338. [CrossRef] [PubMed]
7. Shin, J.-H.; Seo, B.-G.; Lee, I.-W.; Kim, H.-J.; Seo, E.-C.; Lee, K.-M.; Jeon, S.-B.; Baek, S.-K.; Kim, T.-S.; Lee, J.-H.; et al. Functional Characterization of Endothelial Cells Differentiated from Porcine Epiblast Stem Cells. *Cells* **2022**, *11*, 1524. [CrossRef] [PubMed]
8. Petkov, S.; Brenmoehl, J.; Langhammer, M.; Hoeflich, A.; Röntgen, M. Myogenic Precursor Cells Show Faster Activation and Enhanced Differentiation in a Male Mouse Model Selected for Advanced Endurance Exercise Performance. *Cells* **2022**, *11*, 1001. [CrossRef] [PubMed]
9. Zahmatkesh, E.; Ghanian, M.H.; Zarkesh, I.; Farzaneh, Z.; Halvaei, M.; Heydari, Z.; Moeinvaziri, F.; Othman, A.; Ruoß, M.; Piryaei, A.; et al. Tissue-Specific Microparticles Improve Organoid Microenvironment for Efficient Maturation of Pluripotent Stem-Cell-Derived Hepatocytes. *Cells* **2021**, *10*, 1274. [CrossRef] [PubMed]
10. Yan, M.; Fu, L.-L.; Nada, O.A.; Chen, L.-M.; Gosau, M.; Smeets, R.; Feng, H.-C.; Friedrich, R.E. Evaluation of the Effects of Human Dental Pulp Stem Cells on the Biological Phenotype of Hypertrophic Keloid Fibroblasts. *Cells* **2021**, *10*, 1803. [CrossRef] [PubMed]
11. Chrysanthou, S.; Flores, J.C.; Dawlaty, M.M. Tet1 Suppresses p21 to Ensure Proper Cell Cycle Progression in Embryonic Stem Cells. *Cells* **2022**, *11*, 1366. [CrossRef] [PubMed]
12. Dan, J.; Zhou, Z.; Wang, F.; Wang, H.; Guo, R.; Keefe, D.L.; Liu, L. Zscan4 Contributes to Telomere Maintenance in Telomerase-Deficient Late Generation Mouse ESCs and Human ALT Cancer Cells. *Cells* **2022**, *11*, 456. [CrossRef] [PubMed]
13. Ortiz, A.d.C.; Fideles, S.O.M.; Pomini, K.T.; Bellini, M.Z.; Pereira, E.d.S.B.M.; Reis, C.H.B.; Pilon, J.P.G.; de Marchi, M.Â.; Trazzi, B.F.d.M.; da Silva, W.S.; et al. Potential of Fibrin Glue and Mesenchymal Stem Cells (MSCs) to Regenerate Nerve Injuries: A Systematic Review. *Cells* **2022**, *11*, 221. [CrossRef] [PubMed]
14. Sharma, V.; Dash, S.K.; Govarthanan, K.; Gahtori, R.; Negi, N.; Barani, M.; Tomar, R.; Chakraborty, S.; Mathapati, S.; Bishi, D.K.; et al. Recent Advances in Cardiac Tissue Engineering for the Management of Myocardium Infarction. *Cells* **2021**, *10*, 2538. [CrossRef]
15. Messaoudi, O.; Henrionnet, C.; Bourge, K.; Loeuille, D.; Gillet, P.; Pinzano, A. Stem Cells and Extrusion 3D Printing for Hyaline Cartilage Engineering. *Cells* **2021**, *10*, 2. [CrossRef]
16. Nikolits, I.; Nebel, S.; Egger, D.; Kreß, S.; Kasper, C. Towards Physiologic Culture Approaches to Improve Standard Cultivation of Mesenchymal Stem Cells. *Cells* **2021**, *10*, 886. [CrossRef] [PubMed]

cells

MDPI

Article

Functional Characterization of Endothelial Cells Differentiated from Porcine Epiblast Stem Cells

Joon-Hong Shin [1,2,†], Bo-Gyeong Seo [2,3,†], In-Won Lee [1,2], Hyo-Jin Kim [2,3], Eun-Chan Seo [2,3], Kwang-Min Lee [2,3], Soo-Been Jeon [1,‡], Sang-Ki Baek [1,§], Tae-Suk Kim [1], Jeong-Hyung Lee [4], Jung-Woo Choi [5], Cheol Hwangbo [2,3,*] and Joon-Hee Lee [1,6,*]

1 Department of Animal Bioscience, College of Agriculture and Life Sciences, Gyeongsang National University, Jinju 52828, Korea; leonardo6858@gnu.ac.kr (J.-H.S.); yi.innwon@gnu.ac.kr (I.-W.L.); sbub9598@naver.com (S.-B.J.); sangki.beak@kitox.re.kr (S.-K.B.); kts9347@gnu.ac.kr (T.-S.K.)
2 Division of Applied Life Science (BK21), PMBBRC and Research Institute of Life Sciences, Gyeongsang National University, Jinju 52828, Korea; sbk6427@naver.com (B.-G.S.); jin4477@hanmail.net (H.-J.K.); eunchan3927@gmail.com (E.-C.S.); stare4567@naver.com (K.-M.L.)
3 Division of Life Science, College of Natural Sciences, Gyeongsang National University, Jinju 52828, Korea
4 Department of Biochemistry, College of Natural Sciences, Kangwon National University, Chuncheon 24414, Korea; jhlee36@kangwon.ac.kr
5 College of Animal Life Science, Kangwon National University, Chuncheon 24414, Korea; jungwoo.kor@gmail.com
6 Institute of Agriculture & Life Science, College of Agriculture and Life Sciences, Gyeongsang National University, Jinju 52828, Korea
* Correspondence: chwangbo@gnu.ac.kr (C.H.); sbxjhl@gnu.ac.kr (J.-H.L.); Tel.: +82-55-772-1341 (C.H.); +82-55-772-1886 (J.-H.L.)
† These authors contributed equally to this work.
‡ Current address: CHA Advanced Research Institute, Bundang CHA Hospital, CHA University, Seongnam 13488, Korea.
§ Current address: Gyeongsangnamdo Livestock Experiment Station, Sancheong 52263, Korea.

Citation: Shin, J.-H.; Seo, B.-G.; Lee, I.-W.; Kim, H.-J.; Seo, E.-C.; Lee, K.-M.; Jeon, S.-B.; Baek, S.-K.; Kim, T.-S.; Lee, J.-H.; et al. Functional Characterization of Endothelial Cells Differentiated from Porcine Epiblast Stem Cells. *Cells* **2022**, *11*, 1524. https://doi.org/10.3390/cells11091524

Academic Editors: Mehdi Najar and Alexander V. Ljubimov

Received: 14 March 2022
Accepted: 29 April 2022
Published: 2 May 2022

Publisher's Note: MDPI stays neutral with regard to jurisdictional claims in published maps and institutional affiliations.

Abstract: Endothelial cells (ECs), lining blood vessels' lumen, play an essential role in regulating vascular functions. As multifunctional components of vascular structures, pluripotent stem cells (PSCs) are the promising source for potential therapeutic applications in various vascular diseases. Our laboratory has previously established an approach for differentiating porcine epiblast stem cells (pEpiSCs) into ECs, representing an alternative and potentially superior cell source. However, the condition of pEpiSCs-derived ECs growth has yet to be determined, and whether pEpiSCs differentiate into functional ECs remained unclear. Changes in morphology, proliferation and functional endothelial marker were assessed in pEpiSCs-derived ECs in vitro. pEpiSCs-derived ECs were subjected to magnetic-activated cell sorting (MACS) to collect CD-31+ of ECs. We found that sorted ECs showed the highest proliferation rate in differentiation media in primary culture and M199 media in the subculture. Next, sorted ECs were examined for their ability to act as typical vascular ECs through capillary-like structure formation assay, Dil-acetylated low-density lipoprotein (Dil-Ac-LDL) uptake, and three-dimensional spheroid sprouting. Consequently, pEpiSCs-derived ECs function as typical vascular ECs, indicating that pEpiSC-derived ECs might be used to develop cell therapeutics for vascular disease.

Keywords: porcine epiblast stem cells; endothelial cells; magnetic activated cell sorting; functional evaluation

1. Introduction

Endothelial cells (ECs), which constitute the lumen of blood vessels in the body, play a critical role in modulating vascular functions [1]. They are implicated in thrombosis and platelet adhesion, immunological and inflammatory responses, and vascular tone and blood flow regulation [2–4]. Endothelial dysfunction has been linked to a wide range of pathologic events, including atherosclerosis, congestive heart failure, and pulmonary hypertension [5–7]. Hence, ECs have been utilized in a variety of in vitro disease models to explore vascular dysfunction, such as diabetes and atherosclerosis development, coronary artery disease, and COVID-19 virus infection [8–11]. ECs play an essential role in vascular homeostasis by interacting with circulating cells and adjacent cells in the vessel walls. However, functional properties of vascular ECs are gradually reduced due to the loss of homeostasis caused by dietary preferences, smoking, aging and physical inactivity. ECs failure, which results in atherosclerosis, is the leading cause of cardiovascular disease (CVD) and its repercussions, such as heart attacks and strokes [12–15]. Therefore, vascular ECs regeneration avail of pluripotent stem cell (PSCs)-based therapy is an attractive therapeutic approach to the treatment of CVD because ECs essentially participate in any organ regeneration program [16,17].

PSCs retain the pluripotency to differentiate into any cell types in the body and the self-renewal capacity to replicate from mother cells to daughter cells indefinitely [18,19]. In general, PSCs include embryonic stem cells (ESCs) and epiblast stem cells (EpiSCs) derived from particular cell mass (ICM or epiblast) of pre- or post-implantation embryos [20–22]. Furthermore, induced pluripotent stem cells (iPSCs) reprogrammed from terminally differentiated somatic cells by transduction of exogenous reprogramming factors [23–25]. In the mouse, ESCs and EpiSCs established from pre- or post-implantation embryos represented different cytokine dependency to maintain the pluripotent states termed as "naïve" and "primed" [20,26–30]. Depending on the species, it was also that murine ESCs and human ESCs demand different signaling pathways to maintain their pluripotent states [27,30–32]. Although PSCs have significantly different characteristics from sources and species, porcine epiblast stem cells (pEpiSCs) represented a feature of the primed state like murine EpiSCs and human ESCs [33,34].

Since then, ECs were first successfully generated from human ESCs [35], a variety of differentiation protocols have been extensively studied to generate ECs from human PSCs mainly applying sequential addition in culture of growth factors such as bone morphogenetic protein-4 (BMP-4) and vascular endothelial growth factor (VEGF) for the development of vascular disease therapy [36,37]. However, there are highly harsh reactions, such as ethical and stability concerns to apply human ESCs for human disease. As an alternative to theses, we recently reported that an in vitro differentiation of pEpiSCs into ECs is efficiently established for applying the treatment of vascular diseases in humans [38]. The cells differentiated from pEpiSCs, on the other hand, were highly heterogeneous mixed with undifferentiated PSCs and differentiated ECs. It may potentially cause teratoma formation following transplantation into the body, which would exclude direct clinical applications of PSCs. Therefore, there has been a strong emphasis on the purification of contractile ECs from ESCs, such as using Percoll gradient fractionation, fluorescence-activated cell sorting (FACS) separation of CD-31+ cells, surface expression of signal-regulatory protein and activated leukocyte cell adhesion molecule [39–42].

Addressing this current limitation, we describe an efficient procedure for selecting ECs solely from cell mixtures differentiated from pEpiSCs using magnetic beads labeled with an endothelial cell marker CD-31, providing growth conditions for these cells to proliferate effectively. The sorted ECs show typical features of ECs, such as capillary-like structure formation, Dil-labelled acetylated low-density lipoprotein (Dil-Ac-LDL) uptake and sprouting ability in three-dimensional spheroid. Moreover, the cells maintain their functional properties during prolonged culture through at least ten passages.

2. Materials and Methods

2.1. Cells and Cell Culture

Culture of porcine epiblast stem cells (pEpiSCs) were used for this experiment as described in a previous study [34]. Briefly, pEpiSCs were derived from epiblasts that dissociated from in vivo embryos collected nine days after insemination. Mouse embryonic fibroblast cells (MEFs) were derived from fetal mice at 11.0 days in placentas of pregnant female mouse. MEFs were cultured in Dulbecco's modified eagle's medium (DMEM/F12) supplemented with 10% of fetal bovine serum (FBS), 1.5 mM of β-mercaptoethanol, 1% of MEM and 1% of penicillin/streptomycin. When MEFs showed about 80% confluence, MEFs were inactivated by 10 μg/mL mitomycin C for $2\frac{1}{2}$ h. Inactivated MEFs (iMEFs) were treated with 0.05% Trypsin/EDTA, which was used to detach the cells, 1.2×10^6 cells seeded on plate. After 24 h, pEpiSCs cultured on iMEFs with DMEM/F-12 supplemented with 20% of FBS, 1% of GlutaMAX, 1% of MEM, 1% of penicillin/streptomycin and 0.1 mM β-mercaptoethanol and incubated at 39 °C in 5% CO_2. pEpiSCs were passaged using mixture of 1 mg/mL Collagenase IV and 1 mg/mL Dispase for 10 min at 39 °C.

2.2. Reagents and Antibodies

Dulbecco's modified eagle's medium (DMEM/F12; 11995-073), 0.05% Trypsin/EDTA (25300-54), Collagenase IV (17104-019), Dispase (17105-041), M199 (M4530) were purchased from GIBCO (Grand Island, NY, USA). Fetal bovine serum (FBS; S001-01) used for MEF culture was purchased from WELGENE (Gyeongsan, Korea). β-mercaptoethanol (M3148), MEM (M7145), penicillin/streptomycin (P0781), mitomycin C (M4287), bovine serum albumin (BSA, A6003), heparin (H3149), paraformaldehyde (PFA; P6148), methyl cellulose (M7140), HEPES (391340) were purchased from Sigma (St. Louis, MO, USA). Endothelial cell growth basal medium-2 (EBM-2; CC-3156), endothelial cell growth medium-2 SingleQuots® (EGM-2; CC4176) were purchased from Lonza (Basel, Switzerland). Vascular endothelial cell growth factors (VEGF; 293-VE) was purchased from R&D (Minneapolis, MN, USA). Matrigel® growth factor reduced (354230) used for differentiation, matrigel® (354234) used for capillary-like structure formation assay, endothelial cell growth supplement (ECGS; 356006), collagen I (354249) were purchased from Corning (Corning, NY, USA). Dynabeads™ M-280 streptavidin Sheep anti-rabbit (11203D) was purchased from Thermo Fisher Scientific (Grand Island, NY, USA). CD31 antibody (NB100-2284), Ki-67 antibody (NB500-170) were purchased Novusbio (Centennial, CO, USA). Ethylene-diamine-tetraacetic acid (EDTA; EDT001.500) was purchased from BioShop (Ontario, Canada). FBS (TMS-013-BKR) used for solely EC culture was purchased from Merck (Darmstadt, Germany). Phycoerythrin (PE) conjugated CD-31 antibodies (555027) was purchased from BD Pharmingen (Becton Dickinson, NJ, USA). Alexa Fluor® 546 Goat Anti-Rabbit IgG (A11010), Phalloidin (A12379) were purchased from Invitrogen (Carlsbad, CA, USA). Hoechst 33342 was purchased from Life Technologies (Prederick, MD, USA). RNeasy Plus Mini Kit (74134) was purchased from QIAGEN (Valencia, CA, USA). Revoscript™ RT Premix (25087) was purchased from iNtRON Biotechnology Inc (Seongnam, Korea). GoTaq® SYBR Master Mix (QPK201) was purchased from Promega (Madison, WI, USA). NaOH (39155-1250) was purchased from Junsei Chemical (Tokyo, Japan). Dil-Ac-LDL (022K) was purchased from Cell applications Inc. (H-1000; San Diego, CA, USA) was purchased from Vectorlabs (Burlingame, CA, USA).

2.3. In Vitro Differentiation of Endothelial Cells from Porcine Epiblast Stem Cells

pEpiSCs were cultured on iMEFs for 3 days, then detached stem cell colonies using manual picking. The pEpiSCs were differentiated into endothelial cells (ECs) in 50 ng/mL of VEGF included EGM-2 with VEGF excluded endothelial cell growth medium-2 SingleQuots®. Differentiation has proceeded for 8 days on culture plates coated with matrigel® (1:40 dilution with DMEM/F-12 medium) at 39 °C.

2.4. Magnetic-Activated Cells Soring (MACS)

Dynabeads™ M-280 streptavidin Sheep anti-rabbit and CD31 antibody were mixed in 1:10 ratio and incubated at 4 °C for overnight. ECs differentiated from pEpiSCs were washed with Dulbecco's phosphate-buffered saline (D-PBS). The cells were detached in 2 mM of ethylene-diamine-tetraacetic acid (EDTA) for 15 min at 39 °C and then centrifuged at 300× *g* for 3 min. Collected cells were suspended in 1% bovine serum albumin-DMEM medium. Magnetic beads labeled with CD-31 antibody were washed three times with 1%

BSA/DMEM using PolyATtract® system 1000 magnetic separation stand (Promega, Z5410). Suspended cells in 1% BSA/DMEM and magnetic beads labeled with the antibody were mixed and then rotated at room temperature for 1 h. Then, the mixtures were washed with 1% BSA/DMEM to remove unlabeled cells using a magnetic stand five times. Sorted ECs were seeded on culture plates coated with 0.2% gelatin for the primary culture. EGM-2, M199 supplemented with 20% of FBS, 30 μg/mL of endothelial cell growth supplement (ECGS) and 100 μg/mL of heparin were used for the culture medium of sorted ECs.

2.5. Proliferation Assay

Cells were seeded on culture plates coated with 0.2% gelatin and then counted the cell numbers for 5 days. Additionally, cell proliferation rates were examined with the expression of Ki-67 as a representative proliferation nuclear marker [43].

2.6. Flow Cytometry Analysis

Cells were washed with DPBS and then treated with 2 mM EDTA/PBS at 37 °C for 10 min. After centrifugation at 300× *g* for 3 min, collected purified ECs were suspended in stain buffer consisting of 2% BSA in PBS. The cells were stained with phycoerythrin (PE) conjugated CD-31 antibodies for 1 h at room temperature in the dark. The cells were suspended in stain buffer and then analyzed by FACSverse™ (BD Biosciences).

2.7. Immunocytochemistry

Cells were fixed with 4% paraformaldehyde (PFA) for 20 min. The cells were treated with the blocking solution (5% BSA/PBS-T) for 1 h, and then incubated with CD-31 antibody at 4 °C for 16 h, followed by Alexa Fluor® 546 Goat Anti-Rabbit IgG antibody. Hoechst 33342 was used to nucleus counterstain. All images were acquired using the LEICA fluorescence microscope (LEICA, DM 2500) and performed with the Leica Application Suite (LAS; LEICA, version 3.8).

2.8. Quantitative Polymerase Chain Reaction

Total RNA was extracted using RNeasy Plus Mini Kit following the manufacturer's methods. Extracted RNAs were synthesized into cDNA using the Revoscript™ RT Premix. Quantitative real-time polymerase chain reaction (q-PCR) was performed using the GoTaq® SYBR Master Mix with Rotor-Gene Q-Pure Detection system (QIAGEN). The primer list used for quantitative real-time polymerase chain reaction is Table 1. The gene expression was quantified relative to the reference gene (18S).

Table 1. Sequences of primers used in q-RT-PCR.

Gene	Sequence (5′→3′)		References
	Forward	Reverse	
18S	TCG GAA CTG AGG CCA TGA TT	GAA TTT CAC CTC TAG CGG CG	NR_046241.1
OCT-3/4	GGA TAT ACC CAG GCC GAT GT	GTC GTT TGG CTG AAC ACC TT	NM_001113060.1
NANOG	CCC GAA GCA TCC ATT TCC AG	GAT GAC ATC TGC AAG GAG GC	DQ_447201.1
SOX2	CAT GTC CCA GCA CTA CCA GA	GAG AGA GGC AGT GTA CCG TT	NM_001123197.1

2.9. Three-Dimensional Spheroid Sprouting Assay

Three-dimension spheroid sprouting of purified ECs was performed as described in the previous study [44]. Cells were separated into single cells with 0.05% trypsin/EDTA. Spheroids were formed using methocel solutions consisting of 3 g of methyl cellulose in 125 mL of M199. Single cells were counted to 500 cells per 1 spheroid in 25 μL droplet with 20% methocel solutions in each medium. Droplets were formed on the inverted lid of 100 mm culture dishes and then incubated at 37 °C for 24 h. Droplets formation of spheroids were collected from dish lids with PBS containing 10% FBS. Collected spheroids were centrifuged at 100× g for 5 min. For embedding in collagen of spheroids, collagen solution was mixed with acetic acid, 100 mg/mL of collagen I and M199 in a 4:4:1 ratio. Collagen solutions and 80% of methocel were mixed in a 1:1 ratio and then added to spheroids. Mixtures were deposited to 24 well culture plates and then polymerized at 37 °C for 1 h. When mixtures were polymerized, 330 ug/mL ECGS in M199 medium was added to mixtures to induce ECs sprouting. After 24 h, the spheroids in polymerized collagen mixtures were fixed in 4% PFA for 20 min. Phalloidin was stained in mixtures (1:250, Invitrogen, A12379). All images were acquired using OPTIKA fluorescence microscope (OPTIKA, XDS-3FL4) and performed with the software (OPTIKA, vision pro). Sprouts length was calculated using the ImageJ software.

2.10. Capillary-Like Structure Formation Assay

To capillary-like structure formation, cells were cultured on Matrigel, thawed at 4 °C for overnight and 50 μL added to Matrigel on 96 wells plate. Plates coated with Matrigel were incubated at 37 °C for 30 min. Cells were counted to 2×10^4 and then seeded on Matrigel. All images were acquired using an Olympus fluorescence microscope (Olympus, DP70) and performed with the DP manager (Olympus, version 3.1.1.208).

2.11. Dil-Acetylated-LDL Uptake Assay

Cells were added 2 μg/mL of Dil-Ac-LDL and then incubated at 37 °C for $4\frac{1}{2}$ h. The cells were fixed with 4% PFA for 10 min and then washed with PBS. Hoechst 33342 was used for nuclear staining. All images were explored using the LEICA fluorescence microscope (LEICA, DM 2500) and performed with the Leica Application Suite (LAS; LEICA, version 3.8).

2.12. Statistical Analysis

Graph Pad Prism software v7.00 (GraphPad) was used to analysis of data. Relative mRNA levels of OCT-3/4, NANOG, SOX2, quantification of Ki-67 positive cells in culture of differentiation media, sprouts length, branch points and quantification of Dil-Ac LDL uptake assay were analyzed in triplicate and data were presented as means ± SEM. One-way or two-way ANOVA used statistical significance between groups. $P < 0.05$ was considered statistically significant.

3. Results

3.1. Separating Endothelial Cells Differentiated from pEpiSCs

Endothelial cells (ECs) were differentiated from porcine epiblast stem cells (pEpiSCs) using the method described in our previous study [38]. We sought to separate ECs solely from differentiated pEpiSCs by using magnetic-activated cell sorting (MACS) with CD-31 antibody, an endothelial cell marker (Figure 1A), and the expression of CD-31 was analyzed by flow cytometry and immunofluorescence. CD-31 expression was found in about 28% of the unsorted cell population, whereas 100% in the sorted cell population (Figure 1B). Furthermore, unsorted cells showed partial CD-31 expression, but sorted ECs showed most CD-31 expression by immunofluorescence (Figure 1C). pEpiSCs did not express at all. To evaluate pluripotency of sorted ECs, gene expression of pluripotency markers such as OCT-3/4, NANOG and SOX2 were measured. Comparison of the pluripotency in pEpiSCs, unsorted ECs and sorted ECs revealed significant changes. These changes showed that pluripotency markers were significantly decreased in sorted ECs compared to pEpiSCs (Figure 1D). These results indicate that MACS-based cell sorting is sufficient for separating ECs solely and that sorted ECs lost their pluripotency.

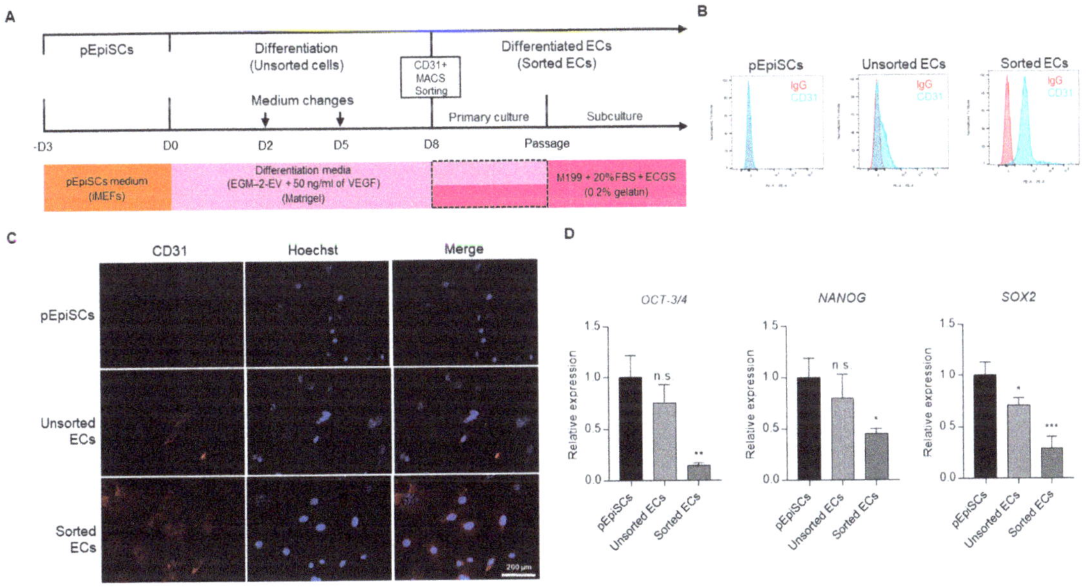

Figure 1. Establishment of differentiated ECs sorting. (**A**) The schematic diagram presented in vitro differentiation of pEpiSCs into ECs in differentiation media on matrigel for eight days. On day 8, ECs were separated from cell mixtures differentiated from pEpiSCs using magnetic beads coated with endothelial cell marker (CD-31). After the cell sorting, ECs were cultured in differentiation media for the primary culture and then proliferated in M199 on 0.2% Gelatin. (**B**) Flow cytometry analysis of CD-31 expression in pEpiSCs, unsorted ECs and sorted ECs. (**C**) Immunofluorescence of CD-31 in pEpiSCs, unsorted ECs and sorted ECs. Blue: staining of Hoechst, Red: staining of CD-31. Scale bar = 200 μm. (**D**) Relative mRNA levels of OCT-3/4, NANOG and SOX2 in pEpiSCs, unsorted ECs and sorted ECs. Values presented as mean SEM. * $p < 0.05$, ** $p < 0.01$, *** $p < 0.001$ vs. pEpiSCs, n.s.: not significant.

3.2. Proliferation of pEpiSCs-Derived ECs

For the primary culture of sorted ECs, these ECs were cultured using various media for a couple of days (Table 2). Sorted ECs were not primarily survived and proliferated in EGM-2 or M199 while growing well in differentiation media (Figure 2A). Photographs were obtained on the day using a phase-contrast microscope, with representative photographs shown. Interestingly, after subculture of these cells, sorted ECs were grown well in M199. The proliferation rate of subculture of ECs were evaluated for five days. The cell growth in each culture condition was compared to differentiation media as the control. As the results, EGM-2 and EGM-2-EV culture showed lower proliferation than control otherwise, M199 showed a significantly increased proliferation rate than control on days four and five. Also, the cells were stained with Ki-67 and showed the highest staining level when cultured in M199 compared to other media (Figure 2B). These results suggested that early passage of sorted ECs are needed differentiation media for stabilization and then, M199 is the best condition for growth and proliferation.

Table 2. Culture conditions for endothelial cells differentiated from porcine epiblast stem cells.

Name	Media	Supplements & Growth Factors
EGM-2	EBM-2	EGM-2 SingleQuot Kit
EGM-2-EV	EBM-2	VEGF excluded EGM-2 SingleQuot Kit
Differentiation Media	EBM-2	50 ng/mL VEGF-165, VEGF excluded EGM-2 SingleQuot Kit
M199	M199	20% PBS, 100 µg/mL heparin, 30 µg/mL ECGS

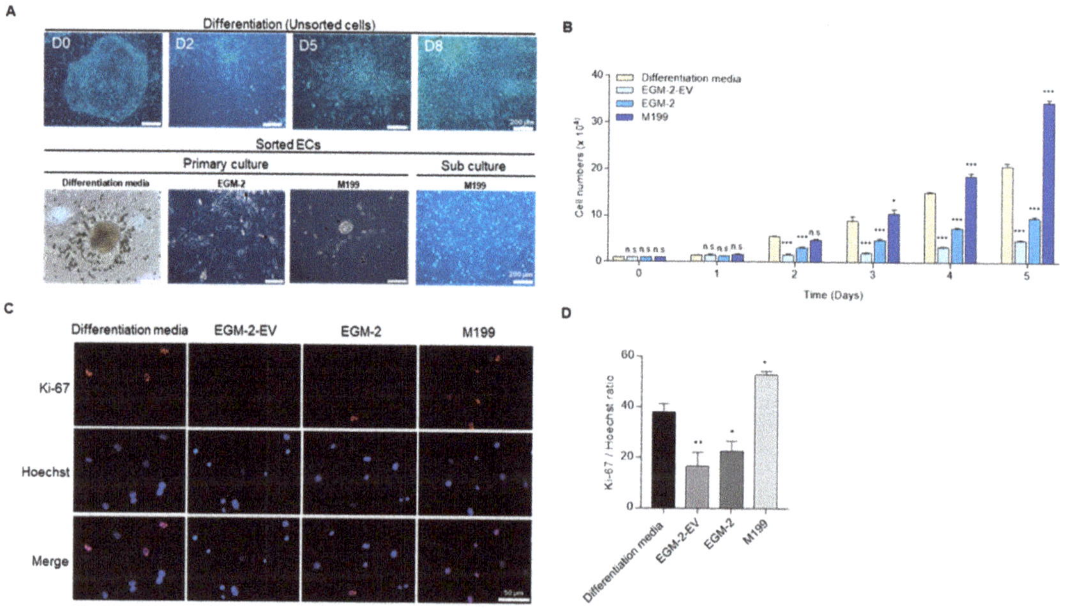

Figure 2. The proliferation of sorted ECs. (**A**) Morphologies of pEpiSCs and differentiated cells in differentiation media on the days 0, 2, 5, 8, respectively. Sorted ECs were cultured in differentiation media, EGM-2, or M199 culture system on 0.2% gelatin for the primary culture. Primary sorted ECs cultured in differentiation media were transferred to M199 culture system for sub-culture. (**B**) Proliferation rates of 1.0×10^4 of sorted ECs were evaluated in four culture conditions. Values presented as mean SEM. * $p < 0.05$, ** $p < 0.01$, *** $p < 0.001$ vs. differentiation media, n.s.: not significant. (**C**) Immunofluorescence of Ki-67 in differentiation media, EBM-2-EV, EGM-2 and M199. Red: staining of Ki-67, blue: staining of Hoechst. Scale bar = 50 µm (**D**) Quantification of Ki-67 positive cells in culture of sorted ECs in differentiation media, EBM-2-EV, EGM-2 and M199. Values presented as mean SEM. * $p < 0.05$, ** $p < 0.01$ vs. differentiation media, n.s.: not significant.

3.3. Angiogenic Function of pEpiSCs-Derived ECs by Three-Dimensional Spheroid Sprouting

Typical ECs have a function of angiogenesis, which is to form new vessels by various signaling. To confirm the angiogenesis capacity of differentiated cells in three-dimensional conditions, spheroids spouting assay derived from three types of cells (pEpiSCs, unsorted ECs and sorted ECs) (Figure 3). pEpiSCs and sorted ECs formed well spheroids, but unsorted ECs did not form spheroids. After allowing the spheroids to sprout, capillary-like structures sprouted out of spheroids were formed in sorted ECs (Figure 3A). However, no branched spheroids were displayed from pEpiSCs and unsorted ECs coexisting with undifferentiated PSCs and differentiated ECs. Additionally, sprouting out of three-dimensional spheroids derived from the sorted ECs were identified with fluorescence staining (Figure 3B,C). Collectively, these results suggested that pEpiSCs-derived ECs have an angiogenetic function as mature vascular endothelial cells in three-dimensional conditions.

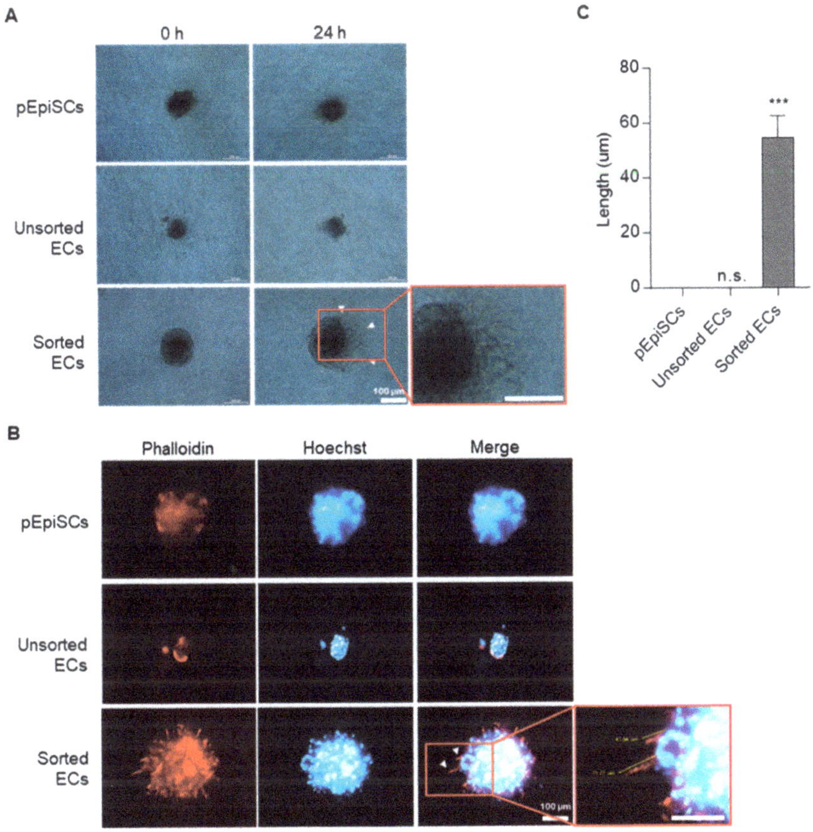

Figure 3. Three-dimensional spheroid sprouting of sorted ECs. (**A**) Three-dimensional spheroids derived from pEpiSCs, unsorted ECs and sorted ECs using the hanging drop protocol were cultured in collagen type I mixtures supplemented with ECGS for 24 h. Arrowheads indicated sprouting (capillary-like) structure out of three-dimensional spheroid derived from the sorted ECs. Scale bar = 100 μm. (**B**) Phalloidin staining in three-dimensional spheroid sprouting. Arrowheads indicated a capillary-like structure (sprouting) out of three-dimensional spheroid derived from the sorted ECs. Blue: staining of Hoechst, Red: staining of phalloidin. Scale bar = 200 μm. (**C**) Sprouts length was calculated by using ImageJ. Values presented as mean SEM. *** $p < 0.001$ vs. pEpiSCs, n.s.: not significant.

3.4. Vessel Organization of pEpiSCs-Derived ECs Using Capillary-like Structure Formation

Capillary-like structure formation was performed to evaluate angiogenesis's reorganization in vascular ECs. This assay was conducted to identify the functional capability of sorted ECs as acts typical ECs. The ability to form a capillary-like structure was assessed by seeding pEpiSCs, differentiated endothelial cells on matrigel-coated plates (Figure 4). As a result, sorted ECs started to form capillary-like structures for two hours. Capillary-like structures from sorted ECs were gradually expanded and then widely broadened. By contrast, unsorted ECs presented unclear capillary-like structure formation because differentiated ECs and undifferentiated pEpiSCs were mixed (Figure 4A,B). Interestingly, single cells derived from pEpiSCs colonies formed partial capillary-like structures. These results show that pEpiSCs-derived ECs have reorganization of vessel tube capability as functional vascular endothelial cells.

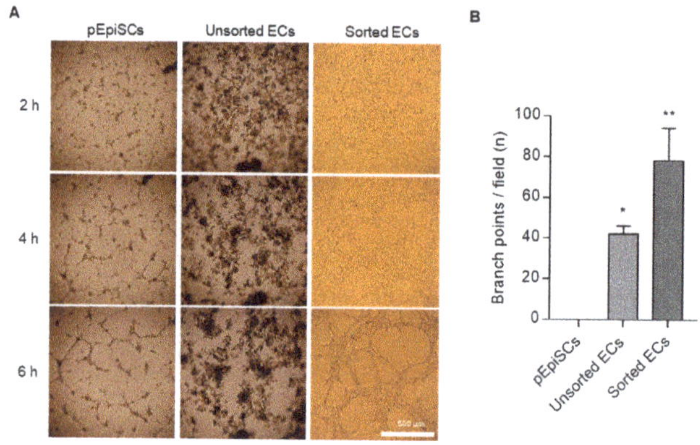

Figure 4. Formation of capillary-like structure of sorted ECs. (**A**) 2.0×10^4 of cells were cultured in 96-well cell culture plates on matrigel. Formation of capillary-like structures of sorted ECs was observed on matrigel for 6 h. Scale bar = 500 µm. (**B**) Branch points of pEpiSCs, unsorted ECs and sorted EC was counted. Values presented as mean SEM. * $p < 0.05$, ** $p < 0.01$ vs. pEpiSCs, n.s.: not significant.

3.5. Acetylated Low Density Lipoprotein Uptake of pEpiSCs-Derived ECs

ECs maintain homeostasis of cholesterol concentration in the blood vessels by uptaking acetylated low-density lipoprotein (Ac-LDL) [45]. To evaluate the function of Ac-LDL uptake in pEpiSCs, unsorted ECs or sorted ECs were examined with Dil-Ac-LDL (Figure 5). As a result, Dil-Ac-LDL uptake was detected only in differentiated endothelial cells (unsorted ECs and sorted ECs). Unsorted ECs showed little uptake of Dil-Ac-LDL, especially sorted ECs were shown the highest uptake of Dil-Ac-LDL. These results revealed that differentiated endothelial cells have a capacity of Ac-LDL uptake as functional property of endothelial cells.

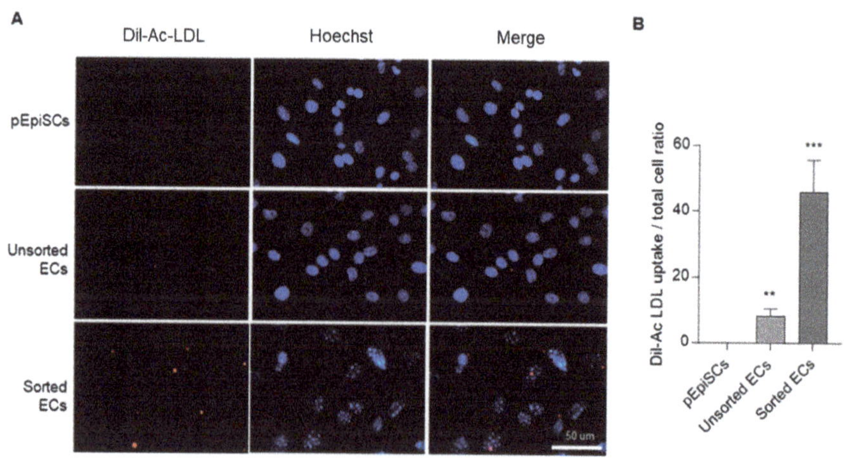

Figure 5. Uptake of Dil-acetylated low-density lipoprotein of Sorted ECs. (**A**) Dil-Ac-LDL uptakes of pEpiSCs, unsorted ECs and sorted ECs were examined at 4.5 h. Blue: staining of Hoechst. Red: staining of Dil-Ac-LDL. Scale bar = 50 µm. (**B**) Quantification of Dil-Ac LDL uptake assay in pEpiSCs, unsorted ECs and sorted EC. Values presented as mean SEM. ** $p < 0.01$, *** $p < 0.001$ vs. pEpiSCs, n.s.: not significant.

4. Discussion

Technologies for the production of a vessel-, tissue-, organ-, disease-, and further patient-specific ECs will become a fundamental necessity for research of molecular target validation, high-throughput drug screening and ECs-based cell therapy, including the engineering of clinically applicable engineered vascular tissue grafts. Although primary ECs have several limitations, such as restricted scalability and high probability of karyotypic defects, ECs have been used in various disease models to explore vascular dysfunction. Because of their indefinite self-renewal and high pluripotency, pluripotent stem cells (PSCs), which include embryonic stem cells (ESCs), epiblast stem cells (EpiSCs), and induced pluripotent stem cells (iPSCs), are the promising therapeutic strategy for human degenerative diseases. However, owing to the harsh ethical and limitation of culture expansion to utilize human ESCs (hECS) for regenerative medicine, many attention has been diverted to non-primate species such as pig retained anatomical and physiological similarities with humans [46]. Accordingly, porcine PSCs were of considerable significance for human degenerative diseases therapy [47].

We recently reported an in vitro differentiation protocol in which pEpiSCs were incubated stem cell culture medium on a feeder layer of mitomycin-treated mouse embryo fibroblasts (MEFs) for three days. Then, the condition changed to EBM-2-EV supplemented with 50 ng/mL VEGF on matrigel for eight days to induce ECs differentiation [38]. We observed differentiation efficiencies of approximately 27% for CD31-positive cells. Such protocols are undoubtedly scalable; however, strategies to select pure ECs with culture conditions that allow pure ECs to proliferate effectively and further functional validation of pEpiSC-derived ECs into typical vascular ECs would be essential for therapeutic application in a clinical trial. Applying the separation of ECs from the heterogeneous mixtures, pEpiSCs-derived cells, including differentiated or undifferentiated ECs were adopted in magnetic-activated cell sorting (MACS) using CD31 antibody. Although the protocol achieved the approach yield of about 27% ECs, 100% purified ECs could be identified through flow cytometry analysis by MACS sorting These were achieved by generating a more efficient amount that is not limited by the current scalable culture conditions of primary ECs.

For the widespread application of ECs, it has been fundamentally important to establish an efficient ECs proliferation system. However, sorted ECs were not proliferated effectively in Endothelial Cell Growth Medium-2 BulletKit (EGM-2) as a specialized ECs growth media. It was reported that adding 50 ng/mL of VEGF to EGM-2-EV supports the ECs survival for the primary culture until the primary sorted cells start growing at the normal proliferation rate [48]. When the required growth factors were not added to the culture medium, proliferation of endothelial progenitor cells (EPCs) proceeded slowly and cell death occurred lastly [49]. For such reasons, it was necessary to culture the sorted ECs in differentiation media until the primary passage. In general, VEGF plays an essential role in the survival and proliferation of ECs by activating of PI3K/Akt/forkhead signaling pathway in scalable suspension culture [50–52]. After the primary culture in differentiation media, the M199 culture system-induced most significant effect on the proliferation of ECs. Therefore, it is reasonable that the requirement of cytokines is different during the ECs proliferation. Collectively, the purified ECs were identified to exhibit the greatest effect on VEGF for the primary culture and the M199 culture system for the proliferation. We further confirmed the expression of Ki67, a marker of cell proliferation, through fluorescence staining.

Considering the requirement of practical grade for the vascular disease therapy, functional assessment of generated pEpiSCs-derived ECs were evaluated applying three different assays. ECs enriched by surface marker selection would provide a safer cell resource. Expression of CD-31 as an ECs-specific surface marker was observed strongly in early vascular development and capillary-like structures derived from ECs on Matrigel [53,54]. Although a previous report described ECs enrichment from human ESCs with a selection of CD-31+ expression [55], the purity of enriched ECs in that study was lower (~20%) than ours (~30%) [38]. On the other hand, PSCs differentiation usually occurs within multicellular, three-dimensional structures called embryoid bodies (EBs). However, in the current study, ECs sorted by selecting CD-31+ expression assembled networks of capillary-like structures, whereas pEpiSCs and unsorted ECs differentiated from pEpiSCs were barely formed. As well as, pEpiSCs-derived ECs solely networks showed much branching points compared with networks from pEpiSCs or unsorted differentiated ECs. The sorted ECs on Matrigel were attached and wrapped around in a way that is reminiscent of angiogenesis.

Additionally, to characterize the phenotypic nature of the ECs derived from pEpiSCs, a functional method that involves measuring Ac-LDL uptake using the fluorescent probe Dil (Dil-Ac-LDL) was performed. The sorted ECs were brilliantly fluorescent, whereas the fluorescent intensity of pEpiSCs and unsorted ECs were barely detectable. Finally, the three-dimensional spheroid sprouting of sorted ECs using the hanging drop protocol cultured in collagen type IV mixtures supplemented with ECGS for one day was examined. Nascent capillary-like structures out of the three-dimensional spheroid were formed in the sorted ECs, elucidating the vessel formation. These suggests that the sorted ECs by selected CD-31+ expression were fully differentiated and functionally competent.

5. Conclusions

The protocol described here offers the first opportunity to generate purified ECs from pEpiSCs with well-set up culture conditions for proliferation, which show the functionality of typical vascular ECs. Functional tests revealed that the generated ECs might be used in vitro assays to examine angiogenesis or cellular responses to various vascular diseases. Additionally, the ability to generate functional-ECs in sufficient quantities for cell therapy techniques may allow these purified pEpiSCs-derived ECs to be used in regenerative treatments.

Author Contributions: Conceptualization, J.-H.S., B.-G.S., C.H. and J.-H.L. (Joon-Hee Lee); investigation and writing, C.H. and J.-H.L. (Joon-Hee Lee); review and editing, I.-W.L., H.-J.K., E.-C.S., K.-M.L., S.-B.J., S.-K.B., T.-S.K., J.-H.L. (Jeong-Hyung Lee) and J.-W.C.; supervision, C.H. and J.-H.L. (Joon-Hee Lee); funding acquisition, C.H. and J.-H.L. (Joon-Hee Lee). All authors have read and agreed to the published version of the manuscript.

Funding: This work was supported by the National Research Foundation of Korea funded by the Korean Government (NRF-2018R1C1B6007728, NRF-2020R1I1A3072689 and NRF-2021R1C1C1006516) Korea. J.-H.S. and B.-G.S. were supported by the scholarship from the BK21 Program, Ministry of Education, Korea.

International Review Board Statement: Not applicable.

Informed Consent Statement: Not applicable.

Data Availability Statement: Data are contained within the article.

Conflicts of Interest: The authors declare no conflict of interest.

References

1. Michiels, C. Endothelial cell functions. *J. Cell. Physiol.* **2003**, *196*, 430–443. [CrossRef] [PubMed]
2. Ignarro, L.J.; Buga, G.M.; Wei, L.H.; Bauer, P.M.; Wu, G.; del Soldato, P. Role of the arginine-nitric oxide pathway in the regulation of vascular smooth muscle cell proliferation. *Proc. Natl. Acad. Sci. USA* **2001**, *98*, 4202–4208. [CrossRef] [PubMed]
3. Pearson, J.D. Endothelial cell function and thrombosis. *Bailliere's Best Pract. Res. Clin. Haematol.* **1999**, *12*, 329–341. [CrossRef]
4. Wang, W.; Deng, M.; Liu, X.; Ai, W.; Tang, Q.; Hu, J. TLR4 activation induces nontolerant inflammatory response in endothelial cells. *Inflammation* **2011**, *34*, 509–518. [CrossRef] [PubMed]
5. Chong, A.Y.; Blann, A.D.; Patel, J.; Freestone, B.; Hughes, E.; Lip, G.Y. Endothelial dysfunction and damage in congestive heart failure: Relation of flow-mediated dilation to circulating endothelial cells, plasma indexes of endothelial damage, and brain natriuretic peptide. *Circulation* **2004**, *110*, 1794–1798. [CrossRef] [PubMed]
6. Gimbrone, M.A., Jr.; Garcia-Cardena, G. Endothelial Cell Dysfunction and the Pathobiology of Atherosclerosis. *Circ. Res.* **2016**, *118*, 620–636. [CrossRef] [PubMed]

7. Ranchoux, B.; Harvey, L.D.; Ayon, R.J.; Babicheva, A.; Bonnet, S.; Chan, S.Y.; Yuan, J.X.; Perez, V.J. Endothelial dysfunction in pulmonary arterial hypertension: An evolving landscape (2017 Grover Conference Series). *Pulm. Circ.* **2018**, *8*, 2045893217752912. [CrossRef]
8. Colmenero, I.; Santonja, C.; Alonso-Riano, M.; Noguera-Morel, L.; Hernandez-Martin, A.; Andina, D.; Wiesner, T.; Rodriguez-Peralto, J.L.; Requena, L.; Torrelo, A. SARS-CoV-2 endothelial infection causes COVID-19 chilblains: Histopathological, immunohistochemical and ultrastructural study of seven paediatric cases. *Br. J. Dermatol.* **2020**, *183*, 729–737. [CrossRef]
9. Munzel, D.; Lehle, K.; Haubner, F.; Schmid, C.; Birnbaum, D.E.; Preuner, J.G. Impact of diabetic serum on endothelial cells: An in-vitro-analysis of endothelial dysfunction in diabetes mellitus type 2. *Biochem. Biophys. Res. Commun.* **2007**, *362*, 238–244. [CrossRef]
10. Schmidt, D.E.; Manca, M.; Hoefer, I.E. Circulating endothelial cells in coronary artery disease and acute coronary syndrome. *Trends Cardiovasc. Med.* **2015**, *25*, 578–587. [CrossRef]
11. Thapa, N.; Hong, H.Y.; Sangeetha, P.; Kim, I.S.; Yoo, J.; Rhee, K.; Oh, G.T.; Kwon, I.C.; Lee, B.H. Identification of a peptide ligand recognizing dysfunctional endothelial cells for targeting atherosclerosis. *J. Control. Release* **2008**, *131*, 27–33. [CrossRef] [PubMed]
12. Gimbrone, M.A., Jr.; Topper, J.N.; Nagel, T.; Anderson, K.R.; Garcia-Cardena, G. Endothelial dysfunction, hemodynamic forces, and atherogenesis. *Ann. N. Y. Acad. Sci.* **2000**, *902*, 230–239; discussion 239–240. [CrossRef] [PubMed]
13. Goya, K.; Otsuki, M.; Xu, X.; Kasayama, S. Effects of the prostaglandin I2 analogue, beraprost sodium, on vascular cell adhesion molecule-1 expression in human vascular endothelial cells and circulating vascular cell adhesion molecule-1 level in patients with type 2 diabetes mellitus. *Metab. Clin. Exp.* **2003**, *52*, 192–198. [CrossRef] [PubMed]
14. Hansson, G.K. Inflammation, atherosclerosis, and coronary artery disease. *N. Engl. J. Med.* **2005**, *352*, 1685–1695. [CrossRef] [PubMed]
15. Vanhoutte, P.M.; Shimokawa, H.; Feletou, M.; Tang, E.H. Endothelial dysfunction and vascular disease—A 30th anniversary update. *Acta Physiol.* **2017**, *219*, 22–96. [CrossRef]
16. Kim, S.; von Recum, H. Endothelial stem cells and precursors for tissue engineering: Cell source, differentiation, selection, and application. *Tissue Eng. Part B Rev.* **2008**, *14*, 133–147. [CrossRef]
17. Reed, D.M.; Foldes, G.; Harding, S.E.; Mitchell, J.A. Stem cell-derived endothelial cells for cardiovascular disease: A therapeutic perspective. *Br. J. Clin. Pharmacol.* **2013**, *75*, 897–906. [CrossRef]
18. Wu, J.; Belmonte, J.C.I. Dynamic pluripotent stem cell states and their applications. *Cell. Stem. Cell* **2015**, *17*, 509–525. [CrossRef]
19. Ying, Q.L.; Wray, J.; Nichols, J.; Batlle-Morera, L.; Doble, B.; Woodgett, J.; Cohen, P.; Smith, A. The ground state of embryonic stem cell self-renewal. *Nature* **2008**, *453*, 519–523. [CrossRef]
20. Brons, I.G.; Smithers, L.E.; Trotter, M.W.; Rugg-Gunn, P.; Sun, B.; Chuva de Sousa Lopes, S.M.; Howlett, S.K.; Clarkson, A.; Ahrlund-Richter, L.; Pedersen, R.A.; et al. Derivation of pluripotent epiblast stem cells from mammalian embryos. *Nature* **2007**, *448*, 191–195. [CrossRef]
21. Evans, M.J.; Kaufman, M.H. Establishment in culture of pluripotential cells from mouse embryos. *Nature* **1981**, *292*, 154–156. [CrossRef] [PubMed]
22. Thomson, J.A.; Kalishman, J.; Golos, T.G.; Durning, M.; Harris, C.P.; Becker, R.A.; Hearn, J.P. Isolation of a primate embryonic stem cell line. *Proc. Natl. Acad. Sci. USA* **1995**, *92*, 7844–7848. [CrossRef]
23. Aoi, T.; Yae, K.; Nakagawa, M.; Ichisaka, T.; Okita, K.; Takahashi, K.; Chiba, T.; Yamanaka, S. Generation of pluripotent stem cells from adult mouse liver and stomach cells. *Science* **2008**, *321*, 699–702. [CrossRef] [PubMed]
24. Hanna, J.; Markoulaki, S.; Schorderet, P.; Carey, B.W.; Beard, C.; Wernig, M.; Creyghton, M.P.; Steine, E.J.; Cassady, J.P.; Foreman, R.; et al. Direct reprogramming of terminally differentiated mature B lymphocytes to pluripotency. *Cell* **2008**, *133*, 250–264. [CrossRef] [PubMed]
25. Takahashi, K.; Yamanaka, S. Induction of pluripotent stem cells from mouse embryonic and adult fibroblast cultures by defined factors. *Cell* **2006**, *126*, 663–676. [CrossRef]
26. Dvorak, P.; Hampl, A. Basic fibroblast growth factor and its receptors in human embryonic stem cells. *Folia Histochem. Cytobiol.* **2005**, *43*, 203–208.
27. Smith, A.G.; Heath, J.K.; Donaldson, D.D.; Wong, G.G.; Moreau, J.; Stahl, M.; Rogers, D. Inhibition of pluripotential embryonic stem cell differentiation by purified polypeptides. *Nature* **1988**, *336*, 688–690. [CrossRef]
28. Tesar, P.J.; Chenoweth, J.G.; Brook, F.A.; Davies, T.J.; Evans, E.P.; Mack, D.L.; Gardner, R.L.; McKay, R.D. New cell lines from mouse epiblast share defining features with human embryonic stem cells. *Nature* **2007**, *448*, 196–199. [CrossRef]
29. Xu, R.H.; Peck, R.M.; Li, D.S.; Feng, X.; Ludwig, T.; Thomson, J.A. Basic FGF and suppression of BMP signaling sustain undifferentiated proliferation of human ES cells. *Nat. Method.* **2005**, *2*, 185–190. [CrossRef]
30. Ying, Q.L.; Nichols, J.; Chambers, I.; Smith, A. BMP induction of Id proteins suppresses differentiation and sustains embryonic stem cell self-renewal in collaboration with STAT3. *Cell* **2003**, *115*, 281–292. [CrossRef]
31. Daheron, L.; Opitz, S.L.; Zaehres, H.; Lensch, M.W.; Andrews, P.W.; Itskovitz-Eldor, J.; Daley, G.Q. LIF/STAT3 signaling fails to maintain self-renewal of human embryonic stem cells. *Stem Cells* **2004**, *22*, 770–778. [CrossRef] [PubMed]
32. Vallier, L.; Alexander, M.; Pedersen, R.A. Activin/Nodal and FGF pathways cooperate to maintain pluripotency of human embryonic stem cells. *J. Cell Sci.* **2005**, *118*, 4495–4509. [CrossRef] [PubMed]
33. Alberio, R.; Croxall, N.; Allegrucci, C. Pig epiblast stem cells depend on activin/nodal signaling for pluripotency and self-renewal. *Stem Cells Dev.* **2010**, *19*, 1627–1636. [CrossRef] [PubMed]

34. Baek, S.K.; Cho, Y.S.; Kim, I.S.; Jeon, S.B.; Moon, D.K.; Hwangbo, C.; Choi, J.W.; Kim, T.S.; Lee, J.H. A Rho-Associated Coiled-Coil Containing Kinase Inhibitor, Y-27632, Improves Viability of Dissociated Single Cells, Efficiency of Colony Formation, and Cryopreservation in Porcine Pluripotent Stem Cells. *Cell. Reprogr.* **2019**, *21*, 37–50. [CrossRef]
35. Levenberg, S.; Golub, J.S.; Amit, M.; Itskovitz-Eldor, J.; Langer, R. Endothelial cells derived from human embryonic stem cells. *Proc. Natl. Acad. Sci. USA* **2002**, *99*, 4391–4396. [CrossRef] [PubMed]
36. Ferrara, N. Molecular and biological properties of vascular endothelial growth factor. *J. Mol. Med.* **1999**, *77*, 527–543. [CrossRef]
37. Olsson, A.K.; Dimberg, A.; Kreuger, J.; Claesson-Welsh, L. VEGF receptor signalling—In control of vascular function. *Nat. Rev. Mol. Cell Biol.* **2006**, *7*, 359–371. [CrossRef]
38. Jeon, S.B.; Seo, B.G.; Baek, S.K.; Lee, H.G.; Shin, J.H.; Lee, I.W.; Kim, H.J.; Moon, S.Y.; Shin, K.C.; Choi, J.W.; et al. Endothelial Cells Differentiated from Porcine Epiblast Stem Cells. *Cell. Reprogr.* **2021**, *23*, 89–98. [CrossRef]
39. Dubois, N.C.; Craft, A.M.; Sharma, P.; Elliott, D.A.; Stanley, E.G.; Elefanty, A.G.; Gramolini, A.; Keller, G. SIRPA is a specific cell-surface marker for isolating cardiomyocytes derived from human pluripotent stem cells. *Nat. Biotechnol.* **2011**, *29*, 1011–1018. [CrossRef]
40. Hattori, F.; Chen, H.; Yamashita, H.; Tohyama, S.; Satoh, Y.S.; Yuasa, S.; Li, W.; Yamakawa, H.; Tanaka, T.; Onitsuka, T.; et al. Nongenetic method for purifying stem cell-derived cardiomyocytes. *Nat. Method.* **2010**, *7*, 61–66. [CrossRef]
41. Rust, W.; Balakrishnan, T.; Zweigerdt, R. Cardiomyocyte enrichment from human embryonic stem cell cultures by selection of ALCAM surface expression. *Regen. Med.* **2009**, *4*, 225–237. [CrossRef]
42. Xu, C.; Police, S.; Rao, N.; Carpenter, M.K. Characterization and enrichment of cardiomyocytes derived from human embryonic stem cells. *Circ. Res.* **2002**, *91*, 501–508. [CrossRef] [PubMed]
43. Sun, X.; Kaufman, P.D. Ki-67: More than a proliferation marker. *Chromosoma* **2018**, *127*, 175–186. [CrossRef] [PubMed]
44. Zahra, F.T.; Choleva, E.; Sajib, M.S.; Papadimitriou, E.; Mikelis, C.M. In Vitro Spheroid Sprouting Assay of Angiogenesis. *Methods Mol. Biol.* **2019**, *1952*, 211–218. [CrossRef] [PubMed]
45. Hassan, H.H.; Denis, M.; Krimbou, L.; Marcil, M.; Genest, J. Cellular cholesterol homeostasis in vascular endothelial cells. *Can. J. Cardiol.* **2006**, *22*, 35B–40B. [CrossRef]
46. Nowak-Imialek, M.; Kues, W.A.; Petersen, B.; Lucas-Hahn, A.; Herrmann, D.; Haridoss, S.; Oropeza, M.; Lemme, E.; Scholer, H.R.; Carnwath, J.W.; et al. Oct4-enhanced green fluorescent protein transgenic pigs: A new large animal model for reprogramming studies. *Stem Cells Develop.* **2011**, *20*, 1563–1575. [CrossRef] [PubMed]
47. Kobayashi, T.; Zhang, H.; Tang, W.W.C.; Irie, N.; Withey, S.; Klisch, D.; Sybirna, A.; Dietmann, S.; Contreras, D.A.; Webb, R.; et al. Principles of early human development and germ cell program from conserved model systems. *Nature* **2017**, *546*, 416–420. [CrossRef] [PubMed]
48. Gu, M. Efficient Differentiation of Human Pluripotent Stem Cells to Endothelial Cells. *Curr. Protoc. Hum. Genet.* **2018**, *98*, e64. [CrossRef]
49. Guan, X.M.; Cheng, M.; Li, H.; Cui, X.D.; Li, X.; Wang, Y.L.; Sun, J.L.; Zhang, X.Y. Biological properties of bone marrow-derived early and late endothelial progenitor cells in different culture media. *Mol. Med. Rep.* **2013**, *8*, 1722–1728. [CrossRef]
50. Abid, M.R.; Guo, S.; Minami, T.; Spokes, K.C.; Ueki, K.; Skurk, C.; Walsh, K.; Aird, W.C. Vascular endothelial growth factor activates PI3K/Akt/forkhead signaling in endothelial cells. *Arterioscler. Throm. Vasc. Biol.* **2004**, *24*, 294–300. [CrossRef]
51. Gerber, H.-P.; McMurtrey, A.; Kowalski, J.; Yan, M.; Keyt, B.A.; Dixit, V.; Ferrara, N. Vascular endothelial growth factor regulates endothelial cell survival through the phosphatidylinositol 3′-kinase/Akt signal transduction pathway: Requirement for Flk-1/KDR activation. *J. Biol. Chem.* **1998**, *273*, 30336–30343. [CrossRef] [PubMed]
52. Olmer, R.; Engels, L.; Usman, A.; Menke, S.; Malik, M.N.H.; Pessler, F.; Göhring, G.; Bornhorst, D.; Bolten, S.; Abdelilah-Seyfried, S. Differentiation of human pluripotent stem cells into functional endothelial cells in scalable suspension culture. *Stem Cell Rep.* **2018**, *10*, 1657–1672. [CrossRef] [PubMed]
53. Albelda, S.M. Endothelial and epithelial cell adhesion molecules. *Am. J. Respir. Cell Mol. Biol.* **1991**, *4*, 195–203. [CrossRef] [PubMed]
54. Gong, T.; Heng, B.C.; Xu, J.; Zhu, S.; Yuan, C.; Lo, E.C.; Zhang, C. Decellularized extracellular matrix of human umbilical vein endothelial cells promotes endothelial differentiation of stem cells from exfoliated deciduous teeth. *J. Biomed. Mater. Res. Part A* **2017**, *105*, 1083–1093. [CrossRef] [PubMed]
55. Figueiredo, L.M.; Costa, E.B.; Orellana, M.D.; Picanco-Castro, V.; Covas, D.T. OP9 Stromal Cells Proteins Involved in Hematoendothelial Differentiation from Human Embryonic Stem Cells. *Cell. Reprogr.* **2015**, *17*, 338–346. [CrossRef] [PubMed]

cells

MDPI

Article

Tet1 Suppresses p21 to Ensure Proper Cell Cycle Progression in Embryonic Stem Cells

Stephanie Chrysanthou [1,2,3], Julio C. Flores [1,2,3] and Meelad M. Dawlaty [1,2,3,*]

1 Ruth L. and David S. Gottesman Institute for Stem Cell and Regenerative Medicine Research, Albert Einstein College of Medicine, 1301 Morris Park Ave, Bronx, NY 10461, USA; stephaniechrys24@gmail.com (S.C.); julio.flores@einsteinmed.edu (J.C.F.)
2 Department of Genetics, Albert Einstein College of Medicine, 1301 Morris Park Ave, Bronx, NY 10461, USA
3 Department of Developmental & Molecular Biology, Albert Einstein College of Medicine, 1300 Morris Park Ave, Bronx, NY 10461, USA
* Correspondence: meelad.dawlaty@einsteinmed.org

Abstract: Ten eleven translocation 1 (Tet1) is a DNA dioxygenase that promotes DNA demethylation by oxidizing 5-methylcytosine. It can also partner with chromatin-activating and repressive complexes to regulate gene expressions independent of its enzymatic activity. Tet1 is highly expressed in embryonic stem cells (ESCs) and regulates pluripotency and differentiation. However, its roles in ESC cell cycle progression and proliferation have not been investigated. Using a series of Tet1 catalytic mutant ($Tet1^{m/m}$), knockout ($Tet1^{-/-}$) and wild type ($Tet1^{+/+}$) mouse ESCs (mESCs), we identified a non-catalytic role of Tet1 in the proper cell cycle progression and proliferation of mESCs. $Tet1^{-/-}$, but not $Tet1^{m/m}$, mESCs exhibited a significant reduction in proliferation and delayed progression through G1. We found that the cyclin-dependent kinase inhibitor p21/*Cdkn1a* was uniquely upregulated in $Tet1^{-/-}$ mESCs and its knockdown corrected the slow proliferation and delayed G1 progression. Mechanistically, we found that p21 was a direct target of Tet1. Tet1 occupancy at the p21 promoter overlapped with the repressive histone mark H3K27me3 as well as with the H3K27 trimethyl transferase PRC2 component Ezh2. A loss of Tet1, but not loss of its catalytic activity, significantly reduced the enrichment of Ezh2 and H3K27 trimethylation at the p21 promoter without affecting the DNA methylation levels. We also found that the proliferation defects of $Tet1^{-/-}$ mESCs were independent of their differentiation defects. Together, these findings established a non-catalytic role for Tet1 in suppressing p21 in mESCs to ensure a rapid G1-to-S progression, which is a key hallmark of ESC proliferation. It also established Tet1 as an epigenetic regulator of ESC proliferation in addition to its previously defined roles in ESC pluripotency and differentiation.

Keywords: Tet1; ESC; cell cycle; p21; proliferation

Citation: Chrysanthou, S.; Flores, J.C.; Dawlaty, M.M. Tet1 Suppresses p21 to Ensure Proper Cell Cycle Progression in Embryonic Stem Cells. *Cells* **2022**, *11*, 1366. https://doi.org/10.3390/cells11081366

Academic Editor: Mehdi Najar

Received: 10 March 2022
Accepted: 14 April 2022
Published: 17 April 2022

Publisher's Note: MDPI stays neutral with regard to jurisdictional claims in published maps and institutional affiliations.

1. Introduction

The ten eleven translocation (Tet) family of proteins (Tet1/2/3) are epigenetic modifiers that promote DNA demethylation through the iterative oxidation of 5-methylcytosine (5mC) to 5-hydroxymethylcytosine (5hmC) and other derivatives [1–3]. They can also form complexes with other epigenetic modifiers and transcription factors to regulate gene expressions independent of their enzymatic activity in DNA demethylation [4–6]. Tet enzymes are dynamically expressed during development and in various embryonic and somatic cell types. In embryonic stem cells (ESCs), Tet1 is a key regulator of pluripotency where its enzymatic activity is required for the proper demethylation of pluripotency genes [2,7] and its non-enzymatic functions promote the stable repression of lineage specifiers [4–6]. The biological relevance of Tet1 in ESC self-renewal and pluripotency is well-investigated in the field, but very little is known about the role of Tet1 in regulating ESC proliferation.

A key feature of ESCs is their rapid proliferation, which is marked by a unique cell cycle structure that has an extended DNA synthesis (S) phase and very short Gap (G1/G2) phases [8]. Mouse ESCs (mESCs) progress through the cell cycle in as fast as 6 h and the S phase constitutes ~65% of the cell cycle [9]. mESCs maintain this cell cycle structure until they differentiate [9,10]. This distinct cell cycle of mESCs is mainly driven by the high expression of cyclins, increased cyclin-dependent kinase (Cdk) activity and the absence of Cdk inhibitors (CKIs) [8,9]. As such, the typical oscillatory activity of Cyclin-Cdk complexes present in a somatic cell cycle is absent in mESCs. For example, Cyclin E/Cdk2 and Cyclin A/Cdk2 activities are so high throughout the ESC cell cycle that they are considered to be cell cycle-independent. A high Cdk2 activity is essential for a rapid G1 phase progression and the establishment of the unique ESC cycle structure [11]. p21/*Cdkn1a* is a key CKI that inhibits Cdk2 and is a negative regulator of the G1/S transition. Therefore, it is suppressed in ESCs to allow for a rapid progression through G1 [8–10]. However, the mechanisms by which p21 is repressed in ESCs have not been well-defined. Although it is widely established that p21 is a downstream target of p53, which can readily activate p21, the p53-independent regulation of p21 has also been reported. For example, in human ESCs, it has been shown that the p21 expression is epigenetically silenced through the deposition of the repressive histone mark H3K27me3 at the p21 promoter, which prevents p21 activation even in the presence of p53 [12]. This suggests that the epigenetic modifications of the p21 promoter override the transcriptional activation by p53 and serve as a safeguard mechanism for proper cell cycle progression.

Recently, we and others have implicated Tet1 in H3K27 trimethylation and the suppression of developmental genes in mESCs. Tet1, independent of its enzymatic activity, facilitates the recruitment of the histone trimethyltransferase PRC2 to deposit H3K27me3 and establish a bivalent state in developmental genes [4,6]. This silences the developmental programs in mESCs, but keeps them poised for activation upon differentiation and is, therefore, important in maintaining the pluripotent state. However, it is not known whether Tet1 can also regulate the cell cycle and proliferation of mESCs through histone modifications and/or DNA demethylation. In this study, we report a catalytic independent role of Tet1 in the regulation of the mESC cell cycle progression through the suppression of p21. We found that Tet1 directly binds to the p21 promoter and, independent of any DNA demethylation activity, recruits PRC2 for H3K27 trimethylation to repress p21. This allows for a rapid G1-to-S progression in mESCs. Consistently, Tet1 knockout, but not catalytic-deficient mESCs, had an extended G1 phase, an increased p21 expression and reduced levels of PRC2 and H3K27me3 at its promoter. The knockdown of p21 or the re-expression of catalytic-dead Tet1 in Tet1 knockout mESCs rescued the cell cycle and proliferation defects. Together, our findings identified Tet1 as an epigenetic regulator of ESC cell cycle progression and proliferation.

2. Materials and Methods

2.1. Embryonic Stem Cell Culture and Proliferation Assays

$Tet1^{-/-}$ and $Tet1^{m/m}$ mESCs were previously generated in our lab [6]. All mESC lines tested negative for mycoplasma by a PCR test and were cultured onto irradiated feeders in a media-containing serum/LIF (DMEM supplemented with 10% FBS, 2 mM glutamine, 1 × non-essential amino acids, 100 U/mL penicillin, 100 μg/mL streptomycin, 0.02 ug/mL LIF, 50 mM b-mercaptoethanol). The cells were passaged onto feeder-coated plates once they reached 70–80% confluency. For RNA and DNA extraction, the mESCs were pre-plated onto gelatin to remove the feeders and then seeded onto gelatin overnight before harvesting. For the differentiation assay, pre-plated mESCs were seeded onto gelatin in mESC media for 12 h and then cultured in mESC differentiation media (without LIF and supplemented with 1 μM retinoic acid) for 3 d. For the proliferation assays, 25,000 cells of each clone were seeded onto gelatin in 12-well plates in triplicate. The viable cells were counted using trypan blue and hemocytometer each day for 4 d. A two-way ANOVA test was used to calculate the statistical significance.

2.2. Cell Cycle Analysis and Apoptosis Assays

The mESCs were cultured onto gelatin and on the following day they were arrested in G2/M by a Nocodazole treatment (M1404, Sigma-Aldrich, Inc., St. Louis, MO, USA, 100 ng/mL for 16 h). The cells were released from the arrest by removing the Nocodazole-treated media, washing with 1× PBS twice and culturing in fresh mESC media for 2, 4, 5 and 7 h. The cells were fixed in 70% ethanol/PBS followed by staining for 30 min in PBS containing 0.1 mg/mL RNase A and 50 μg/mL propidium iodide (PI) (P4864, Sigma-Aldrich, Inc., St. Louis, MO, USA). The cell cycle analysis was performed using a BD LSR II flow cytometer (BD, Biosciences, Franklin Lakes, NJ, USA) and FlowJo software (v.10.8.0). For the quantification of the apoptotic cells, the ESCs were cultured onto gelatin for 24 h and stained with Annexin V and 7AAD using an Annexin V kit (556547, BD Biosciences, Franklin Lakes, NJ, USA) following the manufacturer's guidelines as previously described [13]. Annexin V staining was analyzed on a BD LSR II flow cytometer using FlowJo software (v.10.8.0). Early (7AAD$^-$ Annexin V$^+$) and late (7AAD$^+$ Annexin V$^+$) apoptotic cells were quantified and plotted.

2.3. RT-qPCR and ChIP-qPCR

A total of 2 μg of RNA, extracted from feeder-free mESCs by an Omega E.Z.N.A Total RNA kit (Omega Bio-tek, Inc., Norcross, GA, USA), was used to synthesize cDNA using a Superscript III First-Strand synthesis system (Invitrogen, Waltham, MA, USA). A real-time quantitative PCR was performed using a SYBR green master mix (Applied Biosystems, Waltham, MA, USA) in a QuantStudio 6 Flex Real-Time PCR system following the standard protocols using previously published primers [14–21] as listed in Table S1. The relative gene expression level was analyzed by a comparative Ct method and was normalized to *Gapdh*. A two-way ANOVA test was used to calculate the statistical significance between the three groups. ChIP experiments were performed using feeder-free mESCs cultured onto gelatin following previously published protocols [22] using antibodies against Tet1 (GTX125888, GeneTex, Irvine, CA, USA), Ezh2 (CST 5246, Cell Signaling Technologies, Danvers, MA, USA), and H3K27me3 (07449, Millipore, Burlington, MA, USA). The DNA concentration was measured using a Qubit 2.0 Fluorometer (Invitrogen, Waltham, MA, USA). The protein enrichment at specific loci was quantified by qPCR as mentioned above using the primers in Table S1. ChIP-qPCR signals were calculated as a fold enrichment using a 10% input and IgG as the control. In the case of the Tet1 ChIP-qPCR, the enrichment was also normalized over the *Tet1*$^{-/-}$ controls.

2.4. Western Blotting

The cells were lysed in a Radioimmunoprecipitation assay (RIPA) buffer (50 mM Tris-HCl, pH 7.4, 250 mM NaCl, 2% Nonidet-P40, 2.5 mM EDTA, 0.1% SDS, 0.5% DOC Sigma-Aldrich, Inc., St. Louis, MO, USA) supplemented with PIC and PMSF. The lysates were resolved on an 8–12% SDS-PAGE (Mini-PROTEAN electrophoresis chamber, Bio-Rad, Hercules, CA, USA) and transferred onto PVDF membranes (Mini Trans-Blot apparatus, Bio-Rad, Hercules, CA, USA) following the manufacturer's protocols. The membranes were blocked in 5% milk in PBS with 0.1% Tween-20 (PBST) and incubated overnight at 4 °C or for 1 h at room temperature with primary antibodies (p21: BD Pharmingen 556431; p53: CST2524T; p27: Santa Cruz sc1641; CycD1: CST2978T; Cdk1: Abcam ab18; Cdk2: CST2546T; β-actin: Abcam ab6276). Secondary antibody incubations (HRP-anti-mouse Calbiochem 401253 or HRP-anti-rabbit Calbiochem 401393, 1:3000) were carried out for 1 h at room temperature. In all experiments, β-actin was used as the loading control. The quantification of the Western blot band signal intensities was performed using Image J (v.1.53k) and the data were normalized to the respective actin signal intensities and then plotted.

2.5. Lentivirus Preparation for p21 Knockdown

HEK293T cells were cultured in 10% FBS Dulbecco's Modified Eagle Medium (DMEM) and transfected with a 10 µg pFUGW-H1 empty vector or pFUGW-H1 p21 shRNA1 (Addgene Watertown, MA, USA, plasmid # 25868) [23] along with lentiviral packaging plasmids pPAX2 (7.5 µg) and pMDG (2.5 µg) using an Xtremegene Transfection reagent (06365787001, Roche, Basel, Switzerland). The media were changed after 16 h. The following day, the lentivirus-containing supernatant was collected at 24 h and at 48 h and concentrated by a Lenti-X Concentrator (631232,Takara, Tokyo, Japan), according to the manufacturer's protocol. The mESCs cultured onto the gelatin at a 50% confluency were transduced with a pFUGW-H1 empty vector or a pFUGW-H1 p21 shRNA1 lentivirus supplemented with polybrene. The GFP-positive cells were sorted by flow cytometry and expanded in the culture for use in the experiments.

2.6. Bioinformatic Analysis

The DNA methylation levels and the enrichment of Tet1, H3K27me3 and H3K4me3 peaks at the promoter of p21 were visualized in the integrated genome browser (IGV) using our previously published Tet1 CUT&Tag datasets in *Tet1*$^{+/+}$ and *Tet1*$^{m/m}$ mESCs, H3K4me3 and H3K27me3 CUT&Tag datasets and WGBS datasets in *Tet1*$^{+/+}$, *Tet1*$^{m/m}$ and *Tet1*$^{-/-}$ mESCs (GSE176389) [6]. The genes implicated in the cell cycle progression were identified using our previously published RNA-seq datasets in *Tet1*$^{+/+}$, *Tet1*$^{m/m}$ and *Tet1*$^{-/-}$ ESCs (GSE176389) [6] and their expression was plotted as a heatmap using the pheatmap package in R software (v. 4.1.0).

3. Results

3.1. Deficiency of Tet1, but Not of Its Catalytic Activity, Leads to Reduced Proliferation and Extended G1 Phase in mESCs

To establish the catalytic-dependent and catalytic-independent requirements of Tet1 in mESC proliferation, we examined the proliferation of our previously generated [6] Tet1 knockout (*Tet1*$^{-/-}$), Tet1 catalytic mutant (*Tet1*$^{m/m}$) and wild type (*Tet1*$^{+/+}$) mESCs over 4 d in a culture. We found that *Tet1*$^{-/-}$, but not *Tet1*$^{m/m}$, mESCs had a significantly slower proliferation than the wild type mESCs, with the difference being more prominent on day 4 (Figure 1A). The overexpression of either the wild type or the catalytically inactive Tet1 in *Tet1*$^{-/-}$ mESCs rescued this growth defect (Figure 1B), suggesting that Tet1 non-catalytic functions are required for the proper proliferation of mESCs. To examine whether the reduced proliferation of *Tet1*$^{-/-}$ mESCs was due to increased cell death, we stained *Tet1*$^{-/-}$, *Tet1*$^{m/m}$ and wild type mESCs with Annexin V and quantified the number of apoptotic cells by flow cytometry. Although the number of early apoptotic cells (7AAD^{-} annexin V^{+}) was comparable among all three lines, the number of late apoptotic cells (7AAD^{+} annexin V^{+}) was marginally increased in *Tet1*$^{-/-}$ mESCs (~1% vs. ~0.5% in *Tet1*$^{m/m}$ and *Tet1*$^{+/+}$, respectively) (Figure 1C and Figure S1A). This insignificant increase in apoptosis could not explain the significantly reduced proliferation observed in *Tet1*$^{-/-}$ mESCs. Thus, the slow growth was unlikely due to increased cell death and could involve defects in the cell cycle progression. To this end, we analyzed the cell cycle of *Tet1*$^{-/-}$, *Tet1*$^{m/m}$ and *Tet1*$^{+/+}$ mESCs. First, we synchronized the mESCs by treating them with the microtubule inhibitor Nocodazole for 16 h to arrest the cells in the G2/M phase. We then released the cells to progress through the cell cycle and analyzed the percentage of cells in each phase of the cell cycle at 2, 4, 5 and 7 h post-release. We found that the *Tet1*$^{-/-}$ mESCs exhibited an extended G1 phase (marked by the presence of significantly more cells in the G1 phase and fewer cells in the S phase) compared with the *Tet1*$^{+/+}$ and *Tet1*$^{m/m}$ mESCs, which had comparable cell cycle profiles. This difference was most prominent at 4 h post-release (Figure 1D–E and Figure S1B). These findings suggest that Tet1 non-catalytic functions are important for proper cell cycle progression and proliferation of mESCs.

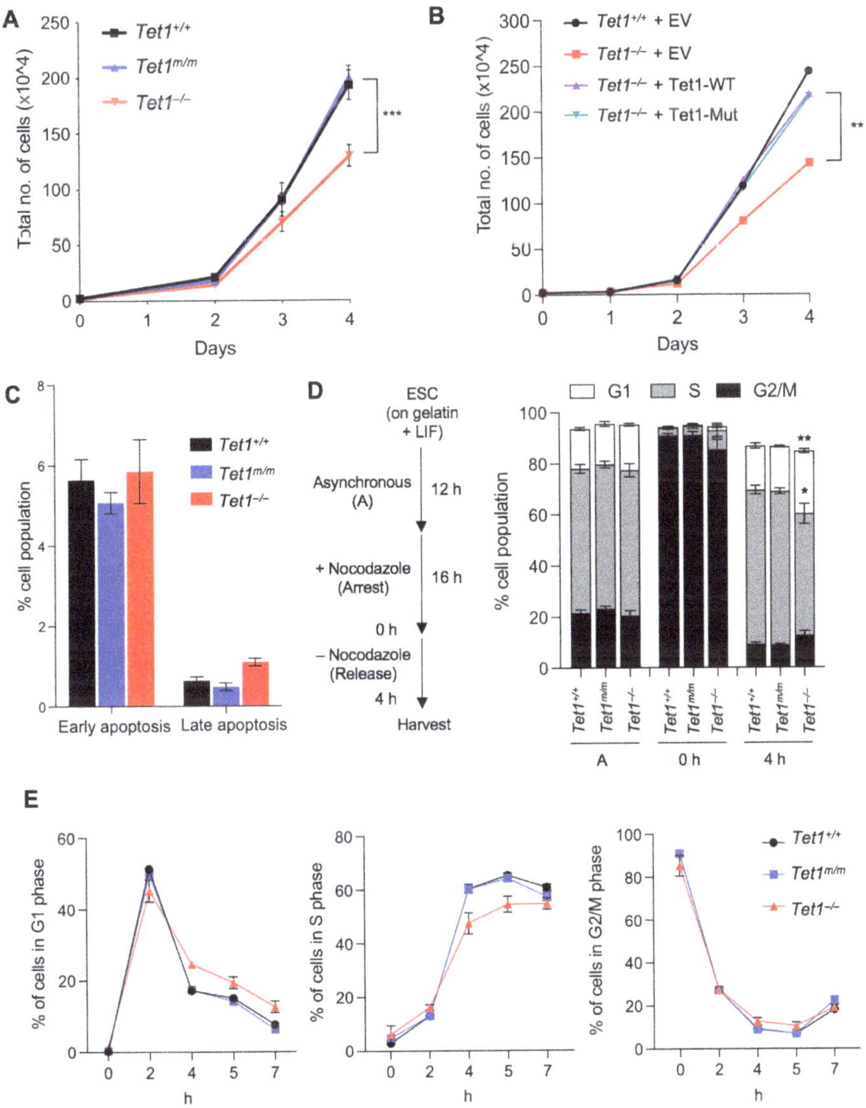

Figure 1. Loss of Tet1, but not the loss of its catalytic activity, reduces mESC proliferation and delays G1/S transition. (**A**) Growth curve of mESCs of indicated genotypes over a 4-day period. Three lines of each genotype were used. (**B**) Growth curve of *Tet1*$^{-/-}$ mESCs stably overexpressing wild type (WT) or catalytic mutant (Mut) Tet1 transgene. *Tet1*$^{+/+}$ and *Tet1*$^{-/-}$ mESCs transfected with an empty vector were used as controls. (**C**) Quantification of early (7AAD^{-} Annexin V^{+}) and late (7AAD^{+} Annexin V^{+}) apoptotic cells by flow cytometry in mESCs of indicated genotypes. Three lines of each genotype were used. Note that the number of apoptotic cells was not increased in *Tet1*$^{-/-}$ mESCs. (**D**) Schematic of cell cycle synchronization strategy (**left**) and percentage of mESCs in each phase of the cell cycle 4 h after release from Nocodazole block (**right**). Three lines of each genotype were used. (**E**) The percentage of mESCs in each phase of the cell cycle at the indicated timepoints shown on the x-axis after release from Nocodazole. Three independent mESC lines of each genotype were used. In all panels, data are presented as ± SEM; statistically significant *** $p < 0.001$, ** $p < 0.01$ and * $p < 0.05$ (two-way ANOVA with Holm–Sidak's multiple comparison test).

3.2. Upregulation of the Cyclin-Dependent Kinase Inhibitor p21 (Cdkn1a) Is Responsible for Delayed Cell Cycle Progression and Reduced Proliferation in $Tet1^{-/-}$ mESCs

To gain a molecular insight into how Tet1 regulates the ESC cell cycle, we examined our previously published transcriptomic data of *Tet1*$^{-/-}$, *Tet1*$^{m/m}$ and *Tet1*$^{+/+}$ mESCs [6] where genes uniquely deregulated in *Tet1*$^{-/-}$ mESCs were enriched for the cell cycle and proliferation gene ontology (GO) terms (Figure S2A). Among these genes, the cyclin-dependent kinase inhibitor (CKI) p21/*Cdkn1a*, which inhibits Cdk2/4 to block the G1/S transition, was a top hit that was significantly upregulated in *Tet1*$^{-/-}$, but not *Tet1*$^{m/m}$ and *Tet1*$^{+/+}$, ESCs. We validated the p21 upregulation at the mRNA level by RT-qPCR (Figure 2A) and at the protein level in both asynchronous and synchronized cultures by a Western blot (Figure 2B and Figure S2B). The levels of other CKIs involved in the G1/S transition—such as p15, p16, p19 and p27—were unaffected (Figure 2A). Likewise, cyclin-dependent kinases Cdk1 and Cdk2 as well as Cyclin D were not deregulated in *Tet1*$^{-/-}$ mESCs (Figure 2C and Figure S2B). This suggested that the upregulation of p21 is a very specific molecular signature of *Tet1*$^{-/-}$ mESCs involved in cell cycle regulation. We found that the levels of p53, the main transcriptional activator of p21, were unaffected in *Tet1*$^{-/-}$ ESCs (Figure 2B and Figure S2B–C). Likewise, the levels of the mir290-95 cluster, which has previously been implicated in the regulation of p21 in ESCs [24–26], was unaffected in *Tet1*$^{-/-}$ mESCs (Figure S2C). Together, these findings suggested that the upregulation of p21 in *Tet1*$^{-/-}$ mESCs is unlikely due to the perturbation of p53 or mir290-95 levels and could involve other transcriptional mechanisms involving Tet1.

Next, we sought to determine whether the p21 upregulation was specifically responsible for the proliferation of and cell cycle defects observed in *Tet1*$^{-/-}$ mESCs. To this end, we used shRNAs to knock down p21 in *Tet1*$^{-/-}$ mESCs equivalent to the levels in *Tet1*$^{m/m}$ and *Tet1*$^{+/+}$ mESCs (Figure 2D–E) and assessed the effects on proliferation and cell cycle progression. We found that reducing the p21 levels in *Tet1*$^{-/-}$ mESCs to near to the wild type levels rescued both the slower proliferation (Figure 2F) and the elongated G1 phase (Figure 2G). This confirmed that the loss of the Tet1-mediated upregulation of p21 was driving the proliferation and cell cycle defects in *Tet1*$^{-/-}$ mESCs. It also suggested that Tet1 was responsible for repressing p21 in the mESCs to allow for a rapid proliferation and cell cycle progression that was unique to the pluripotent state. We consistently found that during differentiation, as mESCs exited the pluripotent state (marked by the downregulation of the pluripotency factors Nanog and Oct4), the Tet1 levels decreased and the p21 levels increased (Figure 2H), corresponding with a longer G1 phase, which is a hallmark of differentiated cells. We also found that the knockdown of p21, which corrected the proliferation and cell cycle progression defects of *Tet1*$^{-/-}$ mESCs (Figure 2F–G), did not rescue the differentiation defects of *Tet1*$^{-/-}$ mESCs (Figure 2I). *Tet1*$^{-/-}$ mESCs expressing either an empty vector or p21 shRNA exhibited an aberrant upregulation of mesoderm marker *Bin1* and trophectoderm marker *Eomes* compared with *Tet1*$^{m/m}$ or *Tet1*$^{+/+}$ mESCs expressing an empty vector. This suggested that the Tet1-mediated suppression of p21 mainly regulated the mESC cell cycle progression and not the differentiation programs.

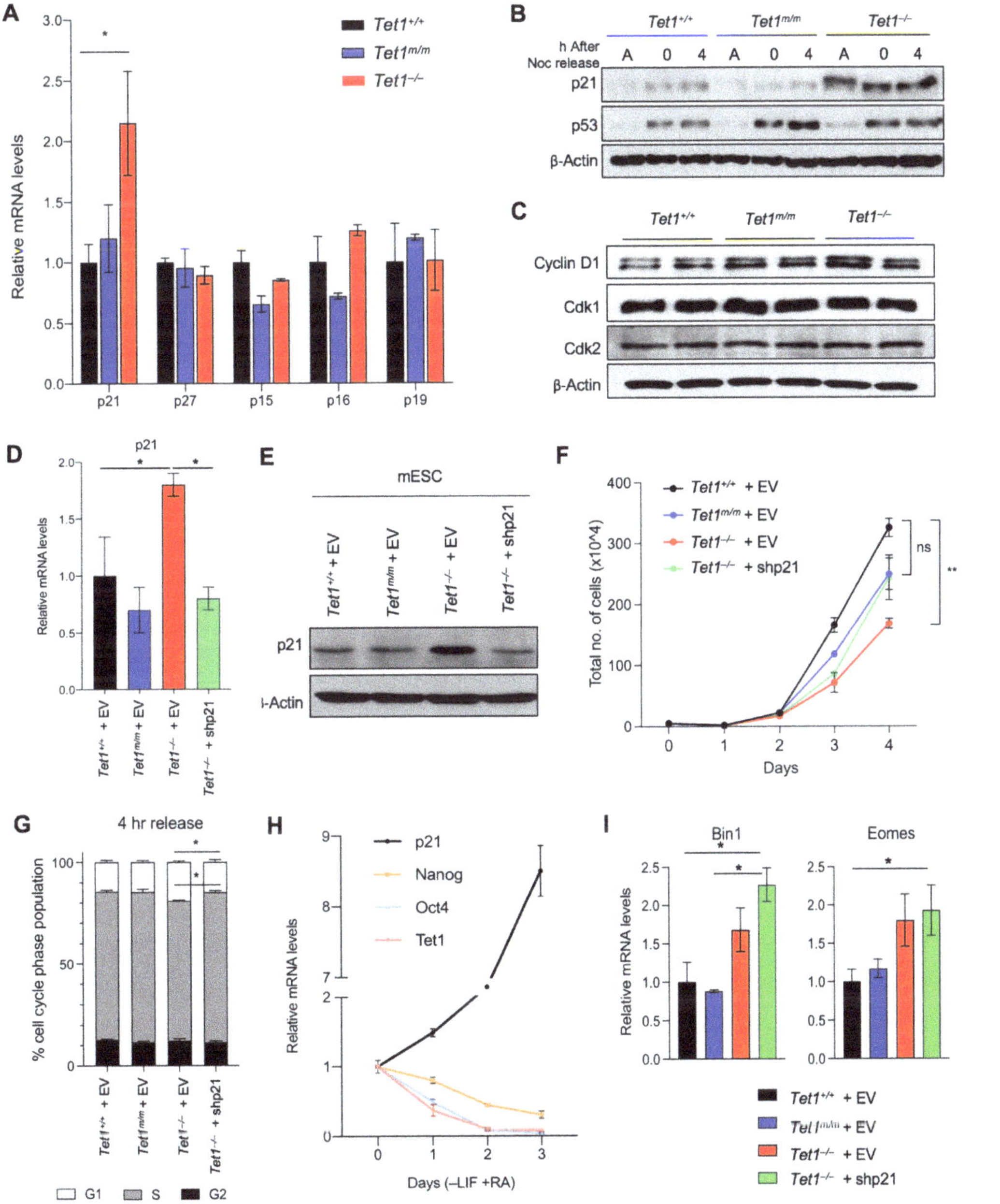

Figure 2. Aberrant upregulation of p21 is responsible for the slow proliferation of *Tet1*$^{-/-}$ mESCs. (**A**) Quantification of *p21, p27, p15, p16* and *p19* mRNA levels in mESCs of indicated genotypes by RT-qPCR. Data normalized to *Gapdh* expression. Three lines of each genotype were used. (**B**) Quantification of p21 and p53 protein levels in asynchronous (**A**), G2/M-arrested (0 h) and released (4 h) mESCs of indicated genotypes by Western blot analysis. β-actin was used as a loading control.

(**C**) Quantification of Cdk1, Cdk2 and Cyclin D1 protein levels in asynchronous mESCs of indicated genotypes by Western blot. β-actin was used as a loading control. (**D**,**E**) Quantification of *p21* mRNA levels by RT-qPCR (**D**) and protein levels by Western blot (**E**) in *Tet1*$^{+/+}$, *Tet1*$^{m/m}$, *Tet1*$^{-/-}$ mESCs transduced with empty vector (EV) or an shRNA against p21 vector (shp21). (**F**) Growth curve of mESCs transduced with an empty vector (EV) or an shRNA against *p21* vector (shp21) over a 4-day period. (**G**) The percentage of mESCs of indicated genotypes expressing empty vector (EV) or an shRNA against *p21* (shp21) in each phase of the cell cycle 4 h after release from Nocodazole block. (**H**) Quantification of mRNA levels of *p21*, *Tet1* and pluripotency markers (Oct4 and Nanog) by RT-qPCR in wild type mESCs differentiated (−LIF +RA) for three days. Three independent mESC lines were used. (**I**) Quantification of mRNA levels of mesoderm markers *Bin1* and *Eomes* by RT-qPCR in *Tet1*$^{+/+}$, *Tet1*$^{m/m}$ and *Tet1*$^{-/-}$ mESCs expressing empty vector (EV) or an shRNA against *p21* (shp21) and cultured in mESC media. Three replicates of each genotype were used. In all panels, data are presented as ± SEM; statistically significant ** $p < 0.01$ and * $p < 0.05$ (two-way ANOVA test).

3.3. Tet1 Suppresses p21 Expression by Binding to Its Promoter and Facilitating PRC2 Recruitment and H3K27 Trimethylation

The upregulation of p21 in *Tet1*$^{-/-}$ mESCs suggested that Tet1 is essential for the suppression of p21 in mESCs. Previously, we and others have implicated Tet1 in the gene suppression of mESCs where Tet1, independent of its catalytic activity, facilitates the recruitment of PRC2 to the gene promoters for H3K27 trimethylation [4,6]. This is essential for establishing the bivalency (H3K4me3^{+}; H3K27me3^{+}) of lineage-specific genes in mESCs, keeping them silenced in the pluripotent state and poised for activation upon differentiation [6]. This, together with the fact that p21 is a known target of PRC2 in human ESCs [12], prompted us to test whether Tet1 suppressed p21 by facilitating PRC2 recruitment and H3K27 trimethylation at its promoter. We examined the enrichment of Tet1 at the *p21* locus and assessed its co-occupancy with the PRC2 component Ezh2 and the bivalent marks using our previously published Tet1 CUT&Tag data in *Tet1*$^{+/+}$ and *Tet1*$^{m/m}$ ESCs as well as H3K4me3 and H3K27me3 CUT&Tag data in *Tet1*$^{+/+}$, *Tet1*$^{m/m}$ and *Tet1*$^{-/-}$ ESCs [6]. Genome browser tracks revealed a significant enrichment of wild type and catalytic mutant Tet1 at the p21 promoter (Figure 3A), which was confirmed by the ChIP-qPCR (Figure 3B). The Tet1 peaks overlapped with the H3K4me3 and H3K27me3 peaks at the *p21* promoter region (Figure 3A). Although the H3K4me3 levels were comparable in *Tet1*$^{+/+}$, *Tet1*$^{m/m}$ and *Tet1*$^{-/-}$ ESCs, the H3K27me3 levels were significantly reduced only in *Tet1*$^{-/-}$ ESCs as shown in the genome browser tracks (Figure 3A) and confirmed by the ChIP-qPCR (Figure 3C). To examine if this corresponded with a reduced PRC2 occupancy, we assessed the enrichment of PRC2 component Ezh2 at the p21 promoter by ChIP-qPCR using the same primers used for assessing the H3K27me3 levels. We found that the Ezh2 levels were significantly reduced in *Tet1*$^{-/-}$, but not in *Tet1*$^{m/m}$, mESCs (Figure 3D). An assessment of the DNA methylation levels at the *p21* locus using our previously published WGBS datasets in *Tet1*$^{+/+}$, *Tet1*$^{m/m}$ and *Tet1*$^{-/-}$ ESCs [6] revealed that the *p21* promoter, as with most bivalent gene promoters, was in a hypomethylated state and remained unchanged in all three lines (Figure 3A). Taken together, these data establish that in mESCs, Tet1 represses the p21 expression independent of its catalytic activity by facilitating PRC2 recruitment and H3K27me3 deposition at its promoter, which allows for the signature rapid cell cycle progression and proliferation of mESCs (Figure 4).

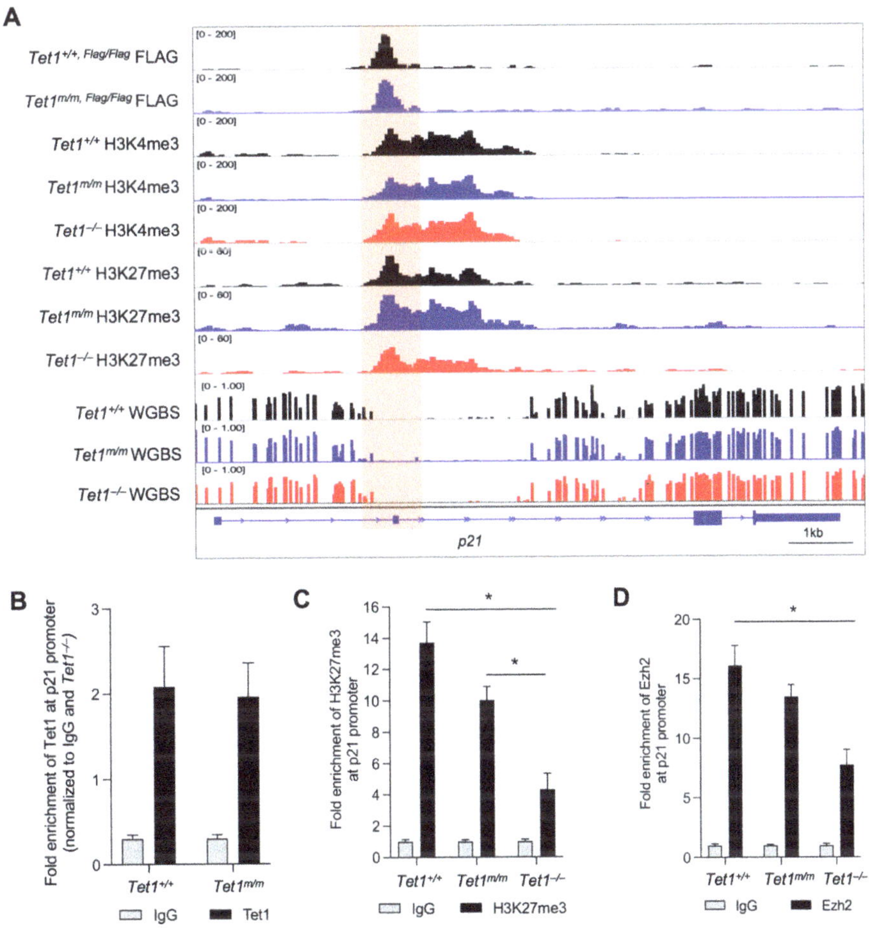

Figure 3. Tet1 silences p21 in mESCs by promoting Ezh2 recruitment to the p21 promoter for H3K27 trimethylation. (**A**) Genome browser tracks displaying co-occupancy of wild type and catalytic mutant Tet1 with bivalent marks (H3K4me3 and H3K27me3) and DNA hypomethylated regions at the p21 locus in mESCs of indicated genotypes using our previously published CUT&Tag and WGBS data in *Tet1*$^{+/+}$, *Tet1*$^{m/m}$ and *Tet1*$^{-/-}$ mESCs (see Methods). Note that the levels of H3K27me3 are reduced in *Tet1*$^{-/-}$ ESCs whereas DNA methylation levels are unaffected at the Tet1-bound p21 locus. (**B–D**) Quantification of enrichment of Tet1 (**B**), H3K27me3 (**C**) and Ezh2 (**D**) at the p21 promoter region in *Tet1*$^{+/+}$, *Tet1*$^{m/m}$ and *Tet1*$^{-/-}$ ESCs (data normalized to 10% input control and IgG). Three replicates of each genotype were used. Note the significant reduction in the levels of H3K27me3 and Ezh2 at the p21 promoter in *Tet1*$^{-/-}$ mESCs. In all panels, data are presented as ± SEM; statistically significant * $p < 0.05$ (two-way ANOVA test).

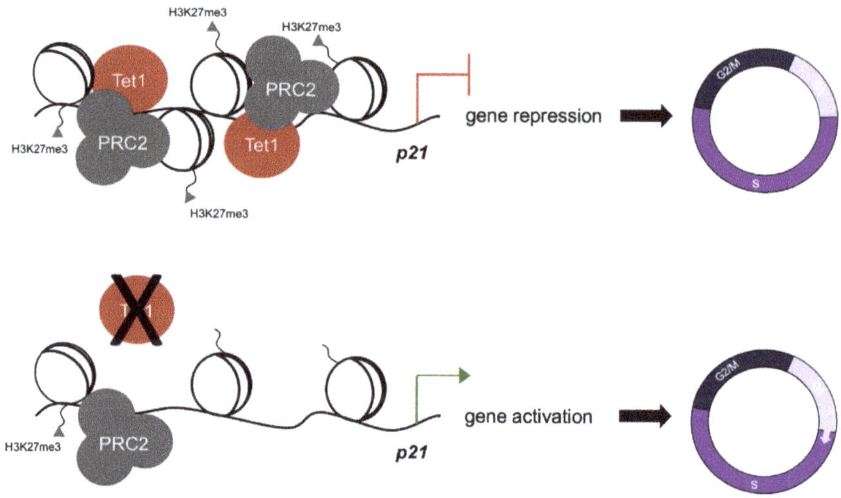

Figure 4. Tet1 regulates mouse ESC cell cycle. Tet1 silences p21 in mESCs by facilitating PRC2 recruitment and H3K27 trimethylation at its promoter to ensure a rapid G1/S transition unique to the pluripotent state of mESCs cultured in serum/LIF.

4. Discussion

Tet1 is highly expressed in mESCs where both its catalytic and non-catalytic functions have been implicated in the regulation of pluripotency and in differentiation programs [2,4,6,14,17]. However, its roles in the regulation of ESC proliferation and cell cycle progression have not been well-studied. mESCs have a unique cell cycle structure consisting of a short G1 phase that allows for a rapid cell division and proliferation [8]. This is, in part, promoted by a lack of expression of CKIs; namely, p21 [8,9]. We provided four lines of evidence that Tet1 regulates ESC proliferation by suppressing p21 independent of its catalytic activity: (1) $Tet1^{-/-}$, but not $Tet1^{m/m}$, ESCs had a reduced proliferation, a delayed G1-S phase progression and an increased expression of p21; (2) the re-expression of catalytic-dead Tet1 or the knockdown of p21 in $Tet1^{-/-}$ ESCs rescued the proliferation defects; (3) Tet1 was enriched at the p21 promoter and its occupancy overlapped with the H3K27me3 repressive mark; and (4) a loss of Tet1, but not a loss of its catalytic activity, reduced H3K27 trimethyltransferase PRC2 occupancy as well as H3K27me3 deposition at the p21 promoter, leading to its aberrant upregulation in ESCs. These findings establish Tet1 as an epigenetic regulator of mESC proliferation in addition to its previously defined roles in mESC pluripotency and differentiation.

It is well-established that p21 expression is largely regulated by a p53-dependent mechanism at the transcriptional level in many cell types [27]. However, Tet1 loss in ESCs did not change the p53 mRNA or protein levels, implying that p21 upregulation is likely to be p53-independent unless a Tet1 deficiency perturbs p53 enrichment at the p21 promoter. Similarly, the mir290-95 cluster—which has been implicated in the regulation of p21 as well as ESC proliferation and pluripotency [24–26]—was not deregulated in $Tet1^{-/-}$ ESCs, suggesting that p21 upregulation is not triggered by the perturbation of this cluster of miRNAs. We found that, in mESCs, p21 is a suppressed bivalent gene, marked by both repressive H3K27me3 and activating H3K4me3 histone modifications, which is consistent with studies in human ESCs where the deposition of the repressive H3K27me3 mark at the p21 promoter has been shown to override activation by p53 [12]. This generates a unique epigenetic strategy to keep certain p53 target genes poised whilst p53 maintains genomic stability in ESCs. Recently, we also showed that Tet1 promotes the establishment of the bivalency of developmental genes, keeping them silenced in mESCs, but poised for activation upon differentiation [6]. The loss of Tet1 leads to an aberrant activation of these genes and differentiation defects [6]. Here, we found that reducing the abnormally high levels of p21 in *Tet1*$^{-/-}$ mESCs by shRNA rescued the cell cycle structure and proliferation defects, but did not correct the aberrant expression of the developmental genes. This is consistent with previous work showing that a p21 overexpression in mESCs does not cause differentiation defects [10]. Together, these suggest that the aberrant differentiation of *Tet1*$^{-/-}$ mESCs is not caused by defects in the cell cycle. We also found that the p21 expression inversely correlated with Tet1 and the pluripotency marker expression. As the mESCs differentiated, the Tet1 levels decreased and the p21 levels increased. This allowed for the establishment of an extended G1 phase pertinent to the differentiated state as the mESCs exited pluripotency. Consistently, p21 expression has been proven to be a roadblock for reprogramming somatic cells to iPSCs and the knockdown of p21 or the selection of cells with a shorter G1 increases reprogramming efficiency [28–30]. Therefore, Tet1, which facilitates reprogramming by promoting pluripotency gene expression programs [31–33], may also play a role in silencing p21 during iPSC generation.

Our findings, implicating Tet1 in the regulation of the mESC cell cycle and proliferation, are in agreement with other studies describing a requirement for Tet1 in the proper regulation of the cell cycle in other cell types, albeit involving different mechanisms. For example, in trophoblast stem cells (TSCs), Tet1 partners with Cyclin B to stabilize its protein levels. The combined loss of Tet1 and Tet2 significantly reduced the proliferation of TSCs [34]. In NIH3T3 cells, a Tet1 loss led to the downregulation of Cyclin D1 and reduced the levels of phosphorylated RB, blocking G1/S entry [35]. We did not observe any deregulation of the key cyclins in *Tet1*$^{-/-}$ mESCs and the p21 knockdown was sufficient to rescue the proliferation defects. Therefore, it is likely that Tet1 influences the cell cycle differently in various cell types. The Tet1-mediated regulation of p21 has also been reported in other cell types. For example, hypercholesterolemia causes Tet1 downregulation in hematopoietic stem cells and leads to the upregulation of p19, p21 and p27 by decreasing the H3K27me3 repressive mark of these genes [19]. Likewise, in human ESCs, p21 is silenced by H3K27 trimethylation at its promoter [12]. In non-small cell lung carcinomas, Tet1 acts as an oncogene and its knockdown leads to the induction of p21 as well as an increased senescence in cancer cells [36]. Finally, we note that the extended G1 phase and reduced proliferation of *Tet1*$^{-/-}$ mESCs, although significant, was not very pronounced, suggesting that other parallel mechanisms may also control the mESC G1/S transition. In conclusion, our findings establish that Tet1, in addition to its previously known function in mESC pluripotency, is an epigenetic regulator of mESC proliferation by suppressing p21 expression. This has implications not only in ESC applications, the reprogramming of iPSCs and development, but also in various cancers where Tet1 is dysregulated.

Supplementary Materials: The following are available online at https://www.mdpi.com/article/10.3390/cells11081366/s1: Figure S1: Apoptosis and cell cycle profiles of *Tet1*$^{+/+}$, *Tet1*$^{m/m}$ and *Tet1*$^{-/-}$ mESCs are analyzed by flow cytometry; Figure S2: mRNA and protein levels of cell cycle regulators are quantified in *Tet1*$^{+/+}$, *Tet1*$^{m/m}$ and *Tet1*$^{-/-}$ mESCs; Table S1: List of oligos used in the study.

Author Contributions: S.C. performed the experiments, analyzed the data and prepared the figures. J.C.F. assisted with the gene expression data analyses. M.M.D. and S.C. designed the study and wrote the manuscript. M.M.D. supervised the study and secured the funding. All authors have read and agreed to the published version of the manuscript.

Funding: This work was supported by NIH R01GM122839 to M.M.D. M.M.D. is also supported by NYSDOH/NYSTEM Contract C32589GG, NIH R01HL148852 and funds from the Albert Einstein College of Medicine Stem Cell Institute and Genetics Department. S.C was in part supported by The Einstein Training Program in Stem Cell Research from the Empire State Stem Cell Fund NYSDOH Contract C30292GG. J.C.F. is supported by NIH F31 predoctoral fellowship award F31GM140554.

Institutional Review Board Statement: Not applicable.

Informed Consent Statement: Not applicable.

Data Availability Statement: This study does not generate datasets that are archived elsewhere. All data generated are shown in this mansucript main figures and supplemental information.

Acknowledgments: We thank the Einstein Flow Cytometry Core for help with the cell cycle and apoptosis analyses and the Dawlaty lab members for helpful discussions and comments on the manuscript.

Conflicts of Interest: The authors declare no conflict of interest.

References

1. Tahiliani, M.; Koh, K.P.; Shen, Y.; Pastor, W.A.; Bandukwala, H.; Brudno, Y.; Agarwal, S.; Iyer, L.M.; Liu, D.R.; Aravind, L.; et al. Conversion of 5-methylcytosine to 5-hydroxymethylcytosine in mammalian DNA by MLL partner TET1. *Science* **2009**, *324*, 930–935. [CrossRef] [PubMed]
2. Ito, S.; D'Alessio, A.C.; Taranova, O.V.; Hong, K.; Sowers, L.C.; Zhang, Y. Role of Tet proteins in 5mC to 5hmC conversion, ES-cell self-renewal and inner cell mass specification. *Nature* **2010**, *466*, 1129–1133. [CrossRef] [PubMed]
3. Pastor, W.A.; Aravind, L.; Rao, A. TETonic shift: Biological roles of TET proteins in DNA demethylation and transcription. *Nat. Rev. Mol. Cell Biol.* **2013**, *14*, 341–356. [CrossRef] [PubMed]
4. Wu, H.; D'Alessio, A.C.; Ito, S.; Xia, K.; Wang, Z.; Cui, K.; Zhao, K.; Sun, Y.E.; Zhang, Y. Dual functions of Tet1 in transcriptional regulation in mouse embryonic stem cells. *Nature* **2011**, *473*, 389–393. [CrossRef] [PubMed]
5. Williams, K.; Christensen, J.; Pedersen, M.T.; Johansen, J.V.; Cloos, P.A.; Rappsilber, J.; Helin, K. TET1 and hydroxymethylcytosine in transcription and DNA methylation fidelity. *Nature* **2011**, *473*, 343–348. [CrossRef] [PubMed]
6. Chrysanthou, S.; Tang, Q.; Lee, J.; Taylor, S.J.; Zhao, Y.; Steidl, U.; Zheng, D.; Dawlaty, M.M. The DNA dioxygenase Tet1 regulates H3K27 modification and embryonic stem cell biology independent of its catalytic activity. *Nucleic Acids Res.* **2022**, *50*, 3169–3189. [CrossRef] [PubMed]
7. Ficz, G.; Branco, M.R.; Seisenberger, S.; Santos, F.; Krueger, F.; Hore, T.A.; Marques, C.J.; Andrews, S.; Reik, W. Dynamic regulation of 5-hydroxymethylcytosine in mouse ES cells and during differentiation. *Nature* **2011**, *473*, 398–402. [CrossRef]
8. White, J.; Dalton, S. Cell cycle control of embryonic stem cells. *Stem Cell Rev.* **2005**, *1*, 131–138. [CrossRef]
9. Ohtsuka, S.; Dalton, S. Molecular and biological properties of pluripotent embryonic stem cells. *Gene Ther.* **2008**, *15*, 74–81. [CrossRef]
10. Li, V.C.; Ballabeni, A.; Kirschner, M.W. Gap 1 phase length and mouse embryonic stem cell self-renewal. *Proc. Natl. Acad. Sci. USA* **2012**, *109*, 12550–12555. [CrossRef]
11. Koledova, Z.; Kramer, A.; Kafkova, L.R.; Divoky, V. Cell-cycle regulation in embryonic stem cells: Centrosomal decisions on self-renewal. *Stem Cells Dev.* **2010**, *19*, 1663–1678. [CrossRef] [PubMed]
12. Itahana, Y.; Zhang, J.; Göke, J.; Vardy, L.A.; Han, R.; Iwamoto, K.; Cukuroglu, E.; Robson, P.; Pouladi, M.A.; Colman, A.; et al. Histone modifications and p53 binding poise the p21 promoter for activation in human embryonic stem cells. *Sci. Rep.* **2016**, *6*, 28112. [CrossRef] [PubMed]
13. Ma, L.; Tang, Q.; Gao, X.; Lee, J.; Lei, R.; Suzuki, M.; Zheng, D.; Ito, K.; Frenette, P.S.; Dawlaty, M.M. Tet-mediated DNA demethylation regulates specification of hematopoietic stem and progenitor cells during mammalian embryogenesis. *Sci. Adv.* **2022**, *8*, eabm3470. [CrossRef] [PubMed]
14. Dawlaty, M.M.; Ganz, K.; Powell, B.E.; Hu, Y.C.; Markoulaki, S.; Cheng, A.W.; Gao, Q.; Kim, J.; Choi, S.W.; Page, D.C.; et al. Tet1 is dispensable for maintaining pluripotency and its loss is compatible with embryonic and postnatal development. *Cell Stem Cell* **2011**, *9*, 166–175. [CrossRef] [PubMed]
15. Dawlaty, M.M.; Breiling, A.; Le, T.; Barrasa, M.I.; Raddatz, G.; Gao, Q.; Powell, B.E.; Cheng, A.W.; Faull, K.F.; Lyko, F.; et al. Loss of Tet enzymes compromises proper differentiation of embryonic stem cells. *Dev. Cell* **2014**, *29*, 102–111. [CrossRef]
16. Vella, P.; Scelfo, A.; Jammula, S.; Chiacchiera, F.; Williams, K.; Cuomo, A.; Roberto, A.; Christensen, J.; Bonaldi, T.; Helin, K.; et al. Tet proteins connect the O-linked N-acetylglucosamine transferase Ogt to chromatin in embryonic stem cells. *Mol. Cell* **2013**, *49*, 645–656. [CrossRef]
17. Koh, K.P.; Yabuuchi, A.; Rao, S.; Huang, Y.; Cunniff, K.; Nardone, J.; Laiho, A.; Tahiliani, M.; Sommer, C.A.; Mostoslavsky, G.; et al. Tet1 and Tet2 Regulate 5-Hydroxymethylcytosine Production and Cell Lineage Specification in Mouse Embryonic Stem Cells. *Cell Stem Cell* **2011**, *8*, 200–213. [CrossRef]
18. Rosello-Diez, A.; Madisen, L.; Bastide, S.; Zeng, H.; Joyner, A.L. Cell-nonautonomous local and systemic responses to cell arrest enable long-bone catch-up growth in developing mice. *PLoS Biol.* **2018**, *16*, e2005086. [CrossRef]
19. Tie, G.; Yan, J.; Khair, L.; Tutto, A.; Messina, L.M. Hypercholesterolemia Accelerates the Aging Phenotypes of Hematopoietic Stem Cells by a Tet1-Dependent Pathway. *Sci. Rep. Investig.* **2020**, *10*, 3567. [CrossRef]
20. Vuong, L.; Brobst, D.E.; Saadi, A.; Ivanovic, I.; Al-Ubaidi, M.R. Pattern of expression of p53, its family members, and regulators during early ocular development and in the post-mitotic retina. *Investig. Ophthalmol. Vis. Sci.* **2012**, *53*, 4821–4831. [CrossRef]
21. Medeiros, L.A.; Dennis, L.M.; Gill, M.E.; Houbaviy, H.; Markoulaki, S.; Fu, D.; White, A.C.; Kirak, O.; Sharp, P.A.; Page, D.C.; et al. Mir-290-295 deficiency in mice results in partially penetrant embryonic lethality and germ cell defects. *Proc. Natl. Acad. Sci. USA* **2011**, *108*, 14163–14168. [CrossRef] [PubMed]

22. Johnson, D.S.; Mortazavi, A.; Myers, R.M.; Wold, B. Genome-wide mapping of in vivo protein-DNA interactions. *Science* **2007**, *316*, 1497–1502. [CrossRef] [PubMed]
23. Fasano, C.A.; Dimos, J.T.; Ivanova, N.B.; Lowry, N.; Lemischka, I.R.; Temple, S. shRNA knockdown of Bmi-1 reveals a critical role for p21-Rb pathway in NSC self-renewal during development. *Cell Stem Cell* **2007**, *1*, 87–99. [CrossRef] [PubMed]
24. Dolezalova, D.; Mraz, M.; Barta, T.; Plevova, K.; Vinarsky, V.; Holubcova, Z.; Jaros, J.; Dvorak, P.; Pospisilova, S.; Hampl, A. MicroRNAs regulate p21(Waf1/Cip1) protein expression and the DNA damage response in human embryonic stem cells. *Stem Cells* **2012**, *30*, 1362–1372. [CrossRef] [PubMed]
25. Wang, Y.; Baskerville, S.; Shenoy, A.; Babiarz, J.E.; Baehner, L.; Blelloch, R. Embryonic stem cell-specific microRNAs regulate the G1-S transition and promote rapid proliferation. *Nat. Genet.* **2008**, *40*, 1478–1483. [CrossRef] [PubMed]
26. Lichner, Z.; Pall, E.; Kerekes, A.; Pallinger, E.; Maraghechi, P.; Bosze, Z.; Gocza, E. The miR-290-295 cluster promotes pluripotency maintenance by regulating cell cycle phase distribution in mouse embryonic stem cells. *Differentiation* **2011**, *81*, 11–24. [CrossRef] [PubMed]
27. Gartel, A.L.; Tyner, A.L. Transcriptional regulation of the p21((WAF1/CIP1)) gene. *Exp. Cell Res.* **1999**, *246*, 280–289. [CrossRef]
28. Roccio, M.; Schmitter, D.; Knobloch, M.; Okawa, Y.; Sage, D.; Lutolf, M.P. Predicting stem cell fate changes by differential cell cycle progression patterns. *Development* **2013**, *140*, 459–470. [CrossRef]
29. Hong, H.; Takahashi, K.; Ichisaka, T.; Aoi, T.; Kanagawa, O.; Nakagawa, M.; Okita, K.; Yamanaka, S. Suppression of induced pluripotent stem cell generation by the p53-p21 pathway. *Nature* **2009**, *460*, 1132–1135. [CrossRef]
30. Hanna, J.; Saha, K.; Pando, B.; van Zon, J.; Lengner, C.J.; Creyghton, M.P.; van Oudenaarden, A.; Jaenisch, R. Direct cell reprogramming is a stochastic process amenable to acceleration. *Nature* **2009**, *462*, 595–601. [CrossRef]
31. Costa, Y.; Ding, J.; Theunissen, T.W.; Faiola, F.; Hore, T.A.; Shliaha, P.V.; Fidalgo, M.; Saunders, A.; Lawrence, M.; Dietmann, S.; et al. NANOG-dependent function of TET1 and TET2 in establishment of pluripotency. *Nature* **2013**, *495*, 370–374. [CrossRef] [PubMed]
32. Chen, J.; Guo, L.; Zhang, L.; Wu, H.; Yang, J.; Liu, H.; Wang, X.; Hu, X.; Gu, T.; Zhou, Z.; et al. Vitamin C modulates TET1 function during somatic cell reprogramming. *Nat. Genet.* **2013**, *45*, 1504–1509. [CrossRef] [PubMed]
33. Gao, Y.; Chen, J.; Li, K.; Wu, T.; Huang, B.; Liu, W.; Kou, X.; Zhang, Y.; Huang, H.; Jiang, Y.; et al. Replacement of Oct4 by Tet1 during iPSC induction reveals an important role of DNA methylation and hydroxymethylation in reprogramming. *Cell Stem Cell* **2013**, *12*, 453–469. [CrossRef] [PubMed]
34. Chrysanthou, S.; Senner, C.E.; Woods, L.; Fineberg, E.; Okkenhaug, H.; Burge, S.; Perez-Garcia, V.; Hemberger, M. A Critical Role of TET1/2 Proteins in Cell-Cycle Progression of Trophoblast Stem Cells. *Stem Cell Rep.* **2018**, *10*, 1355–1368. [CrossRef] [PubMed]
35. Huang, S.; Zhu, Z.; Wang, Y.; Wang, Y.; Xu, L.; Chen, X.; Xu, Q.; Zhang, Q.; Zhao, X.; Yu, Y.; et al. Tet1 is required for Rb phosphorylation during G1/S phase transition. *Biochem. Biophys. Res. Commun.* **2013**, *434*, 241–244. [CrossRef] [PubMed]
36. Filipczak, P.T.; Leng, S.; Tellez, C.S.; Do, K.C.; Grimes, M.J.; Thomas, C.L.; Walton-Filipczak, S.R.; Picchi, M.A.; Belinsky, S.A. p53-Suppressed Oncogene TET1 Prevents Cellular Aging in Lung Cancer. *Cancer Res.* **2019**, *79*, 1758–1768. [CrossRef]

cells

MDPI

Article

Myogenic Precursor Cells Show Faster Activation and Enhanced Differentiation in a Male Mouse Model Selected for Advanced Endurance Exercise Performance

Stefan Petkov [1,†], Julia Brenmoehl [2,*,†], Martina Langhammer [3], Andreas Hoeflich [2] and Monika Röntgen [1,*]

1 Institute of Muscle Biology and Growth, Research Institute for Farm Animal Biology (FBN), Wilhelm-Stahl-Allee 2, 18196 Dummerstorf, Germany; petkov@fbn-dummerstorf.de
2 Institute of Genome Biology, Research Institute for Farm Animal Biology (FBN), Wilhelm-Stahl-Allee 2, 18196 Dummerstorf, Germany; hoeflich@fbn-dummerstorf.de
3 Lab Animal Facility, Institute of Genetics and Biometry, Research Institute for Farm Animal Biology (FBN), Wilhelm-Stahl-Allee 2, 18196 Dummerstorf, Germany; martina.langhammer@fbn-dummerstorf.de
* Correspondence: brenmoehl@fbn-dummerstorf.de (J.B.); roentgen@fbn-dummerstorf.de (M.R.)
† These authors contributed equally to this work.

Abstract: Satellite cells (SATC), the most abundant skeletal muscle stem cells, play a main role in muscle plasticity, including the adaptive response following physical activity. Thus, we investigated how long-term phenotype selection of male mice for high running performance (Dummerstorf high Treadmill Performance; DUhTP) affects abundance, creatine kinase activity, myogenic marker expression (Pax7, MyoD), and functionality (growth kinetics, differentiation) of SATC and their progeny. SATC were isolated from sedentary male DUhTP and control (Dummerstorf Control; DUC) mice at days 12, 43, and 73 of life and after voluntary wheel running for three weeks (day 73). Marked line differences occur at days 43 and 73 (after activity). At both ages, analysis of SATC growth via xCELLigence system revealed faster activation accompanied by a higher proliferation rate and lower proportion of Pax7+ cells in DUhTP mice, indicating reduced reserve cell formation and faster transition into differentiation. Cultures from sedentary DUhTP mice contain an elevated proportion of actively proliferating Pax7+/MyoD+ cells and have a higher fusion index leading to the formation of more large and very large myotubes at day 43. This robust hypertrophic response occurs without any functional load in the donor mice. Thus, our selection model seems to recruit myogenic precursor cells/SATC with a lower activation threshold that respond more rapidly to external stimuli and are more primed for differentiation at the expense of more primitive cells.

Keywords: muscle stem cells; voluntary physical activity; proliferation; myogenic differentiation; Pax7; MyoD

Citation: Petkov, S.; Brenmoehl, J.; Langhammer, M.; Hoeflich, A.; Röntgen, M. Myogenic Precursor Cells Show Faster Activation and Enhanced Differentiation in a Male Mouse Model Selected for Advanced Endurance Exercise Performance. *Cells* **2022**, *11*, 1001. https://doi.org/10.3390/cells11061001

Academic Editors: Mehdi Najar and Michael Deschenes

Received: 11 January 2022
Accepted: 14 March 2022
Published: 16 March 2022

1. Introduction

Skeletal muscle is mainly composed of multinucleated myofibers. In mice, as in other species, the number of myofibers is fixed at birth, and postnatal muscle growth is achieved by hypertrophic fiber growth [1]. This process is most intense during the first three weeks of life and contributes to half of the seven- to eight-fold increase in body weight that occurs during this period in mice [2]. Muscle progenitor cells, so-called satellite cells (SATC), are essentially involved in this most dynamic phase of postnatal growth [3]. After this period, the adult number of SATC and myonuclei is already established in mice [3]. The size and composition of the produced SATC pool greatly affect postnatal muscle growth, maintenance, and repair [4]. Therefore, the number of myogenic precursor cells (MPC)/SATC and their molecular/functional properties play a significant role in muscle development, plasticity, and regeneration [4].

SATC are heterogeneous, and, as in other species, two distinct SATC populations have been identified in mice, including committed progenitors responsible for muscle growth and routine maintenance (≈90%) and reserve SATC (≈10%) [5–7]. The proportion of SATC decreased with age, from about 30% at birth to 13 and 3% at postnatal weeks two and seventeen [8–10]. During the first three postnatal weeks, however, a high percentage (80%) of these SATC is proliferating [1,11], whereas, in the adult, most SATC (97–99%, [12]) are quiescent but poised to activation. SATC determination, functional stages, and fate are regulated by expressing several characteristic myogenic regulatory factors that also participate in cell cycle regulation, hypertrophy, and fiber type determination [13].

For juvenile myogenic progenitors (until about day 21 of life), paired box protein 7 (Pax7) expression is critical for survival, maintenance of their expansive and myogenic capacity, and transition into the adult, quiescent state [14–16]. The latter process achieves the regenerative capacity of muscles [16]. During quiescence, Pax7 and myogenic factor 5 (Myf5) are expressed, whereas myoblast determination protein 1 (MyoD) is induced in activated/proliferating myoblasts [17–19]. Under in vivo conditions, juvenile SATC progenitors of mice are actively proliferating and contribute to the extensive myofiber growth by the net addition of myonuclei. Myoblasts expressing both Pax7 and MyoD form the largest group in the proliferating population [20].

Myf5 and/or MyoD expressing cells can return to quiescence to maintain the SATC pool. Transition into differentiation is initiated by downregulation of Pax7 and increased expression of MyoD, leading to induction of myogenin (MyoG) (early differentiation; myocytes). In vitro, differentiation can be induced by serum starvation and comprises a sequence of highly ordered events: after myogenin expression (mark cells as irreversibly committed to differentiation), permanent cell cycle withdrawal is induced by p21 [21], followed by the production of various contractile proteins including sarcomeric myosin heavy chain (MyHC) proteins (phenotypic differentiation) and fusion [17,22,23]. Fusion consists of two distinct stages, namely, initial myoblast-myoblast fusion to establish nascent myotubes and subsequent myoblast-myotube fusion to increase myotube size [24].

Muscle SATC are essential for muscle repair/regeneration, but their involvement in muscle atrophy and hypertrophy is not fully understood [25]. SATC are known to be activated in muscles to proliferate during muscle hypertrophy and to be incorporated into myofibers, increasing the number of myonuclei. This "myonuclei domain theory" is advocated by many researchers and states that increasing myofiber size requires new myonuclei from SATC to maintain the ratio of myonuclei to cytoplasm, thereby increasing muscle hypertrophy and weight [26,27]. Interestingly, it has been shown in a SATC-depleted mouse model (Pax7-DTA) [20] that SATC-independent muscle fiber hypertrophy is possible [28] without an increase in the number of myonuclei [26,29].

Besides intrinsic signals, skeletal muscle and SATC functionality/phenotype are affected by environmental stimuli such as nutrients, growth factors, injury, and muscle use, e.g., physical exercise [30–32]. Physical activity is critical for muscle mass development and maintenance and, thus, its functionality. By studying SATC in vivo, a positive effect of physical exercise on these cells was demonstrated. Voluntary running wheel (RW) activity for six or eight weeks leads to increased SATC numbers, identified by Pax7 staining, and improves the differentiation capacity in rats' plantaris [33] or vastus lateralis and medial gastrocnemius muscles [34]. Kurosaka et al. could also show a positive correlation between the percentage of SATC and the running distance [33]. Also, enhanced proliferation in murine musculus plantaris and musculus soleus was reported in response to four-week RW activity [35]. The use of genetically labeled SATC impressively visualized in vivo that more SATC fused with fibers in the musculus soleus, musculus plantaris, and musculus gastrocnemius after eight weeks of voluntary RW activity than under sedentary conditions [36]. Long-term moderate-intensity treadmill running (13 weeks) increases the number of SATC per myofiber in the musculus gastrocnemius of female and male rats at the age of 3.5 and 15 months, respectively [37]. Studies on proliferation and the percentage distribution of quiescent, proliferating, and activated SATC isolated from the limb muscles of sedentary mice of different ages or after voluntary activity do not exist to our knowledge.

An excellent model to study the features of SATC in mice is provided by the non-inbred Dummerstorf marathon mouse model DUhTP and its associated control line. Due to paternal phenotype selection over 140 generations, the line DUhTP has a genetically fixed high running ability without previous training [38,39] accompanied by diminished voluntary running wheel activity [40]. A series of experiments [41–43] revealed marked modifications in the energy metabolism in liver and subcutaneous adipose tissue of these DUhTP mice, specifically in fat accumulation and mobilization. In sedentary conditions, they accumulate high amounts of body fat, while voluntary activity in RW completely abolished the obese phenotype, indicating the physiological relevance of prominent mitochondrial oxidation for superior endurance exercise performance of DUhTP mice [42]. Effects of long-term selection for high treadmill performance on MPC/SATC abundance, functional properties and fate, and their role in the muscular adaptive response to exercise have never been investigated in DUhTP mice.

SATC number and, thus, bioavailability is a main determinant of postnatal muscle growth, muscle homeostasis by replenishment of aging myonuclei [4], and for muscle regeneration after injuries. We assume that the number of MPC/SATC is higher in DUhTP mice and determined SATC abundance (as SATC number per gram muscle) to identify possible age-dependent and/or line-dependent differences and the effect of voluntary activity.

Regarding SATC functionality, we hypothesize that MPC from DUhTP mice show an increased ability to proliferate and differentiate, thereby contributing to their superior forced running capacity. To include different stages (juvenile, proliferating; maturation, transition to quiescence; adult, quiescent) [3,44] of postnatal SATC development, cells were isolated from the limb muscles of male mice of different ages (12, 43, 73 d).

2. Materials and Methods

2.1. Mouse Line Establishment

The Dummerstorf mouse lines have been initially generated by systematic crossbreeding of four inbred (CBA/Bln, AB/Bln, C57BL/Bln, and XVII/Bln) and four outbred lines (NMRI orig., Han:NMRI, CFW, and CF1) to obtain a mouse strain with a broad genetic background [45]. From this genetic pool (FztDU), the control line Dummerstorf Control mouse line (DUC) has been established by random mating. The mouse line DUhTP used in the present study had been generated from the same base population by paternal selection for high treadmill performance for 140 generations, four generations per year [38]. Therefore, submaximal high treadmill performance was determined after mating in around 77-day-old male mice, enhancing the selection process's speed. Treadmill performance was determined by a singular submaximal test (0% incline). According to the protocol, mice performed first an initial run for 30 m at a speed of 12 m/min (≙150 s). Then, after stopping the treadmill for 1 min and a second run for 20 m at 12 m/min (≙100 s), the treadmill speed was gradually increased up to a maximum speed of 38 m/min. The stepwise increases consisted of running distances of 50 m each at 22 m/min (≙136 s), 26 m/min (≙115 s), 30 m/min (≙100 s), 34 m/min (≙88 s), and 36 m/min (≙83 s)). Thus, the terminal velocity was reached after 300 m (≙832 s). The test was terminated when the mice repeatedly rested on the stimulation device at the end of the treadmill. The endurance fitness was recorded as running distance in meters on a treadmill. The offspring of the males with the highest running performance were selected as parents for the next generation. By this strategy, a direct effect of the physical performance itself on breeding performance was excluded. Instead, a genetically fixed higher forced running ability without previous training was generated. Both lines, DUhTP and DUC, were maintained by minimizing inbreeding. For this purpose, the control line was bred with 125–200 breeding pairs and the DUhTP line with 60–100 breeding pairs per generation during the whole selection period.

Compared to the animals at the selection start, a selection success of 400% was achieved in DUhTP mice. DUhTP mice run longer than DUC controls, covering a 3.8-fold longer running distance than the unselected control line (DUhTP: 5832.1 ± 838.2 m, DUC:

1537.3 ± 200.5 m). Both lines are non-inbred strains characterized by high variability between individuals of one generation or different generations.

2.2. Animals and Study Design

All in vivo experiments were performed at the Research Institute of Farm Animal Biology (FBN) in Dummerstorf. Animal husbandry in the Laboratory for Innovative Farm Animal Models (LIN) of the FBN and slaughter followed the guidelines of the Animal Care Committee of the State Mecklenburg-Western Pomerania, Germany, based on the German Law of Animal Protection (Animal Welfare Act; TierSchG), and were approved by our internal institutional review board. As animals were not manipulated before slaughter, no animal experiment was conducted according to the German Animal Welfare Act.

Mice of the lines DUhTP and DUC were kept in H-Temp Polysulfon cages (floor area: 370 cm^2; Eurostandard Type II, Tecniplast, Hohenpeißenberg, Germany) in specified pathogen-free conditions and provided with fresh water and autoclaved Ssniff® M-Z feed (Ssniff-Spezialdiäten GmbH, Soest, Germany) ad libitum. After weaning, males of both lines were randomly assigned to different groups and kept in single cages until days 43 and 73 of life, respectively (Figure 1). One half of the latter group had access to an RW in the cage (activity wheel for rats; Tecniplast, Hohenpeißenberg, Germany) from 52 days of age, which they could use voluntarily for three weeks, while the second half was kept as sedentary controls in cages without RW. Voluntary physical activity was registered daily by a wheel counter that recorded all quarter rotations (Figure 1). Based on the RW activity (quarter rounds per day), individual activity was calculated, where one complete revolution of the wheel (diameter = 33.4 cm) corresponded to a running distance of 1 m.

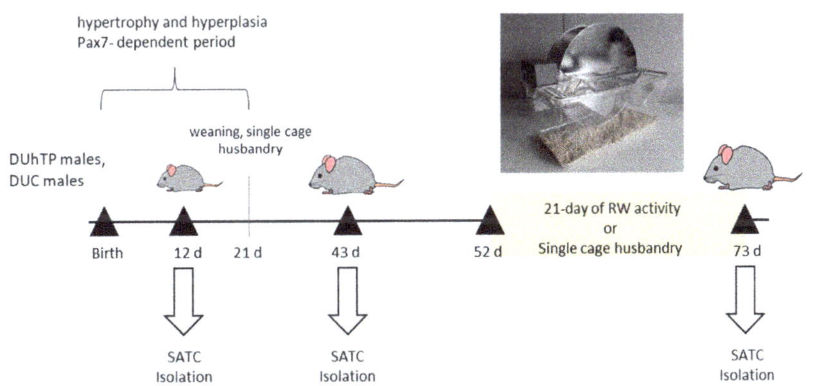

Figure 1. Experimental design. For the isolation of primary myogenic precursor cells/satellite cells from whole limb muscles, male DUC and DUhTP mice of different ages (12, 43, and 73 days) were used. Seventy-three-day-old mice either had access to a running wheel (RW) with a wheel counter (Tecniplast, Hohenpeißenberg, Germany) from day 52 of age or were kept entirely under sedentary conditions.

On days 12, 43, or 73, the mice were weighed, sacrificed, and total limb muscle masses were determined after removing surrounding and intermuscular fat. A total of 94 DUC animals (12 d DUC: 20, 43 d DUC: 22, 73 d DUC: 27, 73 d DUC act: 25) and 79 DUhTP mice (12 d DUhTP: 16, 43 d DUhTP: 25, 73 d DUhTP: 22, 73 d DUhTP act: 16) were included in the experiment and selected for different trials. The mice were from a total of six generations (DUC: 190–193, 195–196; DUhTP: 145–148, 150–151) between May 2018 and November 2019, with three to twelve animals per age or activity group.

2.3. Satellite Cell Isolation and Culture

Total limb muscles (musculus quadriceps femoris, musculus biceps femoris, musculus triceps brachii, extensors, and flexors of the elbow) free of fat and tendons were used for MPC/SATC isolation (Figure 1). Muscles were cut into small pieces, transferred to an enzyme mix containing 2% collagenase Type I (Collagenase from *Clostridium histolyticum*, Sigma-Aldrich, St. Louis, MO, USA) and 2% dispase (Dispase II from *Bacillus polymoxa*, Roche, Mannheim, Germany), and incubated three times for 10 min with intervening mixing/shearing at 37 °C. The suspension was filtered through a 70 µm filter and then incubated at room temperature with a cocktail of monoclonal antibodies against non-target cells (Satellite Cell isolation kit mouse, Miltenyi Biotec, Bergisch Gladbach, Germany) conjugated to MACS® MicroBeads. After 30 min, the cell suspension was filtered through a 50 µm filter. The magnetically labeled non-target cells were depleted by retaining them within the MACS Column in the magnetic field of a MACS Separator, while the unlabeled SATC passed through the column. Cell size and viability were measured (Countess Automatic cell counter, Thermo Fisher Scientific, MA, USA), and the number of live cells was calculated as cells per gram of muscle (10^5). After isolation, cells have spent 40 min in untreated and uncoated dishes (Corning Incorporated, Durham, NC, USA) with growth medium (HAMS F10, 20% Fetal bovine serum (FBS), 2% Penicillin-Streptomycin solution, 1% Amphotericin; PAN Biotech, Aidenbach, Germany) in order to remove fibroblasts (pre-plating). Fibroblasts are heavier than SATC and attach faster to the dish bottom. After pre-plating, floating MPC/SATC were harvested, seeded in collagen type I (Greiner Bio-one, Monroe, NC, USA) coated dishes, and cultured in growth medium under a humidified atmosphere with 5% CO_2 at 37 °C. Twenty-four hours after seeding, the cells were washed with Dulbecco's phosphate-buffered saline (DPBS, PAN Biotech). A fresh growth medium was added, and cells were cultured for three days before renewing the medium. Bacterial and fungal contamination of cells was excluded via inoculation of CASO Bouillon Tryptic Soy Broth and Thioglycolate medium EP (Becton Dickinson, Heidelberg, Germany). For passaging, cultured MPC/SATC were detached by using Accutase cell detachment solution (PAN Biotech), and the reaction was stopped by adding growth media. After centrifugation (5 min, 453 g, 22 °C), cells were re-suspended in the growth medium, and the cell number was determined. The cell morphology was evaluated using a photonic microscope Carl Zeiss Primovert Ser. No. 38420210210.

After the first passage, part of the cells from single mice was used to determine the myogenic markers Pax7 and MyoD and to measure creatine kinase activity (see below). The numbers of animals used per experiment are indicated individually. The magnetic cell separation approach used for SATC isolation in the present study led to a strong 46-fold enrichment of this cell population. Nevertheless, as a control of the myogenic origin of the cells, differentiation assays (see below) were also performed, while the rest of the cells were frozen and stored until making cell pools for further experiments (shown in this manuscript: kinetic growth curves; xCELLigence system, fusion index, areas covered by myotubes, myofiber number, and size distribution). A mixture consisting of 50% growth medium, 30% FBS, and 20% dimethyl sulfoxide (DMSO; Carl Roth, Karlsruhe, Germany) was used as the freezing medium.

After sampling cells over four to five generations, they were thawed to create cell pools. Cells of one age group but from different generations were pooled to obtain 200,000 cells per vial. Care was taken to ensure a similar proportion of cells from individual animals per pool. Therefore, pools consisted of a minimum of three and a maximum of nine different mice. Cell pools were grown to get a maximum of cell yield and, after that, finally seeded for the corresponding assay (e.g., xCELLigence, differentiation) or frozen for biochemical or metabolic studies (results are not part of this manuscript). The numbers of pools and the numbers of individual mice are thus indicated individually.

2.4. xCELLigence

In order to compare the functionality of MPC/SATC isolated from male DUC and DUhTP mice, growth kinetics, adhesion, and proliferation properties were continuously recorded over 72 h using the xCELLigence system (RTCA-SP, ACEA Biosciences Inc., San Diego, CA, USA), which measures impedance changes caused by the attachment and spreading of cells and calculates them as the dimensionless Cell Index (CI).

Therefore, forty thousand cells per well in quadruplicates were used in a 96-well-plate to assess quantitative adhesion and proliferation capabilities. The parameters slope (expressing the adhesion kinetics; AP), maximal cell index (highest CI value, usually associated with the stationary growth phase, here: end point value of the CI; CImax), and doubling time (DT, a measure of proliferation rate) were used and are shown exemplarily in Figure 2 [46,47].

More details on the xCELLigence technology can be found in Atienza et al. (2005) or Ke et al. (2011).

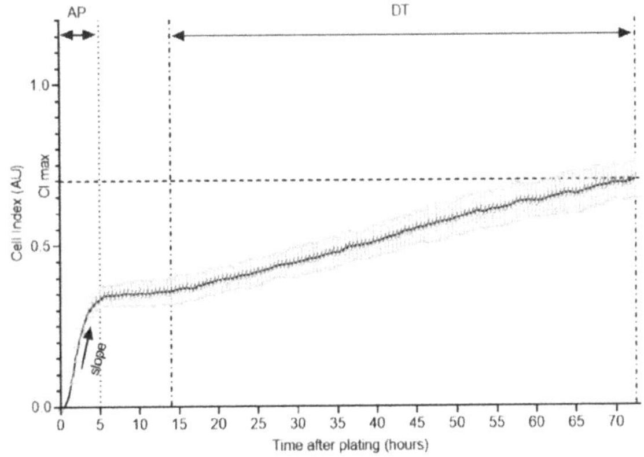

Figure 2. Schematic presentation of xCELLigence system experiment. The growth curves were analyzed within 72 h for slope during the adhesion period (AP), cell index (CI max), and doubling time (DT) during the logarithmic growth phase.

2.5. Immunofluorescence Staining for Myogenic Markers Pax7 and MyoD

MPC/SATC originating from single animals or cell pools were seeded at a density of 50,000 per well on a collagen-coated 24-well plate and cultured in the growth medium with 20% FBS (see Section 2.3) for 24 h. The cells were fixed with 4% Paraformaldehyde (PFA), then permeabilized with 0.1% Triton X100 in DPBS, and blocked with 5% Horse serum (HS, PAN Biotech) at room temperature. The primary anti-MyoD antibody (rabbit, Thermofisher Scientific) was 1:250 diluted in undiluted Hybridoma Mouse anti-Pax7 antibody (Developmental Studies Hybridoma Bank). Cells were incubated with both primary antibodies overnight at 4 °C. The next day, cells were washed with DPBS to remove unbounded primary antibodies and incubated for 45 min at room temperature with the secondary antibodies, diluted 1:1000 in 5% HS in DPBS (MyoD: Alexa 488 donkey anti-rabbit, Pax7: Alexa 546 goat anti-mouse; Thermo Fisher Scientific).

Pictures of the cells were taken with a fluorescent microscope Nikon Diaphot 300 (Nikon Corporation, Tokyo, Japan). To determine the total cell number, six pictures per sample, randomly chosen, were used. The software Photoshop CS6 (Adobe Photoshop Version: 13.0.1, Adobe Inc., San Jose, CA, USA) was used to create the overlay of pictures to adjust the brightness and contrast. Software Image J 2.0.0 Java 1.6 (National Institutes of Health (NIH), Bethesda, MD, USA) was used for quantitative analysis. While the total number of nuclei was counted automatically, nuclei, positive for Pax7 (Pax7+/MyoD−), MyoD (Pax7−/MyoD+), or both (Pax7+/MyoD+), were counted manually.

2.6. Differentiation of SATC

For myogenic differentiation, cells from single animals (passage one) or cell pools were seeded in Matrigel Growth Factor Reduced Basement Membrane Matrix (diluted 1:50, Corning) coated 24-well plates (50,000 cells/well) and cultured in growth medium. When cells started to show signs of spontaneous differentiation, differentiation was induced. After removing the growth medium, cells were washed with 37 °C DPBS, and the differentiation medium was added (PAN Biotech, DMEM with high glucose, 4.5 g/L, 5% HS, 2% Penicillin-Streptomycin). Experiments were performed 72 h after induction.

2.7. Immunofluorescence for Myosin Heavy Chain and Determination of the Fusion Index

For myosin heavy chain (MHC) staining, cells were fixed with 4% PFA and permeabilized with 0.1% TritonX100 in PBS. After blocking with 5% HS for 1 h, myotubes were incubated overnight with a primary mouse anti-skeletal myosin antibody (MF20 hybridoma) in DPBS. Subsequently, primary antibodies were removed by washing with DPBS. Then, myotubes were incubated with an Alexa 488-conjugated rabbit anti-mouse secondary antibody (Life Technologies, MA, USA), diluted 1:1000, and stained with DAPI (Sigma Aldrich). For imaging, the Nikon Diaphot 300 microscope with 10× magnification was used.

The pictures' overlay was created using the software Photoshop CS6 (Version CS6, Adobe Inc., CA, USA); brightness and contrast were adjusted to the same degree in every sample group. MHC positive myotubes containing ≥2 nuclei were encircled to quantify the myotube area using the Software Image J 2.0.0 Java 1.6 (NIH). The total number of cell nuclei was counted automatically. For each experiment, six random sections were analyzed.

Micrographs of the MHC-stained cells were also used to determine the fusion index, according to Miersch et al. [48]. The number of nuclei in the fused multinucleated cells (≥2 nuclei) was divided by the total number of visible nuclei for six randomly chosen pictures per sample and multiplied by 100 to calculate the fusion index.

2.8. Creatine Kinase (CK) Measurement

Frozen MPC/SATC were thawed and differentiated for three days before CK activity was determined using the CK-NAH hit kit (IFCC method, BIOMED Labordiagnostik GmbH, City, Germany) according to the manufacturer's instructions.

2.9. Statistical Analyses

Data from xCELLigence system are presented as means ± standard deviation (SD), data concerning the distribution of myotube area and Pax7 and/or MyoD expressing cells are demonstrated as stacked columns, whereas all other data are shown as Box-Whisker plots with the median. Statistical analyses were performed using Graph Pad Prism (Version 9.2, Graph Pad Software, San Diego, CA, USA) and SigmaPlot 14.0 (Systat Software Inc, Chicago, IL, USA). Data from cell isolation parameters, body weight, muscle mass, recording of kinetic growth curves (xCELLigence system experiment), and immunostaining of myogenic markers were tested with one-way ANOVA, Tukey's multiple comparison test. Data from differentiation assays, quantification of fusion index, sedentary live cell yield comparison, and creatine kinase measurements were tested with one-way ANOVA, Tukey's multiple comparison test and unpaired *t*-test with Welch's corrections. Statistical differences were considered single factorial between the lines at the same age (line-specific effects) or within each mouse line at different ages (age-related effects) or physical activity status (physical activity-related effects). A *p*-value of <0.05 was considered to be statistically significant.

3. Results

3.1. Voluntary Running Performance, Body and Muscle Weight, and Satellite Cell Yield

The voluntary activity level of DUC mice was 2.5-fold higher than that of DUhTP mice (DUC: 5684 ± 2694 m (n = 17), DUhTP: 2205 ± 1279 m (n = 16); $p < 0.001$).

From day 43 on, DUhTP mice had significantly lower body weight and muscle mass than DUC mice (Figure 3a,b). Since the body and muscle weights on day 12 did not differ significantly between the two lines, the DUC phenotypically showed a significantly higher growth capacity by about 30% than the DUhTP line until 43 days of age. After that, there was no difference in growth intensity. In response to voluntary physical activity, bodyand muscle mass were exclusively reduced in DUC mice ($p < 0.05$).

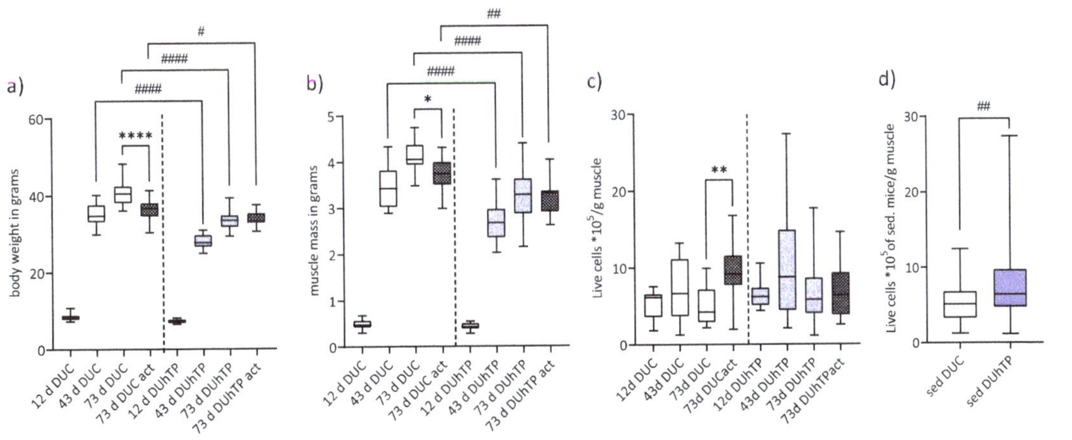

Figure 3. (**a**) Body weight and (**b**) muscle mass of all limbs of 12-, 43-, and 73-day-old male DUC (white bars) and DUhTP mice (blue bars), with older mice either sedentary or active in a running wheel (act). A total of 85 DUC (12 d: 20, 43 d: 22, 73 d: 26, 73 d act: 17) and 72 DUhTP mice (12 d: 16, 43 d: 18, 73 d: 22, 73 d act: 16) were included in the analyses. Satellite cells were isolated from whole limb muscles, and (**c**–**d**) the cell yield per gram of muscle was considered (**c**) for each age (DUC 12 d: 16, 43 d: 22, 73 d: 22, 73 d act: 17 mice; DUhTP 12 d: 16, 43 d: 18, 73 d: 22, 73 d act: 16 mice) and activity group or (**d**) holistically from all sedentary (sed) mice of a mouse line (DUC: $n = 68$ mice; DUhTP: $n = 56$ mice). Results are shown as means and SD, and statistical analyses were performed using one-way ANOVA (**a**–**c**) or unpaired t-test with Welch's correction. Significant line-specific differences (DUC versus DUhTP mice) are marked by hashtags ($^{\#}$ $p < 0.05$; $^{\#\#}$ $p < 0.01$; $^{\#\#\#\#}$ $p < 0.0001$), while activity-induced differences (act versus sedentary) within the mouse lines are indicated by asterisks (* $p < 0.05$; ** $p < 0.01$; **** $p < 0.0001$).

SATC yield obtained from limb muscles was similar in male DUC and DUhTP mice at different ages (Figure 3c). However, a generally higher SATC yield isolated from the muscles of sedentary DUhTP mice was observed compared to sedentary controls (Figure 3d; $p < 0.005$). Furthermore, a strong activity-induced increase in total cell yield was observed in DUC mice (Figure 3c; $p < 0.01$).

3.2. Growth Kinetics of Isolated Satellite Cells

The isolated MPC/SATC were analyzed for their characteristic growth profile, typified by unique morphology, attachment, and proliferative behavior [49], using the xCELLigence system. Continuous real-time growth curves of myoblasts from male DUC and DUhTP mice were recorded over 72 h and are shown in Figure 4a,b, respectively. Kinetic CI traces from myoblasts of all DUC ages were characterized by a rapid adhesion followed by a distinct lag phase and a period of logarithmic growth. No plateau was reached during the measurements. CI curves of myoblasts from 12-day-old and sedentary 73-day-old DUC mice (Figure 4a; green and orange lines) were similar to those of coeval DUhTP mice (Figure 4b; green and orange lines). However, clear differences between lines were found at day 43 (blue lines) and when mice had performed voluntary activity (red lines). Typically, shortening of the lag phase and faster entry into logarithmic growth were observed when myoblasts from DUhTP mice were used (Figure 4b). The ability to proliferate (Figure 4c) and attach (Figure 4d) was clearly higher in myoblast from 43-day-old DUC mice (CImax 2.12 ± 0.02 AU, slope 0.70 ± 0.07 AU/h) compared to cells from 12-day-old (CImax 0.7 ± 0.6 AU, slope 0.30 ± 0.03 AU/h) and 73-day-old (CImax 1.15 ± 0.01 AU, slope 0.45 ± 0.03 AU/h) DUC mice (all $p < 0.0001$) or from myoblasts of 43-day-old DUhTP mice (CI 0.88 ± 0.06 AU, slope 0.10 ± 0.03 AU/h, $p < 0.0001$). DT was slower in 43-day-old DUC than 73-day-old DUC cells ($p < 0.0001$; Figure 4e). In contrast, cells from 43-days-old DUhTP mice had a faster DT than those from 12- or 73-day-old DUhTP mice ($p < 0.05$).

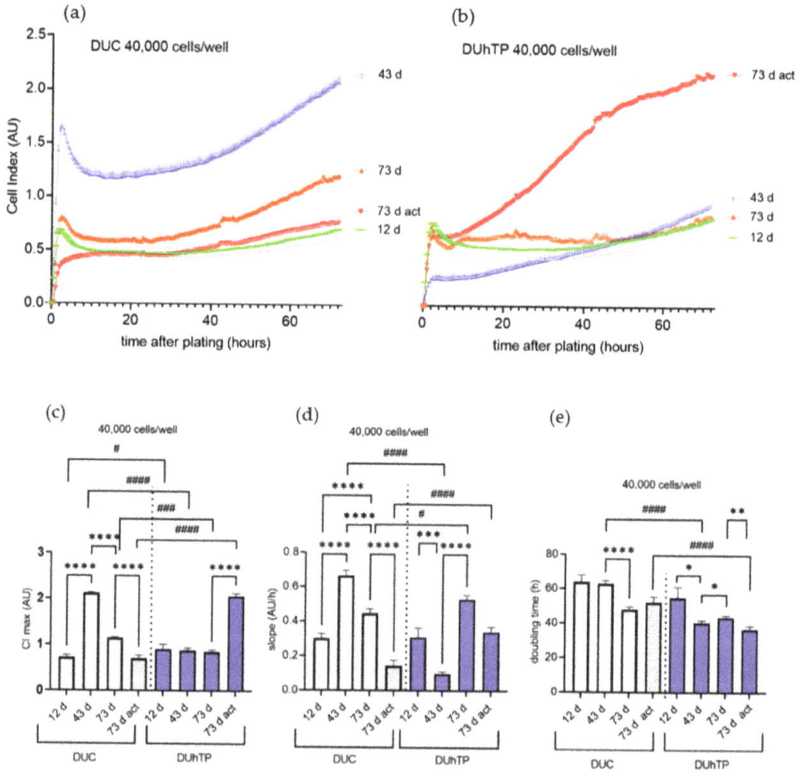

Figure 4. Cultured pools of MPC/SATC from male DUC (left and in white; pools (n = 3–4) represent: 12 d: 6 mice, 43 d: 16 mice, 73 d: 16 mice, 73 d act: 12 mice) and DUhTP animals (right and in blue; pools (n = 3) represent: 43 d: 9 mice, 12 d, 73 d, and 73 d act: 8 mice) were derived from sedentary animals at the age of 12, 43, 73 days (**d**) and physically active (act) animals at the age of 73 days. The impedance-based xCELLigence system was used for real-time monitoring of adhesion and proliferation of SATC populations seeded at densities of 40,000 cells per well. Kinetic growth curves (**a**,**b**) of SATC were recorded for 72 h, and (**c**) the maximum cell index (CI max), (**d**) the slope during the adhesion period, and (**e**) doubling time (DT) were analyzed. DT was obtained during the logarithmic growth phase, while the slope was calculated (ΔCI/Δtime) during the adhesion period. The data are shown in histograms as means and SD. Significant line-specific differences (DUC vs. DUhTP mice) are separated by hashtags ($^{\#}$ $p < 0.05$, $^{\#\#\#}$ $p < 0.001$, $^{\#\#\#\#}$ $p < 0.0001$), while significant age-related or activity-related differences within each line are separated by asterisks (* $p < 0.05$, ** $p < 0.01$, *** $p < 0.001$, **** $p < 0.0001$).

Compared with sedentary 73-day-old mice, voluntary activity led to decreased ($p < 0.0001$) CImax (0.70 ± 0.07 AU) and slope (0.14 ± 0.03 AU/h) in DUC cells, whether CImax increased in DUhTP cells (CImax: 2.10 ± 0.06 vs. 0.84 ± 0.04 AU; $p < 0.0001$). In proliferating cultures from active mice, CImax and slope were higher ($p < 0.0001$) in DUhTP than the DUC line. In addition, the DT was not affected by physical activity in DUC mice (control: 52.07 ± 3.3 h, act: 56.6 ± 2.7 h), but significantly reduced in DUhTP mice (control: 43.7 ± 2.6 h, act: 36.5 ± 2.01 h, $p < 0.01$). Considering both lines, myoblasts from 43-day-old sedentary and 73-day-old active DUhTP mice were characterized by a shorter DT than DUC cells from the same groups ($p < 0.0001$; Figure 4e).

3.3. Pax7 and MyoD Expression in MPC/SATC Cultures from Male DUC and DUhTP Mice

For more detailed characterization, isolated MPC/SATC were cultivated under growth-promoting conditions for three days and, after that, examined for expression of Pax7 and MyoD by immunostaining. While the satellite cell marker Pax7 maintains the stem cell properties of satellite cells, MyoD is expressed in proliferating cells (myoblasts) and is known to activate the transcription of genes responsible for myogenic differentiation [50]. Results are summarized in Figure 5.

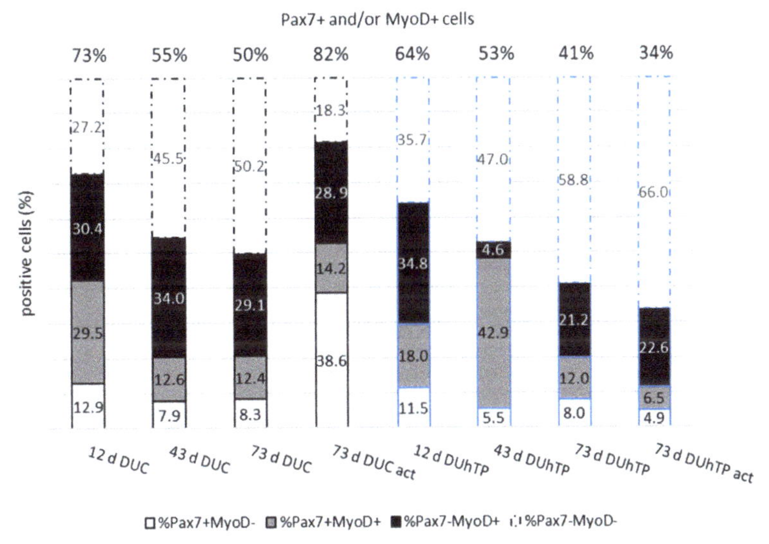

Figure 5. Percentage distribution of Pax7 (white), Pax7 and MyoD (gray), and MyoD (black) positive cells (Pax7+/MyoD−, Pax7+/MyoD+, and Pax7−/MyoD+) derived from sedentary male DUC (left, 12 d: $n = 5$ mice, 43 d: and 73 d: $n = 12$ mice, 73 d act: $n = 7$ mice) and DUhTP mice (right, blue border, 12 d: $n = 8$ mice, 43 d: $n = 4$ mice, 73 d: $n = 5$ mice and 73 d act: $n = 3$ mice) at the age of 12, 43, and 73 days (d) or from physically active (act) 73-day-old mice. At the top of the columns, the percentage of Pax7 and MyoD negative cells (Pax7−/MyoD−), respectively, of all considered cells, is given. For the visualization of Pax7 and MyoD, the cells were cultivated in growth-promoting conditions for three days and subsequently immunostained. At the top of the figure, the percentage of Pax7 and MyoD positive cells, respectively, of all considered cells is shown.

Immunocytochemical staining showed an age-dependent decrease of the proportion of Pax7 and/or MyoD positive cells from 73 to 49% in DUC mice (Figure 5, upper line). This decrease resulted from a reduction of Pax7+/MyoD− and Pax7+/MyoD+ cells. However, the proportion of Pax7−/MyoD+ cells amounted to about 30% over all time points. In 73-day-old sedentary DUC mice, immunostaining revealed that 8% of cells were Pax7+/MyoD−. The voluntary activity resulted in a marked increase in the proportion of Pax7+/MyoD− cells amounting to 39% (not significant). Mice of the high-performance selected DUhTP line also showed the expected age-dependent decrease in the proportion of Pax7 and/or MyoD positive cells. Interestingly, in contrast to cell cultures derived from 12 and 73 d DUhTP mice, proliferating cells from 43-day-old DUhTP were characterized by a high proportion of Pax7+/MyoD+ cells. No significant differences were observed between cells from sedentary and exercising 73-day-old DUhTP mice, although the proportion of Pax7+/MyoD− and Pax7+/MyoD+ myogenic cells was further reduced compared to sedentary 73-day-old DUhTP mice.

However, between the lines, the proportion of Pax7 and/or MyoD positive cells was strongly decreased in active DUhTP mice (35%) compared with the control (82%). Cells from 12-day-old DUhTP mice were characterized by a reduced proportion of Pax7+/MyoD+ cells compared with DUC mice (18 vs. 30%; not significant). On day 43 of age, the percentage of Pax7+/MyoD+ cells was significantly increased in DUhTP compared to DUC mice (43 vs. 13%; $p < 0.001$), whereas the proportion of Pax7−/MyoD+ cells was decreased (5 vs. 34%; not significant).

3.4. MPC/SATC of the DUhTP Line Showed a Higher Ability to Fuse and to Differentiate

After replacing the growth medium with a serum-reduced one, cells isolated from male DUC and DUhTP mice were able to differentiate and form multinucleated myotubes at all time points investigated, as clearly seen in phase-contrast images and images from MHC-stained cultures (Figure S1). The fusion index of myocytes was similar in 12- and 43-day-old DUC mice but increased significantly in cells from 73 day-old DUC mice (43 d vs. 73 d, $p < 0.001$; Figure 6a). In DUhTP mice, the fusion index of myocytes derived from muscles at day 43 was significantly higher than in 12-day-old DUhTP mice ($p < 0.0001$). In response to voluntary activity, a slightly increased fusion index in DUC mice ($p < 0.05$) and a decreased fusion index in DUhTP mice ($p < 0.05$) were found; therefore, the fusion index of the two active groups differed by 11% ($p < 0.0001$). Furthermore, myocytes of 43-day-old DUhTP mice showed a significantly higher fusion capacity than control mice of the same age (DUC: 16 ± 4% vs. DUhTP: 32 ± 4%, $p = 0.005$).

Next, as a measure of the differentiation potential, MHC-positive signals in cultures were quantified to determine the area covered by myotubes (Figure 6b, Table S1) and to analyze the myotube size (Figures 6c and S1). The total myotube-covered area was highest in both mouse lines when MPC/SATC had been isolated at day 43 of life (Figure 6b, Table S1). In cell cultures derived from DUC mice, the proportion of small-sized (<10,000 μm^2) myotubes increased in an age-dependent manner, whereas the proportion of medium-sized (≥10,000 <20,000 μm^2) myotubes decreased (Figure 6c). Thus, the cell cultures from 73-day-old DUC mice formed mainly small-sized myotubes (92%). In response to voluntary physical activity, more middle-, large (≥20,000 <40,000 μm^2)-, and very large-sized (≥40,000 μm^2) myotubes were detectable ($p < 0.04$) than in sedentary littermates at the same age. Differentiated cell cultures from 12-day-old DUhTP mice consisted mainly of small-sized (93%) and some middle-sized myotubes. In cultures of 43-day-old DUhTP mice, only half of the differentiated myotubes were still of small area; the other half was defined by middle-(20%), large-(22%), and very large-area (8%) myotubes. Cell cultures from 73-day-old DUhTP mice were characterized by around 72% small-, 24% medium-, and 4% large-sized myotubes.

Comparing both lines demonstrated an increase in myotubes with a larger myotube area in DUC due to voluntary activity, while in DUhTP, these myotubes were detectable at 43 days of age (Figure 6b,c).

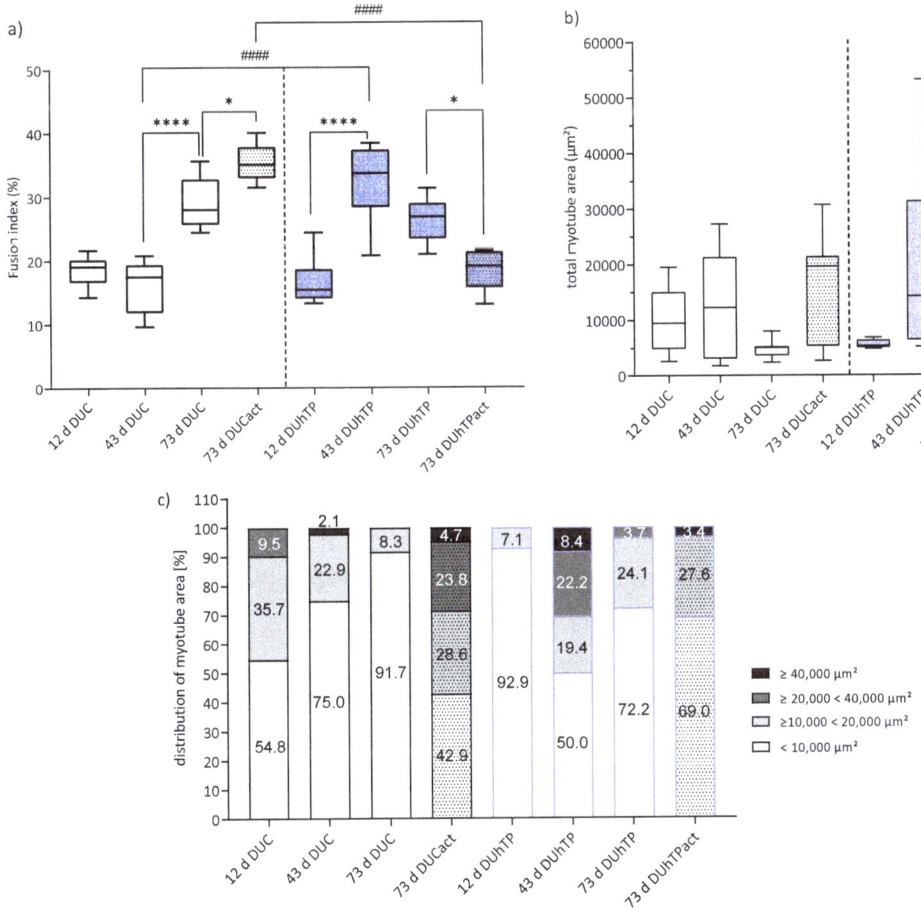

Figure 6. Fusion index and myotube size distribution of differentiated MPC/SATC. Myosin heavy chain immunofluorescence images were performed to quantify (**a**) the fusion index as the percentage of total nuclei being localized in fused multinucleated cells, (**b**) the area covered by myotubes (µm^2), and (**c**) the myotube size distribution (%). All parameters are determined 72 h after induction of differentiation. Myotubes with a size of <10,000 µm^2 were classified as small myotubes, whereas myotubes ≥10,000 <20,000 µm^2 were grouped as middle-sized, ≥20,000 <40,000 µm^2 as large myotubes, and ≥40,000 µm^2 as very large myotubes. Significant line-specific differences between DUC (left; consistent of single animal and pool cultures representing: 12 d: 7 mice, 43 d: 9 mice, 73 d: 27 mice, 73 d act: 25 mice) and DUhTP (right; consistent of single animal and pool cultures representing: 12 d: 15 mice, 43 d: 25 mice, 73 d: 13 mice, 73 d act: 5 mice) lines are separated by hashtags (#### $p < 0.0001$). Significant age-related differences within each line are separated by asterisks (* $p < 0.05$, **** $p < 0.0001$).

3.5. Cellular Creatine Kinase (CK) Activity

Next, we examined total cellular CK activity (mitochondrial and cytosolic) in myotubes after three-day cultivation in a differentiation medium. As shown in Figure 7, differentiated myotubes from DUC mice demonstrated a higher CK activity at the age of 12 days compared to the other age groups ($p < 0.01$). The highest CK activity was observed in cells of 43-day-old DUhTP mice (1.6 IU/mg total protein) compared to cells from 12-day-old (0.4 IU/mg total protein, $p < 0.001$) and 73-day-old DUhTP mice (0.6 IU/mg total protein, $p < 0.01$). Voluntary activity did not affect CK activity in both lines.

Interestingly, myotubes from 12-day-old DUC showed a higher total cellular CK activity than those from 12-day-old DUhTP mice ($p < 0.0001$). However, cells from 43-day-old and physically active 73-day-old DUhTP mice had higher CK activity than those of control mice of the same age ($p < 0.01$, $p < 0.05$, respectively).

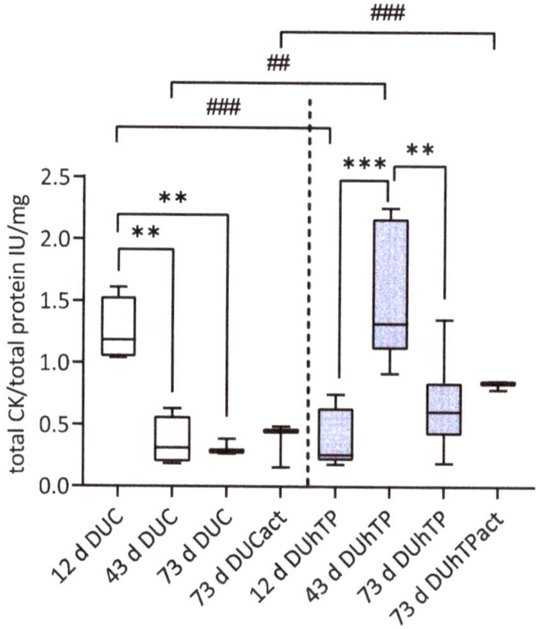

Figure 7. The activity of cellular creatine kinase in myotubes derived from male sedentary DUC (left, 12 d and 43 d: $n = 4$ mice, 73 d and 73 d act: $n = 3$ mice) and DUhTP mice (right, 12 d and 73 d: $n = 7$ mice, 43 d: $n = 5$ mice, and 73 d act: $n = 3$ mice) at the age of 12, 43, and 73 days (d) or physically active (act) 73-day old mice. Creatine kinase activity was determined 72 h after induction of differentiation. Results were obtained from quadruplicates, and total CK activity per total cell protein was calculated and displayed as means ± SD. Statistical significances were calculated by one-way ANOVA. Significant line-specific differences between DUC and DUhTP lines are separated by hashtags (## $p < 0.01$, ### $p < 0.001$), while significant age-related differences within each line are separated by asterisks (** $p < 0.01$, *** $p < 0.001$).

4. Discussion

The present study provides data about MPC/SATC derived from male non-inbred unselected control mice (DUC) and the long-term selected marathon mouse line DUhTP established by paternal phenotypic selection for high treadmill performance without previous training. A single submaximal run on a treadmill after mating served as a selection trait and resulted in genetically fixed high running ability in the offspring [38] without increasing voluntary RW activity [40]. In the context of the superior muscle performance of DUhTP animals, muscle stem cells, namely SATC, are of particular interest.

SATC are 90% progenitor cells responsible for muscle growth and routine maintenance [5–7], usually by fusion with existing myofibers or with each other. In mice, SATC-dependent hypertrophic fiber growth occurs over the first three postnatal weeks [1]. As the proportion of SATC in the total muscle cell population decreases with age, falling from about 30% at birth to 3% at week seventeen [8–10], male mice of both lines were used at different ages to characterize SATC and their progeny in the juvenile proliferative phase (day 12), in transition (day 43), and in the "adult" phase with "quiescent" SATC (day 73). Seventy-three-day-old DUhTP and DUC mice had either access to a RW that they could use voluntarily or were kept in sedentary condition. Although the DUhTP mouse line had been selected for superior running ability, the RW measurements showed an average 2.5-fold reduction in daily voluntary activity in DUhTP mice compared to unselected controls. This difference was already observed with mice of an earlier generation but with less distinctness [40]. The covered running performance of control mice corresponded to published data with a daily RW activity between 4–20 km [51] in the RW, while DUhTP mice ran a daily distance of 2.2 km. This suggests that paternal selection for high treadmill running performance resulted in a mouse line with genetically fixed superior running performance, but that this line exerts its performance strength only under forced conditions such as on a computer-controlled treadmill. Although it is known that there exist sex-specific differences in running capacity [52,53], which we can also confirm in our mouse lines, we have limited our focus of this research to males for comparability with our previous publications.

The muscular proportion to the entire body weight was similar in both lines. However, body and muscle weight (except for day 12) were lower in sedentary DUhTP than sedentary DUC mice at all ages, as previously published for 70-day-old mice [42]. More generally, the occurrence of a small muscle phenotype, also named 'mini muscles', is common in various mouse lines bred for high levels of voluntary wheel running [54]. Among other changes, 'mini muscles' are characterized by a shift to the slower myofiber phenotype, and it is proposed that this may reduce muscle energy use, thereby providing an adaptive advantage regarding endurance-running capacity. Indeed, so far, unpublished results from a study with sedentary 70-day-old DUhTP mice confirm a higher proportion of slow muscle fibers within the musculus rectus femoris. Keefe et al. [4] found a higher number of SATC in slow myofibers of mice limb muscles (5000 per mm^3 versus 2000 per mm^3 in fast myofibers). In accord, our results show that compared with sedentary DUC mice, more MPC/SATC per gram muscle could be isolated from sedentary DUhTP mice, possibly indicating a higher regenerative capacity [16] and an improved ability to maintain muscle homeostasis [4].

Cultivating SATC/myoblasts from male DUC and DUhTP mice under conditions that promote proliferation reveals differences at the functional and regulatory levels. These differences manifest predominantly at day 43 of life in sedentary mice and day 73 after voluntary RW activity.

At all ages, growth profiles of myoblast cultures from DUC mice were characterized by rapid cell attachment/cell spreading (adhesion) followed by a very distinct lag phase before starting the logarithmic growth period. However, growth curves of myoblasts from sedentary 43-day-old or active 73-day-old DUhTP mice showed shortening of the lag phase and a reduced doubling time, reflecting faster activation (G0–G1 transition) and entry into the logarithmic growth period accompanied by a higher proliferation rate compared with controls. Wozniak et al. examined stretch-induced activation of SATC and identified two populations entering proliferation after different time periods (30 min and 2 h) [55]. Thus, it seems possible that SATC/MPC subpopulations with distinct activation thresholds are recruited in DUhTP vs. DUC mice. Our hypothesis is in accord with data from Rodgers et al. [56], showing the existence of two distinct functional phases, G0 and G_{Alert}. In adults, SATC transition into G_{Alert} depends on mTORC1 activity and HGF receptor signaling and allows SATC to respond more rapidly to external stimuli [56,57]. In addition, proliferating cells from 43-day-old sedentary or 73-day-old active mice of the lines differ regarding their ability to attach/spread and the maximum CI. As found by others [48,58,59], a higher adhesion ability (43-day-old DUC > DUhTP; 73-day-old active DUhTP > DUC) is linked to an increased proliferation capacity of SATC descendants in our study. Thus, not only genetic background has an impact on the growth kinetics of SATC populations, but also physical exercise.

Marked differences regarding Pax7 and MyoD expression have been found in myoblast cultures from male DUC and DUhTP mice lines at these two time-points (day 43 and day 73 act) and thus, seem to be at least partly responsible for the functional changes observed.

During the postnatal development of young mice, the SATC marker Pax7 is essential for the enhanced proliferation potential, expansion, and survival of muscle progenitor cells and maintenance of their myogenic potential [5,14,15]. Moreover, during a critical period of postnatal muscle development, Pax7 is required for the transition from muscle progenitor to adult SATC status, which is induced by the withdrawal from differentiation and the transition into quiescence [16,17,60]. Thus, according to other studies [12], the proportion of Pax+/MyoD− cells was highest in day 12 SATC cultures from both lines. MyoD, on the other hand, is expressed in proliferating myoblasts and myotubes [13] and is known to force cells into the differentiation pathway by inducing the cell cycle inhibitors p21 and p57 as well as the early differentiation marker myogenin that marks the irreversible commitment to terminal differentiation [61]. Down-regulation of MyoD in a cohort of cells leads to reserve cells formation, which is important for replenishing the SATC pool [14,21,60,62,63].

It has been shown that Pax7 recognizes the entire SATC population in young mice [64] and that the subpopulation, which is MyoD negative (Pax7+/MyoD−), represents at least near 'quiescent' SATC, whereas Pax7+/MyoD+ cells are activated [65].

Thus, Pax7+/MyoD− cells can be considered a measure of reserve cell formation and maintenance under proliferation-promoting conditions, as in our study. Others have also identified reserve cells in in vitro culture systems and characterized them as a slowly proliferating subpopulation capable of self-renewal and generating fast-dividing progeny that undergoes terminal differentiation [6,21,59,66,67]. Hashimoto et al. (2004) investigated the reserve cell properties of slow and fast myofibers under in vitro conditions [66]. Interestingly, the process of terminal myogenic differentiation starts earlier in SATC progeny from slow myofibers, the dominating fiber type in limb muscles of DUhTP mice, and their reserve cells have a lower potential for self-renewal. Thus, along with the lower activation threshold, this gives a good explanation for the reduced proportion of Pax7+/MyoD− cells in myoblast cultures from 43-day-old sedentary or 73-day-old active DUhTP mice. Physical activity of the donor mice seems to boost this process by further stimulating activation, proliferation rate, and transition of SATC into differentiation. Under in vivo conditions, this could lead to negative effects such as reduction/exhaustion of the reserve cell pool.

In contrast, myoblast cultures from active DUC mice are characterized by a strong, 30% increase of the Pax7+/MyoD− population. This increase in SATC number is in the same order of magnitude (29% or 23%) as found by Chiarifi et al. [68] in response to endurance training (human) and by Kurosaka et al. [33] after voluntary activity in running-wheel cages (8 weeks, rat). Thus, the typical, activity-dependent enlargement of the SATC/reserve cell (Pax7+/MyoD−) compartment observed in muscles (in vivo) was also found in vitro when isolated cells from DUC mice were used. In accord with the expanded SATC pool seen in DUCact-in vitro cultures, more SATC could be isolated from limb muscles of freely active DUC mice, pointing to an increased potential for self-renewal due to activity. In addition, MPC/SATC cultures established from voluntarily active DUC mice display a higher ability to differentiate, mainly reflected by intense fusion processes and formation of more large and very large myotubes, whereas small myotubes clearly dominate in all cultures from sedentary DUC mice. In vivo, activity-induced expansion of the SATC pool and myonuclei accretion via fusion of SATC progeny have been shown to occur during hypertrophic episodes [33–35]. However, as described in various studies using voluntary wheel running [36,51], muscle hypertrophy was not observed in active DUC mice. On the contrary, muscle mass was reduced after voluntary RW activity. Most probably, the strength of the mechanical load was not strong enough to induce a hypertrophic muscle response in our experimental design.

Compared to active DUC mice, voluntary RW activity and the distance run were markedly reduced in active DUhTP and thus, most probably, not supportive of any significant proliferative responses in the SATC compartment. Nevertheless, it is surprising that the proportion of Pax7+ cells (Pax7+/MyoD−, Pax7+/MyoD+) is further reduced (11 vs. 20%) in active compared with sedentary 73-day-old DUhTP mice. A plausible explanation, also proofed by the existing data, could be that the majority of newly generated cells differentiate.

This idea fits the observed differences between cultures of active DUC and DUhTP mice. Active DUC mice show an increased percentage of cells positive for Pax7 and/or MyoD (82 vs. 35%), a higher fusion index (30 vs. 19%), a lower activity of CK (a marker of early differentiation; [69]), and a lower proportion of smaller myotubes than active DUhTP mice. It seems that in contrast to differentiating cultures from DUCact mice, cells from DUhTPact mice have a higher propensity to differentiate but fuse to a lesser extent with existing structures to form more mature myotubes. A deficiency in SATC cell function to supply myonuclei to myofibers accompanied by a 50% reduction of myonuclei number and a smaller fiber size were found in a viable mouse Pax7−/− model [70].

Of greatest relevance to line differences in the myoblast's growth behavior and differentiation potential at day 43 is the ratio between Pax7 expressing (Pax7+/MyoD− and Pax7+/MyoD+) and MyoD+ cells (Pax7−/MyoD+). It amounts to 0.6 in controls and is much higher (10.5) in DUhTP animals. Specifically, the proportion of Pax+/MyoD+ cells is increased in the latter group resulting in prolonged proliferation and provision of more myonuclei. Such an effect has normally been observed after prolonged exercise [33,68,71]. Together with the higher fusion index observed in differentiating cultures from 43-day-old DUhTP mice, this might promote the formation of thicker myotubes reflective of in vitro hypertrophic processes [68]. As the proportion of Pax7−/MyoD+ cells is relatively low (5%) in sedentary 43-day-old DUhTP mice in comparison to age-matched controls (34%), it can be speculated that in controls, the committed differentiated Pax7−/MyoD+ cells persist, whereas in DUhTP mice, the transition of Pax7−/MyoD+ cells into the next steps of differentiation (upregulation of myogenin and myosin expression) is accelerated (as discussed above). This idea is supported by the significant increase in CK activity in cells isolated from sedentary 43-day-old DUhTP mice, which is known to be a useful indicator of early differentiation, also reflecting myogenin induction [69,72]. It also has been shown that the myogenic differentiation factor, MyoDl, binds specifically to two regions of the muscular CK enhancer, which is important for its in vivo expression [73]. In accord, cells isolated from sedentary 43-day-old DUhTP mice have increased proportions of Pax+/MyoD+ SATC and show upregulation of CK activity. In addition, under our in vitro conditions, a greater ability for hypertrophic growth of these cells is shown by the formation of more middle, large, and very large myotubes resulting in a higher area covered by myotubes than controls. Of note, muscular PGC-1alpha has been shown to be upregulated in treadmill-running DUhTP mice [41], and its specific isoform PGC-1alpha4 becomes activated after resistance exercise and promotes hypertrophic myofiber growth [74]. However, voluntary running wheel activity did not increase either muscular Ppargc1a1 or Ppargc1a4 expression or PGC-1alpha protein abundance [41].

Interestingly, a comparable fusion index and distribution of small, middle, large, and very large-sized myotubes were observed in differentiating cells from 73-day-old DUC mice that were physically active and sedentary 43-day-old DUhTP mice. The important role of provision of myonuclei and fusion index for robust myofiber hypertrophy has been shown in other studies [36,75,76]. Our investigation clearly shows that our selection model promotes both factors in sedentary 43-day-old DUhTP mice, but in contrast to active 73-day-old DUC mice, a cell population more primed for differentiation (Pax7+/MyoD+ vs. Pax7+/MyoD− cells in DUCact) predominates. Thus, it seems possible that besides Pax7 and/or MyoD, other genes such as calcitonin receptor, Odz4, DACH1 (a regulator of SIX and CyclinD1), Jagged-1, and CD24 known to regulate quiescence and proliferation/differentiation of SATC and their progeny are differently expressed in the lines [65,77,78]. Specifically, Odz4 has been shown to be important in maintaining the quiescence of muscle SATC and retarded the progression of myogenic differentiation [77,78]. In the study of Ishii et al. [78], Odz4-deficient mice show a reduced number and size of myofibers and reduced size of the SATC pool accompanied by accelerated activation of SATC. These results are similar to our findings with cells from active 73-day-old DUhTP mice pointing to a reduced ability to establish and maintain the quiescent state, which is a prerequisite to preserve maximum regenerative capacity [16].

5. Conclusions

Long-term (140 generations) paternal phenotype selection of mice for high treadmill performance (DUhTP mice) led to marked changes in the functional properties of isolated SATC and their progeny. Differences between male DUhTP and respective controls (DUC) were most prominent in 43-day-old sedentary mice and 73-day-old mice after performing voluntary wheel running for three weeks. At these time points, SATC (Pax7+ cells) from DUhTP mice are characterized by faster activation resulting in higher proliferation rates. Thus, it seems possible that due to selection, a SATC subpopulation with a lower activation threshold is preferentially recruited in male DUhTP mice.

Moreover, the area covered by myotubes and the pattern of myofiber size distribution were similar in differentiating cultures from 43-day-old sedentary DUhTP mice and those from DUC mice that were voluntarily physically active. Under both conditions, a higher proportion of medium, large, and very large myotubes were formed, indicating a robust hypertrophic response under our in vitro conditions. While this was to be expected in active animals (DUC), it shows that the selection model promotes hypertrophic processes (late fusion) in vitro without a previous activity stimulus of the donor animal. In general, the provision of myonuclei and an increased fusion index are the most important factors for a robust hypertrophic response. Consistent with this, the fusion index is high in cultures of physically active 73-day-old DUC and sedentary 43-day-old DUhTP mice. However, whereas the Pax7+/MyoD− SATC population dominates cultures from active controls, Pax+/MyoD+ cells, primed for differentiation, are most prominent in cultures from 43-day-old sedentary DUhTP mice.

Thus, it can be assumed that a combination of lower activation threshold and faster transition into differentiation (observed from day 43 on in DUhTP) could lead to negative effects such as reduction/exhaustion of the reserve cell pool under in vivo conditions. This assumption is supported by the significant reduction in Pax7 positive (Pax7+/MyoD−) cells and the reduced formation of thick fibers in cultures from active DUhTP mice.

Supplementary Materials: The following supporting information can be downloaded at: https://www.mdpi.com/article/10.3390/cells11061001/s1, Figure S1: Myosin heavy chain staining and contrast-phase images of differentiated myogenic progenitor cell cultures derived from 12, 43, and 73 days old sedentary and 73-day-old physically active DUC and DUhTP mice. Table S1: Myotube area (μm^2) of differentiated myogenic progenitor cell cultures derived from 12, 43, and 73 days old sedentary DUC and DUhTP mice and active 73-day-old DUC and DUhTP mice.

Author Contributions: Conceptualization, J.B., A.H. and M.R.; Data curation, S.P. and J.B.; Formal analysis, S.P. and J.B.; Funding acquisition, J.B., A.H. and M.R.; Investigation, S.P., J.B. and M.R.; Methodology, M.L. and M.R.; Resources, M.L.; Supervision, J.B., M.L. and M.R.; Validation, J.B. and M.R.; Visualization, S.P. and J.B.; Writing—original draft, S.P., J.B. and M.R.; Writing—review and editing, M.L. and A.H. All authors have read and agreed to the published version of the manuscript.

Funding: This project and this article's publication were funded by the Future Fund and Open Access Fund of the Research Institute for Farm Animal Biology (FBN), respectively. This research did not receive any grants from the commercial sector.

Institutional Review Board Statement: All in vivo experiments were performed following national and international guidelines and were approved by our internal institutional review board.

Data Availability Statement: The raw data were generated at the FBN Dummerstorf. Derived data supporting the findings of this study are available from the corresponding author on request.

Acknowledgments: The authors thank Angela Steinborn, Kerstin Niemann, Marie Jugert-Lund, and Franziska Feldt for their technical assistance. A special thanks also go to the animal facility of the Research Institute for Farm Animal Biology for taking care of the animals used in this study.

Conflicts of Interest: The authors declare no conflict of interest.

References

1. White, R.B.; Biérinx, A.-S.; Gnocchi, V.F.; Zammit, P.S. Dynamics of Muscle Fibre Growth During Postnatal Mouse Development. *BMC Dev. Biol.* **2010**, *10*, 21. [CrossRef] [PubMed]
2. Gokhin, D.S.; Ward, S.R.; Bremner, S.N.; Lieber, R.L. Quantitative Analysis of Neonatal Skeletal Muscle Functional Improvement in the Mouse. *J. Exp. Biol.* **2008**, *211*, 837–843. [CrossRef] [PubMed]
3. Tajbakhsh, S. Skeletal Muscle Stem Cells in Developmental Versus Regenerative Myogenesis. *J. Intern. Med.* **2009**, *266*, 372–389. [CrossRef]
4. Keefe, A.C.; Lawson, J.A.; Flygare, S.D.; Fox, Z.D.; Colasanto, M.P.; Mathew, S.; Yandell, M.; Kardon, G. Muscle Stem Cells Contribute to Myofibres in Sedentary Adult Mice. *Nat. Commun.* **2015**, *6*, 7087. [CrossRef]
5. Kuang, S.; Kuroda, K.; Le Grand, F.; Rudnicki, M.A. Asymmetric Self-Renewal and Commitment of Satellite Stem Cells in Muscle. *Cell* **2007**, *129*, 999–1010. [CrossRef] [PubMed]
6. Chakkalakal, J.; Christensen, J.; Xiang, W.; Tierney, M.T.; Boscolo, F.S.; Sacco, A.; Brack, A.S. Early Forming Label-Retaining Muscle Stem Cells Require P27kip1 for Maintenance of the Primitive State. *Development* **2014**, *141*, 1649–1659. [CrossRef] [PubMed]
7. Schultz, E. Satellite Cell Proliferative Compartments in Growing Skeletal Muscles. *Dev. Biol.* **1996**, *175*, 84–94. [CrossRef]
8. Schultz, E. A Quantitative Study of the Satellite Cell Population in Postnatal Mouse Lumbrical Muscle. *Anat. Rec.* **1974**, *180*, 589–595. [CrossRef]
9. Ontell, M.; Feng, K.C.; Klueber, K.; Dunn, R.F.; Taylor, F. Myosatellite Cells, Growth, and Regeneration in Murine Dystrophic Muscle: A Quantitative Study. *Anat. Rec.* **1984**, *208*, 159–174. [CrossRef]
10. Allbrook, D.B.; Han, M.F.; Hellmuth, A.E. Population of Muscle Satellite Cells in Relation to Age and Mitotic Activity. *Pathology* **1971**, *3*, 223–243. [CrossRef]
11. Shinin, V.; Gayraud-Morel, B.; Gomès, D.; Tajbakhsh, S. Asymmetric Division and Cosegregation of Template DNA Strands in Adult Muscle Satellite Cells. *Nat. Cell Biol.* **2006**, *8*, 677–687. [CrossRef] [PubMed]
12. Pallafacchina, G.; François, S.; Regnault, B.; Czarny, B.; Dive, V.; Cumano, A.; Montarras, D.; Buckingham, M. An Adult Tissue-Specific Stem Cell in Its Niche: A Gene Profiling Analysis of in Vivo Quiescent and Activated Muscle Satellite Cells. *Stem Cell Res.* **2010**, *4*, 77–91. [CrossRef] [PubMed]
13. Friday, B.B.; Horsley, V.; Pavlath, G.K. Calcineurin Activity Is Required for the Initiation of Skeletal Muscle Differentiation. *J. Cell Biol.* **2000**, *149*, 657–666. [CrossRef] [PubMed]
14. Oustanina, S.; Hause, G.; Braun, T. Pax7 Directs Postnatal Renewal and Propagation of Myogenic Satellite Cells but Not Their Specification. *Embo. J.* **2004**, *23*, 3430–3439. [CrossRef] [PubMed]
15. Relaix, F.; Montarras, D.; Zaffran, S.; Gayraud-Morel, B.; Rocancourt, D.; Tajbakhsh, S.; Mansouri, A.; Cumano, A.; Buckingham, M. Pax3 and Pax7 Have Distinct and Overlapping Functions in Adult Muscle Progenitor Cells. *J. Cell Biol.* **2006**, *172*, 91–102. [CrossRef] [PubMed]
16. Lepper, C.; Conway, S.J.; Fan, C.-M. Adult Satellite Cells and Embryonic Muscle Progenitors Have Distinct Genetic Requirements. *Nature* **2009**, *460*, 627–631. [CrossRef]
17. Olguin, H.C.; Olwin, B.B. Pax-7 up-Regulation Inhibits Myogenesis and Cell Cycle Progression in Satellite Cells: A Potential Mechanism for Self-Renewal. *Dev. Biol.* **2004**, *275*, 375–388. [CrossRef]
18. Liu, W.; Wen, Y.; Bi, P.; Lai, X.; Liu, X.S.; Kuang, S. Hypoxia Promotes Satellite Cell Self-Renewal and Enhances the Efficiency of Myoblast Transplantation. *Development* **2012**, *139*, 2857–2865. [CrossRef]
19. Neal, A.; Boldrin, L.; Morgan, J.E. The Satellite Cell in Male and Female, Developing and Adult Mouse Muscle: Distinct Stem Cells for Growth and Regeneration. *PLoS ONE* **2012**, *7*, e37950. [CrossRef]
20. Dumont, N.A.; Bentzinger, C.F.; Sincennes, M.; Rudnicki, M.A. Satellite Cells and Skeletal Muscle Regeneration. *Compr. Physiol.* **2015**, *5*, 1027–1059.
21. Yoshida, N.; Yoshida, S.; Koishi, K.; Masuda, K.; Nabeshima, Y. Cell Heterogeneity Upon Myogenic Differentiation: Down-Regulation of Myod and Myf-5 Generates' Reserve Cells'. *J. Cell Sci.* **1998**, *111*, 769–779. [CrossRef] [PubMed]
22. Biressi, S.; Molinaro, M.; Cossu, G. Cellular Heterogeneity During Vertebrate Skeletal Muscle Development. *Dev. Biol.* **2007**, *308*, 281–293. [CrossRef] [PubMed]
23. Andrés, V.; Walsh, K. Myogenin Expression, Cell Cycle Withdrawal, and Phenotypic Differentiation Are Temporally Separable Events That Precede Cell Fusion Upon Myogenesis. *J. Cell Biol.* **1996**, *132*, 657–666. [CrossRef]
24. Lehka, L.; Rędowicz, M.J. Mechanisms Regulating Myoblast Fusion: A Multilevel Interplay. *Semin. Cell Dev. Biol.* **2020**, *104*, 81–92. [CrossRef]
25. Fukada, S.I. The Roles of Muscle Stem Cells in Muscle Injury, Atrophy and Hypertrophy. *J. Biochem.* **2018**, *163*, 353–358. [CrossRef] [PubMed]

26. Egner, I.M.; Bruusgaard, J.C.; Gundersen, K. Satellite Cell Depletion Prevents Fiber Hypertrophy in Skeletal Muscle. *Development* **2016**, *143*, 2898–2906. [CrossRef] [PubMed]
27. Goh, Q.; Millay, D.P. Requirement of Myomaker-Mediated Stem Cell Fusion for Skeletal Muscle Hypertrophy. *Elife* **2017**, *6*, e20007. [CrossRef]
28. Murach, K.A.; White, S.H.; Wen, Y.; Ho, A.; Dupont-Versteegden, E.E.; McCarthy, J.J.; Peterson, C.A. Differential Requirement for Satellite Cells During Overload-Induced Muscle Hypertrophy in Growing Versus Mature Mice. *Skelet. Muscle* **2017**, *7*, 14. [CrossRef]
29. McCarthy, J.J.; Mula, J.; Miyazaki, M.; Erfani, R.; Garrison, K.; Farooqui, A.B.; Srikuea, R.; Lawson, B.A.; Grimes, B.; Keller, C.; et al. Effective Fiber Hypertrophy in Satellite Cell-Depleted Skeletal Muscle. *Development* **2011**, *138*, 3657–3666. [CrossRef]
30. Koulmann, N.; Bigard, A.X. Interaction between Signalling Pathways Involved in Skeletal Muscle Responses to Endurance Exercise. *Pflug. Arch.* **2006**, *452*, 125–139. [CrossRef]
31. SShamim, B.; Hawley, J.A.; Camera, D.M. Protein Availability and Satellite Cell Dynamics in Skeletal Muscle. *Sports Med.* **2018**, *48*, 1329–1343. [CrossRef] [PubMed]
32. Bischoff, R.; Heintz, C. Enhancement of Skeletal Muscle Regeneration. *Dev. Dyn.* **1994**, *201*, 41–54. [CrossRef] [PubMed]
33. Kurosaka, M.; Naito, H.; Ogura, Y.; Kojima, A.; Goto, K.; Katamoto, S. Effects of Voluntary Wheel Running on Satellite Cells in the Rat Plantaris Muscle. *J. Sports Sci. Med.* **2009**, *8*, 51–57. [PubMed]
34. Smith, H.K.; Merry, T.L. Voluntary Resistance Wheel Exercise During Post-Natal Growth in Rats Enhances Skeletal Muscle Satellite Cell and Myonuclear Content at Adulthood. *Acta Physiol.* **2012**, *204*, 393–402. [CrossRef]
35. Li, P.; Akimoto, T.; Zhang, M.; Williams, R.S.; Yan, Z. Resident Stem Cells Are Not Required for Exercise-Induced Fiber-Type Switching and Angiogenesis but Are Necessary for Activity-Dependent Muscle Growth. *Am. J. Physiol. Cell Physiol.* **2006**, *290*, C1461–C1468. [CrossRef]
36. Masschelein, E.; D'Hulst, G.; Zvick, J.; Hinte, L.; Soro-Arnaiz, I.; Gorski, T.; Von Meyenn, F.; Bar-Nur, O.; De Bock, K. Exercise Promotes Satellite Cell Contribution to Myofibers in a Load-Dependent Manner. *Skelet. Muscle* **2020**, *10*, 21. [CrossRef]
37. Shefer, G.; Rauner, G.; Yablonka-Reuveni, Z.; Benayahu, D. Reduced Satellite Cell Numbers and Myogenic Capacity in Aging Can Be Alleviated by Endurance Exercise. *PLoS ONE* **2010**, *5*, e13307. [CrossRef]
38. Falkenberg, H.; Langhammer, M.; Renne, U. Comparison of Biochemical Blood Traits after Long-Term Selection on High or Low Locomotory Activity in Mice. *Arch. Anim. Breed.* **2000**, *43*, 513–522. [CrossRef]
39. Brenmoehl, J.; Walz, C.; Spitschak, M.; Wirthgen, E.; Walz, M.; Langhammer, M.; Tuchscherer, A.; Naumann, R.; Hoeflich, A. Partial Phenotype Conversion and Differential Trait Response to Conditions of Husbandry in Mice. *J. Comp. Physiol. B* **2018**, *188*, 527–539. [CrossRef]
40. Brenmoehl, J.; Ohde, D.; Walz, C.; Langhammer, M.; Schultz, J.; Hoeflich, A. Analysis of Activity-Dependent Energy Metabolism in Mice Reveals Regulation of Mitochondrial Fission and Fusion mRNA by Voluntary Physical Exercise in Subcutaneous Fat from Male Marathon Mice (Duhtp). *Cells* **2020**, *9*, 2697. [CrossRef]
41. Brenmoehl, J.; Albrecht, E.; Komolka, K.; Schering, L.; Langhammer, M.; Hoeflich, A.; Maak, S. Irisin Is Elevated in Skeletal Muscle and Serum of Mice Immediately after Acute Exercise. *Int. J. Biol. Sci.* **2014**, *10*, 338–349. [CrossRef] [PubMed]
42. Brenmoehl, J.; Ohde, D.; Walz, C.; Schultz, J.; Tuchscherer, A.; Rieder, F.; Renne, U.; Hoeflich, A. Dynamics of Fat Mass in Duhtp Mice Selected for Running Performance—Fat Mobilization in a Walk. *Obes. Facts* **2015**, *8*, 373–385. [CrossRef] [PubMed]
43. Brenmoehl, J.; Ohde, D.; Albrecht, E.; Walz, C.; Tuchscherer, A.; Hoeflich, A. Browning of Subcutaneous Fat and Higher Surface Temperature in Response to Phenotype Selection for Advanced Endurance Exercise Performance in Male Duhtp Mice. *J. Comp. Physiol. B* **2017**, *187*, 361–373. [CrossRef] [PubMed]
44. Sonia, A.-M.; Rochat, A.; Mademtzoglou, D.; Morais, J.; de Reyniès, A.; Auradé, F.; Chang, T.H.-T.; Zammit, P.S.; Relaix, F. Gene Expression Profiling of Muscle Stem Cells Identifies Novel Regulators of Postnatal Myogenesis. *Front. Cell Dev. Biol.* **2016**, *4*, 58.
45. Dietl, G.; Langhammer, M.; Renne, U. Model Simulations for Genetic Random Drift in the Outbred Strain Fzt:Du. *Arch. Tierz.* **2004**, *47*, 595–604. [CrossRef]
46. Atienza, J.M.; Zhu, J.; Wang, X.; Xu, X.; Abassi, Y. Dynamic Monitoring of Cell Adhesion and Spreading on Microelectronic Sensor Arrays. *J. Biomol. Screen* **2005**, *10*, 795–805. [CrossRef]
47. Ke, N.; Wang, X.; Xu, X.; Abassi, Y.A. The Xcelligence System for Real-Time and Label-Free Monitoring of Cell Viability. *Methods Mol. Biol.* **2011**, *740*, 33–43.
48. Miersch, C.; Stange, K.; Röntgen, M. Separation of Functionally Divergent Muscle Precursor Cell Populations from Porcine Juvenile Muscles by Discontinuous Percoll Density Gradient Centrifugation. *BMC Cell Biol.* **2018**, *19*, 2. [CrossRef]
49. Kirstein, S.L.; Atienza, J.M.; Xi, B.; Zhu, J.; Yu, N.; Wang, X.; Xu, X.; Abassi, Y.A. Live Cell Quality Control and Utility of Real-Time Cell Electronic Sensing for Assay Development. *Assay Drug Dev. Technol.* **2006**, *4*, 545–553. [CrossRef]
50. Weintraub, H.; Dwarki, V.J.; Verma, I.; Davis, R.; Hollenberg, S.; Snider, L.; Lassar, A.; Tapscott, S.J. Muscle-Specific Transcriptional Activation by Myod. *Genes Dev.* **1991**, *5*, 1377–1386. [CrossRef]
51. Manzanares, G.; Brito-Da-Silva, G.; Gandra, P. Voluntary Wheel Running: Patterns and Physiological Effects in Mice. *Braz. J. Med. Biol. Res.* **2018**, *52*, e7830. [CrossRef] [PubMed]
52. Oydanich, M.; Babici, D.; Zhang, J.; Rynecki, N.; Vatner, D.E.; Vatner, S.F. Mechanisms of Sex Differences in Exercise Capacity. *Am. J. Physiol. Regul. Integr. Comp. Physiol.* **2019**, *316*, R832–R838. [CrossRef] [PubMed]

53. Bartling, B.; Al-Robaiy, S.; Lehnich, H.; Binder, L.; Hiebl, B.; Simm, A. Sex-Related Differences in the Wheel-Running Activity of Mice Decline with Increasing Age. *Exp. Gerontol.* **2017**, *87*, 139–147. [CrossRef] [PubMed]
54. McGillivray, D.G.; Garland, T.; Dlugosz, E.M.; Chappell, M.A.; Syme, D.A. Changes in Efficiency and Myosin Expression in the Small-Muscle Phenotype of Mice Selectively Bred for High Voluntary Running Activity. *J. Exp. Biol.* **2009**, *212*, 977–985. [CrossRef] [PubMed]
55. Wozniak, A.C.; Pilipowicz, O.; Yablonka-Reuveni, Z.; Greenway, S.; Craven, S.; Scott, E.; Anderson, J.E. C-Met Expression and Mechanical Activation of Satellite Cells on Cultured Muscle Fibers. *J. Histochem. Cytochem.* **2003**, *51*, 1437–1445. [CrossRef] [PubMed]
56. Rodgers, J.T.; King, K.Y.; Brett, J.O.; Cromie, M.J.; Charville, G.W.; Maguire, K.K.; Brunson, C.; Mastey, N.; Liu, L.; Tsai, C.R.; et al. Mtorc1 Controls the Adaptive Transition of Quiescent Stem Cells from G0 to G(Alert). *Nature* **2014**, *510*, 393–396. [CrossRef]
57. Joseph, R.T.; Schroeder, M.D.; Ma, C.; Rando, T.A. Hgfa Is an Injury-Regulated Systemic Factor That Induces the Transition of Stem Cells into G_{Alert}. *Cell Rep.* **2017**, *19*, 479–486.
58. Sellathurai, J.; Cheedipudi, S.; Dhawan, J.; Schrøder, H.D. A Novel in Vitro Model for Studying Quiescence and Activation of Primary Isolated Human Myoblasts. *PLoS ONE* **2013**, *8*, e64067. [CrossRef]
59. Rouger, K.; Brault, M.; Daval, N.; Leroux, I.; Guigand, L.; Lesoeur, J.; Fernandez, B.; Cherel, Y. Muscle Satellite Cell Heterogeneity: In Vitro and in Vivo Evidences for Populations That Fuse Differently. *Cell Tissue Res.* **2004**, *317*, 319–326. [CrossRef]
60. Zammit, P.S.; Golding, J.P.; Nagata, Y.; Hudon, V.; Partridge, T.A.; Beauchamp, J.R. Muscle Satellite Cells Adopt Divergent Fates: A Mechanism for Self-Renewal? *J. Cell Biol.* **2004**, *166*, 347–357. [CrossRef]
61. Yoshida, K.; Kaji, R.; Kubori, T.; Kohara, N.; Iizuka, T.; Kimura, J. Muscle Afferent Block for the Treatment of Oromandibular Dystonia. *Mov. Disord.* **1998**, *13*, 699–705. [CrossRef] [PubMed]
62. Halevy, O.; Piestun, Y.; Allouh, M.; Rosser, B.W.; Rinkevich, Y.; Reshef, R.; Rozenboim, I.; Wleklinski-Lee, M.; Yablonka-Reuveni, Z. Pattern of Pax7 Expression During Myogenesis in the Posthatch Chicken Establishes a Model for Satellite Cell Differentiation and Renewal. *Dev. Dyn.* **2004**, *231*, 489–502. [CrossRef] [PubMed]
63. Subramaniam, S.; Sreenivas, P.; Cheedipudi, S.; Reddy, V.R.; Shashidhara, L.S.; Chilukoti, R.K.; Mylavarapu, M.; Dhawan, J. Distinct Transcriptional Networks in Quiescent Myoblasts: A Role for Wnt Signaling in Reversible vs. Irreversible Arrest. *PLoS ONE* **2014**, *8*, e65097. [CrossRef] [PubMed]
64. Collins, C.A.; Olsen, I.; Zammit, P.S.; Heslop, L.; Petrie, A.; Partridge, T.A.; Morgan, J.E. Stem Cell Function, Self-Renewal, and Behavioral Heterogeneity of Cells from the Adult Muscle Satellite Cell Niche. *Cell* **2005**, *122*, 289–301. [CrossRef] [PubMed]
65. Gnocchi, V.F.; White, R.B.; Ono, Y.; Ellis, J.A.; Zammit, P.S. Further Characterisation of the Molecular Signature of Quiescent and Activated Mouse Muscle Satellite Cells. *PLoS ONE* **2009**, *4*, e5205. [CrossRef] [PubMed]
66. Hashimoto, N.; Murase, T.; Kondo, S.; Okuda, A.; Inagawa-Ogashiwa, M. Muscle Reconstitution by Muscle Satellite Cell Descendants with Stem Cell-Like Properties. *Development* **2004**, *131*, 5481–5490. [CrossRef]
67. Miersch, C.; Stange, K.; Hering, S.; Kolisek, M.; Viergutz, T.; Röntgen, M. Molecular and Functional Heterogeneity of Early Postnatal Porcine Satellite Cell Populations Is Associated with Bioenergetic Profile. *Sci. Rep.* **2017**, *7*, 45052. [CrossRef]
68. Nadia, C.; Kadi, F.; Féasson, L.; Denis, C. Satellite Cell Frequency in Skeletal Muscle of Old Men: Effects of Endurance Training. *Muscle Nerve* **2003**, *28*, 87–92.
69. John, L.; Bischoff, R. Differentiation of Creatine Phosphokinase During Myogenesis: Quantitative Fractionation of Isozymes. *Dev. Biol.* **1977**, *57*, 330–344.
70. Kuang, S.; Chargé, S.B.; Seale, P.; Huh, M.; Rudnicki, M.A. Distinct Roles for Pax7 and Pax3 in Adult Regenerative Myogenesis. *J. Cell Biol.* **2006**, *172*, 103–113. [CrossRef]
71. Appell, H.-J.; Forsberg, S.; Hollmann, W. Satellite Cell Activation in Human Skeletal Muscle after Training: Evidence for Muscle Fiber Neoformation. *Int. J. Sports Med.* **1988**, *9*, 297–299. [CrossRef] [PubMed]
72. Yutzey, K.E.; Rhodes, S.J.; Konieczny, S.F. Differential Trans Activation Associated with the Muscle Regulatory Factors Myod1, Myogenin, and Mrf4. *Mol. Cell. Biol.* **1990**, *10*, 3934–3944. [PubMed]
73. Lassar, A.B.; Buskin, J.N.; Lockshon, D.; Davis, R.L.; Apone, S.; Hauschka, S.D.; Weintraub, H. Myod Is a Sequence-Specific DNA Binding Protein Requiring a Region of Myc Homology to Bind to the Muscle Creatine Kinase Enhancer. *Cell* **1989**, *58*, 823–831. [CrossRef]
74. Ruas, J.L.; White, J.P.; Rao, R.R.; Kleiner, S.; Brannan, K.T.; Harrison, B.C.; Greene, N.; Wu, J.; Estall, J.; Irving, B.; et al. A Pgc-1alpha Isoform Induced by Resistance Training Regulates Skeletal Muscle Hypertrophy. *Cell* **2012**, *151*, 1319–1331. [CrossRef] [PubMed]
75. Kadi, F.; Charifi, N.; Denis, C.; Lexell, J.; Andersen, J.L.; Schjerling, P.; Olsen, S.; Kjaer, M. The Behaviour of Satellite Cells in Response to Exercise: What Have We Learned from Human Studies? *Pflug. Arch.* **2005**, *451*, 319–327. [CrossRef]
76. Brooks, M.J.; Hajira, A.; Mohamed, J.S.; Alway, S.E. Voluntary Wheel Running Increases Satellite Cell Abundance and Improves Recovery from Disuse in Gastrocnemius Muscles from Mice. *J. Appl. Physiol.* **2018**, *124*, 1616–1628. [CrossRef]
77. Yamaguchi, M.; Ogawa, R.; Watanabe, Y.; Uezumi, A.; Miyagoe-Suzuki, Y.; Tsujikawa, K.; Yamamoto, H.; Takeda, S.; Fukada, S.-I. Calcitonin Receptor and Odz4 Are Differently Expressed in Pax7-Positive Cells During Skeletal Muscle Regeneration. *J. Mol. Histol.* **2012**, *43*, 581–587. [CrossRef]
78. Ishii, K.; Suzuki, N.; Mabuchi, Y.; Ito, N.; Kikura, N.; Fukada, S.-I.; Okano, H.; Takeda, S.; Akazawa, C. Muscle Satellite Cell Protein Teneurin-4 Regulates Differentiation During Muscle Regeneration. *Stem Cells* **2015**, *33*, 3017–3027. [CrossRef]

cells

MDPI

Article

Zscan4 Contributes to Telomere Maintenance in Telomerase-Deficient Late Generation Mouse ESCs and Human ALT Cancer Cells

Jiameng Dan [1,2,*], Zhongcheng Zhou [1], Fang Wang [3], Hua Wang [1], Renpeng Guo [1], David L. Keefe [3] and Lin Liu [1,*]

1 State Key Laboratory of Medicinal Chemical Biology, Department of Cell Biology and Genetics, College of Life Sciences, Nankai University, Tianjin 300071, China; loyal.zhongcheng@163.com (Z.Z.); wangh2@mskcc.org (H.W.); guorp@njau.edu.cn (R.G.)

2 State Key Laboratory of Primate Biomedical Research, Institute of Primate Translational Medicine, Kunming University of Science and Technology, Kunming, Yunnan 650500, China; Yunnan Key Laboratory of Primate Biomedical Research, Kunming, Yunnan 650500, China

3 Department of Obstetrics and Gynecology, New York University Langone Medical Center, New York, NY 10016, USA; fang.wang@nyulangone.org (F.W.); david.keefe@nyulangone.org (D.L.K.)

* Correspondence: danjm@lpbr.cn (J.D.); liulin@nankai.edu.cn (L.L.)

Abstract: Proper telomere length is essential for indefinite self-renewal of embryonic stem (ES) cells and cancer cells. Telomerase-deficient late generation mouse ES cells and human ALT cancer cells are able to propagate for numerous passages, suggesting telomerase-independent mechanisms responding for telomere maintenance. However, the underlying mechanisms ensuring the telomere length maintenance are unclear. Here, using late generation telomerase KO (G4 Terc$^{-/-}$) ESCs as a model, we show that *Zscan4*, highly upregulated in G4 Terc$^{-/-}$ ESCs, is responsible for the prolonged culture of these cells with stably short telomeres. Mechanistically, G4 Terc$^{-/-}$ ESCs showed reduced levels of DNA methylation and H3K9me3 at Zscan4 promoter and subtelomeres, which relieved the expression of Zscan4. Similarly, human ZSCAN4 was also derepressed by reduced H3K9me3 at its promoter in ALT U2 OS cells, and depletion of ZSCAN4 significantly shortened telomeres. Our results define a similar conserved pathway contributing to the telomere maintenance in telomerase-deficient late generation mESCs and human ALT U2OS cancer cells.

Keywords: Zscan4; telomere; ES cells; ALT; DNA methylation; H3K9me3

Citation: Dan, J.; Zhou, Z.; Wang, F.; Wang, H.; Guo, R.; Keefe, D.L.; Liu, L. Zscan4 Contributes to Telomere Maintenance in Telomerase-Deficient Late Generation Mouse ESCs and Human ALT Cancer Cells. *Cells* **2022**, *11*, 456. https://doi.org/10.3390/cells11030456

Academic Editor: Mehdi Najar

Received: 3 January 2022
Accepted: 25 January 2022
Published: 28 January 2022

1. Introduction

Mammalian telomeres consist of repetitive G-rich sequences and associated proteins at the ends of linear chromosomes and function in the maintenance of chromosomal stability and integrity [1,2]. Telomeres are primarily maintained by active telomerase which is composed of telomerase reverse transcriptase (TERT), telomerase RNA (TERC), and dyskerin [2]. Telomerase is expressed highly in a subset of stem cells, as well as in most immortal and cancer cells, presumably to support their indefinite proliferation and self-renewal [3,4]. Most mammalian somatic cell types do not express telomerase activity, such that telomeres shorten progressively with each cell cycle and the end replication problem ensues [5,6].

Previously, we have generated ES cell lines with high efficiency from wild type (WT, Terc$^{+/+}$), heterozygous (Terc$^{+/-}$), and early to late generation of Terc$^{-/-}$ mouse blastocyst (G1, G3 and G4 telomerase RNA null blastocyst) [7]. Interestingly, late generation telomerase-deficient *mTerc*$^{-/-}$ mouse ES cells (G4 Terc$^{-/-}$) show telomere shortening and dysfunction compared with WT ES cells, yet can sustain their proliferation for numerous passages without telomerase activity [7], suggesting the involvement of telomerase-independent alternative lengthening of telomeres (ALT) mechanisms responding for telomere maintenance. Similarly, about 10–15% of human cancer cells lack detectable telomerase activity, and many of these use an ALT mechanism to maintain telomeres in cancer growth (reviewed in [8]). However, the mechanisms ensuring telomere length maintenance in both cases remain elusive. Furthermore, whether late generation telomerase-deficient *mTerc*$^{-/-}$ mouse ES cells and ALT cancer cells share the same or similar mechanisms to maintain telomere homeostasis remains to be determined.

Telomeres and subtelomeres are enriched with repressive DNA and histone modification (e.g., DNA hypermethylation and H3K9me3/H4K20me3 histone modifications) that are important negative regulators of telomere length [9], forming an inhibitive effect to the genes nearby, named Telomere Position Effect (TPE) [10–12]. These DNA and histone modifications at subtelomeric regions were reduced in late generation Terc$^{-/-}$ mouse embryonic fibroblast (MEF) cells [13].

Zscan4, a gene with unique expression pattern, expressed specifically in two-cell embryos and transiently in sporadic ES cells (1–5%) at any given time, marks a transient 2C-like state of mouse ES cells [14–17]. In 2C-like cells (2CLCs), *Zscan4* and many other 2C-specific genes are direct targets of *Dux*, which is necessary and sufficient for the transition of mouse ESCs to 2CLCs [18–20]. Functionally, Zscan4 is required for preimplantation embryonic development, lengthening telomeres promptly by activating telomere recombination and maintaining exceptional genomic stability [16,17,21]. *Zscan4* knockdown in mESCs leads to telomere shortening, abnormal karyotypes (chromosome deletion and fusion), proliferation defects, and, ultimately, apoptosis, while *Zscan4* re-expression could rescue those abnormalities [16]. *Zscan4* knockdown in early embryos delays the progression from two-cell to four-cell stage and produces blastocysts that fail to implant or proliferate in blastocyst outgrowth culture [17]. Most recently, TRF2 was shown to be dispensable for telomere protection in pluripotent stem cells through the upregulation of *Zscan4* [22]. In contrast, TRF2 is critically required for telomere protection in differentiated somatic cells. Depletion of *Zscan4* in TRF2-null ES cells results in re-appearance of telomere fusion that normally seen in TRF2-null differentiated somatic cells, while induced expression of *Zscan4* in TRF2-null differentiated somatic cells prevents telomere fusion [22], supporting the critical roles of Zscan4 in maintaining genomic stability in ES cells. Furthermore, Zscan4 was also found to specifically recognize a subset of $(CA)_n$ microsatellites and protect mouse two-cell embryos from DNA damage, which ensures the successful progression of early embryo development [23]. Surprisingly, 2C-like mESCs were reported to be associated with global DNA hypomethylation [14,15]. Mechanistically, we previously reported that as mESCs proliferate, they sporadically enter a 2C-like totipotent state, resulting in a burst of Zscan4 expression [15]. Zscan4, via its ZNF4 motif, recruits the Uhrf1-Dnmt1 complex and, through self-association mediated by its SCAN domain, brings two or more Uhrf1-Dnmt1 heterodimers into proximity. This facilitates Uhrf1′s E3 ligase-mediated ubiquitination of Uhrf1 and Dnmt1, likely through cross actions between Uhrf1 molecules. Subsequent Uhrf1 and Dnmt1 degradation leads to global DNA demethylation, which facilitates telomere recombination and elongation [15]. This mechanism somehow explains Zscan4′s effects on promoting telomere recombination and elongation through chromatin de-condensation with reduced DNA methylation. However, the exact mechanisms on how Zscan4 promotes telomere recombination and elongation remain to be elucidated.

Interestingly, both mouse *Zscan4* and human ZSCNA4 genes locate at subtelomeric regions, with mouse *Zscan4* at short p arm of chromosome 7 while human ZSCAN4 at short p arm of chromosome 19, indicating that *Zscan4* might be repressed by TPE. Here, we show that reduced repressive epigenetic modifications at *Zscan4* promoters de-repress *Zscan4*, which contributes to telomere maintenance in both late generation Terc$^{-/-}$ ES cells and telomerase-deficient human ALT-U2 OS cancer cells.

2. Materials and Methods

2.1. Cell Culture

Both WT and G4 Terc$^{-/-}$ ES cell lines derived from C57BL/6J mice were isolated and characterized previously in our laboratory [7]. J1 ES cells were cultured without feeder. The

ES cell culture medium consisted of knock-out DMEM (Gibco) with 15% FBS (Hyclone), 1000 U/mL mouse leukemia inhibitory factor (LIF) (ESGRO, Chemicon), 0.1 mM non-essential amino acids, 0.1 mM β-mercaptoethanol, 1 mM L-glutamine, and penicillin (100 U/mL), and streptomycin (100 μg/mL). For culture of all ES cell lines, the medium was changed daily, and cells were routinely passaged every two days. HeLa cells were cultured with high-glucose DMEM (Gibco) with 10% FBS (Hyclone), 0.1 mM non-essential amino acids, 1 mM L-glutamine, and penicillin (100 U/mL), and streptomycin (100 μg/mL). The ALT positive U2 OS cell medium was HeLa medium plus 0.2 mM non-essential amino acids.

2.2. Stable Zscan4 RNAi in G4 Terc$^{-/-}$ ES Cells, ALT-U2 OS Cells, and ZSCAN4 KO in U2 OS Cells

For stable knockdown of mZscan4 in ES cells, control and mZscan4 shRNAs (Table S1) were cloned into pSIREN-RetroQ (Clontech) and introduced into Plat-E cells for packaging of retrovirus. G4 Terc$^{-/-}$ ES cells were infected with indicated retrovirus and selected with 1.5 μg/mL puromycin for about 10 days. The resistant clones were picked and expanded for further culturing. For stable knockdown of human ZSCAN4 in U2 OS cells, initially, EGFP in RNAi lentivirus vector pLL3.7 was replaced with neomycin between NheI/EcoRI sites and named as pLL3.7-neo [24]. After initial test on RNAi efficiency using seven different shRNA sequences against human ZSCAN4, four hZscan4 shRNA and control sequences (Table S1) were cloned into pLL3.7-neo lentivirus vectors. The resultant vectors were introduced into 293 T cells with helper vectors to produce lentivirus, and 48 h later the indicated lentivirus were used to infect U2 OS cells and selected with 700 μg/mL G418 for 10 days. The pLL3.7-EGFP vector was used to monitor the infection efficiency. For ZSCAN4 KO in U2 OS cells, CRISPR-Cas9 technology was used to generate stable KO clones. Guide RNA (gRNA) targeting exon 3 of human ZSCAN4 gene was cloned into a targeting vector, and single clones were picked and expanded for further experiments.

ZSCAN4 gRNA sequence: ACAGCAATAATTCATATGCA.

2.3. Cell Cycle Analysis

Cells were fixed in freshly prepared 70% ethanol at 4 °C overnight, then centrifuged at 1000× *g* for 5 min and stained with propidium iodide (PI) at 37 °C for 30 min in water bath. FACS analysis was used to determine cell cycle progression.

2.4. Gene Expression Analysis by Quantitative Real-Time PCR

Total RNA was isolated from cells using RNeasy mini kit (QIAGEN). Two micrograms of RNA were subject to cDNA synthesis using M-MLV Reverse Transcriptase (Invitrogen). Real-time quantitative PCR reactions were set up in duplicate with the FastStart Universal SYBR Green Master (ROX) (Roche, Hvidovre, Denmark) and run on the iCycler iQ5 2.0 Standard Edition Optical System (Bio-Rad) using primers. Most primers were designed using the IDT DNA website. The primers used are listed in Table S3.

2.5. Immunoblot Analysis

Cells were washed twice in PBS, collected, lysed, and boiled in SDS sample buffer at 99 °C for 5 min; Equal amounts of total proteins of each cell extracts were resolved by 10–12% Bis-Tris SDS-PAGE and transferred to polyvinylidinedifluoride membranes (PVDF, Millipore, Burlington, MA, USA). Nonspecific binding was blocked by incubation in 5% skim milk in TBST at room temperature for 1–2 h. Blots were then probed with various primary antibodies, anti-Zscan4 (AB4340, Millipore, 1:1000), Dnmt1 (sc10221, Santa Cruz, 1:500), Dnmt3a (ab13888, Abcam, 1:1000), Dnmt3b (ab13604, Abcam, 1:1000), Tet2 (Kind gift from Dr. Jinsong Li from SIBS, 1:1000), H3K9me3 (ab8898, Abcam, 1:2000), H3 (ab1791, Abcam, 1:2000), and β-actin ((P30002, Abmart, 1:5000) by overnight incubation in 5% skim milk in TBST at 4 °C. Immunoreactive bands were then probed for 1–2 h at room temperature with the appropriate horseradish peroxidase-conjugated secondary anti-Rabbit IgG-HRP (GE Healthcare, NA934V, 1:5000), or anti-mouse IgG-HRP (Santa Cruz, sc-2031, 1:5000), or anti-goat IgG-HRP (Santa Cruz, sc-2020, 1:5000). The protein bands were detected by Enhanced ECL AmershamTM prime Western blotting detection reagent (GE Healthcare, RPN2232).

2.6. Immunofluorescence Microscopy

Cells were washed twice in PBS, then fixed in freshly prepared 3.7% paraformaldehyde (PFA) in PBS (pH 7.4) for 15 min on ice cube, permeabilized in 0.1% Triton X-100 in blocking solution (3% goat serum plus 0.5% BSA in PBS) for 30 min at room temperature, washed three times (each for 15 min) and left in the blocking solution for 1 h. Cells were then incubated overnight at 4 °C with primary antibodies against Zscan4 (AB4340, Millipore, 1:500), H3K9me3 (ab8898, Abcam, 1:1000), washed three times (each for 15 min), and incubated for 1 h with the secondary antibody Fluor 568 goat anti-rabbit IgG (A11036, MP), diluted 1:200 with blocking solution. Samples were washed, and counterstained with 0.5 μg/mL Hoechst33342 (H1398, MP) in Vectashield mounting medium. Fluorescence was detected and imaged using a Zeiss inverted fluorescence microscope (Axio Imager Z1).

2.7. Telomere Quantitative Fluorescence In Situ Hybridization (Q-FISH)

Telomere length and function (telomere integrity and chromosome stability) was estimated by telomere quantitative FISH. Cells were incubated with 0.5 μg/mL nocodazole for 1.5 h to enrich cells at metaphases. Chromosome spreads were made by a routine method. Metaphase-enriched cells were exposed to hypotonic treatment with 75 mM KCl solution, fixed with methanol: glacial acetic acid (3:1) and spread onto clean slides. Telomere FISH and quantification were performed as described previously [25], except for FITC-labeled (CCCTAA) peptide nucleic acid (PNA) probe used in this study. Telomeres were denatured at 80 °C for 3 min and hybridized with telomere specific PNA probe (0.5 μg/mL) (Panagene, Daejeon, Korea). Chromosomes were counter-stained with 0.5 μg/mL DAPI. Fluorescence from chromosomes and telomeres was digitally imaged on a Zeiss microscope with FITC/DAPI filters, using AxioCam and AxioVision software 4.6. For quantitative measurement of telomere length, telomere fluorescence intensity was integrated using the TFL-TELO program (gift kindly provided by P. Lansdorp), and calibrated using standard fluorescence beads.

2.8. Telomere Measurement by Quantitative Real-Time PCR (T/S Ratio)

Genome DNA was prepared using E.N.Z. DNeasy Blood & Tissue Kit (QIAGEN, Valencia, CA, USA). Average telomere length was measured from total genomic DNA using a real-time PCR assay [26]. PCR reactions were performed on the iCycleriQ real-time PCR detection system (Bio-Rad, Hercules, CA, USA), using telomeric primers, primers for the reference control gene (mouse 36B4 single copy gene) and PCR settings as previously described [27]. For each PCR reaction, a standard curve was made by serial dilutions of known amounts of DNA. The telomere signal was normalized to the signal from the single copy gene to generate a T/S ratio indicative of relative telomere length. Equal amounts of DNA were used for each reaction. Primers for T/S ration assay are shown in Table S2.

2.9. Telomere Chromatid Orientation-Fluorescence In Situ Hybridization (CO-FISH)

CO-FISH assay was performed according to our previous work [24]. Briefly, sub-confluent cells were incubated with 10 μM of 5′-bromo-2′-deoxyuridine (BrdU) for 12 h to allow BrdU incorporation for one cell cycle. Nocodazole was added for 2 h prior to cell harvest, and metaphase spreads were prepared by standard cytogenetic method as described above for Q-FISH. Chromosome slides were treated with RNaseA, fixed with 4% formaldehyde, then stained with Hoechst 33258 (0.5 mg/mL) for 15 min and exposed to 365 nm UV light (Stratalinker 1800 UV irradiator) for 40 min. The BrdU-substituted DNA was digested with Exonuclease III (Promega, Madison, WI, USA). The slides were then dehydrated through cold ethanol series and air dried. PNA-FISH was performed with

Fluorescein-OO-(CCCTAA)$_3$ (Bio-Synthesis, Lewisville, TX, USA). Slides were hybridized, washed, dehydrated, mounted, and counterstained with VectaShield antifade medium (Vector, Peterborough, UK), containing 0.1 mg/mlDAPI. Digital images were captured using a CCD camera on a Zeiss Axio-Imager Z1 microscope equipped with Metasystems Isis software.

2.10. Global DNA Methylation Analysis

DNA methylation was analyzed by immunostaining with 5-MeC (NA81, Calbiochem, San Diego, CA, USA) and 5hmC (39769, active motif) antibodies as we previously described [28]. ES cells were dissociated in trypsin–EDTA and washed in PBS twice and fixed in ice-cold 3.7% PFA for 30 min, washed in 0.05% PBT, permeabilized in 0.2% Triton X-100 solution for 30 min. Then, cells were depurinated in 4 N HCl, 0.1% Triton X-100 for 10 min, washed, and then blocked in 2% BSA in PBS. Samples were incubated with mouse anti-5-MeC or anti-5hmC antibody over-night at 4 °C, washed and then incubated with FITC conjugated goat anti-mouse IgG (A-11001, Molecular Probes, 1:500) or FITC conjugated anti-rabbit IgG (554020, BD Biosciences, 1:500) in blocking solution for 1 h at RT, washed, and then ES cells were placed in the fluorescence-activated cell sorting for analysis and sorting by flow cytometry (BD Biosciences, Franklin Lakes, NJ, USA). Cells with immunostaining only with normal mouse or rabbit IgG served as negative controls. Fluorescence distribution and intensity of cell populations following immunostaining with anti-5-MeC and 5hmC antibodies were quantified by FACS analysis.

2.11. Bisulfate Sequencing

WT, G4 Terc$^{-/-}$ ES cells and WT ES cells treated with 1 μM 5-aza-dC for 2.5 days were analyzed for DNA methylation of mouse *Zscan4c* promoter and *Tcstv1*/*Tcstv3* promoter. HeLa and U2 OS cells were used to analyze for DNA methylation of human ZSCAN4 promoter. Briefly, 1 μg of genomic DNA was subject to sodium bisulfite conversion using the EpiTect Bisulfite kit (QIAGEN). Mouse *Zscan4c* promoter and human ZSCAN4 promoter were amplified using the following bisulfite primers. For mouse *Zscan4c* promoter, forward: GATTAGATTTTGAAGAATATATTTTTTGTG; reverse: CCTCCAATATAACAAAACCCT-TAAC. For mouse *Tcstv1/3* promoter, forward: TTGGTATATTTGGTGGGTTAAGAAG; reverse: AATCCACAATTCTCCTTCAAAAATA. For human ZSCAN4 promoter, forward: TGGTTTTTATAGGTTTTGTATAGATT; reverse: AAAACTAAAATCCCCTACCAACTTC. The PCR products were cloned into the pEasy-T1 simple vector (Transgene) through TA cloning, and 8–15 colonies were randomly picked and sequenced and analyzed for each sample.

2.12. ChIP-qPCR Analysis

WT and G4 Terc$^{-/-}$ ES cells, HeLa and U2 OS cells were used for ChIP experiment as we previously described [24]. Briefly, cells were fixed with freshly prepared 1% paraformaldehyde for 10 min at room temperature. Cells were then harvested and their nuclei extracted, lysed, and sonicated. DNA fragments were then enriched by immunoprecipitation with 4 μg H3K9me3 (ab8898, Abcam) antibody per ChIP. The eluted protein:DNA complex was reverse-crosslinked at 65 °C overnight. DNA was recovered after proteinase and RNase A treatment. In ChIP-qPCR experiments, the ChIP-enriched DNA was analyzed by real-time PCR using primers for mouse and human Zscan4 promoter regions, β-actin served as negative control, and relative fold changes were normalized to IgG controls. The primers used are in Table S4.

2.13. Co-Localization of Telomeresand Human Chromosome 19

Telomeric FISH was performed as previously described [27] with minor modifications. Briefly, after slides were fixed for 2 min in 4% formaldehyde, dehydrated in a series of ethanol (70%, 90%, and 100%) and air dried, they were denatured with 0.5 μg/mL of telomere probe and 2 μL of chromosome 19 probe that specifically hybridizes to 19p13 in

10 μL of hybridization buffer (offered in chr 19 probe kit) (CHR19-10-RE, Empire Genomics) at 73 °C for 3 min and hybridized at 37 °C for 2.5 h. After hybridization, slides were washed twice with 70% formamide, followed by three washes in TBS-Tween (0.08%) buffer. Then slides were dehydrated with a series of ethanol and air dried. Chromosomes were stained with 0.5 μg/mL DAPI. Images were taken at 63× magnification using a Zeiss fluorescence microscope.

2.14. Statistical Analysis

Data were analyzed by ANOVA and means compared by Fisher's protected least-significant difference (PLSD) using the StatView software from SAS Institute Inc. (Cary, NC, USA). Significant differences were defined as $p < 0.05$, 0.01, or lower.

3. Results

3.1. Late Generation Terc$^{-/-}$ ES Cells Maintains Stably Short Telomere Length during Prolonged Passaging

The late fourth generation Terc$^{-/-}$ ES cells (G4) show similar morphological characteristics of WT ES cells, even though they harbor a slower proliferation rate [7]. Short telomeres were maintained in G4 Terc$^{-/-}$ ES cells during prolonged passaging (Figure 1). Telomere quantitative fluorescence in situ hybridization (QFISH) analysis of telomere length dynamics in WT ES cells shows that telomeres elongated dramatically from passage 12 to 20 and to 30 ($p < 0.0001$) (Figure 1A), consistent with previous reports that telomeres lengthened during passaging [29–31]. In contrast, G4 Terc$^{-/-}$ ES cells maintained their telomeres without notable shortening between passages 12 and 22 ($p = 0.2351$, 20.83 ± 14.17 TFU vs. 20.15 ± 16.44 TFU) (Figure 1B), but displayed increased frequency of signal-free ends, indicating telomere loss (Figure 1B). Notably, G4 Terc$^{-/-}$ ES cells showed typical ALT characteristics of heterogeneous telomeres, with some telomeres longer than others (indicated by white arrows), unlike telomeres of WT ES cells (Figure 1C). The telomere QFISH data on telomere elongation in WT and telomere maintenance of G4 Terc$^{-/-}$ ES cells was further validated by quantitative real-time PCR shown as T/S ratio [26] (Figure 1D), suggesting a telomerase-independent mechanism of telomere maintenance might be triggered by short telomeres in G4 Terc$^{-/-}$ ES cells. Indeed, the frequency of telomere sister-chromatid exchange (T-SCE) events, as analyzed by chromosome orientation FISH (CO-FISH), increased remarkably in G4 Terc$^{-/-}$ ES cells (Figure 1E,F). These data suggest that telomerase-independent mechanisms are responsible for rescuing telomeres that would otherwise shorten in G4 Terc$^{-/-}$ ES cells without telomerase.

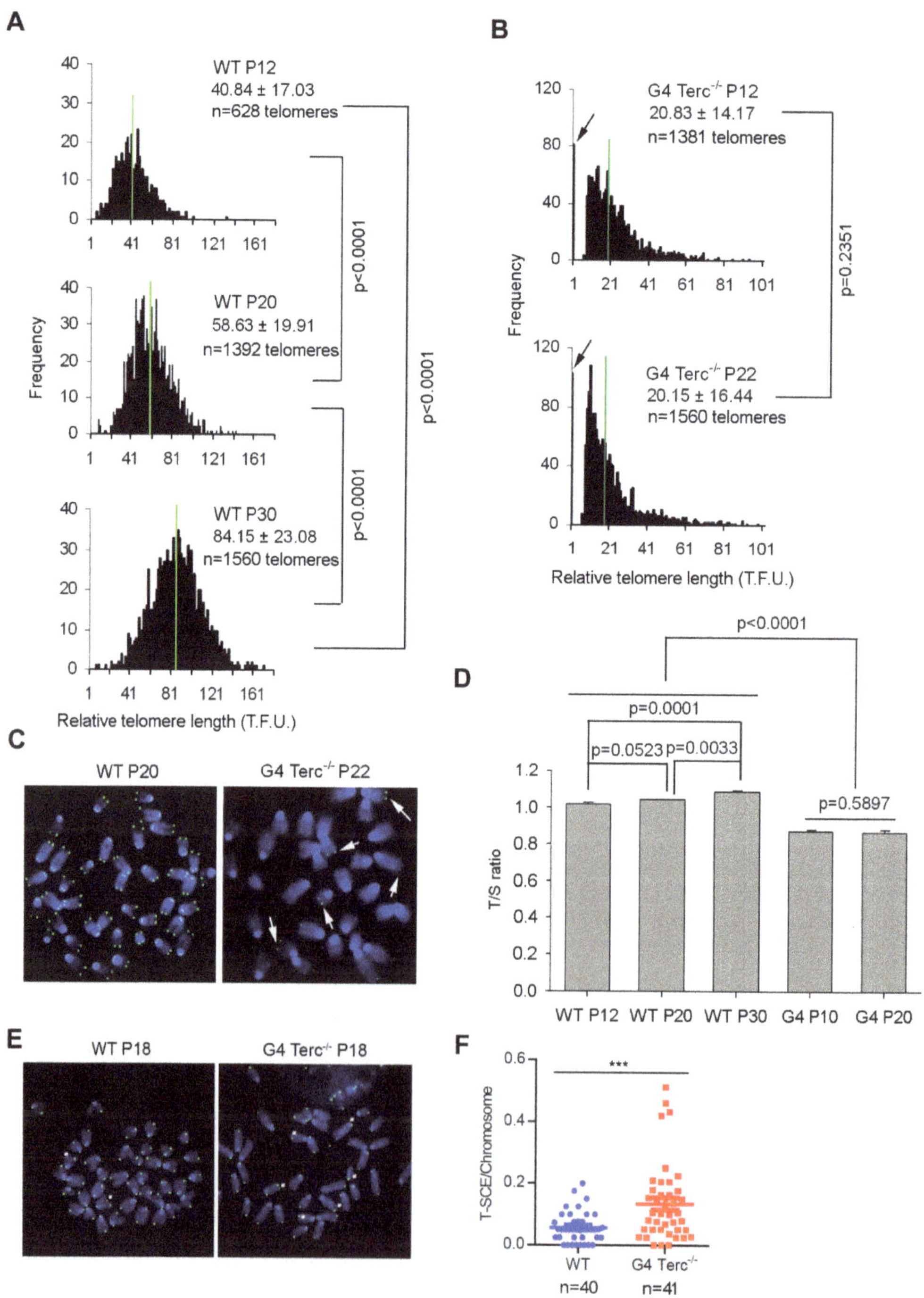

Figure 1. Telomere length dynamics in WT and G4 Terc$^{-/-}$ ES cells at different passages. (**A**,**B**) Distribution histogram of relative telomere length at different passages of WT (**A**) and G4 Terc$^{-/-}$ (**B**) ES cells, as analyzed by telomere Q-FISH and the TFL-TELO software. TFU, arbitrary telomere fluorescence unit.

Green lines indicate medium telomere length. The average length ± SD is given in the upper right corner. Arrows indicate frequency of telomere loss. (**C**) Representative telomere-specific PNA-FISH image of WT P20 and G4 Terc$^{-/-}$ P22 ES cells. The G4 Terc$^{-/-}$ ES cells show typical ALT characteristics, with some telomeres longer than others shown with white arrow. (**D**) Relative telomere length of WT and G4 Terc$^{-/-}$ ES cells at various passages shown as T/S ratio. Error bars indicate mean ± SEM (n = 3). (**E**) Micrographs showing T-SCE (arrows) by CO-FISH analysis. (**F**) Increased frequency of T-SCE per chromosome in G4 Terc$^{-/-}$ ES cells. n, number of metaphase spread images analyzed; ***, $p < 0.001$, compared to controls.

3.2. Up-Regulation of Zscan4 Is Implicated in Telomere Maintenance of G4 Terc$^{-/-}$ ES Cells

Interestingly, in comparison with global gene expression profile of WT ES cells [7], G4 Terc$^{-/-}$ ES cells showed upregulation of two-cell (2C) genes including *Tcstv1, Tcstv3, Gm4340, Dub1, Eif1a-like* (*Gm2022*), and *MuERV-L* [32], and particularly, *Zscan4*, which is involved in telomere recombination and elongation [16] (Figure 2A), validated by qPCR here (Figure 2B). Western blotting analysis also verified the up-regulation of Zscan4 protein level in G4 Terc$^{-/-}$ ES cells (Figure 2C). Notably, Rif1 protein level was not changed noticeably (Figure 2C), which was involved in negative regulation of Zscan4 expression by our previous report [24]. Consistently, *Zscan4* promoter activity analyzed by dual luciferase assay elevated dramatically in G4 Terc$^{-/-}$ ES cells compared with WTES cells ($p < 0.0001$) (Figure 2D), another evidence supporting the elevation of *Zscan4* expression in G4 Terc$^{-/-}$ ES cells with short telomeres. Moreover, the Zscan4$^+$ ES cell population in G4 Terc$^{-/-}$ ES cell cultures increased remarkably (Figure 2E,F). However, Terc$^{+/-}$ ES cells with sufficient telomere lengths and pluripotency [7], expressed *Zscan4* at levels similar to those of WT ES cells, in contrast to much higher *Zscan4* levels of G4 Terc$^{-/-}$ ES cells with short telomere (Figure 2G), indicating highly correlation between Zscan4 expression levels and telomere lengths (Figure 2H). These results together further suggest that short telomeres lead to excessive activation of *Zscan4*. Consistently, a previous report with live cell imaging technique showed that *Zscan4* is activated after telomere shortening in mESCs to recover the shortened telomeres during extended cell cycles [33].

To test whether *Zscan4* is implicated in telomere maintenance of G4 Terc$^{-/-}$ ES cells, we depleted *Zscan4* by specific shRNAs targeting *Zscan4* gene (targeting all *Zscan4* isoforms) (Table S1). *Zscan4* mRNA levels were reduced to minimal after *Zscan4* knockdown (KD) (Figure 2I). Telomeres of *Zscan4* KD G4 Terc$^{-/-}$ ES cells were shorter than those of control KD G4 Terc$^{-/-}$ ES cells analyzed by telomere QFISH (Figure 2J,K). Furthermore, cell cycle progression was reduced as evidenced by shorter S phase but longer G2 phase following *Zscan4* depletion compared to control KD G4 Terc$^{-/-}$ mESCs (Figure 2L). Collectively, *Zscan4* was excessively activated in G4 Terc$^{-/-}$ ES cells with short telomeres in the absence of telomerase activity and likely responsible for ALT-mediated telomere elongation to compensate for telomere shortening.

3.3. DNA Methylation and H3K9me3 Regulate Zscan4 of G4 Terc$^{-/-}$ ES Cells

We searched for the mechanisms underlying up-regulation of *Zscan4* in G4 Terc$^{-/-}$ ES cells with very short telomeres. Further analysis of the microarray data in Figure 2A identified that many of the up-regulated genes in G4 Terc$^{-/-}$ ES cells, including *Zscan4, Tcstv1, Tcstv3,* and *Gm4340,* are located at subtelomeric or pericentromeric regions. *Zscan4* gene clusters are located at subtelomeric regions of short p arm of chromosome 7 (Figure S1A). Telomeres and subtelomeres are densely enriched with repressive DNA and histone modifications (e.g., H3K9me3) that regulate mammalian telomere length negatively [9,34]. We hypothesized that the repressive DNA methylation and histone modifications may inhibit gene expression at subtelomeres including *Zscan4*, and *Tcstv1/3*, and that reduction of the repressive DNA and histone modifications de-represses *Zscan4* and contributes to telomere maintenance by activating recombination-based telomere elongation mechanisms in ES cells with short telomeres.

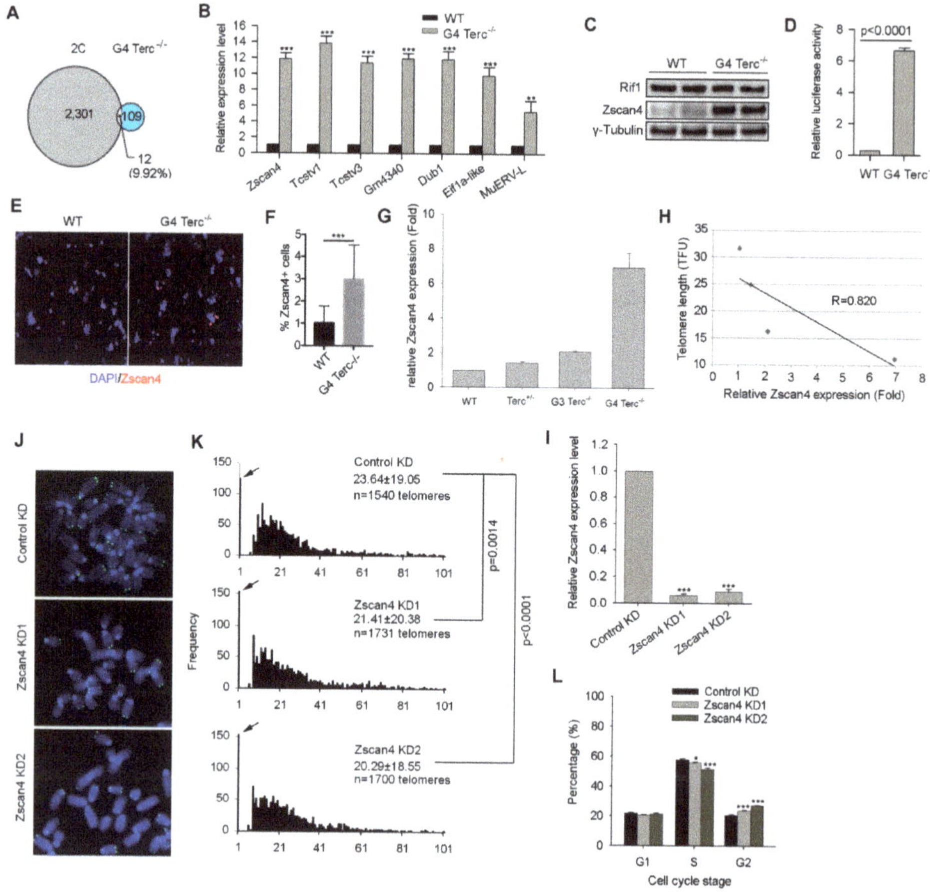

Figure 2. High expression level of Zscan4 contributes to telomere maintenance in G4 Terc$^{-/-}$ ES cells. (**A**) A Venn diagram shows the overlap between 2C-genes and genes upregulated in G4 Terc$^{-/-}$ ES cells (2C-gene data [32] and G4 Terc$^{-/-}$ data [7] were analyzed). (**B**) Two cell-specific genes *Zscan4, Tcstv1, Tcstv3, Gm4340, Dub1, Eif1a-like,* and MuERV-L are all up-regulated in G4 Terc$^{-/-}$ ES cells. *Zscan4, Tcstv1,* and *Tcstv3* are located at sub-telomeric regions and *Gm4340* is located at peri-centromeric region in mouse genome. (**C**) Western blotting analysis of Zscan4 and Rif1 protein levels between WT and G4 Terc$^{-/-}$ ES cells. Two technical repeats were included. (**D**) Relative luciferase activity of Zscan4c promoter between WT and G4 Terc$^{-/-}$ ES cells. (**E**) Representative Zscan4 immunofluorescence images of Zscan4^{+} cells in WT and G4 Terc$^{-/-}$ ES cells. (**F**) Relative percentage of Zscan4^{+} cells. Error bars indicate mean ± SEM. (**G**) Comparison of Zscan4 relative expression levels by qPCR analysis of WT, Terc$^{+/-}$, G3, and G4 Terc$^{-/-}$ ES cells. (**H**) Relative telomere length shown as TFU of ESCs is highly (R = 0.820) correlated with Zscan4 expression levels among different cells lines. The telomere length data used for correlation analysis were showed in our previous paper [7]. (**I**) Zscan4 mRNA levels were reduced by shRNA sequences targeting Zscan4 gene. (**J**) Representative telomere specific PNA-FISH images of control and Zscan4 KD G4 Terc$^{-/-}$ ES cells. (**K**) Histogram shows telomere distribution and relative lengths following Zscan4 KD by telomere QFISH. Frequency of telomere end free-signals is shown on Y-axis indicated by arrows. Upper right-hand corner, the average length ± SD. (**L**) Cell cycle progression analysis shows reduced S phase and increased G2 phase of G4 Terc$^{-/-}$ ES cells following knockdown of *Zscan4*. Error bars indicate mean ± SEM (n = 3). *, $p < 0.05$; **, $p < 0.01$; ***, $p < 0.001$, compared to controls.

We first assessed the global DNA methylation levels between WT and G4 Terc$^{-/-}$ ES cells. FACS analysis showed reduced global 5mC/5hmC relative fluorescence intensity in G4 Terc$^{-/-}$ ES cells compared with WT ES cells (Figure 3A and Figure S1B). DNA methylation levels could be balanced by DNA methyltransferases (DNMT enzymes) and 5-methylcytosine oxidases (TET enzymes). All three DNMT enzymes were decreased while Tet2 was increased in G4 Terc$^{-/-}$ ES cells as analyzed by immunoblot assays (Figure 3B). Quantitative PCR analysis also confirmed the downregulation of *Dnmt3a* and *Dnmt3b*, up-regulation of *Tet2*, in G4 Terc$^{-/-}$ ES cells, respectively (Figure S1C). Interestingly, *Dnmt1* mRNA level was not changed despite significant reduction at protein level in G4 Terc$^{-/-}$ ES cells (Figure 3B and Figure S1C), suggesting post-transcriptional regulation of Dnmt1 involved in G4 Terc$^{-/-}$ ES cells. Consistently, *Tert* KO ES cells with short telomeres also showed the reduction of both Dnmt3a and Dnmt3b protein levels and global DNA methylation [35]. Previously, we have demonstrated that over-expression of Tbx3 in mouse ES cells led to decreased Dnmt3b and increased Tet2 protein levels, respectively, synergistically resulting in reduced DNA methylation level and elevated Zscan4 expression level [28]. Indeed, *Tbx3* was upregulated in G4 Terc$^{-/-}$ ES cells compared to WT ES cells (Figure S1C).

In support of the inhibitory effects of DNA methylation on *Zscan4*, ES cells were treated with DNA methyltransferase inhibitor, 5-aza-deoxycytidine (5-aza-dC), and showed decreased Dnmt1 and Dnmt3b protein levels (Figure 3C). Consistently, the global DNA methylation was decreased significantly following 5-aza-dC treatment (Figure 3A and Figure S1B). Furthermore, bisulfate sequencing analysis showed that *Zscan4c* promoter was heavily methylated in WT ES cells and decreased slightly in G4 Terc$^{-/-}$ ES cells, and decreased further in 5-aza-dC treated ES cells (Figure 3D,E). Moreover, significantly decreased DNA methylation also was found at *Tcstv1/3* promoter (*Tcstv1* and *Tcstv3* genes share the same promoter sequence) in G4 Terc$^{-/-}$ ES cells (Figure S1D,E).

Next, we compared the histone modifications between WT and G4 Terc$^{-/-}$ ES cells. Heterochromatin marked by H3K9me3 foci was reduced in G4 Terc$^{-/-}$ ES cells, compared with WT ES cells as analyzed by both immunofluorescence and immuoblot assays (Figure 3F–H). We have previously demonstrated that *Zscan4* was repressed by H3K9me3 binding at its promoter [24]. Thus, we asked whether the reduced H3K9me3 occupancy at specific genomic loci was within or near the proximity of *Zscan4c* loci in G4 Terc$^{-/-}$ ES cells, which could account for the de-repression of *Zscan4*. We performed ChIP-qPCR analysis to determine the H3K9me3 occupancy at *Zscan4c* loci and found that H3K9me3 level was decreased in six proximal locations upstream to *Zscan4c* promoter and subtelomeres of chromosome 7 further upstream to *Zscan4c* locus (Figure 3I). Moreover, H3K9me3 level was also decreased in *Tcstv1/Tcstv3* proximal promoter at subtelomeres of chromosome 13 (Figure 3I), also in line with the upregulation of *Tcstv1/Tcstv3* expression in G4 Terc$^{-/-}$ ES cells (Figure 2A,B). Taken together, short telomeres reduce DNA methylation and H3K9me3 levels and occupancy at subtelomeres and telomeres of G4 Terc$^{-/-}$ ES cells and thus de-repress *Zscan4* expression at subtelomeres.

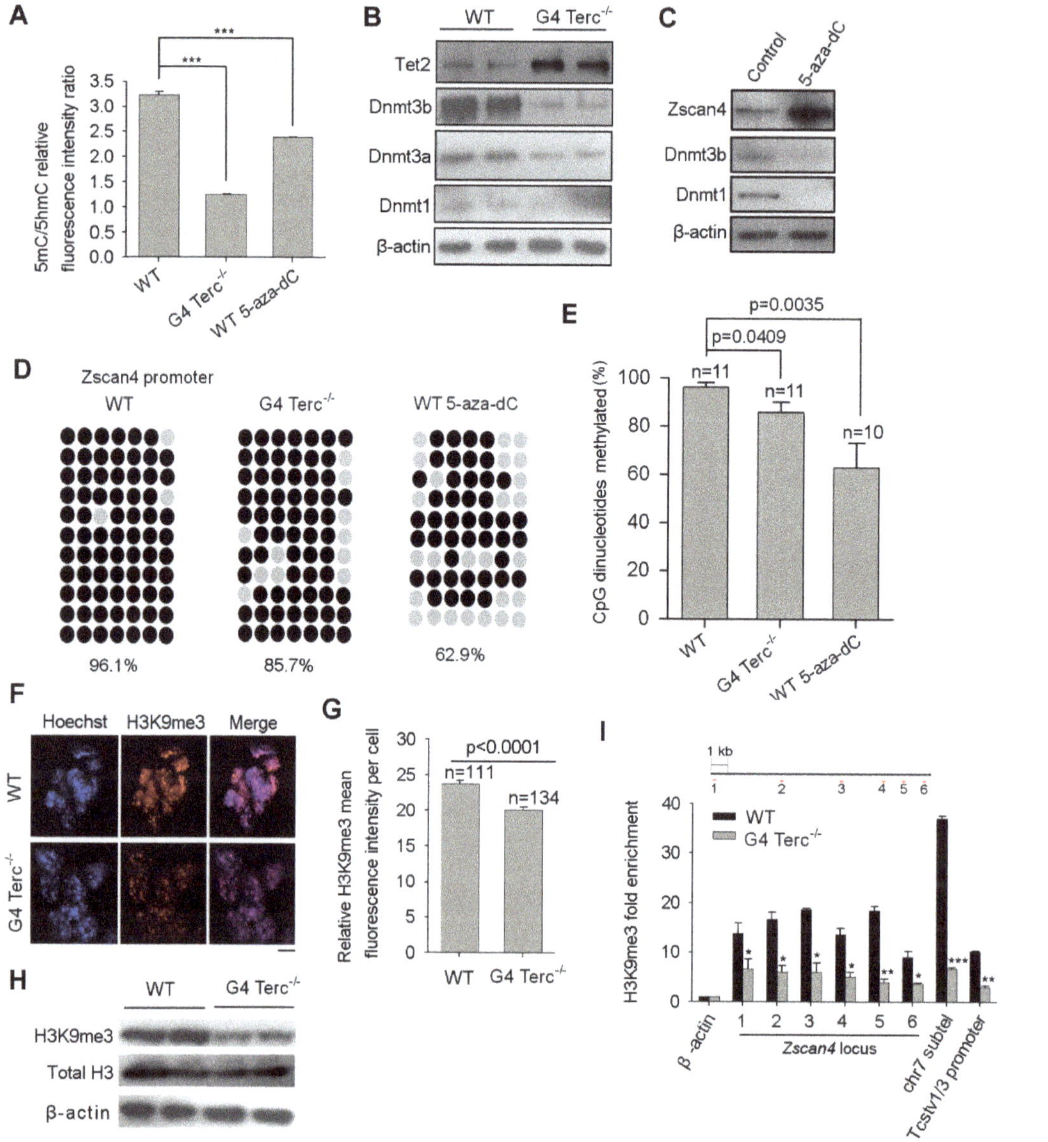

Figure 3. Reduction of DNA methylation and H3K9me3 occupancy at Zscan4 promoter regions in G4 Terc$^{-/-}$ ES cells with short telomeres. (**A**) Relative quantity (RQ) of DNA methylation levels among WT, G4 Terc$^{-/-}$, and 5-aza-dC treated WT ES cells (1 μM for 2.5 days). Quantification was made by quantifying mean fluorescence intensity of cell populations following immunostaining with anti-5mC and 5hmC antibodies and analyzed by FlowJo software. Error bars indicate mean ± SEM (n = 3). ***, $p < 0.001$, compared to WT controls. (**B**) Western blotting shows decreased Dnmt1, Dnmt3a, Dnmt3b, and increased Tet2 protein levels in G4 Terc$^{-/-}$ ES cells, respectively. Two technical repeats were included. (**C**) Western blotting shows decreased Dnmt1, Dnmt3b protein levels after 5-aza-dC treatment (1 μM for 2.5 days) in WT ES cells. (**D**) Bisulfate sequencing analysis of CpG dinucleotides of mouse Zscan4c proximal promoter regions among WT, G4 Terc$^{-/-}$ and 5-aza-dC treated WT ES cells

(1 μM for 2.5 days). Open circles, unmethylated; closed circles, methylated. (**E**) Quantification of bisulfate genomic sequencing results of 10–11 independent clones (n) at Zscan4c proximal promoter. Note the decrease in methylated CpGs at Zscan4c proximal promoter in G4 Terc$^{-/-}$ and 5-aza-dC treated WT ES cells compared with WT ES cells. Error bars indicate mean ± SEM. (**F**) Representative images shows immunofluorescence staining of heterochromatic H3K9me3 (red). Nuclei stained with Hoechst33342 (blue) show brighter fluorescence signal in the heterochromatin. Scale bar, 10 μm. (**G**) Quantification of relative H3K9me3 mean fluorescence intensity between WT and G4 Terc$^{-/-}$ ES cells using ImageJ software. Randomly chosen cells from 9–10 fields of view were counted. n, number of cells counted. (**H**) Western blotting analysis shows the decreased inhibitive histone modification H3K9me3 in G4 Terc$^{-/-}$ ES cells. (**I**) ChIP-qPCR analysis of H3K9me3 occupancy at six proximal locations upstream to *Zscan4c* promoter, sub-telomere further upstream to *Zscan4c* promoter of chr7 and *Tcstv1*/*Tcstv3* promoter at subtelomere of chr13. Relative fold enrichment was normalized to rabbit IgG ChIP signals at the same regions, and the β-actin locus served as negative controls. Error bars indicate mean ± SEM. *, $p < 0.05$; **, $p < 0.01$; ***, $p < 0.001$, compared to controls.

3.4. Inhibition of Repressive DNA and H3K9 Methylation Up-Regulates Zscan4 and Elongates Telomeres

Decreased telomeric repressive DNA and histone modifications (e.g., H3K9me3) relieve the inhibition of *Zscan4* at subtelomeric heterochromatin regions, contributing to telomere maintenance by activating telomere recombination. Next, we tested whether inhibiting repressive DNA and H3K9 methylation by small molecules can also relieve *Zscan4* inhibition and elongate telomere length. Treatment of ES cells with DNA methyltransferase inhibitor 5-aza-dC, or histone H3K9methyltransferase inhibitor BIX-01294, or both, all significantly elevated Zscan4 protein levels (Figure 4A). Consequently, treatments with 5-aza-dC, orBIX-01294, or both, all significantly elongated telomere length as analyzed by QFISH analysis (Figure 4B,C). The telomere QFISH on telomere elongation data was further validated by quantitative real-time PCR shown as T/S ratio (Figure 4D). These data further support that epigenetic modification regulates expression of heterochromatic genes, and *Zscan4* at subtelomeres in this case.

3.5. ZSCAN4 Depletion in Telomerase-Deficient Human ALT U2 OS Cancer Cells Shortens Telomeres

ALT positive U2 OS cancer cells are characterized by promyelocytic leukemia (PML) bodies and high frequency of T-SCE, indicative of telomere recombination [36–38], and show extremely long telomeres: even longer than those of human ES cells. However, these ALT cells have minimal or no telomerase activity, in contrast to hES cells [39,40]. Given the similarity of telomerase-deficient mESCs and ALT human cancer cells in maintaining telomeres independent of telomerase involvement for cellular proliferation, we hypothesized that ZSCAN4 may also be involved in telomere maintenance of human ALT U2 OS cancer cells.

First, we compared the gene expression levels of ZSCAN4 in the well-defined telomerase-positive/ALT negative HeLa cells, telomerase negative/ALT positive U2 OS cells and telomerase/ALT double deficient fibroblast cells (negative control), and found that ZSCAN4 was highly up-regulated in U2 OS cells as analyzed by qPCR (Figure 5A). Western blotting analysis also showed dramatically increased human ZSCAN4 protein level in U2 OS cells (Figure 5B).

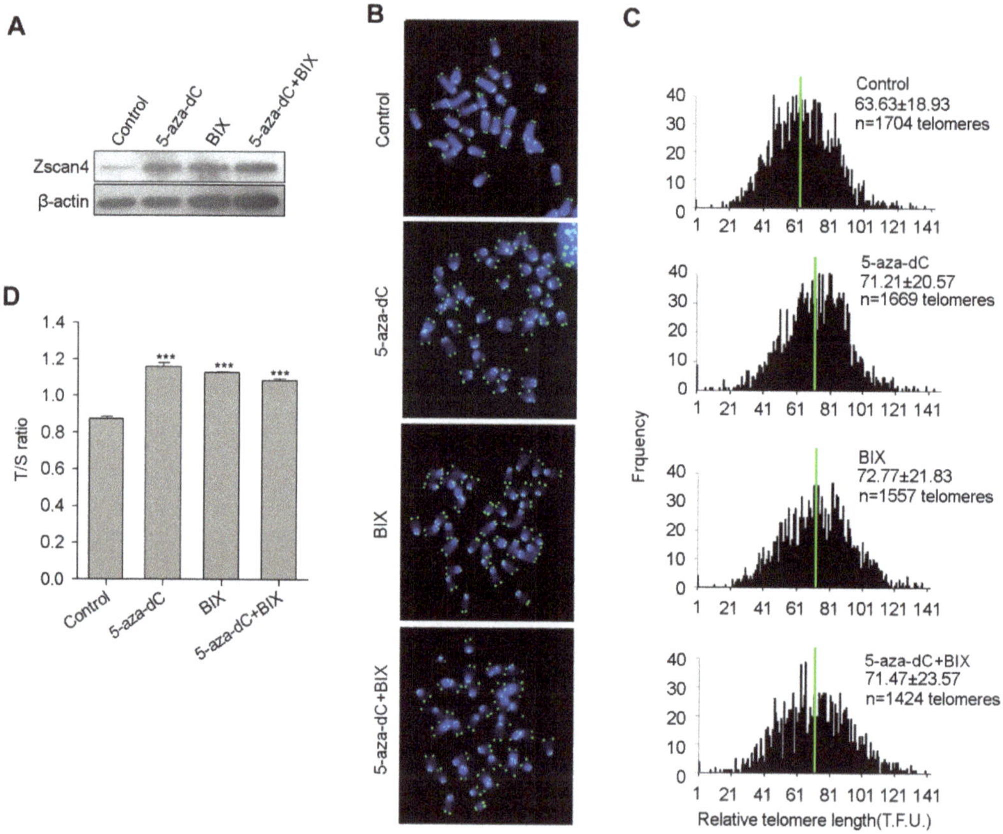

Figure 4. Inhibition of DNA methylation and H3K9 methylation elevates Zscan4 expression levels and elongates telomere lengths. (**A**) Western blotting analysis shows increased Zscan4 protein levels after treatment with small molecules. 5-aza-dC, DNA methyltransferases (DNMTs) inhibitor; BIX-01294, histone lysine methyltransferases (HMTases) inhibitor. WT ES cells were treated with 1 µM 5-aza-dC, or 1 µM BIX-01294, or both for 2.5 days. (**B**) Telomere specific PNA-FISH image of WT ES cells after treatment with 5-aza-dC, or/and BIX-01294 for four passages. (**C**) Distribution histogram of relative telomere length of WT ES cells after different treatments, analyzed by telomere Q-FISH and the TFL-TELO software. Green lines indicate medium telomere length. The average length ± SD is given in the upper right corner. Arrows indicate frequency of telomere loss. (**D**) Relative telomere length of WT ES cells after different treatments shown as T/S ratio. Experiments performed in (**B–D**) were treated with 1 µM 5-aza-dC, or 1 µM BIX-01294, or both, for four passages, 10 days. After the treatments, genomic DNA and chromosome spreads were used for T/S ratio and QFISH assays for telomere length measurement, respectively. Error bars indicate mean ± SEM (n = 3). ***, $p < 0.001$, compared to controls.

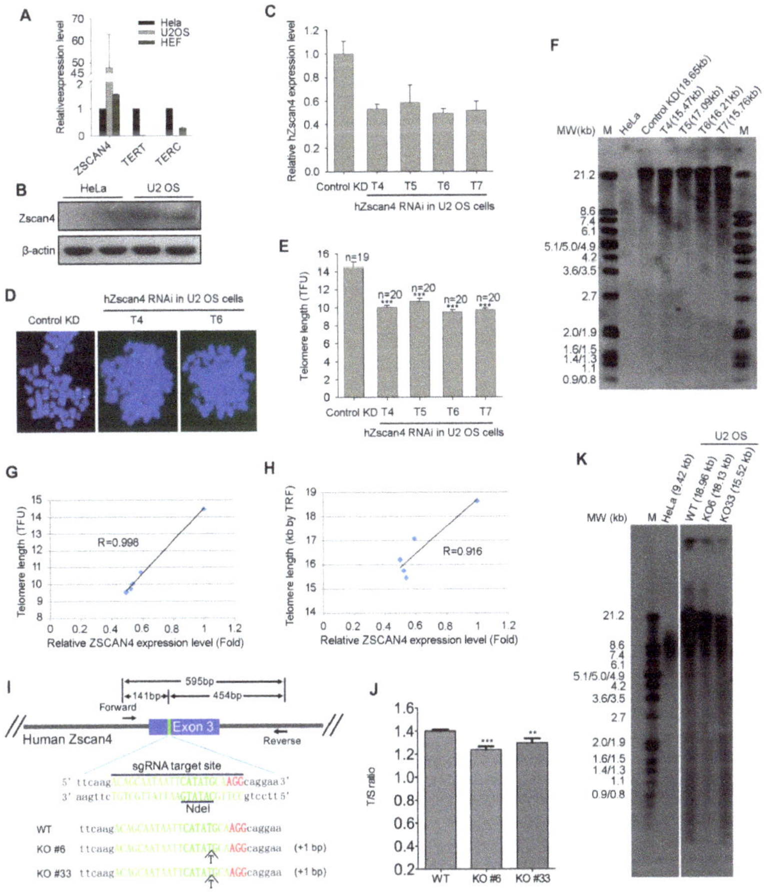

Figure 5. Depletion of ZSCAN4 in ALT U2 OS cells shortens telomere length. (**A**,**B**) Human ZSCAN4 is highly up-regulated in ALT positive U2 OS cells compared to ALT negative HeLa cells, analyzed by qPCR (**A**) and Western blotting (**B**). (**C**) Stable knockdown efficiency of ZSCAN4 analyzed by qPCR with four different ZSCAN4RNAi sequences. (**D**) Representative telomere specific PNA-FISH image of control and stable ZSCAN4 KD U2 OS cells. Image of control KD cells shows telomere length heterogeneity, typical ALT cell characteristics of human cancer cells. (**E**) Quantitative telomere length of control and ZSCAN4 stable KD U2 OS cells. The error bars indicate mean of telomere length ± SEM (n = 19–20 chromosome spreads). TFU, telomere fluorescence unit. ***, $p < 0.0001$, compared to control KD. (**F**) TRF analysis of control and stable ZSCAN4 KD U2 OS cells. A volume of 1.5 µg of DNA was digested and separated on a 0.8% agarose gel. M indicates a set of DNA size markers, and MW refers to DNA sizes in kilobases. Telomere length was measured using Telomeric version 1.2 software, and the median telomere length of each lane was given upper the gel. (**G**,**H**) Human ZSCAN4 expression levels after RNAi with different targeting sequences are highly correlated with relative telomere length analyzed by QFISH (R = 0.998) (**G**) and TRF assay (R = 0.916) (**H**). (**I**) Targeted disruption of ZSCAN4 gene by CRISPR/Cas9 technology in U2 OS cells. The designed sgRNA targeting human ZSCAN4 gene exon 3 is underlined, and the protospacer-adjacent motif (PAM) sequence labeled in red. By sequencing, two independent homologous KO clones were identified by insertion of a T at the restriction site. (**J**,**K**) ZSCAN4 KO in U2 OS cells shortens telomere length as analyzed by real time qPCR shown as T/S ratio (**J**) and TRF assay (**K**). Telomere length shown in (**J**) and (**K**) represents mean telomere length. Error bars indicate mean ± SEM (n = 3). **, $p < 0.01$, ***, $p < 0.001$, compared to controls.

To test whether ZSCAN4 is involved in telomere maintenance of U2 OS cells, we stably knocked-down ZSCAN4 with lentivirus-mediated RNA interference with high efficiency (Figure S2) using four different shRNA sequences targeting ZSCAN4 gene (Table S1). ZSCAN4 expression levels were decreased by the four shRNA sequences as analyzed by qPCR (Figure 5C). Telomeres shortened by telomere Q-FISH analysis of ZSCAN4 knockdown in U2 OS cells, using four different shRNA sequences, compared with knockdown controls (Figure 5D,E). TRF assay also confirmed the telomere shortening following ZSCAN4 KD (Figure 5F). Notably, the telomere shortening rate as analyzed by both Q-FISH and TRF assays is positively correlated with ZSCAN4 KD efficiency in ALT-U2 OS cells (Figure 5G,H). We further confirmed the telomere shortening by CRISPR-Cas9 mediated deletion of ZSCAN4 in U2 OS cells (Figure 5I–K). These data suggest that ZSCAN4-mediated telomerase-independent mechanisms can also be responsible for telomere length maintenance of human ALT positive U2 OS cancer cells.

3.6. Human ZSCAN4 Is Also Repressed by Epigenetic Marks at Its Promoter

Of note, the ZSCAN4 gene also locates at subtelomeric region of the p arm of chromosome 19 (Figure S3A). Interestingly, the telomere signal intensity in the p arm of chromosome 19 where the ZSCAN4 gene located was much weaker compared with the q arm of chromosome 19 in telomerase-deficient U2 OS cells (Figure 6A), which was not seen in telomerase-positive hES cells with similar telomere signal intensity between the p and q arm of chromosome 19 (Figure 6A). It is to note that short telomeres following telomerase-deficiency were coupled with reduced repressive DNA and H3K9me3 methylation marks in $Terc^{-/-}$ MEF cells [13] and in G4 $Terc^{-/-}$ ES cells in the present study (Figure 3), suggesting that ZSCAN4 may also be epigenetic regulated by repressive DNA methylation and H3K9me3 in U2 OS cells.

Bisulfate sequencing analysis showed only minimal difference in DNA methylation levels at ZSCAN4 promoter between HeLa and U2 OS cells (Figure S3B). However, H3K9me3 level was decreased dramatically in U2 OS cells by immunoblot analysis (Figure 6B,C). Furthermore, ChIP-qPCR analysis also showed significant reduced H3K9me3 occupancy at human ZSCAN4 promoter regions (Figure 6D). The above data suggests that human ALT U2 OS cancer cells also may maintain their telomeres in the absence of telomerase activity by reduced epigenetic silencing at subtelomeres and telomeres. Taken together, our results demonstrate that the reduced repressive epigenetic marks-mediated depressed heterochromatin silencing activates Zscan4 expression, which triggers telomere elongation and contributes to telomere maintenance in telomerase-deficient late generation mESCs and ALT U2 OS cancer cells. The positive feedback loop of reduced epigenetic marks/Zscan4 activation/telomere elongation axis ensures telomere homeostasis and the capacity of those cells to proliferate for many passages without telomerase activity (Figure 7).

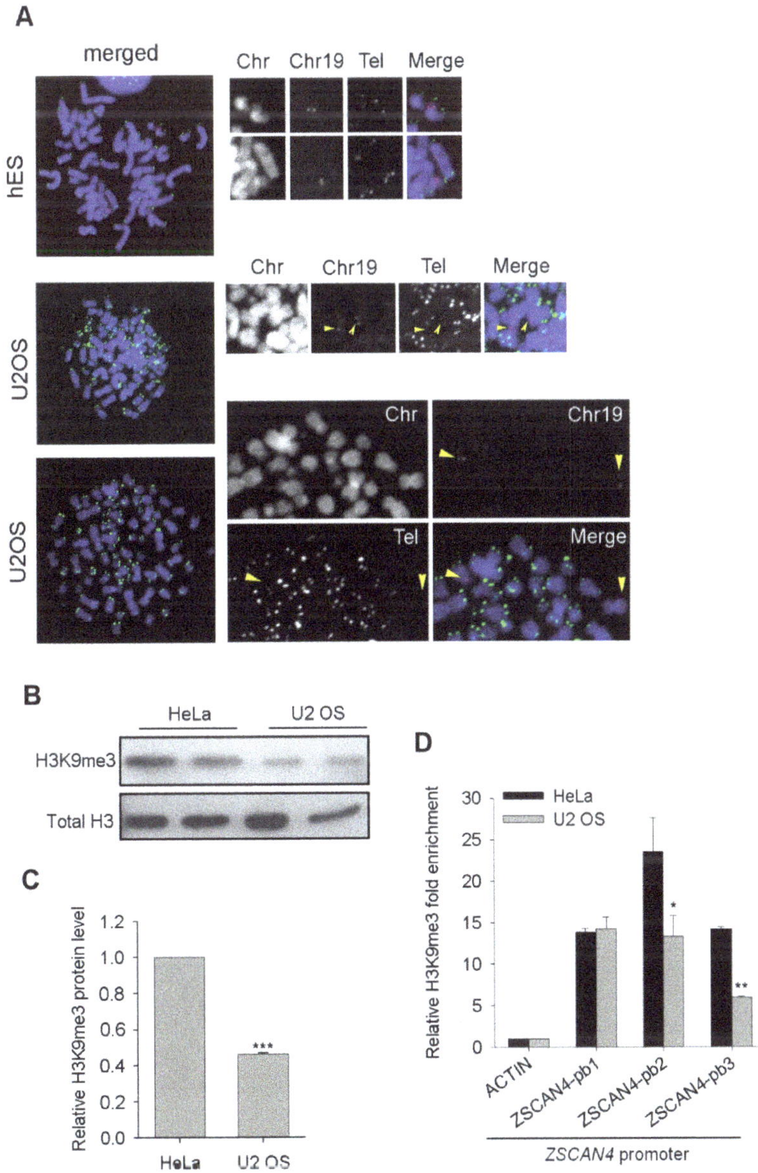

Figure 6. Heterochromatin de-repression activates ZSCAN4 in ALT-U2 OS cells. (**A**) Co-localization of telomeres and ZSCAN4 gene region on chromosome 19 in telomerase-positive hESCs and telomerase negative U2 OS cells. Orange arrowheads indicate telomere signals of chromosome 19 where ZSCAN4 gene locates. (**B**) Western blotting analysis of H3K9me3 levels between HeLa and U2 OS cells. (**C**) Relative protein quantity of H3K9me3 normalized to β-actin by Bio-Rad Quantity One software. Error bars indicate mean ± SEM (n = 2). ***, $p < 0.001$, compared to controls. (**D**) ChIP-qPCR analysis of H3K9me3 occupancy at 3 proximal locations upstream to ZSCAN4 promoter at subtelomeric regions of chromosome 19. Relative fold enrichment was normalized to rabbit IgG-ChIP signals at the same regions, and the β-actin locus served as negative controls. 'pb' stands for promoter binding site. Error bars indicate mean ± SEM (n = 2). *, $p < 0.05$, **, $p < 0.01$, compared to controls.

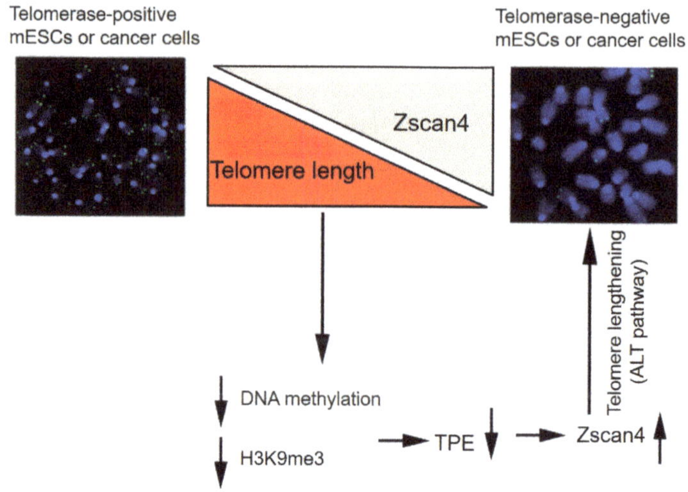

Figure 7. Proposed model of telomere length homeostasis shared in telomerase-deficient late generation mESCs and ALT-U2 OS cells. The telomerase-deficient mouse ES cells and ALT-U2 OS cancer cells share similar mechanisms to maintain telomere homeostasis. *Zscan4* is activated due to reduced TPE associated with shortened telomeres at subtelomeric *Zscan4* loci in those cells. The up-regulation of *Zscan4* activates ALT mechanisms, ensuring telomere homeostasis and the capacity of those cells to proliferate for many passages without telomerase activity.

4. Discussion

Telomere length maintenance is essential for indefinite proliferation of mouse ES cells and cancer cells. Very short telomeres could impair functional cellular activities, leading to cell proliferation crisis, cell senescence, and aging [6]. The late generation telomerase-deficient mouse ES cells and telomerase-deficient human ALT cancer cells are able to propagate for numerous passages [7,41,42]. However, how telomeres are maintained in those cells and the mechanisms responding for the telomere maintenance are unclear. Based on our data, we propose that *Zscan4*, which is highly upregulated in those telomerase-deficient late generation mouse ES cells and human ALT U2 OS cells, contributes to telomere maintenance of those cells without telomerase activities.

Zscan4 was originally found to specifically express in two-cell embryos and transiently in sporadic ES cells (1–5%) at any given time [16,17], and to lengthen telomeres by recombination-based mechanisms in mouse ES cells [16]. In the case of cancer cells, ZSCAN4 was recently demonstrated to highly express in human head and neck squamous cell carcinoma and to play important roles in the maintenance of cancer stem cell phenotype [43], possibly through the enhancement of telomere maintenance. In the current study, we show that telomere lengths of late generation G4 $Terc^{-/-}$ ES cells maintain stable levels while telomeres of WT ES cells lengthen remarkably during prolonged passaging. Shorter telomeres of G4 $Terc^{-/-}$ ES cells coincide with higher expression of *Zscan4*. In contrast, $Terc^{+/-}$ ES cells with telomerase, despite haplo-insufficiency, show much longer telomeres and expression levels of *Zscan4* similar to WT ES cells, supporting the notion that shorter telomeres could reduce TPE and thus activate *Zscan4* expression. Coincidently, *Zscan4* is localized at the subtelomeres of both mouse and human chromosomes, suggesting that activation of *Zscan4* in both species is regulated by TPE. Interestingly, we previously demonstrated that Zscan4 facilitates telomere elongation by promoting global DNA demethylation in 2C-like mouse ES cells [15]. Moreover, we also showed in the current study that *Zscan4* upregulation and telomere elongation were pronounced after treatment with DNA methyltransferase inhibitor 5-aza-dC (Figure 4). Thus, Zscan4 forms a positive feedback loop to activate itself and promote telomere elongation by inducing DNA hypomethylation.

Remarkably, telomeres lengthen rapidly in the early cleavage embryos, progressively in company with global DNA demethylation [27], while *Zscan4* is only expressed in two-cell embryos. Surprisingly, there is no detectable telomerase activity in fertilized zygotes and preimplantation embryos and, telomerase is only reactivated after blastocyst stage [27], suggesting other mechanisms responsible for the telomere elongation beyond the two-cell stage.

Furthermore, specific inhibition of H3K9 methylation results in significant increases in *Zscan4* and telomere elongation, suggesting that loss of repressive histone H3K9me3 may be responsible for ALT cells. By genome-wide epigenetic screen, at the megabase scale, a single feature—levels of H3K9me3—represents major histone modification that can account for more than 40% of mutation-rate variation in human cancer cells [44]. Compared with telomerase-positive HeLa cells, H3K9me3 modification also decreased at human ZSCAN4 promoter regions, located at subtelomeres of the p arm of human chromosome 19 with weak telomere signal intensity, in telomerase-deficient-ALT positive U2 OS cells. Coincidently, ZSCAN4 was highly expressed in telomerase-deficient-ALT positive U2 OS cells compared with telomerase-positive HeLa cells. Moreover, knockdown and Knockout of human ZSCAN4 in U2 OS cells significantly shortened telomere length, suggesting human ZSCAN4 also functions in the contribution of telomere maintenance in human ALT cancer cells without telomerase activity.

However, the exact mechanisms on how Zscan4 facilitates telomere elongation in both mouse ESCs and human ALT-U2 OS cancer cells remain to be elucidated. Interestingly, it has been shown that TERRAs, the telomeric noncoding RNAs transcribed from subtelomeric promoters by RNA polymerase II, destabilize telomere integrity to initiate break-induced replication (BIR) in human ALT cancer cells [45]. Remarkably, there is ample evidence that BIR processes re-elongate damaged telomeres occurring in the G2 and M phases of the cell cycle through DNA polymerase delta accessory subunits POLD3 and POLD4 [46–49]. Thus, TERRA transcription elevation results in alternative telomere lengthening through BIR processes. Notably, TERRA expression is inhibited by the repressive epigenetic marks at subtelomeres and, reduction of repressive H3K9me3 and DNA methylation modifications alleviate the repression of TERRA expression [50,51]. Moreover, telomere length negatively regulates TERRA expression. In somatic and ES cells, short telomeres, which are associated with decreased repressive epigenetic marks [9], promote TERRA expression [50], while telomere elongation decreases TERRA molecule levels [51]. Therefore, it is possible that Zscan4 might promote telomere recombination and elongation by upregulating TERRA expression by modulating subtelomeric DNA demethylation, leading to the activation of BIR processes to elongate telomeres. It will be of great interest to investigate this possibility in the future.

Anti-telomerase therapy provokes ALT and other adaptive mechanisms in telomerase-positive cancers [52,53], which account for 85–90% of human cancers, limiting its potential therapeutic usage in clinical medicine. Telomerase-depletion of telomere short cells indeed can induce cell senescence and cell death, but the prolonged depletion of telomerase leads to resistance and activation of ALT [52]. Thus, ALT has the potential to function as an escape mechanism for cells undergoing anti-telomerase therapies, elucidation of the molecular details of the ALT pathway is critical if we are to target these tumors [54]. Combination of anti-telomerase and anti-ALT therapies may shed promising light to anti-cancer battle. Although the regulators underpinning ALT induction and maintenance are poorly understood, recent evidence suggests that SLX4IP and RAD51AP1 are two essential regulators of ALT telomere maintenance in ALT cancer cells [55,56]. Further elucidation of mechanisms involved in ALT telomere lengthening remains to be determined. Our studies proved that human ZSCAN4 gene was highly upregulated in ALT-U2 OS cells, and depletion of ZSCAN4 by shRNA sequences and CRISPR-Cas9 mediated deletion shorten telomeres in ALT-U2 OS cells.

Together, the telomerase-deficient late generation mESCs and ALT-U2 OS cancer cells harbor a similar mechanism to contribute to telomere maintenance for unlimited self-renewal in the absence of telomerase activity. Our data indicate that ZSCAN4 might be a useful target to complement cancer therapy using telomerase inhibitors.

Supplementary Materials: The following supporting information can be downloaded at: https://www.mdpi.com/article/10.3390/cells11030456/s1, Figure S1: DNA methylation level and occupancy at *Tcstv1/3* promoters in G4 Terc$^{-/-}$ mESCs, Figure S2: ALT U2 OS cells were infected with human ZSCAN4 RNAi lentivirus with high infection efficiency, Figure S3: No noticeable difference in DNA methylation occupancy at human ZSCAN4 promoter between HeLa and U2 OS cells, Table S1: 19 nucleotide sequences for RNA knockdown, Table S2: Primers for qPCR analysis of telomere length, Table S3: Primers used for real-time PCR analysis, Table S4: ChIP-qPCR primers of proximal regions of mouse *Zscan4c* promoter and human ZSCAN4 promoter.

Author Contributions: Conceptualization, J.D. and L.L.; Data curation, J.D., Z.Z., F.W., H.W. and R.G.; Investigation, J.D., Z.Z., F.W., H.W. and R.G.; Methodology, J.D., Z.Z., F.W., H.W. and R.G.; Supervision, D.L.K. and L.L.; Validation, J.D., Z.Z., F.W. and L.L.; Visualization, J.D. and L.L.; Funding acquisition, J.D. and L.L.; Writing—original draft, J.D. and L.L.; Writing—review and editing, J.D. and L.L. All authors have read and agreed to the published version of the manuscript.

Funding: This work was supported by funding from Natural Science Foundation of Yunnan Province (grant number: 202001BC070001, 202102AA100053), China National Key R&D Program (2018YFA0107000), National Natural Science Foundation of China (Grant Nos. 31571546 and 91749129).

Institutional Review Board Statement: Not applicable.

Informed Consent Statement: Not applicable.

Data Availability Statement: The original contributions presented in the study are contained within this article and in the Supplementary Materials, and further inquiries can be directed to the corresponding authors.

Acknowledgments: We thank Peter Lansdorp for the TFL-TELO software, Maja Okuka for help on telomere QFISH assay and Yifei Liu for help on microarray data analysis.

Conflicts of Interest: The authors declare no conflict of interest.

References

1. Blackburn, E.H. Switching and signaling at the telomere. *Cell* **2001**, *106*, 661–673. [CrossRef]
2. Palm, W.; de Lange, T. How shelterin protects mammalian telomeres. *Annu. Rev. Genet.* **2008**, *42*, 301–334. [CrossRef]
3. Hiyama, E.; Hiyama, K. Telomere and telomerase in stem cells. *Br. J. Cancer* **2007**, *96*, 1020–1024. [CrossRef]
4. Kim, N.W.; Piatyszek, M.A.; Prowse, K.R.; Harley, C.B.; West, M.D.; Ho, P.L.; Coviello, G.M.; Wright, W.E.; Weinrich, S.L.; Shay, J.W. Specific association of human telomerase activity with immortal cells and cancer. *Science* **1994**, *266*, 2011–2015. [CrossRef] [PubMed]
5. Allsopp, R.C.; Chang, E.; Kashefi-Aazam, M.; Rogaev, E.I.; Piatyszek, M.A.; Shay, J.W.; Harley, C.B. Telomere shortening is associated with cell division in vitro and in vivo. *Exp. Cell Res.* **1995**, *220*, 194–200. [CrossRef]
6. Blasco, M.A. Telomeres and human disease: Ageing, cancer and beyond. *Nat. Rev. Genet.* **2005**, *6*, 611–622. [CrossRef] [PubMed]
7. Huang, J.; Wang, F.; Okuka, M.; Liu, N.; Ji, G.; Ye, X.; Zuo, B.; Li, M.; Liang, P.; Ge, W.W.; et al. Association of telomere length with authentic pluripotency of ES/iPS cells. *Cell Res.* **2011**, *21*, 779–792. [CrossRef]
8. Shay, J.W.; Reddel, R.R.; Wright, W.E. Cancer. Cancer and telomeres—An ALTernative to telomerase. *Science* **2012**, *336*, 1388–1390. [CrossRef]
9. Blasco, M.A. The epigenetic regulation of mammalian telomeres. *Nat. Rev. Genet.* **2007**, *8*, 299–309. [CrossRef]
10. Ottaviani, A.; Gilson, E.; Magdinier, F. Telomeric position effect: From the yeast paradigm to human pathologies? *Biochimie* **2008**, *90*, 93–107. [CrossRef]
11. Slijepcevic, P. Telomere length regulation—A view from the individual chromosome perspective. *Exp. Cell Res.* **1998**, *244*, 268–274. [CrossRef]
12. Zhimulev, I.F.; Beliaeva, E.S. Heterochromatin, gene position effect and gene silencing. *Genetika* **2003**, *39*, 187–201.
13. Benetti, R.; Garcia-Cao, M.; Blasco, M.A. Telomere length regulates the epigenetic status of mammalian telomeres and subtelomeres. *Nat. Genet.* **2007**, *39*, 243–250. [CrossRef] [PubMed]
14. Eckersley-Maslin, M.A.; Svensson, V.; Krueger, C.; Stubbs, T.M.; Giehr, P.; Krueger, F.; Miragaia, R.J.; Kyriakopoulos, C.; Berrens, R.V.; Milagre, I.; et al. MERVL/Zscan4 Network Activation Results in Transient Genome-wide DNA Demethylation of mESCs. *Cell Rep.* **2016**, *17*, 179–192. [CrossRef] [PubMed]

15. Dan, J.; Rousseau, P.; Hardikar, S.; Veland, N.; Wong, J.; Autexier, C.; Chen, T. Zscan4 Inhibits Maintenance DNA Methylation to Facilitate Telomere Elongation in Mouse Embryonic Stem Cells. *Cell Rep.* **2017**, *20*, 1936–1949. [CrossRef]
16. Zalzman, M.; Falco, G.; Sharova, L.V.; Nishiyama, A.; Thomas, M.; Lee, S.L.; Stagg, C.A.; Hoang, H.G.; Yang, H.T.; Indig, F.E.; et al. Zscan4 regulates telomere elongation and genomic stability in ES cells. *Nature* **2010**, *464*, 858–863. [CrossRef] [PubMed]
17. Falco, G.; Lee, S.L.; Stanghellini, I.; Bassey, U.C.; Hamatani, T.; Ko, M.S. Zscan4: A novel gene expressed exclusively in late 2-cell embryos and embryonic stem cells. *Dev. Biol.* **2007**, *307*, 539–550. [CrossRef]
18. De Iaco, A.; Planet, E.; Coluccio, A.; Verp, S.; Duc, J.; Trono, D. DUX-family transcription factors regulate zygotic genome activation in placental mammals. *Nat. Genet.* **2017**, *49*, 941–945. [CrossRef]
19. Hendrickson, P.G.; Dorais, J.A.; Grow, E.J.; Whiddon, J.L.; Lim, J.W.; Wike, C.L.; Weaver, B.D.; Pflueger, C.; Emery, B.R.; Wilcox, A.L.; et al. Conserved roles of mouse DUX and human DUX4 in activating cleavage-stage genes and MERVL/HERVL retrotransposons. *Nat. Genet.* **2017**, *49*, 925–934. [CrossRef]
20. Whiddon, J.L.; Langford, A.T.; Wong, C.J.; Zhong, J.W.; Tapscott, S.J. Conservation and innovation in the DUX4-family gene network. *Nat. Genet.* **2017**, *49*, 935–940. [CrossRef]
21. Le, R.; Huang, Y.; Zhang, Y.; Wang, H.; Lin, J.; Dong, Y.; Li, Z.; Guo, M.; Kou, X.; Zhao, Y.; et al. Dcaf11 activates Zscan4-mediated alternative telomere lengthening in early embryos and embryonic stem cells. *Cell Stem Cell* **2021**, *28*, 732–747. [CrossRef] [PubMed]
22. Markiewicz-Potoczny, M.; Lobanova, A.; Loeb, A.M.; Kirak, O.; Olbrich, T.; Ruiz, S.; Lazzerini Denchi, E. TRF2-mediated telomere protection is dispensable in pluripotent stem cells. *Nature* **2021**, *589*, 110–115. [CrossRef] [PubMed]
23. Srinivasan, R.; Nady, N.; Arora, N.; Hsieh, L.J.; Swigut, T.; Narlikar, G.J.; Wossidlo, M.; Wysocka, J. Zscan4 binds nucleosomal microsatellite DNA and protects mouse two-cell embryos from DNA damage. *Sci. Adv.* **2020**, *6*, eaaz9115. [CrossRef] [PubMed]
24. Dan, J.; Liu, Y.; Liu, N.; Chiourea, M.; Okuka, M.; Wu, T.; Ye, X.; Mou, C.; Wang, L.; Yin, Y.; et al. Rif1 maintains telomere length homeostasis of ESCs by mediating heterochromatin silencing. *Dev. Cell* **2014**, *29*, 7–19. [CrossRef] [PubMed]
25. Poon, S.S.; Martens, U.M.; Ward, R.K.; Lansdorp, P.M. Telomere length measurements using digital fluorescence microscopy. *Cytometry* **1999**, *36*, 267–278. [CrossRef]
26. Callicott, R.J.; Womack, J.E. Real-time PCR assay for measurement of mouse telomeres. *Comp. Med.* **2006**, *56*, 17–22.
27. Liu, L.; Bailey, S.M.; Okuka, M.; Munoz, P.; Li, C.; Zhou, L.; Wu, C.; Czerwiec, E.; Sandler, L.; Seyfang, A.; et al. Telomere lengthening early in development. *Nat. Cell Biol.* **2007**, *9*, 1436–1441. [CrossRef]
28. Dan, J.; Li, M.; Yang, J.; Li, J.; Okuka, M.; Ye, X.; Liu, L. Roles for Tbx3 in regulation of two-cell state and telomere elongation in mouse ES cells. *Sci. Rep.* **2013**, *3*, 3492. [CrossRef]
29. Wang, F.; Yin, Y.; Ye, X.; Liu, K.; Zhu, H.; Wang, L.; Chiourea, M.; Okuka, M.; Ji, G.; Dan, J.; et al. Molecular insights into the heterogeneity of telomere reprogramming in induced pluripotent stem cells. *Cell Res.* **2012**, *22*, 757–768. [CrossRef]
30. Marion, R.M.; Strati, K.; Li, H.; Tejera, A.; Schoeftner, S.; Ortega, S.; Serrano, M.; Blasco, M.A. Telomeres acquire embryonic stem cell characteristics in induced pluripotent stem cells. *Cell Stem Cell* **2009**, *4*, 141–154. [CrossRef]
31. Varela, E.; Schneider, R.P.; Ortega, S.; Blasco, M.A. Different telomere-length dynamics at the inner cell mass versus established embryonic stem (ES) cells. *Proc. Natl. Acad. Sci. USA* **2011**, *108*, 15207–15212. [CrossRef] [PubMed]
32. Macfarlan, T.S.; Gifford, W.D.; Driscoll, S.; Lettieri, K.; Rowe, H.M.; Bonanomi, D.; Firth, A.; Singer, O.; Trono, D.; Pfaff, S.L. Embryonic stem cell potency fluctuates with endogenous retrovirus activity. *Nature* **2012**, *487*, 57–63. [CrossRef]
33. Nakai-Futatsugi, Y.; Niwa, H. Zscan4 Is Activated after Telomere Shortening in Mouse Embryonic Stem Cells. *Stem Cell Rep.* **2016**, *6*, 483–495. [CrossRef] [PubMed]
34. Wong, L.H. Epigenetic regulation of telomere chromatin integrity in pluripotent embryonic stem cells. *Epigenomics* **2010**, *2*, 639–655. [CrossRef] [PubMed]
35. Pucci, F.; Gardano, L.; Harrington, L. Short telomeres in ESCs lead to unstable differentiation. *Cell Stem Cell* **2013**, *12*, 479–486. [CrossRef]
36. Draskovic, I.; Arnoult, N.; Steiner, V.; Bacchetti, S.; Lomonte, P.; Londono-Vallejo, A. Probing PML body function in ALT cells reveals spatiotemporal requirements for telomere recombination. *Proc. Natl. Acad. Sci. USA* **2009**, *106*, 15726–15731. [CrossRef]
37. Londono-Vallejo, J.A.; Der-Sarkissian, H.; Cazes, L.; Bacchetti, S.; Reddel, R.R. Alternative lengthening of telomeres is characterized by high rates of telomeric exchange. *Cancer Res.* **2004**, *64*, 2324–2327. [CrossRef]
38. Vera, E.; Canela, A.; Fraga, M.F.; Esteller, M.; Blasco, M.A. Epigenetic regulation of telomeres in human cancer. *Oncogene* **2008**, *27*, 6817–6833. [CrossRef]
39. Scheel, C.; Schaefer, K.L.; Jauch, A.; Keller, M.; Wai, D.; Brinkschmidt, C.; van Valen, F.; Boecker, W.; Dockhorn-Dworniczak, B.; Poremba, C. Alternative lengthening of telomeres is associated with chromosomal instability in osteosarcomas. *Oncogene* **2001**, *20*, 3835–3844. [CrossRef]
40. Wang, F.; Pan, X.; Kalmbach, K.; Seth-Smith, M.L.; Ye, X.; Antumes, D.M.; Yin, Y.; Liu, L.; Keefe, D.L.; Weissman, S.M. Robust measurement of telomere length in single cells. *Proc. Natl. Acad. Sci. USA* **2013**, *110*, E1906–E1912. [CrossRef]
41. Niida, H.; Shinkai, Y.; Hande, M.P.; Matsumoto, T.; Takehara, S.; Tachibana, M.; Oshimura, M.; Lansdorp, P.M.; Furuichi, Y. Telomere maintenance in telomerase-deficient mouse embryonic stem cells: Characterization of an amplified telomeric DNA. *Mol. Cell. Biol.* **2000**, *20*, 4115–4127. [CrossRef] [PubMed]
42. Wang, Y.; Erdmann, N.; Giannone, R.J.; Wu, J.; Gomez, M.; Liu, Y. An increase in telomere sister chromatid exchange in murine embryonic stem cells possessing critically shortened telomeres. *Proc. Natl. Acad. Sci. USA* **2005**, *102*, 10256–10260. [CrossRef] [PubMed]

43. Portney, B.A.; Arad, M.; Gupta, A.; Brown, R.A.; Khatri, R.; Lin, P.N.; Hebert, A.M.; Angster, K.H.; Silipino, L.E.; Meltzer, W.A.; et al. ZSCAN4 facilitates chromatin remodeling and promotes the cancer stem cell phenotype. *Oncogene* **2020**, *39*, 4970–4982. [CrossRef] [PubMed]
44. Schuster-Bockler, B.; Lehner, B. Chromatin organization is a major influence on regional mutation rates in human cancer cells. *Nature* **2012**, *488*, 504–507. [CrossRef]
45. Silva, B.; Arora, R.; Bione, S.; Azzalin, C.M. TERRA transcription destabilizes telomere integrity to initiate break-induced replication in human ALT cells. *Nat. Commun.* **2021**, *12*, 3760. [CrossRef]
46. Dilley, R.L.; Verma, P.; Cho, N.W.; Winters, H.D.; Wondisford, A.R.; Greenberg, R.A. Break-induced telomere synthesis underlies alternative telomere maintenance. *Nature* **2016**, *539*, 54–58. [CrossRef]
47. Roumelioti, F.M.; Sotiriou, S.K.; Katsini, V.; Chiourea, M.; Halazonetis, T.D.; Gagos, S. Alternative lengthening of human telomeres is a conservative DNA replication process with features of break-induced replication. *EMBO Rep.* **2016**, *17*, 1731–1737. [CrossRef]
48. Min, J.; Wright, W.E.; Shay, J.W. Alternative Lengthening of Telomeres Mediated by Mitotic DNA Synthesis Engages Break-Induced Replication Processes. *Mol. Cell. Biol.* **2017**, *37*, e00226-17. [CrossRef]
49. Zhang, J.M.; Yadav, T.; Ouyang, J.; Lan, L.; Zou, L. Alternative Lengthening of Telomeres through Two Distinct Break-Induced Replication Pathways. *Cell Rep.* **2019**, *26*, 955–968.e3. [CrossRef]
50. Schoeftner, S.; Blasco, M.A. Developmentally regulated transcription of mammalian telomeres by DNA-dependent RNA polymerase II. *Nat. Cell Biol.* **2008**, *10*, 228–236. [CrossRef]
51. Arnoult, N.; Van Beneden, A.; Decottignies, A. Telomere length regulates TERRA levels through increased trimethylation of telomeric H3K9 and HP1alpha. *Nat. Struct. Mol. Biol.* **2012**, *19*, 948–956. [CrossRef]
52. Hu, J.; Hwang, S.S.; Liesa, M.; Gan, B.; Sahin, E.; Jaskelioff, M.; Ding, Z.; Ying, H.; Boutin, A.T.; Zhang, H.; et al. Antitelomerase therapy provokes ALT and mitochondrial adaptive mechanisms in cancer. *Cell* **2012**, *148*, 651–663. [CrossRef] [PubMed]
53. Bechter, O.E.; Zou, Y.; Walker, W.; Wright, W.E.; Shay, J.W. Telomeric recombination in mismatch repair deficient human colon cancer cells after telomerase inhibition. *Cancer Res.* **2004**, *64*, 3444–3451. [CrossRef]
54. Nittis, T.; Guittat, L.; Stewart, S.A. Alternative lengthening of telomeres (ALT) and chromatin: Is there a connection? *Biochimie* **2008**, *90*, 5–12. [CrossRef] [PubMed]
55. Panier, S.; Maric, M.; Hewitt, G.; Mason-Osann, E.; Gali, H.; Dai, A.; Labadorf, A.; Guervilly, J.H.; Ruis, P.; Segura-Bayona, S.; et al. SLX4IP Antagonizes Promiscuous BLM Activity during ALT Maintenance. *Mol. Cell* **2019**, *76*, 27–43.e11. [CrossRef]
56. Barroso-Gonzalez, J.; Garcia-Exposito, L.; Hoang, S.M.; Lynskey, M.L.; Roncaioli, J.L.; Ghosh, A.; Wallace, C.T.; de Vitis, M.; Modesti, M.; Bernstein, K.A.; et al. RAD51AP1 Is an Essential Mediator of Alternative Lengthening of Telomeres. *Mol. Cell* **2019**, *76*, 11–26.e17. [CrossRef] [PubMed]

cells MDPI

Article

Synaptic Hyaluronan Synthesis and CD44-Mediated Signaling Coordinate Neural Circuit Development

Emily S. Wilson and Karen Litwa *

Department of Anatomy and Cell Biology, Brody School of Medicine, East Carolina University, Greenville, NC 27834, USA; wilsonemi21@ecu.edu
* Correspondence: litwak16@ecu.edu

Abstract: The hyaluronan-based extracellular matrix is expressed throughout nervous system development and is well-known for the formation of perineuronal nets around inhibitory interneurons. Since perineuronal nets form postnatally, the role of hyaluronan in the initial formation of neural circuits remains unclear. Neural circuits emerge from the coordinated electrochemical signaling of excitatory and inhibitory synapses. Hyaluronan localizes to the synaptic cleft of developing excitatory synapses in both human cortical spheroids and the neonatal mouse brain and is diminished in the adult mouse brain. Given this developmental-specific synaptic localization, we sought to determine the mechanisms that regulate hyaluronan synthesis and signaling during synapse formation. We demonstrate that hyaluronan synthase-2, HAS2, is sufficient to increase hyaluronan levels in developing neural circuits of human cortical spheroids. This increased hyaluronan production reduces excitatory synaptogenesis, promotes inhibitory synaptogenesis, and suppresses action potential formation. The hyaluronan receptor, CD44, promotes hyaluronan retention and suppresses excitatory synaptogenesis through regulation of RhoGTPase signaling. Our results reveal mechanisms of hyaluronan synthesis, retention, and signaling in developing neural circuits, shedding light on how disease-associated hyaluronan alterations can contribute to synaptic defects.

Keywords: induced pluripotent stem cells; synapse; hyaluronan; rhogtpases

Citation: Wilson, E.S.; Litwa, K. Synaptic Hyaluronan Synthesis and CD44-Mediated Signaling Coordinate Neural Circuit Development. *Cells* **2021**, *10*, 2574. https://doi.org/10.3390/cells10102574

Academic Editor: Mehdi Najar

Received: 14 August 2021
Accepted: 21 September 2021
Published: 28 September 2021

1. Introduction

The majority of neurodevelopmental disorders exhibit synaptic alterations [1,2]. These synaptic alterations disrupt the balance between excitation and inhibition of developing neural circuits, often leading to hyperexcitability [3–5]. For example, reductions in the brain extracellular matrix glycosaminoglycan, hyaluronan (HA), increase the excitability of neural networks and drive the formation of epileptic seizures [6]. In order to elucidate the role of HA in the emergence of neural circuit disruption, our research focuses on the mechanisms of synapse formation [7–12]. In humans, synapse formation (or synaptogenesis) begins in mid-fetal gestation [12,13]. In contrast to previous studies suggesting that HA is absent from synapses [14,15], our recent work demonstrates that HA is present in the cleft between the pre-and post-synaptic compartments of developing excitatory synapses [11]. To determine whether HA's synaptic localization is developmental specific, our current study confirms that HA is present at the cleft of newly formed synapses of the mouse neonatal brain, but is reduced in the cleft of mature synapses and increases in regions surrounding the synapse, consistent with previous observations of a peri-synaptic extracellular matrix [16,17]. This shift coincides with the formation of perineuronal nets around inhibitory neurons, which form postnatally and are prominent in the adult brain [18]. PNNs signal the closure of developmental plasticity and maintain the established balance between inhibitory and excitatory signaling [19], with the absence of PNNs leading to increased neural network excitability [20]. Prior to PNN formation, our work establishes that HA protects the developing brain from the emergence of hyperexcitability by suppressing excitatory synaptogenesis [11]. However, the mechanisms of HA synthesis and signaling within developing neural circuits have yet to be elucidated.

HA is synthesized by three hyaluronan synthases (HAS 1–3). Mammalian HASs consist of seven transmembrane regions, including a large cytoplasmic loop responsible for glycosyltransferase activity and uridine 5′-diphosphate (UDP) binding [21]. HAS uses UDP-D-glucuronic acid (GlcA) and UDP-*N*-acetyl-D-glucosamine (GlcNac) to synthesize an HA chain, composed of repeating GlcA-GlcNAc units, directly into the extracellular space [21]. While all three HASs utilize this method to produce HA, they have varying enzymatic activities and produce HA of varying sizes [21,22]. HAS2 produces the largest chains compared to HAS1 and HAS3, with a molecular mass greater than 2×10^6 Da [22]. HAS2 is ubiquitous throughout the body and is the most abundant HAS in the cortex [6,14,23]. HAS2 knockout is embryonic lethal due to its roles in cardiac development [14]; thus, conditional knockouts (cKO) are used to evaluate HAS2 in the brain. Brain HAS2 cKO reduces cortical HA levels and results in epileptic seizures [6,23]. In our current study, we sought to elucidate whether HAS2 synthesizes HA at newly formed synapses of developing neural circuits.

In addition to HA synthesis, it is important to understand how HA-mediated signaling regulates synaptic physiology. While HA is a space filling molecule, HA is also known to interact with cell surface receptors to mediate intracellular signaling events. The dominant cell surface receptor for HA is CD44 [24]. CD44-HA interactions regulate diverse cellular functions, including cell migration, proliferation and differentiation [25–31]. In non-neuronal systems, HA binding to CD44 also promotes retention of HA extracellular matrix [32]. CD44 is also known to localize to hippocampal synapses, although it is unclear whether the effects of CD44 on hippocampal synaptic plasticity are HA-dependent [31]. HA binds to *N*-terminal motifs of CD44 that act as loading sites, giving rise to clusters as it binds to HA outside the cell [33]. This alone is hypothesized to cause cellular response through changes in force on the cell surface that are balanced by intracellular attachment of CD44 to the actin cytoskeleton through ERM complexes [34]. Through these interactions, CD44 regulates RhoGTPase activity leading to actin cytoskeletal rearrangements that alter the morphology of hippocampal dendritic spines, astrocytes, breast cancer cells and keratinocytes [25,27,31,35]. Our current study demonstrates that CD44 is necessary for HA retention in developing neural circuits, where it localizes to newly formed excitatory synapses. Furthermore, similar to HAS2 and HA, CD44 suppresses excitatory synaptogenesis by attenuating the activity of the RhoGTPase, Rac1. We have previously demonstrated that elevated Rac1 activity is associated with increased excitatory synapse formation [10]. Our data establishes HA-CD44 interactions as regulators of synaptogenesis in developing neural networks. These mechanistic insights shed light on how synaptic defects may arise in disorders associated with reduced hyaluronan, leading to epilepsy, and may help to explain synaptic reductions associated with elevated hyaluronan in Alzheimer's Disease [36,37].

2. Materials and Methods

2.1. Animals

All procedures using animals adhered to guidelines outlined in the National Research Council Guide for the Care and Use of Laboratory Animals and were approved by the Animal Care and Use Committee at East Carolina University (approval A198a) for the laboratory of Dr. Chris Geyer, from whom the samples were ob. Outbred CD-1 mice (Charles River Laboratories) were used for all studies, and the day of birth was designated as postnatal day (P)0. Mice were humanely euthanized by asphyxiation in CO2 followed by cervical dislocation. Brains from four mice at each P0 and P40 were used for imaging, both males and females were used at P0, but only males were examined at P40.

2.2. Cell Lines

Neurotypic control skin fibroblasts of the cell line 7545 19B were reprogrammed into human-induced pluripotent stem cells (hIPSCs) in the laboratory of Dr. Mike McConnell (University of Virginia) with the addition of Yamanaka transcription factors Oct3/4, Sox2, Klf4, and c-Myc using the CytoTune-iPS 2.0 Sendai Reprogramming Kit (Invitrogen, Waltham, MA, USA). 7545 fibroblasts were obtained under an MTA with the Coriell Institute (Camden, NJ, USA). Control WTC-11-ActBmeGFP and WTC-11-Paxilin-GFP IPSCs were obtained under MTA from the Coriell institute. The parental WTC-11 IPSC line was developed by Bruce Conklin of the Gladstone Institute and was further gene edited by the Allen Institute for Cell Science using CRISPR/Cas9 to tag endogenous β-actin or paxillin with monomeric green fluorescent protein (GFP). hIPSCs were maintained in Essential 8 Medium + E8 supplement (Gibco, Waltham, MA, USA) on hESC Matrigel (Corning, Glendale, AZ, USA) coated plates. Upon splitting, 10 μM of the ROCK inhibitor, Y27632 (Selleck Chemicals, Houston, TX, USA), was added to the cell medium.

Paxillin-GFP IPSCs were used to generate CRISPRi competent cortical spheroids via transduction of lentiviral nuclease-dead Cas9 carrying a C-terminal Krüppel associated box (KRAB) domain. (Applied Biological Materials, Richmond, BC, Canada). Expression of dCas9-KRAB transduced IPSCs was maintained by puromycin selection. Cortical spheroids were produced under puromycin selection resulting in Paxillin-GFP-dCas9-KRAB cortical spheroids. At day 90, spheroids were transduced for 48 h with lentiviral sgRNAs towards CD44 (sgCD44) to selectively repress CD44 or scrambled sgRNAs (sgScrambled) as a control at an MOI of 2.

2.3. Three-Dimensional Cortical Spheroid Culture

Cortical spheroids were produced following the methods described by Pasca et al. [38]. Briefly, enzymatically lifted hIPSCs were transferred to ultra-low attachment plates and cultured in DMEM supplemented with Knockout Serum Replacement (Gibco) supplemented with 5 μM Dorsomorphin (BioVision, Milpitas, CA, USA), 10 μM SB431542 (Miltenyi Biotec, Bergisch Gladbach, Germany), 10 μM Y27632 (Selleck Chemicals) for 6 days. The resulting spheroids were then maintained in neurobasal media until day 90: Neurobasal A medium, 2% B-27 supplement without vitamin A, GlutaMAX L-glutamine supplement (Gibco) penicillin/streptomycin (Gibco). Spheroids were next supplemented with 20 ng/mL of bFGF and EGF (PeproTech, Cranbury, NJ, USA) from day 6 to 25, and 20 ng/mL of BDNF and NT3 (Shenandoah Biotechnology, Warminster, PA, USA) from day 26 to 42. Spheroids were harvested beginning at day 90 for analysis.

2.4. HAS2 and CD44 Manipulation

Adenoviral constructs were used for the overexpression of HAS2 and CD44 in cortical spheroids. The constructs used were prepared in the same manner as Ishizuka et al., 2019 [39]. The human HAS2 ORF (NM_005328) in pCR3.1 (a gift from Dr. Tim Bowen, Cardiff University School of Medicine, Cardiff, UK) was PCR-amplified using AccuPrime Pfx DNA Polymerase (Life Technologies, Waltham, MA, USA) and primers designed specifically for use with the Adeno-X Adenoviral System 3 (Clontech, San Jose, CA, USA). The amplified human HAS2 sequence was ligated into the pAdenoX-CMV-ZsGreen1 linearized vector (Clontech) to form Ad-ZsGreen-hHAS2 and Ad-ZsGreen-hCD44. All PCR- amplified regions were verified by DNA sequencing. In the adeno-ZsGreen-HAS2 and adeno-ZsGreen-CD44 vectors, HAS2 or CD44 and ZsGreen1 sequence are present on the same plasmid but driven by separate CMV-1E promoters. Primers and plasmid for LacZ, included in the Adeno-X Adenoviral System 3 kit (Clontech) were used according to the manufacturer's protocol to generate Ad-ZsGreen-LacZ. Cultures were transduced at 10 IFU/cell for 48 h before media washout and/or subsequent harvesting.

Knockdown of CD44 was accomplished using sgCD44 lentiviral constructs with the KRAB competent Pax-GFP cell line derived cortical spheroids. Spheroids were transduced for 48 h with pools of 3 target sgRNAs toward CD44 or Scrambled controls (Applied Biological Materials) before harvesting.

2.5. Immunohistochemistry

Cortical spheroids and mouse brain samples were fixed in 4% paraformaldehyde for 24 h and placed in 30% sucrose for 24 h. Samples were then embedded in OCT mounting media overnight (Sakura Finetek USA, Torrance, CA, USA), flash frozen, and cryosectioned into 10 μm thick sections. Cryosections were permeabilized with 0.2% TritonX-100 in 1× PBS before immunostaining. Primary antibodies were diluted in 2% normal goat serum in PBS, added to fixed cultures and kept at 4 °C overnight. After three PBS washes, secondary antibodies diluted in 2% normal goat serum in PBS were added to fixed cultures and kept at room temperature for 1 h. Sections fated for dSTORM imaging were additionally post-fixed in 4% paraformaldehyde and 0.1% glutaraldehyde in PBS for 10 min and subsequently washed 3 times in PBS. Cryosections were mounted using Fluoro-gel II with DAPI mounting medium (Electron Microscopy Sciences, Hatfield, PA, USA) for confocal imaging or Vectashield without DAPI (Vector Laboratories, Burlingame, CA, USA, 101098-042) for dSTORM imaging.

Two-dimensional cultures were fixed after 1 week in 4% paraformaldehyde and 4% sucrose in PBS. Cells were permeabilized with 0.2% TritonX-100 in PBS before staining. Primary antibodies were diluted in 2% normal goat serum in PBS, added to fixed cultures and kept at 4 °C overnight. After three PBS washes, secondary antibodies diluted in 2% normal goat serum in PBS were added to fixed cultures and kept at room temperature for 1 h. Cultures were mounted using Fluoro-gel II with DAPI mounting medium (Electron Microscopy Sciences).

HA was visualized using Hyaluronic Acid Binding Protein (HABP). HABP is a biotinylated link protein G1 domain of the proteoglycan versican, which binds selectively to HA. Fluorescently labeled streptavidin was used to detect HABP. See antibody table for further information. After immunostaining, cryosections were subsequently imaged using a ZEISS LSM 700 or LSM 800 confocal or Nikon Ti2-E inverted dSTORM microscope. All immunohistochemistry was performed in triplicate on at least three different sets of spheroids.

2.6. Confocal Microscopy

Cortical spheroid sections were imaged on a ZEISS LSM 700 confocal microscope at 40× total magnification, using the 639, 555, 488 and 405 channels. Z-stacked images were acquired, 5 z-planes with a 1 μm step size, and merged in ImageJ to generate maximum intensity projections. Confocal images were further analyzed using Image J analysis software. Each 4-channel image was analyzed as an 8-bit tiff-file. All samples of the same experiment had equal threshold values for each channel. Ten area samples measuring 100 μm in diameter were used to determine the intensity and density of the fluorescent markers along the edge of the cortical spheroid, coinciding with the cortical plate location. Spheroids were stained with pre- and post-synaptic markers to analyze the total synapse area we used the ImageJ colocalization plug in. This produced an image indicating areas at the threshold of pre- and post-synaptic markers colocalized by at least 10% of their intensities (shown in yellow in figures). The area of each individual marker as well as the colocalized synapse area was measured and normalized to DAPI. Pixel resolution for all confocal images is 0.1563257 μm per pixel.

2.7. Airyscan Microscopy

Three-dimensional reconstruction of HA at excitatory synapses was performed using a ZEISS LSM 800 confocal with Airyscan feature at 63× total magnification using the 639, 555 and 488 channels. Z-stacked images were acquired. Airyscan processed images were then analyzed using the 3D visualization plug-in. Pixel resolution for all Airyscan images is 0.043 μm per pixel.

2.8. Three-Dimensional dSTORM Microscopy

As we have previously described [9–11], 3D direct STochastic Optical Reconstruction Microscopy (dSTORM) images of 10 μm thick cortical spheroid sections were acquired with a Nikon Ti2-E inverted microscope with an L-APPS H-TIRF attachment and 4-line (405, 488, 561, and 640 nm) LUN-F laser module and 100 × 1.49NA Apo TIRF objective. The raw images are collected with the 3D STORM lens to a back-thinned Princeton Instruments Pro- EM-HS EMCCD 512 × 512 camera and acquired and analyzed with NIS-Elements STORM modules. The following secondary fluorophores were used for dSTORM: Atto488, Alexa568, and Alexa647, and the mounting media vectashield was used for its ability to prolong blinking [40–42]. The acquired images were visually inspected for spectral separation that could interfere with analysis of synaptic localization. Processed .nd2 files were analyzed in ImageJ. The .nd2 files were opened as a hyperstack, and a 3D max intensity projection was then used to plot the brightest point with the Y-axis as the axis of rotation. A Gaussian blur filter with a radius of 2.00 pixels was applied to avoid biasing our analysis toward a single bright link. The plot profile function of ImageJ was used to analyze the intensity of synaptic molecules perpendicular to the synaptic cleft. Average pixel resolution for exported dSTORM images of individual synapses prior to image analysis is ~0.002 μm per pixel.

For analysis of dSTORM HA localization, the following classification was used: HA whose centroid distance to that of each synaptic marker was less than 50 nm is referred to as equidistantly located in the synaptic cleft. HA whose centroid distance to that of each synaptic marker greater than 50 nm and less than 100 nm is referred to as equidistantly located near the synaptic cleft. Finally, HA centroid distance difference greater than 100 nm is referred to as preferentially located to either the pre-synaptic or post-synaptic compartment (Supplementary Figure S1).

2.9. Microelectrode Array Analysis

For MEA at analysis, spheroids at day 76 were dissociated using a Primary Neuron Isolation Kit (Pierce Thermo Scientific, Waltham, MA, USA, MAN0016221) and plated onto PEI-coated MEA plates prepared as described below. An Axion BioSystems Maestro Edge Multielectrode Array System (MEA) was used to record spontaneous action potentials in dissociated spheroids. MEA plates were prepared 48 h before cells were plated as follows: Wells were incubated 1 h at 37 °C with 0.1% polyethylenimine in ddH2O and rinsed 4 times with sterile water and left at room temperature overnight. After 24 h, 5 μg/mL laminin was added to wells and left overnight at room temperature. Wells were washed twice with sterile PBS before 250,000 cells/well of dissociated spheroids were added. Cultures were given MEA medium consisting of Neurobasal A medium, 2% B-27 Plus supplement without vitamin A, glutaMAX L-glutamine supplement (Gibco), and penicillin/streptomycin (Gibco). Cells were acclimated to the plates for 2 weeks before treatment and recording. Manipulation of HA levels was conducted for 24 h. Wells were recorded 10 min/h for 24 h. After treatments wells were given fresh media. Washout was recorded for 24 h. (10 min/h). Wells were treated with 4 μM TTX and immediately recorded for 10 min to suppress action potential formation and verify that the recorded electrical currents correspond to spontaneous neural activity. MEA experiments were performed on 6 sets of differentiated spheroids.

2.10. Western Blot Analysis

Total protein was extracted using cell lysis buffer containing protease and phosphatase inhibitor mixtures. Equivalent protein concentrations were loaded into 4–12% NuPAGE Novex Tris acetate gradient gels (ThermoFisher Scientific, Waltham, MA, USA) for electrophoresis. Proteins were subsequently transferred to a nitrocellulose membrane using a Criterion blotter apparatus (Bio-Rad, Hercules, CA, USA). The membrane was incubated in 5% non-fat dry milk in TBST for 1 h. Immunoblots were incubated overnight with primary antibody in TBST at 4 °C. The next day immunoblots receive 3–5 min washes in

TBST before incubation with secondary antibody diluted in TBST for 1 h. Detection of immunoreactive bands was performed using chemiluminescence (Novex ECL, Invitrogen, Waltham, MA, USA). Blots were stripped using Restore Plus Western Stripping Buffer (Thermo Fisher Scientific) for 5–10 min at room temperature and reprobed using another primary antibody. Developed X-ray films were imaged and digitized using a Bio-Rad Gel-Doc with ImageLab software. Pixel intensities for bands were used for quantification after normalization to loading control bands (α-tubulin). Active Rac1-Pulldown was performed as per manufacturer instructions using Rac1 Pull-Down Activation Assay Biochem Kit (Cytoskeleton Inc., Denver, CO, USA).

2.11. *Quantitative Real Time PCR Analysis RT-PCR (qRT-PCR)*

Total RNA was isolated from cortical spheroid or 2D cultures using a Direct-zol RNA Isolation Kit (Zymo Research, Irvine, CA, USA) according to the manufacturer's instructions. Total RNA was reverse-transcribed to cDNA using the iScript cDNA synthesis kit (Bio-Rad). Quantitative PCR was performed using Sso-Advanced SYBR-Green Supermix (Bio-Rad) and amplified on a StepOnePlus real-time PCR system (Applied Biosystems, Waltham, MA, USA) to obtain cycle threshold (Ct) values for target and internal reference cDNA levels. See table for primer sequences.

2.12. *Statistical Analyses*

Statistical analyses were performed using SigmaPlot 13.0 Software. Data sets were first tested for normality using Shapiro–Wilk tests. T-tests were run on parametric data to determine statistical significance. Non-parametric data was analyzed using Mann–Whitney Rank Sum tests to test for statistical significance.

3. Results

3.1. *HA Localizes to the Synaptic Cleft of Excitatory Synapses during Cortical Development*

Our previous work demonstrated that HA localizes to the synaptic cleft of newly formed excitatory synapses in human cortical spheroids that resemble the fetal brain [11]. Given that HA was previously thought to be absent from synapses [14,15], we sought to further clarify the synaptic localization of HA during cortical brain development. Notably, previous studies relied on electron micrographs of isolated adult synaptosomes or isolated synaptic junctions [15,43]. Given the hydrophilic nature of HA, dehydration techniques for EM make it difficult to preserve HA-based structures [44]. However, with the advent of super-resolution fluorescence microscopy techniques, we can now visualize HA without tissue dehydration [11]. We therefore employed both dSTORM super-resolution microscopy and Airyscan processing to visualize HA at developing synapses of the neonatal (P0) mouse cortex and mature synapses in the young adult (P40) mouse cortex (Figure 1). Consistent with our previous studies in human cortical spheroids, we used PSD95 to visualize the post-synaptic compartment and VGLUT1 to visualize the pre-synaptic compartment of excitatory synapses (Figure 1A–E). As with nascent excitatory synapses in the human cortical spheroids, hyaluronan is present in the synaptic cleft of immature excitatory synapses of the neonatal mouse cortex (Figure 1A). However, in the adult mouse brain, we observed decreased HA within the synaptic cleft, corresponding with either no synaptic HA or a redistribution to regions surrounding the synapse (Figure 1B), as has been previously described for peri-synaptic extracellular matrix [16,17]. In certain cases, HA was completely absent from excitatory synapses of the young adult mouse cortex. We quantified the frequency of synaptic localization events in P0 and P40 excitatory synapses. At immature synapses of the neonatal cortex, HA was present at all synapses examined (Figure 1C). In the majority (~80%) of cases, HA was at (51.9%) or near (25.9%) the synaptic cleft. By contrast, HA was only found in the synaptic cleft of approximately one third of mature synapses of the young adult cortex, and there was a ~20% overall reduction in HA at or near the synaptic cleft. Furthermore, HA was completely absent from ~10% of all adult excitatory synapses. To further elucidate this developmental shift from HA cleft localization to peri-synaptic matrix, we used Airyscan imaging and 3D reconstruction to visualize a larger synaptic population (Figure 1D,E). These images corroborate our dSTORM findings, with HA at or near the cleft of most immature synapses but later encapsulating mature synapses. Consistent with these synaptic changes, we also observed the emergence of a condensed hyaluronan matrix in perineuronal nets surrounding parvalbumin interneurons in the young adult mouse cortex (Figure 1F–J). Thus, hyaluronan exhibits a developmentally regulated localization at excitatory synapses, with HA predominantly found within the synaptic cleft of immature synapses but this localization to the synaptic cleft is reduced in mature synapses of the adult brain. Having established that HA exhibits a conserved and developmentally regulated localization to excitatory synapses, we sought to address how HA is synthesized at excitatory synapses, and the mechanisms through which it elicits synaptic signaling events.

3.2. Manipulation of Hyaluronan Synthase 2 Alters HA Synthesis, Synapse Formation, and Excitability of Developing Neural Networks

We previously established that 3D cortical spheroids generate an endogenous HA-based extracellular matrix (ECM) and that they express HAS2, the hyaluronan synthase primarily responsible for cortical HA production [6]. Using monolayer cultures, we demonstrated that HAS2 is expressed by human cortical neurons [11], consistent with research demonstrating that primary rodent cortical neurons express HAS 1–3 [45]. Using confocal image analysis, we observed significant association of HAS2 with excitatory synapses expressing VGLUT1 and PSD95 (Figure 2A–C), while only ~5% of inhibitory synapses exhibited HAS2 expression (Figure 2B). To confirm that HAS2 is directly at excitatory synapses, we used dSTORM super-resolution microscopy to visualize excitatory synapses (Figure 2D–G). We immunostained for PSD95 to visualize the post- synaptic compartment and VGLUT1 to visualize the pre-synaptic compartment of excitatory synapses. dSTORM image analysis reveals that HAS2 directly associates with the synaptic compartment of neurons and is preferentially localized to the pre-synaptic compartment in ~58.3% of HAS2-expressing synapses. To investigate contributions of other HAS isoforms, HAS1 and HAS3, we compared relative mRNA expression of HAS1 and HAS3 to levels of HAS2 (Figure 2H). HAS1 was present at low levels compared to HAS2 expression. HAS3, while more abundant than HAS1, is expressed at lower levels compared to HAS2. To determine if the HAS isoforms were associated with astrocytes rather than neurons, we stained cortical spheroids for GFAP, DAPI and each HAS isoform (Figure 2I). We found that nearly 80% of HAS1 and HAS3 expression is colocalized with astrocyte marker GFAP (Figure 2J), while HAS2 astrocytic association tends to be less. While we cannot exclude the possibility that HAS3 also regulates synaptic HA levels, these data suggest that HAS2 is the predominant HAS in cortical spheroids, where we have found it localized to both astrocytes and neurons. Using dSTORM, we have also now demonstrated that HAS2 localizes directly to excitatory synapses.

Having established that HAS2 localizes to excitatory synapses of developing neural circuits, we sought to determine whether HAS2 is sufficient to increase hyaluronan production and alter synaptogenesis. We used adenoviral constructs to successfully transduce cortical spheroids and drive the expression of HAS2 or a LacZ control (Figure 3). In order to alter HAS2 expression during synaptogenesis, we transduced cortical spheroids at Day 90, when we observe robust synapse formation that it is susceptible to exogenous manipulations [9–11] (see Figure 3A for detailed workflow). RT-PCR confirmed that *HAS2* expression significantly increased 48 h after transduction (Figure 3E). Notably, *HAS1* and *HAS3* were unaltered by exogenous *HAS2* transduction. To visualize the location of increased HAS2 expression, we imaged the adenoviral expression of the virally co-expressed fluorophore, ZsGreen. ZsGreen expression is prominent in outer cortical layers, where synaptogenesis occurs in our model [10–12] (arrows in Figure 3B). Through immunostaining, we confirmed that HAS2 protein expression increased and resulted in increased hyaluronan production in the cortical plate where synaptogenesis occurs (Figure 3C,D). Increased HA levels demonstrate that the expressed HAS2 is functional.

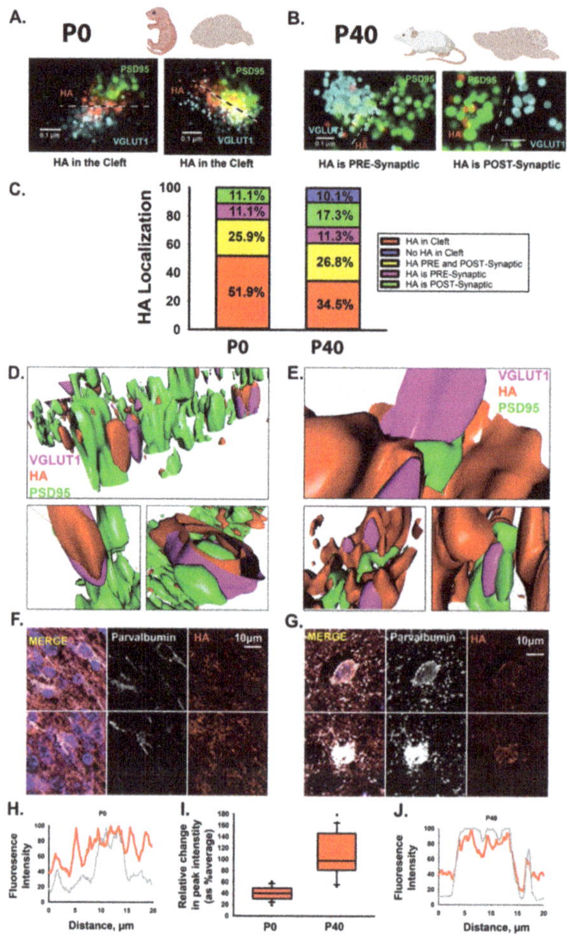

Figure 1. Hyaluronan exhibits a dynamic developmental expression at excitatory synapses. (**A**) Representative dSTORM images of HA within the synaptic cleft between excitatory synaptic markers, VGLUT1 and PSD95, in mouse P0 cortical slices. Cyan: pre-synaptic marker VGLUT1; green: post-synaptic marker PSD95; red: HA. (**B**) Representative dSTORM images of HA outside of the synaptic cleft, yet near synapses, in P40 mouse cortical slices. (**C**) To illustrate the shift in HA localization from P0 to P40 we used percent stacked bar charts to look at the distribution of HA localization at excitatory synapses visualized by dSTORM microscopy. Percentages are based on the distance from HA to each synaptic marker. HA in the cleft between pre- and post- synaptic is shown in red. HA near (within 100 nm) of the pre- and post-synaptic marker is labeled as HA localized to both the pre- and post-synaptic compartment (yellow). HA preferentially localizing to one synaptic compartment is labeled as HA localized to either the pre- or post- synaptic compartment (presynaptic: magenta; postsynaptic: green). Synapses lacking HA are labeled in blue. $n = 168$ synapses from 4 brain samples for each P0 or P40 time point. (**D**,**E**). Three-dimensional reconstruction of mouse excitatory synapses using Airyscan processing reveals that HA is clearly present between synaptic markers on postnatal day 1 (P0) (**D**) but shifts to surround excitatory synapses at P40 (**E**). (**F**–**J**) Inhibitory interneurons in P0 (**F**) and P40 (**G**) mouse cortical brain slices, stained for DAPI in blue, parvalbumin (inhibitory interneurons) in white, and HA in red. While HA is ubiquitous around the cells at P0, condensed perineuronal nets are absent. By contrast, HA is found in the form of a visibly condensed perineuronal net around the soma at P40.

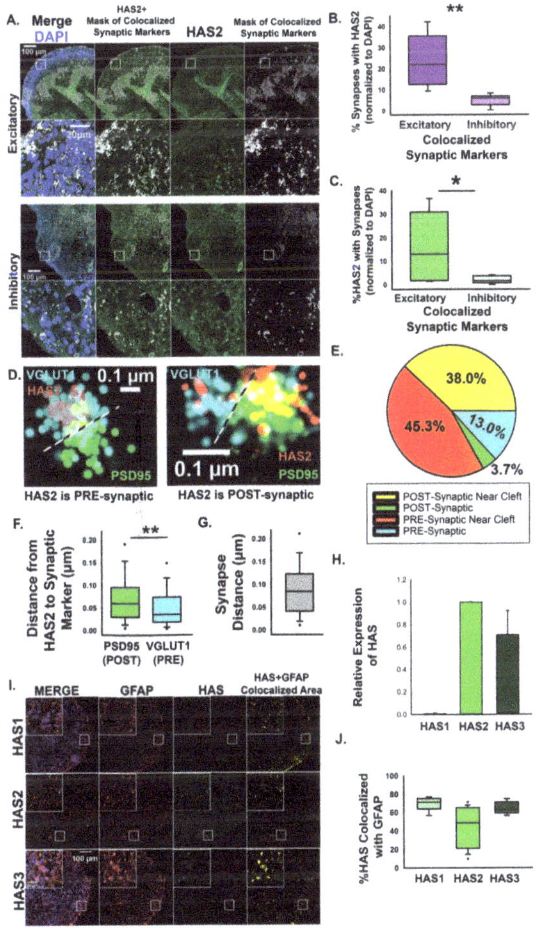

Figure 2. Hyaluronan synthase, HAS2, localizes to excitatory synapses. (**A**) Representative confocal images of HAS2 together with colocalized excitatory synapse markers, VGLUT1 and PSD-95, and colocalized inhibitory synapse markers, VGAT and gephyrin. Panels show DAPI (blue), HAS2 (green) and a mask of colocalized synaptic markers (white). Co-localization analysis between HAS2 and the area defined by the colocalized synapse mask reveals that HA is preferentially enriched at excitatory synapses. (**B**,**C**) Quantification of HAS2 synaptic localization shown in A. $n = 9$ total spheroids derived from 3 separately grown spheroid sets for each set of synaptic markers. * $p = 0.027$ ** $p < 0.001$. (**D**) dSTORM imaging reveals that HAS2 localizes to pre- and post-synaptic compartments of excitatory synapses. The pre-synaptic marker VGLUT1 is shown in blue, HAS2 is shown in red, and post-synaptic PSD95 is shown in green. (**E**) Pie chart showing preferred synaptic localization of HAS2. Within 100 nm of the synaptic cleft is distinguished as 'near cleft'. (**F**) Quantification of the average distance between HAS2 and synaptic markers. $n = 168$ synapses from 9 total spheroids derived from 3 separately grown spheroid sets, ** $p < 0.001$. (**G**) Quantification of the average distance from pre- (VGLUT1) to post-synaptic (PSD95) markers demonstrates that HAS2 is closer to the cleft at the pre-synaptic VGLUT1-postive compartments. (**H**) qRT-PCR relative mRNA expression of HAS1 and HAS3 as compared to HAS2. (**I**) Representative images of cortical spheroids stained for HAS isoforms (green), GFAP (red), and DAPI. Far right panel (yellow) is the area of colocalization between HAS and GFAP. Scale bar 100 µm, inset: 20 µm. (**J**) Quantification of HAS and GFAP colocalization.

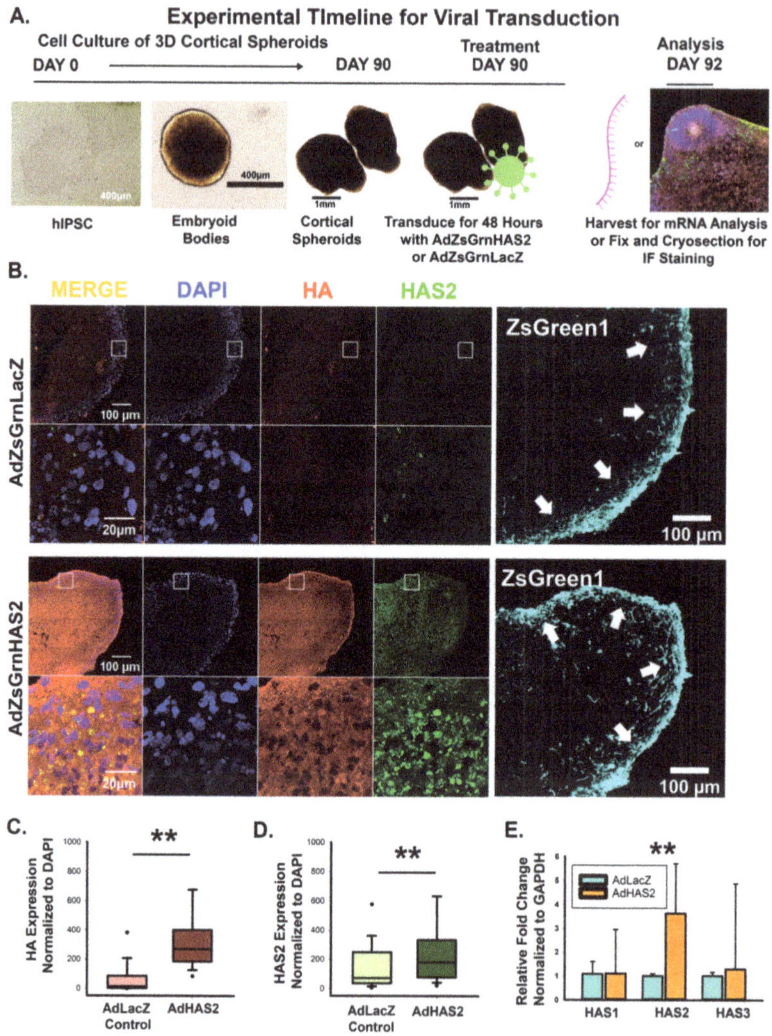

Figure 3. (**A**) Workflow for viral transduction of 3D cortical spheroids for manipulation of HAS2. (**B**) Representative images of cortical spheroids virally transduced to express HAS2. Top panel: adenoviral expression of ZsGreen and LacZ control (AdZsGreenLacZ). Bottom panel: adenoviral expression of ZsGreen and HAS2 (AdZsGreenHAS2), stained for DAPI (blue), HA (red), and HAS2 (green). The highlighted region of the cortical plate is enlarged below. To the right is the viral co-expression of the fluorophore, ZsGreen1. Arrows indicate the cortical plate, where synaptic analysis was performed. Scale bars top: 100 µm, enlarged region: 20 µm. (**C**) Quantified HA expression (via HABP) normalized to DAPI. $n = 9$ total spheroids for each AdZsGrnLacZ and AdZsGrnHAS2 derived from 3 separately grown and transduced spheroid sets ** $p < 0.001$. (**D**) Quantification of HAS2 expression normalized to DAPI. $n = 9$ total spheroids for each transduction (AdZsGrnLacZ and AdZsGrnHAS2). Spheroids were derived from 3 separately grown and transduced spheroid sets ** $p < 0.001$. (**E**) qRT-PCR confirmation of increased HAS2 expression. $n = 20$ total spheroids derived from 4 separately grown and transduced spheroid sets. ** $p < 0.001$.

Given that elevated levels of HA suppress excitatory synapse formation [11], we anticipated that HAS2-mediated increases in HA would similarly impact synaptogenesis. We immunostained for pre- and post-synaptic markers, VGLUT1 and PSD95, of excitatory synapses in cortical spheroid cryosections 48 h after viral transduction, when we observe both increased HAS2 and HA (Figure 3) and analyzed the colocalized area between pre- and post-synaptic markers, to determine the extent of synapse formation (Figure 4A–D). When compared to the LacZ control, HAS2 overexpression significantly decreased the area of individual pre- and post-synaptic excitatory markers as well as their colocalized synaptic area, highlighting a regulatory role for HAS2 in excitatory synapse formation. Notably, our human cortical spheroids resemble dorsal forebrain development and therefore predominantly form excitatory synapses [38,46]. However, we are still able to observe inhibitory synapses, albeit fewer than excitatory synapses [9,11]. To visualize inhibitory synapses, we immunostained for inhibitory pre- and post-synaptic markers, vesicular GABA transporter (VGAT) and gephyrin, respectively (Figure 4E–H). Notably, inhibitory synapses dramatically increase in response to HAS2 overexpression. These data demonstrate that HAS2-mediated HA synthesis at excitatory synapses can significantly alter the emerging balance between excitatory and inhibitory synapses in developing neural networks.

To determine whether these synaptic changes correspond with altered synaptic transmission, we used microelectrode array analysis to measure spontaneous action potential firing rate in LacZ and HAS2-transduced cultures of dissociated cortical spheroids (detailed workflow described in Figure 5A, representative image of transduced dissociated cultures shown in Figure 5B). We have previously established that dissociated cortical spheroid re-secrete and retain an HA-based extracellular matrix that regulates neuronal activity [11]. Control LacZ adenoviral transduction decreases HA levels and increases neuronal activity when compared to untreated spheroids (data not shown). Thus, we compared the effects of adenoviral expression of HAS2 (Figure 5) and later CD44 (Figure 8) with the appropriate LacZ control. Within developing neural circuits of human cortical spheroids, spontaneous action potentials form and increase with further development. We predicted that the decreased excitatory synapse formation resulting from HAS2 overexpression would prevent spontaneous action potential formation. HAS2 overexpression significantly reduced spontaneous activity, decreasing activity by ~50% at both 24 and 72 h after viral transduction (Figure 5C), as illustrated by the raster plots of neural spiking activity in Figure 5D. These results suggest that HAS2 represses the development of spontaneous activity through synaptic HA synthesis. However, whether synaptic HA functions solely as a physical barrier or also interacts with synaptic receptors to alter synapse formation remains to be elucidated.

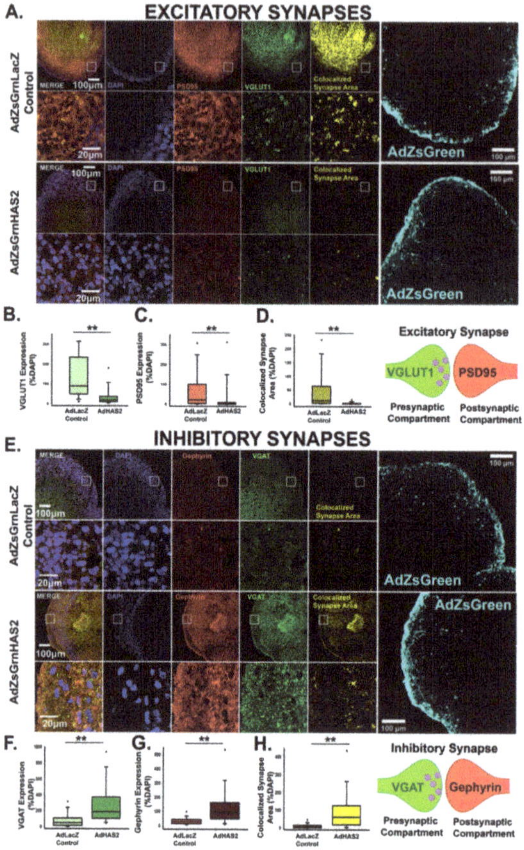

Figure 4. Hyaluronan synthase, HAS2, alters the emerging balance between excitatory and inhibitory synapses in developing neural networks. (**A**) Representative images of cortical spheroids transduced with either LacZ control or HAS2 and stained for DAPI (blue) and excitatory synapses: post-synaptic marker PSD95 (red) and pre-synaptic marker VGLUT1 (green). Right panel shows the excitatory synaptic colocalization of PSD95 and VGLUT1 in yellow. Bottom panel highlights the region of the cortical plate indicated by the white box. Scale bars: top: 100 μm, enlarged region: 20 μm. (**B**) Quantification of the area of pre-synaptic marker VGLUT1 normalized to the area of DAPI. (**C**) Quantification of area of post-synaptic marker PSD95 normalized to the area of DAPI. (**D**) Quantification of colocalized excitatory synapse area normalized to the area DAPI. $n = 9$ total spheroids for each transduction (AdZsGrnLacZ and AdZsGrnHAS2). Spheroids were derived from 3 separately grown and transduced spheroid sets ** $p < 0.001$ for (**B**–**D**). (**E**) Representative images of transduced cortical spheroids stained for DAPI (blue), and inhibitory synapses: post-synaptic marker gephyrin (red) and pre-synaptic maker VGAT (green). Rightmost panel shows the inhibitory synaptic colocalization of gephyrin and VGAT in yellow. Bottom panel highlights the region of the cortical plate indicated by the white box. Scale bars: top: 100 μm, enlarged region: 20 μm. (**F**) Quantification of the area of pre- synaptic marker VGAT normalized to the area of DAPI. (**G**) Quantification of the area of post- synaptic marker gephyrin normalized to DAPI. (**H**) Quantification of colocalized inhibitory synapse area normalized to the area of DAPI. $n = 9$ total spheroids for each transduction (AdZsGrnLacZ and AdZsGrnHAS2). Spheroids were derived from 3 separately grown and transduced spheroid sets ** $p < 0.001$ for (**F**–**H**).

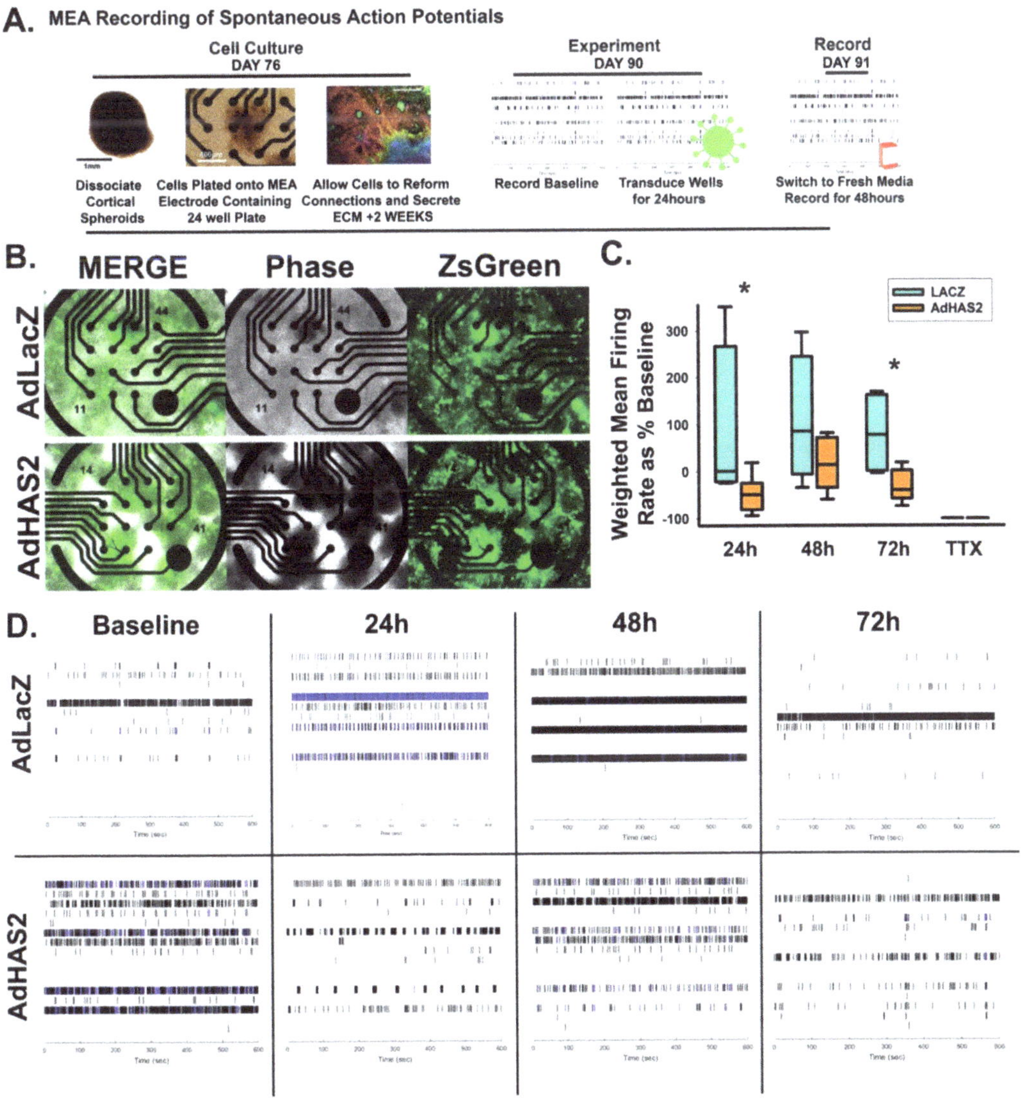

Figure 5. HAS2 suppresses excitability of developing neural networks. (**A**) Workflow of multielectrode array experiments for detection of action potentials propagation in neural networks. (**B**) Representative images of successful viral transduction on MEA plates as indicated by viral expression of ZsGreen in dissociated spheroid transduced with adenovirus constructs for control LacZ (AdZsGrnLacZ) and HAS (AdZsGrnHAS2) expression. (**C**) Quantification of weighted mean firing rate of spontaneous action potentials in HAS2 and LacZ-expressing dissociated spheroids demonstrates that HAS2 suppresses excitability by significantly reducing the firing rate at 24 and 72 h post-transduction. Tetrodotoxin (TTX) was used to inhibit action potential propagation, demonstrating that the firing rate reflects action potentials. $n = 6$ wells of dissociated spheroids per treatment. Spheroids were derived from 3 separately grown and transduced spheroid sets. * $p \leq 0.05$ (**D**) Representative raster plots of spontaneous action potential firing at 0 (baseline), 24, 48, and 72 h post-transduction for AdZsGrnLacZ and AdZsGrnHAS2.

3.3. HA-CD44 Interactions at Developing Excitatory Synapses Attenuate Synaptogenesis through Regulation of RhoGTPase Signaling

To determine whether HA and HAS2 suppress synaptogenesis via synaptic signaling events, we investigated whether the predominant HA receptor, CD44 (Figure 6A), localizes to developing excitatory synapses. Using dSTORM imaging analysis, we observed that CD44 associates with both VGLUT1-positive pre-synaptic and PSD95-positive post-synaptic compartments (Figure 6B–F). Notably, CD44 was found in all examined excitatory synapses. In the majority of excitatory synapses (~66%), CD44 localized to both the pre- and post-synaptic compartment, followed by ~20% solely in post-synaptic compartment and ~14% solely in the pre-synaptic compartment (Figure 6D). Since CD44 promotes HA retention in other tissue systems [32], we sought to determine whether CD44 is responsible for HA retention in developing neural circuits. In order to knockdown CD44 specifically during synaptogenesis, we used human cortical spheroids stably expressing a dead Cas9 enzyme conjugated to the transcriptional repressor, Krüppel-Associated Box (KRAB) for CRISPR interference [47,48]. After 90 days of development, we introduced targeting sgRNA towards the CD44 promoter, which downregulated CD44 (Figure 6G) as confirmed by RT-PCR (Figure 6H). Using immunofluorescence, we observed that this strategy led to global CD44 reductions in our cortical spheroid model (Figure 6G,I). Strikingly, loss of CD44 resulted in significant loss of HA (Figure 6J). Conversely, adenoviral expression of CD44 increased HA levels (Figure 7). These changes in HA levels are unlikely to be driven by increased HA synthesis as we did not observe changes in HAS2 expression resulting from knockdown of CD44 (Figure 6H). Knockdown of CD44 was sufficient to increase excitatory synaptogenesis (Figure 6K–N.) Together, these results demonstrate that CD44 is responsible for HA retention in the developing neural circuits of our model of the human fetal cortex and that loss of HA retention via loss of CD44 alters synaptic networks.

Notably, adenoviral CD44 expression resulted in decreased excitatory synapse formation compared to viral expression of a LacZ control. Viral transduction was confirmed via visualization of viral co-expression of ZsGreen, and the resulting CD44 overexpression was confirmed by qRT-PCR analysis (Figure 7A–D). In addition to regulating the overall area of excitatory synapses (Figure 7E–H), CD44 overexpression also decreased the size and number of individual synapses, which may suggest a role for CD44 in synaptic refinement (Figure 7I,J). Furthermore, CD44 suppressed spontaneous action potential formation similar to HAS2 and HA (Figure 8), suggesting that HAS2 and CD44 cooperatively regulate synapse formation, by HAS2-mediated HA synthesis at developing synapses followed by HA binding to CD44 for retention and potentially influencing intracellular signaling events.

To determine whether HA functions solely in a space-filling physical capacity or whether it can also alter intracellular signaling events, we examined RhoGTPase signaling alterations arising from HA or CD44 manipulations. CD44 is known to be tethered to the actin cytoskeleton through ERM complexes and modulate cell morphology via RhoGTPases [34,49]. We previously established that antagonistic RhoGTPase signaling regulates synaptogenesis [10]. Specifically, RhoA kinase (ROCK)-mediated actin bundling suppresses excitatory synaptogenesis, but ROCK inhibition increased Rac1 activity and excitatory synapse formation [10] (Figure 9A). Based on these data, we hypothesized that HA-CD44 interactions antagonize excitatory synapse formation by promoting RhoA activity and suppressing Rac1 activity (Figure 9A). Consistent with this hypothesis, hyaluronidase degradation of HA, which increases excitatory synapse formation, increases Rac1 activity at synapses (Figure 9B–E). Conversely, HAS2 and CD44 overexpression significantly decreased Rac1 activity (Figure 9F,G), corresponding with the observed decrease in excitatory synapse formation (Figures 4 and 7). Thus, in addition to its physical properties, HA also has the ability to alter intracellular synaptic signaling events through interaction with its receptor, CD44.

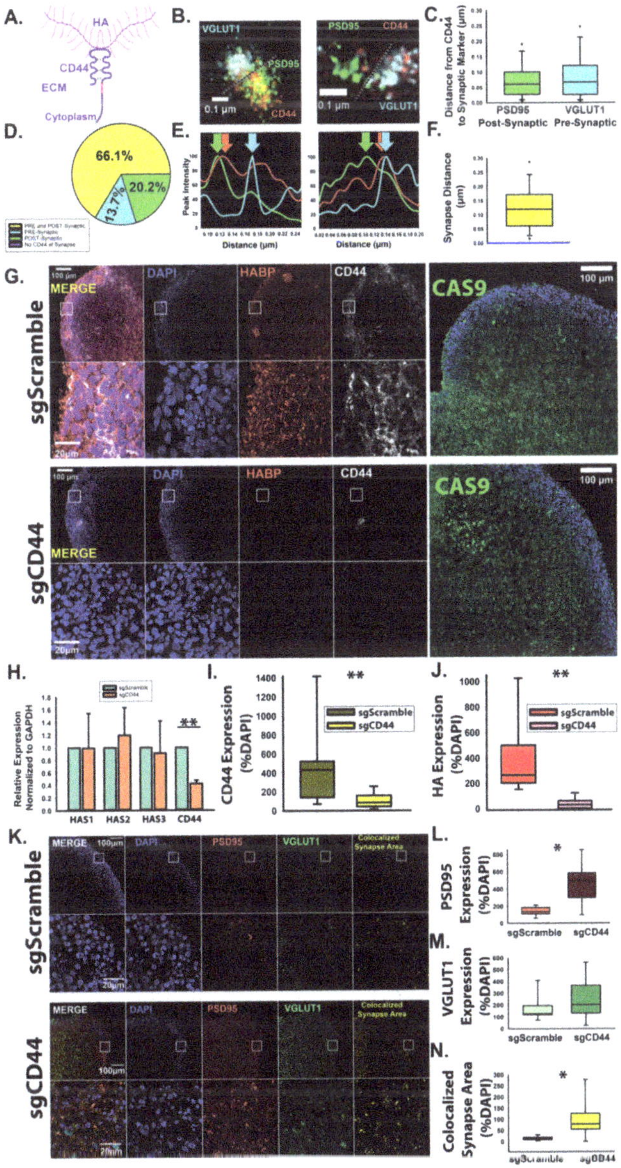

Figure 6. CD44 localizes to excitatory synapses and promotes HA retention in developing neural circuits. (A) Schematic of transmembrane protein CD44 bound to HA. (B) Representative dSTORM images of CD44 at excitatory synapses demonstrates that CD44 localizes to both pre- and post-synaptic compartments. Green: PSD95; red: CD44; cyan: VGLUT1. (C) Quantified distance from CD44 to pre- and post-synaptic markers. $n = 168$ synapses from 9 total spheroids derived from 3 separately grown spheroid sets. (D) Quantification of CD44 localization at individual excitatory synapses reveals that in the majority of cases, CD44 localizes to both the pre- and post-synaptic compartment (yellow) but may also be found solely within the pre- synaptic (blue) or post-synaptic (green) compartment. (E) Synaptic intensity plot profiles corresponding to the synapses shown in B.

Arrow indicate the peak intensity of PSD95, VGLUT1, and CD44 at the synapse. (**F**) Quantification of the synaptic distance between VGLUT1 and PSD95. n = 168 synapses from 9 total spheroids derived from 3 separately grown and transduced spheroid sets. (**G**) Representative images of cortical spheroids stably expressing dCas9-KRAB (green) for CRISPR-mediated interference of gene expression. At day 90, dCAS9-KRAB-expressing cortical spheroids were lentivirally transduced with either non- targeting scrambled sgRNA or sgRNA directed toward the promoter region of CD44. After 48 h, spheroids were fixed and stained for DAPI (blue), HABP (red) and CD44 (white). (**H**) qRT- PCR confirmation of CD44 knockdown. n = 9 total spheroids for each sgScramble and sgCD44 derived from 3 separately grown and transduced spheroid sets ** ≤ 0.001. (**I**) Quantification of area of CD44 expression normalized to the area of DAPI. n = 18 ** $p < 0.001$. (**J**) Quantification of area of HABP expression normalized to DAPI area. n = 9 total spheroids for each sgScramble and sgCD44 derived from 3 separately grown and transduced spheroid sets ** $p < 0.001$. (**K**) Representative images of cortical spheroids stained for excitatory synaptic markers, PSD95 and VGLUT1, and DAPI. Far right panel: colocalized synaptic area. (**L**) Quantification of PSD95 expression normalized to DAPI. * $p < 0.05$. (**M**) Quantification of VGLUT1 expression normalized to DAPI. (**N**) Quantification of colocalized excitatory synapse area normalized to DAPI, * $p < 0.05$.

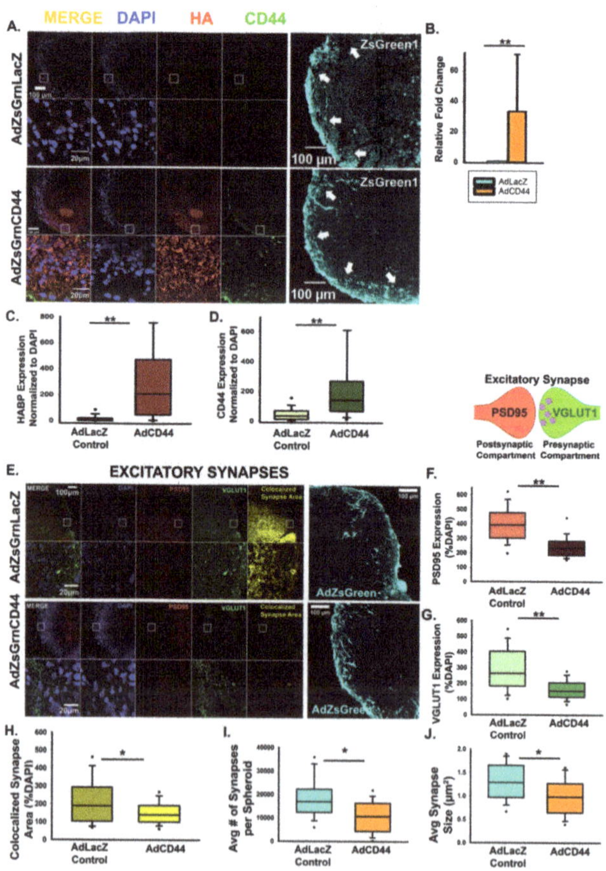

Figure 7. CD44 antagonizes excitatory synapse formation. (**A**) Representative images of cortical spheroids transduced with either LacZ control or CD44 and stained for DAPI (blue), HA (red), and CD44 (green). Highlighted region of the cortical plate is enlarged below. Successful transduction was confirmed via viral expression of the fluorophore, ZsGreen1 (cyan, far right panels). Arrows indicate the cortical plate, where synapse analysis was performed. Scale bars top: 100 µm, enlarged region:

20 μm. (**B**) qRT-PCR confirmation of increased CD44 expression normalized to GAPDH. (**C**) Quantification of the area of HA (HABP) normalized to the area of DAPI reveals that CD44 promotes HA retention. $n = 9$ total spheroids for each transduction (AdZsGrnLacZ and AdZsGrnCD44). Spheroids were derived from 3 separately grown and transduced spheroid sets ** $p < 0.001$. (**D**) Quantification of the area CD44 expression normalized to the area of DAPI. $n = 9$ total spheroids for each transduction (AdZsGrnLacZ and AdZsGrnCD44). Spheroids were derived from 3 separately grown and transduced spheroid sets ** $p < 0.001$. (**E**) Representative images of cortical spheroids transduced with either LacZ control or CD44 and stained for DAPI (blue), and excitatory synapses: post-synaptic marker PSD95 (red), and pre-synaptic maker VGLUT1 (green). Right panel shows the excitatory synaptic colocalization of PSD95 and VGLUT1 in yellow. Bottom panel highlights the region of the cortical plate indicated by the white box. Successful transduction was confirmed via viral co- expression of the fluorophore, ZsGreen1 (cyan, far right panels). Scale bars: top: 100 μm, enlarged region: 20 μm. (**F–H**) Quantification of individual pre- and post-synaptic markers and their co-localization in excitatory synapses reveals that CD44 antagonizes excitatory synapse formation. Quantification of the area of post-synaptic marker PSD95 normalized to the area of DAPI. $n = 9$ total spheroids for each transduction (AdZsGrnLacZ and AdZsGrnCD44). Spheroids were derived from 3 separately grown and transduced spheroid sets. ** $p < 0.001$ (**F**). Quantification of the area of the pre-synaptic marker VGLUT1 normalized to the area of DAPI. $n = 9$ total spheroids for each transduction (AdZsGrnLacZ and AdZsGrnCD44). Spheroids were derived from 3 separately grown and transduced spheroid sets. ** $p < 0.001$ (**G**). Quantification of co-localized excitatory synapse area normalized to the area of DAPI. $n = 9$ total spheroids for each transduction (AdZsGrnLacZ and AdZsGrnCD44). Spheroids were derived from 3 separately grown and transduced spheroid sets. * $p = 0.030$ (**H**). (**I**,**J**) Quantification of the average number of synapses per spheroid (**I**) and individual synapse size (**J**) demonstrates that CD44 promotes the refinement of excitatory synapses into discrete puncta. $n = 9$ total spheroids for each transduction (AdZsGrnLacZ and AdZsGrnCD44). Spheroids were derived from 3 separately grown and transduced spheroid sets. ** $p < 0.001$.

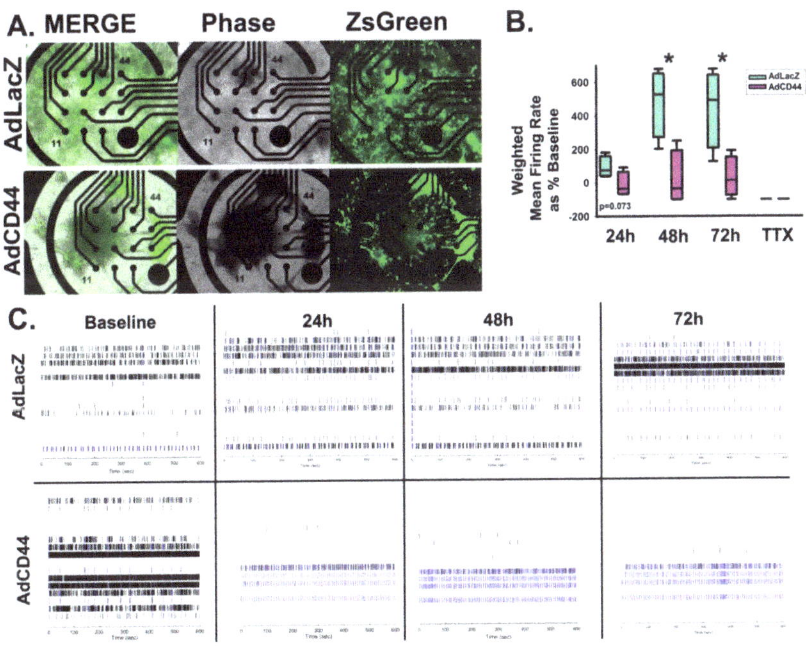

Figure 8. CD44 suppresses excitability of developing neural networks. (**A**) Representative images of successful viral transduction on MEA plates as indicated by viral expression of ZsGreen in dissociated spheroid transduced with adenovirus constructs for control LacZ (AdZsGrnLacZ) and CD44

(AdZsGrnCD44) expression. (**B**) Quantification of weighted mean firing rate of spontaneous action potentials in CD44 and LacZ-expressing dissociated spheroids demonstrates that CD44 suppresses excitability by significantly reducing the firing rate at 48 and 72 h post-transduction. Tetrodotoxin (TTX) was used to inhibit action potential propagation, demonstrating that the firing rate reflects action potentials. m = 6 wells of dissociated spheroids per treatment, spheroids were derived from 3 separately grown and transduced spheroid sets. * $p < 0.05$. (**C**) Representative raster plots of spontaneous action potential firing at 0 (baseline), 24, 48, and 72 h post-transduction for AdZsGrnLacZ and AdZsGrnCD44.

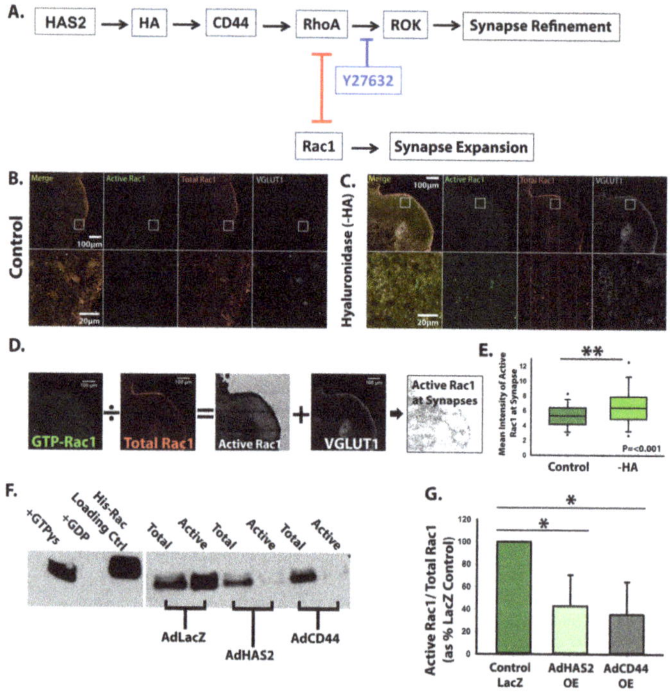

Figure 9. HAS2, HA, and CD44 regulate synapse formation via RhoGTPase signaling. (**A**) Graphic representation of proposed mechanism by which HAS2, HA, and CD44 regulate the antagonistic relationship between RhoGTPases, RhoA and Rac1, to promote refinement of excitatory synapses. (**B**,**C**) Control cortical spheroids (**B**) or cortical spheroids treated with streptomyces hyaluronidase (**C**) to degrade HA were immunostained for active GTP-bound Rac1 and total Rac1 together with VGLUT1 to visualize association with developing excitatory synapses. (**D**) Ratiometric imaging was performed by dividing the fluorescence intensity for GTP-Rac1 by the fluorescence intensity for total Rac1. The resulting ratiometric image intensity was measured in particles corresponding to VGLUT1 area. (**E**) Quantification of Rac1 activity associated with VGLUT1-postive compartments reveals that HA suppresses Rac1 activity, while treatment with hyaluronidase significantly increases Rac1 activity. $n = 9$ total spheroids for each treatment derived from 3 separately grown and treated spheroid sets. ** $p < 0.001$ (**F**) Immunoblot of Active Rac1 pulldown. Glutathione affinity beads were used to pulldown GTP-Rac1. Columns left to right: +GTPγs positive control, +GDP negative control, His-Rac1 antibody loading control, Total (30 μg) and Active (300 μg of lysate incubated with beads) samples for spheroids transduced with AdZsGreenLacZ, AdZsGreenHAS2, or AdZsGreenCD44. (**G**) Quantification of Active Rac1 pulldowns reveal HAS2 and CD44 overexpression decrease Rac1 activity, reported as GTPγs-Rac1/Total-Rac1 as percent of AdZsGreenLacZ control. $n = 3$ spheroids for each AdZsGrnLacZ, AdZsGrnHAS2 and AdZsGrnCD44 derived from 3 separately grown and transduced spheroid sets, ran on 3 separate western blot analyses. * $p < 0.05$.

4. Discussion

While HA is known to be substantially increased during the prenatal developmental window and to impact early events in brain development, such as neurulation [50–52], our research is the first to establish a role for HA in excitatory synapse formation and the emergence of synaptic activity in developing neural circuits [11]. In our current research, we sought to determine the mechanisms underlying the synthesis and signaling of HA at developing excitatory synapses. Given that HA was previously thought to be excluded from the synaptic cleft [15,43], we first demonstrate that this developmental localization is conserved between human brain spheroids and the developing mouse brain. HA localizes to the synaptic cleft of the majority of newly formed excitatory synapses in both human cortical spheroids [11] and the mouse neonatal brain (Figure 1). While human synapse formation begins prenatally during mid-fetal gestation and continues into juvenile development [12], mouse synapse formation occurs predominantly around birth [53]. Thus, we used P0 mouse cortex to capture synapse formation similar to synapse formation in our model of human mid-fetal cortical development [9–12,38]. We compared HA at excitatory synapses of the P0 mouse neonatal brain at the time of synapse formation with the young adult mouse brain (P40) during the maintenance of mature synaptic connections (Figure 1). At immature synapses of the neonate cortex, HA predominantly localizes to the synaptic cleft, corroborating our previous observations in human cortical spheroids that model fetal synaptogenesis [11]. By contrast, in the adult mouse brain, HA largely surrounds synapses, consistent with previous reports of HA in peri-synaptic extracellular matrix [16]. The peri-synaptic extracellular matrix is associated with the formation of a tripartite synapse containing an astrocytic endfoot [53]. In mice, the majority of astrocytic contributions to synaptic development occur postnatally [16,53,54]. For example, during the first and second week of postnatal development, astrocytes segregate into non-overlapping territories, with one astrocyte ensheathing up to 100,000 synapses [54–56]. During this time frame, there is also a robust increase in the number of synapses and a corresponding increase in synaptic activity [13,57]. Astrocytes contribute to synaptic function in part by regulating glutamate uptake from the synaptic cleft, preventing glutamate-induced neurotoxicity [58,59]. Hyaluronan supports the activity of the glial glutamate transporter [60]. Thus, while our data supports a model in which HA is directly synthesized into the developing synaptic cleft, it is interesting to speculate that with further development, HA is synthesized by associated astrocytes, which would surround and stabilize mature synaptic contacts. In addition to the developmental changes in HA synaptic localization, postnatal development also coincides with the emergence of perineuronal nets around inhibitory interneurons [14,19,61]. In early brain development, the HA-based ECM is loosely packed and forms the interstitial ECM [62] (Figure 1G) and lacks the dense meshwork of HA in perineuronal nets around the soma of interneurons (Figure 1H) [14]. Our data thus highlights a critical temporal role for HA at developing synapses, while also elucidating mechanisms of HA synthesis and signaling at these newly formed synapses. It is interesting to speculate that this early localization to synapses may be necessary for initial stages of synaptogenesis. For example, HA binding can lead to clustering of CD44 [33], and CD44 is known to interact with and regulate the expression of post-synaptic density proteins, such as PSD-95 and Bassoon [31,63]. Thus, HA-CD44 interactions may help to initiate post-synaptic density (PSD) formation. Once formed, glutamate receptors clustered within the PSD can readily respond to pre-synaptic glutamate release, leading to synaptic strengthening. Neuronal activity is associated with increased matrix remodeling, which may lead to the observed loss of HA from excitatory synapses of older animals (Figure 1) [42,64]. Future studies are needed to address whether activity impacts HA synaptic synthesis, localization and turnover. However, our current study provides a foundation for such future studies by elucidating mechanisms of HA synthesis, signaling and retention within developing neural circuits.

In cortical brain development, HAS2 is the predominant hyaluronan synthase isoform [6]. Furthermore, data from the Human Brain Transcriptome demonstrates that the

RNA expression of HAS2 is highest during prenatal brain development when synapses are forming (Available online: http://www.hbatlas.org, accessed on 16 September 2021) [65–67]. We establish that HAS2 critically regulates excitatory synapse formation and function through synthesis of HA. Using combined confocal and dSTORM microscopy, we demonstrate that HAS2 preferentially localizes to excitatory synapses, where it is found in both pre- and post-synaptic compartments (Figures 2 and 3). During excitatory synapse formation, pre- and post-synaptic projections contact and adhere to one another [68]. In this developmental context, HAS2 is capable of secreting HA directly into the developing synaptic cleft. Notably, increased HA levels via introduction of high molecular weight HA [11] or overexpression of HAS2 attenuate excitatory synaptogenesis and suppress spontaneous action potential formation (Figures 4 and 5). Thus, within the critical window of synapse formation in early brain development, the synaptic expression of HAS2 uniquely positions it to prevent the emergence of hyperexcitability commonly observed in neurodevelopmental disorders, such as epilepsy and autism spectrum disorders [3–5].

Following HA synthesis, subsequent interaction of HA with its receptor CD44 alters intracellular RhoGTPase signaling pathways that attenuate synapse formation (Figures 6 and 9). dSTORM super-resolution microscopy reveals that CD44 localizes near the synaptic cleft in both the pre- and post-synaptic compartment of excitatory synapses (Figure 6), enabling both sides of the synapse to adhere to the HA-based ECM. In other cells, interactions between HA and CD44 stabilize membrane protrusions leading to directional migration [69]. Thus, we hypothesize that CD44 helps to stabilize newly formed synaptic contacts. Similar to other tissue systems [25,27,70], we also establish that CD44 is necessary for HA retention in developing neural networks (Figure 6). Together with HA retention, CD44 prevents excitatory synaptogenesis and spontaneous action potential formation similar to HAS2 and HA expression [11] (Figures 7 and 8). These effects are driven by regulation of RhoGTPase signaling, as has previously been observed for CD44 in mature hippocampal neuron cultures, although it was not determined whether these effects were dependent on HA [31]. Specifically, HA and CD44 suppress Rac1 activity (Figure 9). We have previously demonstrated that Rac1 activity promotes the formation of dendritic spine precursors [7] and is associated with increased post-synaptic surface area [71] and increased synapse formation [10]. Thus, we propose that CD44 plays a critical role in synaptic refinement, stabilizing newly formed synapses but also restricting their growth and attenuating the formation of new synapses. A possible role for CD44 in synaptic refinement is supported by our observed decrease in the size of individual synapses in response to adenoviral expression of CD44 (Figure 7J). Thus, while previous HA-mediated effects on synaptic function were attributed to changes in the physical space occupied by the complex HA structure [6,23], our data also shows that HA can regulate synaptic activity through cell signaling events mediated by interaction with its receptor CD44 (Figure 9). Thus, our data highlights a critical role for synaptic HAS2-HA-CD44 in protecting developing neural networks from the emergence of hyperexcitability and epilepsy through a RhoGTPase signaling axis, by suppressing aberrant Rac1 hyperactivation [23]. These results are consistent with emerging data suggesting that RhoGTPase signaling pathways may be targeted therapeutically to restore normal synapse development and cognition in neurodevelopmental disorders [72,73]. Furthermore, these results indicate that hyaluronan pathways can be manipulated to restore synaptic function, especially in brain disorders associated with altered hyaluronan levels. For example, elevated HA or CD44-mediated retention of HA may be used to suppress elevated excitatory synaptic transmission in epilepsy [6,23]. Conversely, increased hyaluronan is observed in aging and Alzheimer's Disease [62,74,75]. Together with this increase in hyaluronan, synapse loss is observed prior to neurodegeneration in Alzheimer's Disease [76–79] and correlates with the severity of cognitive decline [80]. It is therefore interesting to speculate that decreased hyaluronan synthesis or CD44 may be used to prevent synapse loss.

Ultimately, our use of dSTORM super-resolution microcopy allowed us to observe a previously unappreciated localization of HA to the synaptic cleft of immature synapses in both a cell culture model of human cortical development and the mouse neonatal brain. Using our genetically tractable and physiologically relevant model of human brain development, we were able to disrupt HA synthesis and signaling specifically during synapse formation, revealing novel roles for HAS2 and CD44 in synapse formation and the emerging balance between excitatory and inhibitory signaling in developing neural circuits. Future modulation of these developing networks and their interactions with the HA-based ECM will allow us to appreciate how changes in HA synaptic localization contribute to synapse maturation and refinement.

Supplementary Materials: The following are available online at https://www.mdpi.com/article/10.3390/cells10102574/s1, Figure S1: HA, Table S1: Primary Antibodies, Table S2: Secondary Antibodies, Table S3. Primer Sequences.

Author Contributions: Conceptualization, K.L. and E.S.W.; methodology, K.L. and E.S.W.; validation, K.L. and E.S.W.; formal analysis, E.S.W.; investigation, E.S.W.; data curation, E.S.W.; writing—original draft preparation, K.L. and E.S.W.; writing—review and editing, K.L. and E.S.W; visualization, E.S.W.; supervision, K.L.; project administration, K.L.; funding acquisition, K.L. All authors have read and agreed to the published version of the manuscript.

Funding: We would like to thank ECU Department of Anatomy and Cell Biology and ECU Research Economic Development and Engagement (REDE) for funding for this project.

Institutional Review Board Statement: Not applicable.

Informed Consent Statement: Not applicable.

Data Availability Statement: Not applicable.

Acknowledgments: We would like to thank the laboratory of Chris Geyer (ECU) for the donation of mouse brain samples, Amanda Petritsch for the development of hIPSCs and subsequent human spheroids stably expressing dCas9-KRAB, and Warren Knudson (ECU) for adenoviruses expressing LacZ, HAS2, and CD44, all of which were used in this study. Portions of Figure 1A,B, Figures 3A, 5A and 6A, and the graphical abstract were created using Biorender under license purchased by the Litwa laboratory.

Conflicts of Interest: The authors declare that they have no conflict of interest.

References

1. Washbourne, P. Synapse Assembly and Neurodevelopmental Disorders. *Neuropsychopharmacology* **2014**, *40*, 4–15. [CrossRef]
2. Zoghbi, H.Y.; Bear, M.F. Synaptic Dysfunction in Neurodevelopmental Disorders Associated with Autism and Intellectual Disabilities. *Cold Spring Harb. Perspect. Biol.* **2012**, *4*, a009886. [CrossRef]
3. Gao, R.; Penzes, P. Common mechanisms of excitatory and inhibitory imbalance in schizophrenia and autism spectrum disorders. *Curr. Mol. Med.* **2015**, *15*, 146–147. [CrossRef]
4. Gandal, M.J.; Haney, J.R.; Parikshak, N.N.; Leppa, V.; Ramaswami, G.; Hartl, C.; Schork, A.J.; Appadurai, V.; Buil, A.; Werge, T.M.; et al. Shared molecular neuropathology across major psychiatric disorders parallels polygenic overlap. *Science* **2018**, *359*, 693–697. [CrossRef] [PubMed]
5. Frega, M.; Selten, M.; Mossink, B.; Keller, J.M.; Linda, K.; Moerschen, R.; Qu, J.; Koerner, P.; Jansen, S.; Oudakker, A.; et al. Distinct Pathogenic Genes Causing Intellectual Disability and Autism Exhibit a Common Neuronal Network Hyperactivity Phenotype. *Cell Rep.* **2020**, *30*, 173–186.e6. [CrossRef]
6. Arranz, A.M.; Perkins, K.L.; Irie, F.; Lewis, D.P.; Hrabe, J.; Xiao, F.; Itano, N.; Kimata, K.; Hrabetova, S.; Yamaguchi, Y. Hyaluronan deficiency due to Has3 knock-out causes altered neuronal activity and seizures via reduction in brain extracellular space. *J. Neurosci.* **2014**, *34*, 6164–6176. [CrossRef]
7. Martin-Vilchez, S.; Whitmore, L.; Asmussen, H.; Zareno, J.; Horwitz, R.; Newell-Litwa, K. RhoGTPase Regulators Orchestrate Distinct Stages of Synaptic Development. *PLoS ONE* **2017**, *12*, e0170464. [CrossRef] [PubMed]
8. Papariello, A.; Newell-Litwa, K. Human-Derived Brain Models: Windows into Neuropsychiatric Disorders and Drug Therapies. *ASSAY Drug Dev. Technol.* **2020**, *18*, 79–88. [CrossRef] [PubMed]
9. Papariello, A.; Taylor, D.; Soderstrom, K.; Litwa, K. CB1 antagonism increases excitatory synaptogenesis in a cortical spheroid model of fetal brain development. *Sci. Rep.* **2021**, *11*, 1–17. [CrossRef] [PubMed]
10. Wilson, E.; Rudisill, T.; Kirk, B.; Johnson, C.; Kemper, P.; Newell-Litwa, K. Cytoskeletal regulation of synaptogenesis in a model of human fetal brain development. *J. Neurosci. Res.* **2020**, *98*, 2148–2165. [CrossRef]

11. Wilson, E.; Knudson, W.; Newell-Litwa, K. Hyaluronan regulates synapse formation and function in developing neural networks. *Sci. Rep.* **2020**, *10*, 1–14. [CrossRef]
12. Wilson, E.S.; Newell-Litwa, K. Stem cell models of human synapse development and degeneration. *Mol. Biol. Cell* **2018**, *29*, 2913–2921. [CrossRef]
13. Semple, B. D.; Blomgren, K.; Gimlina K; Ferriero, D. M; Noble-Haeusslein, L. J. Brain development in rodents and humans: Identifying benchmarks of maturation and vulnerability to injury across species. *Prog. Neurobiol.* **2013**, *106–107*, 1–16. [CrossRef]
14. Frischknecht, R.; Seidenbecher, C.I. The crosstalk of hyaluronan-based extracellular matrix and synapses. *Neuron Glia Biol.* **2008**, *4*, 249–257. [CrossRef]
15. Bondareff, W.; Sjöstrand, J. Cytochemistry of synaptosomes. *Exp. Neurol.* **1969**, *24*, 450–458. [CrossRef]
16. Ferrer-Ferrer, M.; Dityatev, A. Shaping Synapses by the Neural Extracellular Matrix. *Front. Neuroanat.* **2018**, *12*, 40. [CrossRef] [PubMed]
17. Kochlamazashvili, G.; Henneberger, C.; Bukalo, O.; Dvoretskova, E.; Senkov, O.; Lievens, P.M.-J.; Westenbroek, R.; Engel, A.K.; Catterall, W.A.; Rusakov, D.A.; et al. The Extracellular Matrix Molecule Hyaluronic Acid Regulates Hippocampal Synaptic Plasticity by Modulating Postsynaptic L-Type Ca2+ Channels. *Neuron* **2010**, *67*, 116–128. [CrossRef] [PubMed]
18. Rogers, S.; Rankin-Gee, E.; Risbud, R.M.; Porter, B.E.; Marsh, E.D. Normal Development of the Perineuronal Net in Humans; In Patients with and without Epilepsy. *Neuroscience* **2018**, *384*, 350–360. [CrossRef] [PubMed]
19. Fawcett, J.W.; Oohashi, T.; Pizzorusso, T. The roles of perineuronal nets and the perinodal extracellular matrix in neuronal function. *Nat. Rev. Neurosci.* **2019**, *20*, 451–465. [CrossRef]
20. Wen, T.H.; Binder, D.K.; Ethell, I.M.; Razak, K.A. The Perineuronal 'Safety' Net? Perineuronal Net Abnormalities in Neurological Disorders. *Front. Mol. Neurosci.* **2018**, *11*, 270. [CrossRef]
21. Itano, N.; Kimata, K. Mammalian Hyaluronan Synthases. *IUBMB Life* **2002**, *54*, 195–199. [CrossRef]
22. Itano, N.; Sawai, T.; Yoshida, M.; Lenas, P.; Yamada, Y.; Imagawa, M.; Shinomura, T.; Hamaguchi, M.; Yoshida, Y.; Ohnuki, Y.; et al. Three Isoforms of Mammalian Hyaluronan Synthases Have Distinct Enzymatic Properties. *J. Biol. Chem.* **1999**, *274*, 25085–25092. [CrossRef]
23. Perkins, K.L.; Arranz, A.M.; Yamaguchi, Y.; Hrabetova, S. Brain extracellular space, hyaluronan, and the prevention of epileptic seizures. *Rev. Neurosci.* **2017**, *28*, 869–892. [CrossRef] [PubMed]
24. Aruffo, A.; Stamenkovic, I.; Melnick, M. UCB and SB. CD44 is the Principal Cell Surface Receptor for Hyaluronate. *Cell* **1990**, *61*, 1303–1313. [CrossRef]
25. Bourguignon, L.Y. Matrix Hyaluronan-Activated CD44 Signaling Promotes Keratinocyte Activities and Improves Abnormal Epidermal Functions. *Am. J. Pathol.* **2014**, *184*, 1912–1919. [CrossRef]
26. Bourguignon, L.Y.W.; Singleton, P.A.; Zhu, H.; Diedrich, F. Hyaluronan-mediated CD44 Interaction with RhoGEF and Rho Kinase Promotes Grb2-associated Binder-1 Phosphorylation and Phosphatidylinositol 3-Kinase Signaling Leading to Cytokine (Macrophage-Colony Stimulating Factor) Production and Breast Tumor Progression. *J. Biol. Chem.* **2003**, *278*, 29420–29434. [CrossRef] [PubMed]
27. Bourguignon, L.Y.; Wong, G.; Xia, W.; Man, M.-Q.; Holleran, W.M.; Elias, P.M. Selective matrix (hyaluronan) interaction with CD44 and RhoGTPase signaling promotes keratinocyte functions and overcomes age-related epidermal dysfunction. *J. Dermatol. Sci.* **2013**, *72*, 32–44. [CrossRef]
28. Bijata, M.; Labus, J.; Guseva, P.D.; Stawarski, M.; Butzlaff, M.; Dzwonek, J.; Schneeberg, J.; Böhm, K.; Michaluk, P.; Rusakov, D.; et al. Synaptic Remodeling Depends on Signaling between Serotonin Receptors and the Extracellular Matrix. *Cell Rep.* **2017**, *19*, 1767–1782. [CrossRef]
29. Ponta, H.; Wainwright, D.; Herrlich, P. Molecules in focus The CD44 protein family. *Int. J. Biochem. Cell Biol.* **1998**, *30*, 299–305. [CrossRef]
30. Konopka, A.; Zeug, A.; Skupień-Jaroszek, A.; Kaza, B.; Müller, F.; Chwedorowicz, A.; Ponimaskin, E.; Wilczynski, G.; Dzwonek, J. Cleavage of Hyaluronan and CD44 Adhesion Molecule Regulate Astrocyte Morphology via Rac1 Signalling. *PLoS ONE* **2016**, *11*, e0155053. [CrossRef]
31. Roszkowska, M.; Skupień-Jaroszek, A.; Wójtowicz, T.; Konopka, A.; Gorlewicz, A.; Kisiel, M.; Bekisz, M.; Ruszczycki, B.; Doleżyczek, H.; Rejmak-Kozicka, E.; et al. CD44: A novel synaptic cell adhesion molecule regulating structural and functional plasticity of dendritic spines. *Mol. Biol. Cell* **2016**, *27*, 4055–4066. [CrossRef]
32. Knudson, C.B.; Knudson, W. Hyaluronan and CD44: Modulators of chondrocyte metabolism. *Clin. Orthop. Relat. Res.* **2004**, *427*, S152–S162. [CrossRef]
33. Yang, C.; Cao, M.; Liu, H.; He, Y.; Xu, J.; Du, Y.; Liu, Y.; Wang, W.; Cui, L.; Hu, J.; et al. The High and Low Molecular Weight Forms of Hyaluronan Have Distinct Effects on CD44 Clustering. *J. Biol. Chem.* **2012**, *287*, 43094–43107. [CrossRef] [PubMed]
34. Thorne, R.F.; Legg, J.W.; Isacke, C. The role of the CD44 transmembrane and cytoplasmic domains in co-ordinating adhesive and signalling events. *J. Cell Sci.* **2004**, *117*, 373–380. [CrossRef] [PubMed]
35. Bourguignon, L.Y.; Zhu, H.; Shao, L.; Chen, Y.W. Ankyrin–Tiam1 Interaction Promotes Rac1 Signaling and Metastatic Breast Tumor Cell Invasion and Migration. *J. Cell Biol.* **2000**, *150*, 177–192. [CrossRef]
36. Sethi, M.K.; Zaia, J. Extracellular matrix proteomics in schizophrenia and Alzheimer's disease. *Anal. Bioanal. Chem.* **2016**, *409*, 379–394. [CrossRef] [PubMed]

37. Reed, M.J.; Damodarasamy, M.; Pathan, J.L.; Chan, C.K.; Spiekerman, C.; Wight, T.N.; Banks, W.A.; Day, A.J.; Vernon, R.B.; Keene, C.D. Increased Hyaluronan and TSG-6 in Association with Neuropathologic Changes of Alzheimer's Disease. *J. Alzheimer's Dis.* **2019**, *67*, 91–102. [CrossRef]
38. Paşca, A.M.; Sloan, S.A.; Clarke, L.E.; Tian, Y.; Makinson, C.D.; Huber, N.; Kim, C.H.; Park, J.-Y.; O'Rourke, N.A.; Nguyen, K.; et al. Functional cortical neurons and astrocytes from human pluripotent stem cells in 3D culture. *Nat. Methods* **2015**, *12*, 671–678. [CrossRef]
39. Ishizuka, S.; Tsuchiya, S.; Ohashi, Y.; Terabe, K.; Askew, E.B.; Ishizuka, N.; Knudson, C.B.; Knudson, W. Hyaluronan synthase 2 (HAS2) overexpression diminishes the procatabolic activity of chondrocytes by a mechanism independent of extracellular hyaluronan. *J. Biol. Chem.* **2019**, *294*, 13562–13579. [CrossRef]
40. Olivier, N.; Keller, D.; Rajan, V.S.; Gönczy, P.; Manley, S. Simple buffers for 3D STORM microscopy. *Biomed. Opt. Express* **2013**, *4*, 885–899. [CrossRef]
41. Dempsey, G.T.; Vaughan, J.C.; Chen, K.H.; Bates, M.; Zhuang, X. Evaluation of fluorophores for optimal performance in localization-based super-resolution imaging. *Nat. Methods* **2011**, *8*, 1027–1036. [CrossRef]
42. Jourquin, J.; Tremblay, E.; Décanis, N.; Charton, G.; Hanessian, S.; Chollet, A.-M.; Le Diguardher, T.; Khrestchatisky, M.; Rivera, S. Neuronal activity-dependent increase of net matrix metalloproteinase activity is associated with MMP-9 neurotoxicity after kainate. *Eur. J. Neurosci.* **2003**, *18*, 1507–1517. [CrossRef]
43. Bloom, F.E.; Aghajanian, G.K. Fine structural and cytochemical analysis of the staining of synaptic junctions with phosphotungstic acid. *J. Ultrastruct. Res.* **1968**, *22*, 361–375. [CrossRef]
44. Cohen, M.; Klein, E.; Geiger, B.; Addadi, L. Organization and Adhesive Properties of the Hyaluronan Pericellular Coat of Chondrocytes and Epithelial Cells. *Biophys. J.* **2003**, *85*, 1996–2005. [CrossRef]
45. Fowke, T.M.; Karunasinghe, R.N.; Bai, J.-Z.; Jordan, S.; Gunn, A.; Dean, J.M. Hyaluronan synthesis by developing cortical neurons in vitro. *Sci. Rep.* **2017**, *7*, 44135. [CrossRef] [PubMed]
46. Birey, F.; Andersen, J.; Makinson, C.D.; Islam, S.; Wei, W.; Huber, N.; Fan, H.C.; Metzler, K.R.C.; Panagiotakos, G.; Thom, N.; et al. Assembly of functionally integrated human forebrain spheroids. *Nature* **2017**, *545*, 54–59. [CrossRef] [PubMed]
47. Tian, R.; Gachechiladze, M.A.; Ludwig, C.H.; Laurie, M.T.; Hong, J.Y.; Nathaniel, D.; Prabhu, A.V.; Fernandopulle, M.S.; Patel, R.; Abshari, M.; et al. CRISPR Interference-Based Platform for Multimodal Genetic Screens in Human iPSC-Derived Neurons. *Neuron* **2020**, *104*, 239–255. [CrossRef]
48. Deans, P.J.M.; Brennand, K.J. Applying stem cells and CRISPR engineering to uncover the etiology of schizophrenia. *Curr. Opin. Neurobiol.* **2021**, *69*, 193–201. [CrossRef]
49. Su, W.; Foster, S.C.; Xing, R.; Feistel, K.; Olsen, R.H.J.; Acevedo, S.F.; Raber, J.; Sherman, L.S. CD44 Transmembrane Receptor and Hyaluronan Regulate Adult Hippocampal Neural Stem Cell Quiescence and Differentiation. *J. Biol. Chem.* **2017**, *292*, 4434–4445. [CrossRef]
50. Kazanis, I.; Ffrench-Constant, C. Extracellular matrix and the neural stem cell niche. *Dev. Neurobiol.* **2011**, *71*, 1006–1017. [CrossRef]
51. Casini, P.; Nardi, I.; Ori, M. Hyaluronan is required for cranial neural crest cells migration and craniofacial development. *Dev. Dyn.* **2012**, *241*, 294–302. [CrossRef]
52. Preston, M.; Sherman, L.S. Neural Stem Cell Niches: Roles for the Hyaluronan-Based Extracellular Matrix. *Front. Biosci.* **2011**, *3*, 1165–1179. [CrossRef]
53. Farhy-Tselnicker, I.; Allen, N.J. Astrocytes, neurons, synapses: A tripartite view on cortical circuit development. *Neural Dev.* **2018**, *13*, 1–12. [CrossRef]
54. Chung, W.-S.; Allen, N.J.; Eroglu, C. Astrocytes Control Synapse Formation, Function, and Elimination. *Cold Spring Harb. Perspect. Biol.* **2015**, *7*, a020370. [CrossRef]
55. Distler, C.; Dreher, Z.; Stone, J. Contact spacing among astrocytes in the central nervous system: An hypothesis of their structural role. *Glia* **1991**, *4*, 484–494. [CrossRef] [PubMed]
56. Halassa, M.M.; Fellin, T.; Takano, H.; Dong, J.-H.; Haydon, P.G. Synaptic Islands Defined by the Territory of a Single Astrocyte. *J. Neurosci.* **2007**, *27*, 6473–6477. [CrossRef] [PubMed]
57. Kroon, T.; van Hugte, E.; Van Linge, L.; Mansvelder, H.; Meredith, R. Early postnatal development of pyramidal neurons across layers of the mouse medial prefrontal cortex. *Sci. Rep.* **2019**, *9*, 1–16. [CrossRef] [PubMed]
58. Anderson, C.M.; Swanson, R.A. Astrocyte Glutamate Transport: Review of Properties, Regulation, and Physiological Functions. *Glia* **2000**, *32*, 1–14. [CrossRef]
59. Mahmoud, S.; Gharagozloo, M.; Simard, C.; Gris, D. Astrocytes Maintain Glutamate Homeostasis in the CNS by Controlling the Balance between Glutamate Uptake and Release. *Cells* **2019**, *8*, 184. [CrossRef] [PubMed]
60. Hayashi, M.K.; Nishioka, T.; Shimizu, H.; Takahashi, K.; Kakegawa, W.; Mikami, T.; Hirayama, Y.; Koizumi, S.; Yoshida, S.; Yuzaki, M.; et al. Hyaluronan synthesis supports glutamate transporter activity. *J. Neurochem.* **2019**, *150*, 249–263. [CrossRef]
61. Carulli, D.; Rhodes, K.E.; Brown, D.J.; Bonnert, T.P.; Pollack, S.J.; Oliver, K.; Strata, P.; Fawcett, J. Composition of perineuronal nets in the adult rat cerebellum and the cellular origin of their components. *J. Comp. Neurol.* **2006**, *494*, 559–577. [CrossRef] [PubMed]
62. Lau, L.W.; Cua, R.; Keough, M.B.; Haylock-Jacobs, S.; Yong, V.W. Pathophysiology of the brain extracellular matrix: A new target for remyelination. *Nat. Rev. Neurosci.* **2013**, *14*, 722–729. [CrossRef] [PubMed]

63. Kim, H.; Takegahara, N.; Walsh, M.C.; Choi, Y. CD44 Can Compensate for IgSF11 Deficiency by Associating with the Scaffold Protein PSD-95 during Osteoclast Differentiation. *Int. J. Mol. Sci.* **2020**, *21*, 2646. [CrossRef] [PubMed]
64. Wlodarczyk, J.; Mukhina, I.; Kaczmarek, L.; Dityatev, A. Extracellular matrix molecules, their receptors, and secreted proteases in synaptic plasticity. *Dev. Neurobiol.* **2011**, *71*, 1040–1053. [CrossRef]
65. Pletikos, M.; Sousa, A.M.; Sedmak, G.; Meyer, K.A.; Zhu, Y.; Cheng, F.; Li, M.; Kawasawa, Y.I.; Šestan, N. Temporal Specification and Bilaterality of Human Neocortical Topographic Gene Expression. *Neuron* **2014**, *81*, 321–332. [CrossRef] [PubMed]
66. Kang, H.J.; Kawasawa, Y.I.; Cheng, F.; Zhu, Y.; Xu, X.; Li, M.; Sousa, A.M.M.; Pletikos, M.; Meyer, K.A.; Sedmak, G.; et al. Spatio-temporal transcriptome of the human brain. *Nat. Cell Biol.* **2011**, *478*, 483–489. [CrossRef]
67. Johnson, M.B.; Kawasawa, Y.I.; Mason, C.E.; Krsnik, Z.; Coppola, G.; Bogdanović, D.; Geschwind, D.H.; Mane, S.M.; State, M.W.; Šestan, N. Functional and Evolutionary Insights into Human Brain Development through Global Transcriptome Analysis. *Neuron* **2009**, *62*, 494–509. [CrossRef]
68. Yuste, R.; Bonhoeffer, T. Genesis of dendritic spines: Insights from ultrastructural and imaging studies. *Nat. Rev. Neurosci.* **2004**, *5*, 24–34. [CrossRef]
69. Martín-Villar, E.; Fernández-Muñoz, B.; Parsons, M.; Yurrita, M.M.; Megías, D.; Pérez-Gómez, E.; Jones, G.E.; Quintanilla, M. Podoplanin Associates with CD44 to Promote Directional Cell Migration. *Mol. Biol. Cell* **2010**, *21*, 4387–4399. [CrossRef]
70. Huang, Y.; Askew, E.B.; Knudson, C.B.; Knudson, W. CRISPR/Cas9 knockout of HAS2 in rat chondrosarcoma chondrocytes demonstrates the requirement of hyaluronan for aggrecan retention. *Matrix Biol.* **2016**, *56*, 74–94. [CrossRef]
71. Newell-Litwa, K.A.; Badoual, M.; Asmussen, H.; Patel, H.; Whitmore, L.; Horwitz, A.R. ROCK1 and 2 differentially regulate actomyosin organization to drive cell and synaptic polarity. *J. Cell Biol.* **2015**, *210*, 225–242. [CrossRef]
72. Guo, D.; Yang, X.; Shi, L. Rho GTPase Regulators and Effectors in Autism Spectrum Disorders: Animal Models and Insights for Therapeutics. *Cells* **2020**, *9*, 835. [CrossRef]
73. Tejada-Simon, M.V. Modulation of actin dynamics by Rac1 to target cognitive function. *J. Neurochem.* **2015**, *133*, 767–779. [CrossRef]
74. Bonneh-Barkay, D.; Wiley, C.A. Brain Extracellular Matrix in Neurodegeneration. *Brain Pathol.* **2009**, *19*, 573–585. [CrossRef] [PubMed]
75. Reed, M.J.; Vernon, R.B.; Damodarasamy, M.; Chan, C.K.; Wight, T.N.; Bentov, I.; Banks, W.A. Microvasculature of the Mouse Cerebral Cortex Exhibits Increased Accumulation and Synthesis of Hyaluronan With Aging. *J. Gerontol. Ser. A Boil. Sci. Med. Sci.* **2016**, *72*, 740–746. [CrossRef] [PubMed]
76. Koffie, R.M.; Hyman, B.T.; Spires-Jones, T.L. Alzheimer's disease: Synapses gone cold. *Mol. Neurodegener.* **2011**, *6*, 63. [CrossRef] [PubMed]
77. Kashyap, G.; Bapat, D.; Das, D.; Gowaikar, R.; Amritkar, R.E.; Rangarajan, G.; Ravindranath, V.; Ambika, G. Synapse loss and progress of Alzheimer's disease -A network model. *Sci. Rep.* **2019**, *9*, 1–9. [CrossRef]
78. Yoshiyama, Y.; Higuchi, M.; Zhang, B.; Huang, S.-M.; Iwata, N.; Saido, T.C.; Maeda, J.; Suhara, T.; Trojanowski, J.Q.; Lee, V.M.-Y. Synapse Loss and Microglial Activation Precede Tangles in a P301S Tauopathy Mouse Model. *Neuron* **2007**, *53*, 337–351. [CrossRef]
79. Subramanian, J.; Savage, J.C.; Tremblay, M.È. Synaptic Loss in Alzheimer's Disease: Mechanistic Insights Provided by Two-Photon in vivo Imaging of Transgenic Mouse Models. *Front. Cell. Neurosci.* **2020**, *14*, 445. [CrossRef]
80. DeKosky, S.T.; Scheff, S.W. Synapse loss in frontal cortex biopsies in Alzheimer's disease: Correlation with cognitive severity. *Ann. Neurol.* **1990**, *27*, 457–464. [CrossRef]

cells

MDPI

Article

Evaluation of the Effects of Human Dental Pulp Stem Cells on the Biological Phenotype of Hypertrophic Keloid Fibroblasts

Ming Yan [1,2,†], Ling-Ling Fu [1,2,†], Ola A. Nada [1], Li-Ming Chen [3], Martin Gosau [1], Ralf Smeets [1,4], Hong-Chao Feng [3,*] and Reinhard E. Friedrich [1]

1 Department of Oral and Maxillofacial Surgery, University Medical Center Hamburg-Eppendorf, 20246 Hamburg, Germany; cnming.yan@hotmail.com (M.Y.); fu.lingling@hotmail.com (L.-L.F.); ola.abdulmonem@gmail.com (O.A.N.); m.gosau@uke.de (M.G.); r.smeets@uke.de (R.S.); r.friedrich@uke.de (R.E.F.)
2 Department of Oral and Maxillofacial Surgery, Hebei Eye Hospital, Xingtai 054000, China
3 Department of Oral and Maxillofacial Surgery, Guiyang Hospital of Stomatology, Guiyang 050017, China; cnliming.chen@hotmail.com
4 Department of Oral and Maxillofacial Surgery, Division of "Regenerative Orofacial Medicine", University Medical Center Hamburg-Eppendorf, 20246 Hamburg, Germany
* Correspondence: hcfeng@gzu.edu.cn; Tel.: +86-139-8403-0259
† These authors contributed equally to this work.

Abstract: Objective: Despite numerous existing treatments for keloids, the responses in the clinic have been disappointing, due to either low efficacy or side effects. Numerous studies dealing with preclinical and clinical trials have been published about effective therapies for fibrotic diseases using mesenchymal stem cells; however, no research has yet been reported to scientifically investigate the effect of human dental pulp stem cells (HDPSCs) on the treatment of keloids. The objective is to provide an experimental basis for the application of stem cells in the treatment of keloids. Methods: Human normal fibroblasts (HNFs) and human keloid fibroblasts (HKFs) were cultured alone and in combination with HDPSCs using a transwell cell-contact-independent cell culture system. The effects of HDPSCs on HKFs were tested using a CCK-8 assay, live/dead staining assay, quantitative polymerase chain reaction, Western blot and immunofluorescence microscopy. Results: HDPSCs did not inhibit the proliferation nor the apoptosis of HKFs and HNFs. HDPSCs did, however, inhibit their migration. Furthermore, HDPSCs significantly decreased the expression of profibrotic genes (CTGF, TGF-β1 and TGF-β2) in HKFs and KNFs ($p < 0.05$), except for CTGF in HNFs. Moreover, HDPSCs suppressed the extracellular matrix (ECM) synthesis in HKFs, as indicated by the decreased expression of collagen I as well as the low levels of hydroxyproline in the cell culture supernatant ($p < 0.05$). Conclusions: The co-culture of HDPSCs inhibits the migration of HKFs and the expression of pro-fibrotic genes, while promoting the expression of anti-fibrotic genes. HDPSCs' co-culture also inhibits the synthesis of the extracellular matrix by HKFs, whereas it does not affect the proliferation and apoptosis of HKFs. Therefore, it can be concluded that HDPSCs can themselves be used as a tool for restraining/hindering the initiation or progression of fibrotic tissue.

Keywords: human dental pulp stem cells; fibroblast; co-culture; keloid

Citation: Yan, M.; Fu, L.-L.; Nada, O.A.; Chen, L.-M.; Gosau, M.; Smeets, R.; Feng, H.-C.; Friedrich, R.E. Evaluation of the Effects of Human Dental Pulp Stem Cells on the Biological Phenotype of Hypertrophic Keloid Fibroblasts. *Cells* **2021**, *10*, 1803. https://doi.org/10.3390/cells10071803

Academic Editor: Mehdi Najar

Received: 25 May 2021
Accepted: 8 July 2021
Published: 16 July 2021

1. Introduction

During wound healing, a dynamic balance of synthesis and the degradation of collagen usually results in either a physiological or a pathological scar. A pathological scar can be described as a fibroproliferative disease caused by the hyperproliferation of fibroblasts and excessive synthesis of the extracellular matrix (ECM) in the process of dermis wound healing following skin trauma or serious burns [1–3]. Moreover, pathological scars can be classified into the following two types: hypertrophic scar and keloid. The former is predominantly localized above the original wound region, with a reddish or pinkish appearance, which can sometimes be pruritic. Furthermore, a hypertrophic scar usually regresses after years to form a matured scar [4,5], whereas the situation is strikingly different when the latter is involved. Although a keloid is a non-malignant disease, keloids almost always overgrow onto the surrounding skin, where they can often lead to malignant manifestations such as pain, ulceration, secondary infection, active angiogenesis and even carcinogenesis [6]. Such symptoms not only impact the quality of life, but they can also be highly unsightly, contributing further to psychological disturbances and corresponding social distress.

Despite numerous existing treatments for keloids, such as surgical removal, hormonal therapy, laser treatment or radiotherapy, as well as interventional sclerotherapy, the responses in the clinic have been disappointing due to either low efficacy or side effects, or a combination of both [7]. In order to address the aforementioned existing and urgent clinical need, new and more efficient therapeutic strategies are in order.

Numerous studies dealing with preclinical and clinical trials have been published about effective therapies for fibrotic diseases using mesenchymal stem cells, including the following: bone-marrow-derived mesenchymal stem cells (BMSCs) [8–10] and adipose derived mesenchymal stem cells (ADSCs) [11]. Human dental pulp stem cells (HDPSCs) are a type of adult stem cell that possess self-renewal, self-replication and multi-differentiation properties, where they can differentiate into a variety of mesodermal tissue cells, such as chondrocytes, osteoblasts, cardiomyocytes as well as adipocytes [12]. However, no research has yet been reported to scientifically investigate the effect of HDPSCs on the treatment of keloids.

The objective is to provide an experimental basis for the application of stem cells in the treatment of keloids. A co-culture method was set up to investigate the influence and mechanism of dental pulp stem cells on keloid fibroblast properties, such as cell proliferation, migration, collagen synthesis, invasion and apoptosis.

2. Materials and Methods

2.1. Cell Isolation and Cultures

Isolation of human dental pulp stem cells was performed using the explant growth method. Extracted teeth were obtained from patients (aged 18–25 years) with informed consent at the Guiyang Hospital of Stomatology (Guiyang, China). The retrieved dental pulp tissue was cut with sterilized scissors into small pieces and the tissue pieces were then plated onto 100-mm cell culture dishes in DMEM with 10% fetal bovine serum (FBS) and penicillin (100 U/mL)/streptomycin (100 μg/mL) at 37 °C and 5% CO_2. Cells were further grown until confluence was reached. Human dental pulp stem cells were obtained using a limiting dilution method, a method previously established by our group [13]. Only passages 3–5 of dental pulp stem cells were used in this study.

Seven tissue samples (0.5 cm^3) of keloids and seven corresponding normal skin tissue samples were used as a reference (Table 1). All tissue samples were taken from the maxillofacial area of patients in the same hospital with informed consent. The samples were classified into 2 groups (keloid and normal skin) based on clinical and pathological diagnosis. Furthermore, pathological characteristics were diagnosed upon staining with Hematoxylin and Eosin (H&E) with the following findings: (1) the dermal layer of the skin is markedly thickened; (2) disappearance of skin appendages; (3) collagen fibers begin to coarse into thick bundles. Finally, fibroblasts from keloid and normal skin isolation were also performed using the explant growth method. Samples were cut into 0.5 mm^3 tissue pieces. Each tissue piece was then cultivated onto 100-mm cell culture dishes in DMEM with 10% fetal bovine serum (FBS) and penicillin (100 U/mL)/streptomycin (100 μg/mL) at 37 °C and 5% CO_2. Cells were harvested until confluence was reached. Cells from passages 3–5 were used for this study.

Table 1. Patient epidemiological data.

Subject	Age	Sex
1	29	F
2	38	M
3	21	M
4	32	M
5	27	F
6	25	F
7	33	M

2.2. Flow Cytometry Was Used to Detect the Surface Markers of HDPSCs

HDPSCs at passage 3 were trypsinized by 0.25% trypsin, washed twice with PBS and resuspended at a concentration of 1×10^9/L in culture medium, according to the minimum criteria set by The International Society for Cellular Therapy in 2006 for defining multipotent MSCs [14]. The antibodies used in the experiments were CD19 (561295; BD Biosciences, San Jose, CA, USA), CD44 (560531; BD Biosciences, San Jose, CA, USA), CD45 (560368; BD Biosciences, San Jose, CA, USA), CD90 (555595; BD Pharmingen, San Jose, CA, USA), HLA-DR (560652; BD Biosciences, San Jose, CA, USA), CD29 (17-0299; eBiosciences, San Jose, CA, USA), CD73 (561014; BD Biosciences, San Jose, CA, USA), CD105 (560819; BD Biosciences, San Jose, CA, USA), CD73 (561014; BD Biosciences, San Jose, CA, USA), CD34 (560940; BD Biosciences), CD105 (560819; BD Biosciences) and CD11b (11-0113; eBiosciences, San Jose, CA, USA). In flow cytometry tubes, 1 mL of the cell suspension was collected with 5 μL of each of the anti-fluorescein isothiocyanate-conjugated antibodies. Cells were incubated at 4 °C for 1 h in the dark, and the cell surface markers were then analyzed on a BD Fortessa (BD Biosciences) instrument. All data analysis was conducted using FlowJo (v.10) software (FlowJo, Ashland, OR, USA).

2.3. Transwell Co-Culture Systems

In our project, 6-well Transwell systems, where cells shared medium without making any contact with transwell inserts (0.4-micrometer pore size), were employed (Figure 1A). Culture media was composed of DMEM with the following constituents: 10% FBS, 1% glutamine and 1% penicillin/streptomycin. There were the following four culture conditions in this project: (Figure 1B) HNFs monoculture, (Figure 1C) HDPSCs/HNFs co-culture, (Figure 1D) HKFs monoculture and (Figure 1E) HDPSCs/HKFs co-culture. In the monoculture groups, HNFs or HKFs were seeded in the bottom chambers. In the co-culture groups, HNFs or HKFs were seeded into the bottom chambers, and HDPSCs into the top chambers. The ratio of cells in the upper chamber to cells in the lower chamber was 1:1.

2.4. Cell Morphology

Cells were photographed under an inverted microscope (ECLIPSE Ts2-FL, Nikon, Japan) at days 1, 3, 5, and 7.

2.5. Cell Proliferation

Both normal and keloid fibroblasts were each individually seeded into 6-well plates at a concentration of 1×10^4 cells per well. In this study, cellular proliferation was analyzed using the CCK8 assay (Dojindo, Kumamoto, Japan) according to manufacturer's instructions. After removal of the supernatant, 660 μL of CCK8 solution (60 μL CCK8:600 μL medium) was added to each well and incubated in the incubator for 2 h on days 1, 3, 5 and 7. Then, a 100-microliter volume of the supernatant was separated by centrifugation and transferred into a fresh 96-well plate. Readings at wavelengths of 450 nm were recorded using a Fluostar Omega plate reader, and a standard curve was then plotted against the readings of the standards.

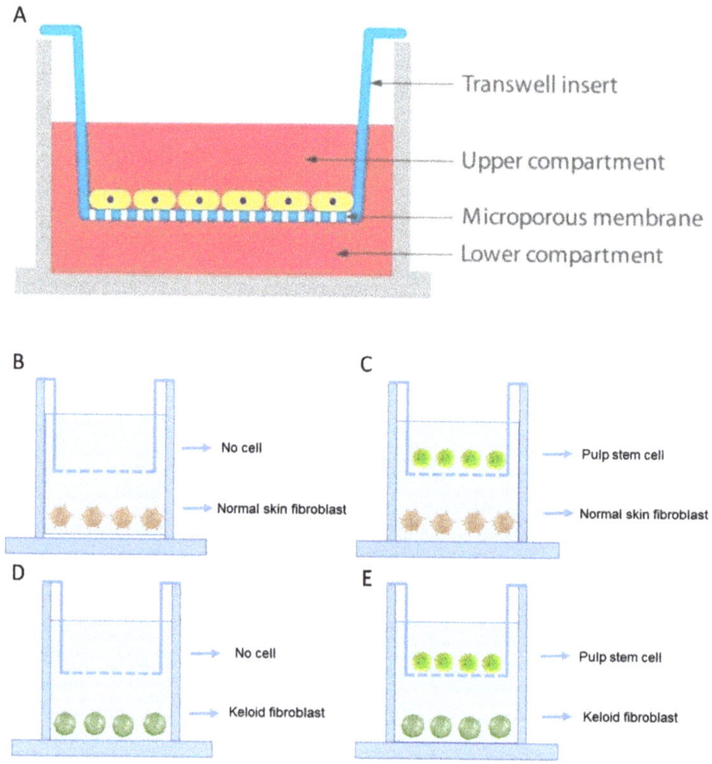

Figure 1. Study design. (**A**): A schematic design of the Transwell co-culture model was established using Transwell chambers with a 0.4-micrometer pore size that allows for the passage of chemical and biochemical molecules. (**B–E**): 4 combinations of the inducing and differentiating cells.

2.6. Collagen Synthesis Detection

Both normal and keloid fibroblasts were each individually seeded into 6-well plates at a concentration of 5×10^5 cells per well. Supernatants were retrieved from 6-well plates after 48 h when cells reached confluence and were handled according to the instructions of Hydroxyproline Assay Kit (MAK008; Sigma Aldrich, St. Louis, MI, USA). A total of 1.5 mL of the supernatant was harvested and resuspended in 0.05 mL of digestion solution at 37 °C for 3 h. Then, 0.5 mL of solution A was transferred into the mixture for 10 min and 0.5 mL of solution B was added and allowed to react for 5 min, followed by the addition of 1 mL of solution C at 60 °C for 15 min. Absorbance was then read at a wavelength of 550 nm. Hydroxyproline concentrations were calculated from a standard curve of hydroxyproline (0–100 mg/mL).

2.7. Wound Scratch Assay to Detect the Cell Migration Ability

Normal and keloid fibroblasts were seeded individually into 6-well plates at a concentration of 5×10^5 cells per well. When the cells reached 90% confluency, 200 μL sterile pipette tips were used to scratch the bottoms of the 6-well plate culture wells. After rinsing out the floating cells using PBS, cells were starved for 24 h in serum-free medium. In the test group, transwell chambers were seeded with a concentration of 5×10^5 of dental pulp stem cells. The wound area was recorded using a light microscope camera after scratching at 0 and 24 h. Finally, the rate of cellular migration was quantified using image J software V1.51 (NIH software, Bethesda, MD, USA). The migration rate of cells was measured using the following formula: $(W_{0h} - W_{24h})/W_{0h} \times 100\%$.

2.8. Live and Dead Staining

Live and dead cells were assessed using LIVE/DEAD® staining kits (Cat# L3224; Molecular Probes, Eugene, OR, USA) according to the manufacturer's instructions, where 500 μL of staining solution was added to each well after the chamber was removed at days 1, 3, 5 and 7, the cells were then incubated in the dark at room temperature for 10 min, and then photographed with the employment of immunofluorescence microscopy. Five images, one from the center and four from the periphery, were obtained from each well using an Olympus inverted fluorescence microscope. Live and dead cells were counted using Image J software, and the rates of cell survival were calculated as Live cells/(Live + Dead cells)%.

2.9. Western Blot

Normal and keloid fibroblasts were seeded in six-well plates at a density of 5×10^5 cells/well and transwell chambers were seeded with a concentration of 5×10^5 of dental pulp stem cells cultured in complete medium for 24 h. The medium was discarded and the fibroblast in the lower chamber washed 3 times with phosphate buffered saline (PBS). Total cell lysates from different groups were obtained by lysing the cells in a RIPA buffer (Beyotime Institute of Biotechnology, Shanghai, China) containing protease inhibitors (100 μM phenylmethylsulfonyl fluoride, 10 μM leupeptin, 10 μM pepstatin and 2 mM EDTA). The protein content was quantitated using a BCA Protein Assay kit (Beyotime Institute of Biotechnology). Proteins (50 μg/lane) were separated via 4–20% SDS-PAGE and transferred to nitrocellulose membranes (Pall Life Sciences, Pensacola, FL, USA). Membranes were blocked with 2% Bovine Serum Albumin (BSA) for 1 h at room temperature and probed overnight at 4 °C with primary antibodies against CTGF (ab6992), α-SMA (ab7817), TGF-β1 (ab64715), TGF-β2 (ab66045) and collagen I (ab292), which were products of Abcam (Cambridge, UK), in a humidified chamber. Membranes were washed and incubated with horseradish peroxidase-conjugated secondary antibodies (1:5000) (cat. nos. AS014 and AS003; ABclonal Biotech Co., Ltd., Woburn, MA, USA) at room temperature for 2 h, and were visualized using an enhanced chemiluminescence system (EMD Millipore, Billerica, MA, USA). Fluorescent secondary antibodies were then added and immunoblots were thereafter imaged with a two-channel (at 700 plus 800 nm) IR fluorescent Odyssey CLx imaging system (LI-COR®, Lincoln, NE, USA). Results were quantified using image J software.

2.10. Real-Time Quantitative PCR (RT-qPCR)

Normal and keloid fibroblasts were seeded in six-well plates at a density of 5×10^5 cells/well and transwell chambers were seeded with dental pulp stem cells at a concentration of 5×10^5 and cultured in complete medium for 24 h. The medium was discarded and the fibroblast in the lower chamber was washed 3 times with phosphate buffered saline (PBS). Thereafter, cells were harvested using a cell scraper after the addition of a lysis buffer. RNA was consequently extracted through the following steps: lysing cells in Trizol reagent (Cat No. 15596-026, Life Technologies, Carlsbad, CA, USA), followed by extracting RNA in trichloromethane, and then precipitating it in isopropanol, and finally resuspending it in RNase-free water. The RNA concentration and purity levels were determined using a Nanodrop2000 Spectrophotometer (Thermo Fischer Scientific, Waltham, MA, USA). Total RNA (2.5 μg) was subjected to cDNA synthesis using a qScript cDNA SuperMix (Quanta BioSciences, Beverly, MA, USA) through the following consequent cycles: firstly at 25 °C for 5 min, followed by 42 °C for 30 min and finally at 85 °C for 5 min. A real-time PCR was performed to determine the mRNA levels of TGF-β1, fibrinogen, α-SMA and GAPDH using SYBR Green Master MIX (ABI, Vernon, CA, USA). For a relative mRNA expression, the $2\Delta\Delta cq$ method, in which ΔCq = each corresponding Cq value − minimum Cq value, was calculated.

2.11. Statistical Analyses

The data are shown as the mean ± standard deviation (SD) from at least three independent repeated experiments. Student *t*-test was used to analyze the differences in mean values between the two groups. Significant differences were defined as $p < 0.05$. All statistical analyses were performed using Graph Pad Prism software (Graph Pad Prism, San Diego, CA, USA; RRID:SCR_002798) version 7.0a.

3. Results

3.1. Clinical and Pathological Characteristics

The samples were classified into two groups (keloid and normal skin). Compared with normal skin tissue, the keloid is harder in texture, possessing either a round or oval structure, and involving excessive growth beyond the boundary of the originally wounded skin, thereby invading neighboring normal tissues. See Figure 2.

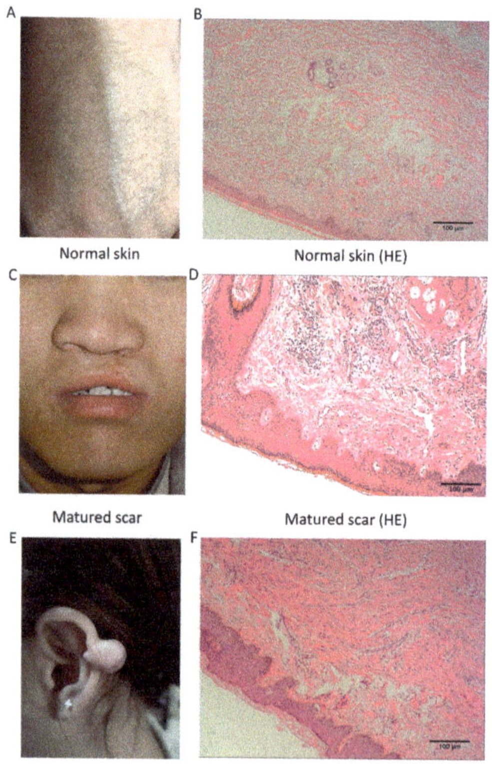

Figure 2. Gross and hematoxylin and eosin (H&E) staining image of normal skin, matured scar and keloid. (**A**) Normal skin. (**B**) H&E staining of normal skin. (**C**) Matured scar. (**D**) H&E staining of matured scar. (**E**) Keloid. (**F**) H&E staining of keloid.

3.2. Identification of Stem Cells

To confirm that these HDPSCs were of mesenchymal origin, cell surface markers were detected using flow cytometry to determine the mesenchymal origin of the HDPSCs. The following markers were positively expressed at the respective percentages: CD29 at 99.30%, CD73 at 99.60%, CD105 at 99.50% and CD90 at 99.60%, whereas the hemopoietic stem cell marker, CD19 at 0.43%, CD44 at 0.17%, CD45 at 0.74%, HLA-DR at 0.14%, CD34 at 0.11% and CD11b at 0.19%. See Figure 3.

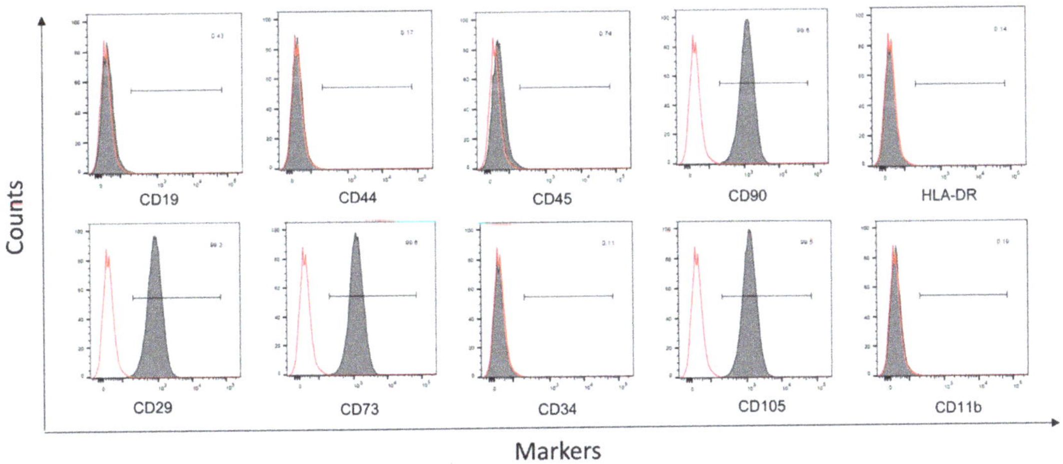

Figure 3. Flow cytometric results of passage 3 HDPSCs. Representative histograms showing antigen expression in bone-marrow-derived MSC. From left to right CD19, CD44, CD45, CD90, HLA-DR, CD29, CD73 CD105, CD73, CD34, CD105 and CD11b. Black-filled histogram: antigen expression; solid red line: auto-fluorescence control.

3.3. HDPSCs Did Not Inhibit the Proliferation of HNFs and HKFs

Compared to the HKFs cultured alone, the CCK-8 assay showed that no difference in the proliferation of HKFs was observed in the HDPSCs/HKFs co-culture group at days 1, 3, 5, and 7 ($p > 0.05$, *t*-test; $n = 7$, mean ± SD). Moreover, as a positive control, there was also no detected difference in the proliferation of HNFs in the DPSCs/HNFs co-culture group at days 1, 3, 5, and 7 ($p > 0.05$, *t*-test; $n = 7$, mean ± SD). However, the proliferation rate of HKFs was much higher than that of HNFs on the third day (Figure 4A). Compared with the control group, the morphology of the fibroblast was still elongated, and there was no variation between the two groups.

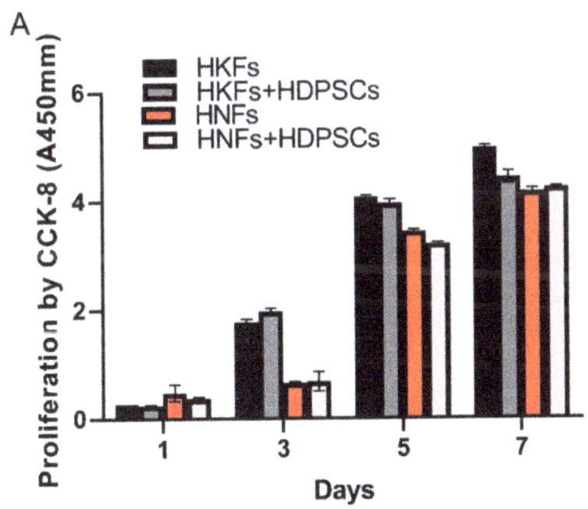

Figure 4. *Cont.*

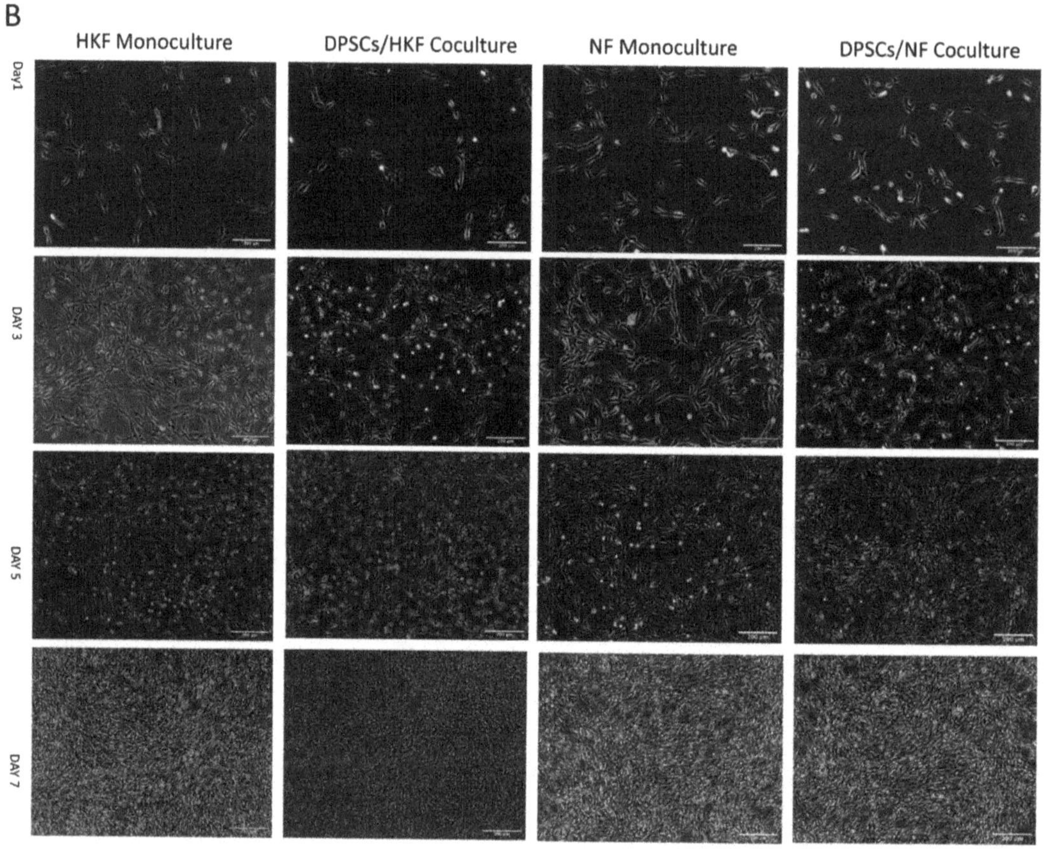

Figure 4. (**A**): The cell counts (proliferation) were analyzed using CCK-8 kits (Dojindo) on days 1, 3, 5, and 7. (**B**): The morphology of the fibroblasts was determined on days 1, 3, 5, and 7 by direct observation with a light microscope. Scale bar: 200 μM. HNFs: human normal fibroblasts, HKFs: human keloid fibroblasts, HDPSCs: dental pulp stem cells.

3.4. HDPSCs Did Not Influence the Apoptosis of HNFs and HKFs

Our results indicate that the cellular apoptosis of HNFs and HKFs was not affected by HDPSCs. As shown in Figure 5, with respect to cell apoptosis, both HKFs and HNFs were not affected through their coculture with HDPSCs ($p > 0.05$). This result further validates the cell proliferation results.

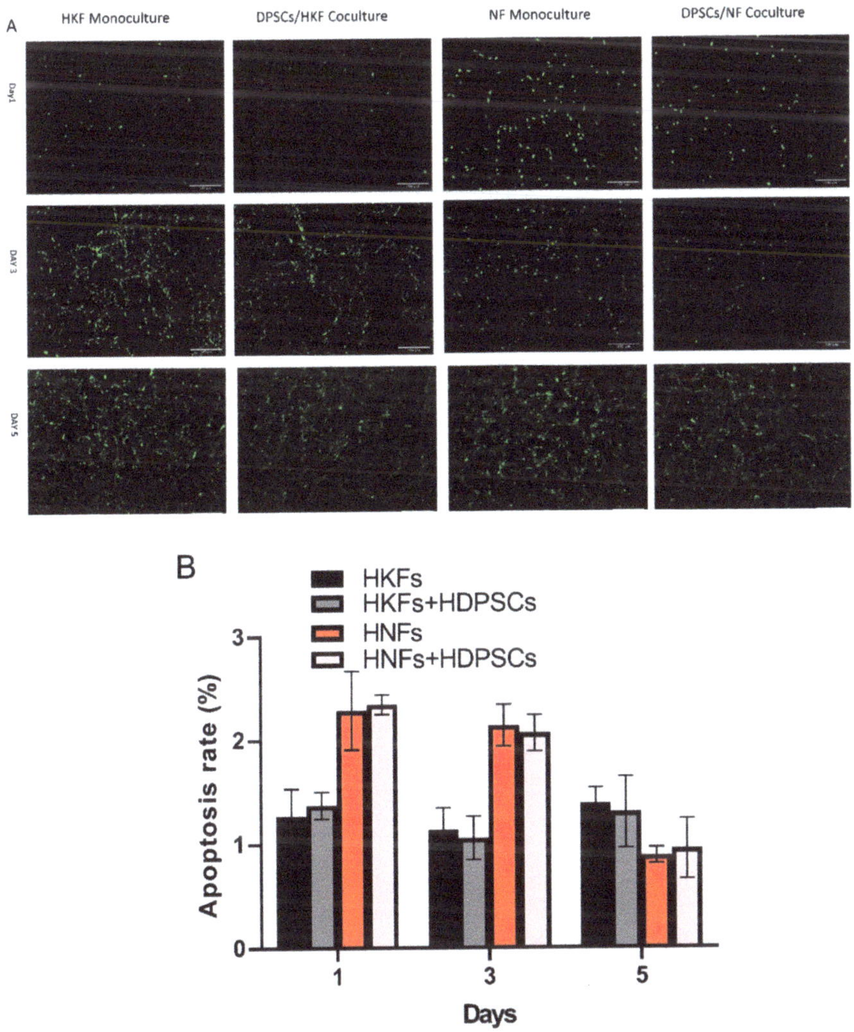

Figure 5. Apoptosis rate of HNFs and HKFs in mono and co-cultures. HDPSCs do not induce apoptosis in HNFs or HKFs. (**A**): Vital cells stain green with Calcein-AM, while dead cells stain red with propidium iodide on days 1, 3, and 5. Scale bar: 200 µm. (**B**): The quantification of cell apoptosis/necrosis using the percentage of PI-positive cells/AM-positive cells. (Mean q-SD, $n = 5$, Student *t*-test, $p < 0.05$).

3.5. The Effects of HDPSCs on the Migration of HKFs and HNFs

The wound scratch assays showed that there was only a difference in the cell migration ability in the co-culture group of the human keloid fibroblast after 24 h of culture ($p < 0.05$). As a positive control, cellular migration of the HNFs remained uninfluenced ($p > 0.05$). See Figure 6.

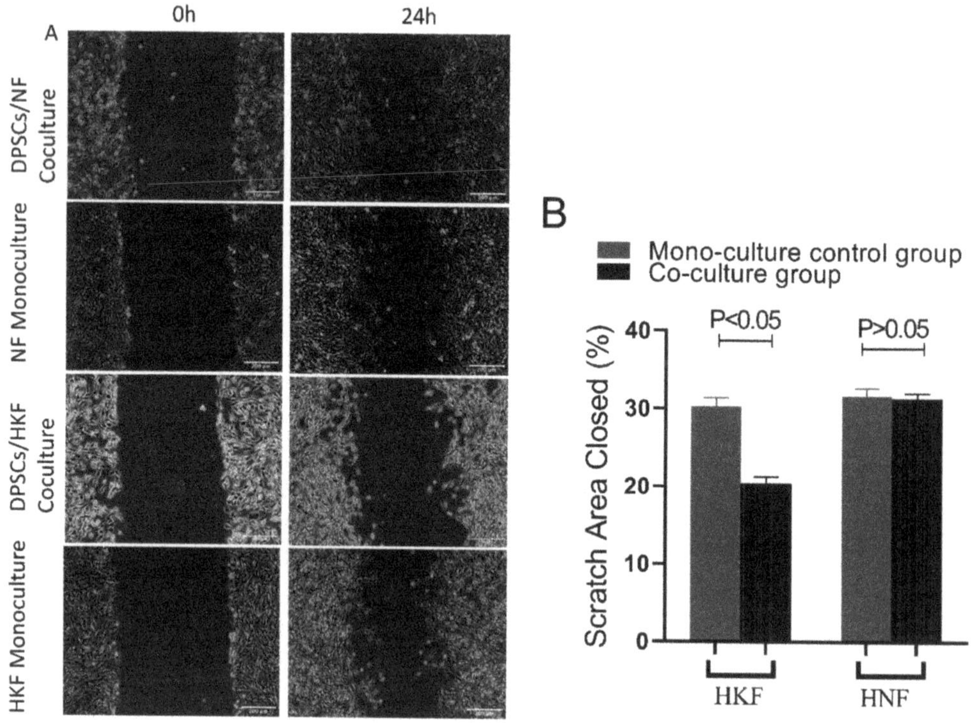

Figure 6. Cellular migration in scratch area. (**A**): Pictures were taken at 0 and 24 h, a magnification of 100× was used. (**B**): Percentage of wound area recovery by migrated cells was quantified by Image J. Significant differences were only detected in the co-culture group of the human keloid fibroblast after 24 h ($p < 0.05$, t-test; $n = 7$, mean ± SD).

3.6. Inhibited Expression of Fibrosis-Associated Gene Phenotype and Protein Expression in HKFs and KNFs

Compared with solely cultured cells, only the expression levels of CTGF in HNFs were observed to be unaffected ($p > 0.05$, t-test; $n = 7$, mean ± SD), whereas the expression levels of CTGF, TGF-β1 and TGF-β2 in HKFs and KNFs at the mRNA and protein levels were significantly inhibited when co-cultured with the HDPSCs ($p < 0.05$, t-test; $n = 7$, mean ± SD). These findings indicate that both the transcriptional and posttranslational levels were inhibited. See Figure 7.

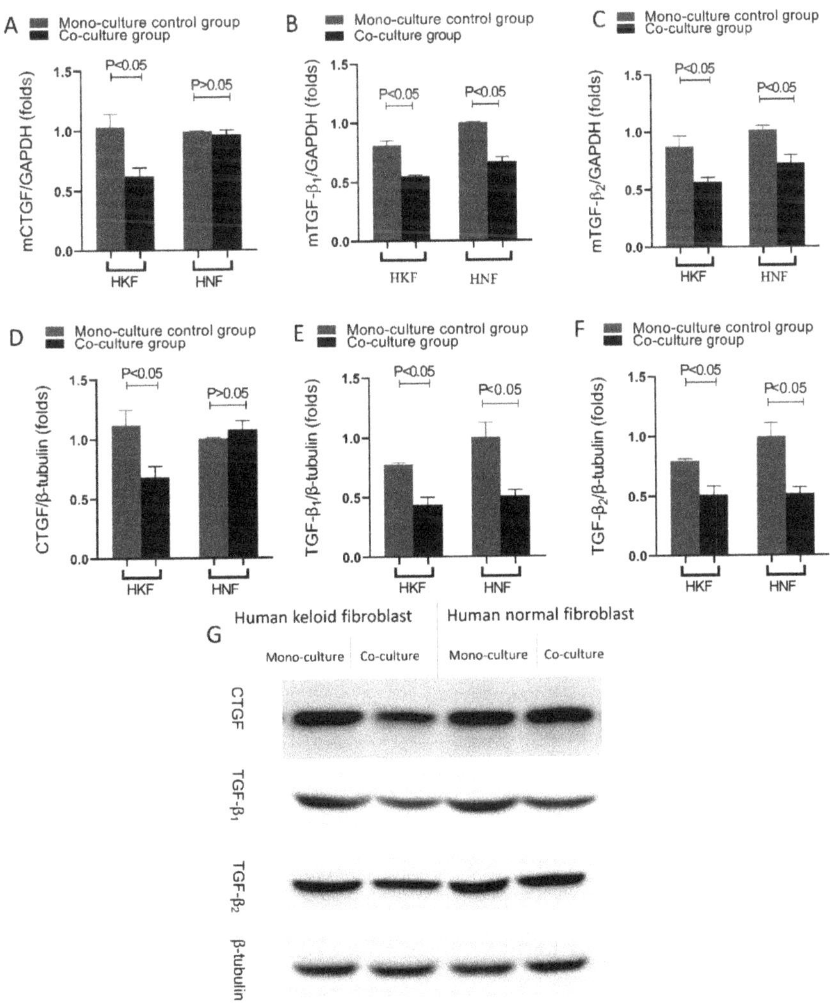

Figure 7. HDPSCs attenuated the pro-fibrotic phenotype of HKFs and HNFs. After 24 h of culture, cells were subjected to RT-qPCR and Western blot. Quantification of (**A**) CTGF, (**B**) TGF-β1 and (**C**) TGF-β2 gene expression, normalized to GAPDH expression (**D**–**F**). Quantification of (**D**) CTGF, (**E**) TGF-β1 and (**F**) TGF-β2 protein levels, normalized to β-tubulin expression. (**G**) Gels were analyzed by immunoblotting using the indicated antiserum.

3.7. HDPSCs Inhibits Extracellular Matrix Synthesis of HKFs and HNFs

Compared with solely cultured cells, the expression levels of collagen I in HNFs were observed to be unaffected ($p > 0.05$, t-test; $n = 7$, mean ± SD), whereas the expression levels of collagen I and α-SMA in HKFs and KNFs at the mRNA and protein levels were significantly inhibited when co-cultured with the HDPSCs ($p < 0.05$, t-test; $n = 7$, mean ± SD). These findings indicate that both the transcriptional and posttranslational levels were inhibited. See in Figure 8A–D,F.

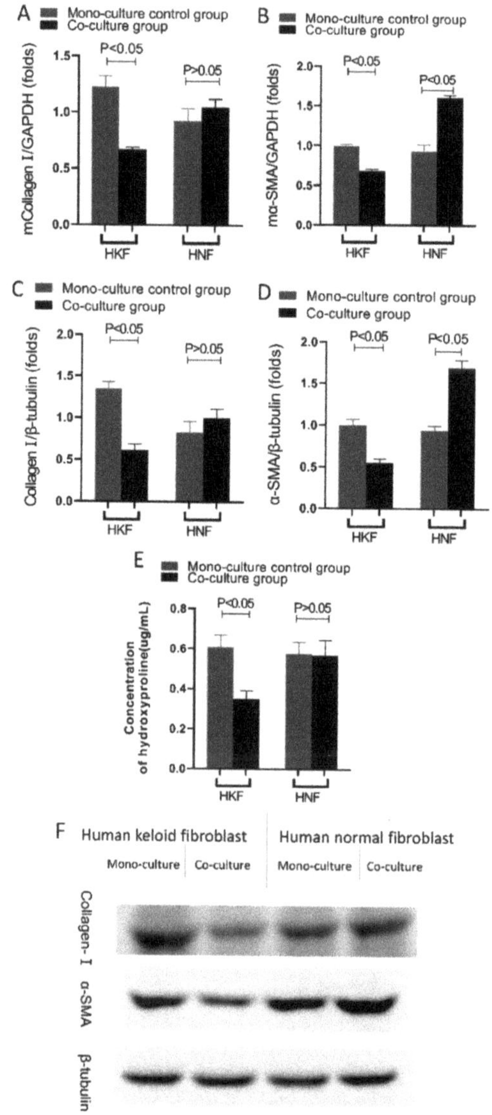

Figure 8. HDPSCs inhibits extracellular matrix synthesis of HKFs and HNFs. (**A**) After 24 h of culture and then subjected to RT-qPCR and Western blot. Quantification of (**A**) Collagen I, and (**B**) α-SMA gene expression, normalized to GAPDH expression. Quantification of (**C**) Collagen I, and (**D**) α-SMA protein levels, normalized to β-tubulin expression levels (**E**) Cell culture supernatants were collected and tested for hydroxyproline content. (**F**) Gels were analyzed by immunoblotting using the indicated antiserum.

There was a significant decrease in the resultant hydroxyproline concentrations of the HKFs co-culture group ($p < 0.05$, t-test; $n = 7$, mean ± SD), whereas no significant changes were detected in the other three groups ($p > 0.05$, t-test; $n = 7$, mean ± SD). See in Figure 8E.

4. Discussion

Wound repair is a complex process that often leads to the formation of scars following traumatic skin injuries. This process is associated with the functions of various cells, such as fibroblasts, endothelial cells, macrophages and lymphocytes [15–17], among which the biological behavior of fibroblasts is considered to be a key factor in the scar formation process. Numerous studies have shown that fibroblasts, which happen to be the main constituents of keloid tissue, have the ability to over proliferate and are accompanied by incomplete apoptosis, together with the abnormal synthesis of collagen, which overall results in the continuous proliferation of keloid tissue [18,19]. Therefore, the inhibition of fibroblast proliferation and induction of apoptosis in keloid tissues can majorly reduce keloid tissue proliferation and thereby delay disease progression, which is important for the improvement and treatment of keloid scars. Therefore, understanding the biology of keloid fibroblasts is important for the treatment of keloids.

Mesenchymal stem cells (MSCs) are an important member of the stem cell family, which have been used to treat scar formation-related diseases such as pulmonary fibrosis. Moreover, great progress has been made where they have also been used to inhibit cardiac scar formation through the secretion of various cytokines [20]. Studies have shown that stem cells have been used in animal models and in a few clinical trials for the regeneration of diseased organs. Furthermore, stem cells have been shown to improve tissue repair by secreting interleukins such as interleukin 6, interleukin 8, interleukin 10 and other proteins that are suitable for inducing tissue regeneration [21,22]. Recent studies have shown that adipose-derived stem cells are able to inhibit mRNA expression levels of COL1A1, transforming growth factor β1, connective tissue growth factor and alpha actin 2 (ACTA2) in renal fibrosis tissues, thereby playing a therapeutic role in renal fibrosis [23]. Whether or not dental pulp stem cells can play a role in the repair of skin scars by inhibiting the proliferation and migration of keloid fibroblasts through the secretion of cytokines has not been clearly reported thus far.

Dental pulp stem cells, human keloid fibroblasts and human normal skin fibroblasts were isolated using the tissue explant method. The molecular mechanisms of action of dental pulp stem cells on the proliferation, migration and apoptosis of keloid fibroblasts were investigated using a transwell co-culture. The histology of keloid scars is highlighted by the secretion and deposition of large amounts of extracellular collagen [24]. Furthermore, both skin keloids and fibrous tumors are pathologically fibrous connective tissue lesions with large extracellular collagen deposition [25]. In this study, we also found that the co-culture of HDPSCs inhibited the migration of HKFs and HNFs but did not inhibit the proliferation of either HKFs and HNFs, nor did it induce the apoptosis of HKFs and HNFs. Our experimental data are similar to the results of a recent study reporting that the conditioned medium and cell lysates of human-derived WJ-MSCs inhibit the migration of human HKFs [26]. In comparison with this report, we used HDPSCs, which have the advantages of being easily accessible with minimal patient harm, and the fact that they can be used autologously. However, studies reporting opposing results also exist [27], in which the conditioned medium of human WJ-MSCs could promote HKFs proliferation in a paracrine manner through an indirect transwell co-culture treatment system. We hypothesize that the reasons for these different results are attributed to the different types of MSCs used, the different cell treatment cultures and different assays, all of which impact the following biological behaviors of keloid fibroblasts: migration, proliferation and collagen secretion, in addition to the apoptosis involved in wound repair after skin injury. Therefore, by inhibiting the migration and proliferation of keloid fibroblasts, keloid formation is also consequently inhibited [28]. Moreover, it has been shown that the inhibition of matrix metalloproteinase expression in keloid fibroblasts can inhibit the migration ability of keloid fibroblasts and, therefore, function as a keloid treatment aid [29]. Furthermore, similar findings have been reported where inhibition of mTOR protein expression in keloid fibroblasts can also inhibit the migratory ability of keloid fibroblasts and thereby further inhibiting the development of keloids [30]. Similarly, and consistent with the aforementioned studies, the results of the present study revealed that dental pulp stem cells are also able to inhibit the migration of keloid fibroblasts in vitro, thereby suggesting that dental pulp-derived stem cells can promote wound healing by inhibiting scar formation.

The expression of anti-fibrotic and pro-fibrotic genes is closely related to the pathogenesis of fibrotic diseases, which in another way confirms that hyperplastic scars and keloids are classified as fibrotic diseases. TGF-β1 and TGF-β2 overexpression is an important cause for excessive scar proliferation and fibrosis, and studies have shown that the targeted reduction in TGF-β1 and TGF-β2 expression in hyperplastic scars and keloids can inhibit scar proliferation and achieve clinical therapeutic effects [29,30]. Therefore, TGF-β1 and TGF-β2 have also become the targets of numerous studies that tackle the treatment of hyperplastic scars and keloids. Moreover, CTGF is a marker protein of fibrotic diseases in which it promotes both the proliferation of fibroblasts as well as the secretion and deposition of extracellular matrix proteins, such as collagen I and fibronectin [31,32]. In the current study, it was found that after 24 h of HDPSC co-culture with HKFs, both the gene and protein expression of TGF-β1, TGF-β2 and CTGF were significantly reduced in HKFs. This is in accordance with studies that have shown MSCs as being capable of secreting cytokines in order to alter some biological phenotypes of fibroblasts, such as fibrotic and proliferative phenotypes, through paracrine functions [33,34]. Increasing evidence thereby suggests that the paracrine function of MSCs is an important potential mechanism for their cellular therapeutic function. Thereupon, after being injected into the body, MSCs can, on the one hand, inhabit the area of tissue damage through the processes of chemotaxis, proliferation and, eventually, differentiation by evolving into the cell type required for the secretion of extracellular matrix that would be needed to repair the damage in the recipient area; on the other hand, MSCs, upon entering the body, can exhibit a paracrine function, which entails the secretion of cytokines and nutrient-active substances required for repairing the damage, and inducing the body's self-generated cells to repair the tissue damage. The SMAD signaling pathway is a downstream mediator of TGF-β. After phosphorylation, the phosphorylation of R-SMAD 3 is upregulated in the keloid, whereas the downregulation of R-SMAD 3 significantly reduces procollagen gene expression in keloid fibroblasts. I-SMAD 6 and I-SMAD 7 inhibit the action of TGF-β. SMAD 6 also inhibits the binding of SMAD 4 and R-SMAD. The expression of I-SMAD 6 and 7 is reduced in keloid fibroblasts. By inhibiting the TGF-β1-SMAD signaling pathway and activating TLR7 or SMAD 7, keloid formation can be significantly reduced. The toll-like receptor signaling pathway plays a protective as well as a destructive role. After skin injury, toll-like receptors (TLR) are combined as damage-related molecular patterns (DAMP) to enable the innate immune system to respond to sterile tissue damage. As the concentrations of several pro-inflammatory and pro-fibrotic cytokines in macrophages increase in response to TLR stimulation in macrophages, fibroblast gene expression and TGF-β responses change, leading to an increased collagen production.

A very distinctive feature of proliferative scars and keloid tissues is the excessive deposition of the extracellular matrix [35]. There are two main mechanisms that can lead to an excessive extracellular matrix deposition, one being an increase in the extracellular matrix synthesis, and the other being a decrease in the extracellular matrix degradation. Compared to normal skin, the amount of collagen synthesis in hyperplastic keloid scars is three times higher, while keloid scars can reach up to 20 times higher [4]. In our experiments, it was further observed that due to the co-culture with DPSCs, both HSFs and HKFs displayed a reduction in extracellular matrix synthesis, as shown by the reduced expression of collagen type I, α-SMA and hydroxyproline. Moreover, in the extracellular matrix of scar tissue, collagen type I is normally the predominantly present collagen type. Furthermore, the detection of hydroxyproline content in cell culture media is recognized as a reliable indicator of the ability of fibroblasts to synthesize collagen. Relevantly, it was previously reported that both bone marrow MSCs and dermal MSCs were able to inhibit collagen synthesis and the expression of α-SMA in keloid fibroblasts [27].

Finally, our findings are consistent with previous studies by showing that dental pulp stem cells were able to inhibit the expression of pro-fibrotic genes and keloid fibroblast collagen synthesis in vitro. Furthermore, this study revealed that dental pulp stem cells inhibit the migration of keloid fibroblasts. Nevertheless, the specifically secreted cytokines that influence the biological behavior of keloid fibroblasts need to be further investigated by subsequent experiments.

5. Conclusions

The co-culture of HDPSCs inhibits the migration of HKFs and the expression of pro-fibrotic genes, while promoting the expression of anti-fibrotic genes. HDPSCs' co-culture also inhibits the synthesis of the extracellular matrix by HKFs, whereas it does not affect the proliferation and apoptosis of HKFs. Therefore, it can be concluded that HDPSCs can be used as a tool for restraining/hindering the initiation or progression of fibrotic tissue.

Author Contributions: M.Y.: conceived the study, supervised the experiments and drafted the manuscript; O.A.N.: conceived the study, supervised the experiments and drafted the manuscript; L.-L.F.: data evaluation and manuscript preparation; L.-M.C.: data evaluation and manuscript preparation; M.G.: analyzed the data and revised the manuscript; R.E.F.: analyzed the data and revised the manuscript; R.S.: performed the data collection; H.-C.F.: conceived the study, designed the data evaluation and prepared the manuscript. All authors have read and agreed to the published version of the manuscript.

Funding: This study was supported in part by grants from the Guiyang Municipal Health Bureau Fund for Science and Technology projects. The funder is in charge of the publication fees.

Institutional Review Board Statement: Administrative permissions were acquired by our team to access the data used in our research. The Ethics approval Research Ethics Committee number: GYSKLL-KY-20161223-01. Accordingly, all teeth were coded with number and all personal identification of the patients were removed.

Informed Consent Statement: All parent or guardian of participants provided written informed consent for using their teeth that otherwise would have been discarded as waste.

Data Availability Statement: The data are available from the corresponding author on reasonable request.

Conflicts of Interest: The authors declare that they have no conflicting interests.

References

1. Bielefeld, K.A.; Amini-Nik, S.; Alman, B.A. Cutaneous wound healing: Recruiting developmental pathways for regeneration. *Cell Mol. Life Sci.* **2013**, *70*, 2059–2081. [CrossRef] [PubMed]
2. McGinty, S.; Siddiqui, W.J. *Keloid*; StatPearls: Treasure Island, FL, USA, 2020.
3. van den Broek, L.J.; Limandjaja, G.C.; Niessen, F.B.; Gibbs, S. Human hypertrophic and keloid scar models: Principles, limitations and future challenges from a tissue engineering perspective. *Exp. Derm.* **2014**, *23*, 382–386. [CrossRef]
4. Ud-Din, S.; Bayat, A. New insights on keloids, hypertrophic scars, and striae. *Derm. Clin.* **2014**, *32*, 193–209. [CrossRef]
5. Armour, A.; Scott, P.G.; Tredget, E.E. Cellular and molecular pathology of HTS: Basis for treatment. *Wound Repair Regen* **2007**, *15* (Suppl. 1), S6–S17. [CrossRef]
6. Andrews, J.P.; Marttala, J.; Macarak, E.; Rosenbloom, J.; Uitto, J. Keloids: The paradigm of skin fibrosis—Pathomechanisms and treatment. *Matrix Biol* **2016**, *51*, 37–46. [CrossRef] [PubMed]
7. Berman, B.; Maderal, A.; Raphael, B. Keloids and Hypertrophic Scars: Pathophysiology, Classification, and Treatment. *Derm. Surg.* **2017**, *43* (Suppl. 1), S3–S18. [CrossRef] [PubMed]
8. Fang, F.; Huang, R.L.; Zheng, Y.; Liu, M.; Huo, R. Bone marrow derived mesenchymal stem cells inhibit the proliferative and profibrotic phenotype of hypertrophic scar fibroblasts and keloid fibroblasts through paracrine signaling. *J. Derm. Sci.* **2016**, *83*, 95–105. [CrossRef]
9. Falanga, V.; Iwamoto, S.; Chartier, M.; Yufit, T.; Butmarc, J.; Kouttab, N.; Shrayer, D.; Carson, P. Autologous bone marrow-derived cultured mesenchymal stem cells delivered in a fibrin spray accelerate healing in murine and human cutaneous wounds. *Tissue Eng.* **2007**, *13*, 1299–1312. [CrossRef] [PubMed]
10. Averyanov, A.; Koroleva, I.; Konoplyannikov, M.; Revkova, V.; Lesnyak, V.; Kalsin, V.; Danilevskaya, O.; Nikitin, A.; Sotnikova, A.; Kotova, S.; et al. First-in-human high-cumulative-dose stem cell therapy in idiopathic pulmonary fibrosis with rapid lung function decline. *Stem Cells Transl. Med.* **2020**, *9*, 6–16. [CrossRef]

11. Xie, F.; Teng, L.; Xu, J.; Lu, J.; Zhang, C.; Yang, L.; Ma, X.; Zhao, M. Adipose-derived mesenchymal stem cells inhibit cell proliferation and migration and suppress extracellular matrix synthesis in hypertrophic-scar and keloid fibroblasts. *Exp. Med.* **2021**, *21*, 139. [CrossRef]
12. Tsutsui, T.W. Dental Pulp Stem Cells: Advances to Applications. *Stem Cells Cloning* **2020**, *13*, 33–42. [CrossRef] [PubMed]
13. Zhao, Y.; Zheng, Y.; Eichhorn, W.; Klatt, J.; Henningsen, A.; Kluwe, L.; Friedrich, R.E.; Gosau, M.; Smeets, R. Enriching Stem/Progenitor Cells from Dental Pulp Cells by Low-density Culturing. *In Vivo* **2019**, *33*, 23–29. [CrossRef]
14. Dominici, M.; Le Blanc, K.; Mueller, I.; Slaper-Cortenbach, I.; Marini, F.; Krause, D.; Deans, R.; Keating, A.; Prockop, D.; Horwitz, E. Minimal criteria for defining multipotent mesenchymal stromal cells. The International Society for Cellular Therapy position statement. *Cytotherapy* **2006**, *8*, 315–317. [CrossRef] [PubMed]
15. Zhang, J.; Liu, Z.; Cao, W.; Chen, L.; Xiong, X.; Qin, S.; Zhang, Z.; Li, X.; Hu, C.A. Amentoflavone inhibits angiogenesis of endothelial cells and stimulates apoptosis in hypertrophic scar fibroblasts. *Burns* **2014**, *40*, 922–929. [CrossRef] [PubMed]
16. He, L.; Marneros, A.G. Macrophages are essential for the early wound healing response and the formation of a fibrovascular scar. *Am. J. Pathol.* **2013**, *182*, 2407–2417. [CrossRef]
17. Bernabei, P.; Rigamonti, L.; Ariotti, S.; Stella, M.; Castagnoli, C.; Novelli, F. Functional analysis of T lymphocytes infiltrating the dermis and epidermis of post-burn hypertrophic scar tissues. *Burns* **1999**, *25*, 43–48. [CrossRef]
18. Gong, Y.F.; Zhang, X.M.; Liu, F.; Wang, Z.Z.; Deng, X.F.; Jiao, Y.; Li, X.J.; Huang, X.Y. Inhibitory effect of recombinant human endostatin on the proliferation of hypertrophic scar fibroblasts in a rabbit ear model. *Eur. J. Pharm.* **2016**, *791*, 647–654. [CrossRef] [PubMed]
19. Wang, H.; Chen, Z.; Li, X.; Tang, Y.; Li, X.; Zhang, S.; Ma, L.; Huang, X. TSG-6 treatment promoted apoptosis in human fibroblasts of pathological scar. *Cell Mol. Biol.* **2016**, *62*, 33–37. [PubMed]
20. Merimi, M.; Lagneaux, L.; Moussa Agha, D.; Lewalle, P.; Meuleman, N.; Burny, A.; Fahmi, H.; Najar, M. Mesenchymal Stem/Stromal Cells in Immunity and Disease: A Better Understanding for an Improved Use. *J. Clin. Med.* **2020**, *9*, 1516. [CrossRef]
21. Cao, J.Q.; Liang, Y.Y.; Li, Y.Q.; Zhang, H.L.; Zhu, Y.L.; Geng, J.; Yang, L.Q.; Feng, S.W.; Yang, J.; Kong, J.; et al. Adipose-derived stem cells enhance myogenic differentiation in the mdx mouse model of muscular dystrophy via paracrine signaling. *Neural Regen. Res.* **2016**, *11*, 1638–1643. [CrossRef]
22. Harn, H.J.; Lin, S.Z.; Hung, S.H.; Subeq, Y.M.; Li, Y.S.; Syu, W.S.; Ding, D.C.; Lee, R.P.; Hsieh, D.K.; Lin, P.C.; et al. Adipose-derived stem cells can abrogate chemical-induced liver fibrosis and facilitate recovery of liver function. *Cell Transplant.* **2012**, *21*, 2753–2764. [CrossRef]
23. Zhang, C.; Wang, T.; Zhang, L.; Chen, P.; Tang, S.; Chen, A.; Li, M.; Peng, G.; Gao, H.; Weng, H.; et al. Combination of lyophilized adipose-derived stem cell concentrated conditioned medium and polysaccharide hydrogel in the inhibition of hypertrophic scarring. *Stem Cell Res.* **2021**, *12*, 23. [CrossRef]
24. Zhao, J.; Shu, B.; Chen, L.; Tang, J.; Zhang, L.; Xie, J.; Liu, X.; Xu, Y.; Qi, S. Prostaglandin E2 inhibits collagen synthesis in dermal fibroblasts and prevents hypertrophic scar formation in vivo. *Exp. Derm.* **2016**, *25*, 604–610. [CrossRef] [PubMed]
25. Tejiram, S.; Zhang, J.; Travis, T.E.; Carney, B.C.; Alkhalil, A.; Moffatt, L.T.; Johnson, L.S.; Shupp, J.W. Compression therapy affects collagen type balance in hypertrophic scar. *J. Surg. Res.* **2016**, *201*, 299–305. [CrossRef] [PubMed]
26. Song, S.Y.; Jung, J.E.; Jeon, Y.R.; Tark, K.C.; Lew, D.H. Determination of adipose-derived stem cell application on photo-aged fibroblasts, based on paracrine function. *Cytotherapy* **2011**, *13*, 378–384. [CrossRef] [PubMed]
27. Abusin, G.A.; Abu-Arja, R.F.; Gingrich, R.D.; Silverman, M.D.; Zamba, G.K.; Schlueter, A.J. An algorithm for utilizing peripheral blood CD34 count as a predictor of the need for plerixafor in autologous stem cell mobilization–cost-effectiveness analysis. *J. Clin. Apher.* **2013**, *28*, 293–300. [CrossRef]
28. Chun, Q.; ZhiYong, W.; Fei, S.; XiQiao, W. Dynamic biological changes in fibroblasts during hypertrophic scar formation and regression. *Int. Wound J.* **2016**, *13*, 257–262. [CrossRef]
29. Kryger, Z.B.; Sisco, M.; Roy, N.K.; Lu, L.; Rosenberg, D.; Mustoe, T.A. Temporal expression of the transforming growth factor-Beta pathway in the rabbit ear model of wound healing and scarring. *J. Am. Coll. Surg.* **2007**, *205*, 78–88. [CrossRef] [PubMed]
30. Chen, W.; Fu, X.; Ge, S.; Sun, T.; Zhou, G.; Jiang, D.; Sheng, Z. Ontogeny of expression of transforming growth factor-beta and its receptors and their possible relationship with scarless healing in human fetal skin. *Wound Repair Regen.* **2005**, *13*, 68–75. [CrossRef]
31. Lee, C.H.; Shah, B.; Moioli, E.K.; Mao, J.J. CTGF directs fibroblast differentiation from human mesenchymal stem/stromal cells and defines connective tissue healing in a rodent injury model. *J. Clin. Investig.* **2010**, *120*, 3340–3349. [CrossRef]
32. Abraham, D. Connective tissue growth factor: Growth factor, matricellular organizer, fibrotic biomarker or molecular target for anti-fibrotic therapy in SSc? *Rheumatology* **2008**, *47* (Suppl. 5), v8–v9. [CrossRef] [PubMed]
33. Ashcroft, K.J.; Syed, F.; Bayat, A. Site-specific keloid fibroblasts alter the behaviour of normal skin and normal scar fibroblasts through paracrine signalling. *PLoS ONE* **2013**, *8*, e75600. [CrossRef] [PubMed]
34. Murao, N.; Seino, K.; Hayashi, T.; Ikeda, M.; Funayama, E.; Furukawa, H.; Yamamoto, Y.; Oyama, A. Treg-enriched CD4+ T cells attenuate collagen synthesis in keloid fibroblasts. *Exp. Derm.* **2014**, *23*, 266–271. [CrossRef]
35. Mukhopadhyay, A.; Wong, M.Y.; Chan, S.Y.; Do, D.V.; Khoo, A.; Ong, C.T.; Cheong, H.H.; Lim, I.J.; Phan, T.T. Syndecan-2 and decorin: Proteoglycans with a difference–implications in keloid pathogenesis. *J. Trauma* **2010**, *68*, 999–1008. [CrossRef]

cells

MDPI

Article

Tissue-Specific Microparticles Improve Organoid Microenvironment for Efficient Maturation of Pluripotent Stem-Cell-Derived Hepatocytes

Ensieh Zahmatkesh [1,2], Mohammad Hossein Ghanian [3], Ibrahim Zarkesh [3], Zahra Farzaneh [1], Majid Halvaei [3], Zahra Heydari [1,2], Farideh Moeinvaziri [1,2], Amnah Othman [4], Marc Ruoß [4], Abbas Piryaei [5,6], Roberto Gramignoli [7], Saeed Yakhkeshi [1], Andreas Nüssler [4], Mustapha Najimi [8,*], Hossein Baharvand [1,2,*] and Massoud Vosough [1,9,*]

1 Department of Stem Cells and Developmental Biology, Cell Science Research Center, Royan Institute for Stem Cell Biology and Technology, ACECR, Tehran 1665659911, Iran; ensieh_zahmatkesh@yahoo.com (E.Z.); zahrafarzaneh2006@yahoo.com (Z.F.); zahrabiology85@gmail.com (Z.H.); f.moeinvaziri@yahoo.com (F.M.); s.yakhkeshi@modares.ac.ir (S.Y.)
2 Department of Developmental Biology, University of Science and Culture, Tehran 1665659911, Iran
3 Department of Cell Engineering, Cell Science Research Center, Royan Institute for Stem Cell Biology and Technology, ACECR, Tehran 1665659911, Iran; biomaterialist@gmail.com (M.H.G.); ibrahimz_edl@yahoo.com (I.Z.); mj.halvaei@yahoo.com (M.H.)
4 Department of Traumatology, Siegfried Weller Institute, University of Tübingen, 72076 Tübingen, Germany; othman.amnah@web.de (A.O.); m.ruoss@hotmail.de (M.R.); andreas.nuessler@googlemail.com (A.N.)
5 Department of Tissue Engineering and Applied Cell Sciences, School of Advanced Technologies in Medicine, Shahid Beheshti University of Medical Sciences, Tehran 1985717443, Iran; piryae_a@yahoo.com
6 Department of Biology and Anatomical Sciences, School of Medicine, Shahid Beheshti University of Medical Sciences, Tehran 1985717443, Iran
7 Division of Pathology, Department of Laboratory Medicine, Karolinska Institutet, 17177 Stockholm, Sweden; roberto.gramignoli@ki.se
8 Laboratory of Pediatric Hepatology and Cell Therapy, Institute of Experimental & Clinical Research, Université Catholique de Louvain, B-1200 Brussels, Belgium
9 Department of Regenerative Medicine, Cell Science Research Center, Royan Institute for Stem Cell Biology and Technology, ACECR, Tehran 1665659911, Iran
* Correspondence: mustapha.najimi@uclouvain.be (M.N.); baharvand@royaninstitute.org (H.B.); masvos@royaninstitute.org (M.V.)

Citation: Zahmatkesh, E.; Ghanian, M.H.; Zarkesh, I.; Farzaneh, Z.; Halvaei, M.; Heydari, Z.; Moeinvaziri, F.; Othman, A.; Ruoß, M.; Piryaei, A.; et al. Tissue-Specific Microparticles Improve Organoid Microenvironment for Efficient Maturation of Pluripotent Stem-Cell-Derived Hepatocytes. *Cells* **2021**, *10*, 1274. https://doi.org/10.3390/cells10061274

Academic Editor: Mehdi Najar

Received: 14 March 2021
Accepted: 19 May 2021
Published: 21 May 2021

Abstract: Liver organoids (LOs) are receiving considerable attention for their potential use in drug screening, disease modeling, and transplantable constructs. Hepatocytes, as the key component of LOs, are isolated from the liver or differentiated from pluripotent stem cells (PSCs). PSC-derived hepatocytes are preferable because of their availability and scalability. However, efficient maturation of the PSC-derived hepatocytes towards functional units in LOs remains a challenging subject. The incorporation of cell-sized microparticles (MPs) derived from liver extracellular matrix (ECM), could provide an enriched tissue-specific microenvironment for further maturation of hepatocytes inside the LOs. In the present study, the MPs were fabricated by chemical cross-linking of a water-in-oil dispersion of digested decellularized sheep liver. These MPs were mixed with human PSC-derived hepatic endoderm, human umbilical vein endothelial cells, and mesenchymal stromal cells to produce homogenous bioengineered LOs (BLOs). This approach led to the improvement of hepatocyte-like cells in terms of gene expression and function, CYP activities, albumin secretion, and metabolism of xenobiotics. The intraperitoneal transplantation of BLOs in an acute liver injury mouse model led to an enhancement in survival rate. Furthermore, efficient hepatic maturation was demonstrated after ex ovo transplantation. In conclusion, the incorporation of cell-sized tissue-specific MPs in BLOs improved the maturation of human PSC-derived hepatocyte-like cells compared to LOs. This approach provides a versatile strategy to produce functional organoids from different tissues and offers a novel tool for biomedical applications.

Keywords: liver organoid; tissue-specific microparticle; pluripotent stem cell; hepatic differentiation; tissue engineering

1. Introduction

End-stage liver diseases account for almost two million deaths per year worldwide [1]. Moreover, drug-induced liver injury, as a public health concern, remains a potential health challenge [2]. It was reported that between 1953 and 2013, over 18% of pharmaceutical withdrawals occurred due to hepatotoxicity [3]. Animal models are used for drug screening, discovery, and toxicity testing, and they offer numerous advantages; however, they are often limited in relevance, time-consuming, and expensive and raise concerns [4,5]. Alternatively, in vitro culture of tissue-specific cells has attracted considerable attention as a promising approach to develop biomimetic systems for the prediction of potential hepatoxicity of drugs [6].

Traditional two-dimensional (2D) monolayer cells, cultured on flat and rigid substrates, are usually used for cell-based assays in drug development. Since almost all cells in the in vivo environment are surrounded by other stromal cells and ECM in a 3D condition, 2D cell culture does not adequately recapitulate the natural 3D environment of cells. In addition, it should be considered that physiological cell–cell interactions that typically happen in a 3D culture influence cell signaling, polarity, viability, and drug response [7,8]. Recently, liver organoids (LOs) were introduced as promising in vitro models for drug discovery, toxicology studies, disease modeling, and cell-based therapy treatments for patients suffering from hepatic failure [9,10]. The main principle of LOs is the recapitulation of major components and developmental steps of organogenesis or tissue repair in a dish [11]. In a pioneer work, Takebe and colleagues showed that the coculture of PSC-derived hepatic endoderm (HE) with mesenchymal stem cells (MSCs) and human umbilical vein endothelial cells (HUVECs) on a 3D Matrigel matrix led to condensation into a multicellular aggregate with liver-mimicking cell organization, called a liver bud organoid [12]. This strategy to develop a liver organoid was not scalable or reproducible to produce cells in clinically relevant numbers. Moreover, homogenous and efficient diffusion of nutrients, wastes, and other soluble factors throughout the aggregates might be restricted due to their large size [8]. Hence, a large number of organoids, similar in size and shape, are required for basic studies and translational applications. To address this requirement, researchers tried to develop liver organoids by coculture in suspension culture setups [13,14]. Those organoid-like spheroids were not functional enough, likely due to the lack of extracellular matrix (ECM) support to provide key cell–ECM interactions in cell-only aggregates. The ECM affects cell behavior and provides a proper niche and the biochemical cues required for the function of various cells in a specific organ [15]. Recent studies on decellularized liver tissue unveiled the role of liver ECM in activating hepatic differentiation and endodermal organoid formation [16–19]. Some studies have demonstrated that liver ECM promotes the maturation of human PSC-derived hepatocytes by downregulation of fetal liver markers (AFP and CYP3A7) and upregulation of other CYP genes as markers of metabolic activity [20–22]. Hence, an innovative strategy to establish liver organoids benefiting from both similar size and cell–ECM interactions is urgently required. In this regard, we previously developed a method for the encapsulation of three cell types in size-controlled alginate microcapsules enriched with sheep liver ECM. Although both coculture and ECM were effective on the functionality of hepatic cells, the migration and aggregation of the cells were restricted in the microcapsules, likely due to hindrances induced by the dense alginate matrix [20]. Here, we have developed an optimized, scalable 3D BLO culture system with a higher level of recapitulation of the liver-specific microenvironment by applying micropatterning technology. This was achieved by the addition of cell-sized MPs derived from liver ECM to the coculture of MSCs, HUVECs, and HE. We found that the expression levels of hepatic-specific genes ALB, G5PC, and CYP3A4 were significantly higher in BLOs compared to

LOs. The maturation was further promoted upon intraperitoneal transplantation of BLOs in an acute liver injury mouse model, leading to an improved survival rate. Furthermore, efficient hepatic maturation was observed after ex ovo transplantation. Therefore, organoid engineering via the incorporation of cell-sized tissue-specific MPs within size-controlled multicellular aggregates provides a versatile strategy for the production of more functional organoids from different sources.

2. Materials and Methods

2.1. Ethical Statement

All experimental protocols using hESCs and hiPSCs primary cells (HUVEC and MSC), and animal studies were done in compliance with guidelines of the "Royan Institute Ethics Committee" (IR.ACECR.ROYAN.REC.1397.104).

2.2. Fabrication of Liver ECM-Derived MPs

Sheep-liver-derived ECM and its pre-gel solution were prepared according to a previously reported protocol, with a few modifications [23,24]. Briefly, after freezing then unfreezing sliced sheep liver was decellularized using 1% sodium dodecyl sulfate and 1% Triton X. Then, the decellularized tissue was lyophilized and milled. After that, the powder was enzymatically digested using 2% pepsin (Sigma, St. Louis, MO, USA) in 0.5 M acetic acid (Sigma) for 48 h at 4 °C to produce a solution of 20 mg/mL pre-gel of liver ECM.

The ECM-derived MPs were fabricated using a water-in-oil (*w/o*) emulsion method. Briefly, 250 mg of gelatin type B was dissolved in 5 mL of liver ECM solution (20 mg/mL) and stirred for 10 min at 40 °C to obtain a homogeneous solution of liver ECM–gelatin. Then, 2 mL of the resulting solution was added dropwise into 80 mL of corn oil and homogenized at 7000 rpm for 5 min to establish a *w/o* emulsion. Afterward, the mixture was chilled on ice for 1 h. For MP cross-linking, a glutaraldehyde solution (0.2% *w/w*) was added to the mixture and stirred at 700 rpm for 12 h. The resulting MPs were retrieved by centrifugation at 4000 rpm, treated with glycine solution (500 mM) to eliminate the residual aldehyde group, washed three times in deionized water, and used.

2.3. Characterization of Liver ECM-Derived MPs

The morphology of MPs was studied by scanning electron microscopy (SEM). The sample was dehydrated and gold-coated using a sputter coater (MSP-1S, Shinku Device, Ibaraki, Japan) and subsequently observed by SEM (SEM; VE-8800, Keyence, Tokyo, Japan). Furthermore, the size of MPs was measured manually for at least 150 MPs using phase-contrast micrographs and ImageJ software. To compare the chemical composition of the MPs with intact and decellularized liver tissues, Fourier-transform infrared spectroscopy (FTIR) (Ettlingen, Germany) was performed. All samples were lyophilized and milled, and FTIR spectra were obtained by a Bruker-Equinox 55 FTIR spectrometer equipped with attenuated total reflectance accessories.

To explore the degradation rate, MPs (10 mg, M1) were incubated at 37 °C in 10 mM Dulbecco's phosphate-buffered saline (DPBS, 21600-010, Gibco, Waltham, MA, USA) (200 µL) at pH 7.4. Next, DPBS was discarded at a regular interval of 24 h, and the MP mass was measured (M2) [25]. Three samples of MPs were analyzed at each time point, and incubation was continued until the MPs were degraded (90% degradation was observed after 12 days). Finally, the percentage of degradation was calculated based on M2/M1 for each time point during the degradation process.

To determine possible cytotoxicity of MPs, human foreskin fibroblast viability was measured using an Orangu kit (OR01-500, Cell guidance systems, St. Louis, MO, USA) Fibroblasts were grown for 24 h under standard cell culture conditions in a 24-well plate at 5×10^4 cells/mL initial concentration in DMEM medium (21331-020, Gibco) supplemented with 10% fetal bovine serum (FBS, 10270, Gibco), 0.1 mM nonessential amino acids (NEAA 35050-038, Gibco), 2 mM L-glutamine (35050-038, Gibco), and 1% penicillin and streptomycin (pen/strep, 15070-063, Gibco) [26,27]. The MPs were sterilized by UV exposure for 45 min, and the cells were cultured for an additional 48 h in medium containing different cell-to-MP ratios (1:1 and 1:2). The medium was then replaced with fresh medium containing Orangu, according to the manufacturer's protocol. After 3–4 h of incubation at 37 °C, solution absorbance was measured at 495 nm.

2.4. Isolation and Expansion of Human-Umbilical-Cord-Derived Endothelial Cells

One donated human umbilical cord was washed with DPBS that contained 3% pen/strep. The umbilical vein was cannulated and washed with DPBS. Next, endothelial growth medium (EGM, CC-4147, Lonza) that contained 1 mg/mL collagenase IV (17104019, Gibco) was injected into the umbilical vein, followed by incubation at 37 °C for 15 min. The obtained cell suspension was sedimented at 1700 rpm for 10 min. The endothelial cells were resuspended in fresh EGM, transferred into a T25 flask, and cultured. The cells were used at passage number 1–3 [21]. All experiments were done in compliance with the guidelines of the Royan Institute Ethics Committee.

2.5. Expansion of Human Bone Marrow-Derived MSCs

The characterized cells were obtained from the Royan Stem Cell Bank (Iran) and cultured in low-glucose DMEM medium (11885-084, Gibco) supplemented with 10% FBS, 2 mM L-glutamine, 0.1 mM NEAA, 1% pen/strep, and 0.1 mM β-mercaptoethanol. The cells were used at passage number 1–4. All experiments were done in compliance with the guidelines of the Royan Institute Ethics Committee.

2.6. hPSCs Expansion and Differentiation towards Hepatic Endoderm in 3D Culture

Expansion and differentiation of human pluripotent stem cells (human embryonic stem cells (RH5) (RSCB0022) and human-induced stem cells (iPS4) (RSCB0082) were performed as previously described [28–30]. Briefly, 2.5×10^5 cells/mL were transferred to a low attachment bacterial plate (628102, Greiner) with hPSC medium that was previously conditioned for 24 h on inactivated human foreskin fibroblasts with mitomycin C (M4287, Sigma-Aldrich). The medium contained DMEM/F12 (21331-020, Gibco), 20% knockout serum replacement (KOSR, 10828028, Gibco), 2 mM L-glutamine, 1% pen/strep, 0.1 mM NEAA, 1% insulin–transferrin–selenium (ITS, 41400045, Gibco), 0.1 mM β-mercaptoethanol (M7522, Sigma-Aldrich), and 100 ng/mL basic fibroblast growth factor (bFGF, GFH28-100, Cell GS). Only on the first day, 10 μM Rho-associated protein kinase (ROCK) inhibitor (Y0503, Sigma-Aldrich) was added to the dispersed cells. The plates were incubated under standard conditions (at 37 °C with 5% CO_2 and saturated humidity), and the medium was replaced every other day.

The 4–5-day hPSC aggregates (average size 150 ± 20 μm) were induced to HE, following a published protocol [31]. Briefly, aggregates were rinsed with PBS plus Ca^{2+} and Mg^{2+} (PBS,14040117, Gibco) and cultured for 3 days in a definitive endoderm induction medium containing RPMI 1640 (5240041, Gibco), 0.1% bovine serum albumin (BSA, A9418, Sigma-Aldrich), and 1X B-27 without insulin (-Ins, 17504044, Gibco). The medium was supplemented with 6 μM CHIR99021 (CHIR, 04-0004-10, Stemgent, Beltsville, WV, USA) for one day. The aggregates were then washed with PBS and treated with 10 ng/mL Activin A (04-0004-10, R&D System) for two days. For HE differentiation, this medium was replaced with DMEM-F12, 2% KOSR, 10 ng/mL hepatocyte growth factor (HGF, 294-HG, R&D Systems, Minneapolis, MN, USA), and 10 ng/mL fibroblast growth factor 4 (FGF4, Royan Biotech, Tehran, Iran) for 4 days. All experiments were done in compliance with the guidelines of the Royan Institute Ethics Committee.

2.7. Formation of Liver Organoids in Microwell Plates

Initially, hESC-derived HE aggregates were dissociated into single cells by 0.05 trypsin-EDTA (25300-062, Gibco). After that, different cell numbers per microwell (500 and 1000) at different cell/MP ratios (1:1, 2:1, and 3:1) were evaluated and set up, respectively (Table S1 and Figure S1). The MPs were mixed with HUVECs, MSCs, and HE at a 2:1 cell-to-MP ratio. After that, single cells at 10:7:2 ratios [27] of HE:HUVECs:MSCs were transferred to a microwell plate (AggreWell™, 34421, STEMCELL, Vancouver, Canada) containing approximately 7000 microwells per well (microwells were 400 μm in size). Before using the AggreWell™ plate, it was rinsed with an antiadherence rinsing solution (07010, STEMCELL) to ensure optimal performance. The cell suspension was centrifuged at 160 g for 5 min, and the resultant pellet was resuspended in a combination of hepatocyte complete medium (HCM, CC-4182, Lonza), DMEM/F12 as a maturation medium, and EGM medium (1:1 *v/v*) and then seeded on an AggreWell™ plate. The medium was supplemented with 2% KOSR, 10 ng/mL HGF, 10 ng/mL FGF4, 10 ng/mL oncostatin M (OSM, Royan Biotech), 0.1 mM dexamethasone (Dex, D-2915, Sigma-Aldrich, St. Louis, MO, USA), 1 mM NEAA, and L-glutamine and changed every other day. Next, the plate was centrifuged at 100 g for 3 min to capture the cells in the microwells. After 24 h, the condensed structures were harvested and treated for an additional 11 days in the same medium, which was changed every other day.

2.8. Incorporation of Liver-ECM-Derived MPs within Liver Organoids

To evaluate the rate of liver-ECM-derived MP incorporation within liver organoids and to visualize the distribution pattern of them in the organoids, a FITC-labeled MP suspension was added to the cell suspension prior to the pellet formation described above. The efficiency of incorporation was calculated by counting free MPs before and after aggregate formation using a hemocytometer. Furthermore, MP incorporation and distribution within the aggregates were traced by fluorescein isothiocyanate (FITC)-labeled MPs. After that, the aggregates were fixed, sectioned, and stained with 4′,6-diamidino-2-phenylindole (DAPI, Sigma-Aldrich, D8417), and micrographs were taken using a fluorescent microscope (IX7, Olympus, Tokyo, Japan).

2.9. RT-Quantitative Polymerase Chain Reaction (qRT-PCR)

Total RNA was isolated using an RNeasy Micro Kit (74004, QIAGEN, Hilden, Germany) according to the manufacturer's instructions. Here, 1 μg RNA was used as a template for complementary DNA (cDNA) synthesis using a cDNA reverse transcription kit (4368813, Thermo Fisher Scientific, Waltham, MA, USA) according to the manufacturer's instructions. For positive controls, we used the cDNA of fetal and adult livers. qRT-PCR was performed with sets of human-specific primers (Table S2) and SYBR Green Master Mix (Takara Bio, Inc., SYBR Premix Ex Taq II RR081Q) using the StepOnePlus™ Real-Time PCR System. The results were normalized against glyceraldehyde 3-phosphate dehydrogenase (*GAPDH*) and calibrated against a combination of the same ratios of HE, HUVECs, and MSCs on Day 0 of the coculture. Data were analyzed and presented using the comparative CT method ($2^{-\Delta\Delta Ct}$). All experiments were done in compliance with the guidelines of the Royan Institute Ethics Committee.

2.10. Tissue Processing and Immunofluorescence Staining

LOs and BLOs derived from hESCs were collected at the end of Day 12 and fixed by 4% paraformaldehyde (at 4 °C, overnight). The agar-embedded structures were processed and embedded in paraffin. After that, 6-μm sections were prepared and treated with antigen retrieval Dako (S2368, Glostrup, Denmark). The sections were incubated with primary antibodies in the blocking buffer overnight at 4 °C and then incubated with secondary antibodies at 37 °C for 1 h. The list of antibodies used in this work is shown in Table S3. Finally, the nuclei were counterstained with DAPI. The micrographs were taken by a fluorescent microscope (IX71; Olympus) equipped with an Olympus DP72 digital camera.

2.11. Live & Dead Assay

Cell viability of the LOs and BLOs was evaluated using a LIVE & DEAD Viability Cytotoxicity Kit (Invitrogen, L3224, Carlsbad, CA, USA) on Day 14, according to instructions provided by the manufacturer. Briefly, the LOs and BLOs were washed with PBS and stained using calcein-AM (2 lM) and ethidium homodimers (4 lM) for 30 min at room temperature. LOs and BLOs were acquired using an inverted fluorescence microscope (Olympus, IX71).

2.12. Periodic Acid-Schiff (PAS) Staining

To evaluate the glycogen storage potential of LOs and BLOs derived from hESCs, periodic acid Schiff (PAS) staining was performed. Briefly, the LO and BLO sections were oxidized in 1% periodic acid for 5 min and then rinsed in dH_2O. After that, the sections were treated with Schiff's reagent for 15 min, followed by color development in dH_2O for 5–10 min. Finally, micrographs were taken using a light microscope.

2.13. Indocyanine Green (ICG) Uptake and Release

For assessment of ICG uptake as a functional proof of hepatic organoid, both groups were incubated with 1 mg/mL ICG (12633, Sigma-Aldrich) for 1 h at 37 °C. Next, the LOs and BLOs derived from hESCs were washed three times with PBS and incubated in fresh medium for another 3 h for ICG release evaluation. The uptake and release of ICG were visualized using a light microscope [32].

2.14. Albumin (ALB) and Fibrinogen Secretion Assay

The ALB and fibrinogen secreted in the culture supernatants were respectively quantified using a human ALB-ELISA kit (Bethyl, E80-129) and a human fibrinogen-ELISA kit (Genway, 10-288-22856), according to the manufacturers' instructions, by a microplate reader (Thermo Scientific, Multiskan Spectrum, 51118650). The results were normalized to the cell number.

2.15. Cytochrome P450 (CYP) Activity and Inducibility

To measure the activity of cytochrome P450 (CYP)-3A4, -2B6, -2C9, and -1A2 in both groups, lytic assays were performed by using a P450-GloTM assay kit (V8802, V8322, and V8792, respectively; Promega) based on the manufacturer's instructions. The LOs and BLOs derived from hESCs were treated with basal media containing rifampicin (25 µM), valproate (100 µM), phenobarbital (100 µM), and omeprazole (100 µM) as CYP-3A4, -2B6, -2C9, and -1A2 inducers, respectively, or DMSO (0.1%) as the inducer solvent. The activity of each enzyme was measured by reading its luminescence using a luminometer (SpectraMax i3x), according to the manufacturer's protocol.

2.16. Acute Toxicity Assay

The viability of cells in the LOs and BLOs derived from hESCs was evaluated by an Orangu kit after 48 h of exposure to four hepatotoxins, namely, paracetamol (APTP; 500, 1000, and 10,000 µM), fluorouracil (5-FU; 50, 100 and 500 µM), busulfan (10, 100 and 1000 µM) and tamoxifen (TAM; 1, 10 and 20 µM). For this purpose, culture media were refreshed every 24 h with serum-free media containing a certain concentration of each compound.

2.17. Ex Ovo Implantation of Bioengineered Liver Organoids on Chorioallantoic Membrane

The ex ovo implantation of BLOs derived from hESCs on CAM was performed based on a previously reported method [28]. The Fertilized Hy-line W-36 haying hens' eggs were supplied by a commercial farm. The eggs were then cracked in a sterile laminar flow hood, and embryos with undamaged yolks were transferred into a surrogate shell that was bigger than the eggshell, sealed with plastic wrap, and incubated in a forced air incubator for 60 h at 37 °C with 60% humidity (Embryonic Day 0 (ED0)). On ED2.5, the yolk-embedded embryo was transferred to a second surrogate shell, which was bigger than the primary surrogate shell, and then sealed and incubated for another 5 days. One-day-old BLOs were implanted onto the surface of the CAM on ED7.5, distal from the embryo and proximal to major blood vessels, by gently scraping the upper CAM layer. Egg windows were sealed with plastic wrap, and the eggs were incubated for 14 additional days (ED 19.5). Finally, to study vasculogenesis, FITC-dextran (SLBT3224, Sigma-Aldrich) was injected into one CAM vessel, and other BLOs were subjected to qPCR and histology analysis. All experiments were done in compliance with the guidelines of the Royan Institute Ethics Committee.

2.18. Fabrication of a 3D-Printed Basket

To have a suitable container for the localization of the transplanted BLOs, a 3D basket device was fabricated using the fused deposition modeling (FDM) method, as described before, with minor modifications [29]. Briefly, the device was designed in CATIA V5 r19 and exported as an "STL" file. Then, it was sliced using Slic3r software. Based on the device dimensions and its vertical porosities, the key parameters were optimized for 3D printing. Printing temperature, nozzle diameter, printing speed, travel speed, and gap speed for this construct were set to 230 °C, 300 μm, 10 mm/min, 60 mm/min, and 60 mm/min, respectively. The porosity of the bottom of the device and the surface of the cap were set to 50% and 70%, respectively. The printing material was transparent polylactic acid (PLA, Befon). All 3D basket devices were printed by the Royan 3D printer (Royan Institute) (Figure S2).

2.19. In Vivo Implantation of Bioengineered Liver Organoids

We used the three following animal groups for the in vivo experiments: healthy mice (normal), acute liver injury model with transplanted empty basket (sham), and acute liver injury model with a transplanted basket containing BLOs derived from hESCs (treated). Acute liver injury in mice was induced by an intraperitoneal injection of 1 mL/kg (sublethal dose) carbon tetrachloride (CCl_4) in 10-week old male C57BL/6 mice [30]. For immunosuppression, mice received a daily injection of cyclosporine (Novartis Pharmaceuticals, 20 mg/kg) for four days before transplantation until one week after transplantation. Next, two-week-old BLOs were transferred to the 3D-printed basket. The mice were then anesthetized by an intraperitoneal injection of 75 mg/kg ketamine (Alfasan) and 0.1 mg/kg medetomidine (Syva). After that, the 3D basket containing either approximately 1000 BLOs and that without BLOs (as BLO-transplanted or sham-transplanted, respectively), were transplanted into the abdominal cavity and sutured to the internal abdominal muscle. The animals were followed for 14 days for determination of survival rate. Furthermore, the sera levels of alanine transaminase (ALT) and aspartate transaminase (AST) on Day 2 after transplantation were evaluated using a commercially available kit (Biorexfars; BXC0215A and BXC0205A). For gene expression analysis, the samples were harvested from the 3D basket on Day 14. Moreover, human ALB in mice sera was quantified by a human ALB-ELISA kit (Bethyl; E80-129). All experiments were done in compliance with the guidelines of the Royan Institute Ethics Committee.

2.20. Chi-Square Analysis

This analysis was performed for the confirmation of survival rate significance. The hypothesis in this analysis was comprised of H0: the survival rate is not different between the two groups and Ha: the survival rate is different between the two groups. The decision rule was as follows: if $X2 > 3.84$, the null hypothesis is rejected (Tables S4–S6).

2.21. Statistical Analyses

Data are presented as mean ± standard deviation (SD) from at least three biological replicates. Based on the normal distribution of data, statistical analyses were performed using the unpaired two-tailed Student's *t*-test or one-way ANOVA to compare two or more groups, respectively. The assessments were followed by Tukey posthoc analysis to show the differences between groups. *P*-values less than 0.05 were considered significant. To evaluate the survival rate, a chi-square analysis was performed.

3. Results

3.1. Fabrication and Characterization of Liver-ECM-Derived MPs

In this study, we applied a *w/o* technique to generate liver-ECM-derived MPs (Figure 1A). The MPs were labeled with FITC for tracking after incorporation within the cell spheroids (Figure 1B). According to SEM findings, the MPs had a round and identical spheroid shape (Figure 1C). Size distribution analysis showed a narrow size distribution for the MPs, with a mean diameter of 7.9 ± 1.0 μm (Figure 1D). FTIR analysis was performed to compare the chemical composition of MPs and intact and decellularized liver tissues (Figure 1E). A comparison of the FTIR data of the MPs, decellularized liver ECM, and natural sheep liver confirmed the presence of collagen and proteoglycan in all of them. Within the region of 1500–1700 cm^{-1}, three major peaks, including amide I (1652 cm^{-1}), amide II (1542 cm^{-1}), and amide III (1236 cm^{-1}) bands, are associated with collagen protein [33]. The amide I band is mainly related to C=O stretching and N-H stretching, while the amide II band derives from N-H bending coupled with C-N stretching. The spectral range of 1000–1200 cm^{-1} mainly represents the C-O-C, C-O-H or C-C and C-C-O, and C-O-C stretches, vibrations of carbohydrate moieties from proteoglycan and polysaccharides, respectively [34]. To explore the stability of MPs under physiologic conditions, degradation over time was analyzed based on weight. The results demonstrated that the MPs were degraded over 11 days in a time-resolved manner (Figure 1F). To evaluate biocompatibility, fibroblasts were cultured on tissue culture plates in the presence of MPs at different cell/MP ratios. The results showed that following cell culture with different ratios of MPs, cell viability remained similar to that of the sham group (without MPs) (Figure 1G).

3.2. Differentiation of hPSCs towards Hepatic Endoderm

First, we optimized the differentiation protocol to generate HE from hESCs in terms of the induction period. Accordingly, the hESC aggregates were differentiated towards hepatoblasts in a 3D condition. Then, the aggregates were harvested on Days 3, 5, 7, and 9 of differentiation and evaluated. The qRT-PCR results demonstrated that the endodermal markers *SOX17, FOXA2,* and *CXCR4* were significantly downregulated ($p < 0.05$, Figure S3A). However, on Days 7 and 9 of differentiation, the expression of HE-specific markers, including *HNF4α, HEX,* and *TBX3,* was significantly higher than that of Days 3 and 5 ($p < 0.05$, Figure S3A). No significant difference was observed in the expression of HE markers between Days 7 and 9 of differentiation. Therefore, we applied a reverse screening experiment by comparing the hepatic markers' gene expression on Days 7 and 9 for HE-derived liver organoids. We observed that the three mature hepatic genes, cytochrome p450 subunit 3A4 (*CYP3A4*), tyrosine aminotransferase (*TAT*), and tryptophan 2,3-dioxygenase (*TDO*), were significantly expressed at higher levels in the LOs derived from Day-7 HE ($p < 0.01$, Figure S3B). Therefore, the PSC-derived HE on Day 7 was selected for the coculture to form LOs and BLOs.

3.3. Formation of Liver Organoids in Microwell Plates

Initially, to optimize condensed aggregate formation, we examined various cell-to-MP ratios. Remarkably, cell condensation occurred only at 500 cell/microwell cell density with cell/MP ratio 2:1 (Table S1 and Figure S1).

Human ESC-derived HE on Day 7 was cocultured with BM-MSCs and HUVECs, with/without MPs, in AggreWell™ plates (Figure 2A). Twenty-four hours after coculture, the condensed aggregates were harvested and treated for further maturation for 12 days in static 3D culture (Figure 2A). Homogenous, round, and dense hepatic-like organoids were observed on Days 2, 7, and 12 post-coculture in the LOs and BLOs groups (Figure 2B). A gradual increase was demonstrated in size for both groups on Days 2, 7, and 12 post-coculture. The mean sizes on Days 2, 7, and 12 post-coculture were 150 ± 48.7, 250 ± 33.2, and 320 ± 45.3 μm, respectively. There was no difference in the mean size between the two groups (Figure S4). The results of MP incorporation indicated that the FITC-labeled MPs were successfully (approximately 80%) incorporated into the aggregates. Fluorescent images and microscopic sections showed that FITC-labeled MPs had a homogenous distribution pattern throughout the aggregates (Figure 2C).

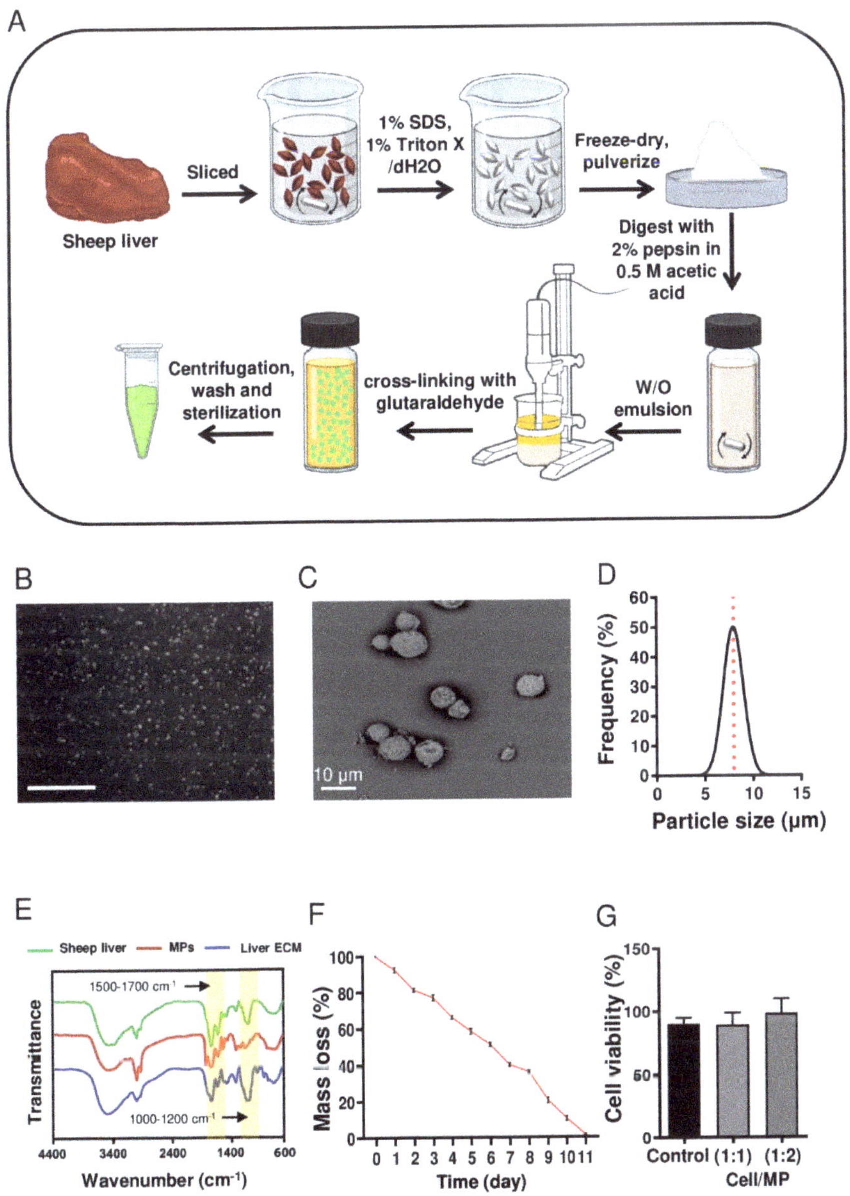

Figure 1. Fabrication and characterization of liver-ECM-derived microparticles (MPs). (**A**) A schematic presentation of liver-ECM-derived MP fabrication. (**B**) Phase-contrast microscopy image of MPs (scale bar: 500 µm). (**C**) A representative scanning electron microscopy (SEM) image of the MPs. (**D**) Size distribution histogram of the MPs. (**E**) Fourier-transform infrared spectroscopy (FTIR) spectra of intact and decellularized liver tissues and liver-ECM-derived MPs. (**F**) Mass loss of MPs in DPBS (pH 7.4, at 37 °C). (**G**) Cytocompatibility analysis of MPs by MTS assay on fibroblast cells cultured for 48 h on tissue culture polystyrene, with/without MPs, at different cell/MP ratios. (ECM: extracellular matrix; MPs: microparticles; SEM: scanning electron microscopy; FTIR: Fourier-transform infrared spectroscopy; DPBS: Dulbecco's phosphate-buffered saline).

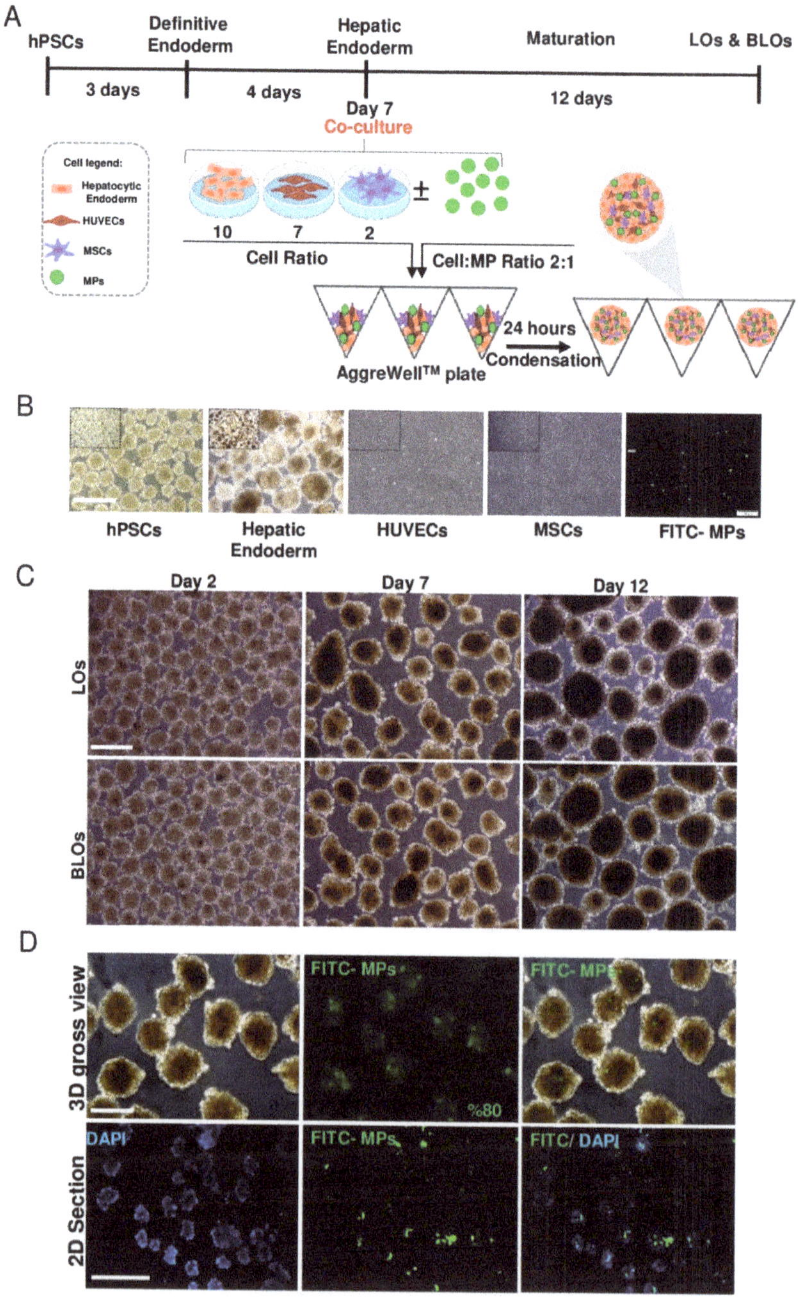

Figure 2. Formation of LOs and BLOs from human pluripotent stem-cell-derived hepatic endoderm. (**A**) A schematic presentation of the stepwise protocol for differentiation of hESCs (RH5 cell line) towards HE and their coculture with HUVECs and MSCs (10:2:7 of HE:MSC:HUVEC), with/without MPs (2:1 cell:MP), in AggreWell™ plates. (**B**) Phase-contrast microscopy images of hPSCs, HE derived from hESCs, HUVECs, MSCs, and FITC-MPs (scale bar: 200 μm) (**C**) Phase-contrast microscopy images

of cell aggregates on Days 2, 7, and 12 post-coculture (scale bar: 200 μm). (**D**) Visual illustration of MP incorporation within cell aggregates composed of three types of cells (HE, HUVECs, and MSCs; 10:7:2, respectively) using fluorescein isothiocyanate (FITC)-labeled MPs. Phase-contrast microscopy images (top left), fluorescence microscopy images (top middle), merged fluorescence microscopy image (top right), and paraffin-embedded cross-section (bottom) of the MP-incorporated cell aggregate. FITC-labeled MPs are visualized in green. The nuclei of cells were counterstained with DAPI. Scale bar: 200 (top) and 500 μm (below). HE: hepatic endoderm; HUVECs: human umbilical vein endothelial cells; MSCs: mesenchymal stem cells; LOs: liver organoids; BLOs: bioengineered liver organoids; hESCs: human embryonic stem cells; MPs: microparticles; FITC: fluorescein isothiocyanate.

3.4. Gene Expression Analysis of the Liver Organoids

To determine the gene expression profile in LOs and BLOs derived from hESC and hiPSC cultures after 12 days, we evaluated the mRNA expression of the following hepatic-specific genes: α-fetoprotein (*AFP*), albumin (*ALB*), tyrosine aminotransferase (*TAT*), tryptophan 2,3-dioxygenase (*TDO*), multidrug resistance-associated protein 2 (*MRP2*), multidrug resistance (*MDR*), asialoglycoprotein receptor1 (*ASGPR1*), carbamoyl phosphate synthase 1 (*CPS1*), glucose-6-phosphatase (*G6PC*) and phase I enzymes, cytochrome p450 subunit 1A2 (*CYP1A2*), subunit 3A4 (*CYP3A4*), subunit 3A7 (*CYP3A7*), subunit 2C9 (*CYP2C9*), and subunit 2B6 (*CYP2B6*). Moreover, we analyzed the gene expression of phase II and III metabolic enzymes, UDP-glucuronosyltransferase 2B15 (*UGT2B15*), UDP-glucuronosyltransferase-2B7 (*UGT2B7*), solute carrier organic anion transporter family member 2B1 (*SLCO2B*), organic anion transporting polypeptide 1B1 (*OATP1B1*), and aquaporin-7 (*AQP7*) (Figure 3). Interestingly, a significant downregulation of immature fetal hepatic markers, *AFP* and *CYP3A7*, was observed in the BLOs group compared to the LOs group ($p< 0.01$; Figure 3). Moreover, the expression of mature hepatic-specific genes (*ALB*, *MRP2*, *MDR*, *ASGPR1*, and *G6PC*) was significantly higher in the BLOs group than the LOs group. In addition, drug-metabolism-related genes, such as phase I enzymes (*CYP1A2*, *CYP3A4*, *CYP2C9*, and *CYP2B6*) as well as phase II and III enzymes (*UGT2B15*, *SLCO2B1*, *OATP1B1*, and *AQP7*), were significantly upregulated in the BLOs group compared to the LOs group (at least $p < 0.05$; Figure 3). We also examined the expression of *ITGa5* and *ITGb1* as genes of ECM receptors and their target genes (*FOXA2*, *HNF4α*, and *PXR*). The results indicated that their expression was significantly upregulated in the BLOs group compared to the LOs group (at least $p < 0.05$; Figure 3). Gene expression analysis was also performed for the liver organoids derived from HE that was differentiated from hiPSCs, and similar results were observed (Figure S5). These data suggest that engineering the microenvironment of liver organoids by the incorporation of liver ECM MPs could improve the maturation and metabolic capacity of the PSC-derived hepatocytes.

3.5. Immunostaining Analysis of the Liver Organoids

The result of live and dead staining indicated that most of the cells inside the LOs and BLOs were alive, and a small number of cells died (Figure S4C). Immunofluorescence staining demonstrated that ALB, CYP3A4, CYP1A2, and E-CAD were expressed in both LOs and BLOs derived from the hESC groups. Z0-1 as an epithelial marker was also well-expressed in both groups. Moreover, vimentin and CD31-positive cells indicated the MSCs and HUVECs in both groups, respectively (Figure 4A). We also quantified and compared the protein expression in BLOs and LOs, respectively, for ALB (57 ± 5.6% vs. 34.3 ± 4.8%), CYP3A4 (51.6 ± 6.2% vs. 31.6 ± 5.1%), ALB/CYP3A4 (58 ± 5.2% vs. 34.3 ± 5.6%), CYP1A1 (33.3 ± 6.1% vs. 15.6 ± 5.3%), ZO-1 (46.6 ± 6.3% vs. 35 ± 5.2%), E-cadherin (44.3 ± 6.1% vs. 15.67 ± 4.5%), vimentin (9 ± 1.1% vs. 9.5 ± 1.01%), and CD31 (26 ± 1.5% vs. 26.5 ± 1.3%). The results indicated that there were more ALB+, CYP3A4+, CYP1A2+, ZO-1+, and E-cadherin+ cells in the BLOs group compared to the LOs group. Additionally, the percentage of vimentin+ cells and CD31+ cells in both groups was almost the same (Figure 4B). Primary human hepatocytes and adult fibroblasts were used as positive and negative controls, respectively (Figure S6A–C).

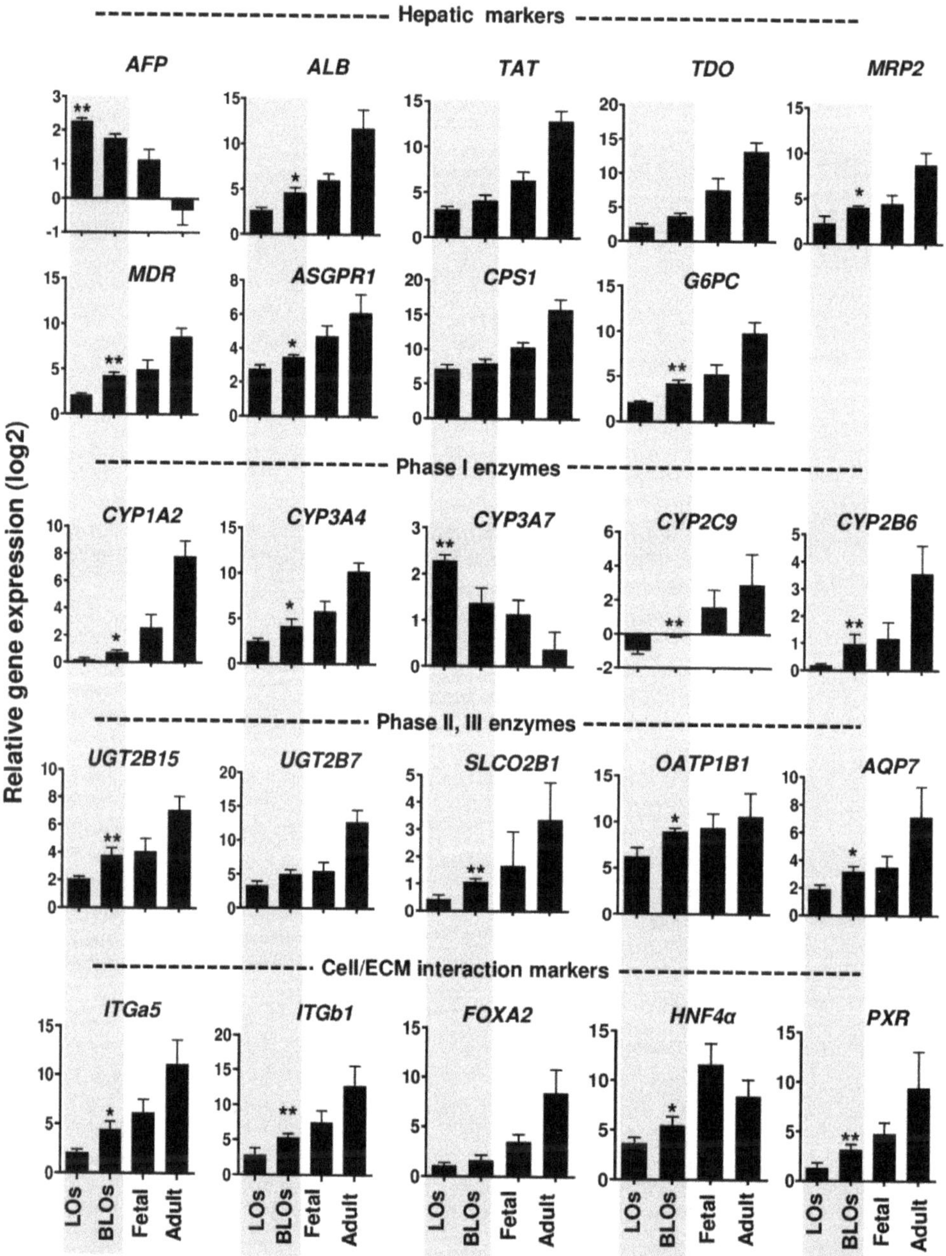

Figure 3. Gene expression of hepatic signatures in the LOs and BLOs derived from hESC-HE. Relative gene expressions of the hepatic-specific genes and the genes related to phases I, II, and III of drug metabolism in the BLOs and LOs groups compared to fetal and adult human livers as control groups (n = 4). Data were normalized against *GAPDH* and are presented as fold changes calibrated to expression values on Day 0. Data are shown as mean ± SD (n = 3). Statistical analysis was performed using unpaired two-tailed Student's t-tests. * $p < 0.05$ and ** $p < 0.01$. LOs: liver organoids; BLOs: bioengineered liver organoids; hESC: human embryonic stem cell (RH5 cell line); HE: hepatic endoderm.

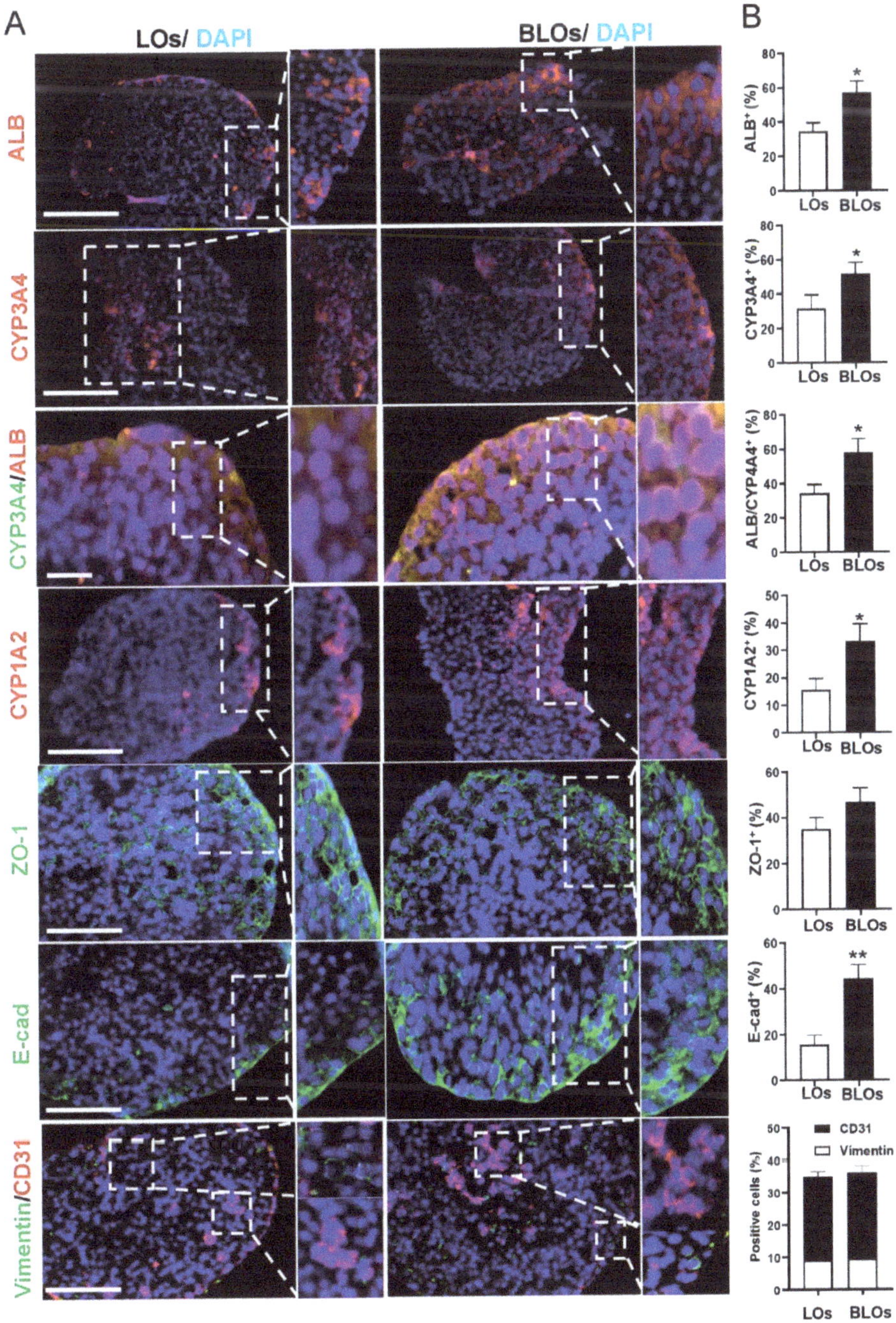

Figure 4. Protein expression profile of the LOs and BLOs groups. (**A**) Immunostaining of BLOs and LOs derived from hESCs for AFP, ALB, CYP3A4, CYP1A2, ZO-1, E-cadherin, and vimentin/CD31. The nuclei were counterstained with DAPI (scale bar: 200 and 100 μm). (**B**) The ratios of ALB^+, $CYP3A4^+$, $ALB/CYP3A4^+$, $CYP1A2^+$, $ZO\text{-}1^+$, $E\text{-}cadherin^+$, $vimentin^+$, and $CD31^+$ cells in all DAPI-stained cells obtained from at least 4 views per section for at least 3 sections * $p < 0.05$ and ** $p < 0.01$. LOs: liver organoids; BLOs: bioengineered liver organoids.

3.6. Functional Analysis of the Liver Organoids

Functional analysis showed that hepatocyte-like cells in both groups (LOs and BLOs derived from hESCs) were able to uptake and release ICG (Figure 5A and Figure S6D). PAS staining showed glycogen storage in both groups (Figure 5B). According to immunostaining results, the hepatic-like cells were localized in the marginal zone, and HUVECs and MSCs were localized in the center of the organoids, which was consistent with the PAS staining results. Moreover, ALB and fibrinogen secretion in the BLOs group was significantly higher than the LOs group ($p < 0.05$; Figure 5C,D). Moreover, our data showed that the activity of CYP3A4, CYP2B6, and CYP2C9 significantly increased ($p < 0.05$) upon treatment with drug inducers in both LOs and BLOs groups; however, CYP1A2 assessment did not show any significant difference (Figure 5E). Notably, higher drug inducibility was observed in the BLOs group compared to the LOs group (at least $p < 0.05$; Figure 5E). To evaluate their applicability in drug toxicity assessments, the BLOs and LOs groups were treated with different concentrations of various hepatotoxic compounds. The results showed that the susceptibility and sensitivity of the BLOs group towards the hepatotoxic compounds were significantly higher than the other group (at least $p < 0.05$; Figure 5F). Altogether, the BLOs group showed higher maturity and functionality than the LOs group; hence, we selected this group to perform subsequent experiments for further evaluations.

3.7. Further Maturation of Bioengineered Liver Organoids by Ex Ovo Transplantation

We next explored whether the highly vascularized microenvironment of CAM could improve the maturation of the BLOs derived from hESCs. To show the potential of vascularization and integration with the host vasculature, the chick was injected with dextran-FITC for visualization. The comparison of fluorescence microscopy images, taken before and after dextran-FITC injection, showed multiple blood vessels in the close vicinity of the BLOs after 14 days (Figure S7A). Moreover, H&E staining showed the penetration of capillaries into the implanted BLOs (Figure S7B). Gene expression analyses demonstrated that at 14 days after transplantation, early hepatic development genes were significantly downregulated while late hepatic genes were significantly upregulated, and the BLOs were closer to the adult liver in terms of gene expression patterns (Figure S7C). Thus, the transplantation of BLOs into a highly vascularized microenvironment could promote their maturation. Therefore, CAM could enhance the maturation of BLOs by providing a highly vascularized microenvironment.

3.8. Ectopic Transplantation of Bioengineered Liver Organoids in Mice with Acute Liver Injury

Acute liver injury and transplantation procedures are illustrated in Figure 6A. BLOs derived from hESCs were transplanted in mice in a 3D-printed basket. The baskets were highly porous in order to facilitate mass transfer and engraftment between the transplanted organoids and the host (Figure 6B). In order to investigate the capacity of BLOs to improve survival rates in the mouse model, the baskets with/without BLOs were transplanted into the internal abdominal muscles of cyclosporine-treated mice with acute liver injury. Animal survival rate was evaluated for 14 days post-transplantation; the results demonstrated that in the BLO-transplanted group, all animals survived, while 60% in the sham group were alive (Figure 6C). The result of chi-square analysis indicated that X2 = 5.48, which exceeds the critical value of 3.84. Therefore, we reject H0. We have significant evidence, $\alpha = 0.05$, to show that the two survival curves are different.

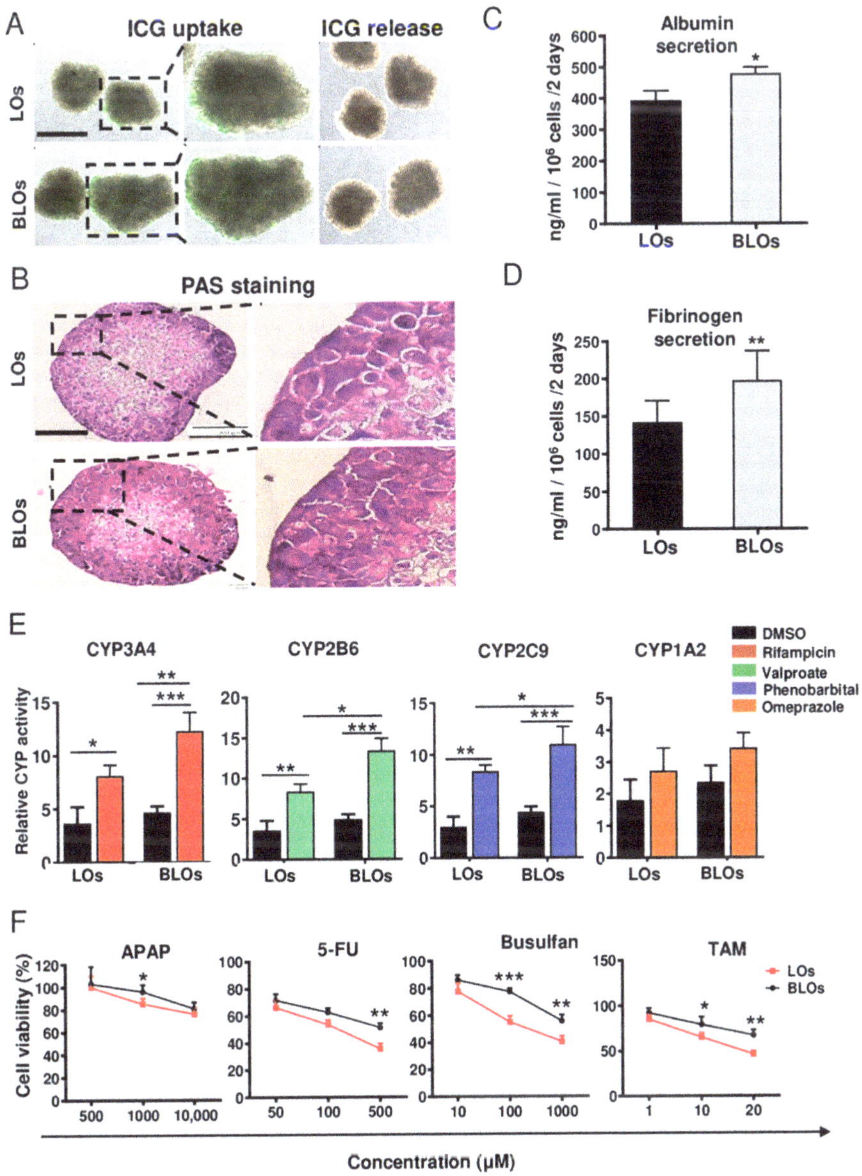

Figure 5. Functional assays of the LOs and BLOs derived from the hESC groups. (**A**) The BLOs and LOs groups were examined for their ability to take up and release indocyanine green (ICG). The ICG uptake and release took 1 and 3 h, respectively. Scale bar: 200 μm. (**B**) Periodic acid Schiff (PAS) staining was done to analyze glycogen storage. Scale bar: 200 μm. (**C**,**D**) Quantification of albumin (ALB) and fibrinogen secretion after 48 h. Data are shown as mean ± SD (n = 3). Statistical analysis was performed by using unpaired two-tailed Student's t-tests. * $p < 0.05$ and ** $p < 0.01$. (**E**) Induction of CYP3A4, CYP2B6, CYP2C9, and CYP1A2 by rifampicin, valproate, phenobarbital, and omeprazole, respectively, after 72 h of treatment. (**F**) Cell viability in the BLOs and LOs groups was assessed by Orangu® kit after 48 h of exposure to different concentrations of hepatotoxic compounds, paracetamol (ATAP), fluorouracil (5-FU), busulfan, and tamoxifen (TAM). Data are shown as mean ± SD (n = 3). Statistical analysis was performed using one-way ANOVA, followed by Tukey posthoc. * $p < 0.05$, ** $p < 0.01$, and *** $p < 0.001$. LOs: liver organoids; BLOs: bioengineered liver organoids.

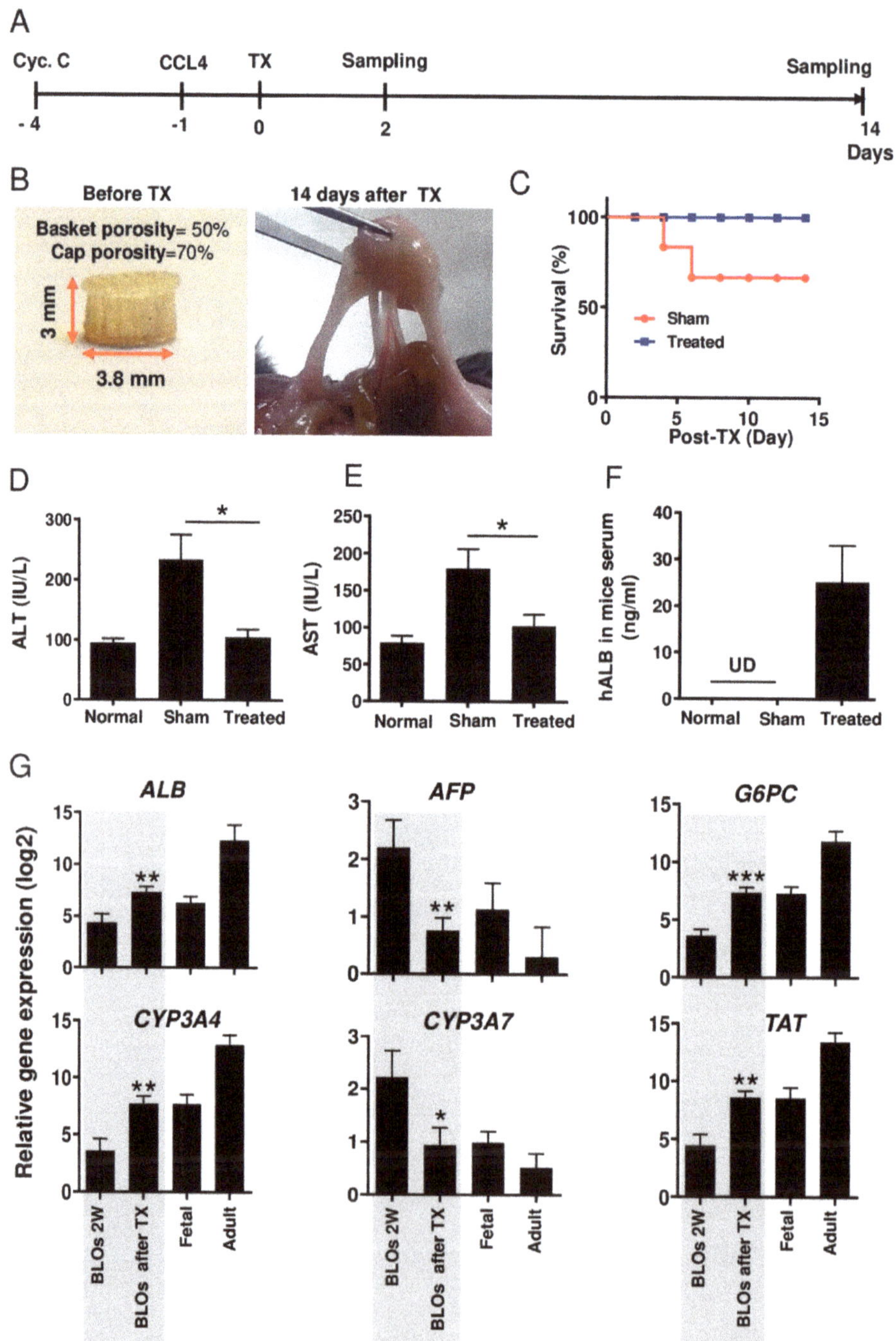

Figure 6. Ectopic implantation of the BLOs in mice with acute liver injury. (**A**) The timeline representing the time points in the figure. (**B**) 3D-printed basket containing the BLOs, before (left) and after (right) transplantation. (**C**) A graphic presentation of the survival rate of the control and treated groups. (**D**,**E**) Biochemical assessment of blood sera from mice with acute liver injury treated with the BLOs. BLO transplantation led to significantly decreased alanine transaminase (ALT)

and aspartate transaminase (AST) levels. (**F**) Human ALB was measured by ELISA in mice sera. Normal: normal mice; Control: CCl_4-treated mice transplanted with empty baskets; Treated: CCl_4-treated mice transplanted with 3D baskets containing approximately 1000 BLOs (n =8). Data are shown as mean ± SD (n = 3). Statistical analysis was performed by using unpaired two-tailed Student's t-tests. * $p < 0.05$. (**G**) Relative gene expression of hepatic-specific genes (*ALB*, *AFP*, *G6PC*, *CYP3A4*, *CYP3A7*, and *TAD*) in the BLOs, before and after transplantation, and fetal and adult liver tissues as control groups. Data were normalized against *GAPDH* and are presented as fold changes compared with data obtained on coculture Day 0 as the calibrator. Data are shown as mean ± SD (n = 8). Statistical analysis was performed by using unpaired two-tailed Student's t-tests. * $p < 0.05$, ** $p < 0.01$, and *** $p < 0.001$. BLOs: bioengineered liver organoids; 3D: three-dimensional; ALT: alanine transaminase; AST: aspartate transaminase; ALB: albumin; CCL4: carbon tetrachloride.

In the next step, we measured changes in liver injury indicators, ALT and AST, two days after transplantation. Data showed significant decreases in the plasma levels of ALT (Figure 6D) and AST (Figure 6E) in mice transplanted with baskets containing BLOs compared to the sham group transplanted with empty baskets. Moreover, to explore BLO functionality, we measured human ALB in mice sera. Using an ELISA test, 20–35 ng/mL human ALB was detected in the sera of mice transplanted with baskets containing BLOs (Figure 6F). Relative gene expression analysis showed that two immature assigned genes (*AFP* and *CYP3A7*) were significantly decreased compared to the in vitro cultured, two-week-old BLOs (at least $p < 0.05$). In contrast, the expression of many mature genes was significantly increased to near that of the adult liver compared to the in vitro cultured, two-week-old BLOs (Figure 6G). All primers used in this section were human-specific.

4. Discussion

Bioengineered organoids have broad applications in experimental medical research, and they are progressively opening up new avenues in drug screening, toxicology, and disease modeling. Nowadays, only 10% of new components that enter phase I clinical trials proceed to the next phases. This emphasizes the need to develop and optimize methodologies that reduce the expenses, time, and preclinical animal studies needed to identify a suitable compound and perform drug screening [35,36]. Recently, advances in technologies involving the production of organoids from human PSCs have presented a great opportunity for developing a more reliable, rapid, and cost-effective drug-screening platform compared to the currently used animal and in vitro models [37–39].

Liver development is a complex process in which distinct microenvironmental and biophysical signals are crucial. Human PSC technology has promoted in vitro recreation of processes that occur during liver organogenesis [35]. Takebe et al. took advantage of this approach and generated liver organoids using a coculture system and a Matrigel matrix, a solubilized basement membrane preparation extracted from Engelbreth–Holm–Swarm mouse sarcoma cells [12]. However, current organoid fabrication technologies have faced some drawbacks, such as the nonspecific matrix provided by Matrigel, uncontrolled size, and poor reproduction [36]. Bioengineering approaches using liver-specific ECM and micropatterning systems have been proposed as novel strategies to overcome these limitations [37]. Herein, we addressed this challenge by incorporation of cell-sized sheep liver ECM-derived MPs within cell aggregates during aggregate formation in microwell plates. The MPs were fabricated by chemical cross-linking of a water-in-oil dispersion of digested decellularized liver tissue. Then, HE cells were cocultured with MSCs and HUVECs in AggreWell™ plates, with or without the MPs, to form 3D structures named bioengineered liver organoids (BLOs) and LOs, respectively. Furthermore, advances in the micropatterning of liver organoids can potentially open an avenue to screen thousands of compounds/drugs to narrow down potential candidates. This model is more suitable for in vitro drug screening and disease modeling rather than cell-based therapy.

In this study, we successfully developed a simple and efficient method to produce liver-ECM-derived MPs with a round shape and homogenous size in the range of the typical size of single cells (~8 µm). Moreover, FTIR assessment of the LEM gel, MPs, and liver ECM verified the preservation of the key components of liver tissue after processing into the liver ECM solution and hydrogel MPs, which was consistent with previously reported data [1,23].

Gene expression analysis showed significant upregulation of specific hepatic genes (*ALB, TAT, MRP2, TDO, MDR, ASGPR1, CPS1,* and *G6PC*) as well as phase I (*CYP1A2, CYP3A4, CYP2C9,* and *CYP2B6*), phase II, and phase III enzyme genes (*UGT2B15, UGT2B7, SLCO2B, OATP1B1,* and *AQP7*) in the BLOs compared to the LOs group. In parallel, the hepatic immature genes (*AFP* and *CYP3A7*) in the BLOs group were downregulated compared to the LOs group. These results agree with other reports that have demonstrated that liver ECM successfully improves hepatocyte differentiation. In this context, mouse liver scaffolds could enhance the differentiation of hiPSC-derived hepatocytes [16]. Huanjing and colleagues showed that porcine liver ECM promoted the hepatic differentiation of BM-MSC [38]. Moreover, our results showed that hPSC-derived hepatocytes in the BLOs group expressed epithelial membrane protein markers like E-cadherin and ZO–1, implying that compared to hepatocytes in the LOs group, they were more polarized. This observation is consistent with previous studies, which have shown that liver ECM and supportive cells enhance the epithelialization and polarity of hepatocytes and other epithelial cells [40–43]. Moreover, based on our findings, the BLOs group was more functionally mature than the LOs group, as shown by higher glycogen storage, ability to uptake and release ICG, and CYP induction. We also found significant upregulations of ALB and fibrinogen secretion (as two important factors secreted by mature hepatocytes) in the BLOs group compared to the LOs group. This was comparable to ALB secretion by hepatospheres in the study of Pettinato and colleagues (450 vs. 230 ng/mL, respectively) [44]. Moreover, ICG uptake by hepatocytes happens through organic anion-transporting polypeptides (OATP1B1 and OATP1B3), which are expressed on the sinusoidal membrane of hepatocytes and ICG and then excreted into bile canaliculi via multidrug resistance associated protein 2 (MRP2). The successful ICG uptake and release indicated that ICG absorption and clearance occurred after 1 and 3 h, respectively. The results were consistent with the gene expression of transporters. In contrast to previous studies, ICG release took a shorter time (i.e., 3 h versus approximately 6 h in other studies) [45,46]. Previous studies have shown that recapitulation of the native microenvironment can improve the maturation and functionality of different cells [47–49]. Our finding shows that BLOs are more sensitive in the evaluation of drug-induced cytotoxicity compared to LOs. These data suggest that the BLOs could detect the toxicity of the reactive metabolites generated by drug-metabolizing enzymes such as CYP enzymes. This is important because, in many cases, drug-induced hepatotoxicity is caused by the reactive metabolites produced by drug-metabolizing enzymes [42]. Likewise, ex ovo and in vivo ectopic transplantations of the BLOs resulted in their further maturation and functionality. We found that the application of chick CAM provided a highly vascularized microenvironment. Ectopic transplantation also enhanced the BLOs' maturation, as indicated by the markedly increased expression of hepatic maturation genes. We assumed that the vascular structures that surrounded the BLOs through the highly porous 3D basket enhanced the maturation of the BLOs. Some studies have demonstrated the effect of in vivo and ex ovo vascularizations on the maturation of organoids or other tissues [33,34,50–52]. Here, we found that the transplanted BLOs could rapidly perform hepatic functions such as ALB secretion; in addition, ALT and AST levels in BLO-transplanted mice were significantly decreased compared to the sham group. These findings demonstrate that the BLO-transplanted group successfully performed metabolic detoxification and biosynthesis of albumin and other crucial biomolecules in acute liver injury and eliminated toxic metabolites to provide the healthy microenvironment required for liver regeneration, resulting in an improved survival rate. The ability of BLOs to quickly perform hepatic functions may be attributed to the presence of liver-ECM-derived MPs and the fact that the BLOs were surrounded by vessels.

Liver-specific ECM components provide a natural microenvironment that promotes and mimics cell–cell and cell–matrix interactions similar to the liver [45]. Cells interact with ECM via integrins, which are cell adhesion receptors that regulate cellular behavior [53]. The ECM provides guiding cues for better phenotype maintenance, differentiation, proliferation, and polarization of hepatocytes both in vitro and in vivo [47]. Studies have demonstrated that ECM hydrogels derived from different organs, such as heart [48,49], liver [54], skeletal muscles, and kidney [51], could serve as a proper cell culture platform to improve cell functionality and maturity. In our previous studies, liver ECM was used as a hydrogel [21] and patch [22] to improve the functionality of hepatic cells. The reconstruction of an ECM in a 3D culture of microscale cell spheroids is a challenging subject. The collagen-cell-sized MPs can be incorporated into the inner space of the cell spheroids and work as microenvironment regulators due to their physicochemical properties [55].

There is evidence that integrins have other functions besides acting as cell adhesion molecules, and it is likely that interactions between specific receptors and distinct ligand binding sites can convey unique information to the cells [56]. Guo and colleagues demonstrated that the endothelial cell matrix could upregulate the expression of *FOXA2*, *HNF4α*, and *PXR* and, thus, modulate the expression of several downstream functional targets, including CYP450 enzymes and transporters, promoting the functional and metabolic maturation of hepatic-like cells [57]. Therefore, in the present attempt to understand the liver ECM–hepatocytes cross-talk and the molecular mechanism through which it could promote the functionality and maturation of BLOs, we examined the expression of integrin α5 (*ITGa5*) and β1 (*ITGb1*), the genes of the ECM receptors in the cells, and their target genes (*FOXA2*, *HNF4α*, and *PXR*). We found that their expression significantly increased in the BLOs group compared to the LOs group. Therefore, we assumed that the effects of the liver-ECM-derived MPs on the functionality and maturation of the BLOs could be a result of α5β1 integrin and natural biomolecule upregulation in MP-mediated signaling pathways.

Despite the promising results, the present work still has to consider and address additional issues in future studies, with the aim to consolidate the relevance of the data presented here. The size of the organoids should be controlled to avoid impairing the quality of cells that are located in the core. Hence, in vitro culture conditions should be improved to limit the merging of LOs and BLOs that we noticed after two weeks in 3D static-suspension conditions. An advanced analysis of the molecular interactions between MPs and cells is mandatory to appraise the beneficial experimental conditions on both the morphology and physiology of the cells. Additional investigations into the bioengineered liver organoids matured subsequently to ex ovo transplantation should help in getting deep information on the microstructure and maturation aspects of the organoids as well as the quality of anastomosis. Finally, for in vivo transplantation studies, the engraftment level of those organoids should be evaluated in order to appraise if the effects noticed on the markers of the liver injury induced in the mouse model are related to the implanted differentiated cells or their paracrine activity.

To the best of our knowledge, this is the first study reporting the generation of cell-sized organ-specific MPs derived from the native ECM and their application in engineering an organoid microenvironment. We tried to recapitulate the original niche for HE by using MPs derived from liver ECM and a 3D coculture system to generate more functional and mature hepatocyte-like cells in a BLO manner. These liver-ECM-derived MPs also provide a scalable platform for the mass production of BLOs. Therefore, organoid engineering via the incorporation of cell-sized tissue-specific MPs within size-controlled multicellular aggregates provides a versatile strategy for the production of more functional organoids from different sources for biomedical applications.

Supplementary Materials: The following are available online at https://www.mdpi.com/article/10.3390/cells10061274/s1, Figure S1: Optimization of cell/MP and cell/microwell ratio for successful cell aggregate formation. The cell combination consisted of HE:HUVEC:MSC (10:7:2), Figure S2:

Fabrication of 3D basket by 3D printing, Figure S3: Determination of optimal conditions for differentiation of human embryonic stem cells into hepatic endoderm by gene expression analysis, Figure S4: Size distribution of the cell aggregates on different days post-co-culture, immunostaining and live and dead staining of LOs and BLOs, Figure S5: Gene expression profile in the LOs and BLOs derived from 7-day HE differentiated from induced pluripotent stem cells (iPSC4 cell line), Figure S6: Immunostaining of primary human hepatocyte and fibroblast and ICG assay of BLOs, Figure S7: Ex ovo implantation of the BLOs, Table S1: Optimization of cell/MP and cell/microwell ratios for successful cell aggregate formation, Table S2: Primer sequences used in qRT-PCR, Table S3: List of antibodies used for immunoflourescent staining, Table S4: Data for Log Rank Test to Compare Survival Curves, Table S5: Expected Numbers of Events in Each Group, Table S6: Total Observed and Expected Numbers of Observed in each Group.

Author Contributions: E.Z. performed all the experiments and drafted the manuscript. M.H.G., I.Z., M.H., and S.Y. helped in the bioengineering part, the concept and bioprinting, and reviewed the manuscript. Z.F., Z.H., F.M., A.O., and M.R. were involved in the bioassays and analysis and discussion of data. A.P., R.G., A.N., and M.N. helped in the conception and development of the proposal, the review of data and analysis, and the revision of the manuscript. H.B. and M.V. developed the idea, supported the project, and revised and approved the manuscript. All authors have read and agreed to the published version of the manuscript.

Funding: This project was financially supported by grants from the Royan Institute (96000182), Bahar Tashkhis Teb Co. (BTT, 9701, and 9801), and the National Institute for Medical Research Development (NIMAD, 940231 and 982403), Tehran, Iran, to M.V. and the Royan Institute (96000182), the Iran National Science Foundation (INSF, grant No.96001316), and the Ministry of Health and Medical Education (No. 56700/147) to H.B.

Institutional Review Board Statement: The study was conducted according to the guidelines of the "Royan Insti-tute Ethics Committee" (IR.ACECR.ROYAN.REC.1397.104).

Informed Consent Statement: Not applicable.

Data Availability Statement: All data supporting the findings of this study are available within thearticle and its supplemental data file or from the corresponding author upon reasonable request.

Acknowledgments: We would like to express our sincere gratitude to Hassan Ansari, Paria Poyan, and Mehdi Alikhani from the Royan Institute core facilities in the stem cell department and Poya Tavakol from the preclinical study department for their technical assistance.

Conflicts of Interest: The authors declare no conflict of interest.

Abbreviations

ALB: Albumin, ALT: Alanine Transaminase, AST: Aspartate Transaminase, bFGF: Basic Fibroblast Growth Factor, BLO: Bioengineered Liver Organoid, BSA: Bovine Serum Albumin, CYP: Cytochrome P450, CAM: Chorioallantoic Membrane, CCl_4: Carbon Tetrachloride, cDNA: complementary DNA, dECM: Decellularized Extra Cellular Matrix, DPBS: Dulbecco's Phosphate-Buffered Saline minus Ca^{2+} and Mg^{2+}, Dex: Dexamethasone, ECM: Extra Cellular Matrix, EGM: Endothelial Growth Medium, FTIR: Fourier-Transform Infrared Spectroscopy, FBS: Fetal Bovine Serum, FGF4: Fibroblast Growth Factor4, FITC: Fluorescein isothiocyanate, 5-FU: Fluorouracil, GPDH: Glyceraldehyde 3-phosphate Dehydrogenase, HE: Hepatic Endoderm, HGF: Hepatocyte Growth Factor, hESCs: Human Embryonic Stem Cells, hiPSCs: Human Induced Pluripotent Stem Cells, HBM: Hepatocyte Basal Medium, HUVECs: Human Umbilical Vein Endothelial Cells, ICG: Indocyanine Green, KOSR: Knockout Serum Replacement, LOs: Liver Organoids, Liver ECM: Liver Extra Cellular Matrix, Liver ECM-G: Liver ECM-gelatin, MPs: Microparticles, MSCs: Mesenchymal Stem Cells, NEAA: Nonessential Amino Acids:, OSM: Oncostatin M, PSCs: Pluripotent Stem Cells, PHH: Primary Human Hepatocyte, Pen/Step: Penicillin and Streptomycin, PAS: Periodic Acid Schiff, APTP: Paracetamol, PBS: Phosphate-Buffered Saline plus Ca^{2+} and Mg^{2+}, qRT-PCR: Quantitative Real-Time PCR, SEM: Scanning Electron Microscopy, TAM: Tamoxifen, w/o: Water-in-Oil.

References

1. Obiri-yeboa, I.; Mbundi, L.; Sumasondaram, M.; Sibbons, P.D.; Campus, R.F. Development and characterisation of a Porcine Liver Scaffold. for Clinical Application. *Stem Cell Dev.* **2019**, *44*, 1–32. [CrossRef]
2. Raschi, E.; Ponti, F. De Strategies for Early Prediction and Timely Recognition of Drug-Induced Liver Injury: The Case of Cyclin-Dependent Kinase 4 / 6 Inhibitors. *Front. Pharacology* **2019**, *10*, 1–15. [CrossRef]
3. Walker, P.A.; Ryder, S.; Lavado, A.; Dilworth, C.; Riley, R.J. The evolution of strategies to minimise the risk of human drug-induced liver injury (DILI) in drug discovery and development. *Arch. Toxicol.* **2020**, *94*, 2559–2585. [CrossRef]
4. Rowe, R.G.; Daley, G.Q. Induced pluripotent stem cells in disease modelling and drug discovery R. *Nat. Rev. Genet.* **2019**, *20*, 377–388. [CrossRef]
5. Hansel, M.C.; Davila, J.C.; Vosough, M.; Gramignoli, R.; Skvorak, K.J.; Dorko, K.; Marongiu, F.; Blake, W.; Strom, S.C. The Use of Induced Pluripotent Stem Cells for the Study and Treatment of Liver Diseases. *Curr. Protoc. Toxicol.* **2016**, *67*, 1–27. [CrossRef]
6. Vunjak-Novakovic, G.; Ronaldson-Bouchard, K. Organs on a chip: A fast-track for engineered human tissues in drug development. *Physiol. Behav.* **2016**, *176*, 139–148. [CrossRef]
7. Kitaeva, K.V.; Rutland, C.S.; Rizvanov, A.A.; Solovyeva, V.V. Cell Culture Based in vitro Test Systems for Anticancer Drug Screening. *Front. Bioeng. Biotechnol.* **2020**, *8*, 1–9. [CrossRef]
8. Langhans, S.A. Three-Dimensional in Vitro Cell Culture Models in Drug Discovery and Drug Repositioning. *Front. Pharmacol.* **2018**, *9*, 1–14. [CrossRef] [PubMed]
9. Cotovio, P.; Fernandes, T.G. Production of human pluripotent stem cell-derived hepatic cell lineages and liver organoids: Current status and potential applications. *Bioengineering* **2020**, *7*, 36. [CrossRef]
10. Zahmatkesh, E.; Vosough, M. A quick update from the past to current status of human pluripotent stem cell-derived Hepatocyte culture systems. *Mod. Med. Lab. J.* **2018**, *2*, 110–112. [CrossRef]
11. Akbari, S.; Arslan, N.; Senturk, S.; Erdal, E. Next-Generation Liver Medicine Using Organoid Models. *Front. Cell Dev. Biol.* **2019**, *7*, 1–15. [CrossRef] [PubMed]
12. Takebe, T.; Sekine, K.; Enomura, M.; Koike, H.; Kimura, M.; Ogaeri, T.; Zhang, R.; Ueno, Y.; Zheng, Y.; Koike, N.; et al. Vascularized and functional human liver from an iPSC-derived organ bud transplant. *Nature* **2013**, *499*, 481–484. [CrossRef] [PubMed]
13. Zhu, Y.; Shi, Q.; Peng, Q.; Gao, Y.; Yang, T.; Cheng, Y.; Wang, H.; Luo, Y.; Huang, A.; He, T.C.; et al. A simplified 3D liver microsphere tissue culture model for hepatic cell signaling and drug-induced hepatotoxicity studies. *Int. J. Mol. Med.* **2019**, *44*, 1653–1666. [CrossRef] [PubMed]
14. Wang, Z.; Li, W.; Jing, H.; Ding, M.; Fu, G.; Yuan, T.; Huang, W.; Dai, M.; Tang, D.; Zeng, M.; et al. Generation of hepatic spheroids using human hepatocyte-derived liver progenitor-like cells for hepatotoxicity screening. *Theranostics* **2019**, *9*, 6690–6705. [CrossRef] [PubMed]
15. Agmon, G.; Christman, K.L.; Diego, S. Controlling stem cell behavior with decellularized extracellular matrix scaffolds. *Curr. Opin. Solid State Mater. Sci.* **2017**, *20*, 193–201. [CrossRef] [PubMed]
16. Scottoni, F.; Crowley, C.; Fiadeiro, R.; Maghsoudlou, P.; Pellegata, A.F.; Mazzacuva, F.; Gjinovci, A.; Lyne, A.; Zulini, J.; Little, D.; et al. Mouse decellularised liver scaffold improves human embryonic and induced pluripotent stem cells differentiation into hepatocyte-like cells. *PLoS ONE* **2017**, *12*, 1–23. [CrossRef]
17. Jiang, W.; Cheng, Y.; Yen, M.; Chang, Y.; Yang, V.W.; Lee, O.K. Cryo-chemical decellularization of the whole liver for mesenchymal stem cells-based functional hepatic tissue engineering. *Biomaterials* **2015**, *35*, 3607–3617. [CrossRef] [PubMed]
18. Giobbe, G.G.; Crowley, C.; Luni, C.; Campinoti, S.; Khedr, M.; Kretzschmar, K.; De Santis, M.M.; Zambaiti, E.; Michielin, F.; Meran, L.; et al. Extracellular matrix hydrogel derived from decellularized tissues enables endodermal organoid culture. *Nat. Commun.* **2019**, *10*. [CrossRef] [PubMed]
19. Carolina, N.; Thomas, E.; Brown, J.; Affairs, V. Functional Maturation of Induced Pluripotent Stem Cell Hepatocytes in Extracellular Matrix — A Comparative Analysis of Bioartificial Liver Microenvironments. *Stem Cells Transl. Med.* **2016**, *5*, 1257–1267. [CrossRef]
20. Darakhshan, S.; Bidmeshki Pour, A.; Kowsari-Esfahan, R.; Vosough, M.; Montazeri, L.; Ghanian, M.H.; Baharvand, H.; Piryaei, A. Generation of Scalable Hepatic Micro-Tissues as a Platform for Toxicological Studies. *Tissue Eng. Regen. Med.* **2020**, *17*, 459–475. [CrossRef]
21. Saheli, M.; Sepantafar, M.; Pournasr, B.; Farzaneh, Z.; Vosough, M.; Piryaei, A.; Baharvand, H. Three-Dimensional Liver-derived Extracellular Matrix Hydrogel Promotes Liver Organoids Function. *J. Cell. Biochem.* **2018**, *119*, 1–34. [CrossRef] [PubMed]
22. Nobakht, F.; Saheli, M.; Farzaneh, Z.; Taheri, P.; Dorraj, M.; Baharvand, H.; Vosough, M.; Piryaei, A. Generation of Transplantable Three-Dimensional Hepatic- Patch to Improve the Functionality of Hepatic Cells in vitro and in vivo. *Stem Cell Dev.* **2019**, *29*, 1–30. [CrossRef] [PubMed]
23. Batista, P.; Castro, P.; Madureira, A.R.; Sarmento, B.; Pintado, M. Development and characterization of chitosan microparticles-in-films for buccal delivery of bioactive peptides. *Pharmaceuticals* **2019**, *12*, 32. [CrossRef] [PubMed]
24. McConnell, K.I.; Shamsudeen, S.; Meraz, I.M.; Mahadevan, T.S.; Ziemys, A.; Rees, P.; Summers, H.D.; Serda, R.E. Reduced cationic nanoparticle cytotoxicity based on serum masking of surface potential. *J. Biomed. Nanotechnol.* **2016**, *12*, 154–164. [CrossRef]
25. Zustiak, S.P.; Leach, J.B. Hydrolytically degradable poly(ethylene glycol) hydrogel scaffolds with tunable degradation and mechanical properties. *Biomacromolecules* **2010**, *11*, 1348–1357. [CrossRef]

26. Heidariyan, Z.; Ghanian, M.H.; Ashjari, M.; Farzaneh, Z.; Najarasl, M.; Larijani, M.R.; Piryaei, A.; Vosough, M.; Baharvand, H. Efficient and cost-effective generation of hepatocyte-like cells through microparticle-mediated delivery of growth factors in a 3D culture of human pluripotent stem cells. *Biomaterials* **2018**, *159*, 174–188. [CrossRef] [PubMed]
27. Takebe, T.; Zhang, R.; Koike, H.; Kimura, M.; Yoshizawa, E.; Enomura, M.; Koike, N.; Sekine, K.; Taniguchi, H. Generation of a vascularized and functional human liver from an iPSC-derived organ bud transplant. *Nat. Protoc.* **2014**, *9*, 396–409. [CrossRef]
28. Varzideh, F.; Pahlavan, S.; Ansari, H.; Halvaei, M.; Kostin, S. Human cardiomyocytes undergo enhanced maturation in embryonic stem cell-derived organoid transplants Fahimeh. *Biomaterials* **2018**, *192*, 537–550. [CrossRef]
29. Soltanian, A.; Mardpour, S.; Ghezelayagh, Z.; Baharvand, H. Generation of functional human pancreatic organoids by transplants of embryonic stem cell derivatives in a 3D - printed tissue trapper. *J. Cell. Physiol.* **2018**, 1–13. [CrossRef]
30. Cong, M.; Zhao, W.; Liu, T.; Wang, P.; Fan, X.; Zhai, Q.; Bao, X.; Zhang, D.; You, H.; Kisseleva, T.; et al. Protective effect of human serum amyloid P on CCl4-induced acute liver injury in mice. *Int. J. Mol. Med.* **2017**, *40*, 454–464. [CrossRef]
31. Farzaneh, Z.; Najarasl, M.; Abbasalizadeh, S.; Vosough, M. Developing a Cost-Effective and Scalable Production of Human Hepatic Competent Endoderm from Size-Controlled Pluripotent Stem Cell Aggregates. *Stem Cell Dev.* **2017**, *27*, 1–37. [CrossRef] [PubMed]
32. Pettinato, G.; Ramanathan, R.; Fisher, R.A.; Mangino, M.J.; Zhang, N.; Wen, X. Scalable Differentiation of Human iPSCs in a Multicellular Spheroid-based 3D Culture into Hepatocyte-like Cells through Direct Wnt/β-catenin Pathway Inhibition. *Sci. Rep.* **2016**, *6*, 1–17. [CrossRef] [PubMed]
33. Ahn, G.; Min, K.; Kim, C.; Lee, J.; Kang, D.; Won, J.; Cho, D.; Kim, J.; Jin, S.; Yun, W. Precise stacking of decellularized extracellular matrix based 3D cell-laden constructs by a 3D cell printing system equipped with heating modules. *Sci. Rep.* **2017**, 1–11. [CrossRef] [PubMed]
34. Seif-naraghi, S.B.; Horn, D.; Schup-magoffin, P.A.; Christman, K.L. Injectable extracellular matrix derived hydrogel provides a platform for enhanced retention and delivery of a heparin- binding growth factor. *Acta Biomater* **2013**, *8*, 3695–3703. [CrossRef]
35. Simunovic, M.; Brivanlou, A.H. Embryoids, organoids and gastruloids: New approaches to understanding embryogenesis. *Development* **2017**, *144*, 976–985. [CrossRef]
36. Peng, W.; Datta, P.; Wu, Y.; Dey, M.; Ayan, B.; Dababneh, A.; Ozbolat, I.T. Challenges in Bio-fabrication of Organoid Cultures. *Adv. Exp. Med. Biol.* **2018**, *1107*, 53–71.
37. Rossi, E.A.; Quintanilha, L.F.; Kymie, C.; Nonaka, V.; Solano, B.; Souza, D.F. Advances in Hepatic Tissue Bioengineering with Decellularized Liver Bioscaffold. *Stem Cell Int.* **2019**, *10*, 1–13. [CrossRef]
38. Bi, H.; Ming, L.; Cheng, R.; Luo, H.; Zhang, Y.; Jin, Y. Liver extracellular matrix promotes BM-MSCs hepatic differentiation and reversal of liver fi brosis through activation of integrin pathway. *J. Tissue Eng. Regen. Med.* **2016**, *11*, 2685–2698. [CrossRef]
39. Zahmatkesh, E.; Khoshdel-Rad, N.; Mirzaei, H.; Shpichka, A.; Timashev, P.; Mahmoudi, T.; Vosough, M. Evolution of organoid technology: Lessons learnt in Co-Culture systems from developmental biology. *Dev. Biol.* **2021**, *21*, 0012–0056. [CrossRef]
40. Rijal, G. Native-mimicking in vitro microenvironment: An elusive and seductive future for tumor modeling and tissue engineering. *J. Biol. Eng.* **2018**, *12*, 1–22. [CrossRef]
41. Chaudhuri, R.; Ramachandran, M.; Moharil, P.; Harumalani, M.; Jaiswal, A.K. Biomaterials and cells for cardiac tissue engineering: Current choices: Current Choices. *Mater. Sci. Eng. C* **2017**, *79*, 950–957. [CrossRef] [PubMed]
42. Nell, P. Highlight report: Spheroids from stem cell-derived hepatocyte-like cells. *Arch. Toxicol.* **2018**, *92*, 3603–3604. [CrossRef]
43. Takebe, T.; Enomura, M.; Yoshizawa, E.; Kimura, M.; Koike, H.; Ueno, H.; Matsuzaki, T.; Yamazaki, T.; Toyohara, T.; Osafune, K.; et al. Vascularized and Complex Organ Buds from Diverse Short Article Vascularized and Complex Organ Buds from Diverse Tissues via Mesenchymal Cell-Driven Condensation. *Stem Cell* **2015**, *16*, 556–565. [CrossRef]
44. Pettinato, G.; Lehoux, S.; Ramanathan, R.; Salem, M.M.; He, L.; Muse, O.; Flaumenhaft, R.; Thompson, M.T.; Rouse, E.A.; Cummings, R.D.; et al. Generation of fully functional hepatocyte-like organoids from human induced pluripotent stem cells mixed with endothelial Cells. *Sci. Rep.* **2019**, *9*, 1–21. [CrossRef] [PubMed]
45. Deng, J.; Wei, W.; Chen, Z.; Lin, B.; Zhao, W. Engineered Liver-On-A-Chip Platform to Mimic Liver Functions and Its Biomedical Applications: A Review. *micromachines* **2019**, *10*, 676. [CrossRef] [PubMed]
46. Legate, K.R.; Wickström, S.A.; Fässler, R. Genetic and cell biological analysis of integrin outside-in signaling. *Genes Dev.* **2009**, *23*, 397–418. [CrossRef] [PubMed]
47. Jain, E.; Damania, A.; Kumar, A. Biomaterials for liver tissue engineering. *Hepatol. Int.* **2014**, *8*, 185–197. [CrossRef] [PubMed]
48. Ungerleider, J.L.; Johnson, T.D.; Rao, N.; Christman, K.L. Fabrication and characterization of injectable hydrogels derived from decellularized skeletal and cardiac muscle. *Methods* **2016**, *84*, 53–59. [CrossRef]
49. Leong, J.Y.; Amir, S.; Patel, R.R. Evidence for mechanisms underlying the functional benefits of a myocardial matrix hydrogel for post-MI treatment. *J. Am. Coll. Cardiol.* **2017**, *176*, 139–148. [CrossRef]
50. Miranda, C.; Fernandes, T.G.; Diogo, M.M.; Cabral, J.M.S. Towards Multi-Organoid Systems for Drug Screening Applications. *Bioeng.* **2018**, *5*, 49. [CrossRef]
51. Su, J.; Satchell, S.C.; Shah, R.N.; Wertheim, J.A. Kidney decellularized extracellular matrix hydrogels: Rheological characterization and human glomerular endothelial cell response to encapsulation. *J. Biomed. Mater. Res. A* **2019**, *106*, 2448–2462. [CrossRef] [PubMed]
52. Khoshdel-rad, N.; Zahmatkesh, E.; Moeinvaziri, F. Promoting maturation of human pluripotent stem cell-derived renal micro-tissue by incorporation of endothelial and. *Stem Cell Dev.* **2021**, 1–36. [CrossRef]

53. Sainio, A.; Järveläinen, H. Extracellular matrix-cell interactions: Focus on therapeutic applications. *Cell. Signal.* **2020**, *66*, 109487. [CrossRef]
54. Loneker, A.E.; Faulk, D.M.; Hussey, G.S.; Amore, A.D.; Badylak, S.F. Solubilized liver extracellular matrix maintains primary rat hepatocyte phenotype in-vitro. *Soc. biomatherials* **2016**, 957–965. [CrossRef] [PubMed]
55. Yamada, M.; Hori, A.; Sugaya, S.; Yajima, Y.; Utoh, R.; Yamato, M.; Seki, M. Cell-sized condensed collagen microparticles for preparing microengineered composite spheroids of primary hepatocytes. *Lab Chip* **2015**, *15*, 3941–3951. [CrossRef] [PubMed]
56. Kechagia, J.Z.; Ivaska, J. Integrins as biomechanical sensors of the microenvironment. *Nat. Rev. Mol. Cell Biol.* **2019**, *20*, 457–473. [CrossRef] [PubMed]
57. Guo, X.; Li, W.; Ma, M.; Lu, X.; Zhang, H. Endothelial cell-derived matrix promotes the metabolic functional maturation of hepatocyte via integrin-Src signalling. *J. Cell. Mol. Med.* **2017**, *21*, 2809–2822. [CrossRef] [PubMed]

cells MDPI

Review

Potential of Fibrin Glue and Mesenchymal Stem Cells (MSCs) to Regenerate Nerve Injuries: A Systematic Review

Adriana de Cássia Ortiz [1], Simone Ortiz Moura Fideles [1], Karina Torres Pomini [1,2], Márcia Zilioli Bellini [3], Eliana de Souza Bastos Mazuqueli Pereira [2], Carlos Henrique Bertoni Reis [1,4], João Paulo Galletti Pilon [4,5], Miguel Ângelo de Marchi [6], Beatriz Flavia de Moraes Trazzi [7], Willian Saranholi da Silva [8], Marcelo Rodrigues da Cunha [9,10], Daniela Vieira Buchaim [2,11] and Rogerio Leone Buchaim [1,12,*]

1 Department of Biological Sciences, Bauru School of Dentistry (FOB/USP), University of São Paulo, Bauru 17012-901, Brazil; adrianaortiz@usp.br (A.d.C.O.); simoneortiz@usp.br (S.O.M.F.); karinatp@usp.br (K.T.P.); dr.carloshenriquereis@usp.br (C.H.B.R.)
2 Postgraduate Program in Structural and Functional Interactions in Rehabilitation, Postgraduate Department, University of Marilia (UNIMAR), Marília 17525-902, Brazil; elianabastos@unimar.br (E.d.S.B.M.P.); danibuchaim@alumni.usp.br (D.V.B.)
3 Pro-Rectory of Research and Graduate Studies, University Center of Adamantina (UniFAI), Adamantina 17800-000, Brazil; mzbellini@fai.com.br
4 UNIMAR Beneficent Hospital (HBU), Medical School, University of Marilia (UNIMAR), Marília 17525-160, Brazil; joao.pilon@unesp.br
5 Postgraduate Program in Speech Therapy, Sao Paulo State University (UNESP—Univ Estadual Paulista), Marília 17525-900, Brazil
6 Coordination of the Medical School, University Center of Adamantina (UniFAI), Adamantina 17800-000, Brazil; coordmedicina@fai.com.br
7 Coordination of the Dentistry School, University of Marilia (UNIMAR), Marília 17525-902, Brazil; flavia.odonto@unimar.br
8 Dentistry School, University of Marilia (UNIMAR), Marília 17525-902, Brazil; williansaranholi@unimar.br
9 Interunit Postgraduate Program in Bioengineering (EESC/FMRP/IQSC), University of São Paulo (USP), São Carlos 13566-590, Brazil; cunhamr@hotmail.com
10 Department of Morphology and Pathology, Jundiaí Medical School, Jundiaí 13202-550, Brazil
11 Teaching and Research Coordination of the Medical School, University Center of Adamantina (UniFAI), Adamantina 17800-000, Brazil
12 Graduate Program in Anatomy of Domestic and Wild Animals, Faculty of Veterinary Medicine and Animal Science, University of São Paulo (FMVZ/USP), São Paulo 05508-270, Brazil
* Correspondence: rogerio@fob.usp.br; Tel.: +55-14-3235-8220

Citation: Ortiz, A.d.C.; Fideles, S.O.M.; Pomini, K.T.; Bellini, M.Z.; Pereira, E.d.S.B.M.; Reis, C.H.B.; Pilon, J.P.G.; de Marchi, M.Â.; Trazzi, B.F.d.M.; da Silva, W.S.; et al. Potential of Fibrin Glue and Mesenchymal Stem Cells (MSCs) to Regenerate Nerve Injuries: A Systematic Review. *Cells* **2022**, *11*, 221. https://doi.org/10.3390/cells11020221

Academic Editor: Mehdi Najar

Received: 3 December 2021
Accepted: 20 December 2021
Published: 10 January 2022

Abstract: Cell-based therapy is a promising treatment to favor tissue healing through less invasive strategies. Mesenchymal stem cells (MSCs) highlighted as potential candidates due to their angiogenic, anti-apoptotic and immunomodulatory properties, in addition to their ability to differentiate into several specialized cell lines. Cells can be carried through a biological delivery system, such as fibrin glue, which acts as a temporary matrix that favors cell-matrix interactions and allows local and paracrine functions of MSCs. Thus, the aim of this systematic review was to evaluate the potential of fibrin glue combined with MSCs in nerve regeneration. The bibliographic search was performed in the PubMed/MEDLINE, Web of Science and Embase databases, using the descriptors ("fibrin sealant" OR "fibrin glue") AND "stem cells" AND "nerve regeneration", considering articles published until 2021. To compose this review, 13 in vivo studies were selected, according to the eligibility criteria. MSCs favored axonal regeneration, remyelination of nerve fibers, as well as promoted an increase in the number of myelinated fibers, myelin sheath thickness, number of axons and expression of growth factors, with significant improvement in motor function recovery. This systematic review showed clear evidence that fibrin glue combined with MSCs has the potential to regenerate nervous system lesions.

Keywords: nerve regeneration; fibrin glue; fibrin sealant; scaffolds; stem cells; systematic review

1. Introduction

The nervous system is a highly specialized tissue, responsible for the functional sensory and motor activity of the organism. As other tissues, the nervous system is susceptible to trauma by mechanical, electrical and thermal action, as well as by ischemic compression or drug injection [1,2]. The severity of the sequelae and the tissue recovery ability depends on the nature and extent of the injury. Physiologically, the nervous system has a spontaneous self-repair capacity, with a regenerative rate of 1 to 3 mm per day [2]. However, adequate functional recovery of this tissue is still a major challenge, especially in large-scale injuries [2,3]. Thus, in most cases, therapeutic and surgical interventions are essential to encourage the regeneration of the nervous system [2,4]. Although, the late or inadequate interventions can cause several disorders, such as chronic pain, neuropathies, muscle atrophy, paralysis and severe functional incapacity with compromised quality of life [2,5].

Several strategies have been proposed to favor the regeneration of nervous tissue injuries, such as epineural suture, autogenous nerve grafts, allografts, nerve conduits, cell therapy and biological molecules, such as growth factors [5]. All these strategies have advantages and disadvantages, with limited potential to regenerate the damaged nerve [5]. Among the available techniques, autogenous nerve grafts are considered the gold standard, especially in cases of large nerve gaps [1,3–9]. Autogenous nerve grafts constitute a biological scaffold that contains native cells, neurotrophic growth factors, blood vessels, extracellular matrix proteins, adhesion molecules and neuronal cells, such as Schwann cells [5–8]. However, some disadvantages are associated with autogenous grafts, such as sensory-motor loss, morbidity, scarring and the possibility of neuroma formation in the donor area, in addition to the limited supply of donor nerve [4,7].

Allografts, in turn, provide availability of material and avoid the lack of donor área [5]. Despite the absence of viable cells, allografts maintain a native matrix that supports axonal growth, cell migration and angiogenesis [4,10]. Although, the required immunosuppression and inherent complications are disadvantages of allografts [3,4,7,8]. Nerve conduits are an option for the treatment of peripheral nerve injuries [5]. Conduits are tubular structures that surround the ends of the nerve and favor the direction of axonal growth [4]. The conduits can be constructed of synthetic materials, biodegradable or of natural components, such as veins, arteries and tendons. Nonetheless, an ideal conduit must have some fundamental properties, such as biocompatibility, biodegradability, low immunogenicity, mechanical strength and adequate permeability to allow the exchange of nutrients and biological molecules [1]. Additionally, the success of this technique depends on the formation of a fibrin matrix in the conduit lumen, which will support vascular infiltration and cell migration [5]. Furthermore, the use of conduits should only be indicated for the treatment of small nerve gaps (<3 cm) [1,3–5,7].

In addition to these strategies, cell therapy emerges as an interesting alternative for treating nervous tissue injuries. Cellular components play an important role in overall tissue healing.In the same way, nerve regeneration is a complex process that requires the interaction of various types of cells, extracellular matrix components, blood vessels, cytokines and growth factors [1]. Besides that, it is also critical that an endogenous fibrin matrix forms at the injury site in order to support axonal growth, vascular infiltration and immune system cell migration [4,8]. Nerve regeneration process involves a sequence of steps. After the injury, morphological and molecular changes occur in nerve cells, transport of proteins and influx of ions, such as calcium [4]. These events signal a disturbance in the homeostasis of the microenvironment that leads to changes in cell metabolism, with consequent activation of several signaling pathways and regulation of gene expression [4]. Schwann cells, macrophages and fibroblasts act in nerve regeneration from the initial stages. Schwann cells act together with macrophages in the Wallerian degeneration process, phagocytizing fragmented axons and myelin debris [4].

Throughout the regenerative process, the cells present in the microenvironment, mainly Schwann cells, overexpress neurotrophins and others growth factors involved with neuronal regeneration. Neurotrophins include nerve growth factor (NGF), brain derived neurotrophic factor (BDNF), neurotrophin-3 (NT-3) and neurotrophin-4/5 (NT-4/5), which are responsible for neuron growth and survival [8,11]. Other factors stimulate cell proliferation, differentiation and survival, such as ciliary neurotrophic factor (CNTF), glial cell line-derived growth factor (GDNF) and fibroblast growth factor (FGF) [8]. Among them, NGF is the most commonly characterized growth factor to assess the neurotrophic potential of cells [11]. Thus, nerve regeneration process is marked by overexpression of genes related to the inflammatory response, angiogenesis, synthesis of neurotrophic factors and expression of proteins involved in axonal growth and nerve fiber myelination [4]. Considering these issues, Schwann cells constitutes one of the most promising strategies to favor the regeneration of nervous tissue [5,12–17].

However, certain limitations make it difficult to use Schwann cells, such as limited availability of donor tissue and difficulties related to the cultivation and isolation of these cells [4,5]. As a result of these limitations, other cell lines have been investigated for use in regenerative therapy, such as embryonic stem cells (ESCs), induced pluripotent stem cells (iPSCs) and mesenchymal stem cells (MSCs) [5,12,17]. Among the candidate cells, MSCs stand out as a viable alternative for use in regenerative medicine, considering the ethical and genetic manipulation issues that limit the use of ESCs and iPSCs, respectively [12,17].

MSCs are undifferentiated cells that can be obtained from different sources, such as bone marrow, periosteum, adipose tissue, skin, muscle, tendons, umbilical cord, peripheral circulation and dental tissue [18,19]. Due to their adherence to plastic, MSCs can be expanded in culture and characterized by the expression of specific surface antigens, such as CD29, CD73, CD90 and CD109 [17,18,20–22]. MSCs have several biological properties that favor tissue regeneration. MSCs have the capacity for self-renewal, proliferation and differentiation in different cell lines, depending on the stimulus from the microenvironment [17,18,20,21]. Under neural induction, MSCs can differentiate into neuronal cells and express neurotrophic growth factors, such as nerve growth factor (NGF) and brain-derived neurotrophic factor (BDNF) [22–26]. MSCs also have anti-apoptotic, angiogenic and immunomodulatory potential, and act by regulating the production of inflammatory cytokines through several signaling pathways [19–21,25,27].

Modulation of the inflammatory response is essencial to limit tissue destruction, to reduce the formation of fibrous scars and to favor the regeneration of the injured area [12]. Additionally, MSCs express angiogenic factors, such as vascular endothelial growth factor (VEGF), which contribute to vascular neoformation [22,28,29]. MSCs are still able to migrate to the site of injury and recruit other cells through paracrine mechanisms [21].In nervous system injuries, MSCs can be used to favor axonal regeneration and myelinization of the myelin sheath [22–24,26,30]. Together, these characteristics make MSCs promising candidates for use in cell therapy strategies.

For use in cell therapy strategies, MSCs can be harvested from autologous tissues and implanted at the injury site through a cell delivery vehicle, such as fibrin glue or fibrin sealant. In regenerative medicine, fibrin glue can be used as a delivery system for drugs, biomolecules, growth factors and cells [31–34]. Fibrin is a natural polymer formed by the combination of thrombin and fibrinogen, which are components of the blood coagulation system [33,35,36]. Fibrin glue had its use approved by the FDA and since 1976 it has been widely used as a hemostatic agent to treat coagulopathies and in several surgical specialties, such as neurological, gastrointestinal and cardiovascular surgeries, among others [31,35,37]. In medical practice, fibrin glue can be obtained from autologous blood components, which reduces the risk of immune reactions [31,35,38].

Fibrin glue has several other interesting biological properties for use in regenerative therapies. Fibrin is a biocompatible matrix that can be naturally degraded by the action of fibrinolytic enzymes, such as metalloproteinases [33,39–41]. When used as a cell delivery vehicle, fibrin glue has the advantage of allowing a uniform distribution of cells in the matrix and it can also be injected into the lesion area through less invasive procedures [34,41]. Furthermore, fibrin glue provides a bioactive matrix that favors cell adhesion, viability, proliferation and differentiation [33,34,42]. The fibrin glue also contains numerous binding sites with cells, molecules and growth factors that act in the tissue regeneration process [34,41]. Thus, the fibrin glue favors cell-matrix interactions and supports axonal growth. Additionally, the porosity of the fibrin matrix favors angiogenesis and vascular infiltration, which are essential for the restoration of the injured area [34,41].

The viability of growing MSCs in fibrin glue has been reported in several studies [23,24,26,30,40,43]. Kalbermatten et al. (2008) reported that fibrin glue improved the adhesion of MSCs and Schwann cells in bioresorbable poly-3-hydroxybutyrate nerve conduits [43]. This study showed that cells seeded in fibrin glue were optimally distributed along the conduit, better than those seeded in growth medium [43]. In the study by Gardin et al. (2011), adult stem cells organized in neurospheres and seeded in fibrin glue meshes were able to grow and differentiate into glial/neuron cells under neural induction, without any chromosomal alteration [30]. Likewise, Park et al. (2012) showed that MSCs were able to differentiate into neuronal cells, to exhibit neuron-like cell morphology and to express various neural markers and transcription factors [23] (Figure 1).

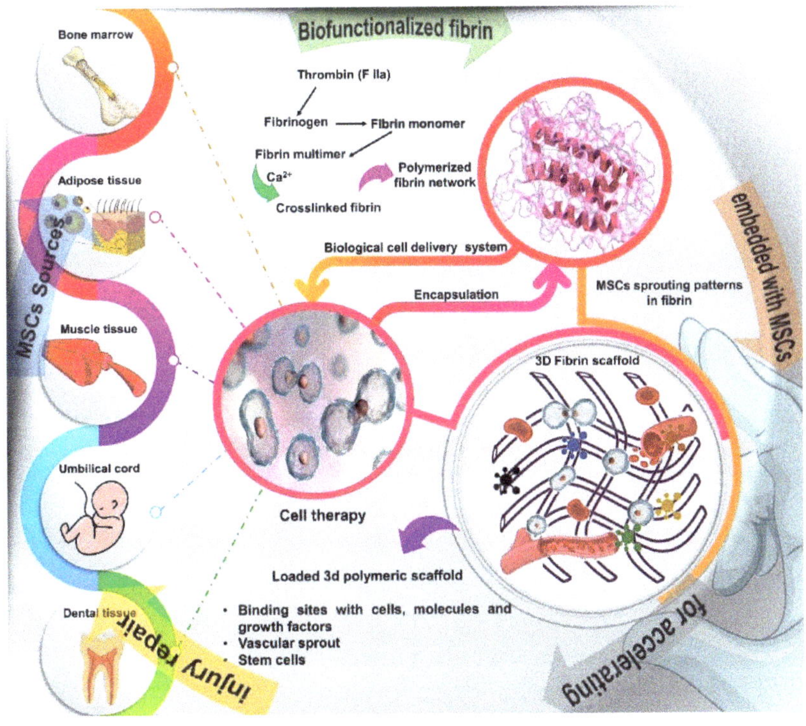

Figure 1. Schematic overview of the different sources of stem cells, such as bone marrow, adipose tissue, muscle, tendons, umbilical cord, and dental tissue and description of how to obtain the fibrin network, a natural polymer formed by the combination of thrombin and fibrinogen, which are components of the blood coagulation system. Fibrin glue provides a 3D bioactive matrix that favors cell adhesion, viability, proliferation and differentiation, in addition to containing numerous binding sites with cells, molecules and growth factors that act in the tissue regeneration process.

Further studies also demonstrated that MSCs cultivated in fibrin glue were able to differentiate into neuronal cells and to express neurogenic marker proteins, such as microtubule-associated protein 2 (MAP2), nerve growth factor (NGF), brain-derived neurotrophic factor (BDNF) and Tau protein [24,26]. The compatibility and bioactivity of MSCs in fibrin glue was also reported by Krug et al. (2018) [40]. This study showed that the fibrin glue favored adhesion, migration, proliferation and metabolic activity of MSCs in culture [40]. These authors also reported that MSCs were uniformly distributed in the fibrin glue and strongly interacted with matrix components [40]. Another relevant data addressed in this study refers to the fact that, after 14 days of culture, fibrin degradation was accompanied by overexpression of metalloproteinases by MSCs, which could favor axonal growth and regeneration in clinical situations [40].

Thus, considering the several biological properties of MSCs and fibrin matrix, this systematic literature review aimed to investigate the potential of therapy with MSCs associated with fibrin glue on the regeneration of the central or peripheral nervous system.

2. Materials and Methods

2.1. Study Design and Bibliographic Search Strategy

This systematic review was conducted according to the PRISMA guidelines and the PICO strategy (Patient, Intervention, Comparison and Outcomes). The electronic bibliographic search was carried out in August–September/2021 in the PubMed/MEDLINE, Embase and Web of Science databases, combining the descriptors ("fibrin sealant" OR "fibrin glue") AND "stem cells" AND "nerve regeneration". The design of this review involved studies published up to the year 2021 that evaluated the effect of therapy with MSCs associated with fibrin glue on central and peripheral nervous system regeneration.

2.2. Study Eligibility

The eligibility criteria involved studies that compared the effect of fibrin glue alone or associated with MSCs as regenerative therapy for the treatment of nerve injuries, with or without the use of nerve conduits. Studies that used different analysis methods were considered. In vitro studies and literature reviews were not included within the eligibility criteria. Studies that used only differentiated or genetically modified MSCs to express growth factors were excluded from this review. Likewise, studies that evaluated the regenerative effect of therapy with MSCs and fibrin glue, through generalized exposure of experimental groups to immunosuppressive or others drugs, were also excluded.

For the eligibility of the experiments included in Tables 1 and 2 for detailed analysis, manuscripts with methodology and main results and their conclusions on neuroregeneration were also considered, in relation to morphophysiological and functional reparative aspects. Furthermore, the quality of the strategy for selecting studies included in this review was confirmed by the analysis of independent reviewers. The selection of studies was carried out carefully following the eligibility criteria in order to minimize bias.

3. Results

The results of the bibliographic research are outlined in the Prism Flow Diagram (Figure 2). The search strategy showed 32 articles in the PubMed/MEDLINE database, of which 19 were excluded because they were outside the eligibility criteria. In the Web of Science and Embase databases, 29 and 17 articles were found, respectively, which were not included due to duplicity or because they were outside the eligibility criteria. We selected 13 studies from the PubMed/MEDLINE database to compose this review.

The literature search strategy resulted in the compilation of studies with animal models that used MSCs and fibrin glue to treat peripheral nerve and spinal cord injuries. Tables 1 and 2 summarizes the main results of the studies included in this review, containing in its columns the manuscript reference, the original source of the stem cells, the treatment groups used in each of the studies, the location of implantation of the stem cells, which were the analyzes performed in the experiment and, in the last column, the main results and conclusions. Table 1 shows the experiments with peripheral nerve injuries and Table 2 shows central nerve injuries.

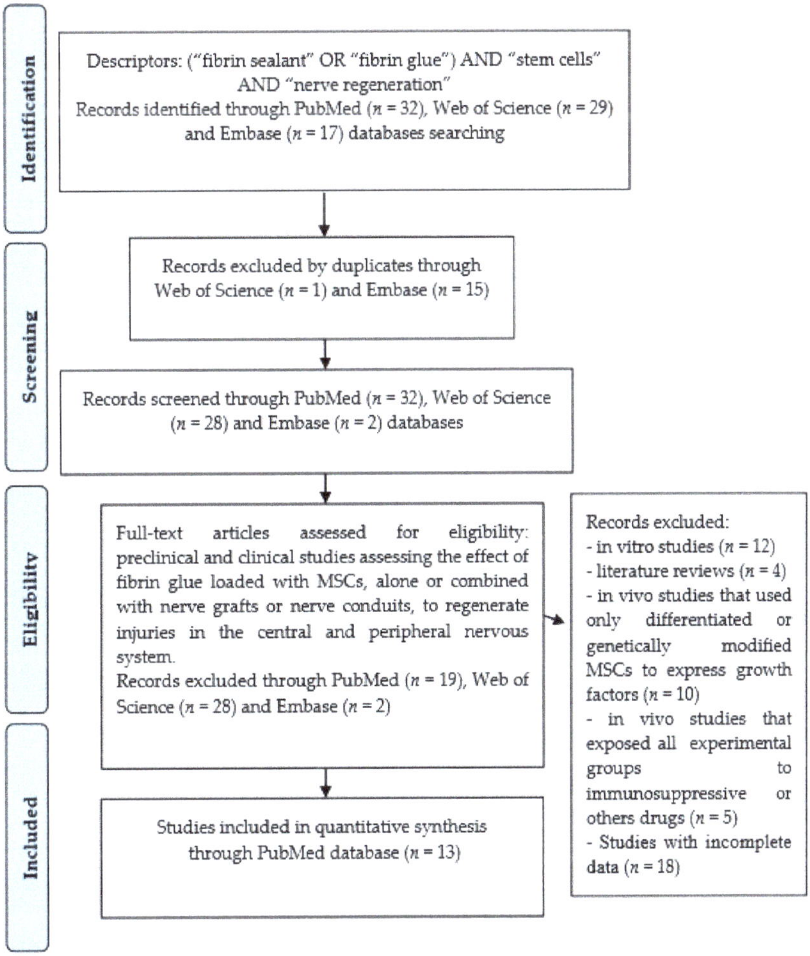

Figure 2. Prisma flow diagram resulted from an electronic bibliographic search for the potential of MSCs associated with fibrin glue in central and peripheral nervous system regeneration.

In the studies included in this review, therapy with MSCs and fibrin glue was used to treat sciatic [24,26,44–51], femoral [23], mandibular [52] and spinal cord nerve injuries [53]. MSCs were harvested from several sources, such as bone marrow from the tibia/femurs [26,44,45], iliac crest [47], adipose tissue [46,51], aminiotic fluid [48,50,54], fetal brain [52,53], dental pulp [24] and skin [23]. Among studies, the effect of therapy on nerve regeneration was evaluated by one or more methods, such as physical examination, electromyography, electrophysiology, immunohistochemistry, immunofluorescence, enzyme-linked immunosorbent and tunnel assays, histology and morphometric analysis.

Table 1. Studies selected according to eligibility criteria, in peripheral nerve injuries, in chronological order.

Reference	Stem Cells Source	Treatment Groups Delivery System	Intervention Implantation Site	Analysis	Main Outcomes Conclusions
Pan et al. (2006) [50]	Allogeneic MSCs (amniotic fluid)	G1: Fibrin glue (FG) + Surgicel® G2: FG + MSCs + Surgicel®	Rats sciatic nerve injuries were sutured, the gap (5 mm) was filled with MSCs + FG + Surgicel or FG + Surgicel (*n* = 10). Analyses were performed after 8 weeks of the procedures.	Electrophysiological and immunohistochemical analysis.	MSCs + FG + Surgicel® significantly improved muscle action potential amplitude and axonal growth. Compound muscle action potential values for G1 and G2 were 28.5 ± 1.3% and 42.5 ± 1.25%, respectively. Amniotic MSCs significantly improved nerve regeneration in a sciatic nerve gap.
Pan et al. (2007) [48]	Allogeneic rats MSCs (amniotic fluid)	G1: Fibrin glue (FG) + Surgicel® G2: FG + MSCs + Surgicel®	Rats sciatic nerve injuries (crush site) were sutured and wrapped with FG (*n* = 20) or FG + MSCs (*n* = 30). Analyzes were performed after 4 weeks of the procedures.	Enzyme-linked immunosorbent assay (ELISA), immunocytochemistry, motor function, electrophysiology, histology and immunocytochemistry	FG + MSCs significantly improved the motor function recovery, reduced the fibrosis at the injury site and favored the expression of neurotrophic factors, such as CNTF and NT-3.Compound muscle action potential values for G1 and G2 were 27.8 ± 4.22% and 67 ± 6.98%, with conduction latency of 3.91 ± 0.303 and 1.33 ± 0.048 msec, respectively. MSCs favored the sciatic nerve regeneration after crush injury.
Pan et al. (2009) [54]	Human amniotic fluid MSCs	G1: Fibrin glue (FG) + Surgicel® G2: FG + Surgicel® + Natto G3: MSCs + FG + Surgicel® G4: MSCs + FG + Surgicel® + Natto	Rats sciatic nerve injuries (crush site) were sutured and wrapped with FG or FG + MSCs (*n* = 6). G2/G4 were fed with Natto extract (16 mg/day) for 7 days. Analyzes were performed after 7 and 28 days of the procedures.	Electrophysiological, immunohistochemical, histological, cell apoptosis (Tunel assay) and pro-inflammatory cytokines analysis.	MSCs + FG + Surgicel® promoted better nerve regeneration than the FG + Surgicel® alone. Compound muscle action potential values for G1, G2, G3 and G4 were 0.25 ± 0.04%, 0.47 ± 0.03%, 0.51 ± 0.02% and 0.68 ± 0.02%, with conduction latency of 3.92 ± 0.31%, 1.85 ± 0.07%, 1.84 ± 0.08%, and 1.38 ± 0.11% respectively. Natto, alone or combined with MSCs, reduced cell apoptosis and proinflammatory cytokines levels, such as *TNF-α* and *IL-1β*. Combined treatment of MSCs+Natto showed the most beneficial effects.

Table 1. Cont.

Reference	Stem Cells Source	Treatment Groups Delivery System	Intervention Implantation Site	Analysis	Main Outcomes Conclusions
McGrath et al. (2012) [47]	Human BM-MSCs (iliac crest)	G1:fibrin conduit (FC) + fibrin matrix (FG, Tisseel®) G2: FC + FG + cyclosporine G3: FC + FG + BM-MSCs G4: FC + FG + BM-MSCs + cyclosporine	Implantation of the fibrin conduit (14 mm) in rats sciatic nerve injuries (gap 10 mm). The conduct was sutured and filled with FG containing BM-MSCs. G2/G4 received daily intraperitoneally injections of cyclosporine A. Analyses were performed after 3 weeks of the procedures.	Immunohistochemistry	FC + FG + MSCs + cyclosporine was the most effective treatment to increase axonal regeneration and to reduce the macrophage-mediated inflammatory response.
Park et al. (2012) [23]	Autologous porcine SMSCs (ear skin)	G1: Collagen membrane (Lyoplant®) + fibrin glue (FG) + SMSCs G2: FG	Implantation of the collagen membrane in porcine femoral nerve injuries (gap 10 mm). The membrane was sutured and filled with FG containing SMSCs (n = 4). Analyses were performed after 2 and 4 weeks of the procedures.	Immunohistochemical and histological analysis.	G1 showed remarkable nerve regeneration with complete nerve bundles and higher expression of S-100 protein and p75 nerve growth factor. Autologous SMSCs and fibrin glue may act as an available substitute for nerve conduit material in peripheral nerve defect sites.
Zhao et al. (2014) [26]	Allogeneic rats BM-MSCs (femurs/tibias)	G1: Autograft G2: Acellular allograft (AL) + Fibrin glue (FG) + BM-MSCs (MSCs injected inside the graft) G3: AL + FG + BM-MSCs (MSCs injected around the graft) G4: AL + FG	Implantation of the graft in rats sciatic nerve injuries (gap 12 mm), suture and injection of FG containing BM-MSCs into/around the graft. Analyzes were performed after 2 weeks (n = 5) and 12 weeks (n = 8) of the procedures.	Muscle weight, histological, histomorphometric, sensory and motor functional analysis.	G1, G2 and G3 groups showed greater potential for nerve regeneration with greater axonal growth and myelination of nerve fibers. The graft implant with FG + BM-MSCs is successful in maintaining nerve structure and may support nerve regeneration.

Table 1. *Cont.*

Reference	Stem Cells Source	Treatment Groups Delivery System	Intervention Implantation Site	Analysis	Main Outcomes Conclusions
Kurwale et al. (2015) [45]	Allogeneic rats BM-MNCs (femurs/tibias)	G1: Fibrin glue (FG) (Tisseel®) G2: FG + BM-MNCs	Rats sciatic nerve injuries were microsutured, the gap (2 mm) was filled with BM-MNCs and covered with FG (n = 5, per group/period). Analyses were performed after 15 and 60 days of the procedures.	Histological, immunohistochemical and morphometric analysis	FG + BM-MNCs showed a significant increase in axon diameter, nerve and myelin thickness at the repair site and at the distal most sites on early and late phases, respectively. Transplantation of BM-MNCs to the site of peripheral nerve injury leads to a significantly better recovery.
Reichenberger et al. (2015) [51]	Allogeneic rats AD-MSCs (inguinal tissue)	G1: Fibrin glue (FG) (Beriplast®) G2:FG + AD-MSCs	Sciatic nerve injuries of 50 rats were microsutured and the gap was filled with FG or FG containing AD-MSCs. Analyses were performed after 7, 21, 35 and 63 days of the procedures.	Immunofluorescence, muscle weight, histological and histomorphometric analysis.	FG + AD-MSCs showed better nerve regeneration with significant increase in muscle weight, axon number and myelin sheath thickness. AD-MSCs significantly enhanced the regeneration of peripheral nerve injuries after primary coaptation.
Ullah et al. (2017) [24]	Human dental pSCs (pulp-derived stem cells)	G1: Sham (n = 4) G2: Collagen membrane (Lyoplant®) + fibrin glue (FG) G3:Lyoplant® + FG + pSCs G4:Lyoplant® + FG + DpSCs	Implantation of the collagen membrane in rats sciatic nerve injuries (gap 5 mm). The membrane was sutured and filled with FG containing pSCs or differentiated neuronal cells (DpSCs) (n = 6/per group). Analyzes were performed after 12 weeks of the procedures.	Muscle contraction activity, immunohistochemical and histological analysis.	G3 and G4 showed considerable and similar regenerative potential, better muscle activity and greater expression of specific markers for angiogenesis, axonal fiber and myelin sheath, such as VEGFR-1, GFAP, S-100 protein, MBP-2 and p75NGFR. pSCs could exhibit excellent peripheral nerve regeneration potential.
Goel et al. (2019) [44]	Allogeneic rats BM-MNCs (femurs/tibias)	G1: Fibrin glue (FG) (Tisseel®) G2:FG + BM-MNCs	Rats' sciatic nerve injuries were microsutured, filled with BM-MNCs and covered with FG (n = 5, per group/period). Analyses were performed after 30 and 60 days of the procedures.	Histological analysis	FG + BM-MNCs presented better axonal regeneration and remyelination with greater density of myelinated fibers. Local delivery of BM-MNCs improved peripheral nerve regeneration.

Table 1. *Cont.*

Reference	Stem Cells Source	Treatment Groups Delivery System	Intervention Implantation Site	Analysis	Main Outcomes Conclusions
Masgutov et al. (2019) [46]	Allogeneic rats AD-MSCs (inguinal tissue)	G1:autologous nerve graft (AG) + fibrin glue (FG) + AD-MSCs Tissucol-Kit® G2: AG + FG G3: AG G4: Intact animals	Rats' sciatic nerve injuries (gap 5 mm) were filled with AG, sutured and covered with FG containing AD-MSCs (n = 5). Analyses were performed after 30 and 60 days of the procedures.	Functional motor test, electromyography, EasyLDI laser Doppler and morphological analysis	AG + FG + AD-MSCs increased the number of myelinated fibers and neurons of the L5 spinal ganglion, with improved motor activity and angiogenesis. AG + FG + AD-MSCs resulted in motor function recovery after injury.
Bayir et al. (2021) [52]	Allogeneic rats NMSCs (fetal brain)	G1: No treatment G2: NMSCs G3: Fibrin glue (FG) Tisseel® G4: NMSCs + FG	Injection of NMSCs or NMSCs + FG in rats mandibular facial nerve injuries(crush). Analyses were performed immediately after surgery and after 3, 5 and 8 weeks of the procedures.	Physical examination, TUNEL assay and histochemical analysis.	NMSCs + FG showed a statistically significant functional improvement. NMSCs + FG may play a promising role as adjuvant regenerative therapy in traumatic facial paralysis.

Table 2. Studies selected according to eligibility criteria, in central nerve injuries.

Reference	Stem Cells Source	Treatment Groups Delivery System	Intervention Implantation Site	Analysis	Main Outcomes Conclusions
Pan et al. (2008) [53]	Allogeneic rats NSCs (fetal brain)	G1: Fibrin glue (FG) G2: FG + granulocyte colony-stimulating factor (G-CSF) G3: FG + NSCs G4: FG + NSCs + G-CSF	Injection of NSCs into the rats spinal cord (T8-T9) gap (2 mm) and sealing with FG + gelfoam. Subcutaneous injection of *G-CSF* for 5 days in G2/G3 groups (n = 10). Analyses were performed after 3 months of the procedures.	Electrophysiological, hind-limb motor function, histological and immunohisto-chemical analysis.	FG + NSCs + G-CSF significantly improved clinical motor activity, conduction latency and spinal cord regeneration scores. Motor evoked potential values for G1, G2, G3 and G4 were 24.69 ± 3.51, 31.64 ± 3.06, 38.97 ± 2.30 and 47.7 ± 3.17 mV, with conduction latency of 1.54 ± 0.04, 1.39 ± 0.03, 1.39 ± 0.04 and 1.29 ± 0.02 msec, respectively. Therapy associated with NSCs and C-CSF promoted better spinal cord regeneration.

Overall, the studies showed that fibrin glue combined with MSCs significantly favored nerve regeneration when compared to the isolated use of fibrin glue. Therapy with MSCs favored axonal regeneration and remyelination of nerve fibers, as well as increased myelinated fiber thickness, axon number and expression of neurotrophic factors. Additionally, treatment with MSCs improved muscle weight and motor function recovery, while reducing fibrosis at the injury site. In the study by Pan et al. (2006), treatment with fibrin glue and MSCs significantly improved the compound muscle action potential (42.5 ± 1.25%)

compared to the isolated use of fibrin glue (28.5 ± 1.3%) [50]. Likewise, Pan et al. (2007) reported that groups treated with fibrin glue alone or associated with MSCs presented compound muscle action potential of 27.8 ± 4.22% and 67 ± 6.98%, with conduction latency of 3.91 ± 0.303 and 1.33 ± 0.048 msec, respectively [48]. When used with autologous nerve graft, fibrin glue with adipose MSCs (AD-MSCs) promoted a significant increase in the number of myelinated fibers with improved motor activity and angiogenesis, compared to the use of autologous graft alone or associated with fibrin glue without cells [46].

Likewise, the use of acellular allograft associated with fibrin glue and bone marrow MSCs (BM-MSCs) presented regenerative potential similar to the isolated use of autologous graft and superior to the use of allograft combined with fibrin glue without cells [26]. Studies that administered agents with anti-inflammatory or immunomodulatory properties, such as Natto extract [54], granulocyte colony-stimulating factor (G-CSF) [53] and cyclosporine A [47], concurrently with therapy with fibrin glue and MSCs, reported an additional beneficial effect on certain parameters. These studies showed a reduction in cell apoptosis, in the inflammatory response mediated by macrophages and in the levels of pro-inflammatory cytokines, such as TNF-α and IL-1β, in addition to an improvement in axonal regeneration and functional motor activity.

Nerve regeneration was also achieved in studies that enveloped the resected nerve ends with absorbable bovine collagen dura mater (Lyoplant®) [23,24] or fibrin conduit [47] and implanted fibrin glue with MSCs into the biodegradable nerve tubule. Finally, the study that compared the effect of implantation of fibrin glue containing undifferentiated or differentiated MSCs in neuronal cells related that nerve regeneration was not affected by the cellular differentiation stage, as the two types of cells presented considerable and similar regenerative potential [24], exhibiting greater expression of specific markers for angiogenesis, axonal fiber and myelin sheath, such as vascular endothelial growth factor receptor-1 (VEGFR-1), glial fibrillary acid protein (GFAP), S-100 protein, myelin basic protein-2 (MBP-2) and p75 nerve growth factor receptor (p75NGFR).

4. Discussion

The studies included in this systematic review used animal models to assess the regenerative potential of therapy with fibrin glue and MSCs in the treatment of nervous system injuries. Of the studies included in this review, six of them compared the effect of fibrin glue alone or associated with MSCs to treat sciatic [44,45,48,50,51] and mandibular nerve injuries [52]. The outcomes of these studies showed that the association of fibrin glue with MSCs significantly favored axonal regeneration and myelination of nerve fibers compared to the use of fibrin glue without cells [44,45,48,50–52]. Additionally, treatment with fibrin glue and MSCs resulted in a significant increase in the levels of neurotrophic factors [48] and less formation of scar tissue at the site of nerve injury [46,48], which accelerated reinnervation and recovery of functional activity [46]. Alterations in the levels of neurotrophic growth factors are indicators of nerve regeneration process, since low levels of these factors are found in intact peripheral nerves, whereas both the expression of mRNA and the synthesis of the corresponding proteins significantly increase when the nerve is transsected or crushed [55].

In the study by Pan et al. (2007), ELISA measurement and immunocytochemical analyzes revealed increased levels of ciliary neurotrophic factor (CNTF) and neurotrophin-3 (NT-3) in the regenerating sciatic nerve and positive staining for these factors in transplanted MSCs, respectively [48]. Kurwale et al. (2015) detected changes in the microstructure of the injured sciatic nerve in the early postoperative phase (15 days), so that treatment with fibrin glue and MSCs promoted a significant increase in the mean diameter of the axon, in the mean nerve thickness and in the myelin thickness at the repair site compared to the isolated use of fibrin glue [45]. Similar results were obtained by Goel et al. (2019), who reported that sciatic nerve regeneration was more significant in the group treated with fibrin glue and MSCs at 30 and 60 days after surgery, while degenerative changes, such as ballooning of axons and degeneration of the sheath of myelin, were more prominent in the group treated only with fibrin glue [44].

Nervous system regeneration is a complex biological process and the isolated use of fibrin glue is not sufficient to promote significant regeneration. The regenerative process takes place through the interaction of cells, matrix components, growth factors and cytokines. When implanted at the site of nerve injury, MSCs are able to differentiate into neuronal cells and express neurotrophic factors, which are essential for axonal growth and nerve fiber remyelination [23,24,26]. In addition, MSCs express angiogenic factors and several other growth factors that, through various signaling pathways, act on cells present in the tissue, such as fibroblasts, macrophages and endothelial cells, modulating the inflammatory response, reducing scar formation and favoring vascularization of the regenerating nerve [56,57].

It is important to emphasize, however, that an adequate regeneration of the nervous system depends on the correct growth orientation of the injured axons. Thus, analyzses of postoperative motor function are important to assess whether there was an adequate regeneration of the injured nerve extremities [55]. Considering these issues, in addition to histological analyses, some studies performed physical [52], electrophysiological [48,50] and muscle weight [51] examinations to analyze the recovery of sensory and motor functions of the injured nerve. These analyses showed that therapy with fibrin glue and MSCs significantly improved functional activity [52], muscle action potential amplitude and motor function recovery [48,50], with significant increase in muscle weight [51] and satisfactory nerve regeneration even in gap of 5 mm [50].

In the study by Pan et al. (2006), electrophysiological analyzes showed that the group treated with fibrin glue and MSCs had an average action potential of 42.5%, differing statistically from the group treated only with fibrin glue (28.5%) [50]. The outcomes of these studies showed that the interactions that MSCs established in the damaged tissue microenvironment contributed not only to favor axonal growth, but to modulate the inflammatory response and reduce the formation of scar tissue, facilitating the reorganization of nerve fibers and orientation of axonal growth. Another relevant issue is that, in these studies, therapy with fibrin glue and MSCs had the potential to regenerate both crush and transection injuries, whereas axons regenerate and interconnect more accurately in crush injuries, in which often Schwann cells are maintained viable, rather than by transection lesions [55].

The regenerative potential of MSCs was also observed in studies that used nerve conduits to guide axonal growth [23,24,47]. The implantation of conduits that act as guide tubules is a strategy used to support and adequately direct axonal growth in peripheral nerve injuries [56]. In transsection injuries in which the nerve fiber is disruption, the use of conduits favors the correct orientation of growth between the extremities of the injured nerve, in addition to reducing fibroblast infiltration and the formation of scar tissue [56]. The success of this technique can be improved by filling the conduit with a bioactive matrix that provides binding sites with cells, molecules and growth factors, considering that nerve regeneration requires the formation of an endogenous supporting matrix [56]. In the three studies included in this review that used nerve conduits, the implant of fibrin glue in the conduit achieved even better results when associated with MSCs [23,24,47].

Thus, McGrath et al. (2012) obtained better axonal regeneration of the sciatic nerve in the groups treated with MSCs [47]. Corroborating these data, Ullah et al. (2017) obtained better muscle activity, greater expression of specific markers for angiogenesis, axonal fiber and myelin sheath in groups treated with MSCs associated with fibrin glue to regenerate sciatic nerve [24]. Additionally, this study also evaluated the rate of cellular apoptosis at the site of injury and found that fibrin glue minimized the effect of cytokines on implanted MSCs, keeping these cells viable in the damaged area [24]. Despite the inherent regenerative properties of MSCs, fibrin glue also played an important role in the tissue regeneration process, providing a bioactive matrix that supported the MSCs to perform their functions. Thus, these studies showed that fibrin glue loaded with MSCs can act as a biological substitute for filling nerve conduits in peripheral nerve injury regeneration strategies [23]. In addition to the filling components, the biological composition of the conduit is one of the factors that influence the success of this technique.

Some of the studies in this review used biodegradable bovine collagen duramater (Lyoplant®) filled with fibrin glue and MSCs to regenerate gaps of 5 mm [24] and 10 mm [23] in the sciatic and femoral nerves, respectively. After 4 weeks, histological and immunohistochemical analyses showed remarkable nerve regeneration with complete nerve bundles in gaps of 10 mm with higher expression of S-100 protein and p75 nerve growth factor (p75NGFR) [23]. Collagen conduits allow the transport of nutrients and growth factors to the area of injury, favoring nerve regeneration. In addition to being semipermeable, collagen is a natural component of the extracellular matrix, which favors cell proliferation and migration [1,56]. Similar outcomes were also obtained by McGrath et al. (2012), who used fibrin conduit filled with fibrin glue and MSCs to regenerate nerve injury with a gap of 10 mm [47]. Like collagen, fibrin is an excellent biomaterial capable of supporting axonal growth, in addition to having a porous structure that allows the passage of molecules into the conduit [11,22,56]. Additionally, fibrin reduces the formation of scar tissue, favoring nerve regeneration [11,56].

However, although the use of biocompatible and biodegradable conduits constitutes an interesting strategy to direct growth and favor axonal regeneration, autogenous grafts are still the gold standard, especially in lesions with extensive loss of nervous tissue, with gaps greater than 3 cm [57]. Two studies included in this review investigated the regenerative potential of fibrin glue and MSCs associated with allograft or autologous nerve graft [26,46]. The study by Masgutov et al. (2019) used autologous nerve graft, fibrin glue and MSCs to treat a gap of 5 mm in a sciatic peripheral nerve [46]. In this study, the use of autologous nerve graft associated with fibrin glue and MSCs (AG + FG + AD-MSCs) resulted in better motor function recovery after injury compared to the use of autologous nerve graft isolated (AG) or combined with fibrin glue (AG + FG) [46]. Sixty days after injury, AG + FG + AD-MSCs showed a significant increase of 26% and 28% in motor activity when compared to the AG and AG + FG groups, respectively [46]. Additionally, AG + FG + AD-MSCs promoted an increase in the number of myelinated fibers and improved angiogenesis [46]. Laser Doppler analysis also showed that the vascular supply at the lesion site was reestablished after 14 days post-operatively in the AG + FG + AD-MSCs group, at levels similar to those found in intact animals [46]. However, no vascular difference between the different groups was found after 30 days [46]. Additionally, after 30 days, AG + FG + AD-MSCs showed a significant increase around 18% in the number of myelinated fibers in the distal segment of the nerve compared to AG + FG [46]. Both groups showed an additional increase of 20% in the number of myelinated fibers after 60 days postoperatively [46].

Corroborating these data, Zhao et al. (2014) showed that the allogeneic graft associated with fibrin glue and MSCs had a similar potential to the isolated use of autogenous graft to regenerate a gap of 12 mm in the peripheral sciatic nerve [26]. In this study, three months after the injury, both groups had greater regenerative potential with greater axonal growth and myelination of nerve fibers [26].

Some studies in this review investigated whether the regenerative potential of fibrin glue and MSCs could be improved by administration of drugs or extracts with an immunoregulatory or anti-inflammatory effects [47,49,53]. Pan et al. (2008) used harvested fetal brain stem cells (NSCs) to regenerate a gap of 2 mm in the spinal cord [53]. The cells were implanted and the lesion area was sealed with fibrin glue (FG) and gelfoam [53]. Two experimental groups also received subcutaneous injection of granulocyte colony-stimulating factor (G-CSF) for 5 days [53]. G-CSF has immunoregulatory, anti-inflammatory and anti-apoptotic effects on neuronal cells. Three months after the injury, the analyses showed that the therapy associating FG + NSCs + G-CSF promoted better spinal cord regeneration and improved clinical motor activity, conduction latency and spinal cord regeneration scores [53].

Posteriorly, these authors investigated the effects of Natto extract administration on the regenerative potential of therapy with fibrin glue and MSCs (FG + MSCs + Surgicel®) to treat sciatic nerve crush and showed that, 28 days after the injury, FG + MSCs + Surgicel® promoted better nerve regeneration than the FG + Surgicel® alone [49]. Additionally, animals fed with Natto extract had a lower rate of cellular apoptosis and lower levels of pro-inflammatory cytokines at the site of injury [49]. Nerve crush injury induces an increase in the number of vacuoles and a reduction in the expression of the S-100 protein, which is related to the nerve fiber myelination process [49]. Thus, the therapy combined with Natto extract showed other additional benefits, reducing the number of vacuoles at the injury site and increasing the expression of S-100 protein [49].

The study by McGrath et al. (2012) that used fibrin conduit and implantation of fibrin glue and MSCs to regenerate a gap of 10 mm in the sciatic nerve obtained better results in the groups in which intraperitoneal injections of cyclosporin A were administered [47]. Cyclosporine A is an immunosuppressive drug that has neuroprotective potential. The groups treated with FG + MSCs + cyclosporine showed better axon regeneration with reduced inflammatory response mediated by macrophages [47]. Cyclosporin A regulates the expression of matrix metalloproteinases, which are responsible for accelerating the degradation of the fibrin matrix. Thus, groups treated with cyclosporine showed faster degradation of fibrin glue and less macrophage infiltration [47]. The administration of drugs as a complementary therapy to surgical interventions for the treatment of injuries in the nervous system can have beneficial effects. However, the risks-benefits and possible side effects associated with drug therapies must be considered, especially with regard to the use of immunosuppressants [4,5]. Furthermore, research must evolve to assess the effect and safety of drug therapies, considering that there is a lack of clinical evidence about the best therapy to treat injuries in the nervous system [5].

Regardless of the technique used, the main outcomes of these in vivo studies showed that therapy with fibrin glue and MSCs had considerable potential to regenerate lesions in the nervous system. The implantation of fibrin glue and MSCs reduced degenerative alterations, macrophage infiltration, destruction of myelin fibers and necrosis of small blood vessels [46]. Regarding MSCs, some important issues must be addressed, such as the tissue of origin and the stage of cell differentiation. In the studies included in this review, MSCs were harvested from different tissues, such as bone marrow, adipose tissue, amniotic fluid, fetal brain, skin and dental pulp. It is important to report that MSCs tend to keep peculiar characteristics of their tissue of origin, which can influence their regenerative properties [18,58,59] (Figure 3).

However, considering the main outcomes of the studies in this review, there was no deficit in terms of nerve tissue regeneration resulting from differences between the MSCs obtained from different sources. Regarding the stage of cell differentiation, only one of the studies included in this review compared the regenerative potential of undifferentiated MSCs and MSCs differentiated in neuronal cells [24]. In this study, the two groups of cells showed considerable and similar potential to regenerate sciatic nerve damage. These cells showed greater expression of specific markers for angiogenesis, axonal fiber and myelin sheath, in addition to better muscle activity, compared to the group treated only with collagen membrane and fibrin glue. Corroborating these data, the regenerative potential of MSCs differentiated into neuronal cells has been reported in several studies with animal models [13–16]. Therefore, there is evidence in the literature that therapy with MSCs has the potential to favor nerve regeneration, regardless of the stage of cell differentiation [60]. Some review studies have demonstrated the potential of stem cells in different areas of neuroregeneration, for example, in the treatment of traumatic brain injury [61,62].

Mesenchymal/stromal stem cells (MSCs) have translational potential in regenerative medicine, moving from researchers' benches to clinical application, and have aroused interest as cell-based therapies for a variety of autoimmune and inflammatory diseases [63]. Furthermore, MSCs are environmentally responsive and they are able to adapt their behavior according to tissue challenges [64].

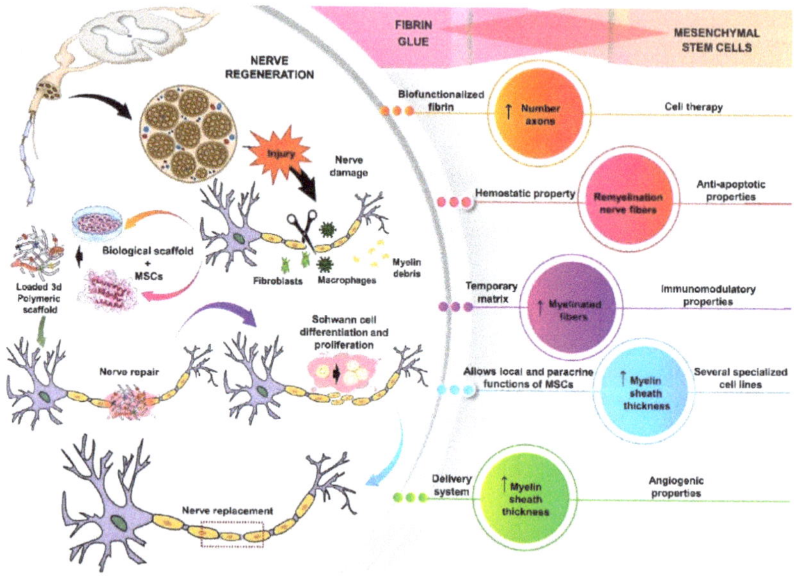

Figure 3. Schematic illustration of central or peripheral nervous system regeneration through cell therapy associated with a 3D polymeric matrix of fibrin. Furthermore, it explains the intrinsic characteristics of each, a proportionally synergistic effect when associated. The mesenchymal stem has angiogenic, anti-apoptotic and immunomodulatory properties, in addition to its ability to differentiate into several specialized cell lines. Fibrin glue has hemostatic properties and acts as a temporary matrix that favors cell-matrix interactions and allows local and paracrine functions of MSCs. In the nervous repair process, events occur that signal a disturbance in the homeostasis of the microenvironment, which leads to changes in cell metabolism, with consequent activation of several signaling pathways and regulation of gene expression. At this stage, the MSCs differentiate into Schwann cells, they overexpress neurotrophins and other growth factors involved in neuronal regeneration and the binding sites in the fibrin network provide interaction with macrophages and fibroblasts, aiding Wallerian degeneration, phagocytizing fragmented axons and myelin debris and sequentially regeneration nervous.

In summary, the methods of analysis of the different studies in this review provided relevant information about the effect of therapy with fibrin glue and MSCs, not only with regard to the histological aspects of the treated area, but also concerning the quality of the regeneration of the injured nerve, evaluated by the recovery of functional motor activity. In these studies, additional assays provided data about the alterations in the levels of neurotrophic growth factors, expression of specific markers of angiogenesis and pro-inflammatory cytokines at the site of injury. Taken together, these results showed that therapy with fibrin glue and MSCs had considerable regenerative potential and could be an advantageous strategy for treating nervous system injuries. Despite the promising results of these in vivo studies, research must advance in the field of clinical trials in order to assess the potential of cell-based therapy to regenerate nerve lesions with different patterns of severity.

5. Conclusions

The combined use of fibrin glue and MSCs has been used as a less invasive strategy to regenerate lesions in different types of tissues. In the nervous system, fibrin glue loaded with MSCs had significant potential to regenerate transection or crush injuries in peripheral nerves or spinal cord. Therapy with fibrin glue and MSCs favored axonal regeneration and remyelination of nerve fibers, as well as increased the number of myelinated fibers, myelin sheath thickness, number of axons and the expression of neurotrophic and angiogenic factors, with improved motor function recovery. Fibrin glue and MSCs associated with nerve grafts or immunoregulatory/anti-inflammatory drug therapies showed an additional beneficial effect on nerve regeneration. Likewise, the use of nerve conduits with implant of fibrin glue and MSCs also had considerable potential to regenerate peripheral nerve injuries.

In short, this systematic review based on in vivo studies showed clear evidence that therapy with fibrin glue and MSCs has the potential to regenerate nervous system damage. However, advances in research are still required to investigate the clinical efficacy of therapy based on combined use of fibrin glue and MSCs as a strategy for treating nervous system injuries, taking into account the regeneration of the damaged area and the recovery of functional motor activity.

Author Contributions: Conceptualization, A.d.C.O., S.O.M.F. and R.L.B.; methodology, S.O.M.F., A.d.C.O. and K.T.P.; formal analysis, S.O.M.F., A.d.C.O., C.H.B.R. and M.Z.B.; investigation, S.O.M.F., A.d.C.O. and J.P.G.P.; data curation, M.Â.d.M. and M.R.d.C.; writing—original draft preparation, S.O.M.F. and A.d.C.O.; writing—review and editing, S.O.M.F., A.d.C.O. and R.L.B.; visualization, B.F.d.M.T., W.S.d.S. and E.d.S.B.M.P.; supervision, R.L.B. and D.V.B. All authors have read and agreed to the published version of the manuscript.

Funding: This study was financed in part by the Coordenação de Aperfeiçoamento de Pessoal de Nível Superior—Brasil (CAPES)—Finance Code 001.

Institutional Review Board Statement: Not applicable.

Informed Consent Statement: Not applicable.

Data Availability Statement: Not applicable.

Acknowledgments: We are grateful to Isabela Maria dos Santos, a scholarship holder of the Unified Scholarship Program at the University of São Paulo for collaborating in the preliminary phase of the study.

Conflicts of Interest: The authors declare no conflict of interest.

References

1. Wang, Y.; Zhang, Y.; Li, X.; Zhang, Q. The progress of biomaterials in peripheral nerve repair and regeneration. *J. Neurorestoratol.* **2020**, *8*, 252–269. [CrossRef]
2. Yi, S.; Xu, L.; Gu, X. Scaffolds for peripheral nerve repair and reconstruction. *Exp. Neurol.* **2019**, *319*, 112761. [CrossRef]
3. Siemionow, M.; Brzezicki, G. Chapter 8: Current techniques and concepts in peripheral nerve repair. *Int. Rev. Neurobiol.* **2009**, *87*, 141–172. [CrossRef]
4. Pan, D.; Mackinnon, S.E.; Wood, M.D. Advances in the repair of segmental nerve injuries and trends in reconstruction. *Muscle Nerve* **2020**, *61*, 726–739. [CrossRef] [PubMed]
5. Fernandez, L.; Komatsu, D.E.; Gurevich, M.; Hurst, L.C. Emerging strategies on adjuvant therapies for nerve recovery. *J. Hand Surg. Am.* **2018**, *43*, 368–373. [CrossRef]
6. Isaacs, J. Major peripheral nerve injuries. *Hand Clin.* **2013**, *29*, 371–382. [CrossRef] [PubMed]
7. Ray, W.Z.; Mackinnon, S.E. Nerve problems in the lower extremity. *Foot Ankle Clin.* **2011**, *16*, 243–254. [CrossRef] [PubMed]
8. Rivlin, M.; Sheikh, E.; Isaac, R.; Beredjiklian, P.K. The role of nerve allografts and conduits for nerve injuries. *Hand Clin.* **2010**, *26*, 435–446. [CrossRef]
9. Sinis, N.; Kraus, A.; Papagiannoulis, N.; Werdin, F.; Schittenhelm, J.; Meyermann, R.; Haerle, M.; Geuna, S.; Schaller, H.E. Concepts and developments in peripheral nerve surgery. *Clin. Neuropathol.* **2009**, *28*, 247–262.
10. Safa, B.; Buncke, G. Autograft substitutes: Conduits and processed nerve allografts. *Hand Clin.* **2016**, *32*, 127–140. [CrossRef]
11. Schmidt, C.E.; Leach, J.B. Neural tissue engineering: Strategies for repair and regeneration. *Annu. Rev. Biomed. Eng.* **2003**, *5*, 293–347. [CrossRef] [PubMed]
12. Assinck, P.; Duncan, G.J.; Hilton, B.J.; Plemel, J.R.; Tetzlaff, W. Cell transplantation therapy for spinal cord injury. *Nat. Neurosci.* **2017**, *20*, 637–647. [CrossRef]
13. di Summa, P.G.; Kingham, P.J.; Raffoul, W.; Wiberg, M.; Terenghi, G.; Kalbermatten, D.F. Adipose-derived stem cells enhance peripheral nerve regeneration. *J. Plast. Reconstr. Aesthet. Surg.* **2010**, *63*, 1544–1552. [CrossRef] [PubMed]
14. Di Summa, P.G.; Schiraldi, L.; Cherubino, M.; Oranges, C.M.; Kalbermatten, D.F.; Raffoul, W.; Madduri, S. Adipose derived stem cells reduce fibrosis and promote nerve regeneration in rats. *Anat. Rec.* **2018**, *301*, 1714–1721. [CrossRef] [PubMed]
15. Schaakxs, D.; Kalbermatten, D.F.; Pralong, E.; Raffoul, W.; Wiberg, M.; Kingham, P.J. Poly-3-hydroxybutyrate strips seeded with regenerative cells are effective promoters of peripheral nerve repair. *J. Tissue Eng. Regen. Med.* **2017**, *11*, 812–821. [CrossRef]

16. Schiraldi, L.; Sottaz, L.; Madduri, S.; Campisi, C.; Oranges, C.M.; Raffoul, W.; Kalbermatten, D.F.; di Summa, P.G. Split-sciatic nerve surgery: A new microsurgical model in experimental nerve repair. *J. Plast. Reconstr. Aesthet. Surg.* **2018**, *71*, 557–565. [CrossRef] [PubMed]
17. Wang, Y.; Pan, J.; Wang, D.; Liu, J. The use of stem cells in neural regeneration: A review of current opinion. *Curr. Stem Cell Res.* **2018**, *13*, 608–617. [CrossRef]
18. Brown, C.; McKee, C.; Bakshi, S.; Walker, K.; Hakman, E.; Halassy, S.; Svinarich, D.; Dodds, R.; Govind, C.K.; Chaudhry, G.R. Mesenchymal stem cells: Cell therapy and regeneration potential. *J. Tissue Eng. Regen. Med.* **2019**, *13*, 1738–1755. [CrossRef]
19. Murphy, M.B.; Moncivais, K.; Caplan, A.I. Mesenchymal stem cells: Environmentally responsive therapeutics for regenerative medicine. *Exp. Mol. Med.* **2013**, *45*, e54. [CrossRef]
20. Han, Y.; Li, X.; Zhang, Y.; Han, Y.; Chang, F.; Ding, J. Mesenchymal stem cells for regenerative medicine. *Cells* **2019**, *8*, 886. [CrossRef]
21. Squillaro, T.; Peluso, G.; Galderisi, U. Clinical trials with mesenchymal stem cells: An update. *Cell Transplant.* **2016**, *25*, 829–848. [CrossRef]
22. Wu, X.; Ren, J.; Li, J. Fibrin glue as the cell-delivery vehicle for mesenchymal stromal cells in regenerative medicine. *Cytotherapy* **2012**, *14*, 555–562. [CrossRef]
23. Park, B.W.; Kang, D.H.; Kang, E.J.; Byun, J.H.; Lee, J.S.; Maeng, G.H.; Rho, G.J. Peripheral nerve regeneration using autologous porcine skin-derived mesenchymal stem cells. *J. Tissue Eng. Regen. Med.* **2012**, *6*, 113–124. [CrossRef]
24. Ullah, I.; Park, J.M.; Kang, Y.H.; Byun, J.H.; Kim, D.G.; Kim, J.H.; Kang, D.H.; Rho, G.J.; Park, B.W. Transplantation of human dental pulp-derived stem cells or differentiated neuronal cells from human dental pulp-derived stem cells identically enhances regeneration of the injured peripheral nerve. *Stem Cells Dev.* **2017**, *26*, 1247–1257. [CrossRef]
25. Yousefi, F.; Arab, F.L.; Nikkhah, K.; Amiri, H.; Mahmoudi, M. Novel approaches using mesenchymal stem cells for curing peripheral nerve injuries. *Life Sci.* **2019**, *221*, 99–108. [CrossRef] [PubMed]
26. Zhao, Z.; Wang, Y.; Peng, J.; Ren, Z.; Zhang, L.; Guo, Q.; Xu, W.; Lu, S. Improvement in nerve regeneration through a decellularized nerve graft by supplementation with bone marrow stromal cells in fibrin. *Cell Transplant.* **2014**, *23*, 97–110. [CrossRef] [PubMed]
27. Xiao, L.; Tsutsui, T. Human dental mesenchymal stem cells and neural regeneration. *Hum. Cell* **2013**, *26*, 91–96. [CrossRef]
28. Chun, S.Y.; Lim, J.O.; Lee, E.H.; Han, M.H.; Ha, Y.S.; Lee, J.N.; Kim, B.S.; Park, M.J.; Yeo, M.G.; Jung, B.; et al. Preparation and Characterization of Human Adipose Tissue-Derived Extracellular Matrix, Growth Factors, and Stem Cells: A Concise Review. *Tissue Eng. Regen. Med.* **2019**, *16*, 385–393. [CrossRef]
29. Man, A.J.; Kujawski, G.; Burns, T.S.; Miller, E.N.; Fierro, F.A.; Leach, J.K.; Bannerman, P. Neurogenic potential of engineered mesenchymal stem cells overexpressing VEGF. *Cell Mol. Bioeng.* **2016**, *9*, 96–106. [CrossRef]
30. Gardin, C.; Vindigni, V.; Bressan, E.; Ferroni, L.; Nalesso, E.; Puppa, A.D.; D'Avella, D.; Lops, D.; Pinton, P.; Zavan, B. Hyaluronan and fibrin biomaterial as scaffolds for neuronal differentiation of adult stem cells derived from adipose tissue and skin. *Int. J. Mol. Sci.* **2011**, *12*, 6749–6764. [CrossRef] [PubMed]
31. Ahmed, T.A.; Dare, E.V.; Hincke, M. Fibrin: A versatile scaffold for tissue engineering applications. *Tissue Eng. Part B Rev.* **2008**, *14*, 199–215. [CrossRef] [PubMed]
32. Breen, A.; O'Brien, T.; Pandit, A. Fibrin as a delivery system for therapeutic drugs and biomolecules. *Tissue Eng. Part B Rev.* **2009**, *15*, 201–214. [CrossRef] [PubMed]
33. Brower, J.; Blumberg, S.; Carroll, E.; Pastar, I.; Brem, H.; Chen, W. Mesenchymal stem cell therapy and delivery systems in nonhealing wounds. *Adv. Skin Wound Care* **2011**, *24*, 524–532. [CrossRef] [PubMed]
34. Heher, P.; Mühleder, S.; Mittermayr, R.; Redl, H.; Slezak, P. Fibrin-based delivery strategies for acute and chronic wound healing. *Adv. Drug Deliv. Rev.* **2018**, *129*, 134–147. [CrossRef] [PubMed]
35. Montana, M.; Tabélé, C.; Curti, C.; Terme, T.; Rathelot, P.; Gensollen, S.; Vanelle, P. Organic glues or fibrin glues from pooled plasma: Efficacy, safety and potential as scaffold delivery systems. *J. Pharm. Pharm. Sci.* **2012**, *15*, 124–140. [CrossRef] [PubMed]
36. Valbonesi, M. Fibrin glues of human origin. *Best Pract. Res. Clin. Haematol.* **2006**, *19*, 191–203. [CrossRef]
37. Whelan, D.; Caplice, N.M.; Clover, A.J. Fibrin as a delivery system in wound healing tissue engineering applications. *J. Control. Release* **2014**, *196*, 1–8. [CrossRef]
38. Khodakaram-Tafti, A.; Mehrabani, D.; Shaterzadeh-Yazdi, H. An overview on autologous fibrin glue in bone tissue engineering of maxillofacial surgery. *Dent. Res. J.* **2017**, *14*, 79–86.
39. Jeschke, M.G.; Finnerty, C.C.; Shahrokhi, S.; Branski, L.K.; Dibildox, M. Wound coverage technologies in burn care: Novel techniques. *J. Burn Care Res.* **2013**, *34*, 612–620. [CrossRef]
40. Krug, C.; Beer, A.; Hartmann, B.; Prein, C.; Clause-Schaumann, H.; Holzbach, T.; Aszodi, A.; Giunta, R.E.; Saller, M.M.; Volkmer, E. Fibrin glue displays promising in vitro characteristics as a potential carrier of adipose progenitor cells for tissue regeneration. *J. Tissue Eng. Regen. Med.* **2019**, *13*, 359–368. [CrossRef]
41. Noori, A.; Ashrafi, S.J.; Vaez-Ghaemi, R.; Hatamian-Zaremi, A.; Webster, T.J. A review of fibrin and fibrin composites for bone tissue engineering. *Int. J. Nanomed.* **2017**, *12*, 4937–4961. [CrossRef] [PubMed]
42. Kim, B.S.; Sung, H.M.; You, H.K.; Lee, J. Effects of fibrinogen concentration on fibrin glue and bone powder scaffolds in bone regeneration. *J. Biosci. Bioeng.* **2014**, *118*, 469–475. [CrossRef]

43. Kalbermatten, D.F.; Kingham, P.J.; Mahay, D.; Mantovani, C.; Pettersson, J.; Raffoul, W.; Balcin, H.; Pierer, G.; Terenghi, G. Fibrin matrix for suspension of regenerative cells in an artificial nerve conduit. *J. Plast. Reconstr. Aesthet. Surg.* **2008**, *61*, 669–675. [CrossRef] [PubMed]
44. Goel, R.K.; Suri, V.; Suri, A.; Sarkar, C.; Mohanty, S.; Sharma, M.C.; Yadav, P.K.; Srivastava, A. Effect of bone marrow-derived mononuclear cells on nerve regeneration in the transection model of the rat sciatic nerve. *J. Clin. Neurosci.* **2009**, *16*, 1211–1217. [CrossRef]
45. Kurwale, N.S.; Suri, V.; Srivastava, A.; Suri, A.; Mohanti, S.; Yadav, P.; Sharma, M.C.; Sarkar, C. Role of bone marrow derived pluripotent stem cells in peripheral nerve repair in adult rats: A morphometric evaluation. *J. Neurosci. Rural Pract.* **2015**, *6*, 152–159. [CrossRef] [PubMed]
46. Masgutov, R.; Masgutova, G.; Mullakhmetova, A.; Zhuravleva, M.; Shulman, A.; Rogozhin, A.; Syromiatnikova, V.; Andreeva, D.; Zeinalova, A.; Idrisova, K.; et al. Adipose-Derived Mesenchymal Stem Cells Applied in Fibrin Glue Stimulate Peripheral Nerve Regeneration. *Front. Med.* **2019**, *6*, 68. [CrossRef]
47. McGrath, A.M.; Brohlin, M.; Kingham, P.J.; Novikov, L.N.; Wiberg, M.; Novikova, L.N. Fibrin conduit supplemented with human mesenchymal stem cells and immunosuppressive treatment enhances regeneration after peripheral nerve injury. *Neurosci. Lett.* **2012**, *516*, 171–176. [CrossRef]
48. Pan, H.C.; Cheng, F.C.; Chen, C.J.; Lai, S.Z.; Lee, C.W.; Yang, D.Y.; Chang, M.H.; Ho, S.P. Post-injury regeneration in rat sciatic nerve facilitated by neurotrophic factors secreted by amniotic fluid mesenchymal stem cells. *J. Clin. Neurosci.* **2007**, *14*, 1089–1098. [CrossRef]
49. Pan, H.C.; Chin, C.S.; Yang, D.Y.; Ho, S.P.; Chen, C.J.; Hwang, S.M.; Chang, M.H.; Cheng, F.C. Human amniotic fluid mesenchymal stem cells in combination with hyperbaric oxygen augment peripheral nerve regeneration. *Neurochem. Res.* **2009**, *34*, 1304–1316. [CrossRef] [PubMed]
50. Pan, H.C.; Yang, D.Y.; Chiu, Y.T.; Lai, S.Z.; Wang, Y.C.; Chang, M.H.; Cheng, F.C. Enhanced regeneration in injured sciatic nerve by human amniotic mesenchymal stem cell. *J. Clin. Neurosci.* **2006**, *13*, 570–575. [CrossRef]
51. Reichenberger, M.A.; Mueller, W.; Hartmann, J.; Diehm, Y.; Lass, U.; Koellensperger, E.; Leimer, U.; Germann, G.; Fischer, S. ADSCs in a fibrin matrix enhance nerve regeneration after epineural suturing in a rat model. *Microsurgery* **2016**, *36*, 491–500. [CrossRef] [PubMed]
52. Bayır, Ö.; Karagöz, T.; Pınarlı, F.A.; Sarıbaş, G.S.; Özoğul, C.; Keseroğlu, K.; Saylam, G.; Tatar, E.Ç.; Karahan, S.; Öcal, B.; et al. Impact of fetal brain tissue derived mesenchymal stem cell and fibrin glue on facial nerve crash injury. *Turk. J. Med. Sci.* **2021**, *51*, 1481–1490. [CrossRef]
53. Pan, H.C.; Cheng, F.C.; Lai, S.Z.; Yang, D.Y.; Wang, Y.C.; Lee, M.S. Enhanced regeneration in spinal cord injury by concomitant treatment with granulocyte colony-stimulating factor and neuronal stem cells. *J. Clin. Neurosci.* **2008**, *15*, 656–664. [CrossRef]
54. Pan, H.C.; Chen, C.J.; Cheng, F.C.; Ho, S.P.; Liu, M.J.; Hwang, S.M.; Chang, M.H.; Wang, Y.C. Combination of G-CSF administration and human amniotic fluid mesenchymal stem cell transplantation promotes peripheral nerve regeneration. *Neurochem. Res.* **2009**, *34*, 518–527. [CrossRef]
55. Fawcett, J.W.; Keynes, R.J. Peripheral nerve regeneration. *Annu Rev. Neurosci.* **1990**, *13*, 43–60. [CrossRef]
56. Fornasari, B.E.; Carta, G.; Gambarotta, G.; Raimondo, S. Natural-Based Biomaterials for Peripheral Nerve Injury Repair. *Front. Bioeng. Biotechnol.* **2020**, *8*, 554257. [CrossRef] [PubMed]
57. Grinsell, D.; Keating, C.P. Peripheral nerve reconstruction after injury: A review of clinical and experimental therapies. *Biomed. Res. Int.* **2014**, *2014*, 698256. [CrossRef]
58. Heo, J.S.; Choi, Y.; Kim, H.S.; Kim, H.O. Comparison of molecular profiles of human mesenchymal stem cells derived from bone marrow, umbilical cord blood, placenta and adipose tissue. *Int. J. Mol. Med.* **2016**, *37*, 115–125. [CrossRef] [PubMed]
59. Ortiz, A.d.C.; Fideles, S.O.M.; Pomini, K.T.; Reis, C.H.B.; Bueno, C.R.d.S.; Pereira, E.d.S.B.M.; Rossi, J.d.O.; Novais, P.C.; Pilon, J.P.G.; Rosa Junior, G.M.; et al. Effects of Therapy with Fibrin Glue combined with Mesenchymal Stem Cells (MSCs) on Bone Regeneration: A Systematic Review. *Cells* **2021**, *10*, 2323. [CrossRef]
60. Dekmak, A.; Mantash, S.; Shaito, A.; Toutonji, A.; Ramadan, N.; Ghazale, H.; Kassem, N.; Darwish, H.; Zibara, K. Stem cells and combination therapy for the treatment of traumatic brain injury. *Behav. Brain Res.* **2018**, *340*, 49–62. [CrossRef]
61. Zibara, K.; Ballout, N.; Mondello, S.; Karnib, N.; Ramadan, N.; Omais, S.; Nabbouh, A.; Caliz, D.; Clavijo, A.; Hu, Z.; et al. Combination of drug and stem cells neurotherapy: Potential interventions in neurotrauma and traumatic brain injury. *Neuropharmacology* **2019**, *145 Pt B*, 177–198. [CrossRef]
62. Nasser, M.; Bejjani, F.; Raad, M.; Abou-El-Hassan, H.; Mantash, S.; Nokkari, A.; Ramadan, N.; Kassem, N.; Mondello, S.; Hamade, E.; et al. Traumatic Brain Injury and Blood-Brain Barrier Cross-Talk. *CNS Neurol. Disord. Drug Targets* **2016**, *15*, 1030–1044. [CrossRef] [PubMed]
63. Merimi, M.; Lewalle, P.; Meuleman, N.; Agha, D.M.; El-Kehdy, H.; Bouhtit, F.; Ayoub, S.; Burny, A.; Fahmi, H.; Lagneaux, L.; et al. Mesenchymal Stem/Stromal Cell Therapeutic Features: The Bridge between the Bench and the Clinic. *J. Clin. Med.* **2021**, *10*, 905. [CrossRef] [PubMed]
64. Merimi, M.; Lagneaux, L.; Agha, D.M.; Lewalle, P.; Meuleman, N.; Burny, A.; Fahmi, H.; Najar, M. Mesenchymal Stem/Stromal Cells in Immunity and Disease: A Better Understanding for an Improved Use. *J. Clin. Med.* **2020**, *9*, 1516. [CrossRef] [PubMed]

cells

MDPI

Review

Roles of Mesenchymal Cells in the Lung: From Lung Development to Chronic Obstructive Pulmonary Disease

Amel Nasri [1], Florent Foisset [1], Engi Ahmed [2,3], Zakaria Lahmar [2,3], Isabelle Vachier [2], Christian Jorgensen [1], Said Assou [1], Arnaud Bourdin [2,3] and John De Vos [1,4,*]

1 Institute for Regenerative Medicine and Biotherapy, Université de Montpellier, INSERM, Centre Hospitalier Universitaire de Montpellier, 34000 Montpellier, France; amel.nasri@inserm.fr (A.N.); florent.foisset@inserm.fr (F.F.); christian.jorgensen@inserm.fr (C.J.); said.assou@inserm.fr (S.A.)
2 Department of Respiratory Diseases, Université de Montpellier, INSERM, Centre Hospitalier Universitaire de Montpellier, 34090 Montpellier, France; noussa31@gmail.com (E.A.); zakaria.lahmar@gmail.com (Z.L.); isabelle.vachier@medbiomed.fr (I.V.); a-bourdin@chu-montpellier.fr (A.B.)
3 PhyMedExp, Université de Montpellier, INSERM, Centre Hospitalier Universitaire de Montpellier, 34295 Montpellier, France
4 Department of Cell and Tissue Engineering, Université de Montpellier, Centre Hospitalier Universitaire de Montpellier, 34000 Montpellier, France
* Correspondence: john.devos@inserm.fr; +33-467337888

Abstract: Mesenchymal cells are an essential cell type because of their role in tissue support, their multilineage differentiation capacities and their potential clinical applications. They play a crucial role during lung development by interacting with airway epithelium, and also during lung regeneration and remodeling after injury. However, much less is known about their function in lung disease. In this review, we discuss the origins of mesenchymal cells during lung development, their crosstalk with the epithelium, and their role in lung diseases, particularly in chronic obstructive pulmonary disease.

Keywords: mesenchyme; lung; development; COPD

Citation: Nasri, A.; Foisset, F.; Ahmed, E.; Lahmar, Z.; Vachier, I.; Jorgensen, C.; Assou, S.; Bourdin, A.; De Vos, J. Roles of Mesenchymal Cells in the Lung: From Lung Development to Chronic Obstructive Pulmonary Disease. *Cells* **2021**, *10*, 3467. https://doi.org/10.3390/cells10123467

Academic Editor: Mehdi Najar

Received: 6 November 2021
Accepted: 7 December 2021
Published: 9 December 2021

1. Introduction

The lung is a complex organ that carries out the vital task of blood oxygenation. To offer the surface required for this process, the lung gas exchange units (i.e., the alveoli) and the corresponding airways have been multiplied through iterative, fractal branching. This comes at a cost: the direct contact with the external environment of a surface larger than the skin. This contact surface is estimated to be bigger than 100 square meters. The bronchial epithelium is composed of polarized cells attached to a basement membrane and is closely interconnected through junctions, such as adherent junctions, tight junctions, gap junctions, desmosomes and hemi-desmosomes. The bronchial epithelium is composed of different cell types, including ciliated cells, club cells and goblet cells. It functions as a protective barrier and maintains the tissue homeostasis [1–3]. The main protection of the epithelium against external aggression (dusts, airborne particulate matters, noxious compounds, and microbes) is the mucociliary clearance system [4]. Goblet cells secrete mucin glycoproteins to form the mucus that covers bronchioles with a gel layer to trap the inhaled physical matter. Then, cilia on the ciliated cell surface beat to remove these pollutants from the airways.

Like all epithelia, the bronchial epithelium relies on the lung mesenchyme for physical support, nutrient supply [5–8], and key differentiation cues during development. The composition of the lung mesenchyme is still not fully known, but it comprises a large array of cells: endothelial cells, lymphatic cells, pericytes, fibroblasts, mesenchymal stromal cells, smooth muscle cells, and myofibroblasts [9–13]. Mesenchymal stromal cells and fibroblasts are morphologically indistinguishable, and they are differentiated mainly using functional assays [14]. In this review, these different cell types are collectively called "mesenchymal cells". They have critical supporting roles and display specific features, such as migration and invasion [15]. Gene expression analysis and more recently single-cell mRNA sequencing have been useful to explore the mesenchymal cell diversity. For instance, their transcriptomic signature varies according to their localization in the body [11,14,16–19]. Moreover, a comprehensive catalogue of all mesenchymal cells has been proposed for human [20], and mouse [21] lungs. This list includes already known cell types (airway smooth muscle cells, adventitial fibroblasts, lipofibroblasts, mesothelial cells, myofibroblasts, pericytes, vascular smooth muscle cells), and also novel cell types identified on the basis of their gene expression profile, such as fibromyocytes and alveolar fibroblasts. All these mesenchymal cell types harbor a distinct gene expression signature (Figure 1 shows some of these cell types and their gene expression signature). Discrepancies among studies may reflect differences linked to the exact tissue localization or the species investigated (proximal versus distal pattern species, human versus mouse, etc.), and suggest that the different mesenchymal cell types could represent different steps in a continuous process of differentiation that varies also between different cell types but also between healthy and diseased lung. As the lung mesenchymal compartment contributes to lung homeostasis and repair after injury [22,23], mesenchymal cells are implicated in many lung diseases, particularly asthma, idiopathic pulmonary fibrosis, and chronic obstructive pulmonary disease (COPD). In this review, we highlight the importance of the crosstalk between epithelium and mesenchyme during lung development, adult life, and diseases, with a focus on COPD.

Human			
Authors	**References**	**Cell Types**	**Gene Expression profile**
Travaglini et al	[20]	Fibroblasts	BSG, **COL1A2**
		Myofibroblasts	**ACTA2,** ASPN, BSG, **COL1A2**, CREB5, FGF18, TAGLN, WIF1
		Others	
		Alveoli Fibroblasts	**COL1A2,** DKK3, FGFR4 GPC3, MYC, SPINT2, TCF21
		Adventitial Fibroblasts	COL1A1, **COL1A2**, C3, IGFBP4, SERPINF1, SFRP2, PI16, PDGFRL
		Fibromyocytes	ACTA2, ACTG2, MYH11, **COL1A2**, CNN1, CREB5, TAGLN, TBX2
		Lipofibroblasts	APOE, CREM, FST, MAFF, MYC, NFIL3, NR4A3, **PLIN2,** SERPINF1
Habermann et al	[24]	Fibroblasts	**COL14A1**, **COL1A1**, COL3A1, COL6A1, COL6A2, **DCN**, EFEMP1, EFEMP2, FBN1, **FBLN1**, FBLN2, FBLN5, ITBP1, **LAMA2**, MAFP2, MAFP5, MGP, PLIN2
		Myofibroblasts	**ACTA2**, **COL1A1**, COL3A1, COL6A2, **LTBP2**, MGP, **PDGFRA**
		Others	
		Mesothelial Cells	MYC, **WT1**
Reyfman et al	[25]	Fibroblasts	**ACTA2**, CAV1, **COL1A1**, COL3A1, HIF1A, NOTCH2, NOTCH3, PDGFRA, PDGFRB, POSTN, TAGLN, TWIST2, **WNT2**, YAP1, ZEB2
Hein et al	[26]	Smooth Muscle Cell types	**ACTA2**, ACTG2, SM22, MYL9, FOXF1, HHIP, MYH11, WNT5A
		Others	
		Non Smooth Muscle Cell types	EGR1, WNT2

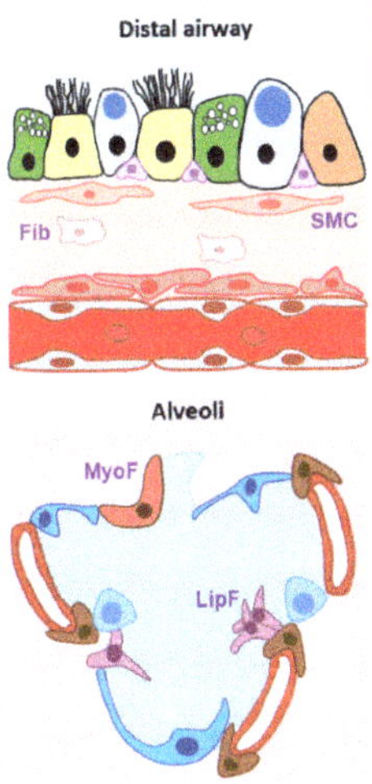

Mouse			
Authors	**References**	**Cell Types**	**Gene Expression profile**
Negretti et al	[21]	Fibroblasts	**Acta2**, **Col1a1**, Gyg, Macf1, Mki67, **Wnt 2**
		Myofibroblasts	**Acta2**, Ednrb, Enpp2, Eln, **Pdgfra,** Tgfbi, Wnt5a
		Smooth Muscle Cell types	**Acta2**, Eln, Tgfb2
		Others	
		Adventitial Fibroblasts	Col1a1, Decorin, Lumican, Eln
		Mesothelial Cells	Col1a1, **Dcn**, Eln, Upk3b, **Wt1**
		Pericytes	Acta2, Cox4i2, Col4a1, Pdgfrb
		Lipofibroblasts	Gyg, **Plin2**, Tcf21,
Guo et al	[27]	Fibroblasts	**Col1a1**, **Col1a2,** Fgf10, Fn1, Vcam1, Tcf21,
		Myofibroblasts	**Col1a1**, **Col1a2**, **Pdgfra**, Rgs2, Vim
		Smooth Muscle Cell types	**Acta2**, **Col1a2**, **Col1a1**, Des, Rgs2
		Others	
		Pericytes	**Acta2**, Notch3, Pdgfrb
Strunz et al	[28]	Fibroblasts	Abi3bp, Adamts2, Angptl, **Col14a1**, Clec3b, C1qtnf7, Cxcl12, **Dcn**, Efemp1, Efemp2, **Fbln1**, Figf, Gas6, Gpc3, Igfbp4, Lama2, Lamc, Ltbp3, Igfbp6, Mdk, Mfap4, Mmp3, Npnt, Serpina3n, Serpina3m, Serpinf1, Pcolce2, Plxdc2
		Myofibroblasts	**Acta2**, Anxa1, Col5a3, Col12a1, Csf1, Ctgf, Cthrc1, Grem1, Inhba, Lgals1, Lman1, Loxl2, Ltbp1, **Ltbp2**, Serpine1, Srpx2, Spp1, S100a6, Tnc, Thbs1, Wisp1
		Smooth Muscle Cell types	Aspn, Crim1, Pgf
Angelidis et al	[29]	Smooth Muscle Cell types	**Acta2**, **Col1a2**
		Others	
		Mesothelial Cells	**Dcn**, Msln
		Intersitial Fibroblasts	**Col1a2**, Dcn, Inmt
		Lipofibroblasts	**Col1a2**, Inmt

Figure 1. Summary of genes expressed in the different lung mesenchymal cell types (distal airways and alveoli), based on different recent studies in human (**left**) [20,24–26] and mouse (**right**) [21,27–29] samples. For fibroblasts cells (Fib), smooth muscle cell types (SMC), myofibroblasts (MyoF) and lipofibroblasts (LipF), genes present in at least two different studies are listed in bold. Distal airway: goblets, ciliated, club and basal cells are green, yellow, blue and lila, respectively. Alveoli: alveolar type 1, alveolar type 2 and pericytes are dark blue, light blue and brown, respectively.

2. Origin of Pulmonary Mesenchymal Cells

At the bilaminar disc stage in the second week of human development, gastrulation starts with the formation of the primitive streak [30]. Cells forming the epiblast undergo epithelial to mesenchymal transition (EMT) and migrate through the primitive streak to form the endoderm and mesoderm cell layers (Figure 2A) [31]. Specifically, the first wave of cells integrates the hypoblast layer and forms the endoderm [30], from which the lung will derive. A second group of cells migrates between the epiblast and endoderm layers and constitutes the mesoderm layer that will give rise to a large variety of tissues, such as skeletal muscle, bone, cartilage, and many mesenchymal cell types (e.g., fibroblasts, chondroblasts, osteoblasts, blood cells). The mesoderm is a major contributor to trunk and limb stromal cells, but neural crest cells also contribute to mesenchymal cell lineages [32,33], although their specific role in lung mesenchyme is still poorly defined.

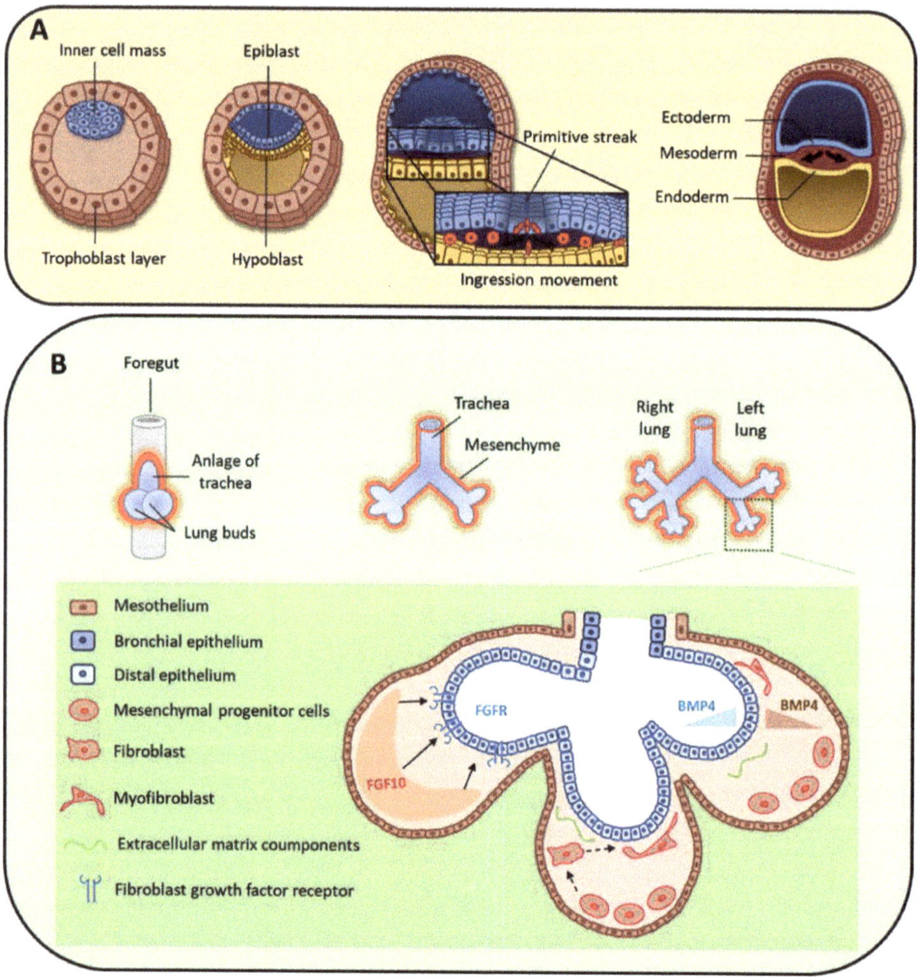

Figure 2. (**A**) During early human embryo development, the inner cell mass differentiates and becomes organized in the epiblast and hypoblast layers. Primitive streak formation leads to an ingression movement of epiblastic cells that elongate and detach from each other via a critical epithelial–mesenchymal transition [31]. This results in the formation of the three germ layers: ectoderm, mesoderm and endoderm. (**B**) On day 22, the foregut forms a ventral outgrowth leading to the formation of larynx and trachea in its proximal part, and lung buds in the distal part. Bifurcation and splitting of the lung buds give rise to the future right and left lungs. These structures grow ventrally to caudally through the surrounding mesenchyme. Mesenchymal progenitor cells secrete many factors, including fibroblast growth factor 10 (FGF10) that interacts with fibroblast growth factor receptor (FGFR) expressed by distal epithelial cells. Moreover, some cytokines, such as bone morphogenic protein 4 (BMP4), are secreted by both epithelial and mesenchymal cells.

3. Mesenchymal Cells during Lung Organogenesis

At the beginning of the fourth week of development, the anterior foregut endoderm develops at the cranial extremity after the cephalocaudal folding of the embryo. The foregut produces a ventral evagination that leads to lung bud development. These buds are surrounded by the splanchnopleuric mesoderm that is part of the lateral mesoderm and will contribute to lung vascularization, cartilage, muscles and conjunctive tissue (Figure 2B). In addition, the embryonic lateral splanchnic mesoderm generates mesothelial cells that form a thin layer of squamous-like cells lining the visceral pleura (mesothelium) [34–36]. Then, at day 26, the lung buds divide into right and left primitive bronchial buds, which are the precursors of the two lungs (Figure 2B).

During the pseudo-glandular stage, a second division at week 5 of development leads to the formation of the future pulmonary lobes by creating three secondary bronchial buds on the right and two on the left side. Each lung bud and the surrounding splanchnopleuric mesoderm grow, elongate and branch until the formation of the terminal bronchioles (17th branching generation) to create the respiratory tree (Figure 2B) [37]. At this stage, the tracheobronchial tree is coated by prismatic epithelial cells, the precursors of ciliated and secretory cells. Bronchioles appears during the canalicular stage (week 16 to 25), forming the basis of the gas exchange units. This is accompanied by geometric modifications of epithelial cells that flatten and by the appearance of capillaries throughout the mesenchyme that surround the bronchioles. Finally, at the saccular stage (week 24 to 40), alveolar ducts start to form. Their formation will continue after birth, and will terminate only in adulthood [7]. Of note, the development of the gas exchange units in utero and during early childhood is critical for achieving full adult lung function [38]. Lung development is the consequence of an interweaved relationship between embryonic lung epithelial and mesenchymal cells, through direct interactions and also indirectly via the secretion of extracellular matrix (ECM) components and growth factors [39–41]. Moreover, the mesothelium plays an important role during lung development [36,42,43], partly by secreting fibroblast growth factor (FGF) 9 [35,44,45].

Fetal airway smooth muscle (ASM) development begins early during gestation (from week 5–6 in human airways) [46]. Fetal ASM surrounds the airways and guides lung development and branching. ASM cells spontaneously contract early in fetal life, with proximal to distal peristaltic-like contractions that displace the amniotic liquid along the lumen [47,48]. At the pseudo-glandular and canalicular stages in pigs [48], the mechanical distention and stretching of the developing lungs, produced by ASM contractions, influence lung growth via mechanotransduction, through the pressure exerted across the airway wall and the surrounding parenchyma. Furthermore, the transmural pressure regulates the rate of airway epithelial bud branching [49].

The next part of this review will focus on some of the most relevant factors implicated in the epithelium–mesenchyme interactions during lung development.

3.1. Peptide Growth Factors

3.1.1. Fibroblast Growth Factors

The large FGF family plays an important role in the regulation of cell differentiation, proliferation and development, including lung branching [50]. Several studies have identified FGFs implicated in the bidirectional signaling between epithelium and mesenchyme during lung development [44,50,51]. For example, FGF10 is expressed in the distal submesothelial mesenchyme and activates FGF receptor 2b (FGFR2b) in the adjacent epithelial cells (Figure 3) to induce lung budding, epithelial cell expansion and migration, and ECM organization [52–54]. Indeed, in *Fgf10*$^{-/-}$ mice, the lung does not develop below the trachea [55,56]. Moreover, Bellusci et al. have identified, in mice, a subtype of Axin2+/FGF10+-resident mesenchymal alveolar niche cells that are close to alveolar type 2 (AT2) stem cells and that control the proliferation and differentiation of AT2 cells [57].

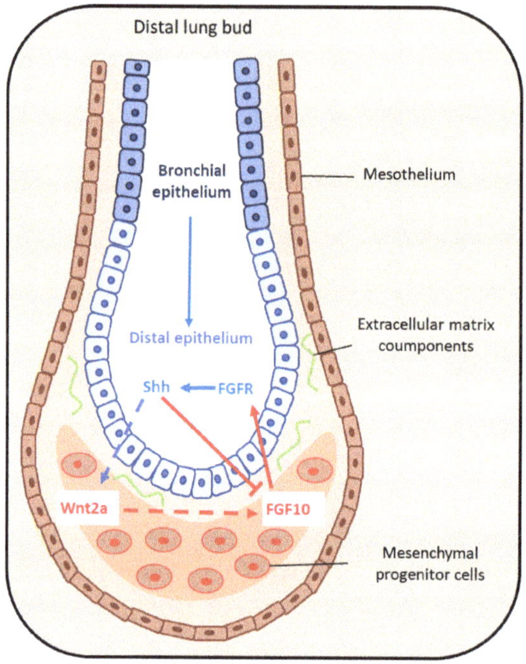

Figure 3. Representative schema of epithelium-mesenchyme interactions during lung budding. Many factors are secreted and exchanged between the lung epithelium (blue arrows) and mesenchyme (red arrows). For instance, fibroblast growth factor 10 (FGF10) is locally expressed by distal submesothelial mesenchymal cells and interacts with its receptor expressed in the distal epithelium. Sonic hedgehog (SHH) is expressed by epithelial cells, and downregulates FGF10 expression through its receptor Patched 1, but also activates FGF10 through the Wnt pathway.

FGF9 is another FGF family member that is expressed in the mesothelium from the pseudo-glandular stage (week 5 to 17) onwards, regulating the local activation of Wnt signaling to promote mesenchymal cell proliferation [35,44] (Figure 3). FGF9 also plays a critical role in lung development, as indicated by the finding that *Fgf9*$^{-/-}$ mice die at the neonatal stage due to lung hypoplasia caused by very reduced mesenchymal cell expansion [58].

3.1.2. Bone Morphogenic Protein 4 (BMP4)

BMP4 belongs to the transforming growth factor superfamily. *Bmp4*$^{-/-}$ mice die early during development, mainly due to an absence of mesoderm differentiation [59]. During lung development, BMP4 expression is detected in the distal epithelium buds and in the adjacent mesenchyme already at the pseudo-glandular stage (week 5 to 17) [60,61]. Conversely, BMP type I receptor (BMPR1) is expressed in both epithelium and mesenchyme (Figure 2B) [62]. In association with sonic hedgehog (SHH), BMP4 antagonizes FGF10 that is expressed in the surrounding mesenchyme [63]. Conditional knock-out of *Bmpr 1a*, the gene encoding the BMP4 receptor in the epithelium, leads to abnormal lung development with reduced cell proliferation, increased apoptosis and abnormal lung morphogenesis. This indicates that BMP4 plays important roles in lung development [64]. Moreover [65], in cultured mouse embryonic lung, reduction of gremlin expression, a BMP4 antagonist, using antisense oligonucleotides promotes epithelial cell proliferation and abnormal lung formation/function. Finally, BMP4 overexpression in the distal bud tips leads to lung hypoplasia, reduction of AT2 cells, and enlargement of the terminal buds [62]. However, the exact role of this signaling pathway during lung development remains debated, because mathematical models to mimic the FGF10-SHH interaction accurately model bronchial branching independently of BMP4 expression [66,67].

3.1.3. Sonic Hedgehog

SHH also is part of a key developmental signaling pathway. It is implicated in central nervous system patterning, and limb, digit and lung development [68]. In *Shh*$^{-/-}$ mice, a single lobe, lung hypoplasia with absence of left and right asymmetry, enhanced cell death and decreased lung mesenchymal cell proliferation are observed [69]. SHH is expressed with BMP4 in the distal bud epithelium during lung development. It binds to and activates its receptor Patched 1 (PTCH1) that is localized in the adjacent mesenchyme. Patched 1 activation downregulates FGF10 expression [44,52,70]. Indeed, in *Shh*$^{-/-}$ mice, FGF10 expression in the mesenchyme immediately adjacent to the epithelium is increased [71]. In addition, BMP4 is overexpressed and wingless-related integration site family member 2 (WNT2) is downregulated in the mesenchyme. In 2013, Peng et al. [72] identified a cardiopulmonary mesoderm progenitor population that is defined by the expression of WNT2, glioma-associated oncogene 1 (Gli1) and Islet 1, and gives rise to the lung mesenchyme and cardiac lineage in the mouse. This population is regulated by Hedgehog signaling because Gli proteins are the main transcriptional effectors of this pathway, and start to be expressed at the lung organogenesis step [72]. This suggests that SHH is broadly involved in mesenchymal signaling in the developing lung.

3.1.4. Epidermal Growth Factor (EGF)

EGF and its tyrosine kinase receptor EGF-R are expressed in the epithelial and mesenchymal compartments during lung development. EGF stimulates lung branching in fetal mice [40]. In agreement, in *Egfr*$^{-/-}$ mice, neonatal lethality is high and epithelial cell development is impaired in several organs, including the lung [73]. Importantly, lung branching is reduced and alveolarization and septation are deficient in *Egfr*$^{-/-}$ mice [74]. Similarly, mouse lung branching can be inhibited in ex-vivo cultures using antisense oligonucleotides against EGF [40,75]. Furthermore, the interplay between retinoid acid (RA) and EGFR during fetal lung development stimulates lung branching [76].

3.1.5. Retinoic Acid (RA)

RA is essential for normal embryo development, including lung development [77]. In the mouse, retinoic acid receptor (RAR) double knock-out (RARα-/-/RARβ2-/-) leads to agenesis of the left lung and hypoplasia of the right lung [78]. RA is produced by the mesenchyme surrounding the lung primordia, and also by the epithelial compartment of the proximal bronchi [79]. Importantly, a RA-transforming growth factor (TGF)-β-FGF10 interaction has been described during lung bud induction, where RA downregulates TGF-β to allow FGF10 expression [80].

3.1.6. TGF-β

TGF-β, a pleiotropic growth factor and a key EMT inducer, is another modulator of epithelium–mesenchyme interactions in the developing lung [81]. Its three isoforms play a crucial role during lung organogenesis [82]. Their expression is well characterized in the mouse where TGF-β1 is expressed in the mesenchyme, TGF-β2 in the distal epithelium, and TGF-β3 in the proximal mesenchyme and pulmonary mesothelium [83,84]. In *Tgf-β2*$^{-/-}$ mice, the distal airways are collapsed and the proximal airways are dilated [82], whereas in *Tgf-β3*$^{-/-}$ mice lung development is delayed, causing their death [85]. Conversely, *Tgfr-β2* ablation in mesodermal tissue results in abnormal lung branching and lung development, while its ablation in epithelial cells that produce surfactant protein C results in a decrease of alveolar type I (AT1) epithelial cells during post-natal alveolarization [86]. TGF-β also plays a role in FGF10 regulation by RA [80] as illustrated by the finding that ectopic TGF-β1 expression inhibits FGF10-induced lung morphogenesis in cultured embryonic lung endodermal explants [87].

3.1.7. The Hippo Pathway

The Hippo pathway and its downstream targets Yes-associated protein (YAP) and transcriptional coactivator with PDZ-binding motif (TAZ) play major roles in tissue homeostasis, organ development, and organ size [88,89] by regulating various processes, such as cell proliferation/survival, the response to mechanical stress and cell geometry. This pathway involves a kinase cascade and adaptors that ultimately regulate YAP/TAZ activities. The name comes from the Drosophila kinase Hippo, the mammalian orthologs of which are the kinases MST1/2. Upon activation of the Hippo signaling pathway, MST1/2 phosphorylates LATS1/2, thereby activate these kinases that in turn phosphorylate YAP and TAZ. Phosphorylated YAP and TAZ are retained in the cytoplasm or degraded by the ubiquitin system. Conversely, when the Hippo pathway is inactive, YAP and TAZ shuttle to the nucleus where they bind to TEA domain (TEAD) transcription factors and regulates the transcription of many downstream genes [89]. *Taz* knockout mice show abnormal alveolarization, leading to airway enlargement that mimics human emphysema [90,91]. Conditional *Yap* knockout in lung epithelium causes disruption of bronchial morphogenesis [91]. During alveologenesis, YAP/TAZ inactivation by cell crowding orients NKX2.1 activity towards AT2 cells, a mechanism thought to be a negative feedback to limit AT1 cell expansion [92]. In bronchia, YAP is activated in distal airways, and its induction prevents multi-ciliated cell differentiation [93]. These observations suggest a general role of YAP/TAZ in favoring the stem cell compartment at the expenses of terminal differentiation. In line, YAP overexpression in adult tracheal cells results in basal cell hyperplasia and stratification [94]. Interestingly, injury in airway epithelial cells leads to downregulation of hippo signaling that increases the concentration of nuclear YAP [38], and induces FGF10 secretion by the adjacent mesenchymal cells [95]. Moreover, in the absence of *Yap*, epithelial progenitors cannot respond to local TGF-β signaling [93]. Overall, the Hippo pathway plays a critical role in lung development and response to injury directly in epithelial cells or indirectly through epithelium/mesenchyme signaling.

Other factors also are involved in lung development, such as components of the vascular endothelial growth factor (VEGF) and WNT signaling pathways. Their expression is highly regulated in space and time, allowing optimal lung development, at least partly through epithelium-mesenchyme interactions. Single-cell transcriptomic analyses will help to identify all the factors involved in lung development.

3.2. Extracellular Matrix Compounds

The ECM exerts functions of support, structure and stabilization that are essential for organ development and homeostasis. By directly binding to growth factors, it can also modulate the activity of secreted factors [96]. At the end of the human pseudo-glandular stage, proteoglycans, such as decorin and lumican, are located between the epithelium and the mesenchyme compartments, along with collagen type I, III and VI [97]. In ex-vivo cultures of mouse embryonic lungs, decorin binds to and neutralizes exogenous TGF-β with high affinity, thus inhibiting TGF-β signaling [98]. Furthermore, heparan sulfates present in the EMC can stabilize the interaction between growth factors and receptors. For example, Izvolsky et al. [99] showed that endogenous gradients of heparan sulfates, especially highly sulfated heparan sulfates, help lung budding induced by FGF10.

4. Physiological Roles of Mesenchymal Cells in Bronchioles

In the airways, epithelial cells are directly exposed to the outside environment and form a protective barrier against pathogens and toxic particles. Some bronchial epithelial cell populations contribute to epithelium repair, such as basal cells [100], one of the main airway stem cell sources [101], and club cells that can trans-differentiate for epithelium renewal [102–104]. Mesenchymal cells also participate by supporting lung epithelium homeostasis and repair [105,106].

Bronchial epithelium maintains the quiescence of mesenchymal cells through the Hedgehog pathway paracrine signaling [105]. Upon epithelial injury, SHH is downregulated to promote proliferation of peribronchial mesenchymal cells and the repair of the damaged tissue [105]. Peng et al. [72] showed that loss of SHH expression in airway epithelial cells leads to expansion of the surrounding mesenchymal Gli1 population that promotes the increase of the total club cell number and consequently bronchial hyperplasia.

Recently, several new lung mesenchyme cell types that can self-renew and contribute to mesenchyme self-maintenance have been described. Zepp et al. [107] characterized Axin2-Pdgfr+ mesenchymal cells that promote the regeneration and growth of alveolar cells, and Axin2+ myofibrogenic progenitors that contribute to the pathological myofibroblast response after lung injury. The origin of myofibroblasts in the lung is still debated because it requires multiple signals from fibroblasts, pericytes and other cell types [108,109]. Although it is still difficult to distinguish the different mesenchymal cell types, presumably because of an underlying continuum of differentiation, new tools such as single-cell RNA-sequencing will provide new data to expand the taxonomy of mesenchymal cells (Figure 4). In an analysis of mesenchymal cells from multiple tissues, Buechler et al. identified a dermatopontin-positive cell fibroblast population, possibly representing a universal fibroblast cell population that can give rise to other fibroblast subsets in the different tissues [110].

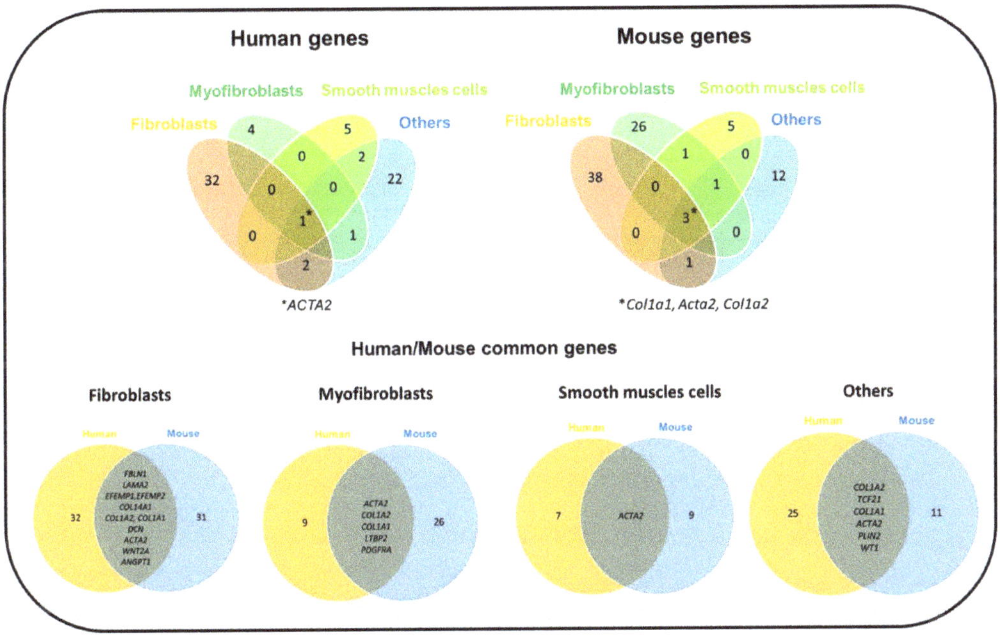

Figure 4. Venn diagrams showing the number of differentially expressed genes in each comparison (between human and mouse) and the overlaps between the four main compared mesenchymal cell types (fibroblasts, myofibroblasts, smooth muscle cells and other cell types). 1* and 3*, number of genes shared by the four cell types. This analysis is based on data from eight single-cell RNA sequencing studies (see Figure 1). The Venn diagrams were generated using the Venny tool.

More recently, Fang et al. [106] tagged with a green fluorescent protein the basic helix-loop-helix transcription factor TWIST2/Dermo1 that is strongly expressed in mouse mesodermal tissues [111]. They found that after lung injury using lipopolysaccharide and naphthalene (a polycyclic aromatic hydrocarbon), the Dermo+ mesenchymal population differentiated into club cells, ciliated cells, goblet cells and neuroendocrine cells. In mice, exposure to naphthalene specifically depletes club cells by binding to CYP2F2 enzymes. These results suggest that besides their pleiotropic (e.g., anti-inflammatory and trophic) effects [112], mesenchymal cells may also directly contribute to epithelium regeneration.

5. Mesenchymal Cells in Chronic Obstructive Pulmonary Disease

Increased environmental susceptibility, as observed in asthma, might be seen as a failure of the innate immune system to prevent the adaptive immunity engagement and the subsequent airway inflammation. Ultimately, this might lead to cell loss and alterations that cause respiratory functional disruption. The lung physiological defense systems and regeneration capacities could be overwhelmed. In some individuals, this can lead to chronic lung diseases, such as COPD, pulmonary fibrosis and lung cancer (see Wolters et al. and Gohy et al. [113,114]). This review will focus on the role of mesenchymal cells in COPD.

5.1. Genetic Contribution

COPD is the third leading cause of death worldwide: more than 3.2 million of deaths in 2017 that should increase to more than 4.4 million by 2040 [115,115]. This chronic airway disease is associated with inflammation and structural changes, leading to permanent bronchial obstruction [116–118]. The insidious progression of the disease might explain its frequent underdiagnosis and late diagnosis [119]. Moreover, COPD is associated with many comorbidities, such as cardiac, gastrointestinal, cerebral and muscular diseases [120–125]. Tobacco smoking is the first cause of COPD in western countries [126–128]. Air pollution, including by biomass combustion [129–131], occupational and non-occupational exposure to dust and chemical agents [132], and repeated airway infections during childhood [133] are now recognized as contributing causes of COPD. Genetic and environmental factors could influence the susceptibility to COPD. Multiple large-cohort genome-wide association studies to understand the link between loci associated with lung function impairment and COPD found many polymorphisms near the Hedgehog Interacting Protein (*HHIP*) gene [134–137] that encodes a physiological inhibitor of SHH [138]. Furthermore, several studies have highlighted a relationship between susceptibility to COPD, *FGF*10 gene variants, and human airway branching variations. For instance, Smith et al., using computed tomography observed greater central airway bifurcation density, bronchial anatomic variations and narrower airway lumens in all lobes of patients with COPD than in controls. They found that these changes were significatively associated with *FGF*10 variants [139]. Gene mutations in components of key signaling pathways involved in lung development may promote COPD development.

The huge heterogeneity in COPD triggers and clinical expression may be explained by the different underlying mechanisms. Indeed, COPD should not be seen as a unique entity, but as a syndrome [140].

5.2. The Epithelial–Mesenchymal Crosstalk

The release of inflammatory mediators, such as TNF-α, interleukin (IL)-6 and IL-8, promotes chronic airway inflammation and ECM deposition [140,141]. In the epithelial cell compartment, goblet and basal cell hyperplasia, squamous metaplasia, mucus hypersecretion and altered cilia beating are the classical structural changes observed in patients with COPD and in smokers [4,142–144]. The contribution of the bronchial mesenchymal compartment to COPD progression is strongly suggested by its role in bronchial injury repair, but this is still debated. Many studies have identified an increase in the expression of mesenchymal markers, such as vimentin and the fibroblast protein S100A4, in COPD lung samples [145]. Myofibroblasts and fibroblasts could remodel the ECM by releasing matrix metalloproteinases, such as MMP9 [146,147]. Interestingly, a recent study identified different fibroblast subtypes that are localized in the lung subepithelial, subpleural and parenchymal regions and that contribute to ECM expansion in pulmonary fibrosis [24]. Similar injury-response mechanisms could be involved in COPD. These data suggest that the epithelial-mesenchymal crosstalk or the "epithelial-mesenchymal trophic unit" plays crucial roles in driving lung pathology [148].

5.3. Peribronchiolar Fibrosis

In COPD, the major site of obstruction is located in the small conducting airways (i.e., bronchioles with a diameter < 2 mm) [149,150]. During COPD, small airways become narrower due to the airway wall thickening and peribronchiolar fibrosis. The mechanisms of peribronchiolar fibrosis are poorly understood, and small airway fibroblasts have not been well characterized. Senescent fibroblasts could have a role in small airway fibrosis [151] due to their increased secretion of collagen 1A1 and 3A1 and increased expression of matrix metalloproteinases (MMP2, MMP9).

Cigarette smoke and oxidative stress may stimulate the release of profibrotic mediators, such as TGF-β and IL-1 beta, by airway epithelial cells [152]. Increased TGF-β1 expression (mRNA and protein) has been observed in epithelial and endothelial cells from small airways of patients with COPD compared with controls [153]. Air–liquid interface culture of airway epithelial cells from patients with COPD showed increased EMT and increased release of TGF-β that were correlated with the degree of peribronchiolar fibrosis and airway obstruction [154]. These profibrotic growth factors may induce a profibrotic phenotype in adjacent airway fibroblasts or promote the differentiation of bronchial smooth muscle cells into myofibroblasts. Indeed, a recent study showed an increase of the αSMA+ myofibroblast population in small airways of patients with COPD compared with controls [155].

5.4. Extracellular Matrix Composition

In COPD, ECM is degraded by enzymes, such as neutrophil elastase, metalloproteinases, hyaluronidases and chondroitinases. It has been shown that in different lung compartments, elastic fibers, elastin, glycosaminoglycans (e.g., hyaluronic acid) and type I collagen decrease drastically, but not fibronectin, tenascin and other collagens. These changes contribute to peri-bronchial fibrosis and progressive emphysema that profoundly impact the respiratory functions (Figure 5) [156–160].

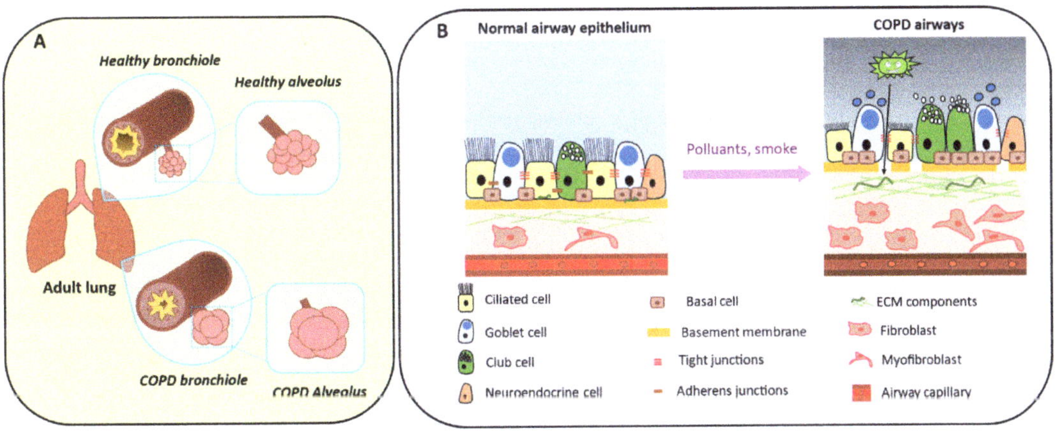

Figure 5. Aspects of COPD development. (**A**) (**Top panel**) Healthy bronchioles present a thin wall and a lumen where mucus is secreted to externalize pathogens. Alveoli contain elastin fibers and regulate gas exchanges between lung and blood. (**Bottom panel**) Over time, chronic exposure to tobacco, air pollution and/or recurrent infections lead to a thickened bronchiole wall associated with airflow obstruction. Emphysema is caused by gas trapping that damages the alveoli, followed by progressive peribronchiolar fibrosis. (**B**) (**Right panel**) In the alveolar-capillary unit, epithelial cells are interconnected and attached to the basement membrane above the mesenchymal compartment. (**Left panel**) Chronic exposure to pollutants damages ciliated cells (leading to impaired mucociliary clearance), alters the epithelial cell barrier and the basement membrane with loss of tight junctions and pathogen penetration in the lower layers. This is associated with mucus plugging and goblet cell hypersecretion. Pro-proliferative fibroblasts and myofibroblasts contribute to extracellular matrix (ECM) deposition that impairs injury repair.

5.5. Epithelial–Mesenchymal Transition

Cigarette smoke also induces EMT in bronchial epithelial cells of smokers and patients with COPD, thus altering epithelial function and contributing to defective lung remodeling [161]. EMT is a morphogenetic cell conversion program of epithelial cells to mesenchymal cells. EMT and mesenchymal–epithelial transition are key processes during embryogenesis and organogenesis [31]. Normally, epithelial cells are polarized and attached to the basement membrane via their basal surface. Depending on the microenvironment conditions and the cell physiological state [162], these cells can lose many epithelial characteristics, such as cell–cell adhesion (Figure 5B) and cell polarity, and acquire features of mesenchymal cells, such as migration and invasion [163–165]. EMT illustrates the epithelial cell phenotype plasticity. Of note, an intermediate "metastable" state where cells co-express markers of both epithelial and mesenchymal cells has been described [166,167]. In chronic airway diseases, EMT is mainly promoted through TGF-β/SMAD signaling [168] that with other signaling pathways promotes myofibroblast proliferation [169,170] and activates several ECM components. By acquiring mesenchymal cell properties, such as migration through the basement membrane, these cells could contribute to peribronchial fibrosis [145,171]. In COPD, EMT could also be promoted by the bronchial epithelium basement membrane fragmentation that facilitates pathogen penetration in the subepithelium and increases local inflammation [172,173]. Hence, mesenchymal cell proliferation and EMT might contribute to COPD pathophysiology and explain some of the observed architectural changes, such as bronchial lumen thickening (Figure 5).

5.6. Airway Smooth Muscle

ASM cells are the main pharmacological target in COPD through inhaled drugs, such as long-acting β-agonists, anticholinergic and corticosteroids. However, little is known about ASM cell changes in COPD. Differently from asthma, no ASM cell alteration (morphology and size) has been observed in the large airways of patients with COPD, and the ASM cell number does not correlate with airflow limitation [174]. Similarly, the proliferation rate of cultured ASM cells from patients with COPD is not increased [175]. Conversely, in small airways, ASM mass is significantly increased in COPD and is inversely correlated with lung function [176].

Some mesenchymal cell populations, such as mesenchymal stromal cells, could have a therapeutic potential in COPD. Indeed, administration of mesenchymal stromal cells in mouse [177] and in rat [178] models of COPD attenuates lung damage. However, it does not improve COPD outcomes [179,180] and therefore, its use is still debated [181–183]. Recently, in patients with COVID-19-associated pneumonia, the systemic delivery of allogeneic mesenchymal stromal cells (MSC) has been proposed. MSC anti-inflammatory properties and their ECM remodeling capacity suggest that MSC- based therapy may prevent fibrosis. Several clinical trials have shown the feasibility of this approach but preliminary data are mixed, possibly due to bronchial inflammation which may dampen their therapeutic efficacy [184]. Therefore, randomized trials are needed before drawing definitive conclusions [185].

6. Conclusions

Lung mesenchyme plays an important role in lung development and homeostasis, and an imbalance or a defect in the mesenchymal cell response to chronic injury could contribute to lung diseases. However, the exact role of the mesenchymal compartment in lung homeostasis and disease is still largely unknown. As a first step to improve this knowledge, it is essential to precisely determine the taxonomy of lung mesenchymal stromal cells during lung development and disease (a task accelerated by single-cell RNA sequencing) and the composition of that niche.

Author Contributions: A.N. wrote the review and made all the figures. F.F., E.A., Z.L., I.V., C.J., S.A., A.B. and J.D.V. have written, reviewed and edited the manuscript. All authors have read and agreed to the published version of the manuscript.

Funding: This work was supported by a grant to F.F. from la Foundation du Souffle.

Institutional Review Board Statement: Not applicable.

Informed Consent Statement: Not applicable.

Data Availability Statement: Not applicable.

Acknowledgments: We thank E. Andermarcher for critical reading of this manuscript.

Conflicts of Interest: The authors declare no conflict of interest.

References

1. Gohy, S.; Hupin, C.; Ladjemi, M.Z.; Hox, V.; Pilette, C. Key role of the epithelium in chronic upper airways diseases. *Clin. Exp. Allergy* **2020**, *50*, 135–146. [CrossRef] [PubMed]
2. Wittekindt, O.H. Tight junctions in pulmonary epithelia during lung inflammation. *Pflug. Arch.* **2017**, *469*, 135–147. [CrossRef]
3. Powell, D.W. Barrier function of epithelia. *Am. J. Physiol.-Gastrointest. Liver Physiol.* **1981**, *241*, G275–G288. [CrossRef] [PubMed]
4. Bustamante-Marin, X.M.; Ostrowski, L.E. Cilia and Mucociliary Clearance. *Cold Spring Harb. Perspect. Biol.* **2017**, *9*, a028241. [CrossRef] [PubMed]
5. Loh, K.M.; Chen, A.; Koh, P.W.; Deng, T.Z.; Sinha, R.; Tsai, J.M.; Barkal, A.A.; Shen, K.Y.; Jain, R.; Morganti, R.M.; et al. Mapping the Pairwise Choices Leading from Pluripotency to Human Bone, Heart, and Other Mesoderm Cell Types. *Cell* **2016**, *166*, 451–467. [CrossRef]
6. Delorme, B.; Chateauvieux, S.; Charbord, P. The concept of mesenchymal stem cells. *Regen. Med.* **2006**, *1*, 497–509. [CrossRef]
7. Bishop, A.E. Pulmonary epithelial stem cells. *Cell Prolif.* **2004**, *37*, 89–96. [CrossRef]
8. Mammoto, A.; Mammoto, T. Vascular Niche in Lung Alveolar Development, Homeostasis, and Regeneration. *Front. Bioeng. Biotechnol.* **2019**, *7*, 318. [CrossRef]
9. Hogan, B.L.M.; Barkauskas, C.E.; Chapman, H.A.; Epstein, J.A.; Jain, R.; Hsia, C.C.W.; Niklason, L.; Calle, E.; Le, A.; Randell, S.H.; et al. Repair and Regeneration of the Respiratory System: Complexity, Plasticity, and Mechanisms of Lung Stem Cell Function. *Cell Stem Cell* **2014**, *15*, 123–138. [CrossRef]
10. McCulley, D.; Wienhold, M.; Sun, X. The Pulmonary Mesenchyme Directs Lung Development. *Curr. Opin. Genet. Dev.* **2015**, *32*, 98–105. [CrossRef] [PubMed]
11. Barron, L.; Gharib, S.A.; Duffield, J.S. Lung Pericytes and Resident Fibroblasts. *Am. J. Pathol.* **2016**, *186*, 2519–2531. [CrossRef]
12. Lee, J.-H.; Tammela, T.; Hofree, M.; Choi, J.; Marjanovic, N.D.; Han, S.; Canner, D.; Wu, K.; Paschini, M.; Bhang, D.H.; et al. Anatomically and Functionally Distinct Lung Mesenchymal Populations Marked by Lgr5 and Lgr6. *Cell* **2017**, *170*, 1149–1163.e12. [CrossRef] [PubMed]
13. Sveiven, S.N.; Nordgren, T.M. Lung-resident mesenchymal stromal cells are tissue-specific regulators of lung homeostasis. *Am. J. Physiol.-Lung Cell. Mol. Physiol.* **2020**, *319*, L197–L210. [CrossRef]
14. Soundararajan, M.; Kannan, S. Fibroblasts and Mesenchymal Stem Cells: Two Sides of the Same Coin? *J. Cell. Physiol.* **2018**, *233*, 9099–9109. [CrossRef]
15. Hay, E.D. The mesenchymal cell, its role in the embryo, and the remarkable signaling mechanisms that create it. *Dev. Dyn.* **2005**, *233*, 706–720. [CrossRef] [PubMed]
16. da Silva Meirelles, L. Mesenchymal stem cells reside in virtually all post-natal organs and tissues. *J. Cell Sci.* **2006**, *119*, 2204–2213. [CrossRef]
17. Rinn, J.L.; Bondre, C.; Gladstone, H.B.; Brown, P.O.; Chang, H.Y. Anatomic Demarcation by Positional Variation in Fibroblast Gene Expression Programs. *PLoS Genet.* **2006**, *2*, e119. [CrossRef]
18. Spadafora, R.; Lu, J.; Khetani, R.S.; Zhang, C.; Iberg, A.; Li, H.; Shi, Y.; Lerou, P.H. Lung Resident Mesenchymal Stromal Cells Reveal Transcriptional Dynamics of Lung Development in Preterm Infants. *Am. J. Respir. Crit. Care Med.* **2018**, *198*, 961–964. [CrossRef]
19. Abreu, S.C.; Rolandsson Enes, S.; Dearborn, J.; Goodwin, M.; Coffey, A.; Borg, Z.D.; dos Santos, C.C.; Wargo, M.J.; Cruz, F.F.; Loi, R.; et al. Lung inflammatory environments differentially alter mesenchymal stromal cell behavior. *Am. J. Physiol. Lung Cell. Mol. Physiol.* **2019**, *317*, L823–L831. [CrossRef] [PubMed]
20. Travaglini, K.J.; Nabhan, A.N.; Penland, L.; Sinha, R.; Gillich, A.; Sit, R.V.; Chang, S.; Conley, S.D.; Mori, Y.; Seita, J.; et al. A molecular cell atlas of the human lung from single-cell RNA sequencing. *Nature* **2020**, *587*, 619–625. [CrossRef]
21. Negretti, N.M.; Plosa, E.J.; Benjamin, J.T.; Schuler, B.A.; Christian, A.; Jetter, C.; Gulleman, P.; Taylor, C.J.; Nichols, D.; Matlock, K.; et al. A Single Cell Atlas of Lung Development. *bioRxiv* **2021**. [CrossRef]
22. Slavkin, H.C.; Snead, M.L.; Zeichner-David, M.; Jaskoll, T.F.; Smith, B.T. Concepts of epithelial-mesenchymal interactions during development: Tooth and lung organogenesis. *J. Cell. Biochem.* **1984**, *26*, 117–125. [CrossRef]

23. Agha, E.E.; Herold, S.; Alam, D.A.; Quantius, J.; MacKenzie, B.; Carraro, G.; Moiseenko, A.; Chao, C.-M.; Minoo, P.; Seeger, W.; et al. Fgf10-positive cells represent a progenitor cell population during lung development and postnatally. *Development* **2014**, *141*, 296–306. [CrossRef] [PubMed]
24. Habermann, A.C.; Gutierrez, A.J.; Bui, L.T.; Yahn, S.L.; Winters, N.I.; Calvi, C.L.; Peter, L.; Chung, M.-I.; Taylor, C.J.; Jetter, C.; et al. Single-Cell RNA Sequencing Reveals Profibrotic Roles of Distinct Epithelial and Mesenchymal Lineages in Pulmonary Fibrosis. *Sci. Adv.* **2020**, *6*, eaba1972. [CrossRef] [PubMed]
25. Reyfman, P.A.; Walter, J.M.; Joshi, N.; Anekalla, K.R.; McQuattie-Pimentel, A.C.; Chiu, S.; Fernandez, R.; Akbarpour, M.; Chen, C.-I.; Ren, Z.; et al. Single-Cell Transcriptomic Analysis of Human Lung Provides Insights into the Pathobiology of Pulmonary Fibrosis. *Am. J. Respir. Crit. Care Med.* **2019**, *199*, 1517–1536. [CrossRef]
26. Hein, R.F.C.; Wu, J.H.; Tsai, Y.-H.; Wu, A.; Miller, A.J.; Holloway, E.M.; Frum, T.; Conchola, A.S.; Szenker-Ravi, E.; Reversade, B.; et al. R-SPONDIN2^{+} Mesenchymal Cells Form the Bud Tip Progenitor Niche During Human Lung Development. *bioRxiv* **2021**. [CrossRef]
27. Guo, M.; Du, Y.; Gokey, J.J.; Ray, S.; Bell, S.M.; Adam, M.; Sudha, P.; Perl, A.K.; Deshmukh, H.; Potter, S.S.; et al. Single Cell RNA Analysis Identifies Cellular Heterogeneity and Adaptive Responses of the Lung at Birth. *Nat. Commun.* **2019**, *10*, 37. [CrossRef]
28. Strunz, M.; Simon, L.M.; Ansari, M.; Kathiriya, J.J.; Angelidis, I.; Mayr, C.H.; Tsidiridis, G.; Lange, M.; Mattner, L.F.; Yee, M.; et al. Alveolar regeneration through a Krt8+ transitional stem cell state that persists in human lung fibrosis. *Nat. Commun.* **2020**, *11*, 3559. [CrossRef]
29. Angelidis, I. An Atlas of the Aging Lung Mapped by Single Cell Transcriptomics and Deep Tissue Proteomics. *Nat. Commun.* **2019**, *10*, 963. [CrossRef]
30. Solnica-Krezel, L.; Sepich, D.S. Gastrulation: Making and Shaping Germ Layers. *Annu. Rev. Cell Dev. Biol.* **2012**, *28*, 687–717. [CrossRef]
31. Thiery, J.P.; Acloque, H.; Huang, R.Y.J.; Nieto, M.A. Epithelial-Mesenchymal Transitions in Development and Disease. *Cell* **2009**, *139*, 871–890. [CrossRef] [PubMed]
32. Douarin, N.M.L.; Creuzet, S.; Couly, G.; Dupin, E. Neural crest cell plasticity and its limits. *Development* **2004**, *131*, 4637–4650. [CrossRef] [PubMed]
33. Takashima, Y.; Era, T.; Nakao, K.; Kondo, S.; Kasuga, M.; Smith, A.G.; Nishikawa, S.-I. Neuroepithelial Cells Supply an Initial Transient Wave of MSC Differentiation. *Cell* **2007**, *129*, 1377–1388. [CrossRef]
34. Que, J.; Wilm, B.; Hasegawa, H.; Wang, F.; Bader, D.; Hogan, B.L.M. Mesothelium contributes to vascular smooth muscle and mesenchyme during lung development. *Proc. Natl. Acad. Sci. USA* **2008**, *105*, 16626–16630. [CrossRef] [PubMed]
35. Yin, Y.; Wang, F.; Ornitz, D.M. Mesothelial- and epithelial-derived FGF9 have distinct functions in the regulation of lung development. *Development* **2011**, *138*, 3169–3177. [CrossRef]
36. Batra, H.; Antony, V.B. The pleural mesothelium in development and disease. *Front. Physiol.* **2014**, *5*, 284. [CrossRef]
37. Schittny, J.C. Development of the lung. *Cell Tissue Res.* **2017**, *367*, 427–444. [CrossRef]
38. Lange, P.; Celli, B.; Agustí, A.; Boje Jensen, G.; Divo, M.; Faner, R.; Guerra, S.; Marott, J.L.; Martinez, F.D.; Martinez-Camblor, P.; et al. Lung-Function Trajectories Leading to Chronic Obstructive Pulmonary Disease. *N. Engl. J. Med.* **2015**, *373*, 111–122. [CrossRef]
39. Nogawa, H.; Ito, T. Branching Morphogenesis of Embryonic Mouse Lung Epithelium in Mesenchyme-Free Culture. *Development* **1995**, *121*, 1015–1022. [CrossRef]
40. Warburton, D.; Seth, R.; Shum, L.; Horcher, P.G.; Hall, F.L.; Werb, Z.; Slavkin, H.C. Epigenetic role of epidermal growth factor expression and signalling in embryonic mouse lung morphogenesis. *Dev. Biol.* **1992**, *149*, 123–133. [CrossRef]
41. Warburton, D.; El-Hashash, A.; Carraro, G.; Tiozzo, C.; Sala, F.; Rogers, O.; De Langhe, S.; Kemp, P.J.; Riccardi, D.; Torday, J.; et al. Lung Organogenesis. *Curr. Top. Dev. Biol.* **2010**, *90*, 73–158. [CrossRef]
42. Koopmans, T.; Rinkevich, Y. Mesothelial to mesenchyme transition as a major developmental and pathological player in trunk organs and their cavities. *Commun. Biol.* **2018**, *1*, 1–14. [CrossRef]
43. Ariza, L.; Carmona, R.; Cañete, A.; Cano, E.; Muñoz-Chápuli, R. Coelomic epithelium-derived cells in visceral morphogenesis. *Dev. Dyn.* **2016**, *245*, 307–322. [CrossRef] [PubMed]
44. White, A.C.; Xu, J.; Yin, Y.; Smith, C.; Schmid, G.; Ornitz, D.M. FGF9 and SHH signaling coordinate lung growth and development through regulation of distinct mesenchymal domains. *Development* **2006**, *133*, 1507–1517. [CrossRef] [PubMed]
45. Yin, Y.; Ornitz, D.M. FGF9 and FGF10 activate distinct signaling pathways to direct lung epithelial specification and branching. *Sci. Signal.* **2020**, *13*, 621. [CrossRef]
46. Nakamura, K.T.; McCray, P.B. Fetal airway smooth-muscle contractility and lung development. A player in the band or just someone in the audience? *Am. J. Respir. Cell Mol. Biol.* **2000**, *23*, 3–6. [CrossRef]
47. Fayon, M.; Andrieux, A.; Bara, I.; Rebola, M.; Labbé, A.; Marthan, R.; Berger, P. An Age-Wise Comparison of Human Airway Smooth Muscle Proliferative Capacity. *PLoS ONE* **2015**, *10*, e0122446. [CrossRef]
48. Bokka, K.K.; Jesudason, E.C.; Lozoya, O.A.; Guilak, F.; Warburton, D.; Lubkin, S.R. Morphogenetic Implications of Peristalsis-Driven Fluid Flow in the Embryonic Lung. *PLoS ONE* **2015**, *10*, e0132015. [CrossRef]
49. Nelson, C.M.; Gleghorn, J.P.; Pang, M.-F.; Jaslove, J.M.; Goodwin, K.; Varner, V.D.; Miller, E.; Radisky, D.C.; Stone, H.A. Microfluidic chest cavities reveal that transmural pressure controls the rate of lung development. *Development* **2017**, *144*, 4328–4335. [CrossRef] [PubMed]

50. Ornitz, D.M.; Itoh, N. Fibroblast Growth Factors. *Genome Biol.* **2001**, *2*, 3005.1–3005.12. [CrossRef]
51. Shannon, J.M.; Hyatt, B.A. Epithelial-Mesenchymal Interactions in the Developing Lung. *Annu. Rev. Physiol.* **2004**, *66*, 625–645. [CrossRef]
52. Bellusci, S.; Grindley, J.; Emoto, H.; Itoh, N.; Hogan, B.L. Fibroblast Growth Factor 10 (FGF10) and Branching Morphogenesis in the Embryonic Mouse Lung. *Development* **1997**, *124*, 4867–4878. [CrossRef]
53. Park, W.Y.; Miranda, B.; Lebeche, D.; Hashimoto, G.; Cardoso, W.V. FGF-10 Is a Chemotactic Factor for Distal Epithelial Buds during Lung Development. *Dev. Biol.* **1998**, *201*, 125–134. [CrossRef] [PubMed]
54. Lü, J.; Izvolsky, K.I.; Qian, J.; Cardoso, W.V. Identification of FGF10 Targets in the Embryonic Lung Epithelium during Bud Morphogenesis. *J. Biol. Chem.* **2005**, *280*, 4834–4841. [CrossRef]
55. Sekine, K.; Ohuchi, H.; Fujiwara, M.; Yamasaki, M.; Yoshizawa, T.; Sato, T.; Yagishita, N.; Matsui, D.; Koga, Y.; Itoh, N.; et al. Fgf10 is essential for limb and lung formation. *Nat. Genet.* **1999**, *21*, 138–141. [CrossRef]
56. Min, H.; Danilenko, D.M.; Scully, S.A.; Bolon, B.; Ring, B.D.; Tarpley, J.E.; DeRose, M.; Simonet, W.S. Fgf-10 is required for both limb and lung development and exhibits striking functional similarity to Drosophila branchless. *Genes Dev.* **1998**, *12*, 3156–3161. [CrossRef] [PubMed]
57. Taghizadeh, S.; Heiner, M.; Vazquez-Armendariz, A.I.; Wilhelm, J.; Herold, S.; Chen, C.; Zhang, J.S.; Bellusci, S. Characterization in mice of the resident mesenchymal niche maintaining AT2 stem cell proliferation in homeostasis and disease. *Stem Cells* **2021**, *39*, 1382–1394. [CrossRef] [PubMed]
58. Colvin, J.S.; Green, R.P.; Schmahl, J.; Capel, B.; Ornitz, D.M. Male-to-Female Sex Reversal in Mice Lacking Fibroblast Growth Factor 9. *Cell* **2001**, *104*, 875–889. [CrossRef]
59. Winnier, G.; Blessing, M.; Labosky, P.A.; Hogan, B.L. Bone morphogenetic protein-4 is required for mesoderm formation and patterning in the mouse. *Genes Dev.* **1995**, *9*, 2105–2116. [CrossRef]
60. Weaver, M.; Dunn, N.R.; Hogan, B.L. Bmp4 and Fgf10 Play Opposing Roles during Lung Bud Morphogenesis. *Development* **2000**, *127*, 2695–2704. [CrossRef]
61. Weaver, M.; Batts, L.; Hogan, B.L.M. Tissue interactions pattern the mesenchyme of the embryonic mouse lung. *Dev. Biol.* **2003**, *258*, 169–184. [CrossRef]
62. Bellusci, S.; Henderson, R.; Winnier, G.; Oikawa, T.; Hogan, B.L. Evidence from Normal Expression and Targeted Misexpression That Bone Morphogenetic Protein (Bmp-4) Plays a Role in Mouse Embryonic Lung Morphogenesis. *Development* **1996**, *122*, 1693–1702. [CrossRef]
63. Herriges, M.; Morrisey, E.E. Lung development: Orchestrating the generation and regeneration of a complex organ. *Development* **2014**, *141*, 502–513. [CrossRef]
64. Eblaghie, M.C.; Reedy, M.; Oliver, T.; Mishina, Y.; Hogan, B.L.M. Evidence that autocrine signaling through Bmpr1a regulates the proliferation, survival and morphogenetic behavior of distal lung epithelial cells. *Dev. Biol.* **2006**, *291*, 67–82. [CrossRef] [PubMed]
65. Shi, W.; Zhao, J.; Anderson, K.D.; Warburton, D. Gremlin negatively modulates BMP-4 induction of embryonic mouse lung branching morphogenesis. *Am. J. Physiol.-Lung Cell. Mol. Physiol.* **2001**, *280*, L1030–L1039. [CrossRef] [PubMed]
66. Hirashima, T.; Iwasa, Y.; Morishita, Y. Mechanisms for split localization of Fgf10 expression in early lung development. *Dev. Dyn.* **2009**, *238*, 2813–2822. [CrossRef]
67. Menshykau, D.; Kraemer, C.; Iber, D. Branch Mode Selection during Early Lung Development. *PLoS Comput. Biol.* **2012**, *8*, e1002377. [CrossRef]
68. Carballo, G.B.; Honorato, J.R.; de Lopes, G.P.F.; de Sampaio e Spohr, T.C.L. A highlight on Sonic hedgehog pathway. *Cell Commun. Signal.* **2018**, *16*, 11. [CrossRef]
69. Litingtung, Y.; Lei, L.; Westphal, H.; Chiang, C. Sonic hedgehog is essential to foregut development. *Nat. Genet.* **1998**, *20*, 58–61. [CrossRef]
70. Cardoso, W.V.; Lü, J. Regulation of early lung morphogenesis: Questions, facts and controversies. *Development* **2006**, *133*, 1611–1624. [CrossRef]
71. Pepicelli, C.V.; Lewis, P.M.; McMahon, A.P. Sonic hedgehog regulates branching morphogenesis in the mammalian lung. *Curr. Biol.* **1998**, *8*, 1083–1086. [CrossRef]
72. Peng, T.; Tian, Y.; Boogerd, C.J.; Lu, M.M.; Kadzik, R.S.; Stewart, K.M.; Evans, S.M.; Morrisey, E.E. Coordination of heart and lung co-development by a multipotent cardiopulmonary progenitor. *Nature* **2013**, *500*, 589–592. [CrossRef] [PubMed]
73. Miettinen, P.J.; Berger, J.E.; Meneses, J.; Phung, Y.; Pedersen, R.A.; Werb, Z.; Derynck, R. Epithelial immaturity and multiorgan failure in mice lacking epidermal growth factor receptor. *Nature* **1995**, *376*, 337–341. [CrossRef] [PubMed]
74. Miettinen, P.J.; Warburton, D.; Bu, D.; Zhao, J.-S.; Berger, J.E.; Minoo, P.; Koivisto, T.; Allen, L.; Dobbs, L.; Werb, Z.; et al. Impaired Lung Branching Morphogenesis in the Absence of Functional EGF Receptor. *Dev. Biol.* **1997**, *186*, 224–236. [CrossRef] [PubMed]
75. Seth, R.; Shum, L.; Wu, F.; Wuenschell, C.; Hall, F.L.; Slavkin, H.C.; Warburton, D. Role of Epidermal Growth Factor Expression in Early Mouse Embryo Lung Branching Morphogenesis in Culture: Antisense Oligodeoxynucleotide Inhibitory Strategy. *Dev. Biol.* **1993**, *158*, 555–559. [CrossRef] [PubMed]
76. Schuger, L.; Varani, J.; Mitra, R.; Gilbride, K. Retinoic acid stimulates mouse lung development by a mechanism involving epithelial-mesenchymal interaction and regulation of epidermal growth factor receptors. *Dev. Biol.* **1993**, *159*, 462–473. [CrossRef]
77. Mark, M.; Ghyselinck, N.B.; Chambon, P. Function of retinoid nuclear receptors: Lessons from Genetic and Pharmacological Dissections of the Retinoic Acid Signaling Pathway During Mouse Embryogenesis. *Annu. Rev. Pharmacol. Toxicol.* **2006**, *46*, 451–480. [CrossRef]

78. Mendelsohn, C.; Lohnes, D.; Décimo, D.; Lufkin, T.; LeMeur, M.; Chambon, P.; Mark, M. Function of the retinoic acid receptors (RARs) during development (II). Multiple abnormalities at various stages of organogenesis in RAR double mutants. *Development* **1994**, *120*, 2749–2771. [CrossRef]
79. Fernandes-Silva, H.; Araújo-Silva, H.; Correia-Pinto, J.; Moura, R.S. Retinoic Acid: A Key Regulator of Lung Development. *Biomolecules* **2020**, *10*, 152. [CrossRef]
80. Chen, F.; Desai, T.J.; Qian, J.; Niederreither, K.; Lü, J.; Cardoso, W.V. Inhibition of Tgfβ signaling by endogenous retinoic acid is essential for primary lung bud induction. *Development* **2007**, *134*, 2969–2979. [CrossRef]
81. Kahata, K.; Dadras, M.S.; Moustakas, A. TGF-β Family Signaling in Epithelial Differentiation and Epithelial–Mesenchymal Transition. *Cold Spring Harb. Perspect. Biol.* **2018**, *10*, a022194. [CrossRef] [PubMed]
82. Saito, A.; Horie, M.; Nagase, T. TGF-β Signaling in Lung Health and Disease. *Int. J. Mol. Sci.* **2018**, *19*, 2460. [CrossRef] [PubMed]
83. Bragg, A.D.; Moses, H.L.; Serra, R. Signaling to the epithelium is not sufficient to mediate all of the effects of transforming growth factor β and bone morphogenetic protein 4 on murine embryonic lung development. *Mech. Dev.* **2001**, *109*, 13–26. [CrossRef]
84. Warburton, D.; Bellusci, S.; De Langhe, S.; Del Moral, P.-M.; Fleury, V.; Mailleux, A.; Tefft, D.; Unbekandt, M.; Wang, K.; Shi, W. Molecular Mechanisms of Early Lung Specification and Branching Morphogenesis. *Pediatr. Res.* **2005**, *57*, 26–37. [CrossRef] [PubMed]
85. Kaartinen, V.; Voncken, J.W.; Shuler, C.; Warburton, D.; Bu, D.; Heisterkamp, N.; Groffen, J. Abnormal lung development and cleft palate in mice lacking TGF-beta 3 indicates defects of epithelial-mesenchymal interaction. *Nat. Genet.* **1995**, *11*, 415–421. [CrossRef] [PubMed]
86. Chen, H.; Zhuang, F.; Liu, Y.-H.; Xu, B.; del Moral, P.; Deng, W.; Chai, Y.; Kolb, M.; Gauldie, J.; Warburton, D.; et al. TGF-β receptor II in epithelia versus mesenchyme plays distinct roles in the developing lung. *Eur. Respir. J.* **2008**, *32*, 285–295. [CrossRef]
87. Xing, Y.; Li, C.; Hu, L.; Tiozzo, C.; Li, M.; Chai, Y.; Bellusci, S.; Anderson, S.; Minoo, P. Mechanisms of TGFβ Inhibition of Lung Endodermal Morphogenesis: The role of TβRII, Smads, Nkx2.1 and Pten. *Dev. Biol.* **2008**, *320*, 340–350. [CrossRef]
88. Edgar, B.A. From Cell Structure to Transcription: Hippo Forges a New Path. *Cell* **2006**, *124*, 267–273. [CrossRef]
89. Meng, Z.; Moroishi, T.; Guan, K.-L. Mechanisms of Hippo pathway regulation. *Genes Dev.* **2016**, *30*, 1–17. [CrossRef]
90. Makita, R.; Uchijima, Y.; Nishiyama, K.; Amano, T.; Chen, Q.; Takeuchi, T.; Mitani, A.; Nagase, T.; Yatomi, Y.; Aburatani, H.; et al. Multiple renal cysts, urinary concentration defects, and pulmonary emphysematous changes in mice lacking TAZ. *Am. J. Physiol. Ren. Physiol.* **2008**, *294*, F542–F553. [CrossRef] [PubMed]
91. Isago, H.; Mitani, A.; Mikami, Y.; Horie, M.; Urushiyama, H.; Hamamoto, R.; Terasaki, Y.; Nagase, T. Epithelial Expression of YAP and TAZ Is Sequentially Required in Lung Development. *Am. J. Respir. Cell Mol. Biol.* **2020**, *62*, 256–266. [CrossRef] [PubMed]
92. Little, D.R.; Lynch, A.M.; Yan, Y.; Akiyama, H.; Kimura, S.; Chen, J. Differential chromatin binding of the lung lineage transcription factor NKX2-1 resolves opposing murine alveolar cell fates in vivo. *Nat. Commun.* **2021**, *12*, 2509. [CrossRef] [PubMed]
93. Mahoney, J.E.; Mori, M.; Szymaniak, A.D.; Varelas, X.; Cardoso, W.V. The Hippo Pathway Effector Yap Controls Patterning and Differentiation of Airway Epithelial Progenitors. *Dev. Cell* **2014**, *30*, 137–150. [CrossRef]
94. Zhao, R.; Fallon, T.R.; Saladi, S.V.; Pardo-Saganta, A.; Villoria, J.; Mou, H.; Vinarsky, V.; Gonzalez-Celeiro, M.; Nunna, N.; Hariri, L.P.; et al. Yap Tunes Airway Epithelial Size and Architecture by Regulating the Identity, Maintenance, and Self-renewal of Stem Cells. *Dev. Cell* **2014**, *30*, 151–165. [CrossRef]
95. Volckaert, T.; Yuan, T.; Chao, C.-M.; Bell, H.; Sitaula, A.; Szimmtenings, L.; El Agha, E.; Chanda, D.; Majka, S.; Bellusci, S.; et al. Fgf10-Hippo epithelial mesenchymal crosstalk maintains and recruits lung basal stem cells. *Dev. Cell* **2017**, *43*, 48–59.e5. [CrossRef]
96. McGowan, S.E. Extracellular matrix and the regulation of lung development and repair1. *FASEB J.* **1992**, *6*, 2895–2904. [CrossRef] [PubMed]
97. Godoy-Guzmán, C.; San Martin, S.; Pereda, J. Proteoglycan and collagen expression during human air conducting system development. *Eur. J. Histochem.* **2012**, *56*, e29. [CrossRef]
98. Zhao, J.; Sime, P.J.; Bringas, P.; Gauldie, J.; Warburton, D. Adenovirus-mediated decorin gene transfer prevents TGF-β-induced inhibition of lung morphogenesis. *Am. J. Physiol.-Lung Cell. Mol. Physiol.* **1999**, *277*, L412–L422. [CrossRef] [PubMed]
99. Izvolsky, K.I.; Shoykhet, D.; Yang, Y.; Yu, Q.; Nugent, M.A.; Cardoso, W.V. Heparan sulfate–FGF10 interactions during lung morphogenesis. *Dev. Biol.* **2003**, *258*, 185–200. [CrossRef]
100. Kotton, D.N.; Morrisey, E.E. Lung regeneration: Mechanisms, applications and emerging stem cell populations. *Nat. Med.* **2014**, *20*, 822–832. [CrossRef]
101. Rock, J.R.; Onaitis, M.W.; Rawlins, E.L.; Lu, Y.; Clark, C.P.; Xue, Y.; Randell, S.H.; Hogan, B.L.M. Basal cells as stem cells of the mouse trachea and human airway epithelium. *Proc. Natl. Acad. Sci. USA* **2009**, *106*, 12771–12775. [CrossRef] [PubMed]
102. Rawlins, E.L.; Okubo, T.; Xue, Y.; Brass, D.M.; Auten, R.L.; Hasegawa, H.; Wang, F.; Hogan, B.L.M. The role of Scgb1a1+ Clara cells in the long-term maintenance and repair of lung airway, but not alveolar, epithelium. *Cell Stem Cell* **2009**, *4*, 525–534. [CrossRef]
103. Tata, P.R.; Mou, H.; Pardo-Saganta, A.; Zhao, R.; Prabhu, M.; Prabhu, M.; Law, B.M.; Vinarsky, V.; Cho, J.L.; Breton, S.; et al. Dedifferentiation of committed epithelial cells into stem cells in vivo. *Nature* **2013**, *503*, 218–223. [CrossRef]
104. Lafkas, D.; Shelton, A.; Chiu, C.; de Leon Boenig, G.; Chen, Y.; Stawicki, S.S.; Siltanen, C.; Reichelt, M.; Zhou, M.; Wu, X.; et al. Therapeutic antibodies reveal Notch control of transdifferentiation in the adult lung. *Nature* **2015**, *528*, 127–131. [CrossRef] [PubMed]

105. Peng, T.; Frank, D.B.; Kadzik, R.S.; Morley, M.P.; Rathi, K.S.; Wang, T.; Zhou, S.; Cheng, L.; Lu, M.M.; Morrisey, E.E. Hedgehog actively maintains adult lung quiescence and regulates repair and regeneration. *Nature* **2015**, *526*, 578–582. [CrossRef]
106. Fang, S.; Zhang, S.; Dai, H.; Hu, X.; Li, C.; Xing, Y. The role of pulmonary mesenchymal cells in airway epithelium regeneration during injury repair. *Stem Cell Res. Ther.* **2019**, *10*, 366. [CrossRef]
107. Zepp, J.A.; Zacharias, W.J.; Frank, D.B.; Cavanaugh, C.A.; Zhou, S.; Morley, M.P.; Morrisey, E.E. Distinct Mesenchymal Lineages and Niches Promote Epithelial Self-Renewal and Myofibrogenesis in the Lung. *Cell* **2017**, *170*, 1134–1148.e10. [CrossRef]
108. Barnes, J.L.; Gorin, Y. Myofibroblast differentiation during fibrosis: Role of NAD(P)H oxidases. *Kidney Int.* **2011**, *79*, 944–956. [CrossRef] [PubMed]
109. El Agha, E.; Kramann, R.; Schneider, R.K.; Li, X.; Seeger, W.; Humphreys, B.D.; Bellusci, S. Mesenchymal Stem Cells in Fibrotic Disease. *Cell Stem Cell* **2017**, *21*, 166–177. [CrossRef] [PubMed]
110. Buechler, M.B.; Pradhan, R.N.; Krishnamurty, A.T.; Cox, C.; Calviello, A.K.; Wang, A.W.; Yang, Y.A.; Tam, L.; Caothien, R.; Roose-Girma, M.; et al. Cross-tissue organization of the fibroblast lineage. *Nature* **2021**, *593*, 575–579. [CrossRef]
111. Yu, K.; Xu, J.; Liu, Z.; Sosic, D.; Shao, J.; Olson, E.N.; Towler, D.A.; Ornitz, D.M. Conditional inactivation of FGF receptor 2 reveals an essential role for FGF signaling in the regulation of osteoblast function and bone growth. *Development* **2003**, *130*, 3063–3074. [CrossRef]
112. Harrell, C.R.; Jankovic, M.G.; Fellabaum, C.; Volarevic, A.; Djonov, V.; Arsenijevic, A.; Volarevic, V. Molecular mechanisms responsible for anti-inflammatory and immunosuppressive effects of mesenchymal stem cell-derived factors. In *Tissue Engineering and Regenerative Medicine*; Pham, P.V., Ed.; Springer International Publishing: Cham, Switzerland, 2019; pp. 187–206.
113. Wolters, P.J.; Collard, H.R.; Jones, K.D. Pathogenesis of Idiopathic Pulmonary Fibrosis. *Annu. Rev. Pathol.* **2014**, *9*, 157–179. [CrossRef] [PubMed]
114. Gohy, S.T.; Hupin, C.; Pilette, C.; Ladjemi, M.Z. Chronic inflammatory airway diseases: The central role of the epithelium revisited. *Clin. Exp. Allergy* **2016**, *46*, 529–542. [CrossRef] [PubMed]
115. Rabe, K.F.; Watz, H. Chronic obstructive pulmonary disease. *Lancet* **2017**, *389*, 1931–1940. [CrossRef]
116. Barnes, P.J. Chronic Obstructive Pulmonary Disease. *N. Engl. J. Med.* **2000**, *343*, 1969–1971. [CrossRef] [PubMed]
117. Barnes, P.J.; Burney, P.G.J.; Silverman, E.K.; Celli, B.R.; Vestbo, J.; Wedzicha, J.A.; Wouters, E.F.M. Chronic obstructive pulmonary disease. *Nat. Rev. Dis. Primers* **2015**, *1*, 15076. [CrossRef] [PubMed]
118. Celli, B.R.; Wedzicha, J.A. Update on Clinical Aspects of Chronic Obstructive Pulmonary Disease. *N. Engl. J. Med.* **2019**, *381*, 1257–1266. [CrossRef] [PubMed]
119. Perez-Padilla, R.; Thirion-Romero, I.; Guzman, N. Underdiagnosis of chronic obstructive pulmonary disease: Should smokers be offered routine spirometry tests? *Expert Rev. Respir. Med.* **2018**, *12*, 83–85. [CrossRef] [PubMed]
120. Rabe, K.F.; Hurst, J.R.; Suissa, S. Cardiovascular disease and COPD: Dangerous liaisons? *Eur. Respir. Rev.* **2018**, *27*, 180057. [CrossRef]
121. Lee, A.L.; Goldstein, R.S. Gastroesophageal reflux disease in COPD: Links and risks. *Int. J. Chronic Obstr. Pulm. Dis.* **2015**, *10*, 1935–1949. [CrossRef]
122. Lahousse, L.; Tiemeier, H.; Ikram, M.A.; Brusselle, G.G. Chronic obstructive pulmonary disease and cerebrovascular disease: A comprehensive review. *Respir. Med.* **2015**, *109*, 1371–1380. [CrossRef] [PubMed]
123. Gea, J.; Sancho-Muñoz, A.; Chalela, R. Nutritional status and muscle dysfunction in chronic respiratory diseases: Stable phase versus acute exacerbations. *J. Thorac. Dis.* **2018**, *10*, S1332–S1354. [CrossRef]
124. Jaitovich, A.; Barreiro, E. Skeletal Muscle Dysfunction in Chronic Obstructive Pulmonary Disease. What We Know and Can Do for Our Patients. *Am. J. Respir. Crit. Care Med.* **2018**, *198*, 175–186. [CrossRef] [PubMed]
125. Smith, M.C.; Wrobel, J.P. Epidemiology and clinical impact of major comorbidities in patients with COPD. *Int. J. Chronic Obstr. Pulm. Dis.* **2014**, *9*, 871–888. [CrossRef]
126. Anderson, D.O. Smoking And Respiratory Disease. *Am. J. Public Health Nations Health* **1964**, *54*, 1856–1863. [CrossRef]
127. Fletcher, C.; Peto, R. The natural history of chronic airflow obstruction. *Br. Med. J.* **1977**, *1*, 1645–1648. [CrossRef]
128. Laniado-Laborín, R. Smoking and Chronic Obstructive Pulmonary Disease (COPD). Parallel Epidemics of the 21st Century. *Int. J. Environ. Res. Public Health* **2009**, *6*, 209–224. [CrossRef]
129. Fullerton, D.G.; Bruce, N.; Gordon, S.B. Indoor air pollution from biomass fuel smoke is a major health concern in the developing world. *Trans. R. Soc. Trop. Med. Hyg.* **2008**, *102*, 843–851. [CrossRef] [PubMed]
130. Salvi, S.S.; Barnes, P.J. Chronic obstructive pulmonary disease in non-smokers. *Lancet* **2009**, *374*, 733–743. [CrossRef]
131. Jiang, X.-Q.; Mei, X.-D.; Feng, D. Air pollution and chronic airway diseases: What should people know and do? *J. Thorac. Dis.* **2016**, *8*, E31–E40. [CrossRef]
132. Cohen, R.; Patel, A.; Green, F. Lung Disease Caused by Exposure to Coal Mine and Silica Dust. *Semin. Respir. Crit. Care Med.* **2008**, *29*, 651–661. [CrossRef] [PubMed]
133. Martinez, F.D. Early-Life Origins of Chronic Obstructive Pulmonary Disease. *N. Engl. J. Med.* **2016**, *375*, 871–878. [CrossRef]
134. Wilk, J.B.; Chen, T.-H.; Gottlieb, D.J.; Walter, R.E.; Nagle, M.W.; Brandler, B.J.; Myers, R.H.; Borecki, I.B.; Silverman, E.K.; Weiss, S.T.; et al. A genome-wide association study of pulmonary function measures in the Framingham Heart Study. *PLoS Genet.* **2009**, *5*, e1000429. [CrossRef] [PubMed]
135. Pillai, S.G.; Ge, D.; Zhu, G.; Kong, X.; Shianna, K.V.; Need, A.C.; Feng, S.; Hersh, C.P.; Bakke, P.; Gulsvik, A.; et al. A genome-wide association study in chronic obstructive pulmonary disease (COPD): Identification of two major susceptibility loci. *PLoS Genet.* **2009**, *5*, e1000421. [CrossRef] [PubMed]

136. Repapi, E.; Sayers, I.; Wain, L.V.; Burton, P.R.; Johnson, T.; Obeidat, M.; Zhao, J.H.; Ramasamy, A.; Zhai, G.; Vitart, V.; et al. Genome-wide association study identifies five loci associated with lung function. *Nat. Genet.* **2010**, *42*, 36–44. [CrossRef]
137. Soler Artigas, M.; Loth, D.W.; Wain, L.V.; Gharib, S.A.; Obeidat, M.; Tang, W.; Zhai, G.; Zhao, J.H.; Smith, A.V.; Huffman, J.E.; et al. Genome-wide association and large-scale follow up identifies 16 new loci influencing lung function. *Nat. Genet.* **2011**, *43*, 1082–1090. [CrossRef] [PubMed]
138. Chuang, P.-T.; McMahon, A.P. Vertebrate Hedgehog signalling modulated by induction of a Hedgehog-binding protein. *Nature* **1999**, *397*, 617–621. [CrossRef]
139. Smith, B.M.; Traboulsi, H.; Austin, J.H.M.; Manichaikul, A.; Hoffman, E.A.; Bleecker, E.R.; Cardoso, W.V.; Cooper, C.; Couper, D.J.; Dashnaw, S.M.; et al. Human airway branch variation and chronic obstructive pulmonary disease. *Proc. Natl. Acad. Sci. USA.* **2018**, *115*, E974–E981. [CrossRef]
140. Agusti, A.; Faner, R. Lung function trajectories in health and disease. *Lancet Respir. Med.* **2019**, *7*, 358–364. [CrossRef]
141. Barnes, P.J. Inflammatory mechanisms in patients with chronic obstructive pulmonary disease. *J. Allergy Clin. Immunol.* **2016**, *138*, 16–27. [CrossRef]
142. Shaykhiev, R.; Crystal, R.G. Early Events in the Pathogenesis of Chronic Obstructive Pulmonary Disease. Smoking-induced Reprogramming of Airway Epithelial Basal Progenitor Cells. *Ann. Am. Thorac. Soc.* **2014**, *11*, S252–S258. [CrossRef]
143. Baraldo, S.; Turato, G.; Badin, C.; Bazzan, E.; Beghé, B.; Zuin, R.; Calabrese, F.; Casoni, G.; Maestrelli, P.; Papi, A.; et al. Neutrophilic infiltration within the airway smooth muscle in patients with COPD. *Thorax* **2004**, *59*, 308–312. [CrossRef]
144. Leopold, P.L.; O'Mahony, M.J.; Lian, X.J.; Tilley, A.E.; Harvey, B.-G.; Crystal, R.G. Smoking Is Associated with Shortened Airway Cilia. *PLoS ONE* **2009**, *4*, e8157. [CrossRef]
145. Mahmood, M.Q.; Sohal, S.S.; Shukla, S.D.; Ward, C.; Hardikar, A.; Noor, W.D.; Muller, H.K.; Knight, D.A.; Walters, E.H. Epithelial mesenchymal transition in smokers: Large versus small airways and relation to airflow obstruction. *Int. J. Chronic Obs. Pulm. Dis.* **2015**, *10*, 1515–1524. [CrossRef]
146. Higham, A.; Quinn, A.M.; Cançado, J.E.D.; Singh, D. The pathology of small airways disease in COPD: Historical aspects and future directions. *Respir. Res.* **2019**, *20*, 1–11. [CrossRef]
147. Yang, J.; Antin, P.; Berx, G.; Blanpain, C.; Brabletz, T.; Bronner, M.; Campbell, K.; Cano, A.; Casanova, J.; Christofori, G.; et al. Guidelines and definitions for research on epithelial–mesenchymal transition. *Nat. Rev. Mol. Cell Biol.* **2020**, *21*, 341–352. [CrossRef]
148. Osei, E.T.; Hackett, T.-L. Epithelial-mesenchymal crosstalk in COPD: An update from in vitro model studies. *Int. J. Biochem. Cell Biol.* **2020**, *125*, 105775. [CrossRef]
149. Hogg, J.C. Pathophysiology of airflow limitation in chronic obstructive pulmonary disease. *Lancet* **2004**, *364*, 709–721. [CrossRef]
150. McDonough, J.E.; Yuan, R.; Suzuki, M.; Seyednejad, N.; Elliott, W.M.; Sanchez, P.G.; Wright, A.C.; Gefter, W.B.; Litzky, L.; Coxson, H.O.; et al. Small-Airway Obstruction and Emphysema in Chronic Obstructive Pulmonary Disease. *N. Engl. J. Med.* **2011**, *365*, 1567–1575. [CrossRef] [PubMed]
151. Barnes, P.J. Small airway fibrosis in COPD. *Int. J. Biochem. Cell Biol.* **2019**, *116*, 105598. [CrossRef] [PubMed]
152. Araya, J.; Cambier, S.; Markovics, J.A.; Wolters, P.; Jablons, D.; Hill, A.; Finkbeiner, W.; Jones, K.; Broaddus, V.C.; Sheppard, D.; et al. Squamous metaplasia amplifies pathologic epithelial-mesenchymal interactions in COPD patients. *J. Clin. Investig.* **2007**, *117*, 3551–3562. [CrossRef] [PubMed]
153. de Boer, W.I.; van Schadewijk, A.; Sont, J.K.; Sharma, H.S.; Stolk, J.; Hiemstra, P.S.; van Krieken, J.H. Transforming growth factor beta1 and recruitment of macrophages and mast cells in airways in chronic obstructive pulmonary disease. *Am. J. Respir. Crit. Care Med.* **1998**, *158*, 1951–1957. [CrossRef] [PubMed]
154. Gohy, S.T.; Hupin, C.; Fregimilicka, C.; Detry, B.R.; Bouzin, C.; Gaide Chevronay, H.; Lecocq, M.; Weynand, B.; Ladjemi, M.Z.; Pierreux, C.E.; et al. Imprinting of the COPD airway epithelium for dedifferentiation and mesenchymal transition. *Eur. Respir. J.* **2015**, *45*, 1258–1272. [CrossRef]
155. Eapen, M.S.; Lu, W.; Hackett, T.L.; Singhera, G.K.; Mahmood, M.Q.; Hardikar, A.; Ward, C.; Walters, E.H.; Sohal, S.S. Increased myofibroblasts in the small airways, and relationship to remodelling and functional changes in smokers and COPD patients: Potential role of epithelial-mesenchymal transition. *ERJ Open Res.* **2021**, *7*, 00876-2020. [CrossRef]
156. Karakioulaki, M.; Papakonstantinou, E.; Stolz, D. Extracellular matrix remodelling in COPD. *Eur. Respir. Rev.* **2020**, *29*, 190124. [CrossRef]
157. Annoni, R.; Lanças, T.; Tanigawa, R.Y.; Matsushita, M.D.M.; Fernezlian, S.D.M.; Bruno, A.; da Silva, L.F.F.; Roughley, P.J.; Battaglia, S.; Dolhnikoff, M.; et al. Extracellular matrix composition in COPD. *Eur. Respir. J.* **2012**, *40*, 1362–1373. [CrossRef]
158. Eurlings, I.M.; Dentener, M.A.; Cleutjens, J.P.; Peutz, C.J.; Rohde, G.G.; Wouters, E.F.; Reynaert, N.L. Similar matrix alterations in alveolar and small airway walls of COPD patients. *BMC Pulm. Med.* **2014**, *14*, 90. [CrossRef] [PubMed]
159. Bidan, C.M.; Veldsink, A.C.; Meurs, H.; Gosens, R. Airway and Extracellular Matrix Mechanics in COPD. *Front. Physiol.* **2015**, *6*, 346. [CrossRef]
160. Hedström, U.; Hallgren, O.; Öberg, L.; DeMicco, A.; Vaarala, O.; Westergren-Thorsson, G.; Zhou, X. Bronchial extracellular matrix from COPD patients induces altered gene expression in repopulated primary human bronchial epithelial cells. *Sci. Rep.* **2018**, *8*, 3502. [CrossRef]
161. Milara, J.; Peiró, T.; Serrano, A.; Cortijo, J. Epithelial to mesenchymal transition is increased in patients with COPD and induced by cigarette smoke. *Thorax* **2013**, *68*, 410–420. [CrossRef]

162. Nieto, M.A.; Cano, A. The epithelial–mesenchymal transition under control: Global programs to regulate epithelial plasticity. *Semin. Cancer Biol.* **2012**, *22*, 361–368. [CrossRef]
163. Kalluri, R.; Neilson, E.G. Epithelial-mesenchymal transition and its implications for fibrosis. *J. Clin. Investig.* **2003**, *112*, 1776–1784. [CrossRef] [PubMed]
164. Kalluri, R.; Weinberg, R.A. The basics of epithelial-mesenchymal transition. *J. Clin. Investig.* **2009**, *119*, 1420–1428. [CrossRef] [PubMed]
165. Zeisberg, M.; Neilson, E.G. Biomarkers for epithelial-mesenchymal transitions. *J. Clin. Investig.* **2009**, *119*, 1429–1437. [CrossRef] [PubMed]
166. Lee, J.M.; Dedhar, S.; Kalluri, R.; Thompson, E.W. The epithelial–mesenchymal transition: New insights in signaling, development, and disease. *J. Cell Biol.* **2006**, *172*, 973–981. [CrossRef]
167. Nieto, M.A.; Huang, R.Y.-J.; Jackson, R.A.; Thiery, J.P. EMT: 2016. *Cell* **2016**, *166*, 21–45. [CrossRef]
168. Soltani, A.; Sohal, S.S.; Reid, D.; Weston, S.; Wood-Baker, R.; Walters, E.H. Vessel-Associated Transforming Growth Factor-Beta1 (TGF-β1) Is Increased in the Bronchial Reticular Basement Membrane in COPD and Normal Smokers. *PLoS ONE* **2012**, *7*, e39736. [CrossRef]
169. Eapen, M.S.; Sharma, P.; Gaikwad, A.V.; Lu, W.; Myers, S.; Hansbro, P.M.; Sohal, S.S. Epithelial–mesenchymal transition is driven by transcriptional and post transcriptional modulations in COPD: Implications for disease progression and new therapeutics. *Int. J. Chronic Obstr. Pulm. Dis.* **2019**, *14*, 1603–1610. [CrossRef]
170. Baarsma, H.A.; Spanjer, A.I.R.; Haitsma, G.; Engelbertink, L.H.J.M.; Meurs, H.; Jonker, M.R.; Timens, W.; Postma, D.S.; Kerstjens, H.A.M.; Gosens, R. Activation of WNT/β-Catenin Signaling in Pulmonary Fibroblasts by TGF-β1 Is Increased in Chronic Obstructive Pulmonary Disease. *PLoS ONE* **2011**, *6*, e25450. [CrossRef]
171. Sohal, S.S.; Reid, D.; Soltani, A.; Ward, C.; Weston, S.; Muller, H.K.; Wood-Baker, R.; Walters, E.H. Evaluation of epithelial mesenchymal transition in patients with chronic obstructive pulmonary disease. *Respir. Res.* **2011**, *12*, 130. [CrossRef]
172. Heijink, I.H.; Brandenburg, S.M.; Postma, D.S.; van Oosterhout, A.J.M. Cigarette smoke impairs airway epithelial barrier function and cell-cell contact recovery. *Eur. Respir. J.* **2012**, *39*, 419–428. [CrossRef]
173. Sohal, S.S.; Reid, D.; Soltani, A.; Ward, C.; Weston, S.; Muller, H.K.; Wood-Baker, R.; Walters, E.H. Reticular basement membrane fragmentation and potential epithelial mesenchymal transition is exaggerated in the airways of smokers with chronic obstructive pulmonary disease. *Respirology* **2010**, *15*, 930–938. [CrossRef] [PubMed]
174. Johnson, P.R.; Roth, M.; Tamm, M.; Hughes, M.; Ge, Q.; King, G.; Burgess, J.K.; Black, J.L. Airway smooth muscle cell proliferation is increased in asthma. *Am. J. Respir. Crit. Care Med.* **2001**, *164*, 474–477. [CrossRef] [PubMed]
175. Trian, T.; Benard, G.; Begueret, H.; Rossignol, R.; Girodet, P.-O.; Ghosh, D.; Ousova, O.; Vernejoux, J.-M.; Marthan, R.; Tunon-de-Lara, J.-M.; et al. Bronchial smooth muscle remodeling involves calcium-dependent enhanced mitochondrial biogenesis in asthma. *J. Exp. Med.* **2007**, *204*, 3173–3181. [CrossRef] [PubMed]
176. Saetta, M.; Di Stefano, A.; Turato, G.; Facchini, F.M.; Corbino, L.; Mapp, C.E.; Maestrelli, P.; Ciaccia, A.; Fabbri, L.M. CD8+ T-lymphocytes in peripheral airways of smokers with chronic obstructive pulmonary disease. *Am. J. Respir. Crit. Care Med.* **1998**, *157*, 822–826. [CrossRef]
177. Antunes, M.A.; Abreu, S.C.; Cruz, F.F.; Teixeira, A.C.; Lopes-Pacheco, M.; Bandeira, E.; Olsen, P.C.; Diaz, B.L.; Takyia, C.M.; Freitas, I.P.; et al. Effects of different mesenchymal stromal cell sources and delivery routes in experimental emphysema. *Respir. Res.* **2014**, *15*, 1–14. [CrossRef]
178. Gu, W.; Song, L.; Li, X.-M.; Wang, D.; Guo, X.-J.; Xu, W.-G. Mesenchymal stem cells alleviate airway inflammation and emphysema in COPD through down-regulation of cyclooxygenase-2 via p38 and ERK MAPK pathways. *Sci. Rep.* **2015**, *5*, 8733. [CrossRef]
179. Weiss, D.J.; Casaburi, R.; Flannery, R.; LeRoux-Williams, M.; Tashkin, D.P. A Placebo-Controlled, Randomized Trial of Mesenchymal Stem Cells in COPD. *Chest* **2013**, *143*, 1590–1598. [CrossRef]
180. Le Thi Bich, P.; Nguyen Thi, H.; Dang Ngo Chau, H.; Phan Van, T.; Do, Q.; Dong Khac, H.; Le Van, D.; Nguyen Huy, L.; Mai Cong, K.; Ta Ba, T.; et al. Allogeneic umbilical cord-derived mesenchymal stem cell transplantation for treating chronic obstructive pulmonary disease: A pilot clinical study. *Stem Cell Res. Ther.* **2020**, *11*, 60. [CrossRef]
181. Antunes, M.A.; Lapa e Silva, J.R.; Rocco, P.R. Mesenchymal stromal cell therapy in COPD: From bench to bedside. *Int. J. Chronic Obs. Pulm. Dis.* **2017**, *12*, 3017–3027. [CrossRef]
182. Broekman, W.; Khedoe, P.P.S.J.; Schepers, K.; Roelofs, H.; Stolk, J.; Hiemstra, P.S. Mesenchymal stromal cells: A novel therapy for the treatment of chronic obstructive pulmonary disease? *Thorax* **2018**, *73*, 565–574. [CrossRef] [PubMed]
183. Ikonomou, L.; Wagner, D.E.; Turner, L.; Weiss, D.J. Translating Basic Research into Safe and Effective Cell-based Treatments for Respiratory Diseases. *Ann. ATS* **2019**, *16*, 657–668. [CrossRef] [PubMed]
184. Najar, M.; Johanne, M.P.; Jean Pierre, P.; Fahmi, H. Novel Insights for Improving the Therapeutic Safety and Efficiency of Mesenchymal Stromal Cells. *World J. Stem Cells* **2020**, *12*, 1474–1491. [CrossRef] [PubMed]
185. Paris, G.C.; Azevedo, A.A.; Ferreira, A.L.; Azevedo, Y.M.A.; Rainho, M.A.; Oliveira, G.P.; Silva, K.R.; Cortez, E.A.C.; Stumbo, A.C.; Carvalho, S.N.; et al. Therapeutic potential of mesenchymal stem cells in multiple organs affected by COVID-19. *Life Sci.* **2021**, *278*, 119510. [CrossRef]

cells

MDPI

Review

Tissue-Nonspecific Alkaline Phosphatase, a Possible Mediator of Cell Maturation: Towards a New Paradigm

Masahiro Sato [1], Issei Saitoh [2], Yuki Kiyokawa [2], Yoko Iwase [3], Naoko Kubota [4], Natsumi Ibano [2], Hirofumi Noguchi [5], Youichi Yamasaki [4] and Emi Inada [4,*]

1 Department of Genome Medicine, National Center for Child Health and Development, 2-10-1 Okura, Setagaya, Tokyo 157-8535, Japan; sato-masa@ncchd.go.jp
2 Department of Pediatric Dentistry, Asahi University School of Dentistry, Gifu 501-0296, Japan; isaitoh@dent.asahi-u.ac.jp (I.S.); ykiyokawa@dent.asahi-u.ac.jp (Y.K.); bano@dent.asahi-u.ac.jp (N.I.)
3 Department of Dentistry for the Disabled, Asahi University School of Dentistry, Gifu 501-0296, Japan; iwase@dent.asahi-u.ac.jp
4 Department of Pediatric Dentistry, Graduate School of Medical and Dental Sciences, Kagoshima University, Kagoshima 890-8544, Japan; k1744111@dent.kagoshima-u.ac.jp (N.K.); yamasaki@dent.kagoshima-u.ac.jp (Y.Y.)
5 Department of Regenerative Medicine, Graduate School of Medicine, University of the Ryukyus, Okinawa 903-0215, Japan; noguchih@med.u-ryukyu.ac.jp
* Correspondence: inada@dent.kagoshima-u.ac.jp; Tel.: +81-3-3416-0181

Abstract: Alkaline phosphatase (ALP) is a ubiquitous membrane-bound glycoprotein capable of providing inorganic phosphate by catalyzing the hydrolysis of organic phosphate esters, or removing inorganic pyrophosphate that inhibits calcification. In humans, four forms of ALP cDNA have been cloned, among which tissue-nonspecific ALP (TNSALP) (TNSALP) is widely distributed in the liver, bone, and kidney, making it an important marker in clinical and basic research. Interestingly, TNSALP is highly expressed in juvenile cells, such as pluripotent stem cells (i.e., embryonic stem cells and induced pluripotent stem cells (iPSCs)) and somatic stem cells (i.e., neuronal stem cells and bone marrow mesenchymal stem cells). Hypophosphatasia is a genetic disorder causing defects in bone and tooth development as well as neurogenesis. Mutations in the gene coding for TNSALP are thought to be responsible for the abnormalities, suggesting the essential role of TNSALP in these events. Moreover, a reverse-genetics-based study using mice revealed that TNSALP is important in bone and tooth development as well as neurogenesis. However, little is known about the role of TNSALP in the maintenance and differentiation of juvenile cells. Recently, it was reported that cells enriched with TNSALP are more easily reprogrammed into iPSCs than those with less TNSALP. Furthermore, in bone marrow stem cells, ALP could function as a "signal regulator" deciding the fate of these cells. In this review, we summarize the properties of ALP and the background of ALP gene analysis and its manipulation, with a special focus on the potential role of TNSALP in the generation (and possibly maintenance) of juvenile cells.

Keywords: alkaline phosphatase; tissue-nonspecific alkaline phosphatase; pluripotent stem cells; somatic stem cells; signal regulator; juvenile cells; reprogramming; induced pluripotent stem cells

Citation: Sato, M.; Saitoh, I.; Kiyokawa, Y.; Iwase, Y.; Kubota, N.; Ibano, N.; Noguchi, H.; Yamasaki, Y.; Inada, E. Tissue-Nonspecific Alkaline Phosphatase, a Possible Mediator of Cell Maturation: Towards a New Paradigm. *Cells* **2021**, *10*, 3338. https://doi.org/10.3390/cells10123338

Academic Editor: Mehdi Najar

Received: 12 October 2021
Accepted: 25 November 2021
Published: 28 November 2021

1. General Property of Alkaline Phosphatase (ALP)

Alkaline phosphatase (ALP; EC 3.1.3.1) is a ubiquitous membrane-bound glycoprotein found in many organisms, from bacteria to humans. In most cases, they are homodimeric enzymes, and each catalytic site contains three metal ions, i.e., two Zn and one Mg. The enzymes catalyze the hydrolysis by catalyzing the hydrolysis of organic phosphate esters, or removing inorganic pyrophosphate (PPi), an inhibitor of biomineralization [1]. As physiological substrates for ALPs, PPi, phosphoethanolamine (PE), and pyridoxal 5′-phosphate (PLP) are generally known [1]. For example, ALP (which is specifically known as tissue-nonspecific alkaline phosphatase (TNSALP), as shown later) isolated from human SAOS-2 osteosarcoma cells hydrolyzes PE and PLP at physiologic pH [2]. ALPs also appear to be involved in the metabolism of nucleotides. For instance, Say et al. [3] reported that purified TNSALP could hydrolyze adenosine triphosphate (ATP), adenosine diphosphate (ADP), adenosine monophosphate (AMP), PPi, glucose-1-phosphate, glucose-6-phosphate, fructose-6-phosphate, β-glycerophosphate, bis-(p-nitrophenyl)-phosphate and p-nitrophenyl phosphate.

ALP was originally described in a histochemical study as a marker for various tissues, especially bone and tooth formation-related tissues [4]. Additionally, ALP is reportedly highly expressed in juvenile cells, such as early preimplantation mouse embryos (cleavage stage embryos and inner cell mass (ICMs)) [5,6], pluripotent stem cells (PSCs)/embryonic stem cells (ESCs)/induced pluripotent stem cells (iPSCs) [7,8], primordial germ cells (PGCs) [9] and spermatogonia [10], stem cells for spermatogenesis, and some types of somatic stem cells, such as neuronal stem cells (NSCs) [11], and bone marrow mesenchymal stem cells (BMMSCs) [12]. Moreover, ALP is known to be highly expressed in the intestine, placenta, adipose, B lymphocytes, colon T lymphocytes, and osteoblasts (precursors for bone formation) [13]. Besides the normal cells, ALP expression is correlated with the progression of some types of cancers (i.e., colon cancer, osteosarcomas, neuroblastoma, and leukemia), making it a clinical marker for these cancer cells [13].

Clinically, a mutation in the gene coding for TNSALP has been closely associated with a severe skeletal deformity disease termed "hypophosphatasia (HPP)," which is characterized by several pathological abnormalities, including rickets, osteomalacia, epilepsy-like seizures associated with vitamin B6 deficiency, muscle weakness, and respiratory disturbance [14,15].

These findings suggest that ALP has some roles in various biological systems.

2. ALP Isoforms and Their Detailed Properties

In humans, at least four forms of *ALP* cDNA have been cloned: intestinal ALP (IALP or IAP; restricted to the intestine), placental ALP (PLALP, PLAP or Regan isozyme; restricted to the placenta), germ cell ALP (GCALP, GCAP or NAGAO isozyme; restricted to early embryonic cells), and liver/bone/kidney ALP (L/B/K ALP; widely distributed) [1,16]. The last form is generally called TNSALP or TNAP [1,16]. Similarly, at least four forms of ALP (i.e., embryonic ALP (EAP), IALP, a putative pseudogene, and TNSALP) have been identified in mice [1,16]. In humans, the gene for TNSALP is located on chromosome 1, and the genes for the other three isoforms (IALP, PLALP, and GCALP) are located on chromosome 2 [16].

Notably, PLALP and GCALP have approximately 98% homology, PLALP and IALP have approximately 87% homology, and IALP and TNSALP have only approximately 57% homology [17]. According to Whyte [18], the gene encoding TNSALP seems to be the ancestral gene. PLALP, GCALP, IALP, and TNSALP belong to the ALP family.

2.1. PLALP

PLALP is a polymorphic heat-stable enzyme present at high levels in the placenta, with up to 18 allozymes resulting from point mutations, in contrast to the other ALP isoenzymes [16]. The gene coding for PLALP is defined as *ALPP* (Table 1) [17], and the gene can be re-expressed by cancer cells as the Regan isoenzyme [16].

The biological functions of PLALP have been extensively determined by reverse genetics-based analysis. For example, Skynner et al. [20] demonstrated that systemic overexpression of human PLALP has no adverse effects on mouse development or viability using transgenic (Tg) mouse lines. Based on these findings, they suggested that PLALP could be used as a reporter gene in conjunction with, or as an alternative to ß-galactosidase (ß-gal; encoded by *lacZ*). A similar conclusion was also drawn by DePrimo et al. [21] (Table 2). The results of knock-out (KO), knock-in (KI), and Tg mouse models for assessing the gene function of the ALP family are summarized in Table 2.

Table 1. Summary [1] of the gene nomenclature, accession number, and common names for the human and mouse alkaline phosphatase (ALP).

Genes	Accession Numbers	Protein Names	Tissue Distribution	Function
Human genes *ALPL*	NM_000478	TNAP, TNSALP, L/B/K ALP	Developing nervous system, skeletal tissues, liver and kidney	Skeletal mineralization
ALPP	NM_001632	PLAP, PLALP, Regan isoenzyme	Syncytiotrophoblast, a variety of tumors	Unknown
ALPP2	NM_031313	GCAP, GCALP, NAGAO isozyme	Testis, malignant trophoblasts, testicular cancer	Unknown
ALP1	NM_001631	IAP, IALP, Kasahara isoenzyme	Gut	Fat absorption, Detoxification of lipopolysaccharide
Mouse genes *Akp2 (Alpl)*	NM_007431	TNAP, TNSALP, L/B/K ALP	Developing nervous system, skeletal tissues and kidney	Skeletal mineralization
Akp3	NM_007432	IAP, IALP, dIALP	Gut	Fat absorption, detoxification of lipopolysaccharide
Akp5 (Alpl2)	NM_007433	EAP	Preimplantation embryo, testis, gut	Early embryogenesis
Akp-ps1	NG 001340	ALP pseudogene, pseudoALP	Not described	
Akp6 (Alpi)	AK008000	gIALP	Gut	Under investigation

[1] Created based on the data of Linder et al. [19]. Abbreviations: dIALP, duodenum-specific intestinal ALP; EAP, embryonic ALP; GCALP (or GCAP), germ cell ALP; gIALP, global IALP; IALP (or IAP), intestinal ALP; L/B/K ALP, liver/bone/kidney ALP; PLALP (or PLAP), placental ALP; TNSALP (or TNAP), tissue-nonspecific ALP.

Table 2. Summary of knock-out (KO), knock-in (KI) or transgenic (Tg) mice for assessing gene function of alkaline phosphatase (ALP) family.

Methods	Target Gene	Outcome (Note)	Reference
Overexpression	*ALPP*	Tg mice overexpressing human *PLALP* systemically were produced. There were no adverse effects on mouse development or viability.	[20,21]
Gene targeting for KO	*ALP1*	Homozygous mutants (called $Akp3^{-/-}$) were histologically normal and fertile. However, accelerated transport of fat droplets through the intestinal epithelium and elevation of serum triglyceride levels were discernible, which was associated with higher intestinal Ca^{2+} uptake.	[22,23]
Overexpression	*ALPP2*	Tg mouse lines harboring human *ALPP2* linked to short or long promoter region of *ALPP2* were produced. A 450 bp promoter sequence directs *GCALP* expression to the intestine and endothelial cells, while a 1.7 kb promoter sequence directs *GCALP* expression to the spermatogenic lineage and to the eight-cell embryos through the blastocysts.	[24]
Gene targeting for KO	*Akp5 (Alpl2)*	No obvious phenotypic abnormalities. Normal reproductive activity with acquisition of live offspring, indicating the nonessential role of EAP during embryonic development.	[25]
Gene targeting for KO	*Akp5 (Alpl2)*	In homozygous mutants (called *EAP.ko*), preimplantation embryos exhibited slower development and higher rates of degeneration, delayed parturition, and reduced litter size.	[26]

Table 2. *Cont.*

Methods	Target Gene	Outcome (Note)	Reference
Gene targeting for KI	*ALPL*	In homozygous mutants (called *Alpl*^{tm1Sor}), primordial germ cells appear unaffected indicating that *ALPL* is not essential for their development or migration. At first, the mice exhibited normal skeletal development; however, homozygous mutant mice developed seizures and apnea at approximately two weeks after birth, and died before weaning. Rescued animals subsequently develop defective dentition. TNSALP modulates T lymphocyte function (specifically T cell-dependent colitis) when examined using heterozygous *Alpl*^{tm1Sor} mice.	[27–29]
Gene targeting for KO	*ALPL*	In homozygous mutants (called *Alpl*^{tm1Jlm} or *Akp2*$^{-/-}$), abnormal bone mineralization was evident. Morphological changes in the osteoblasts, aberrant development of the lumbar nerve roots, disturbances in intestinal physiology, increased apoptosis in the thymus, and abnormal spleens are also discernible. Loss of *ALPL* causes myelin abnormalities and synaptic dysfunction.	[25,30,31]
Gene targeting for KI	*ALPL*	KI of Cre expression unit was carried out into the region between exons 6 and 7 of *ALPL* gene. The resulting line was called *Alpl*$^{tm1(cre)Nagy}$. After crossed *Alpl*$^{tm1(cre)Nagy}$ line with the double-reporter line, Z/AP, human ALP expression is discernible in PGCs at E9.5–10.5. After mid-gestation, however, it was also expressed in the labyrinthine region of the placenta, the intestine and the neural tube.	[32]
Gene targeting for KI	*ALPL*	KI of an around 12 kb genomic sequence of *Akp2* in which two *loxP* sites (located in introns 2 and 4, respectively) and a cassette containing neomycin resistance gene expression unit into the endogenous *Akp2* locus. This floxed mouse (called *Alpl*$^{flox/flox}$) is normal in the absence of Cre expression. However, in the presence of Cre, the deletion of exons 3 and 4 should occur, which may result in the ablation of endogenous TNSALP expression.	[33]
Overexpression	*ALPL*	Tg mouse line (called "*Col1a1-Tnap*") expressing human *ALPL* under control of an osteoblast-specific collagen type I α1 chain (*Col1a1*) promoter was produced. This line is healthy and exhibits increased bone mineralization.	[34]
Overexpression	*ALPL*	Tg mice carrying human *ALPL* under the vascular smooth muscle cell-specific transgelin (*Tagln*) promoter were produced. They developed severe arterial medial calcification and reduced viability.	[35]
Overexpression	*ALPL*	Tg mice (celled "*Endothelial TNSALP mice*") carrying *ALPL* under the endothelial-specific tunica intima endothelial kinase 2 (*Tie2*) promoter were produced. They survived well into adulthood and displayed generalized arterial calcification together with elevated blood pressure and compensatory left ventricular hypertrophy.	[36]

Abbreviations: *ALPL*, tissue-nonspecific ALP (TNSALP); E, embryonic day; *EAP*, embryonic ALP; *GCALP*, germ cell ALP; PGC, primordial germ cell; *PLALP*, placental ALP.

2.2. IALP

IALP is a partially heat-stable isozyme present at high levels in intestinal tissue. In contrast to the other ALP isoenzymes, the carbohydrate side-chains of IALP are not terminated by sialic acid [16].

Although trace expression of IALP has been detected in the thymus, IALP expression in mice and humans is largely restricted to the gut during late embryonic development and adult stages [37]. Notably, IALP can be re-expressed in cancer cells as a modified form, designated as Kasahara isoenzyme [16,38]. The gene encoding IALP is defined as *ALP1* in humans [17] (Table 1) and Akp3 in mice (Table 1) [1,17], which encodes duodenum-specific intestinal IALP (dIALP) [19]. Notably, Akp6 encodes global IALP (gIALP) [19].

To examine the biological functions of IALP, Narisawa et al. [22] generated *Akp3* null-mutated mice, called *Akp3*$^{-/-}$ mice, which were histologically normal and fertile. However, long-term maintenance on a high-fat diet resulted in higher body weight gain compared with wild-type (WT) animals. Histological examination revealed the accelerated transport of fat droplets through the intestinal epithelium and elevated serum triglyceride levels in the *Akp3*$^{-/-}$ mice compared to that in WT mice. Narisawa et al. [22] suggested that IALP participates in a rate-limiting step, regulating fat absorption (Table 2).

Notably, Lynes et al. [39] demonstrated that efficient fat (long-chain fatty acid, LCFA) transport across the small intestinal epithelium is mediated by IALPs and the putative fatty acid translocase/CD36. According to them, phosphorylated CD36 (pCD36) exists in mouse enterocytes, and pCD36 is a substrate of gIALP. gIALP-mediated dephosphorylation results in increased LCFA transport capability.

2.3. GCALP

GCALP is a heat-stable isozyme present at low levels in germ cells, embryonal, and some neoplastic tissues [16]. Similar to IALP, it can be re-expressed as a NAGAO isozyme in cancer cells [16]. The gene encoding GCALP is defined as *ALPP2* (Table 1) [1,17].

To examine the properties of the *ALPP2* promoter, Narisawa et al. [24] generated several Tg mouse lines harboring the entire human *ALPP2* gene. The results demonstrated that the 450 bp promoter sequence directed human *ALPP2* expression to the intestine and endothelial cells, whereas the 1.7 kb promoter sequence directed human *ALPP2* expression to the spermatogenic lineage and to the eight-cell embryos through the blastocysts (Table 2). More importantly, *ALPP2* expression in Tg mice induced cellular immune tolerance to GCALP. Narisawa et al. [24] suggested that these *ALPP2* Tg mice could be useful for studying immune responses associated with repeated administration of conjugated or derivatized anti-GCALP monoclonal antibodies.

2.4. EAP (in Mice)

EAP is first expressed during early embryogenesis (from the 2- to 8-cell stages to the blastocyst stage), but is not detectable in embryos older than 7.5 days post coitum (dpc) [37]. EAP is subsequently re-expressed in the thymus, intestine, and testis in adult mice [6]. The gene encoding EAP in mice is defined as Akp5 (Table 1) [1,17].

To examine the biological functions of EAP, Narisawa et al. [25] generated Akp5 null-mutated mice via homologous recombination, and the resulting mice had no obvious phenotypic abnormalities, indicating the nonessential role of EAP during embryonic development (Table 2). Dehghani et al. [26] independently created *Akp5* null-mutated mice called *EAP.ko* mice. *EAP.ko* preimplantation embryos had slower development and higher rates of degeneration in vitro, leading to fewer blastocysts. In vivo, *EAP.ko* mice had delayed parturition and reduced litter size. Furthermore, there was no compensation for the absence of *Akp5* in the embryos by other isozymes of ALP (Table 2). Overall, the study's findings showed that the presence of an active *Akp5* is beneficial for mouse preimplantation development.

2.5. TNSALP

TNSALP is a heat-labile isozyme [4,13], which is expressed in the ICM of blastocysts and in migrating PGCs [28,37]. Additionally, TNSALP is expressed in developing neural tubes of mouse embryos between 8.5 dpc and 13.5 dpc [40]; however, it is expressed in skeletal tissues at later stages of embryogenesis. In adult mice, TNSALP is expressed in a wide variety of cell types, such as osteoblasts, neutrophils, renal tubules, capillaries in the brain, and myoid cells in the testis [25].

The gene encoding TNSALP is defined as *ALPL* in humans and as *Akp2* in mice (Table 1) [1,17]. ALPL consists of 12 exons, with the coding sequence beginning within the second exon [40–42]. Eleven exons are translated to form 507-amino-acid residues comprising TNSALP [40–42]. Exons 2–12 are contained within 25 kb of the DNA. *ALPL* has two promoters and two corresponding 5′ noncoding exons 1a and 1b, resulting in two different mRNAs. Transcription from the upstream promoter (1a) is used in osteoblasts, whereas the downstream promoter (1b) is used in the liver and kidney [43]. That *ALPL* has a dual-function promoter is also shown in humans [44], rats [45,46], and mice [47].

As already described, TNSALP is a zinc-containing metalloenzyme and functions as a dimeric molecule [16]. Based on the analysis of the *ALPL* cDNA, this gene encodes a preproprotein that is proteolytically cleaved to yield a signal peptide (comprising 17 amino acids) and a proprotein that is subsequently processed to generate the active mature peptide. TNSALP has five potential *N*-linked glycosylation sites that are essential for full activity [17]. The deduced active site of TNSALP is encoded by six exons comprising 15 amino acid residues and shares the same nucleotide sequence with other ALPs. There is a C-terminal hydrophobic domain that is replaced by a glycosylphosphatidylinositol (GPI) anchor, which is involved in inserting protein into the lipid bilayer of cells and liposomes [17].

Regarding the biological role of TNSALP, Kotobuki et al. [48] suggested an important role of TNSALP in osteoblastic function. They introduced small interfering RNAs (siRNAs) of the osteogenic-related genes (including runt-related transcription factor 2 (*RUNX2*), collagen type I α1 chain (*COL1A1*), *ALPL*, and osteocalcin (*OCN*)) into human ALP high-expressing osteoblast-like cells. They found that only *ALPL* siRNA inhibited matrix mineralization. In this instance, the expression of not only *ALPL* but also *COL1A1* and *RUNX2* were influenced by the inhibition of *ALPL*. In contrast, *OCN* expression was not affected by the inhibition of *ALPL*. Kotobuki et al. [48] concluded that TNSALP is a strong candidate for promoting matrix mineralization. Furthermore, as mentioned below, TNSALP promotes bone mineralization mainly by hydrolyzing extracellular inorganic PPi, which is a strong mineralization inhibitor.

2.5.1. Creation of Akp2 KI or KO Mice

To examine the biological functions of TNSALP, Soriano and Millán's groups generated *Akp2* KO mice using traditional gene targeting methods, and MacGregor et al. [27] generated KI mice (called $Alpl^{tm1Sor}$). In these mice, the genomic *Akp2* sequence spanning from the middle of exon 2 to the middle of exon 6 was replaced with a *βgeo* cassette (an expression unit for both *lacZ* and neomycin resistance gene (neo^r)) to enable the expression of β-gal under the control of the endogenous *TNSALP* promoter. Assessment of β-gal activity in the heterozygous progeny of $Alpl^{tm1Sor}$ showed that TNSALP was not expressed in PGC progenitors prior to gastrulation. The ß-gal activity was first found in an extraembryonic lineage destined to form the chorion. In homozygous null mutants, PGCs appeared unaffected, indicating that TNSALP is not essential for their development or migration. In contrast, $Alpl^{tm1Sor}$ mice had elevated plasma PLP levels and died from seizures caused by PL deficiency in cells of the central nervous system (CNS), resulting from a decrease in the hydrolysis of PLP to pyridoxal (PL) (Table 2) [31]. Additionally, Waymire et al. [28] reported that although $Alpl^{tm1Sor}$ mice had normal skeletal development, they exhibited elevated PLP levels, resulting in death approximately two weeks after birth (Table 2). However, the seizure phenotype can be rescued by the administration of PL and a semi-solid diet, but the rescued animals subsequently manifested dental dysplasia.

Narisawa et al. [25] generated *Akp2* KO mice (called *Alpl*^{tm1Jlm} or *Akp2*$^{-/-}$) by inserting a *neo*^{r} cassette into exon 6 of the *Akp2* gene. The *Akp2* KO mice exhibited impaired growth, vitamin-B6-dependent seizures, impaired bone mineralization, and apnea, and died before weaning (Table 2) [25]. The plasma of the KO mice contained low ALP activity (Table 2) [22], whereas *Akp2* heterozygous mice had approximately 50% of the plasma ALP activity of WT mice and were healthy and fertile [25]. Narisawa et al. [25] speculated that TNSALP appears not to be essential for the initial events leading to bone mineral deposition but might play a role in maintaining the bone mineralization process after birth.

Notably, *Akp2* KO lines, *Alpl*^{tm1Sor} and *Alpl*^{tm1Jlm}, were compared in detail by Fedde et al. [30], who reported an increase in putative natural substrates (phosphoethanolamine, PPi, and PLP) for TNSALP in both lines. Bone abnormality was first observed around 10 days after birth, and osteopenia and fractures worsened at later stages, similar to the characteristics of infantile HPP.

2.5.2. Creation of Conditional Akp2 KO Mice

Narisawa [33] generated a floxed mouse line *Alpl*$^{flox/flox}$, by KI of an around 12 kb genomic sequence of *Akp2* in which two *loxP* sites (located in introns 2 and 4, respectively) and a cassette containing *neo*^{r} expression unit were transfected into the endogenous *Akp2* locus (Table 2). The results show that the floxed mouse was normal in the absence of Cre expression. However, in the presence of Cre, deletion of exons 3 and 4 occurred, which resulted in the inhibition of endogenous TNSALP expression. Narisawa [33] generated a Cre Tg mouse line, called *Col1a1-Cre*, in which bone-specific expression of Cre was driven under the control of *Col1a1* promoter. There was around 30% reduction in the plasma ALP levels in the bigenic offspring of *Alpl*$^{flox/flox}$ and *Col1a1-Cre* compared with that in the WT mice. The *Alpl*$^{flox/flox}$ line is thus useful for examining the function of TNSALP in any cell or tissue if appropriate Cre Tg lines are available.

Notably, Lomeli et al. [32] generated a Cre Tg mouse line, called *Alpl*$^{tm1(cre)Nagy}$, by inserting *Cre* into the endogenous *Akp2* locus (Table 2), indicating that Cre expression is controlled by the *Akp2* promoter. When the *Alpl*$^{tm1(cre)Nagy}$ mice were crossed with the double-reporter line, Z/AP [49], carrying a floxed sequence containing *loxP* site, *lacZ* gene, *neo*^{r}, transcription stopper, *loxP* site, and a gene encoding human alkaline phosphatase (hAP), the resulting bigenic offspring exhibited expression of reporter genes in PGCs at E9.5–10.5. At mid-gestational stages, it was expressed in the labyrinthine region of the placenta, the intestine, and the neural tube. Crossing *Alpl*$^{flox/flox}$ with *Alpl*$^{tm1(cre)Nagy}$ would result in the tissue-specific (i.e., PGCs, labyrinthine region of the placenta, and embryonic intestines and neural tubes) inhibition of TNSALP expression.

2.5.3. Creation of Akp2 Mutant Mice through N-Ethyl-N-Nitrosourea (ENU) Mutagenesis

ENU mutagenesis is the forward genetics or phenotype-driven approach (phenotype to the gene) involving the screening of mice for mutant phenotypes without previous knowledge of the genetic basis of the mutation [50]. This approach is used when there is a need for highly efficient induction of point mutations in mice [51]. To perform ENU mutagenesis, male mice (aged 6- to 8-week old) are usually intraperitoneally injected with a solution containing ENU, an alkylating agent, which acts as a powerful mutagen [51].

In 2007, the Gena 328 mouse (called *Alpl*^{Hpp}) exhibited low plasma ALP activity and late-onset skeletal abnormalities but a normal life span and no epilepsy [52]. A point mutation at the splice site for exon 8 of *Akp2* produced a truncated, inactive *TNSALP* having 276 residues rather than the 525 residues of WT *TNSALP*.

In 2009, the mouse mutant (called *Alpl*^{ALPLD1}) exhibited low plasma ALP activity [53]. An A to G point mutation in exon 5 of *Akp2* caused an Asp to Gly change at residue 109 of the TNSALP protein.

In 2012, the same group reported new mouse mutants, called *Alpl*^{BAP023}, *Alpl*^{BAP026}, and *Alpl*^{BAP027}, produced by ENU mutagenesis [54]. *Alpl*^{BAP023} carries a missense T to G mutation in exon 7 at nucleotide (nt) 755 (Leu to Pro at residue 251). *Alpl*^{BAP026} has a splice site mutation in intron 9. *Alpl*^{BAP027} carries a T to A mutation in exon 10 at nt 1194 (Ile to Asn at residue 395).

In 2015, a murine model for odontohypophosphatasia (odonto-HPP) was first generated by Foster et al. [55]. Odonto-HPP is a mild form of HPP characterized only by oral manifestations, including premature exfoliation of deciduous teeth. The odonto-HPP mouse model is characterized by a missense mutation that changes codon 116 from Ala to Thr, which has been identified similarly with autosomal dominant (AD) odonto-HPP. Overall, the odonto-HPP mouse model has around 50% WT plasma ALP activity of WT mice, with similar litter size, survival, and body weight.

2.5.4. Creation of Tg Mice Overexpressing TNSALP

Furthermore, studies have examined the effect of the overexpression of TNSALP in various tissues on their behavior and function. Narisawa et al. [34] generated a Tg mouse line (called "*Col1a1-Tnap*") expressing human *ALPL* under the control of an osteoblast-specific *Col1a1* promoter. The offspring of this line was healthy but exhibited increased mineralization of vertebrae bones (Table 2). Additionally, there was higher mineralization of osteoblasts from the Tg mice compared with that from non-Tg controls.

Sheen et al. [35] demonstrated that Tg mice carrying human *ALPL* and overexpressing TNSALP in vascular smooth muscle cells under the smooth muscle cell-specific transgelin (*Tagln*) promoter developed severe arterial medial calcification and reduced viability (Table 2).

Savinov et al. [36] determined the effect of TNSALP overexpression on vascular calcification and cardiovascular function using the endothelial-specific tunica intima endothelial kinase 2 (*Tie2*) promoter in mice. Mice with endothelial TNSALP overexpression, called "endothelial *TNSALP* mice," survived well into adulthood and displayed generalized arterial calcification (Table 2). Genes associated with osteochondrogenesis (*Runx2*, bone γ-carboxyglutamate protein 2 (*Bglap2*), osteopontin (*Opn*), osteoprotegerin (*Opg*), and collagen type II α1 chain (*Col2a1*)) were upregulated in the aortas of the Tg offspring. At 23 weeks of age, endothelial *TNSALP* mice developed elevated blood pressure and compensatory left ventricular hypertrophy with normal ejection fraction. Savinov et al. [36] concluded that TNSALP has osteogenic potential in endothelial cells, promoting vascular calcification.

3. Various Biological Roles of TNSALP

As mentioned above, TNSALP is expressed in a variety of soft tissues, which do not undergo mineralization (e.g., kidney and liver), thereby casting some doubt on previous thought that TNSALP is involved only in osteoblastic function related to mineralization. The various roles of TNSALP are discussed below.

3.1. Osteogenic System

Mesenchymal stem cells (MSCs) are precursors of bone-producing osteoblasts, and the protein encoded by ALPL is enriched in the stem cell membrane, with involvement in ATP metabolism during cell differentiation. In Figure 1, the mechanism underlying the involvement of TNSALP in osteogenesis is shown. TNSALP in MSCs is tightly associated with AMP-activated protein kinase (AMPK) activation during osteogenesis [56,57]. In this system, AMPK is a key regulator to maintain cellular energy homeostasis. For example, Liu et al. [56] suggested that TNSALP deficiency in MSCs enhances ATP release and reduces ATP hydrolysis. The excessive extracellular ATP is, in turn, internalized by MSCs and causes an elevation in the intracellular ATP level, which consequently inactivates the AMPK pathway. However, AMPK is activated by the increased AMP, which is generated through the sequential degradation/hydrolysis reactions of polyP catalyzed by TNSALP.

RUNX2 is a substrate of AMPK, which directly phosphorylates serine 118 in the DNA-binding domain of RUNX2 [57]. The resulting phosphorylated RUNX2 may activate bone matrix protein genes, leading to enhanced osteogenesis.

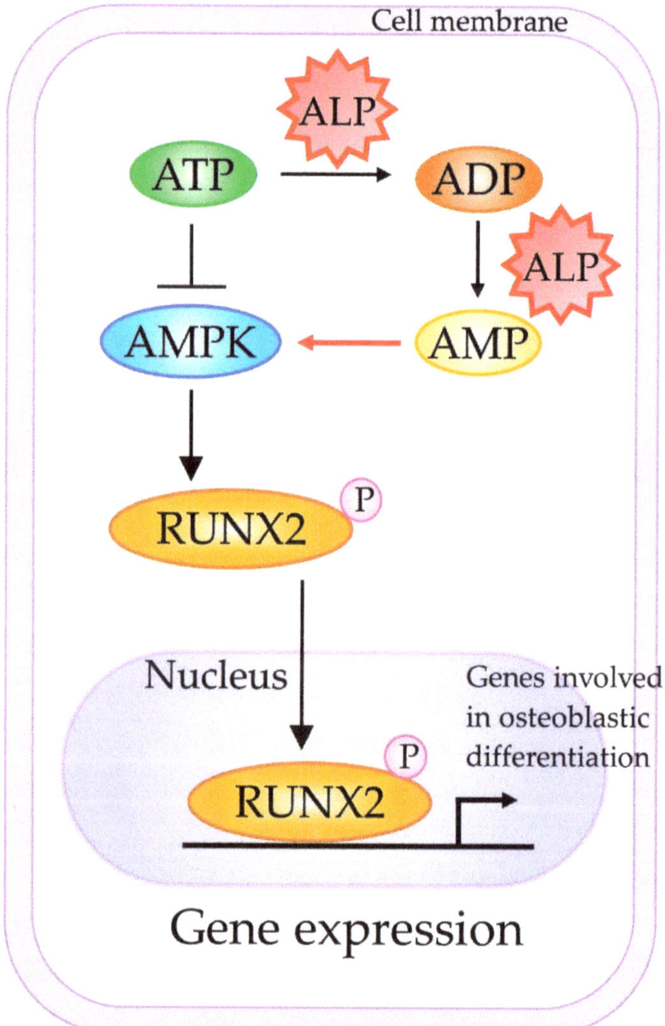

Figure 1. A schematic representation of the molecular mechanism underlying TNSALP-mediated activation of osteogenesis. Active AMPK and RUNX2 phosphorylation are preferentially associated with osteogenesis. Abbreviations: AMP, adenosine monophosphate; ADP, adenosine diphosphate; AMPK, AMP-activated protein kinase; ATP, adenosine triphosphate; MSCs, mesenchymal stem cells; RUNX2, RUNX family transcription factor 2. This figure was drawn in-house, based on the data shown in the paper of Chava et al. [57].

3.2. *Lipid and Energy Metabolism of Fat Cells*

According to Hernández-Mosqueira et al. [58], TNSALP is expressed in adipose tissue and in 3T3-F442A adipocytes. Moreover, there was an upregulation of glycerophosphate dehydrogenase, adiponectin (a recently described adipokine), and fatty acid-binding protein 4 (FABP4 levels; which is also called adipocyte protein 2 (aP2)) genes, but decreased expression of leptin in TNSALP knockdown cells. The latter three proteins are important in adipocyte systemic signaling and insulin sensitivity. Furthermore, inhibition of ALP activity in adipocytes by levamisole (a reversible inhibitor for TNSALP) reduced lipolysis and the expression of various lipogenic genes. Hernández-Mosqueira et al. [58] suggest that the activity of TNSALP might have a critical role in the energy balance of the adipocyte, probably participating in obesity and metabolic syndrome.

3.3. Neuronal System

Hanics et al. [31] assessed the function of TNSALP in a neuronal system in vivo using *Akp2* KO mice ($Akp2^{-/-}$) and reported that TNSALP is expressed in the synaptic cleft and the node of Ranvier in normal adults. Ablation of TNSALP function resulted in a significant decrease in the white matter of the spinal cord accompanied by cellular degradation around the paranodal regions and a decreased ratio and diameter of the myelinated axons (Table 2). Hanics et al. [31] concluded that loss of TNSALP causes myelin abnormalities and synaptic dysfunction.

As previously shown, NSCs express TNSALP more abundantly. To gain insight into the functional role of TNSALP, Kermer et al. [59] applied a knockdown protocol using short hairpin RNA (shRNA) and retroviral infection toward cultured NSCs. TNSALP knockdown reduced the proliferation and differentiation of NSCs into neurons or oligodendrocytes, suggesting an important role of TNSALP in NSC proliferation and differentiation. Interestingly, the RNAi-mediated effect was abrogated by adding ALP (derived from a commercially available frozen product) to the culture medium, suggesting that TNSALP may act on the cell membrane. Kermer et al. [59] provided some reasons for this: (1) TNSALP may be involved in the metabolism of extracellular nucleotides through coupling with purine receptor function, (2) TNSALP may function as an ecto-phosphoprotein phosphatase with phosphodiesterase activity or mediate the hydrolysis of PPi to Pi, or (3) TNSALP may interact with extracellular matrix proteins, such as collagen, through a loop region of TNSALP.

Graser et al. [60] reported an increase in the expression of neuronal marker genes, such as RNA binding protein, fox-1 homolog (*C. elegans*) 3 (NEUN), contactin-associated protein-like 2 (CNTNAP2), neurexin 1 (NRXN1), doublecortin (DCX), and protein kinase C, alpha (PRKCA), as well as microtubule-binding proteins such as microtubule-associated protein 2 (MAP2) and microtubule-associated protein tau (TAU) during neurogenic differentiation in human SH-SY5Y neuroblastoma cell line overexpressing TNSALP. Graser et al. [60] concluded that TNSALP is capable of modulating intercellular communication in the CNS.

3.4. Immune System

TNSALP is known to be highly expressed in murine B lymphocytes when they are activated by either polyclonal mitogens or T helper cells (Th cells) [61]. Marquez et al. [62] extended this notion further and demonstrated that TNSALP expression is correlated with B cell differentiation into Ig-secreting cells. Similar results were obtained by another group [63], who found a parallel increase in electrophoretic mobility and TNSALP expression during B cell progression from the proliferative to the immunoglobulin (Ig) secretion stage. According to Marquez et al. [62], the phosphorylation–dephosphorylation mechanism may play a role in controlling the growth/differentiation rate in the B cell lineage.

Two ALP isoforms, IALP and TNSALP, are co-expressed in the mouse colon, with the latter predominating in colitis. Hernández-Chirlaque et al. [29] examined the role of TNSALP in T lymphocytes, using heterozygous *Akp2* KO mice called *B6.129S7-Akp2^{tm1Sor}/J* (as homozygous *Akp2* KO mice are non-viable). In vitro primary cultures from these mice demonstrated that stimulated splenocytes and T lymphocytes showed decreased cytokine production and expression compared with WT cells. Decreased T cell activation was reproduced by levamisole in WT cells. Hernández-Chirlaque et al. [29] concluded that TNSALP modulates T lymphocyte function (specifically T-cell-dependent colitis).

3.5. Vascular Calcification

Ectopic vascular calcification (VC) is associated with aging, atherosclerosis, diabetes, and/or chronic kidney disease [64]. As mentioned previously, endothelium-specific expression of TNSALP in Tg mice resulted in VC [35,36]. Savinov et al. [36] suggested that TNSALP possesses osteogenic potential in endothelial cells, thus promoting VC.

Notably, Goettsch et al. [65] recently suggested that TNSALP can be a therapeutic target for cardiovascular calcification (CVC), which is associated with increased morbidity and mortality. According to them, CVC develops in several diseases and locations, such as in the tunica intima in atherosclerosis plaques, in the tunica media in type 2 diabetes and chronic kidney disease, and in aortic valves. At present, no treatment is yet available. Moreover, most CVC occurs in a mode similar to skeletal/dental mineralization caused by the overexpression of TNSALP. Thus, tools are now being developed to inhibit TNSALP activity using animal models of CVC. For instance, Andleeb et al. [66] discovered small drug-like molecules (a series of novel triazolyl pyrazole derivatives (10a-y)) as potent and selective inhibitors of human TNSALP. If these drugs are proven useful using animal models, they may be candidate compounds to target VC. Furthermore, Millan's group [67] found a potent and orally bioavailable TNSALP inhibitor, called SBI-425, to inhibit pathological soft-tissue calcification in vivo.

3.6. Role of TNSALP in Renal Cell Carcinoma (RCC)

According to Sharma et al. [68], decreased activity of TNSALP is remarkable in renal cell carcinoma (RCC). Forced expression of *ALPL* cDNA in renal cell carcinoma (RCC) cell lines resulted in decreased migratory property and cell viability compared with the controls. Furthermore, the transfected cells exhibited increased apoptosis, suggesting the role of TNSALP in cell viability and apoptosis during renal tumorigenesis.

3.7. Role of TNSALP Fibroblastic-Like Cell Lines

Hui et al. [69] examined the effects of TNSALP on various aspects of cellular activity by transferring the *ALPL* into three fibroblastic-like cell lines: Chinese Hamster ovary (CHO), R1610, and Rat-2. The expression of TNSALP under the control of simian virus 40 (SV40) promoter (but not R1610) caused a decrease in the proliferation and migration of CHO and Rat-2 cells, indicating the cell proliferation and migration inhibitory effect of TNSALP.

3.8. Role of TNSALP in the Hair-Inductive Capacity of Human Dermal Papilla Spheres

Human dermal papilla (DP) regulates the overlying epithelial cells and plays a key role in regulating hair growth and regeneration. To examine the effect of TNSALP in human DP cells, Kwack et al. [70] overexpressed TNSALP in DP spheres and carried out a hair reconstitution assay. Overexpression of TNSALP significantly increased hair follicle induction, which was closely associated with the Wnt/β-catenin signaling pathway. Additionally, there was a significant increase in the expression levels of Wnt/β-catenin target genes, such as axis inhibition protein 2 (AXIN2), *versican* (VCAN), and lymphoid enhancer-binding factor 1 (LEF1), in DP cells overexpressing TNSALP (Table 3). Moreover, overexpression of TNSALP significantly affected the expression level of bone morphogenetic protein 4 (BMP4) but did not affect the expression profiles of bone morphogenetic protein 2 (BMP2), *noggin* (NOG), and fibroblast growth factor 7 (FGF7) (Table 3). Based on these findings, it was concluded that TNSALP augments the hair-inductive capacity of human DP spheres by regulating the Wnt/β-catenin signaling pathway and possibly by maintaining the characteristics of the DP.

Table 3. Summary of candidate genes whose expression fluctuates after overexpression of exogenous *ALPL* or suppression of endogenous *ALPL* in mammals.

Method [1]	Type of Cells Used	Genes Upregulated	Genes Downregulated	Genes Unaltered	References
siRNA	High-ALP-expressing cells derived from human osteoblast-like cells (HOS)	*COL1A1* *RUNX2*		*OCN*	[48]
siRNA	3T3-F442A adipocytes	*GAPDH* *ADIPOQ* *FABP4*	*LEP*		[58]
Revamisol	T3-F442A adipocytes		*PPARG* *LEP* *ADIPOQ* *ATGL*	*TIP47* *SCARB1*	[58]
Revamisol	Murine osteoblast precursor cells		*RUNX2* *SP7* *BGLAP2* *DMP1*		[71]
Overexpression	Tg mice overexpressing *ALPL* gene	*OPN* *BMP2* *MGP* *SOX9* *SLC20A1*	*ACTA2* *TAGLN*	*ANK*	[35]
Overexpression	Dermal papilla	*BMP4* *AXIN2* *VCAN* *LEF1* *SOX2*		*BMP2* *NOGFGF7*	[70]

[1] Suppression of endogenous *ALPL* expression was carried out by transfection with small interfering RNA (siRNA) or by treating cells with revamisol, a reversible inhibitor for TNSALP. Additionally, an increased level of TNSALP was achieved by overexpression of exogenous *ALPL* gene. Then, genes upregulated, downregulated or unaltered after these treatments were examined through comparison with untreated parental cells. Abbreviations: ACTA2, actin α2; ADIPOQ, adiponectin; ALPL, a gene encoding tissue non-specific alkaline phosphatase (TNSALP); ANK, progressive ankylosis protein; ATGL, adipose triglyceride lipase; AXIN2, axin 2; BGLAP2, bone γ-carboxyglutamate protein 2; BMP2, bone morphogenetic protein 2; BMP4, bone morphogenetic protein 4; COL1A1, collagen type I α1 chain; FABP4, fatty acid-binding protein 4; FGF7, fibroblast growth factor 7; GAPDH, glycerophosphate dehydrogenase; LEF1, lymphoid enhancer-binding factor 1; LEP, leptin; MGP, matrix Gla protein; NOG, noggin; OCN, osteocalcin; OPN, osteopontin; PPARG, peroxisome proliferator activated receptor gamma; RUNX2, runt-related transcription factor 2; SCARB1, scavenger receptor class B member 1; SLC20A1, solute carrier family 20 member 1; SP7, Sp7 transcription factor; SOX2, SRY-box transcription factor 2; SOX9, SRY-box transcription factor 9; TAGLN, transgelin; Tg, transgenic; TIP47, tail-interacting protein of 47 kD; VCAN, versican.

3.9. TNSALP May Be Involved in Premature Bone Aging

As mentioned previously, Liu et al. [56] determined the mechanism of bone aging and found that TNSALP deficiency in stem cells enhanced the release of ATP and reduced its hydrolysis to cause extracellular ATP boost. When internalized by MSCs, ATP inactivated the AMPK α subunit (AMPKα) cell signaling pathway (master regulator of cellular energy homeostasis), contributing to MSC cell fate switch by impairing their ability to grow and proliferate. The work was conducted in vitro and in a mutant mouse model (*Alpl*$^{+/-}$) exhibiting premature aging, followed by metformin treatment to improve the function of endogenous MSCs by reactivating the cell signaling pathway. Liu et al. [56] demonstrated that knockdown of *Akp2* induced premature bone aging, which was characterized by bone mass loss and marrow fat gain. Liu et al. [56] then reactivated the pathway using the diabetes drug metformin to restore the ability of stem cells to grow and differentiate into bone-producing osteoblasts and prevent bone aging.

3.10. Mitochondrial TNSALP Controls Thermogenesis

Adaptive thermogenesis has been defined as the change in energy expenditure following acute and/or long-term overfeeding and underfeeding. Recent data have indicated that thermogenic fat cells use creatine to stimulate futile substrate cycling, dissipating chemical energy as heat based on the super-stoichiometric relationship between the amount of creatine added to mitochondria and the quantity of oxygen consumed. Sun et al. [72] recently provided direct evidence for the critical role of TNSALP as a phosphocreatine phosphatase in the futile creatine cycle using mice. The thermogenic fat cells have a high degree of phosphocreatine phosphatase activity, caused by the action of mitochondrial TNSALP capable of hydrolyzing phosphocreatine to initiate a futile cycle of creatine dephosphorylation and phosphorylation. The expression of mitochondrial TNSALP is reportedly triggered in mice exposed to cold conditions, and its inhibition in isolated mitochondria leads to a loss of futile creatine cycling. Genetic ablation of TNSALP in adipocytes reduces whole-body energy expenditure, leading to rapid-onset obesity in mice.

4. Possible Involvement of TNSALP in Differentiation and Maintenance in Juvenile State

Although there is a strong correlation between TNSALP expression and undifferentiated juvenile cells, the role of TNSALP in the induction, maintenance, and differentiation of these cells remains unknown. For example, TNSALP expression is induced during the reprogramming of somatic cells to iPSCs [73]. TNSALP expression is rapidly reduced when juvenile cells, such as ESCs or iPSCs, are induced to differentiate [74]. Based on these findings, we hypothesized that TNSALP expression might cause changes in cellular behavior, which may be associated with cellular differentiation or di-differentiation. Therefore, the role of TNSALP in cell differentiation is discussed in detail below.

4.1. TNSALP as an Early Marker Expressed in Intermediate Cells towards iPSCs

During reprogramming, candidate iPSC colonies generate approximately 20 days after transfection with reprogramming factors and are generally examined for the expression of pluripotent markers, such as octamer-binding transcription factor-3/4 (*OCT-3/4*), sex-determining region Y-box 2 (*SOX2*), stage-specific embryonic antigen 4 (SSEA-4), tumor rejection antigens (TRA)-1-60 (TRA-1-60), and TRA-1-80, using molecular methods [73]. Moreover, ALP expression is discernible in the colonies as early as 14 days post-transduction and can be used as a marker for the early identification of iPSCs. Brambrink et al. [73] examined the timing of known pluripotency marker activation during mouse iPS cell derivation and observed that ALP was activated first, followed by stage-specific embryonic antigen 1 (SSEA-1). Expression of the homeobox-containing transcription factor Nanog, whose name stems from the Celtic/Irish mythical word Tír na nÓg, and the endogenous *OCT-3/4* gene, marking fully reprogrammed cells, were only observed late in the process. According to David and Polode [75], ALP-positive cells are considered as "intermediate cells" being reprogrammed into undifferentiated cells, such as iPSCs (Figure 2).

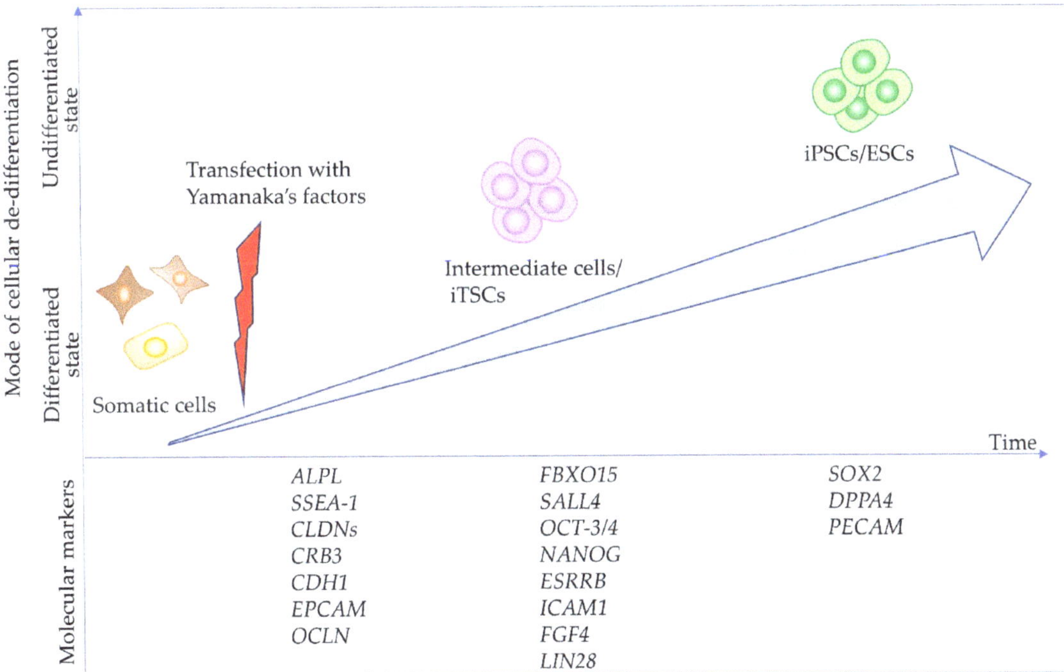

Figure 2. Cell state and molecular events during the reprogramming of somatic cells into induced pluripotent cells (iPSCs). When complete reprogramming occurs, somatic cells are successfully converted into iPSCs. The resulting iPSCs can be further reprogrammed into naïve iPSCs through transfection with vectors carrying Yamanaka's factors or via treatment with chemicals. Additionally, somatic cells are converted into "intermediate cells" called iTSCs, when partial reprogramming occurs. There are at least three phases with respect to (de-)differentiation ("somatic state" which may correspond to the initiation stage, "intermediate state" which may correspond to the maturation stage, and "pluripotent state" which may correspond to the stabilization stage), according to the paper of Samavarchi-Tehrani et al. [76]. Importantly, several molecular markers define each of the above phases, as per Samavarchi-Tehrani et al. and Polo et al. [76,77]. This figure was drawn in-house, based on the data shown in the paper of Adachi et al. [78]. Abbreviations: ALPL, gene encoding tissue-nonspecific alkaline phosphatase (TNSALP or TNAP); CDH1, E-cadherin; CLDNs, claudins; CRB3, crumbs cell polarity complex component 3; EPCAM, epithelial cell adhesion molecule; ESRRB, estrogen-related receptor Beta; FBXO15, F-box protein 15; ESCs, embryonic stem cells; FGF4, fibroblast growth factor 4; DPPA4, developmental pluripotency associated 4; ICAM1, intercellular adhesion molecule 1; LIN28, Lin-28 homolog; NANOG, Nanog homeobox; OCLN, occludin; OCT-3/4, octamer-binding transcription factor-3/4; PECAM, platelet endothelial cell adhesion molecule; SALL4, spalt-like transcription factor 4; SOX2, sex-determining region Y-box 2; SSEA-1, stage-specific embryonic antigen 1.

Generally, stem cells included in the adult organs/tissues (which are called "somatic stem cells" or "adult stem cells") express certain stemness markers (such as Oct-3/4 and ALP). These stem cells are more easily reprogrammed than the other somatic cells because adult neural stem cells (which show endogenous high expression of pluripotent genes, such as Sox2 and c-Myc) are known to be reprogrammed with only two factors [79]. These suggest that ALP-positive somatic cells may be easily reprogrammed to iPSCs by reprogramming factors other than ALP-negative cells. To confirm this point, Inada et al. [80] examined the potential of five isolated primary human deciduous tooth-derived dental pulp cells (HDDPCs) to induce iPSCs after reprogramming. They demonstrated that two lines highly enriched with ALP-positive cells were successfully reprogrammed. However, three ALP-negative lines failed to be reprogrammed. Notably, HDDPCs with a high degree of ALP activity (also associated with increased expression of ALPL mRNA) exhibited an active proliferation rate and expression of stemness factors, such as OCT-3/4, SOX2, and NANOG.

In contrast, HDDPCs with a low degree of ALP activity did not show such properties. Based on these findings, Inada et al. [80] speculated that ALP-positive cells included in HDDPCs may be stem cells or intermediate cells between fully differentiated cells and undifferentiated cells such as iPSCs. Similarly, Soda et al. [81] reported that only one cell line exhibiting ALP activity among six cell lines was successfully reprogrammed into iPSCs by reprogramming factors, with the other five lines remaining negative for ALP activity after 10 days of transfection. However, when the five lines were transfected again with the reprogramming factors, four exhibited ALP activity approximately 10 days after the second transfection (shown in Figure 3), and were successfully reprogrammed to form iPSCs. Based on these findings, it can be concluded that the appearance of intermediates cells that express TNSALP is necessary to form iPSCs. Notably, according to Štefková et al. [82], little is known about the inability to isolate ESCs from *Akp2* KO embryos.

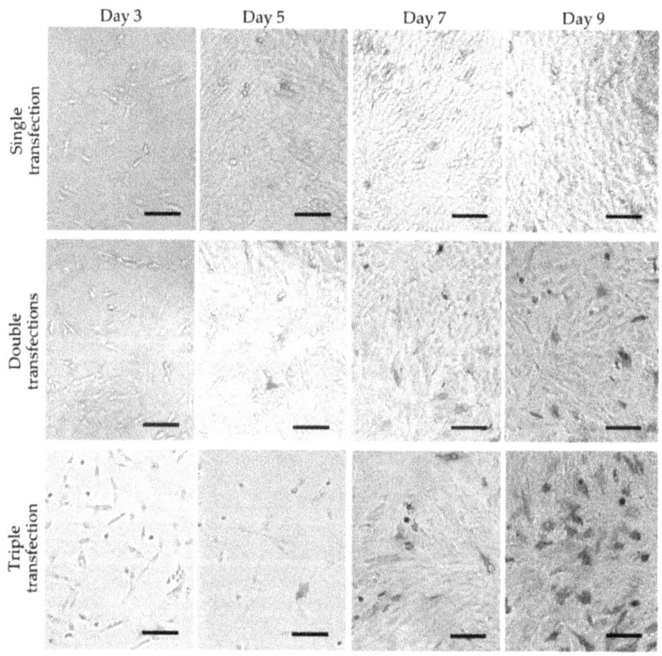

Figure 3. Cytochemical evaluation of ALP activity in HDDPCs after repeated transfections with the reprogramming factors. HDDPCs (P05 line) were transfected with Yamanaka's four reprogramming factors once, twice, or three times. The treated cells were subjected to cytochemical staining for ALP activity at 3, 5, 7, and 9 days after the final transfection. These photographs were originally constructed using data used in the paper of Soda et al. [80]. Bar = 500 μm.

4.2. Possible Involvement of Wnt/β-Catenin Signaling Pathway in Generation of iPSCs

As previously mentioned, the successful development of somatic cells (potentially ALP-negative like fibroblastic cells) to iPSCs is always associated with the appearance of ALP-positive cells ("intermediate cells"). However, it is necessary to examine the mechanism responsible for changing ALP-negative cells to ALP-positive cells after transfection with reprogramming factors. It has been speculated that several signaling pathways may involve in this process, among which the Wnt/β-catenin signaling has been examined.

The Wnt family consists of a large number of secreted glycoproteins that are involved in multiple cellular events, such as cell proliferation, differentiation, and apoptosis through β-catenin-dependent canonical and β-catenin-independent noncanonical pathways [83,84]. In the canonical Wnt/β-catenin pathway, Wnt proteins bind to the seven transmembrane receptors, called Frizzled (FZD), and the low-density lipoprotein receptor-related protein (LRP5/6) co-receptors located on the cell surface. The binding of Wnt protein to its receptors leads to the phosphorylation of the disheveled (Dsh) protein (Dvl) through its association with Axin and the adenomatous polyposis coli tumor suppressor (APC). Additionally, this binding inhibits the phosphorylation of β-catenin by glycogen synthase kinase-3β (GSK-3β) and the cytosolic accumulation of β-catenin, resulting in the translocation of unphosphorylated β-catenin to the nucleus. In the nucleus, β-Catenin interacts with members of the T cell factor-4 (TCF-4)/lymphoid enhancer factor (LEF) family of transcription factors to activate the expression of downstream genes involved in proliferation (e.g., *MYC*, cyclin D1 (*CCND1*), peroxisome proliferator-activated receptor δ (*PPARD*)), stem cell fate (e.g., achaete-scute family bHLH transcription factor 2 (*ASCL2*)), survival (e.g., ATP binding cassette subfamily B member 1 (*ABCB1*), baculoviral IAP repeat-containing 5 (*BIRC5*) (also known as survivin)), differentiation (e.g., inhibitor of DNA-binding 2 (*ID2*), transcription factor 4 (*TCF4*; also known as *ITF2*), ectoderm-neural cortex protein 1 (*ENC1*)), migration (e.g., matrix metallopeptidase 7 (*MMP7*), matrix metallopeptidase 14 (*MMP14*)), and angiogenesis (e.g., vascular endothelial growth factor (*VEGF*)) [85]. The pathway is schematically shown in Figure 4.

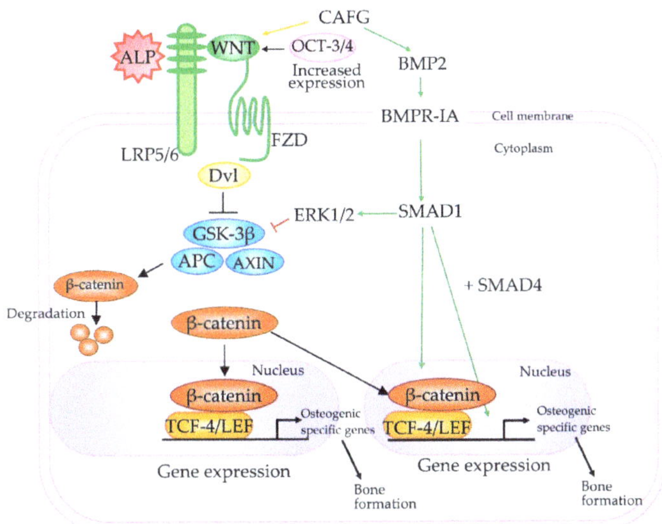

Figure 4. Molecular mechanisms underlying TNSALP-mediated bone marrow stem cells (BMMSC) lineage switching. According to Liu et al. [86], overexpressed TNSALP interacts with low-density lipoprotein-related receptors 5 and 6 (LRP5/6) molecules, one of the important elements of the canonical Wnt/β-catenin pathway, to inhibit phosphorylation of glycogen synthase kinase-3β (GSK-3β). As a result, the nuclear location of β-catenin is accelerated, leading to activation of downstream genes that are involved in osteogenesis and controlled by T cell factor-4 (TCF-4)/lymphoid enhancer factor (LEF) (TCF-4/LEF) proteins. These osteogenesis-related downstream genes may also be regulated by the BMP2-related signaling pathway. β-Catenin can also interact with pluripotency-related genes, such as Krüppel-like factor 4 (KLF4), octamer-binding transcription factor-3/4 (OCT-3/4), and sex-determining region Y-box 2 (SOX2). Abbreviations: ALP, alkaline phosphatase; APC, adenomatous polyposis coli tumor suppressor; AXIN, axis inhibition protein; BMP2, bone morphogenetic protein 2; BMPR-IA, bone morphogenetic protein receptor type IA; CAFG, caviunin 7-O-[β-D-apiofuranosyl-(1-6)-β-D-glucopyranoside]; Dvl, dishevelled (Dsh) protein; ERK1/2, extracellular signal-regulated kinase (ERK)1/2; FZD, Frizzled; SMAD 1, SMAD family member 1; SMAD4, SMAD family member 4. This figure was drawn in-house, based on the data shown in the paper of Liu et al. [86].

The importance of the Wnt/β-catenin signaling pathway in somatic cell reprogramming was confirmed by Kimura et al. [87], who observed high expression of Wnt2 in the early stage of reprogramming. Furthermore, *Wnt2* knockdown suppressed the nuclear accumulation of β-catenin and reduced reprogramming efficiency, whereas *Wnt2* overexpression promoted nuclear accumulation of β-catenin and enhanced reprogramming efficiency. Moreover, experiments using drugs that regulate the Wnt pathway confirmed the importance of the nuclear accumulation of β-catenin in reprogramming. Overall, it can be concluded that the upregulation of *Wnt2* and subsequent accumulation of β-catenin in the nucleus are key events in reprogramming.

Notably, in mouse embryonic stem (ES) cells, stabilized β-catenin forms a complex with Oct-3/4 and enhances the activity of Oct-3/4, thus increasing pluripotency through a TCF-4/LEF-independent mechanism [88]. Furthermore, β-catenin and TCF3 target Nr5a2 (also known as liver receptor homolog-1 (Lrh-1)) and Nr5a2, in turn, directly activating the expression of Tbx3, Nanog, and Oct-3/4 in mouse ES cells (Figure 5) [89]. Moreover, Nr5a2 can replace Oct-3/4 in the reprogramming of mouse somatic cells [90]. According to Tanaka et al. [91], activation of β-catenin targets can maintain pluripotency and enhance cell reprogramming.

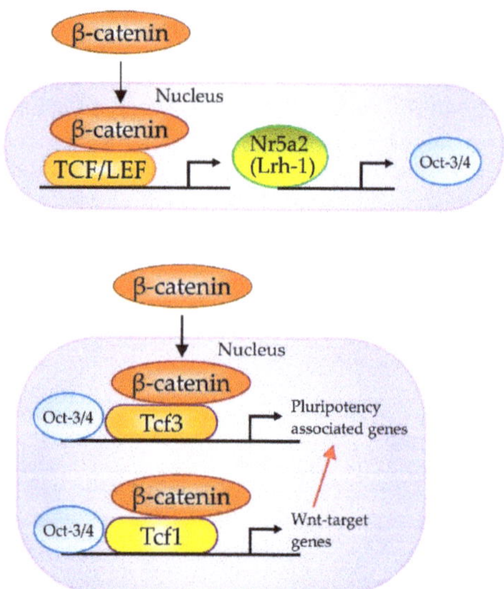

Figure 5. In murine embryonic stem (ES) cells, WNT/β-catenin/NR5A2 (LRH-1) is known to modify pluripotency genes expression. According to Tanaka et al. [91], WNT3A activates the WNT/β-catenin pathway and increases the expression of Nr5a2, which could directly enhance the expression of core pluripotency factors T-box transcription factor 3 (Tbx3), Nanog and Oct-3/4; however, the activation of this pathway is limited when the cells are treated with WNT3A alone. Abbreviations: NR5A2, nuclear receptor subfamily 5 group A member 2; LRH-1, liver receptor homolog-1. This figure was drawn in-house, based on the data shown in the paper of Tanaka et al. [91].

Furthermore, the relationship between the Wnt/β-catenin signaling pathway and upregulation of TNSALP expression has been examined. Si et al. [92] reported that Wnt3A (a representative canonical Wnt ligand) induced ALP activity in MSCs, which was inhibited by Dickkopf WNT signaling pathway inhibitor 1 (*Dkk1*) and dominant-negative *Tcf4*. Moreover, silencing *Wnt5a* expression enhanced the Wnt3a-mediated increase in *ALPL* expression in a dental follicle cell line [93]. Treatment with 6-bromoindirubin-3′-oxime (BIO), a drug capable of inhibiting GSK-3β phosphorylation and its activity, increased *ALPL* expression in cells, thereby activating the Wnt/β-catenin signaling pathway [94].

BIO enhanced *ALPL* mRNA expression in canine bone MSCs, indicating the importance of Wnt/β-catenin signaling pathway during somatic cells reprogramming to iPSCs.

4.3. Possible Involvement of BMP-2 Signaling Pathway in the Generation of iPSCs

BMP-2 is one of the most potent bone-inducing agents in osteoblast differentiation. Moreover, it induces osteogenic trans-differentiation of fibrogenic, myogenic, and adipogenic cells both in vitro and in vivo [95]. A Wnt autocrine loop (a Wnt autocrine/paracrine loop) mediates the induction of *ALPL* and mineralization by BMP-2 in pre-osteoblastic cells [96,97]. Because LRP5/6 acts as a coreceptor for Wnt proteins, loss of function of LRP5/6 leads to osteoporosis (osteoporosis-pseudoglioma syndrome (OPPG)), and a specific point mutation in this same receptor results in high bone mass.

As shown in Figure 4, signaling pathways related to osteogenesis seemingly overlap with those related to iPSCs genesis. TCF-4/LEF proteins, to which activated β-catenin binds, activate downstream genes of the Wnt/β-catenin signaling pathway, including osteogenesis-related proteins, such as RUNX2, osterix (OSX), OPN, and TNSALP [98]. Additionally, BMP-2 stimulation activated RUNX2 and TNSALP expression, thus increasing the proliferation of osteoblastic cells. Zhang et al. [99] reported that the Wnt/β-catenin signaling pathway enhances iPSCs induction at the early stage of reprogramming through the interaction of β-catenin with Yamanaka's reprogramming factors (Krüppel-like factor 4 (KLF4), OCT-3/4, and SOX2), further enhancing the expression of pluripotency circuitry genes. These findings suggest that Wnt signaling may be implicated in somatic cell reprogramming.

Notably, as already shown by Kimura et al. [87], transfection with shRNAs against *Wnt2* caused a reduction in the level of Wnt2, reducing the number of cells showing nuclear localization of β-catenin and reducing the number of ALP-positive or NANOG-positive colonies. These findings suggest that Wnt2 is required for both the nuclear localization of β-catenin and the initiation of reprogramming. Moreover, Wnt2 promotes tumor progression and is involved in epithelial–mesenchymal transition (EMT) events [100]. Mesenchymal–epithelial transition (MET) occurs at the initial step of fibroblast to iPS cell reprogramming. Prolonged activation of the Wnt2-mediated Wnt/β-catenin signaling pathway may inhibit MET in fibroblasts. Thus, for proper fibroblasts to iPS cell reprogramming, the Wnt2-mediated Wnt/β-catenin signaling pathway has to be "on" in the initial step of reprogramming, but "off" in the later steps.

Samavarchi-Tehrani et al. [76] reported that treatment with Yamanaka's factors increased the expression of several epithelial cell-related genes during the "initiation phase" of reprogramming (first 5 days). Specifically, there was an increase in the expression of epithelial cell markers, such as E-cadherin (*Cdh1*), Claudins-2, -4, and -11 (*Cldns-3, -4, -7, and -11*), occludin (*Ocln*), epithelial cell adhesion molecule (*Epcam*), and Crumbs homolog (*Crb3*), a decrease in the expression of mesenchymal-specific genes, such as zinc finger protein SNAI1 (*Snail*), *Slug*, zinc finger E-box binding homeobox 1 (*Zeb1*), and zinc finger E-box binding homeobox 2 (*Zeb2*), and a loss of fibroblastic marker genes during the initiation of MET. siRNAs targeting *Cdh1* are known to inhibit the appearance of ALP-positive colonies, suggesting that MET is a key event that controls the success of cell reprogramming. To examine the role of the BMP signaling pathway in MET, knockdown of survival of motor and autonomic neurons 1 (*Sman1*) suppressed the formation of ALP-positive cells. Similarly, knockdown of SMAD family member 4 (*Smad4*), BMP type II receptor (*BMPRII*), or BMP type I receptor activin receptor-like kinase 3 (*Alk3*) suppressed the de-differentiation of ALP-positive cells. Based on these findings, MET and BMP signaling could be necessary at the initiation phase of reprogramming.

5. TNSALP as a Signal Regulator

As previously mentioned, TNSALP is involved in a variety of cell behavior, including cellular proliferation, cell movement, and cell differentiation/di-differentiation. Evidence suggesting that TNSALP may function as a signal regulator is discussed in the preceding sections.

5.1. TNSALP May Be Involved in Lineage Switching

BMMSCs isolated from HPP patients exhibit decreased capacity to differentiate into osteocytes and increased capacity to differentiate into adipocytes [101]. Specifically, in vitro assay showed a decrease in osteogenic differentiation as evidenced by a decrease in the expression of RUNX2 and OCN, and an increase in adipogenic differentiation as evidenced by positive Oil Red staining and an increase in peroxisome proliferator-activated receptor gamma (PPARγ) expression [101]. Furthermore, in vivo transplantation assay showed a decrease in the differentiation of BMMSCs isolated from HPP patients into the osteogenic lineage, similar to results obtained in BMMSC isolated from *Akp2* KO mice.

Liu et al. [86] examined the regulatory role of TNSALP in BMMSC lineage switching and observed a decrease in the expression of Frizzled class receptor 2 (FZD2), Frizzled class receptor 9 (FZD9), and LRP5/6. Overexpression of TNSALP in BMMSC from HPP patients caused an increase in the level of LRP6 alone. Downregulation of TNSALP in normal BMMSC caused the partial inhibition of the WNT3A-mediated activation of the canonical Wnt/β-catenin pathway. The downregulation of LRP6 caused a similar phenomenon. Anti-TNSALP antibody successfully immunoprecipitated LRP5/6. Based on these findings, it was speculated that TNSALP does not interact with GSK-3β or β-catenin (*CTNNB1*) directly, but with LRP5/6 molecules to inhibit phosphorylation of GSK-3β, accelerate nuclear location of β-catenin, and activate genes controlled by TCF-4/LEF proteins. This process is shown in a schematic diagram in Figure 4. The findings of Liu et al. [86] appear to be the first to reveal the signal regulator role of TNSALP, which regulates the lineage switching of BMMSCs by regulating the LRP5/6/GSK-3β cascade. However, studies are yet to identify the site of TNSALP capable of interacting with LRP5/6.

Notably, Najar et al. [71] demonstrated that adipose tissue-derived MSCs contained two populations, so-called ALDH+ and ALDH-, based on the aldehyde dehydrogenase (ALDH) activity known to be a classical feature of stem cells. When transcriptome analysis of both cell populations was carried out, ALDH+ cells exhibited higher expression of osteogenic differentiation-related genes such as *RUNX2*, *OSX* and *OPG* than ALDH- cells. According to Najar et al. [71], the potential of differentiation towards the osteogenic lineages seems to be not equal and ALDH- cells present a more differentiated state than ALDH+ cells. In this context, it will be of interest to examine which types of adipose tissue-derived MSCs are related to TNSALP-mediated lineage switching.

5.2. Overexpression or Suppression of ALPL May Affect the Expression of Some Genes

Overexpression of *ALPL* was associated with decreased expression of smooth muscle aortic alpha-actin (*ACTA2*) and *TAGLN* [35]. Furthermore, siRNA-mediated suppression of *ALPL* mRNA resulted in suppression of *COL1A1* and *RUNX2* expression [48]. Nakamura et al. [102] reported that treatment of murine osteoblast precursor cells with revamisol decreased the expression of osteogenesis-related proteins: *Runx2*, Sp7 transcription factor (*Sp7*), *Bglap2*, and dentin matrix protein 1 (*Dmp1*). In contrast, overexpression of TNSALP in osteoblasts isolated from *Akp2* KO mice caused increased expression of *Runx2*, *Bglap2*, and *Dmp1*. The genes affected by the level of *TNSALP* are listed in Table 3.

5.3. Expression of ALPL May Be Affected by Some Genes

The mechanism underlying the regulation of TNSALP remains unclear. Previous studies identified transcription factor binding motifs on the *ALPL* promoter, such as TATA box, Sp1 binding site, E-box-like sequences, and vitamin D response element-like motifs in humans [103] and mice [104]. Notably, Štefková et al. [81] performed in silico analysis and identified several binding sites for transcription factors associated with pluripotency, such as OCT-3/4, NANOG, transcription factor 3 (TCF3), and Forkhead box D3 (FOXD3) in the ALPL promoter.

To date, several transcription factors (differentiation-inducing factor-1 (DIF-1) [105], forkhead transcription factor FOXO1 (previously known as *FKHR*) [106], distal-less homeobox 5 (*DLX5*)-binding cis-acting element [107], and p107 retinoblastoma family transcription factor [108]) have been shown to bind the *ALPL* promoter. For example, Hatta et al. [109] examined the expression of FOXO1, a regulator of hepatic glucose metabolic and proapoptotic genes, in osteogenic cells and the effect of FOXO1 on transcription of the *ALPL* gene. RT-PCR and immunoblot analyses revealed the expression of FOXO1 in osteogenic cells, such as MC3T3-E1, SaOS2, and UMR 106. Consequently, it was demonstrated that overexpression of FOXO1 stimulated *ALPL* promoter activity through the forkhead response element in its promoter. These results suggest that *ALPL* is a target gene regulated by FOXO1 and that FOXO1 contributes to osteoblast maturation and osteogenesis. Yusa et al. [108], using hematopoietic cells, demonstrated that Sp3 transcription factor (Sp3) could bind to the fragment spanning around 150 bp upstream from the transcription initiation site of *ALPL*, suggesting that Sp3 activates the *ALPL* promoter in hematopoietic cells.

As previously described, TNSALP expression is stimulated by BMP-2 treatment. However, how BMP-2 induces TNSALP expression is not clearly understood. Kim et al. [107] attempted to determine the mechanism of BMP-2 on TNSALP using the murine *Akp2* promoter, which contains a *Dlx5*-binding cis-acting element. They demonstrated that Dlx5 transactivates *Akp2* expression directly by binding to the *Dlx5*-binding cis-acting element.

6. Therapeutic Aspect of TNSALP

Increase in serum ALP activity commonly originates from liver and bone. Consequently, the examination of serum ALP activity is of particular important in assessing possible hepatobiliary disease. The response of the liver to any form of biliary obstruction induces the synthesis of ALP by hepatocytes, which results in canalicular leakage of ALP into the hepatic sinusoid and subsequent inflow into blood vessels to increase serum ALP activity. A similar increase is seen in patients with advanced liver cancer or widespread secondary hepatic metastases [110]. The increased level of serum ALP has also been attributed to an increased activity of ALP, which is localized in the plasma membrane of osteoblasts before extracellular release, and also correlates with the extent of the bone diseases (i.e., Paget disease or rickets/osteomalasia). Notably, ALP is normally elevated in growing children and adults over the age of fifty [111].

As already described, serum ALP activity and increased amount of PPi are known to be associated with patients with HPP, an inheritable disease caused by mutations in the *ALPL* gene [112,113]. HPP has a broad-range of severity from stillbirth to pathological fractures in adulthood, depending on the degree of *ALPL* deficiency [112–114]. Patients with the most severe type of HPP experience respiratory failure soon after birth, thus requiring respiratory support [14]. To date, more than 350 mutations have been reported in the *ALPL* gene mutations database (www.sesep.uvsq.fr/03_hypo_mutations.php (accessed on 24 November 2021)).

Enzyme replacement therapy (ERT) using bone-targeting recombinant ALP Strensiq®(asfotase alfa) that is comprised by catalytic domain of TNSALP, Fc domain of human immunoglobulin, and 10 asparagine peptides is now available in several countries and has improved the prognosis of patients [115,116]. Although ERT is expensive, subcutaneous injections three or six times a week can improve the medical management of HPP patients [117,118].

As an alternative to ERT in HPP treatment, gene-engineered MSC-based transplantation therapies are considered [99]. For example, Nakano et al. [119] introduced a gene correction targeting vector into iPSCs isolated from two HPP patients by TALENS to correct c.1559delT mutation, the most frequent mutation in Japanese HPP patients. After selection with antibiotics, some clones with successful gene correction were obtained. These clones exhibited ALP activity. Osteoblasts differentiated from the corrected iPSCs exhibited high ALP activity and some calcification in vitro. Overall, the gene-corrected iPSCs can be used as a source for cell replacement therapy for HPP patients. On the other hand, viral gene delivery-based therapy is also considered as an alternative to ERT. For example, Yamamoto et al. [120] tested the possible use of a lentiviral vector (carrying human *ALPL* with deca-aspartate motif (D_{10}) at the C-terminus (TNALP-D_{10}), which is hereinafter referred to as TNALP-D_{10}). They injected the vector into the jugular vein of 1-day-old *Akp2*$^{-/-}$ mice, a murine model for severe infantile HPP, to rescue the HPP phenotype. The injected mice exhibited no epilepsy and survived more than 10 months with an improved bone phenotype. Matsumoto et al. [121] attempted to use a recombinant adeno-associated virus (rAAV) because rAAV is thought to be safer than a lentiviral vector that potentially integrates their genome into host chromosomes. They performed an intravenous injection of rAAV expressing bone-targeted TNALP-D_{10} into *Akp2*$^{-/-}$ mice [25]. To develop a safer and more clinically applicable transduction strategy for HPP gene therapy, the same group [122] recently examined the efficacy of muscle-directed expression of TNALP-D10 using an rAAV with a serotype of 8 (which is hereinafter referred to as rAAV-8). Injection of this vector into the bilateral quadriceps of neonatal *Akp2*$^{-/-}$ mice resulted in healthy mice with more than 3 months of survival. Recently, Kinoshita et al. [123] further extended the work of Nakamura-Takahashi [122] by performing a single intramuscular administration of rAAV-8 encoding TNALP-D10 to increase the life span and improve the skeletal and dentoalveolar phenotypes in *Akp2*$^{-/-}$ mice within 5 days after birth. Treated mice exhibited elevated serum ALP activity, suppressed plasma PPi, extended life span, no sign of rickets, normal-like bone microstructure, and no ectopic calcifications in the kidneys, aorta, coronary arteries, or brain. They suggest that rAAV-mediated, muscle-specific expression of TNALP-D_{10} through intramuscular administration of therapeutic rAAV-8 may be a promising alternative to ERT to treat severe infantile forms of HPP.

As mentioned above, the most frequently used animal model for HPP is *Akp2*$^{-/-}$ mice. Williams et al. [124] recently generated another animal model, namely, a sheep model for HPP ("HPP sheep"), by introducing a single point mutation (1077 C > G) in *ALPL* using a CRISPR/Cas9 system. The results show that HPP sheep exhibited reduced serum ALP activity, decreased tail vertebral bone size, and metaphyseal flaring, consistent with the mineralization deficits observed in human HPP patients. Overall, the animal model would be beneficial for developing treatment strategies for HPP.

Molecular analysis of HPP patients revealed the presence of numerous mutations in the *ALPL* gene, as previously mentioned. To examine the function of these mutations in a short time, transfection of mutated *ALPL* into cultured cells, such a COS-1 cells, would be very convenient. For example, Takinami et al. [125] transfected F310L or F310L and V365I (F310L/V365I) in *ALPL* from HPP patients into COS-1 cells and observed a 67% and 31% reduction, respectively, in ALP activity compared with that in WT mice.

As previously mentioned, for treating HPP patients, administration of drugs, such as Strensiq®(asfotase alfa), is now being frequently used as ERT. However, due to drug dependence and the Quality of Life (QOL) problem, an alternative to ERT has been explored. In this context, cell-based transplantation of gene-corrected cells (i.e., gene-corrected BMMSCs derived from HPP patients) or local (muscle-targeted) administration of therapeutic rAAV would be the best option. Particularly, in the former case, gene correction of the mutated *ALPL* would be highly accelerated because new genome editing technologies (as exemplified by the CRISPR/Cas9 system) are being updated daily.

7. ALP as Possible New Tools for Cell Research

7.1. Usefulness of Live Staining Kit for Isolating Live ALP-Expressing Cells without Fixation

During a reprogramming protocol, candidate iPSC colonies are generally examined for the expression of pluripotent markers, such as SSEA-4, TRA-1-60, and TRA-1-80, using immunocytochemical analyses. This method, however, is typically used after distinct colonies emerge, probably at 21 days post transduction [126]. ALP expression is discernible in the colonies as early as 14 days post transduction and can be used as a marker for the early identification of iPSCs. Singh et al. [127] employed Alkaline Phosphatase Live Stain

Kit (provided from ThermoFisher) to label early intermediates during iPSC generation or clonal populations of ESCs/iPSCs for further selection and expansion.

7.2. Usefulness of Ecto-Alkaline Phosphatase-Mediated System to Eliminate ALP-Positive Cells

The incomplete differentiation of human iPSCs poses a serious safety risk owing to their potential tumorigenicity, thus hindering their clinical application. Kuang et al. [128] explored the potential of phospho-D-peptides as novel iPSC-eliminating agents. They reported that overexpression of ALP in iPSCs dephosphorylated phospho-D-peptides into hydrophobic peptides that aggregate and induce cell death. They isolated a peptide candidate, D-3, that selectively and rapidly induced toxicity in iPSCs within 1 h, but had little influence on various non-iPSCs, including primary hepatocytes and iPSC-derived cardiomyocytes. Additionally, D-3 prevented residual iPSC-induced teratoma formation in a mouse tumorigenicity assay. Kuang et al. [128] concluded that D-3 is a low-cost and effective anti-iPSC agent for both laboratory use and the safe clinical application of iPSC-derived cells in regenerative medicine.

8. Conclusions

TNSALP is a ubiquitous membrane-bound glycoprotein that catalyzes the hydrolysis of phosphate monoesters at basic pH values. It is highly expressed in juvenile cells, such as stem cells (or precursor cells) and PSCs such as ESCs/iPSCs, and is thus considered as one of the markers for defining cells with stemness properties. Additionally, TNSALP is a useful marker for identifying cancerous states in leukemia and some types of cancers. Furthermore, TNSALP is known to be one of the early markers for reporting the presence of intermediate cells, which are reprogrammed from somatic cells toward juvenile cells after transfection with reprogramming factors (also called "Yamanaka's factors").

Unfortunately, little is known about the biological role of TNSALP. Analysis of patients with HPP and KO mice, in which *ALPL* has been specifically mutated, suggests the critical role of TNSALP in bone calcification and the prevention of calcification of skeletal and neuronal systems. In experimental systems, the knockdown of *ALPL* expression reduces cell proliferation and differentiation into neurons or oligodendrocytes in NSCs. The overexpression of TNSALP caused the calcification of skeletal muscle cells and endothelial cells. *Akp2* KO mice models often died from seizures, with surviving animals manifesting dental dysplasia. In some cases, the *Akp2*-null mice exhibited impaired growth, vitamin B6-dependent seizures, impaired bone mineralization and apnea, and died before weaning. Thus, improving the understanding of the role of TNSALP in osteogenesis and neurogenesis is important.

However, the mechanisms of TNSALP pluripotency and differentiation of PSCs remains poorly understood. Recent studies clarified the possibility that TNSALP has multiple functions. According to Liu et al. [100], TNSALP may act as a signal regulator by binding to LRP5/6, which activates a canonical Wnt/β-catenin-dependent increase in somatic cell differentiation to osteogenic lineage cells. Furthermore, activation of this system leads to the activation of TNSALP. Therefore, Wnt/β-catenin system-mediated TNSALP loop may exist in some instances, especially when cells are forced to differentiate into the osteogenic lineage.

Notably, expression of TNSALP can trigger the nuclear localization of β-catenin, which is one of the key regulators accelerating cell differentiation to PSCs. This means that forced expression of *ALPL* in ALP-negative cells might be the first step through which cells are reprogrammed to juvenile PSCs. Supporting this hypothesis, we demonstrated that ALP-positive HDDPCs are more amenable to be reprogrammed into iPSCs than ALP-negative HDDPCs when cells are forced to be reprogrammed by transfection with Yamanaka's factors [79]. In this regard, it would be interesting to test whether the forced expression of *ALPL* in ALP-negative HDDPCs triggers somatic cell differentiation to iPSCs. These recently developed gene-engineering technologies improve the understanding of the biological roles of TNSALP.

Author Contributions: Conceptualization, M.S.; writing—original draft preparation, M.S., I.S., Y.K., Y.I., N.K., N.I., E.I.; writing—review and editing, M.S., I.S., H.N., Y.Y., E.I. All authors have read and agreed to the published version of the manuscript.

Funding: This study was partly supported by grants from the Ministry of Education, Science, Sports, and Culture, Japan (No. 19K06372 for M.S.; No. 21K10165 for E.I. and No. 245804124 for I.S.).

Conflicts of Interest: The authors declare no conflict of interest.

References

1. Millán, J.L. Alkaline Phosphatases: Structure, substrate specificity and functional relatedness to other members of a large superfamily of enzymes. *Purinergic Signal.* **2006**, *2*, 335–341. [CrossRef] [PubMed]
2. Fedde, K.N.; Lane, C.C.; Whyte, M.P. Alkaline phosphatase is an ectoenzyme that acts on micromolar concentrations of natural substrates at physiologic pH in human osteosarcoma (SAOS-2) cells. *Arch. Biochem. Biophys.* **1988**, *264*, 400–409. [CrossRef]
3. Say, J.C.; Ciuffi, K.; Furriel, R.P.; Ciancaglini, P.; Leone, F.A. Alkaline phosphatase from rat osseous plates: Purification and biochemical characterization of a soluble form. *Biochim. Biophys. Acta* **1991**, *1074*, 256–262. [CrossRef]
4. Buchet, R.; Millán, J.L.; Magne, D. Multisystemic functions of alkaline phosphatases. *Methods Mol. Biol.* **2013**, *1053*, 27–51. [CrossRef]
5. Mulnard, J.; Huygens, R. Ultrastructural localization of non-specific alkaline phosphatase during cleavage and blastocyst formation in the mouse. *J. Embryol. Exp. Morphol.* **1978**, *44*, 121–131. [CrossRef] [PubMed]
6. Narisawa, S.; Hofmann, M.C.; Ziomek, C.A.; Millán, J.L. Embryonic alkaline phosphatase is expressed at M-phase in the spermatogenic lineage of the mouse. *Development* **1992**, *116*, 159–165. [CrossRef]
7. Pease, S.; Braghetta, P.; Gearing, D.; Grail, D.; Williams, R.L. Isolation of embryonic stem (ES) cells in media supplemented with recombinant leukemia inhibitory factor (LIF). *Dev. Biol.* **1990**, *141*, 344–352. [CrossRef]
8. Takahashi, K.; Tanabe, K.; Ohnuki, M.; Narita, M.; Ichisaka, T.; Tomoda, K.; Yamanaka, S. Induction of pluripotent stem cells from adult human fibroblasts by defined factors. *Cell* **2007**, *131*, 861–872. [CrossRef]
9. Heath, J.K. Characterization of a xenogeneic antiserum raised against the fetal germ cells of the mouse: Cross-reactivity with embryonal carcinoma cells. *Cell* **1978**, *15*, 299–306. [CrossRef]
10. Wang, P.; Suo, L.-J.; Shang, H.; Li, Y.; Li, G.-X.; Li, Q.-W.; Hu, J.-H. Differentiation of spermatogonial stem cell-like cells from murine testicular tissue into haploid male germ cells in vitro. *Cytotechnology* **2014**, *66*, 365–372. [CrossRef] [PubMed]
11. Mishra, S.K.; Braun, N.; Shukla, V.; Füllgrabe, M.; Schomerus, C.; Korf, H.W.; Gachet, C.; Ikehara, Y.; Sévigny, J.; Robson, S.C.; et al. Extracellular nucleotide signaling in adult neural stem cells: Synergism with growth factor-mediated cellular proliferation. *Development* **2006**, *133*, 675–684. [CrossRef] [PubMed]
12. Bianco, P.; Cao, X.; Frenette, P.S.; Mao, J.J.; Robey, P.G.; Simmons, P.J.; Wang, C.Y. The meaning, the sense and the significance: Translating the science of mesenchymal stem cells into medicine. *Nat. Med.* **2013**, *19*, 35–42. [CrossRef]
13. Rico, L.G.; Juncà, J.; Ward, M.D.; Bradford, J.; Petriz, J. Is alkaline phosphatase the smoking gun for highly refractory primitive leukemic cells? *Oncotarget* **2016**, *7*, 72057–72066. [CrossRef] [PubMed]
14. Salles, J.P. Hypophosphatasia: Biological and clinical aspects, avenues for therapy. *Clin. Biochem. Rev.* **2020**, *41*, 13–27. [CrossRef]
15. Villa-Suárez, J.M.; García-Fontana, C.; Andújar-Vera, F.; González-Salvatierra, S.; de Haro-Muñoz, T.; Contreras-Bolívar, V.; García-Fontana, B.; Muñoz-Torres, M. Hypophosphatasia: A unique disorder of bone mineralization. *Int. J. Mol. Sci.* **2021**, *22*, 4303. [CrossRef]
16. Sharma, U.; Pal, D.; Prasad, R. Alkaline phosphatase: An overview. *Indian J. Clin. Biochem.* **2014**, *29*, 269–278. [CrossRef]
17. Zimmermann, H.; Zebisch, M.; Strater, N. Cellular function and molecular structure of ecto-nucleotidases. *Purinergic Signal.* **2012**, *8*, 437–502. [CrossRef]
18. Whyte, M.P. Hypophosphatasia. In *The Metabolic and Molecular Bases of Inherited Disease*, 8th ed.; Scriver, C.R., Beaudet, A.L., Sly, W.S., Valle, D., Eds.; McGraw-Hill: New York, NY, USA, 2001; pp. 5313–5329.
19. Linder, C.H.; Englund, U.H.; Narisawa, S.; Millán, J.L.; Magnusson, P. Isozyme profile and tissue-origin of alkaline phosphatases in mouse serum. *Bone* **2013**, *53*, 399–408. [CrossRef] [PubMed]
20. Skynner, M.J.; Drage, D.J.; Dean, W.L.; Turner, S.; Watt, D.J.; Allen, N.D. Transgenic mice ubiquitously expressing human placental alkaline phosphatase (PLAP): An additional reporter gene for use in tandem with beta-galactosidase (lacZ). *Int. J. Dev. Biol.* **1999**, *43*, 85–90.
21. DePrimo, S.E.; Stambrook, P.J.; Stringer, J.R. Human placental alkaline phosphatase as a histochemical marker of gene expression in transgenic mice. *Transgenic Res.* **1996**, *5*, 459–466. [CrossRef]
22. Narisawa, S.; Huang, L.; Iwasaki, A.; Hasegawa, H.; Alpers, D.H.; Millán, J.L. Accelerated fat absorption in intestinal alkaline phosphatase knockout mice. *Mol. Cell Biol.* **2003**, *23*, 7525–7530. [CrossRef] [PubMed]
23. Brun, L.R.; Lombarte, M.; Roma, S.; Perez, F.; Millán, J.L.; Rigalli, A. Increased calcium uptake and improved trabecular bone properties in intestinal alkaline phosphatase knockout mice. *J. Bone Min. Metab.* **2018**, *36*, 661–667. [CrossRef]
24. Narisawa, S.; Smans, K.A.; Avis, J.; Hoylaerts, M.F.; Millán, J.L. Transgenic mice expressing the tumor marker germ cell alkaline phosphatase: An in vivo tumor model for human cancer antigens. *Proc. Natl. Acad. Sci. USA* **1993**, *90*, 5081–5085. [CrossRef] [PubMed]

25. Narisawa, S.; Fröhlander, N.; Millán, J.L. Inactivation of two mouse alkaline phosphatase genes and establishment of a model of infantile hypophosphatasia. *Dev. Dyn.* **1997**, *208*, 432–446. [CrossRef]
26. Dehghani, H.; Narisawa, S.; Millán, J.L.; Hahnel, A.C. Effects of disruption of the embryonic alkaline phosphatase gene on preimplantation development of the mouse. *Dev. Dyn.* **2000**, *217*, 440–448. [CrossRef]
27. MacGregor, G.R.; Zambrowicz, B.P.; Soriano, P. Tissue non-specific alkaline phosphatase is expressed in both embryonic and extraembryonic lineages during mouse embryogenesis but is not required for migration of primordial germ cells. *Development* **1995**, *121*, 1487–1496. [CrossRef]
28. Waymire, K.G.; Mahuren, J.D.; Jaje, J.M.; Guilarte, T.R.; Coburn, S.P.; Coburn, S.P. Mice lacking tissue non-specific alkaline phosphatase die from seizures due to defective metabolism of vitamin B-6. *Nat. Genet.* **1995**, *11*, 45–51. [CrossRef]
29. Hernández-Chirlaque, C.; Gámez-Belmonte, R.; Ocón, B.; Martínez-Moya, P.; Wirtz, S.; Sánchez de Medina, F.; Martínez-Augustin, O. Tissue non-specific alkaline phosphatase expression is needed for the full stimulation of T cells and T cell-dependent colitis. *J. Crohns Colitis* **2017**, *11*, 857–870. [CrossRef]
30. Fedde, K.N.; Blair, L.; Silverstein, J.; Coburn, S.P.; Ryan, L.M.; Weinstein, R.S.; Waymire, K.; Narisawa, S.; Millán, J.L.; MacGregor, G.R.; et al. Alkaline phosphatase knock-out mice recapitulate the metabolic and skeletal defects of infantile hypophosphatasia. *J. Bone Miner. Res.* **1999**, *14*, 2015–2026. [CrossRef]
31. Hanics, J.; Barna, J.; Xiao, J.; Millán, J.L.; Fonta, C.; Négyessy, L. Ablation of TNAP function compromises myelination and synaptogenesis in the mouse brain. *Cell Tissue Res.* **2012**, *349*, 459–471. [CrossRef]
32. Lomelí, H.; Ramos-Mejía, V.; Gertsenstein, M.; Lobe, C.G.; Nagy, A. Targeted insertion of Cre recombinase into the TNAP gene: Excision in primordial germ cells. *Genesis* **2000**, *26*, 116–117. [CrossRef]
33. Narisawa, S. Genetically modified mice for studying TNAP function. *Subcell. Biochem.* **2015**, *76*, 45–57. [CrossRef] [PubMed]
34. Narisawa, S.; Yadav, M.C.; Millán, J.L. In vivo overexpression of tissue-nonspecific alkaline phosphatase increases skeletal mineralization and affects the phosphorylation status of osteopontin. *J. Bone Miner. Res.* **2013**, *28*, 1587–1598. [CrossRef] [PubMed]
35. Sheen, C.R.; Kuss, P.; Narisawa, S.; Yadav, M.C.; Nigro, J.; Wang, W.; Chhea, T.N.; Sergienko, E.A.; Kapoor, K.; Jackson, M.R.; et al. Pathophysiological role of vascular smooth muscle alkaline phosphatase in medial artery calcification. *J. Bone Miner. Res.* **2015**, *30*, 824–836. [CrossRef]
36. Savinov, A.Y.; Salehi, M.; Yadav, M.C.; Radichev, I.; Millán, J.L.; Savinova, O.V. Transgenic overexpression of tissue-nonspecific alkaline phosphatase (TNAP) in vascular endothelium results in generalized arterial calcification. *J. Am. Heart Assoc.* **2015**, *4*, e002499. [CrossRef]
37. Hahnel, A.C.; Rappolee, D.A.; Millán, J.L.; Manes, T.; Ziomek, C.A.; Theodosiou, N.G.; Werb, Z.; Pedersen, R.A.; Schultz, G.A. Two alkaline phosphatase genes are expressed during early development in the mouse embryo. *Development* **1990**, *110*, 555–564. [CrossRef]
38. Manes, T.; Glade, K.; Ziomek, C.A.; Millán, J.L. Genomic structure and comparison of mouse tissue-specific alkaline phosphatase genes. *Genomics* **1990**, *8*, 541–554. [CrossRef]
39. Lynes, M.; Narisawa, S.; Millán, J.L.; Widmaier, E.P. Interactions between CD36 and global intestinal alkaline phosphatase in mouse small intestine and effects of high-fat diet. *Am. J. Physiol. Regul. Integr. Comp. Physiol.* **2011**, *301*, R1738–R1747. [CrossRef]
40. Weiss, M.J.; Ray, K.; Henthorn, P.S.; Lamb, B.; Kadesch, T.; Harris, H. Structure of the human liver/bone/kidney alkaline phosphatase gene. *J. Biol. Chem.* **1988**, *263*, 12002–12010. [CrossRef]
41. Weiss, M.J.; Henthorn, P.S.; Lafferty, M.A.; Slaughter, C.; Raducha, M.; Harris, H. Isolation and characterization of a cDNA encoding a human liver/bone/kidney-type alkaline phosphatase. *Proc. Natl. Acad. Sci. USA* **1986**, *83*, 7182–7186. [CrossRef] [PubMed]
42. Misumi, Y.; Tashiro, K.; Hattori, M.; Sakaki, Y.; Ikehara, Y. Primary structure of rat liver alkaline phosphatase deduced from its cDNA. *Biochem. J.* **1988**, *249*, 661–668. [CrossRef]
43. Abdul, R.B.; Shafquat, A.; Mohammad, T. Differentially expressed three non-coding alternate exons at 5′ UTR of regulatory type I beta subunit gene of mouse. *Mol. Biol. Rep.* **2012**, 3375–3383. [CrossRef]
44. Matsuura, S.; Kishi, F.; Kajii, T. Characterization of a 5′-flanking region of the human liver/bone/kidney alkaline phosphatase gene: Two kinds of mRNA from a single gene. *Biochem. Biophys. Res. Commun.* **1990**, *168*, 993–1000. [CrossRef]
45. Toh, Y.; Yamamoto, M.; Endo, H.; Misumi, Y.; Ikehara, Y. Sequence divergence of 5′ extremities in rat liver alkaline phosphatase mRNAs. *J. Biochem.* **1989**, *105*, 61–65. [CrossRef] [PubMed]
46. Zernik, J.; Thiede, M.A.; Twarog, K.; Stover, M.L.; Rodan, G.A.; Upholt, W.B.; Rowe, D.W. Cloning and analysis of the 5′ region of the rat bone/liver/kidney/placenta alkaline phosphatase gene. A dual-function promoter. *Matrix* **1990**, *10*, 38–47. [CrossRef]
47. Terao, M.; Studer, M.; Gianní, M.; Garattini, E. Isolation and characterization of the mouse liver/bone/kidney-type alkaline phosphatase gene. *Biochem. J.* **1990**, *268*, 641–648. [CrossRef] [PubMed]
48. Kotobuki, N.; Matsushima, A.; Kato, Y.; Kubo, Y.; Hirose, M.; Ohgushi, H. Small interfering RNA of alkaline phosphatase inhibits matrix mineralization. *Cell Tissue Res.* **2008**, *332*, 279–288. [CrossRef] [PubMed]
49. Lobe, C.G.; Koop, K.E.; Kreppner, W.; Lomeli, H.; Gertsenstein, M.; Nagy, A. Z/AP, a double reporter for cre-mediated recombination. *Dev. Biol.* **1999**, *208*, 281–292. [CrossRef]
50. Acevedo-Arozena, A.; Wells, S.; Potter, P.; Kelly, M.; Cox, R.D.; Brown, S.D.M. ENU mutagenesis, a way forward to understand gene function. *Annu. Rev. Genom. Hum. Genet.* **2008**, *9*, 49–69. [CrossRef]
51. Stottmann, R.; Beier, D.R. ENU mutagenesis in the mouse. *Curr. Protoc. Hum. Genet.* **2014**, *82*, 15.4.1–15.4.10. [CrossRef]

52. Hough, T.A.; Polewski, M.; Johnson, K.; Cheeseman, M.; Nolan, P.M.; Vizor, L.; Rastan, S.; Boyde, A.; Pritzker, K.; Hunter, A.J.; et al. Novel mouse model of autosomal semidominant adult hypophosphatasia has a splice site mutation in the tissue nonspecific alkaline phosphatase gene *Akp*J. *Bone Miner. Res.* **2007**, *22*, 1397–1407. [CrossRef] [PubMed]
53. Aigner, B.; Rathkolb, B.; Klaften, M.; Sedlmeier, R.; Klempt, M.; Wagner, S.; Michel, D.; Mayer, U.; Klopstock, T.; de Angelis, M.H. Generation of N-ethyl-N-nitrosourea-induced mouse mutants with deviations in plasma enzyme activities as novel organ-specific disease models. *Exp. Physiol.* **2009**, *94*, 412–421. [CrossRef] [PubMed]
54. Sabrautzki, S.; Rubio-Aliaga, I.; Hans, W.; Fuchs, H.; Rathkolb, B.; Calzada-Wack, J.; Cohrs, C.M.; Klaften, M.; Seedorf, H.; Eck, S. New mouse models for metabolic bone diseases generated by genome-wide ENU mutagenesis. *Mamm. Genome* **2012**, *23*, 416–430. [CrossRef] [PubMed]
55. Foster, B.L.; Sheen, C.R.; Hatch, N.E.; Liu, J.; Cory, E.; Narisawa, S.; Kiffer-Moreira, T.; Sah, R.L.; Whyte, M.P.; Somerman, M.J.; et al. Periodontal defects in the A116T knock-in murine model of odontohypophosphatasia. *J. Dent. Res.* **2015**, *94*, 706–714. [CrossRef]
56. Liu, W.; Zhang, L.; Xuan, K.; Hu, C.; Liu, S.; Liao, L.; Jin, F.; Shi, S.; Jin, Y. *Alpl* prevents bone ageing sensitivity by specifically regulating senescence and differentiation in mesenchymal stem cells. *Bone Res.* **2018**, *6*, 27. [CrossRef] [PubMed]
57. Chava, S.; Chennakesavulu, S.; Gayatri, B.M.; Reddy, A.B.M. A novel phosphorylation by AMP-activated kinase regulates RUNX2 from ubiquitination in osteogenesis over adipogenesis. *Cell Death Dis.* **2018**, *9*, 754. [CrossRef]
58. Hernández-Mosqueira, C.; Velez-delValle, C.; Kuri-Harcuch, W. Tissue alkaline phosphatase is involved in lipid metabolism and gene expression and secretion of adipokines in adipocytes. *Biochim. Biophys. Acta* **2015**, *1850*, 2485–2496. [CrossRef]
59. Kermer, V.; Ritter, M.; Albuquerque, B.; Leib, C.; Stanke, M.; Zimmermann, H. Knockdown of tissue non-specific alkaline phosphatase impairs neural stem cell proliferation and differentiation. *Neurosci. Lett.* **2010**, *485*, 208–211. [CrossRef]
60. Graser, S.; Mentrup, B.; Schneider, D.; Klein-Hitpass, L.; Jakob, F.; Hofmann, C. Overexpression of tissue-nonspecific alkaline phosphatase increases the expression of neurogenic differentiation markers in the human SH-SY5Y neuroblastoma cell line. *Bone* **2015**, *79*, 150–161. [CrossRef]
61. García-Rozas, C.; Plaza, A.; Díaz-Espada, F.; Kreisler, M.; Martínez-Alonso, C. Alkaline phosphatase activity as a membrane marker for activated B cells. *J. Immunol.* **1982**, *129*, 52–55.
62. Marquez, C.; Toribio, M.L.; Marcos, M.A.; de la Hera, A.; Barcena, A.; Pezzi, L.; Martinez, C. Expression of alkaline phosphatase in murine B lymphocytes. Correlation with B cell differentiation into Ig secretion. *J. Immunol.* **1989**, *142*, 3187–3192. [PubMed]
63. Bauer, J.; Kachel, V. The increase of electrophoretic mobility and alkaline phosphatase activity are parallel events during B-cell maturation. *Immunol. Invest.* **1990**, *19*, 57–68. [CrossRef]
64. Palit, S.; Kendrick, J. Vascular calcification in chronic kidney disease: Role of disordered mineral metabolism. *Curr. Pharm. Des.* **2014**, *20*, 5829–5833. [CrossRef]
65. Goettsch, C.; Strzelecka-Kiliszek, A.; Bessueille, L.; Quillard, T.; Mechtouff, L.; Pikula, S.; Canet-Soulas, E.; Millan, J.L.; Fonta, C.; Magne, D. TNAP as a therapeutic target for cardiovascular calcification: A discussion of its pleiotropic functions in the body. *Cardiovasc. Res.* **2020**, cvaa299. [CrossRef] [PubMed]
66. Andleeb, H.; Hussain, M.; Ejaz, S.A.; Sevigny, J.; Farman, M.; Yasinzai, M.; Zhang, J.; Iqbal, J.; Hameed, S. Synthesis and computational studies of highly selective inhibitors of human recombinant tissue non-specific alkaline phosphatase (h-TNAP): A therapeutic target against vascular calcification. *Bioorganic Chem.* **2020**, *101*, 103999. [CrossRef]
67. Pinkerton, A.B.; Sergienko, E.; Bravo, Y.; Dahl, R.; Ma, C.-T.; Jackson, M.R.; Cosford, N.D.P.; Millán, J.L. Discovery of 5-((5-chloro-2-methoxyphenyl)sulfonamido)nicotinamide (SBI-425), a potent and orally bioavailable tissue-nonspecific alkaline phosphatase (TNAP) inhibitor. *Bioorg. Med. Chem. Lett.* **2018**, *28*, 31–34. [CrossRef]
68. Sharma, U.; Pal, D.; Singh, S.K.; Kakkar, N.; Prasad, R. Reduced L/B/K alkaline phosphatase gene expression in renal cell carcinoma: Plausible role in tumorigenesis. *Biochimie* **2014**, *104*, 27–35. [CrossRef] [PubMed]
69. Hui, M.Z.; Sukhu, B.; Tenenbaum, H.C. Expression of tissue non-specific alkaline phosphatase stimulates differentiated behaviour in specific transformed cell populations. *Anat. Rec.* **1996**, *244*, 423–436. [CrossRef]
70. Kwack, M.H.; Jang, Y.J.; Won, G.H.; Kim, M.K.; Kim, J.C.; Sung, Y.K. Overexpression of alkaline phosphatase improves the hair-inductive capacity of cultured human dermal papilla spheres. *J. Dermatol. Sci.* **2019**, *95*, 126–129. [CrossRef]
71. Najar, M.; Crompot, E.; van Grunsven, L.A.; Dollé, L.; Lagneaux, L. Aldehyde dehydrogenase activity in adipose tissue: Isolation and gene expression profile of distinct sub-population of mesenchymal stromal cells. *Stem Cell Rev. Rep.* **2018**, *14*, 599–611. [CrossRef]
72. Sun, Y.; Rahbani, J.F.; Jedrychowski, M.P.; Riley, C.L.; Vidoni, S.; Bogoslavski, D.; Hu, B.; Dumesic, P.A.; Zeng, X.; Wang, A.B.; et al. Mitochondrial TNAP controls thermogenesis by hydrolysis of phosphocreatine. *Nature* **2021**, *593*, 580–585. [CrossRef]
73. Brambrink, T.; Foreman, R.; Welstead, G.G.; Lengner, C.J.; Wernig, M.; Suh, H.; Jawnisch, R. Sequential expression of pluripotency markers during direct reprogramming of mouse somatic cells. *Cell Stem Cell* **2008**, *2*, 151–159. [CrossRef]
74. O'Connor, M.D.; Kardel, M.D.; Iosfina, I.; Youssef, D.; Lu, M.; Li, M.M.; Vercauteren, S.; Nagy, A.; Eaves, C. Alkaline phosphatase-positive colony formation is a sensitive, specific, and quantitative indicator of undifferentiated human embryonic stem cells. *Stem Cells* **2008**, *26*, 1109–1116. [CrossRef]
75. David, L.; Polode, J.M. Phases of reprogramming. *Stem Cell Res.* **2014**, *12*, 754–761. [CrossRef] [PubMed]
76. Samavarchi-Tehrani, P.; Golipour, A.; David, L.; Sung, H.K.; Beyer, T.A.; Datti, A.; Woltjen, K.; Nagy, A.; Wrana, J.L. Functional genomics reveals a BMP-driven mesenchymal-to-epithelial transition in the initiation of somatic cell reprogramming. *Cell Stem Cell* **2010**, *7*, 64–77. [CrossRef] [PubMed]

77. Polo, J.M.; Anderssen, E.; Walsh, R.M.; Schwarz, B.A.; Nefzger, C.M.; Lim, S.M.; Borkent, M.; Apostolou, E.; Alaei, S.; Cloutier, J.; et al. A molecular roadmap of reprogramming somatic cells into iPS cells. *Cell* **2012**, *151*, 1617–1632. [CrossRef] [PubMed]
78. Adachi, K.; Kopp, W.; Wu, G.; Heising, S.; Greber, B.; Stehling, M.; Araúzo-Bravo, M.J.; Boerno, S.T.; Timmermann, B.; Vingron, M.; et al. Esrrb unlocks silenced enhancers for reprogramming to naive pluripotency. *Cell Stem Cell* **2018**, *23*, 266–275.e6. [CrossRef] [PubMed]
79. Kim, J.B.; Zaehres, H.; Wu, G.; Gentile, L.; Ko, K.; Sebastiano, V.; Araúzo-Bravo, M.J.; Ruau, D.; Han, D.W.; Zenke, M. Pluripotent stem cells induced from adult neural stem cells by reprogramming with two factors. *Nature* **2008**, *454*, 646–650. [CrossRef] [PubMed]
80. Inada, E.; Saitoh, I.; Kubota, N.; Murakami, T.; Soda, M.; Matsueda, K.; Murakami, T.; Sawami, T.; Kagoshima, A.; Yamasaki, Y. Alkaline phosphatase and OCT-3/4 as useful markers for predicting susceptibility of human deciduous teeth-derived dental pulp cells to reprogramming factor-induced iPS cells. *J. Investig. Clin. Dent.* **2017**, 8. [CrossRef]
81. Soda, M.; Saitoh, I.; Murakami, T.; Inada, E.; Iwase, Y.; Noguchi, H.; Shibasaki, S.; Kurosawa, M.; Sawami, T.; Terunuma, M.; et al. Repeated human deciduous tooth-derived dental pulp cell reprogramming factor transfection yields multipotent intermediate cells with enhanced iPS cell formation capability. *Sci. Rep.* **2019**, *9*, 1490. [CrossRef]
82. Štefková, K.; Procházková, J.; Pacherník, J. Alkaline phosphatase in stem cells. *Stem Cells Int.* **2015**, 628368. [CrossRef]
83. Cadigan, K.M.; Nusse, R. Wnt signaling: A common theme in animal development. *Genes Dev.* **1997**, *11*, 3286–3305. [CrossRef] [PubMed]
84. Wodarz, A.; Nusse, R. Mechanisms of Wnt signaling in development. *Annu. Rev. Cell Dev. Biol.* **1998**, *14*, 59–88. [CrossRef]
85. Barker, N.; Clevers, H. Mining the Wnt pathway for cancer therapeutics. *Nat. Rev. Drug Discov.* **2006**, *5*, 997–1014. [CrossRef] [PubMed]
86. Liu, W.; Zhang, L.; Xuan, K.; Hu, C.; Li, L.; Zhang, Y.; Jin, F.; Jin, Y. Alkaline phosphatase controls lineage switching of mesenchymal stem cells by regulating the LRP6/GSK3β complex in hypophosphatasia. *Theranostics* **2018**, *8*, 5575–5592. [CrossRef]
87. Kimura, M.; Nakajima-Koyama, M.; Lee, J.; Nishida, E. Transient expression of WNT2 promotes somatic cell reprogramming by inducing β-catenin nuclear accumulation. *Stem Cell Rep.* **2016**, *6*, 834–843. [CrossRef] [PubMed]
88. Kelly, K.F.; Ng, D.Y.; Jayakumaran, G.; Wood, G.A.; Koide, H.; Doble, B.W. β-catenin enhances Oct-4 activity and reinforces pluripotency through a TCF-independent mechanism. *Cell Stem Cell* **2011**, *8*, 214–227. [CrossRef]
89. Wagner, R.; Xu, X.; Yi, F.; Merrill, B.; Cooney, A. Canonical Wnt/ß-catenin regulation of liver receptor homolog-1 mediates pluripotency gene expression. *Stem Cells* **2010**, *28*, 1794–1804. [CrossRef]
90. Heng, J.; Feng, B.; Han, J.; Jiang, J.; Kraus, P.; Orlov, Y.; Huss, M.; Yang, L.; Lufkin, T.; Lim, B.; et al. The nuclear receptor Nr5a2 can replace Oct4 in the reprogramming of murine somatic cells to pluripotent cells. *Cell Stem Cell* **2010**, *6*, 167–174. [CrossRef]
91. Tanaka, S.S.; Kojima, Y.; Yamaguchi, Y.L.; Nishinakamura, R.; Tam, P.P. Impact of WNT signaling on tissue lineage differentiation in the early mouse embryo. *Dev. Growth Differ.* **2011**, *53*, 843–856. [CrossRef]
92. Si, W.; Kang, Q.; Luu, H.H.; Park, J.K.; Luo, Q.; Song, W.-X.; Jiang, W.; Luo, X.; Li, X.; Yin, H.; et al. CCN1/Cyr61 is regulated by the canonical Wnt signal and plays an important role in Wnt3A-induced osteoblast differentiation of mesenchymal stem cells. *Mol. Cell Biol.* **2006**, *26*, 2955–2964. [CrossRef]
93. Sakisaka, Y.; Tsuchiya, M.; Nakamura, T.; Tamura, M.; Shimauchi, H.; Nemoto, E. Wnt5a attenuates Wnt3a-induced alkaline phosphatase expression in dental follicle cells. *Exp. Cell Res.* **2015**, *336*, 85–93. [CrossRef]
94. Zhao, X.-E.; Yang, Z.; Gao, Z.; Ge, J.; Wei, Q.; Ma, B. 6-Bromoindirubin-3′-oxime promotes osteogenic differentiation of canine BMSCs through inhibition of GSK3β activity and activation of the Wnt/β-catenin signaling pathway. *An. Acad. Bras. Cienc.* **2019**, *91*, e20180459. [CrossRef]
95. Katagiri, T.; Yamaguchi, A.; Ikeda, T.; Yoshiki, S.; Wozney, J.M.; Rosen, V.; Wang, E.A.; Tanaka, H.; Omura, S.; Suda, T. The non-osteogenic mouse pluripotent cell line, C3H10T1/2, is induced to differentiate into osteoblastic cells by recombinant human bone morphogenetic protein-2. *Biochem. Biophys. Res. Commun.* **1990**, *172*, 295–299. [CrossRef]
96. Takuwa, Y.; Ohse, C.; Wang, E.A.; Wozney, J.M.; Yamashita, K. Bone morphogenetic protein-2 stimulates alkaline phosphatase activity and collagen synthesis in cultured osteoblastic cells, MC3T3-E. *Biochem. Biophys. Res. Commun.* **1991**, *174*, 96–101. [CrossRef]
97. Rawadi, G.; Vayssiere, B.; Dunn, F.; Baron, R.; Roman-Roman, S. BMP-2 controls alkaline phosphatase expression and osteoblast mineralization by a Wnt autocrine loop. *J. Bone Miner. Res.* **2003**, *18*, 1842–1853. [CrossRef] [PubMed]
98. Gaur, T.; Lengner, C.J.; Hovhannisyan, H.; Bhat, R.A.; Bodine, P.V.N.; Komm, B.S.; Javed, A.; van Wijnen, A.J.; Stein, J.L.; Stein, G.S.; et al. Canonical WNT signaling promotes osteogenesis by directly stimulating Runx2 gene expression. *J. Biol. Chem.* **2005**, *280*, 33132–33140. [CrossRef]
99. Zhang, P.; Chang, W.-H.; Fong, B.; Gao, F.; Liu, C.; Al Alam, D.; Bellusci, S.; Lu, W. Regulation of induced pluripotent stem (iPS) cell induction by Wnt/β-catenin signaling. *J. Biol. Chem.* **2014**, *289*, 9221–9232. [CrossRef] [PubMed]
100. Jiang, H.; Li, Q.; He, C.; Li, F.; Sheng, H.; Shen, X.; Zhang, X.; Zhu, S.; Chen, H.; Chen, X.; et al. Activation of the Wnt pathway through Wnt2 promotes metastasis in pancreatic cancer. *Am. J. Cancer Res.* **2014**, *4*, 537–544.
101. Katsube, Y.; Kotobuki, N.; Tadokoro, M.; Kanai, R.; Taketani, T.; Yamaguchi, S.; Ohgushi, H. Restoration of cellular function of mesenchymal stem cells from a hypophosphatasia patient. *Gene Ther.* **2010**, *17*, 494–502. [CrossRef]

102. Nakamura, T.; Nakamura-Takahashi, A.; Kasahara, M.; Yamaguchi, A.; Azuma, T. Tissue-nonspecific alkaline phosphatase promotes the osteogenic differentiation of osteoprogenitor cells. *Biochem. Biophys. Res. Commun.* **2020**, *524*, 702–709. [CrossRef] [PubMed]
103. Kiledjian, M.; Kadesch, T. Analysis of the human liver/bone/kidney alkaline phosphatase promoter in vivo and in vitro. *Nucleic Acids Res.* **1990**, *18*, 957–961. [CrossRef] [PubMed]
104. Kobayashi, T.; Sugimoto, T.; Kanzawa, M.; Chihara, K. Identification of an enhancer sequence in 5′-flanking region of 1A exon of mouse liver/bone/kidney-type alkaline phosphatase gene. *IUBMB Life* **1998**, *44*, 683–691. [CrossRef] [PubMed]
105. Matsuzaki, E.; Takahashi-Yanaga, F.; Miwa, Y.; Hirata, M.; Watanabe, Y.; Sato, N.; Morimoto, S.; Hirofuji, T.; Maeda, K.; Sasaguri, T. Differentiation-inducing factor-1 alters canonical Wnt signaling and suppresses alkaline phosphatase expression in osteoblast-like cell lines. *J. Bone Miner. Res.* **2006**, *21*, 1307–1316. [CrossRef]
106. Hatta, M.; Daitoku, H.; Matsuzaki, H.; Deyama, Y.; Yoshimura, Y.; Suzuki, K.; Matsumoto, A.; Fukamizu, A. Regulation of alkaline phosphatase promoter activity by forkhead transcription factor FKHR. *Int. J. Mol. Med.* **2002**, *9*, 147–152. [CrossRef]
107. Kim, Y.-J.; Lee, M.-H.; Wozney, J.M.; Cho, J.-Y.; Ryoo, H.-M. Bone morphogenetic protein-2-induced alkaline phosphatase expression is stimulated by Dlx5 and repressed by Msx2. *J. Biol. Chem.* **2004**, *279*, 50773–50780. [CrossRef] [PubMed]
108. Flowers, S.; Patel, P.J.; Gleicher, S.; Amer, K.; Himelman, E.; Goel, S.; Moran, E. p107-dependent recruitment of SWI/SNF to the alkaline phosphatase promoter during osteoblast differentiation. *Bone* **2014**, *69*, 47–54. [CrossRef] [PubMed]
109. Yusa, N.; Watanabe, K.; Yoshida, S.; Shirafuji, N.; Shimomura, S.; Tani, K.; Asano, S.; Sato, N. Transcription factor Sp3 activates the liver/bone/kidney-type alkaline phosphatase promoter in hematopoietic cells. *J. Leukoc. Biol.* **2000**, *68*, 772–777. [CrossRef]
110. Lammers, W.J.; van Buuren, H.R.; Hirschfield, G.M.; Janssen, H.L.A.; Invernizzi, P.; Mason, A.L.; Ponsioen, C.Y.; Floreani, A.; Corpechot, C.; Mayo, M.J.; et al. Levels of alkaline phosphatase and bilirubin are surrogate end points of outcomes of patients with primary biliary cirhosis: An international follow-up study. *Gastroenterology* **2014**, *147*, 1338–1349.e5. [CrossRef]
111. Saraç, F.; Saygılı, F. Causes of high bone alkaline phosphatase. *Biotechnol. Biotechnol. Equip.* **2007**, *21*, 194–197. [CrossRef]
112. Whyte, M.P. Hypophosphatasia and the role of alkaline phosphatase in skeletal mineralization. *Endocr. Rev.* **1994**, *15*, 439–461. [CrossRef]
113. Rader, B.A. Alkaline phosphatase, an unconventional immune protein. *Front. Immunol.* **2017**, *8*, 897. [CrossRef]
114. Whyte, M.P.; Wenkert, D.; Zhang, F. Hypophosphatasia: Natural history study of 101 affected children investigated at one research center. *Bone* **2016**, *93*, 125–138. [CrossRef]
115. Whyte, M.P.; Greenberg, C.R.; Salman, N.J.; Bober, M.B.; McAlister, W.H.; Wenkert, D.; Van Sickle, B.J.; Simmons, J.; Edgar, T.S.; Bauer, M.L.; et al. Enzyme-replacement therapy in life-threatening hypophosphatasia. *N. Engl. J. Med.* **2012**, *366*, 904–913. [CrossRef] [PubMed]
116. Scott, L.J. Asfotase Alfa in perinatal/infantile-onset and juvenile-onset hypophosphatasia: A guide to its use in the USA. *BioDrugs* **2016**, *30*, 41–48. [CrossRef] [PubMed]
117. Kishnani, P.S.; Rush, E.T.; Arundel, P.; Bishop, N.; Dahir, K.; Fraser, W.; Harmatz, P.; Linglart, A.; Munns, C.F.; Nunes, M.E.; et al. Monitoring guidance for patients with hypophosphatasia treated with asfotase alfa. *Mol. Genet. Metab.* **2017**, *122*, 4–17. [CrossRef] [PubMed]
118. Whyte, M.P. Hypophosphatasia: Enzyme replacement therapy brings new opportunities and new challenges. *J. Bone Miner. Res.* **2017**, *32*, 667–675. [CrossRef] [PubMed]
119. Nakano, C.; Kitabatake, Y.; Takeyari, S.; Ohata, Y.; Kubota, T.; Taketani, K.; Kogo, M.; Ozono, K. Genetic correction of induced pluripotent stem cells mediated by transcription activator-like effector nucleases targeting ALPL recovers enzyme activity and calcification in vitro. *Mol. Genet. Metab.* **2019**, *127*, 158–165. [CrossRef]
120. Yamamoto, S.; Orimo, H.; Matsumoto, T.; Iijima, O.; Narisawa, S.; Maeda, T.; Millán, J.L.; Shimada, T. Prolonged survival and phenotypic correction of *Akp2*$^{-/-}$ hypophosphatasia mice by lentiviral gene therapy. *J. Bone Miner. Res.* **2011**, *26*, 135–142. [CrossRef] [PubMed]
121. Matsumoto, T.; Miyake, K.; Yamamoto, S.; Orimo, H.; Miyake, N.; Odagaki, Y.; Adachi, K.; Iijima, O.; Narisawa, S.; Millán, J.L.; et al. Rescue of severe infantile hypophosphatasia mice by AAV-mediated sustained expression of soluble alkaline phosphatase. *Hum. Gene Ther.* **2011**, *22*, 1355–1364. [CrossRef] [PubMed]
122. Nakamura-Takahashi, A.; Miyake, K.; Watanabe, A.; Hirai, Y.; Iijima, O.; Miyake, N.; Adachi, K.; Nitahara-Kasahara, Y.; Kinoshita, H.; Noguchi, T.; et al. Treatment of hypophosphatasia by muscle-directed expression of bone-targeted alkaline phosphatase via self-complementary AAV8 vector. *Mol. Ther. Methods Clin. Dev.* **2016**, *3*, 15059. [CrossRef]
123. Kinoshita, Y.; Mohamed, F.F.; de Oliveira, F.A.; Narisawa, S.; Miyake, K.; Foster, B.L.; Millán, J.L. Gene therapy using adeno-associated virus serotype 8 encoding TNAP-D10 improves the skeletal and dentoalveolar phenotypes in *Alpl−/−* mice. *J. Bone Miner. Res.* **2021**, *36*, 1835–1849. [CrossRef] [PubMed]
124. Williams, D.K.; Pinzón, C.; Huggins, S.; Pryor, J.H.; Falck, A.; Herman, F.; Oldeschulte, J.; Chavez, M.B.; Foster, B.L.; White, S.H.; et al. Genetic engineering a large animal model of human hypophosphatasia in sheep. *Sci. Rep.* **2018**, *8*, 16945. [CrossRef] [PubMed]
125. Takinami, H.; Goseki-Sone, M.; Watanabe, H.; Orimo, H.; Hamatani, R.; Fukushi-Irie, M.; Ishikawa, I. The mutant (F310L and V365I) tissue-nonspecific alkaline phosphatase gene from hypophosphatasia. *J. Med. Dent. Sci.* **2004**, *51*, 67–74. [CrossRef] [PubMed]

126. Sun, N.; Panetta, N.J.; Gupta, D.M.; Wilson, K.D.; Lee, A.; Jia, F.; Hu, S.; Cherry, A.M.; Robbins, R.C.; Longaker, M.T.; et al. Feeder-free derivation of induced pluripotent stem cells from adult human adipose stem cells. *Proc. Natl. Acad. Sci. USA* **2009**, *106*, 15720–15725. [CrossRef]
127. Singh, U.; Quintanilla, R.H.; Grecian, S.; Gee, K.R.; Rao, M.S.; Lakshmipathy, U. Novel live alkaline phosphatase substrate for identification of pluripotent stem cells. *Stem Cell Rev. Rep.* **2012**, *8*, 1021–1029. [CrossRef]
128. Kuang, Y.; Miki, K.; Parr, C.J.C.; Hayashi, K.; Takei, I.; Li, J.; Iwasaki, M.; Nakagawa, M.; Yoshida, Y.; Saito, H. Efficient, selective removal of human pluripotent stem cells via ecto-alkaline phosphatase-mediated aggregation of synthetic peptides. *Cell Chem. Biol.* **2017**, *24*, 685–694.e4. [CrossRef]

cells

MDPI

Review

Recent Advances in Cardiac Tissue Engineering for the Management of Myocardium Infarction

Vineeta Sharma [1,†], Sanat Kumar Dash [2,†], Kavitha Govarthanan [1], Rekha Gahtori [3], Nidhi Negi [4], Mahmood Barani [5], Richa Tomar [6], Sudip Chakraborty [7], Santosh Mathapati [8], Dillip Kumar Bishi [9], Poonam Negi [10], Kamal Dua [11,12], Sachin Kumar Singh [13], Rohit Gundamaraju [14], Abhijit Dey [15], Janne Ruokolainen [16], Vijay Kumar Thakur [17,18], Kavindra Kumar Kesari [16,19], Niraj Kumar Jha [20], Piyush Kumar Gupta [21,*] and Shreesh Ojha [22,*]

1 Stem Cell and Molecular Biology Laboratory, Department of Biotechnology, Indian Institute of Technology Madras, Bhupat and Jyoti Mehta School of Biosciences, Chennai 600036, India; vineetabt17@gmail.com (V.S.); govarthanan_kavitha@yahoo.co.in (K.G.)
2 Heat Transfer and Thermal Power Laboratory, Department of Mechanical Engineering, Indian Institute of Technology Madras, Chennai 600036, India; 12345sanat@gmail.com
3 Department of Biotechnology, Sir J. C. Bose Technical Campus, Kumaun University, Nainital 263136, India; rekhagahtori11@gmail.com
4 Department of Chemistry, DSB Campus, Kumaun University, Nainital 263001, India; nidhi.negi2@gmail.com
5 Medical Mycology and Bacteriology Research Center, Kerman University of Medical Sciences, Kerman 7616913555, Iran; mahmoodbarani7@gmail.com
6 Department of Chemistry and Biochemistry, School of Basic Sciences and Research, Sharda University, Knowledge Park III, Greater Noida 201310, India; richa.tomar@sharda.ac.in
7 School of Chemistry, University of New South Wales, Anzac Parade, Kensington, NSW 2033, Australia; sudipchakraborty3@gmail.com
8 Translational Health Science and Technology Institute, NCR Biotech Science Cluster, 3rd Milestone, Faridabad-Gurugram Expressway, Faridabad 121001, India; santosh@thsti.res.in
9 Department of Biotechnology, Rama Devi Women's University, Bhubaneswar 751022, India; dillipkumar.bishi@gmail.com
10 School of Pharmaceutical Sciences, Shoolini University of Biotechnology and Management Sciences, Solan 173212, India; poonam.546@shooliniuniversity.com
11 Discipline of Pharmacy, Graduate School of Health, University of Technology Sydney, Ultimo, Sydney, NSW 2007, Australia; Kamal.Dua@uts.edu.au
12 Australian Research Centre in Complementary and Integrative Medicine, Faculty of Health, University of Technology Sydney, Ultimo, Sydney, NSW 2007, Australia
13 School of Pharmaceutical Sciences, Lovely Professional University, Phagwara 144001, India; singhsachin23@gmail.com
14 ER Stress and Mucosal Immunology Laboratory, School of Health Sciences, University of Tasmania, Launceston, TAS 7248, Australia; rohit.gundamaraju@utas.edu.au
15 Department of Life Sciences, Presidency University, College Street, Kolkata 700073, India; abhijit.dbs@presiuniv.ac.in
16 Department of Applied Physics, School of Science, Aalto University, 00076 Espoo, Finland; janne.ruokolainen@aalto.fi (J.R.); kavindra.kesari@aalto.fi or kavindra_biotech@yahoo.co.in (K.K.K.)
17 Biorefining and Advanced Materials Research Centre, Scotland's Rural College (SRUC), Kings Buildings, Edinburgh EH9 3JG, UK; Vijay.Thakur@sruc.ac.uk
18 School of Engineering, University of Petroleum and Energy Studies (UPES), Dehradun 248007, India
19 Department of Bioproducts and Biosystems, School of Chemical Engineering, Aalto University, 00076 Espoo, Finland
20 Department of Biotechnology, School of Engineering and Technology, Sharda University, Knowledge Park III, Greater Noida 201310, India; nirajkumarjha2011@gmail.com
21 Department of Life Sciences, School of Basic Sciences and Research, Sharda University, Knowledge Park III, Greater Noida 201310, India
22 Department of Pharmacology and Therapeutics, College of Medicine and Health Sciences, United Arab Emirates University, Al Ain P.O. Box 17666, United Arab Emirates
* Correspondence: dr.piyushkgupta@gmail.com (P.K.G.); shreeshojha@uaeu.ac.ae (S.O.)
† Both authors contributed equally as first author.

Citation: Sharma, V.; Dash, S.K.; Govarthanan, K.; Gahtori, R.; Negi, N.; Barani, M.; Tomar, R.; Chakraborty, S.; Mathapati, S.; Bishi, D.K.; et al. Recent Advances in Cardiac Tissue Engineering for the Management of Myocardium Infarction. *Cells* **2021**, *10*, 2538. https://doi.org/10.3390/cells10102538

Academic Editors: Mehdi Najar and Wayne Carver

Received: 3 August 2021
Accepted: 21 September 2021
Published: 25 September 2021

Abstract: Myocardium Infarction (MI) is one of the foremost cardiovascular diseases (CVDs) causing death worldwide, and its case numbers are expected to continuously increase in the coming years. Pharmacological interventions have not been at the forefront in ameliorating MI-related morbidity and mortality. Stem cell-based tissue engineering approaches have been extensively explored for their regenerative potential in the infarcted myocardium. Recent studies on microfluidic devices employing stem cells under laboratory set-up have revealed meticulous events pertaining to the pathophysiology of MI occurring at the infarcted site. This discovery also underpins the appropriate conditions in the niche for differentiating stem cells into mature cardiomyocyte-like cells and leads to engineering of the scaffold via mimicking of native cardiac physiological conditions. However, the mode of stem cell-loaded engineered scaffolds delivered to the site of infarction is still a challenging mission, and yet to be translated to the clinical setting. In this review, we have elucidated the various strategies developed using a hydrogel-based system both as encapsulated stem cells and as biocompatible patches loaded with cells and applied at the site of infarction.

Keywords: myocardial infarction; stem cells; regeneration; biomaterial; cardiomyocytes; tissue engineering

1. Introduction

Cardiovascular disease, predominantly MI, is attributed the highest mortality rate worldwide [1]. Reduced contractility and function, irregular left ventricle remodeling, and uneven stress distribution in the heart muscle are among the complications occurring post-MI, eventually resulting in catastrophic heart failure. According to the American Heart Association (AHA)'s "Heart Disease and Stroke Statistics—2021", the prevalence of CVD (including heart failure, hypertension, and stroke) in the US population is 49.2% in the age range of 20 years and above [2]. In 2014, 150,000 people died due to MI; thus an estimated approximately 14% of global death occurs mainly due to MI. Furthermore, MI survivors are also 15 times more likely to develop post-disease complications that lead to heart failure, and are prone to die sooner rather than later compared to the normal population [1]. Cardiac ischemia-related deaths have also ascended to the top of the list of causes of death in India, the United States, and Europe, apart from MI [1,3,4]. After MI incidence, male and female patients above 45 years of age have a lower life expectancy, of 8.2 and 5.5 years, respectively [1]. Socioeconomic burdens such as health care infrastructure and treatment costs (USD 11.5 billion) have made MI one of the top ten most expensive illnesses in the United States [1,5]. Researchers and clinicians around the world have been working extensively to reduce the global incidence of MI and develop significant cost-effective treatment strategies to reduce the mortality rate from MI.

The human heart is a complex organ composed of various types of cells such as cardiomyocytes (CM), fibroblasts, endothelial cells, valve interstitial cells, and resident cardiac stem cells. The cells of the heart are very active metabolically, as it physiologically requires adenosine tri-phosphate (ATP) for its function. Nonetheless, the heart lacks endogenous repair or regeneration potential, thus it remains devoid of regenerative capacity. Any defect in size or deficiency in cardiomyocyte numbers leads to life-threatening MI-related cardiovascular complications [6]. Currently, mitigation of CVDs by pharmaceutical drugs and other clinical practices have effectively improved the patient's survival and quality of life after tissue damage [7]. However, this remains only a short-term solution of temporary duration; the permanent curative would be via heart transplant. Severe shortage of donor organs, post-graft complications, and the limited efficacy of pharmacological interventions has placed the emphasis on cell-loaded scaffold-based therapeutic approaches for cardiovascular complications (CVDs).

The emergence of cardiac tissue engineering (CTE) has not only given substantial hope for resolving or rescuing the damaged heart after MI but also for prompting the regeneration of the damaged myocardium, thus providing a permanent curative. The idea of CTE was first impelled in 1995 by in vitro-generated cardiac tissue obtained from embryonic chicken CMs. This further ushered in the prospect of new research areas around CTE, mainly idealised to translate the bench to bedside. CTE primarily aims to recapitulate the in vivo cardiac niche under in vitro conditions. Therefore, the long-term goals of CTE are considered the construction of in vitro-fabricated tissues for in vivo cardiac repair and regeneration, in vitro preclinical models for evaluation of drug toxicity, and disease models for understanding the development and pathophysiology of heart-related disorders [8]. With the global rise in CVD cases, it is essential to reinforce the treatment modalities for better disease management. In the current scenario, the previously mentioned CTE is considered as at the forefront; however, the conducting of numerous clinical trials is crucial in prioritizing CTE in clinical practice.

Delivery of an engineered scaffold loaded with stem cells, CM mitogens, or pharmacological molecules directly to the infarcted site via either the intra-coronary or intramyocardial mode leads to the relatively prompt recovery of the infarcted tissue, followed by regeneration and regaining of functional significance. Still to be addressed are the current roadblocks to successive clinical utilization, such as an optimized protocol for stem cell-derived CMs and their source, biomaterials for CM cultures, as well as their delivery strategies [9]. Advanced delivery approaches using injectable or patch-based methods are recently gaining significant attention due to their complexity in design and versatility in application. The most advanced technology, using iPSC-derived CM-loaded microfluidic devices, has now been providing unprecedented opportunities to understand the mechanisms of MI development. This technology can also be employed to study the effects of drugs in the preclinical drug screening phase [10].

In this review, we elaborate the recent technological advances in cardiac tissue engineering, particularly the therapeutic approach to regenerating the infarcted myocardium. This review emphasizes mainly cell-based therapy, patch-based therapy, and microfluidics for studying micro-tissue physiology under laboratory set-up. The overall representation of the aforesaid approaches has been elucidated in Figure 1.

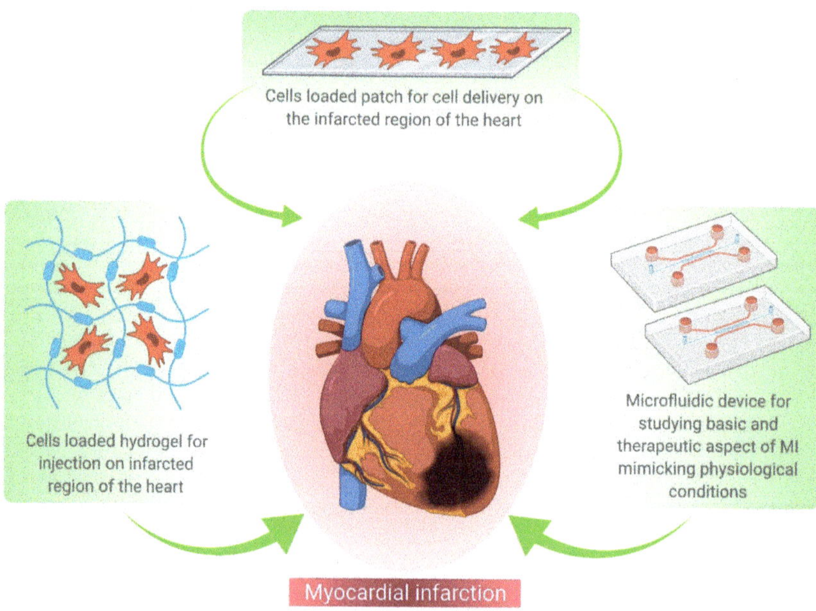

Figure 1. Schematic representation of various tissue engineering approaches for MI treatment. These approaches include hydrogel-based cell delivery (left hand corner), patch-based cell delivery (middle panel), and microfluidics-based drug screening (right corner) during the regenerative therapy of damaged heart tissue.

2. Regenerative Therapy

Heart failure due to the progressive complications of MI mainly occurs due to the limited intrinsic regenerative potential of the myocardium. A left-ventricular assist device (LVAD), fixed internally to relieve the pressure on the heart left region, is the only available medication for the management of post-MI complications. This temporarily delays post MI complications; meanwhile, in such cases, a heart transplant can be the only permanent solution. However, the lack of organ donors has led researchers and clinicians to contemplate alternative therapies available immediately after MI for minimizing cardiomyocyte damage and thereby preventing subsequent heart failure. Under these conditions, cell-based therapy is ideal for repairing the initial injury, restoring lost cardiomyocytes, and preventing the development of a scar (which impairs cardiac function) (Figure 2). Many research groups are currently investigating the possibility of restoring cardiac function by replacing lost cardiomyocytes or via rejuvenating the resident cardiac stem cell population to counterbalance the lost CMs. Cell therapy has conventionally been a modest procedure in which cells are directly injected into the myocardium [11–20]. These cells may function in various ways, including differentiation into cardiomyocytes, support for endogenous regeneration, and/or protection of the affected cells. Several cell types have been investigated for their regenerative ability, each with its own set of benefits and side effects [13]; among these are mesenchymal stem cells, bone marrow cells, and cardiac progenitor cells. These have been shown to improve heart function in preclinical studies and are currently being studied in clinical trials [21]. Cell therapy using mesenchymal stem cells and bone marrow cells has been effective to some extent, but these cells are unable to differentiate into cardiomyocytes due to their restricted differentiation potential.

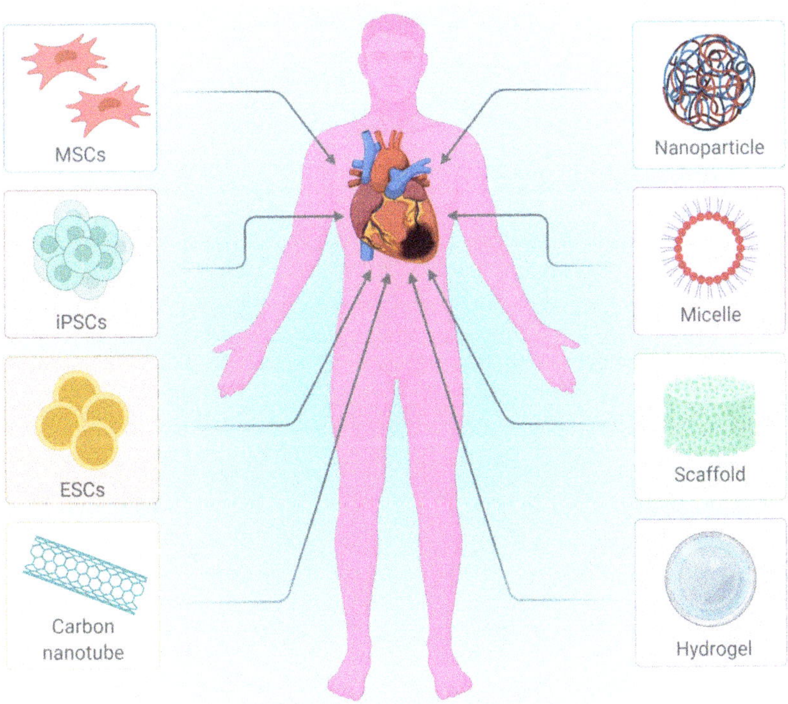

Figure 2. Illustration of stem cell-based regenerative medicine in MI treatment. Various stem cells like MSCs, ESCs, and iPSCs have been employed in MI treatment. Stem cells are delivered through engineered novel biomaterials (via encapsulation) that mimic the native niche. The stem cell-based regenerative therapy is beneficial to either replace the injured area or the whole organ.

The pluripotent properties of induced pluripotent stem cells do improve cardiac function and vascularization, but they still harbor the risk of teratoma formation [13]. Cardiomyocytes, smooth muscle cells, and endothelial cells can all be differentiated from cardiovascular lineage-specific progenitor cells which reside in the heart. Injection after MI demonstrated in vivo regenerative capacity, but progenitor cells were rejected by the host. Cell viability preservation and survival in the hostile environment of the infarcted heart, and further effective coupling to the existing myocardium, remain significant challenges.

Another technique has recently emerged in which new cardiomyocytes are produced by inducing fibroblasts to transdifferentiate. Although the initial success rate of transdifferentiation was poor, it has the ability to introduce new cardiomyocytes even if the scar has completely developed [11]. Despite the fact that cell therapy has not yet proven to be as effective as anticipated, the findings obtained so far have provided further insights into how the damage in the infarcted heart can be handled. While mesenchymal stem cells and bone marrow cells do not contribute to the myocardium, they can still benefit the heart and its function via its secretive nature. Furthermore, cardiomyocyte progenitor cells positively impact the heart even though they do not differentiate into cardiomyocytes. As a result, other cell factors have a positive impact on the regenerative processes. Understanding regenerative processes and their underlying signaling mechanisms, as well as how cell therapy can affect these processes, is critical to reaping the benefits of cell therapy.

3. Cell Based Therapy

In the tissue engineering and regeneration process, various types of cells are involved (Figure 3). Before using any cells, the key issues such as administration of immunosuppression and disease transmission to the host have to be addressed. Although autologous cell transplantation circumvents the use of immunosuppressants and holds a lower risk of disease transmission, the restricted supply hinders its application. Allogenic cells can also be used, but they require immunosuppression and pose a danger of disease transmission. Other disadvantages include the difficulty of collecting cells from donor sources, and of expanding their prior integration into the host. Furthermore, depending on the source of extraction (e.g., elderly persons or diabetic patients), autologous cells may have limited proliferation and differentiation [13,22,23]. Pluripotent stem cells such as embryonic stem cells (ESCs) and induced pluripotent stem cells (iPSCs) are cells that have the ability to self-renew and to give rise to any of the three primary germ cell layers, but not extra-embryonic tissues [24]. Studies employing stem cells in an animal model of cardiac injury and their outcomes have been summarized in Table 1.

a. Embryonic stem cells (ESCs)

ESCs are isolated from the inner mass cells of an embryo at the blastocyst stage. ESCs have the ability to proliferate for an infinite number of passages and can be differentiated into any cell type. ESC treatments with multiple induction cues can differentiate into cardiomyocytes or cardiac progenitor cells. Due to ethical concerns and the generation of teratoma, the use of ESC in clinical trials is ethically restricted. Some of the most challenging aspects of ESC research are teratogenic potential, obtaining pure lineage, and guiding differentiation to a particular lineage type [25–27]. To circumvent these constraints, genetic modifications, biological factor treatment, and diverse cultural approaches are applied. Chong et al. were the first to obtain a large number of cardiomyocytes from ESCs and use them to repair injured myocardium [24].

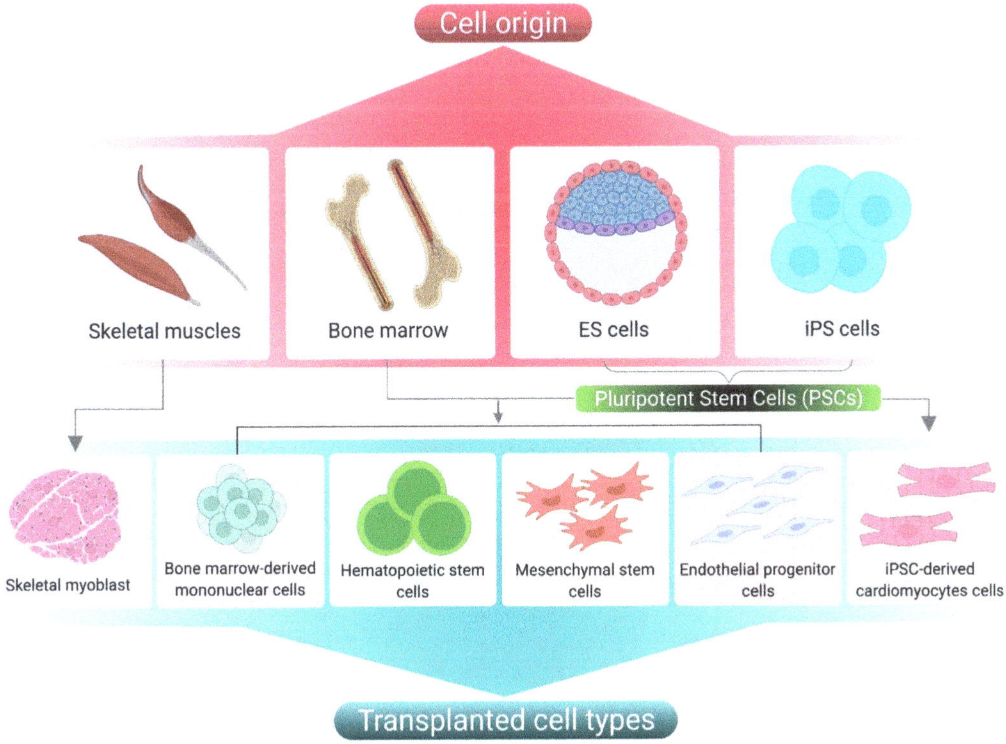

Figure 3. Distinct cells of different origins which are used in the regenerative medicine depicted herein. Depending on the pluripotency, the transplanted cell types can differentiate into various other cells such as skeletal myoblasts, endothelial progenitor cells, chondrocytes, adipocytes, or cardiomyocytes.

b. Induced pluripotent stem cells (iPSCs)

Takahashi and Yamanaka used viral vectors to generate induced pluripotent cells from somatic cells for use as personalized medicine. Since then, several studies have been published that have explained the ability of induced pluripotent stem cells (iPSCs) to differentiate into endoderm (e.g., hepatocytes), mesoderm (e.g., cardiomyocytes) and ectoderm (e.g., neurons), excepting extraembryonic tissues. Martens et al. and Yu et al. identified distinct cardiac phenotypes in infarcted mouse hearts using iPSCs [28,29]. These can replace the fundamental program settings through reprogramming with genetic factors and signaling molecules. As a result, a question about the experimental efficiency of iPSCs arises. Malignancies and oncogenes can also be induced in the host when employing viral vectors [23,30]. Maza et al. found that lowering the Mbd3 gene can enable all cells to acquire pluripotency, which is a key hurdle to employing iPSCs in clinical practice [31].

c. Adult stem cells (ASCs)

Autologous cells can be isolated from various sources (e.g., bone marrow, adipose tissue, etc.). Orlic et al. used transplanted bone marrow-derived cells (BMCs) to regenerate infarcted myocardium [32]. Several other investigations, however, cast doubt on this finding. Clinical experiments utilizing BMCs have shown short-term benefits as well as a higher survival rate [33]. When BMCs are coupled with growth factors, the benefits can last longer [34–36]. BMCs must be cultivated in vitro, but ASCs do not need to be grown. Cells are isolated from human fat tissue. The PRECISE, APOLLO, and RECATABI projects have conducted clinical trials [37]. A 3D polymer scaffold printed using a peptide gel and loaded with ASCs imparts mechanical strength to the already-dilated ventricle, according to the findings [38].

d. Cardiac stem cells (CSCs)

Cardiac stem cells can be isolated from a biopsy and then cultured in the lab. Smooth muscle cells, endothelial cells, and cardiomyocytes can be differentiated from undifferentiated CSCs [39–43]. Lineage tracing studies without a specific cardiac marker showed the existence of endogenous CSCs in the fetal heart; however, it was also pointed out in the study that there is a lack of data supporting the existence of CSCs in the adult heart. Moreover, recent studies revealed progenitors supporting regeneration of the damaged heart via secreting factors that rejuvenate the resident CSCs in order to counter balance the lost cells. However, massive damage requires a high number of cells to maintain homeostasis. Recent clinical studies conducted with CADUCEUS, using autologous cardiosphere-derived cells (CDCs), showed improved heart function [44–46].

Table 1. Description of the cells delivered to the heart by injection. This cell delivery approach has used various cell types including ESCs, iPSCs, MSCs, and CSCs.

Initial Cell Type	Target Cell Type	Composition of Delivery Vehicle	Mode of Delivery	Animal Models	Outcomes	Limitations	References
iPSCs	CMs	Polyethylene glycol hydrogel	Trans-epicardial	MI in nude rats	Increased infarct thickness and improved muscle content	No donor cell engraftment was observed	[47]
Mouse ESCs	CMs	PA-RGDS based gel	Trans-epicardial	Mice	Engraftment and integration of mESC-CMs into host myocardium improved cardiac function	No information available on cardiac remodelling	[12]
iPSCs	CMs	PBS solution	Trans-epicardial	Post-infarcted swine	Enhanced angiogenesis, reduced apoptosis, and blunted cardiac remodelling	No detailed information available on the engraftment of donor cell	[48]
MSCs	****	Self-assembling peptide hydrogels (3-D Matrix, Ltd.)	Surface immobilization by spreading	Lewis rats	Augmented microvascular formation and reduced interstitial fibrosis	No detailed information available on the engraftment of donor cell and CMs differentiation from MSC	[49]
MSCs	****	Si-HPMC	Trans-epicardial	Lewis rats	Short-term recovery of ventricular function and attenuated mid-term remodelling	No detailed information available on the engraftment of donor cell and CMs differentiation from MSC	[50]
c-Kit overexpressing CSCs	****	PBS solution	Intracoronary	Fischer 344 rats	Preserved LV function and structure	Increased cell dose was found to be harmful. Cell tracing or engraftment were not available in detail	[50]
CSCs	****	Matrigel and dimethylpolysiloxane mixture gel	Trans-epicardial	NOD-SCID mice	Improved long-term retention of CSCs, cardiac structure and function	Cell tracing or engraftment were not available	[51]

**** Studies were carried out to observe the improvement in cardiac function. The available report did not present specifics on the engraftment and in situ differentiation of delivered cells to mature CMs.

e. Skeletal myoblast cells (SMs)

SM cells can live in a low-oxygen environment better than other cells [52]. The ability to contract is the most crucial feature of these cells, as it allows them to contribute and adhere to beating cardiomyocytes [53]. These cells, however, are unable to integrate electromechanically with host cardiomyocytes due to a deficiency of connexin 43, a gap junctional protein. The results of the Myoblast Autologous Grafting in Ischemic Cardiomyopathy (MAGIC) clinical trials demonstrated that a pacemaker or a defibrillator with the incorporation of cells is needed for reducing arrhythmias. Phase II clinical trials of MAGIC involved the implantation of a cardioverter defibrillator along with SM via coronary artery bypass grafting [54]. Modifications based on the expression of connexin 43 are also being studied for avoiding myofiber arrhythmogenicity. Gap junction protein modification cannot tolerate arrhythmogenicity [55].

f. Umbilical cord blood cells (UCBC)

Umbilical Cord Blood Cells were isolated from the umbilical cord and have been extensively used for scientific research. UCBCs do not need ethical clearance, which allows for extreme flexibility when studying their biological dimensions. Even though these cells are less immunogenic, they show better reversing ventricular function ability in animal models [56]. Umbilical cord Wharton's jelly-derived MSCs could also be considered a candidate cell type in cardiac tissue engineering, as they were shown to express cardiac-specific genes inherently without any manipulation [57]. Studies demonstrating the use of small molecule inhibitors to reinforce the enhanced differentiation potential of WJ-MSCs also highlighted the abundant supply of cells necessary for transplantation without any invasive procedures [58].

g. Amniotic fluid stem Cells (AFSCs)

AFSCs are prenatal stem cells having the potential to differentiate into cardiac cells or endothelial cells in vitro. These cells pose no risk of tumorigenicity or ethical concerns. In an immunosuppressed rat model, Yeh et al. demonstrated that these cells conserved ventricular wall thickness and improved heart functionality [59].

h. Cells Aggregates

Although stem cell transplantation is currently implemented clinically, it is difficult to accomplish minimally invasive injectable cell delivery while retaining high cell retention and animal survival. Strategies involving stem cell retention in the infarct region such as patch-based therapy and delivery of cell aggregates are being studied. Tang et al. demonstrated the safety and efficacy of encapsulating human cardiac stem cells (hCSCs) in thermosensitive poly (N-isopropylacrylamineco-acrylic acid) or P(NIPAM-AA) nanogel in mouse and pig models of MI. Unlike xenogeneic hCSCs injected in saline, injection of nanogel-encapsulated hCSCs did not elicit systemic inflammation or local T cell infiltration in immunocompetent mice. The developed thermosensitive nanogels can be used as a stem cell carrier: the porous and convoluted inner structure not only allows nutrient, oxygen, and secretion diffusion but also prevents the stem cells from being attacked by immune cells [60]. Compared to the traditional approaches of single cell injection, cell aggregate deliveries have demonstrated higher retention of cells and prevention of teratoma development [61]. Another such study on cell aggregates by Bauer et al. showed that the better survivability of these aggregates could be attributed to the imitation of the endogenous state by ensuring adequate cell-cell interaction [62]. A bioengineered 3D framework which enhances cellular contact while still allowing for certain cell ratios was developed by Monsanto et al. These injectable cardio clusters enhance adhesions and reduce cell loss [63].

4. Patch Based Cell Therapy Development

Various cell-based experiments have been conducted to treat MI. However, very few cells get migrated and incorporated in the infarct region owing to its hostile environment. Low oxygen supply to the infarct area prevents cells from being incorporated. A few pioneering studies have investigated the engineering of sheet-based cardiac patches constructed to harbor well-aligned and interconnected cardiomyocytes for successful implants to regenerate the myocardium [64]. Zimmermann et al. constructed myocardial tissue employing 3D models to mimic the native heart muscles, resulting in the restoration and improvement of cardiac function. Helfer & Bursac demonstrated a versatile framed hydrogel methodology to generate engineered cardiac tissue with enhanced mature functional properties. Therefore, it is well understood that in order to engineer a cardiac patch, the most important prerequisite is the cells that proliferate and gain functionality in the infarcted region. Different strategies involved in patch design for treating injured myocardium are shown in Figure 4 and described in the following section.

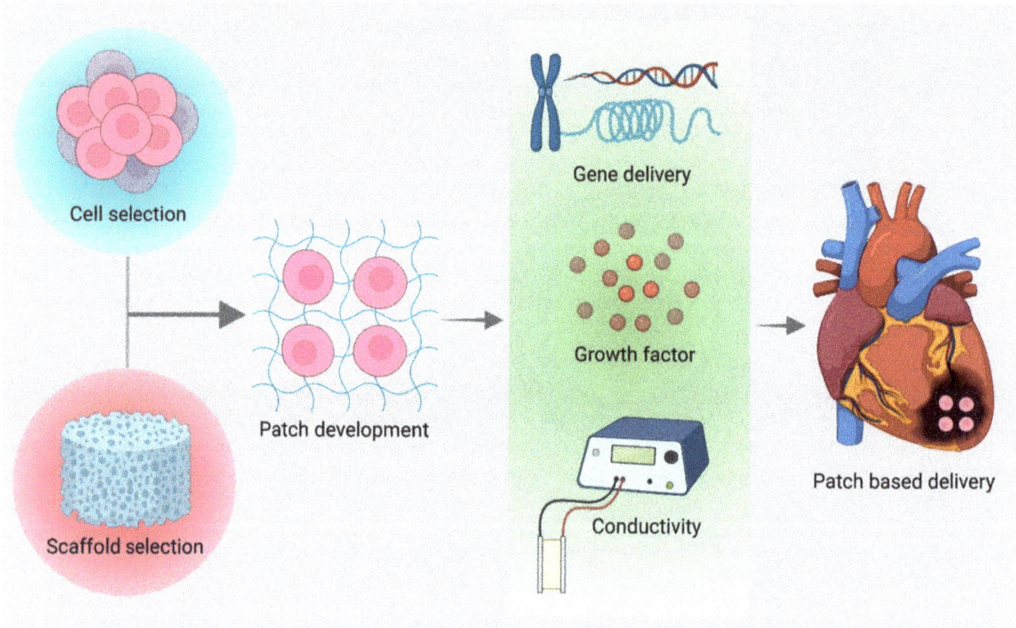

Figure 4. Patch-based cell therapy development started with the fabrication of different patches made up of carbon nanostructures, conductive or non-conductive polymers, hydrogel, etc. followed by patch-based stem cell delivery for MI disease treatment.

4.1. Properties for Patch Design

Scaffold materials are classified into synthetic or biologic and degradable or non-degradable conditional to their usage. Polymers used in the scaffold are subject to the composition, structure, and arrangement of their constituent macromolecules, which can be characterized into different types such as structural, chemical, and biological. Naturally occurring polymers, synthetic biodegradables, and synthetic non-biodegradable polymers are the leading types used as biomaterials. The successful fabrications of the 3D scaffolds utilized in MI can be broadly categorized into four groups based on their properties: chemical, electrical, mechanical, and biological.

Chemical: Surface properties (e.g., surface energy, chemistry, charge, surface area)
Electrical: Conductivity
Physical: Mechanical competence (e.g., compressive and tensile strength), External geometry (e.g., macrostructure, microstructure, interconnectivity), porosity, and pore size
Biological: Interface adherence, biocompatibility, biodegradation

The scaffold's potential lies in its ability to serve as a cell surface receptor for cell differentiation, tissue formation, homeostasis, and regeneration, mimicking the natural ECM. Scaffold geometry plays a vital role in upholding highly interconnected porous fabrics of high surface density, thereby providing an increased surface-to-volume ratio, favoring cell attachment and proliferation [65].

4.1.1. Chemical Properties

Cellular adhesion and proliferation are surface-dependent properties, and the fabricated scaffold should facilitate the attachment of the cells. Altering the surface functionality by thin film deposition can aid in the better anchorage of cells wherein biomolecules viz. collagen, fibronectin, RGD peptides, and growth factors like bFGF, EGF, insulin, etc., are employed in scaffold design [66]. Microfabrication methods include manipulating topographic cues using lithography, which promotes cell organization into anisotropic 2D tissue layers [67]. The Angio-Chip scaffolds are formed with the help of POMaC incorporated nanopores. The microholes in the vessel walls help to increase permeability and permit intercellular crosstalk [68].

Scaffold degradation occurs as a result of physical, chemical, or biological processes. Enzymes involved in tissue remodeling also participate in the degradation of a scaffold. Consequently, scaffold dismantling and material dissolution occur through bulk or surface degradation. The polymeric biomaterials are degraded as a result of hydrolytic or enzymatic cleavage. The factors responsible for the degradation of polymeric biomaterials are the intrinsic properties of the polymer, chemical structure, the presence of hydrolytically unstable bonds, the level of hydrophilicity/hydrophobicity, morphology, glass transition temperature, the copolymer ratio, and the molecular weight of the polymer [65].

4.1.2. Electrical Properties

The lack of electrical conductivity between the cardiac cells, i.e., synchronous beating between different parts of the patch, also needs to be addressed while designing biomaterials for cardiac tissue engineering. The nanostructures can be incorporated within the biomaterials to improve scaffold conductivity. This electrically designed bionic cardiac patch makes it possible to monitor and control engineered tissue functions after implantation [66]. Pairing tissue building blocks impregnated with specific DNA strands and complementary sequences with other building blocks governs their assembly to thicker tissues, while utilization of nanowired FETs allows for high sensitivity. Gold electrodes with a nanometric layer of titanium nitride have been used to increase the surface area, thereby improving cell adherence. The multifunctional electronic cardiac patch can monitor tissue activity in multiple locations. The electroactive polymers can release both positively and negatively charged molecules in response to the current where incorporated [67]. Incorporation of electrically conductive GNRs to the GelMA hydrogels has been designed to implicate the electrical conductivity of GelMA hydrogel constructs. GelMA–GNR hydrogel has an electrical impedance (1.35–0.36 kU) close to the physiological range [69]. It has been observed that conductive biomaterials, whether or not coupled with external electrical stimulation, enhanced the outcome of current tissue engineering strategies by improving cells or biomaterials' native myocardium electromechanical integration [70].

4.1.3. Physical Properties

Mimicking the mechanical properties of cardiac ECM is essential in order to provide biophysical cues to the cells. Acting as the natural cardiac microenvironment, the engineered patch produces the proper contractions. Thus, biomaterials with cardiomimetic mechanical properties such as PGS are used to promote the assembly of functional tissues with more native-like properties [66]. Rigidity and rheological parameters contribute to the mechanical stability of the scaffolds. It becomes crucial to retain the mechanical properties of the scaffold for its biological applications. These in turn help the regeneration of tissue. Parameters such as elastic modulus, flexural modulus, tensile strength, and maximum strain impart good mechanical strength to the scaffold and enhance its potential for several applications. The porous structure of the scaffold is entirely interconnected geometry. It is essential for cell ingrowth, uniform cell distribution, and enables the neovascularization of the construct. The crucial parameters when designing a scaffold are average pore size, pore size distribution, pore volume, pore interconnectivity, pore shape, pore throat size, and pore wall roughness. The pore size varies in different scaffold types, such as 5 μm for neovascularization, 5–15 µm for fibroblast ingrowth, 20 µm for the ingrowth of hepatocytes, 200–350 µm for osteoconduction, and 20–125 µm for regeneration of adult mammalian skin. It is essential to monitor the scaffold-pore interconnectivity to ensure oxygen mass transfer and nutrient transfer between cells [65]. PEG scaffolds have better elastomeric mechanical properties, with extended-release behavior and prolonged transgene expression [71]. GelMA hydrogel was engineered for surface micro-topographies that help to mimic the physiologically relevant myocardium function. Furthermore, it allows the cells to form uniformly dense and highly aligned cardiac tissues on GelMA–GNR hydrogels [72]. Collagen hydrogels, owing to their mechanical stiffness and high biocompatibility, are preferred candidates for cardiac tissue engineering. Further conjugating CNTs at subtoxic levels to collagen hydrogels demonstrate higher toughness, tensile stress, tensile strain, and electrical conductivity [73].

4.1.4. Biological Properties

Tissue engineering is promising with biocompatible materials, and the scaffold or matrix should support cellular activity without hindering the signaling cascade. The chemistry and morphology of the materials used in the scaffold design as well as polymer synthesis and scaffold processing affect their biocompatibility. According to their biocompatibility various polymers are used in general medical applications, including PLA, PGA, PLGA, PDO, and PTMC [65]. Synthetic polymers possess advantages as their properties can be tailored to a specific application. They can be produced in large quantities, less cheaply compared to than biologic scaffolds though with enhanced shelf life. They demonstrate characteristic behavior in terms of tensile strength, elastic modulus, and degradation rate. PLA, PGA, and PLGA copolymers are widely used synthetic polymers in tissue engineering [65].

4.2. Biomaterials Used for Cardiac Tissue Engineering

Almost all normal cells in human tissue except for blood cells either reside in or adhere to the extracellular matrix (ECM). ECM provides structural support to the cells and contributes to the mechanical properties of the tissue. It also regulates cell behavior by influencing homeostasis, cell proliferation, cell shape, cell survival, differentiation, and migration. ECM also acts as a reservoir of growth factors and potentiates their actions while allowing remodeling during development and assisting differentiation [74–81]. Tissue engineering employs various cells, temporary scaffolds, and growth-promoting signals for achieving tissue regeneration and tissue repair [80,82–85]. Essentially, the scaffold should be an ECM analogue unique to the tissue of interest [80,82]. Although many types of ECM exist, all of them are primarily made up of a complicated assemblage of various polysaccharides and proteins that differ by tissue [86]. Several requirements should be met by cardiac tissue-specific construction. The constructs should mimic the native environment of heart muscles and should be viable during and after implantation. They should also help in the improvement of the systolic and diastolic functions of the injured myocardium [76,86–88]. The elastic and electrical properties of the scaffolds should be kept in consideration while designing the construct in order to achieve impulse conduction and contractile properties [86]. As a result, the ideal construct for cardiac regeneration should be mechanically robust and have high contractility and flexibility. It should also have angiogenic potential and induce vascularization after implantation. The constructs should also be non-immunogenic and electrophysiologically stable. There have been reports of different cardiac constructs based on various techniques. Biomaterial gels encapsulating cells, scaffolds encapsulating cells, cell films, and fibrous or porous sheets are among them [89]. The scaffold material chosen is a significant aspect that can influence the regeneration strategy's success. The key polymer biomaterials used in cardiac tissue engineering are described in the sections below Table 2.

Table 2. Types of natural and synthetic polymers used during cardiac tissue engineering.

Natural Polymers	
Chitosan	Hyaluronic acid
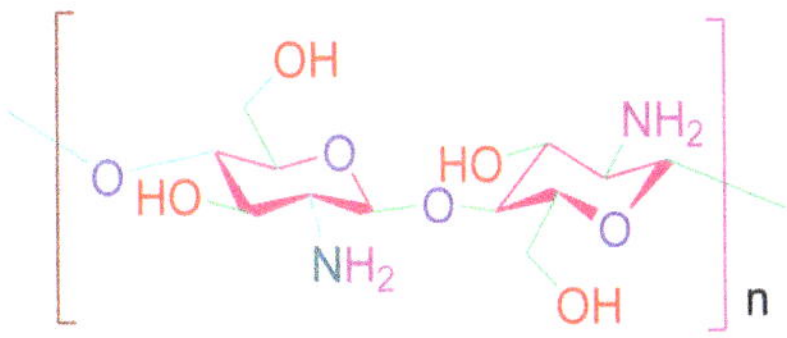	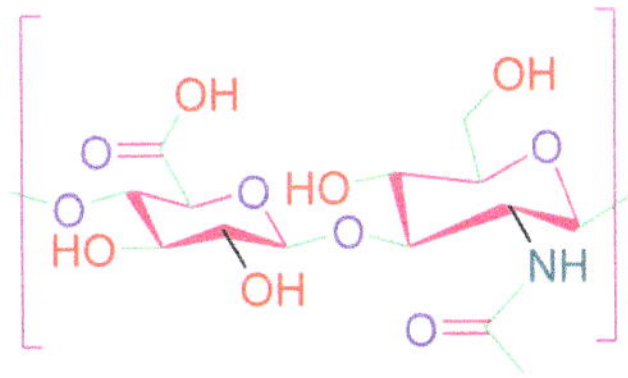
Alginate	Fibrin
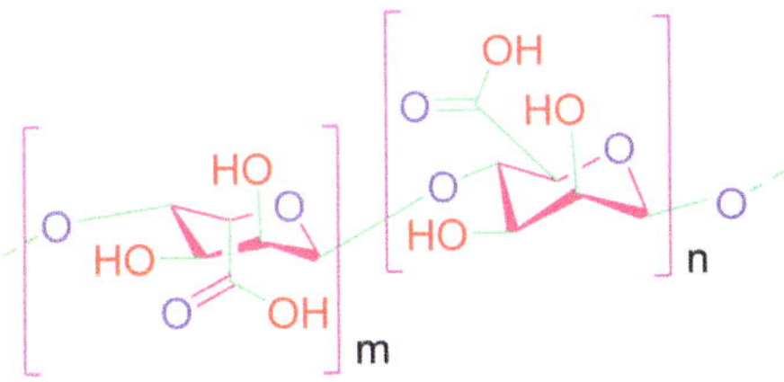	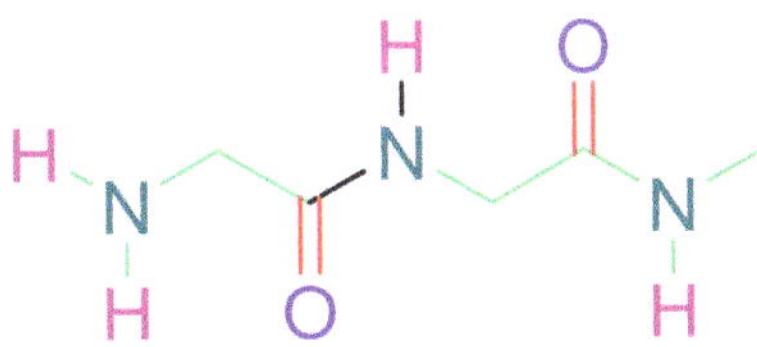
Synthetic Polymers	
Poly(glycolic acid)	Poly(ε-caprolactone)
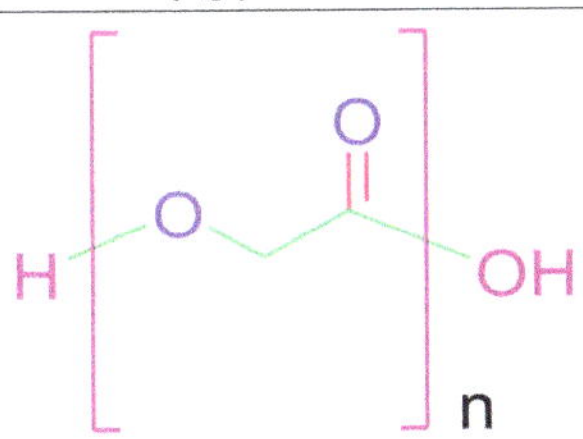	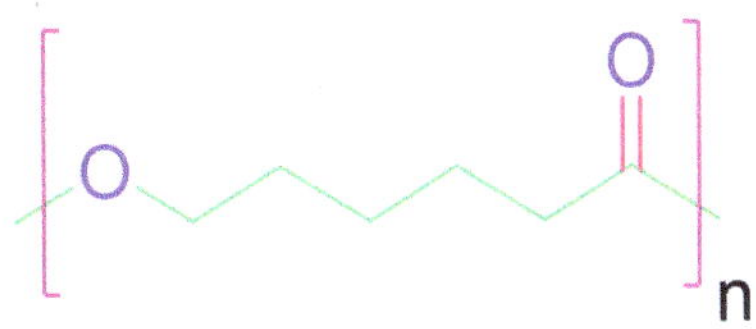
Poly(N-isopropylacrylamide)	Poly(ethylene glycol) 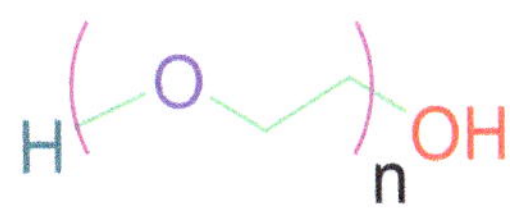
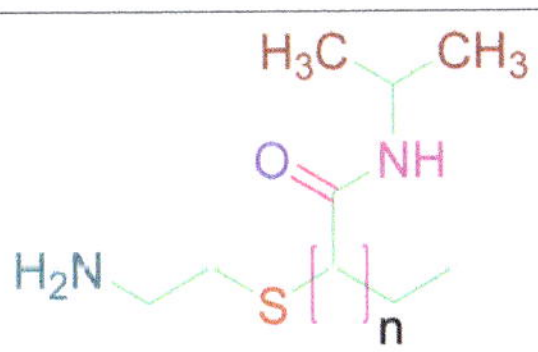	Poly(lactic acid) 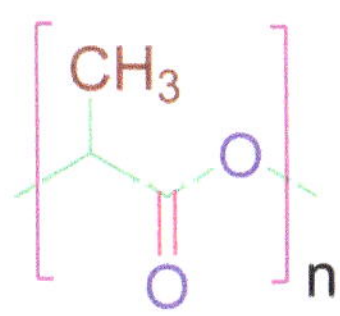

Table 2. *Cont.*

Gelatin methacryloyl

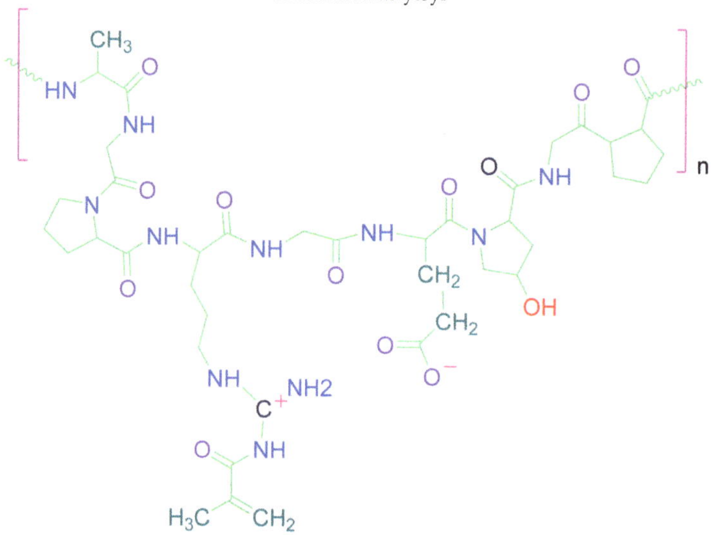

4.2.1. Natural Polymers

The cell-biomaterial interaction is a crucial factor to consider when choosing a biomaterial for cellular delivery. Primary cells are adherent-dependent cells that die if the cell-matrix connection is disrupted. Many naturally occurring biomaterials exhibit the motif that is required for cellular contact and adhesion in this regard. Collagen, gelatin, decellularized tissue/organ, ECM, and silk fibroin are examples of such biomaterials [90,91]. These materials contain peptides which enable integrin mediated contact, leading to crosstalk between cell-matrix and cellular adherence via multimodal signaling pathways. Thus, in the case of natural biomaterials, cell-biomaterial interactions are more prevalent, which results in cell proliferation and differentiation. Since of intrinsic proteins, growth factors, and glycosaminoglycans (GAGs) are present, hydrogels obtained from decellularized ECM are a prominent natural biomaterial that demonstrates outstanding angiogenesis and cardiac regeneration abilities [92,93]. Hydrogels produced from decellularized hearts were used in a rodent model that exhibited better angiogenesis and myocardial function. Silk fibroin is a natural protein-based biomaterial generated from silkworms. Silk fibroin generated from non-mulberry silkworms contains an innate RGD peptide sequence that stimulates the adhesion of stem cells, proliferation, and differentiation [92].

(a) Fibrin

Fibrin is a natural polymer that FDA has approved for clinical use. It contains the amino acids arginine, glycine, and asparagine (RGD), which can help cells adhere together [94–97]. Fibrin can be made from fibrinogen monomers and thrombin polymerization, a proteolytic enzyme. Fibrin gels, on the other hand, have weak mechanical qualities and can shrink when injected into the heart. The injection of fibrin can cause intravascular thrombosis [98,99]. Using a blend of injectable fibrin with multiple types of cells (e.g., bone marrow cells, myoblasts, autologous endothelium cells, etc.) can outperform the efficacy of using just one type of cell [98,100].

(b) Chitosan

Chitosan is a carbohydrate that occurs naturally in chitin [101]. It is more accepted in the tissue engineering sector because of its biocompatibility and antifungal and antibacterial qualities. Because of its great temperature sensitivity, bioactive compounds are easily integrated into chitosan-based hydrogels [101–104]. Cardiomyocyte metabolic function can be improved by a thermoresponsive hybrid hydrogel made of chitosan, collagen, and QHREDGS (a peptide derived from angiopoietin 1) [105]. The hybrid hydrogel promotes cell survival, cell proliferation, and angiogenesis while enhancing myocardial wall thickness [105].

(c) Alginate

Alginate is a polysaccharide that can be made from both seaweed and bacteria. It is biocompatible to a high degree [86,106,107]. Its qualities can be tweaked by adjusting the concentration or molecular weight. Alginate was administered into the zone of infarction of the rat heart, resulting in a reduction in both scar thickness and systolic and diastolic cycle dysfunction. It also necessitates a purifying process prior to use in tissue engineering to ensure that alginate impurities do not create any adverse effects in humans. Clinical experiments with alginate hydrogels were reported by Anker et al., with patients having a higher mortality rate [108].

(d) Hyaluronic acid

Hyaluronic acid (HA) is a polysaccharide that is present in practically every cell in the body. It aids in the transport of nutrients into cells, maintains homeostasis, is nonimmunogenic, anti-inflammatory, and has numerous other beneficial effects on cell regeneration and repair [109]. The FDA approved it for research and certain human uses, and HA is commercially accessible in a cross-linkable form. Depending on its molecular weight, HA can initiate and contribute to a variety of biological processes. During degradation, low molecular weight HA that induces cell proliferation and angiogenesis is generated [110]. PEG-Sh4 biological development can also be functionalized [111]. Shen et al. evaluated various HA hydrogels with commercial chitosan, fibrin, and elastin hydrogels and observed that HA hydrogels were more biocompatible, less immunogenic and cytotoxic, and possessed better angiogenic properties than other hydrogels [112]. For cell attachment and proliferation, HA alone is inadequate. Cell adhesion can be achieved using hyaluronan hydrogels crosslinked with thiol-reactive poly(ethylene glycol) diacrylate [113,114].

(e) Collagen

Collagen is a vital component of a matured heart extracellular matrix (ECM), and it can help cardiomyocytes grow and survive in a natural way [115]. In animal models, commercially available collagen alone can improve heart function [116]. Collagen was administered using a catheter in a pig model, along with several types of cells, to demonstrate the possibility of a noninvasive delivery system using collagen [116]. Despite their advantages, collagen-derived gels are mechanically weak [117]. Using collagen matrix with induced carbon nanotubes (CNTs) helped to enhance rigidity, electrical, and mechanical abilities [118].

(f) ECM

The decellularized tissue scaffolds closely resemble natural ECM since they already possess its physiological and environmental characteristics [75]. Every tissue's ECM has its own set of characteristics and components, such as proteins and proteoglycans. If the decellularized matrix is accessible, it is one of the finest choices for cardiac repair and regeneration [119]. The two most commonly available types of small intestine submucosa (SIS) derived injectable gels were investigated for heart repair in a mouse model. The gels had varied concentrations of fibroblast growth factors, and larger concentrations were more effective for heart regeneration [120]. Slow gelation and rapid breakdown are disadvantages of ECM-derived hydrogels [121,122]. Jefford et al. employed genipin for crosslinking porcine ECM hydrogels. Analysis showed that the degradation rate of genipin-crosslinked hydrogel was slower in vitro than gels without a crosslinker [123]. In an MI rat model, Efraim et al. investigated a functionalized chitosan decellularized porcine cardiac ECM using genipin as a crosslinker. The results demonstrated that this substance could dramatically improve heart function [124]. Gelatin, Matrigel, hair keratin, and laminin are examples of natural polymers. Natural polymers are biocompatible with host tissue and exhibit biological properties. On the other hand, natural polymers have some disadvantages, such as reproducibility concerns. Synthetic polymers are being generated to address the disadvantages of natural polymers while also improving the scaffold's suitability for tissue engineering. ECM provides structural, mechanical, and biochemical signals to govern cellular processes, and the relationship between cells and their ECM has been widely investigated [125].

Scientists have been able to identify essential ECM components as well as the mechanisms by which ECM regulates normal cellular activities like migration and differentiation, as well as pathological events like cancer [119,121,126], fibrosis [127,128], and wound healing [78]. There is mounting evidence that changes in ECM mechanical properties significantly impact on cell structure and function. Excessive collagen-I deposition/crosslinking orchestrated by activated fibroblasts during fibrosis and tumor growth, in particular, is thought to have a role in aberrant mechano-sensing and atypical cell behaviors [129].

4.2.2. Synthetic Polymers

Synthetic biomaterials have a number of benefits over their natural equivalents, including the ability to modify chemical, mechanical, and biological properties. Various synthetic biomaterials help in improving infarcted heart ventricular function when injected into the infarcted region [130]. When compared to sham, different types of poly(N-isopropylacrylamide) (PNIPAm) showed improvement in left ventricular end diastolic diameter (LVEDD) and decreased EF after injection [131]. Similarly, researchers found that injecting a poly (ethylene glycol) (PEG) hydrogel into saline-injected hearts reduced dilation by inhibiting an increase in LVED [132]. Synthetic biomaterials can be modified to fine-tune the target environment, resulting in improved bioavailability, cellular proliferation, and differentiation. A cell binding motif does not exist in pure synthetic biomaterials. Thus, it is necessary that cell binding motifs are introduced in these biomaterials covalently, which helps these synthetic biomaterials with cell adhesion. Similarly, covalent grafting can be used to introduce various functional groups on these polymers to link multiple pharmaceutical and biological molecules that would promote cellular proliferation and have differentiation and angiogenic properties [130,132].

Natural polymers have a wide range of compositions but are mechanically weak. As a result, a composite method has been applied to generate a composite biomaterial for cell transport made up of natural and synthetic polymers. Biocompatibility, mechanical characteristics, and the possibility of grafting molecules onto hydrogels generated from such composite materials are all improved [133]. An injection of fibrin and collagen combined with alginate, for example, hindered the migration of the infarct zone [134].

(a) Poly(ethylene glycol)

PEG is a frequently utilized synthetic polymer, attributed to its biocompatible nature [117]. PEG is a benign, nonimmunogenic polymer that may be customized by adding functional groups to its backbone [135]. It is soluble in water or organic solvents. It is nontoxic and non-immunogenic in nature. As a result, PEG is an appropriate polymer for heart regeneration. Since PEG is bioinert in nature, it does not mimic the microenvironment for cell survival. This constraint can be circumvented by crosslinking PEG hydrogels with natural polymers or employing bioactive compounds in the gels. PEG hydrogel containing cyclodextrin/MPEG–PCL–MPEG was developed by Wang et al. and supplied with erythropoietin (EPO) [136]. EPO is an antioxidant hormone that can protect the infarcted myocardium and minimize cell death. These developed gels demonstrated neovascularization when tested on rats, and resulted in a reduction in infarct size. PEG nanoparticles were also administered intravenously in the form of PEGylated liposomes (142 nm in size). This vehicle transports therapeutic compounds and releases them in a regulated manner [137]. For the binding of nanoparticles to the infarct region, overexpression of AT1 receptor (angiotensin II type1) was used, although the results were not very promising.

(b) Poly(glycolic acid) & Poly (lactic acid)

To tune the desirable qualities, polylactic acid (PLA) and polyglycolic acid (PGA) were combined with poly (lactic–co-glycolic acid) (PLGA). PLA and PGA are both FDA-approved and biocompatible materials. PLA is a non-cytotoxic suture material, and its degradation component, lactic acid, is also non-cytotoxic. Although PLA breakdown makes the microenvironment slightly acidic [138], PGA is a non-cytotoxic thermoplastic. Neither PLA nor PGA, on the other hand, can equal the flexibility of heart tissue. As a result, they are mixed with other polyesters. Using a fibrous membrane made of electrospun PLGA, cardiomyocytes could be oriented to the direction of nanofibers [139]. The delivery of porous PLGA beads seeded with hAFSCs to a rat infarct model using a vehicle or "Cellularized Micro scaffolds" resulted in good cell retention [140]. To increase biological characteristics, PLGA could be mixed with natural polymer laminin or with carbon nanofibers (CNF) to induce conductivity [141,142].

(c) Poly(ε-caprolactone)

At body temperature, poly(ε-caprolactone) has a low glass transition temperature and behaves like rubber or elastic [143]. A 3D structure made up of up to five layers of electrospun PCL nanofibrous mats were evaluated for new-born cardiomyocyte culture, and the layers were able to form electrical connections and beat in time [144]. It is usually mixed with PLA or PGA copolymers. A biodegradable porous scaffold made of polyglycolide-co-caprolactone (PGCL) was employed in a rat infarcted myocardium model to distribute bone marrow-derived mononuclear cells (BMMNC). BNMC was transported from the scaffold to the implant, and neovascularization was seen [145].

(d) N-isopropylacrylamide (poly(N-isopropylacrylamide)) (PNIPAAm)

Thermosensitive polymer PNIPAAm at 32 °C possesses a reversible transition point. It is suited for biomedical applications because of its solution-to-gelation (sol-to-gel) transition point [146,147]. Hydrogels based on PNIPAAm can support cell co-cultures, which are believed to aid cardiac tissue regeneration [148]. Navaei et al. created a hydrogel containing 3D PNIPAAm-gelatin and co-cultured neonatal rat ventricular myocytes (NRVMs) and cardiac fibroblasts (CFs) [149]. As a result, they discovered that co-culturing improved cell contact and homogenous beating when compared to monoculture. Although PNIPAAm has a number of benefits for cardiac tissue engineering, its biodegradability is a concern [150]. Scientists have devised a number of solutions to this problem. With poly (NIPAAmco-2-hydroxyethyl methacrylate (HEMA), Fan et al. created an acrylate oligolactide (AOLA) degradable hydrogel [151]. They discovered that adding HEMA into poly (NIPAAm) caused the hydrogel to break down into a byproduct that is water soluble at body temperature. The most intriguing aspect of PNIPAAm is that it can be conjugated with carbon nanotubes (CNTs) to make the scaffold conduct [152,153]. In a rat MI model, Li et al. employed PNIPAAm in an injectable form in combination with single-walled carbon nanotubes (SWCNTs) to produce brown adipose-derived stem cells (BASCs) [154]. As a result, there was a noticeable increase in cell integration. Materials made of aniline, due to its electroactive and antioxidant capabilities, are a desirable material for cardiac tissue engineering [155,156]. A chitosan graft-aniline tetramer (CS-AT) and poly (ethylene glycol) (PEG-DA) hydrogel were developed by Dong et al. [157]. Electrical cues were transmitted due to the presence of polyaniline in the polymer backbone. Murine myoblasts and adipose-derived MSCs (ADMSCs) displayed vitality and proliferation in vitro after being encased in the hydrogel.

(e) Hybrid gelatin methacryloyl (GelMA)

The superior biocompatibility and controlled biodegradability of hybrid gelatin methacryloyl (GelMA) hydrogels makes them ideal for tissue engineering [158,159]. GelMA is made by combining gelatin and methacrylic anhydride. Hydrogel strength and stiffness can be modified by altering the amount of methacrylic anhydride [158]. GelMA/PEGDMA (PEG di-methacrylate) was used to encapsulate C2C12 myoblasts with stiffness ranging from 12 to 42 kPa by Li et al. [160]. This combination has the potential to cause muscular myofiber development. This crosslinking was created by exposure to UV radiation, and it was then employed to build blood vessels [160]. UV light, on the other hand, can be harmful to the heart [161]. To circumvent this constraint, Noshadi et al. created a cross-linkable GelMA hydrogel that can be exposed to visible light. Neonatal rat ventricular myocytes (NRVMs) were cultivated on top of this hydrogel for at least seven days, and the cells maintained their cardiac phenotype [161].

4.3. Delivery Strategies of Cells from Patch

VEGF-encapsulated MSCs are used to treat MI tissue, which helps in the improvement of cardiac function by angiogenesis based on the tropism of the MSCs to the MI area [162]. When encumbered with stem cells, hdECM is used as bio-ink in the 3D printing of pre-vascularized and functional multimaterial structures. The printed structure is comprised of the spatial patterning of dual stem cells, which are associated with enhanced cardiac function, decreased cardiac hypertrophy and fibrosis, elevated migration from the patch to the infarct area, neo-muscle, and capillary formation with the improvement in cardiac function. The hdECM potentiates in epicardial-mediated cardiac tissue. It is regenerated following the migration of WT1 positive progenitor cells using the EMT process [163].

Polymeric scaffold-mediated viral delivery showed that the release period could be extended from several days to one month if the molecular weights and concentration of the PEG varied. Sustained viral delivery from the core sheath fibers is facilitated by the solid fiber sheath that uses PEG as a porogen [71].

(a) Invasive method

The cardiac patch developed requires open-heart surgery to implant the patch, either by suturing [164,165] or by applying bioglue to the patch [166]. A PCL/gelatin patch incorporated with MSCs activated endogenous cardiac repair by enhancing the survival of MSCs and their HIF-1a, Tb4, VEGF, and SDF-1 expression and decreased CXCL14 expression in hypoxic and serum-deprived conditions. The engrafted MSCs were examined for survival and distribution, and it was observed that the engrafted MSCs migrated across the epicardium and into the myocardium. The epicardium was activated, and EDCs migrated into the deep tissue, which was differentiated into ECs and SMCs, with a few differentiated into CMs [167].

Early vascularization in the ischemic heart is critical for a better outcome. The delivered iVPCs were grown on polymer microbundle scaffolds made up of PLGA, showing beneficial effects on cardiac repair and recovery. When the iVPCs were integrated into a micro-bundle scaffold of PLGA that served as a carrier, the treatment efficacy improved [168].

(b) Minimum Invasive Method

The application of the cardiac patch to the heart requires open-chest surgery, which is traumatic. To overcome this issue, different groups came up with different ideas. Tang et al. designed a biospray using platelet-fibrin gel "paint" which polymerizes in situ with a minimally invasive procedure. They demonstrated that spray treatment improved cardiac repair and weakened cardiac dysfunction after MI [169]. Miles et al. engineered an elastic and microfabricated scaffold that could be delivered via injection. The scaffold material could recover its initial shape while maintaining the viability and function of the cells, as it has occurred after being delivered through an orifice as small as 1 mm. They demonstrated significant improvement in cardiac function in an MI rat model [170]. Brisa et al. designed an injectable RTG and functionalized it with CNTs. It transitions from a solution at room temperature to a 3D gel-based matrix shortly after reaching body temperature, which supports long-term CM survival, promotes CM alignment and proliferation, and improves CM function compared to control. Functionalized RTG with CNT renders both topographical and electrophysiological cues for native CMs. It promotes long-term survival with a more aligned cell organization, and suppresses fibroblast proliferation [171]. VEGF-encapsulated MSCs are prepared by self-assembling gelatin and alginate polyelectrolytes. It is a minimally invasive therapy to treat MI [162].

4.4. Advantages and Disadvantages of Patch

4.4.1. Advantages

The cardiac patch designed from natural or synthetic polymers provides a niche to stem cells [66,67]. The cardiac patch acts as a reservoir for growth factors and viruses for gene therapy [71]. The electroconductive cardiac patch helps in the proper alignment of cardiac cells while aiding in vitro monitoring [68]. The cellular component incorporated in the patch system is anticipated to ameliorate the pre-existing inflammatory milieu and subside the at the tissue damage site. Particularly, MSCs are reported to exist in the various tissue sources, having distinct proliferation and differentiation potential. Earlier studies were reported about the repopulation of worn-out populations at the tissue damage site; however, more recent studies have shown that the paracrine secretory profiles of MSCs are responsible for their beneficial effects, such as anti-inflammatory and immunomodulatory features. Therefore, the environmental sensing potential and its further responsive actions via the secretion of membrane-bound vesicles known as exosomes (which harbor growth factors, cytokines, and miRNAs), have contributed to establishing physiological homeostasis at the site of tissue damage [172].

4.4.2. Disadvantages

Under significant stimulatory conditions, the secretory profile of MSC-derived exosomes have not been completely streamlined for immediate harnessing from bench to bedside. Systemic targeted delivery of MSC-derived exosomes for the amelioration of hypoxia, apoptosis, and inflammatory milieu for accelerated myocardial regeneration may warrant its clinical utility. Therefore, we recommend in-depth research on the patch-based cell and cell-free approaches using humanized models and clinical trials, and highlighting its importance in improving the quality, safety and efficiency of future cardiac therapy [172]. Despite successes in constructing different types of tissue-engineered cardiac patches, clinical use has yet to be achieved due to a lack of in vivo verification of feasibility.

5. Microfluidics Based MI Research

Microfluidics deals with the handling of very small volumes of liquid, typically in a length scale of a few micrometers. Since its development in the early 1980s, it has been discovered to be an effective tool with a wide range of applications, from inkjet printers to LOC systems. Over the last two decades, microfluidics has been used in many biological fields such as genomics and proteomics, point-of-care diagnostic systems, biohazard detection, and more. Its recent developments have enabled tissue engineering research to be done in a more cost-effective and highly sensitive way. Research related to MI has been explored using microfluidic devices as it becomes feasible to create a 3D tissue structure and study it in a dynamic condition mimicking both healthy and pathophysiological states of the heart [173].

Microfluidics emerged from the conventional silicon and glass micro-machining process. With the advent of photo-lithography and other BioMEMS fabrication techniques, the whole process has become more user-friendly. The range of materials used for fabrication is vast, and includes different polymers, silicon, silicon-based materials, metals, etc. Materials are selected based on properties like rigidity or flexibility, optical transparency, biocompatibility, and reactivity to reagents. PDMS is a common material used for microfluidics because of properties like high oxygen permeability, ease of fabrication, and biocompatibility. As cell behavior also depends upon the topography of its environment, the surface chemistry of these materials plays a vital role in cell culture. The biocompatibility of these surfaces can also be increased via processes such as plasma deposition [174].

Cell culture work close to the physiologic environment is possible in a microfluidic chip, with better control over the process parameters. Such work has led to the use of human samples for research and reduced dependence on animal models for drug discovery and therapeutics. The small sample size requirement for microfluidic devices helps in handling costly samples and reagents. Miniaturized 3D tissue models can produce better studies on physiological systems and their behavior for toxicity assay studies. The flow conditions in a microfluidic device are very much essential in cellular studies, as the physiological environment is seldom static.

The microfluidic LOC device has also been designed to easily generate different physiological, mechanical, or electrical forces on the culture, which are complicated in macro bioreactors. The whole system can be designed and optimized through different available computational software (e.g., Comsol, Ansys, etc.). These devices can also be linked to external analytical devices in order to study real-time cell behavior (Figure 5) [175].

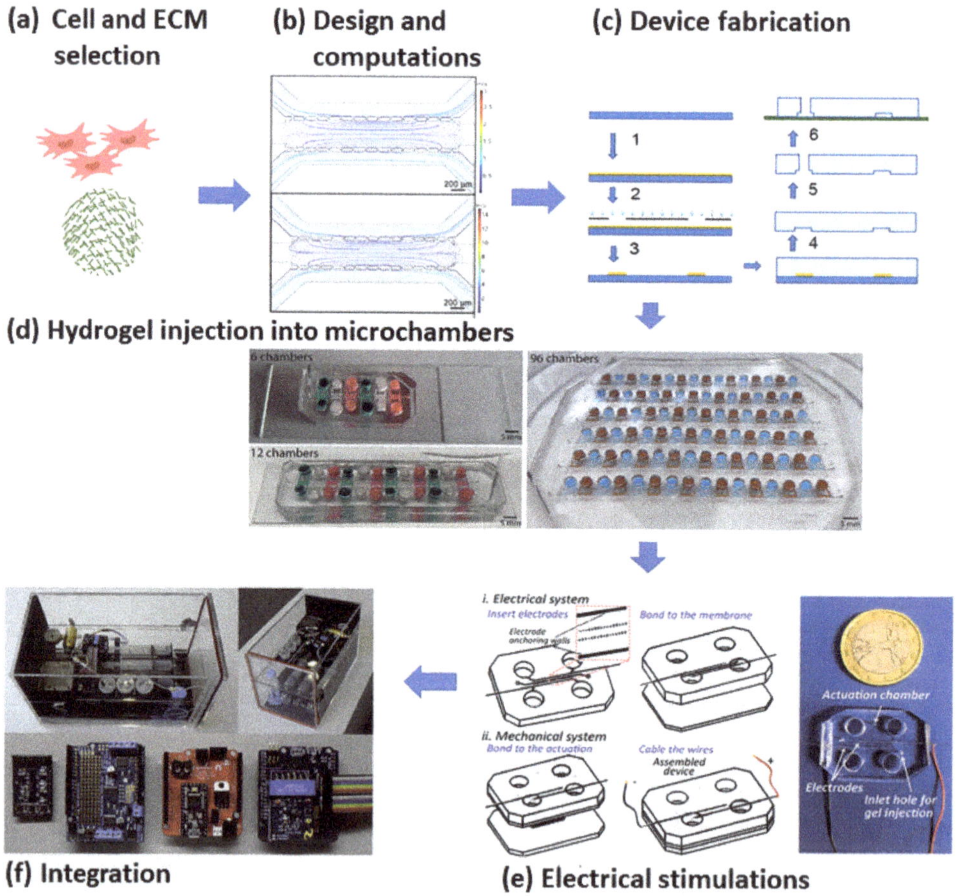

Figure 5. Steps in microfluidic cardiac model generation—(**a**) Selection of cells and extracellular matrix are performed based on the physiology to be studied; (**b**) this is followed by design of the chip through different computational software to achieve the desired flow contours; (**c**) based on the design, the device is fabricated by various microfabrication procedures (the most common being photolithography; the steps of which are 1. spin coating of clean silicon wafer, 2. UV exposure with a mask, 3. dissolution of unwanted resist with developer solution to generate the master pattern, 4. PDMS mold creation from the master pattern, 5. punching of required inlet and outlet holes in the PDMS mold, and 6. bonding of the PDMS mold with a glass plate or another PDMS slab to close the device); (**d**) introduction of cell-laden hydrogel into the device for 3D culture (its selection is based on the mechanical properties needed for the micro tissue under study; (**e**) completion of the electrical circuit required for stimulation of the cardiac cells; (**f**) integration of the device with external circuitry and pumping mechanism for seamless operation of the chip [176–179].

Various peculiar properties come into existence when the fluid flows in a microfluidic network, which are generally suppressed in macrosystems. The fluid flow is always in the laminar regime, as the Reynolds number stays very low. Essentially, this prevents the mixing of two adjacent layers of fluid. Diffusion becomes a major cause for mass transfer, which is used for easily generating a stable spatiotemporal chemical gradient on LOC devices; as the surface-to-volume ratio is very high, surface forces such as capillary forces become dominant. These properties come into play in the efficient designing and control of microfluidic devices.

5.1. Microfluidics for Cardiac Cell Biology

Among all CVDs, MI is the primary cause of mortality in the world. MI is caused due to occlusion in the coronary artery, leading to reduced or no blood flow to cardiac tissues. It causes reduced functionality of CMs, interstitial fibrosis, and ultimately cell death. Current therapeutic approaches to treat MI include the use of drugs to alleviate the symptoms or the use of stents to increase blood flow that can extend the heart's life. However, these approaches do not consider the repair of tissue or compensate for the loss of cardiac cells [46]. Recently developed cell-based and patch-based methods try to fill this gap, and concentrate more on developing techniques to repair and regenerate the infarcted area with a healthy network of blood vessels. To design a patch efficiently, there is a need to closely understand cardiac physiology. The heart is a very dynamic system, with different physical and chemical cues that control its functions. At this stage the role of the microfluidic system becomes important, as this novel tool can be efficiently utilized in MI diagnosis and the study of differentiation of stem cells to cardiac cells. Furthermore, it helps to generate close to physiological conditions in cell culture models, which can produce highly applicable results in tissue engineering research. Co-culture of different cardiac cells, which is necessary in order to understand cell-to-cell interaction, becomes easy with microfluidic devices. The use of human cells in these kinds of micro biomimetic environments makes them replicate the myocardium of an individual human patient. While such a myocardium-on-a-chip is currently intended as a tool for in vitro analyses, the platform may provide the foundation for future development of superior tissue engineering constructs for cardiac tissue replacement.

5.2. Application of Microfluidics in MI Research

Despite the scarce publications relating to the study of MI in microfluidics, some significant research is being done in microfluidic devices to explore the basic as well as the applied aspects. Recently, it was revealed by a single-cell microfluidic study that human iPSC-derived CMs secreted cytokines that ameliorate conditions of MI [180]. Such results were further confirmed by an in vivo study, where the role of cytokine secretion was established in a mouse model of acute MI. This opens up the opportunity to explore and study the role and effect of various cell types which are potential candidates for differentiating into mature CMs and/or have a therapeutic paracrine effect. Another use of microfluidics related to MI research was discovered when the role of inflammatory monocytes was studied in MI patients.

Phonotypic study of monocytes revealed that CD11c/CD18 was an inducible integrin whose expression correlated with a monocyte inflammatory state in MI patients. It was observed that under shear conditions simulating the low dynamic of blood, $CD14^{++}$ $CD16^{+}$ monocyte adhesion was double that of the healthy human subject. It is essential to recruit monocytes or inflammatory cells in ischemic tissue, which is done with the WBC components from circulation. Such cells are attracted when they sense chemicals secreted in response to injury, tissue necrosis, significant cell death, or another inflammatory cellular/tissue response. This dynamic could be easily simulated by a microfluid-based LOC platform [181]. Cellular analysis in a microfluidic platform is more similar to the dynamic conditions of the physiological state. Cell encapsulation in hydrogels enhances their survival in transplanted tissue and makes regeneration easier, and also aids in the production of prohealing cytokines and extracellular vesicles. Microfluidic cocooning has several advantages over vortex cocooning, such as limited consumption of biomaterials, precise cocoon size and cell number, extremely limited shear stresses on cells during cocooning, high individual unit throughput, etc. [182]. Another key benefit of the microfluidic system is the ability to pattern cell and extracellular matrix (ECM) at the cell length scale. Patient-derived cells can also be used, as they require fewer numbers of cells than conventional methods. Both 2D and 3D structures can be effectively generated in the microfluidic chip with the help of surface coating or the use of hydrogels [183]. Such devices can be made to mimic cardiovascular physiology (including ECM structure, cell composition, electrophysiology, heart mechanics, vascularization, etc.) for drug discovery and interventional strategies. Ischemia in the myocardium develops due to thrombus formation inside the coronary artery. Sono-thrombolysis is an effective method for the dissolution of a thrombus in blood vessels. Flow focusing techniques in microfluidics can form microbubbles that can be used for mechanically disrupting the thrombus. Such a technique has been used in a rat model to treat cerebral infarct with Nitrogen microbubbles, and shows a 50% reduction in infarct after thrombus dissolution [184].

Following MI, there is an irreversible loss of CMs that results in the alteration of electrical propagation in the heart. Stem cells are seeded onto the heart directly or through a patch to compensate for the loss of CMs, which helps repair the infarcted area. Mesenchymal stem cells are one of the important cell sources for cardiac tissue engineering. It has been found that their harvesting efficiency is higher when they are cultured in dissolvable gelatin-based microcarriers [185]. The surrounding microenvironment of cardiomyocytes needs to be controlled for appropriate growth and regeneration of the ischemic tissue. Different factors like cell seeding density, type of hydrogel, the percentage and type of non-myocytes, and electrical conditioning can be studied in detail on a microfluidic platform [186].

Different research groups have developed many animal models to test the efficacy of cell types and drugs in myocardial regeneration. However, the results from these animal models cannot be replicated in a patient heart. The development of a cardiac muscle model in microfluidics eliminates all these obstacles. These chips can incorporate on-chip analyses of the electrical conductivity of different cell types. The Laser patterning technique comes in handy as it can arrange cells in desired directions to increase the contact between the stem cells and CMs. Such chips can replicate the in vivo conditions very well. Along with this, it is also possible to test the importance of cell alignment and cell-to-cell contact in stem cell delivery on cardiac tissues [187–189]. The applications of microfluidic techniques for CVD research are shown in Table 3.

5.3. *Microfluidic 3D Culture Models*

Cells in our body reside in a complex 3D environment made out of cells and the ECM. Crucial processes such as nutrient transportation and cell-to-cell interaction depend upon the spatial arrangement of cells and the properties of the ECM. The spatiotemporal gradient of nutrient and oxygen concentration is important in tissue maintenance and regulation. Moreover, the heart is such a complex organ, having multiple cell types and specific cell arrangements, that it becomes challenging to recapitulate the in vivo physiology of the heart in 2D cultures. To obtain culture conditions close to physiological ones, the generation of a 3D environment becomes essential. In conventional culture conditions there are few models which can capture the physiology of infarcted heart tissue using cardiac organoids that mimic the oxygen gradient and adrenergic stimulus in MI [190]. Microfluidics give a vast range of options in which a micro 3D environment can be created. Hydrogel-based micro 3D tissues can be formed inside a microfluidic chip. The 3D model can have a perfusion-based distribution of media or can have a diffusion-based media flow that mimics the endothelial barrier in cardiac tissue [46,191]. Cardiac bodies are other options for microfluidic cell culture models, which do not require a hydrogel mixture [192]. Due to in vivo-like environment generation 3D culture promotes more interactions between cells, allowing them to remain in optimal conditions for proliferation. Drug cytotoxicity is also significantly lower for 3D than for 2D culture.

The efficiency of reprogramming human somatic cells to iPSCs can be improved up to 50-fold in a microfluidic environment [193]. It also consumes fewer reagents than a six-well plate culture. Patient-specific hiPSCs can be differentiated without cell expansion to produce functional CMs in as little as a month's time using such devices. Differentiation and proliferation from cells like MSCs, ESCs, or iPSCs to CMs can be achieved very economically and with better efficiency.

Cellular scaffold generation is an important part of developing a patch for MI or studying cell behavior in 3D constructs. The strength of the scaffold, along with its porosity and biodegradability, are essential elements in its design. Micro-bioprinting techniques help in generating the desired scaffold preparation using different polymeric materials and extracellular proteins. Precise loading of scaffolds becomes easy with this technique allowing the formation of highly viable and functional in vitro constructs with excellent resolution. This technique also allows the use of heterogeneous bioink [194]. Easy fabrication methods make this scaffold matrix highly tunable in forming a network of blood vessels [68]. Researchers have also combined microfluidic technology with 3D micropatterning to produce unique cell-laden structures, which can acquire contractile stresses generated by cardiac cells [195]. Biopolymers have also been used as scaffolds that have similar biochemical components as native tissues [196].

5.4. Implementation of Physical Forces

Cells in the myocardium are always subject to physical forces like hydraulic pressure, mechanical and fluid stresses, electrical field forces, etc. To develop a treatment method for the CVDs, the effect of these forces on cardiac cells and stem cells used for treatment needs to be studied extensively. A mechanical cue like cyclic stretch is a fundamental aspect of heart tissue that can vary from normal to pathological. Stretch at 1 Hz and 10% strain has been shown to yield significantly less cardiomyogenic differentiation of ESC culture [197]. This may be caused by disruption of the cardiomyogenic differentiation process prior to the expression of an MHC, a late-stage marker for cardiogenesis. It can be concluded that the strain magnitude, frequency, direction of strain, duration of stretch application, and the stage of differentiation at which the stretch is applied are all variables that are needed to be systematically investigated. Except for cyclic stretch, other mechanical factors such as hydrostatic pressure and fluid shear stress also play an important role in cardiac physiology. An important thing to note is that these variables change drastically in pathological conditions. Recently, a device to recreate the mechanical loading conditions of the left ventricle (LV) was developed [198]. This device could produce different pressure conditions that simulate normal and pathological conditions like hypertension, hypotension, tachycardia, bradycardia etc. After occlusion of the coronary artery, ischemia develops in the myocardium. While it is necessary to restore blood flow to the ischemic tissue, rapid perfusion can also cause tissue damage called ischemic reperfusion injury (IRI). It is difficult to study IRI in human models, although there are in vitro models which could recapitulate the oxygen concentration and pH condition in IRI [199]. Such models also show that cardioprotective therapeutics and ischemic preconditioning have the ability to reduce IRI [200]. Microfluidic models, with their greater flexibility and control, can model the in vivo environment and reperfusion issues in a better way. Results from such a model show that reperfusion indeed activates endothelial cells with a higher expression of I-CAM 1. Reperfusion can also lead to endothelial injury [201].

Embryonic cardiomyoblast line (H9c2 cells) studied on microfluidic devices showed organized F actin alignment similar to in vivo conditions with cyclic strain. This device could also be used for the co-culture of different cells of cardiac tissue that helps to study their physiology and cell-to-cell interactions. Uniaxial cyclic strain could also induce better electrical and mechanical coupling between cells, with a remarkable increase in junction complexes [202]. It is necessary to understand the behavior of all types of cells present in the heart in order to develop fully functional cardiac tissue. Cardiac fibroblasts (CF) are one type of cell which are primarily involved in cardiac tissue remodeling. CFs have shown increased elongation with stretching [203]. Their proliferation behavior can also be peculiar at different strain intensities. 2% stretching has shown higher proliferation, whereas 8% stretching gives rise to time-dependent proliferative behavior with a drop in the number of mitotic cells after 72 h. Studies like this can give insight into the pathological evolution of fibrotic cardiac disease. The extracellular environment can also direct cell behavior; this is true for both cardiac cells and the pluripotent stem cells from which they are differentiated. Microfluidics makes an excellent tool for such studies. It is very easy to pattern the cell attachment surfaces through micropatterning technologies [204]. Microfluidics offers numerous substrate choices, as mentioned earlier, which makes the study of substrate stiffness on cell physiology convenient. The cellular environment can also be changed by using different hydrogels and ECM proteins in micro cardiac tissues.

Through microfluidic techniques, the contractile efficiency of micro heart tissue can be measured by the deformation of a micropillar array on a chip [205]. Different alignments of cardiomyocytes can be tested in order to understand the contractile efficiency with cell orientation [206]. Electrical stimulation is another cue that stimulates CM function and the differentiation of pluripotent stem cells into CMs, and also helps in the maturation of iPSC-derived CMs. Micro-patterning of nontoxic electrodes like ITO in microfluidic devices offers an excellent platform for such studies. These electrodes have excellent electrical conductivity along with optical transparency. Therefore, they can help in designing a highly efficient system that can conveniently provide on-chip imaging [207]. In another work, a microelectrode array platform was fabricated for recording the beating rate and conductive velocity of cardiomyocytes. In this device, the cardiac monolayer beat in a synchronized fashion, and the conduction velocity was close to the physiological value [208]. Different methods can measure the beating of heart tissue. A cell culture substrate was generated in a novel microfluidic device based on the adhesive properties of polyethylene glycol diacrylate (PEGDA) and gelatin acryloyl (GelMA). Owing to this adhesive property, the cardiomyocytes adhered only to a specific region. During beating, the stretching of PEGDA substrate triggered a shift in its color, which can be visualized through optics. Reduced graphene oxide (rGO) was doped on this substrate to increase the contrast of light [209].

Table 3. Use of microfluidics platforms in the study of various aspects of CVD, and applications in cardiac tissue engineering.

Device Function	Cell Source	Techniques Used	Chemical or Physical Cues Studied	Scaffold Used	Fabrication Technique	Important Observations	References
Differentiation to CMs	ESCs	External motor for stretching the microfluidic device	Uniaxial cyclic mechanical stretch	2D culture	Lithography	Reduction in cardiogenesis	[197]
	hESCs	Micropatterned surface generation through direct micro contact printing	—	Micropatterned fibronectin hydrogel	—	Display of beating foci earlier than non-patterned substrates	[204]
Drug toxicity testing	Human iPSC-CBs	Micro niches to trap CBs in microchannel, Perfusion based system	Veparamil, Quinidine, Doxorubicin	No external scaffold	Standard photo lithography	3D environment showed different effect on beating frequency of cells	[193]
	Human CMs	Micropillar based system to prohibit direct contact between 3D cell matrix from media flow, diffusion-based transport	Isoproterenol	Puramatrix hydrogel	PMMA micromilling	Cell viability appeared better in 3D culture	[192]

Table 3. *Cont.*

Device Function	Cell Source	Techniques Used	Chemical or Physical Cues Studied	Scaffold Used	Fabrication Technique	Important Observations	References
Contractile stress measurement	Neonatal rat ventricular myocytes and human iPSC derived CMs	Electronic quantification of stress through Cantilever deflection measurement	Isoproterenol	3D printed matrix of PDMS with polyamide electrical network	Multimaterial 3D printing	Positive chronotropic response to drug similar to engineered NRVM microtissues and ESC-derived CM tissue	[196]
	Neonatal mouse CMs	Stress measurement by use of PIV technique to capture nanoparticle displacement coupled with finite element method.	Epinephrine	Sandwich of GelMA hydrogel and polyacrylamide hydrogels	3D patterning	Increased frequency and amplitude of contraction cycles	[195]
Generation of in vitro constructs for tissue engineering application	Neonatal rat CMs	Coaxial needle extrusion system	—	GelMA	3D printing	Generated complex heterogenous structures with single bioink extruder	[194]
Hydraulic pressure and mechanical strain condition generation	H9c2 cells	Use of peristaltic pump coupled with pneumatically actuated valve to generate pathological heart conditions	—	—	PDMS molding	Organized F actin alignment similar to in vivo	[198]
	Neonatal rat CM	Pneumatic deflection of thin PDMS membrane to generate stretch	Uniaxial cyclic stretch	Cell laden gel	Lithography	Superior cardiac differentiation with better electrical and mechanical coupling	[202]
Effect of electrical field on proliferation and differentiation	Neonatal rat CM	—	Square monophasic electrical pulses	2D cell culture on collagen coated matrix	Laser ablation of ITO coated glass slides to generate electrodes	Cell aligned in the direction perpendicular to the electric field	[207]
3D environment mimicking shear protection from endothelial barrier	hiPSC derived CMs	—	Verapamil, Isoproterenol, Metoprolol, E-4031	—	Two step photolithography process	IC50 and EC50 values were more consistent with the data on tissue-scale references	[191]

5.5. Drug Discovery and Disease Modelling

Diseases related to cardiovascular tissues are the most prevalent diseases worldwide. The major goals of tissue engineering are to develop, screen, and test drugs that can either cure these ailments or at least maintain healthy cardiovascular function. Conventionally, a 2D culture system is used to study the effects of drugs on cardiac cells, but this system fails more often than not owing to the reason that 3D in vivo physiological conditions change cell behavior. Because of this, many of drugs show negative results and are rejected in preclinical models. Microfluidic devices are the most useful tool in modern-day drug discovery research for generating 3D cell culture conditions and easily introducing in vivo dynamics. These microfluidic models can be used for drug discovery and for toxicity testing of different drugs in cardiac tissues. These microtissue models can be generated by different techniques. In recent work, such a model of cardiac bodies (CBs) was developed from induced pluripotent stem cell (iPSC) clusters [192]. Injection of hydrogels inside the infarct region has been used for myocardial regeneration, but hydrogels lack a porous structure that can support the regenerating tissue. Through microfluidic manufacturing, microporous particles can be generated with microgel building blocks, which will contain drugs to induce tissue generation and support tissue growth with their porous structure [210]. Numerous micro niches can be formed inside the device that can accommodate numerous tissue samples for drug testing studies, reducing the number of experiments compared to the macro system. Researchers have studied the effects of different drugs such as verapamil, quinidine, and doxorubicin, epinephrine, etc., on the function of cardiac cells.

Microdevices can be useful in producing a 3D network of cardiac cells with proper in vivo like alignment and cell-to-cell interactions. By mimicking the cardiac microenvironment closely, a microphysiological system can also be designed to establish in vivo conditions like cell alignment in the 3D microtissue, diffusion-mediated nutrient transportation, and micro-circulatory network. Mathur et al. could generate endothelial barrier-like physiology inside their device by patterning micropillar-like structures in it [191]. The introduction of all these features has shown significantly improved drug response results compared to conventional 2D cultures. Treatment with pharmacological agents like isoproterenol and clinical drugs like verapamil showed comparable beating behavior of these cardiac tissues to that seen in acute experiments performed on human ventricular heart tissue slices.

Production of drug encapsulated microspheres can be formed using microfluidics, with controllable size distribution. Proteins like ACE2 with a short half-life and poor stability can be encapsulated in alginate microspheres with microfluidic electrospray techniques [211]. Droplet formation techniques in microfluidics can be used to form microcapsules with a drug in the inner core and a polymeric shell to control its release. Drugs with a higher concentration can be loaded in the core for sustained delivery. In the acute myocardial infarction model of a rat, drugs like VEGF and PDGF have been delivered through this method. It was found that the repair of the infraction was significant with respect to the reduction in ventricular wall thickness and fibrosis, and the percentage of scar tissue was reduced to 11.2%, compared to 21.4% with saline [212].

Another aspect of drug discovery is cytotoxicity studies on other tissues in the body apart from for the primarily targeted tissue. After the metabolism of the drug molecules, byproducts are released which can affect the functioning of other tissues. Microfluidics allows the formation of a close circulation loop between different tissues on the same chip that can predict the cytotoxic effect of drugs and byproducts, along with their efficacy. Researchers have used cells from tissues such as the heart, liver, and kidney on the same chip to predict the fate of the drug in the human body and its metabolism and excretion mechanism. The cardiotoxicity of other drugs such as doxorubicin can be studied on a heart and liver organ on a chip [213]. The investigation of cytotoxicity and efficacy can be precisely predicted on such platforms to achieve personalized medicine and patient-specific care.

The bioelectrical activity of the heart is a measure of healthy cardiac function. During myocardial infarction due to hypoxia and nutrient deprivation, the electrical physiology varies. In a recently developed heart-on-a-chip system, both extracellular and intracellular potential have been measured in hypoxic conditions with the help of top-down lithography and nanofabricated platinum electrodes, respectively [214]. It was found that initial hypoxia caused tachycardia which then transformed to beat rate reduction and eventually arrhythmia. Intracellular bioelectronics showed narrowed action potential after the introduction of hypoxia.

5.6. Point of Care Devices and Disease Diagnosis

Disease diagnosis is an important stage in cardiac tissue engineering, in order to find out the type and nature of the ailment. Microfluidics allows a simple and fast method to detect diseases. The devices can be more cost-effective than conventional systems, and they have the added benefit of less sample volume usage. The use of nanoparticles as a signal label for the detection of biomarkers such as NT-proBNP and cTnI in microfluidic devices is highly suitable for immunoassays [215]. Although the commercial availability of such devices is very low (e.g., Alere Triage R system, Abbott i-STAT R system), the technology can be optimized considering the type (magnetic or nonmagnetic) and size of nanoparticles, as well as methods to reduce noise due to nonspecific adsorption, etc.

CVDs are the main contributor to global deaths per year. This cause surpasses the sum of deaths due to cancer and infectious diseases. Most of the deaths occur in low-income countries where disease diagnosis and treatment are costlier. Microfluidic technology can bridge this gap and help everybody get proper medical attention. Microfluidic chip-based immunoassays are more sensitive and can give better results. Currently available centrifugal microfluidic devices on the market, like Abaxis Piccolo Xpress, can use capillary and centrifugal action together and can separate plasma and perform fluid mixing and spectrometric measurement right on the microfluidic compact disk [216]. Biomarker detection through chemiluminescent, electrochemical, and optical methods is currently in the research stage and can enrich microfluidic technology for cardiac disease detection. Paper-based microfluidics has inherent microfluidic benefits along with the performances of lateral flow strips that have been proven to be a novel platform for such studies. Multiple marker detection for acute myocardial infarction is necessary for its proper diagnosis. Paper microfluidics-based chemiluminescence techniques can perform multiplexed analysis of antibodies like cardiac troponin I (cTnI), heart-type fatty acid-binding protein (H-FABP), and copeptin [217]. Such devices with a photomultiplier tube (PMT) have great potential for point-of-care analysis. Other techniques like radioimmunoassay, enzyme-linked immunosorbent assay, fluorescence, electrochemistry, and electrochemiluminescence (ECL) can also be used on a chip for diagnostic studies. Early myocardial ischemia stages could be detected with biomarkers such as glycogen phosphorylase isoenzyme BB (GPBB) with a paper microfluidic platform, which is difficult to perform by a conventional method. On the same platform, late markers like cardiac troponin T (cTnT) and CK-MB were also detected [218]. Detailed reviews of microfluidic diagnostic tools can be found elsewhere [219–221], as it is not the central part of this work.

6. Future Directions

In spite of technological advances in the study of cell biology, cellular behavior, and the pathophysiology of MI, there are many of unconnected questions that have to be solved in order to achieve a more efficient and effective stem cell-based therapy for MI. Understandably, no single therapeutic approach will suit all, but to do what can be done to narrow down stem cells, delivery approaches, and biomaterials to the minimal and most efficient formation is the major challenge requiring a detailed and thorough exploration. However, it is still not clear which stem cell is ideal for treating acute or chronic MI. Which mode of delivery is best or better than others, and which biomaterial is most suitable for stem cell delivery to the infarcted region? Many clinical trials have been completed so far, and many will be conducted in the upcoming years. However, a clear strategy has to be evolved in order to find the best combination of stem cells, delivery approach, and biomaterial components to meet the requirement in most cases, if not all.

Microfluidics is another major research area that has helped in the resolution of many complicated questions such as the selection of biomaterials and the mechanisms involved in transplanted stem cell-mediated MI heart regeneration, as well as the basic pathophysiology of MI. Most importantly, microfluidics can be used to more efficiently understand the angiogenesis process and the therapeutic interventions required to increase the vasculature density in the infarcted region. Eventually, it could help to improve the functionality of the infarcted heart.

7. Conclusions

Stem cell delivery to the infarcted myocardium has become indispensable for regenerating the heart so as to regain functionality similar to native cardiac function. To this end, various stem cells have been used for delivery to the infarcted heart. Processes used include direct injection into the myocardium and using a 3D biocompatible scaffold (commonly called a patch) to deliver a higher dose of cells with uniform distribution across the infarcted region. Several mechanisms have been documented where either the transplanted stem cells differentiate into functional CMs and integrate with the heart to improve cardiac function or act through a paracrine manner to induce the regeneration of CMs, increase neovascularization, reduce scar formation, increase ventricular function and decrease remodeling of the myocardium. Hence, the forthcoming mechanistic studies on exosomes and direct tagging of stem cells for monitoring the homing studies after transplantation may further result in translating bench studies to the bedside.

Author Contributions: Conceptualization, writing—original draft preparation, writing—review and editing, supervision, project administration, P.K.G.; writing—original draft preparation, V.S., S.K.D.; writing—review and editing, K.G., R.G. (Rekha Gahtori), N.N., M.B., R.T., S.C., S.M., D.K.B., S.K.S., R.G. (Rohit Gundamaraju), A.D., J.R., V.K.T.; supervision, project administration, K.D., P.N., K.K.K.; artwork and schemes, N.K.J.; funding acquisition, supervision, project administration, S.O. All authors have read and agreed to the published version of the manuscript.

Funding: The authors sincerely acknowledge the research fund support from ADEK award for Research Excellence (AARE18-194). The authors also acknowledge the research grant (21R075) from Zayed Bin Sultan Al Nahyan Center for Health Sciences (ZCHS), UAE University, UAE.

Institutional Review Board Statement: Not Applicable.

Informed Consent Statement: Not Applicable.

Data Availability Statement: Not Applicable.

Conflicts of Interest: The authors declare no competing interests with the work presented in the manuscript.

References

1. Benjamin, E.J.; Blaha, M.J.; Chiuve, S.E.; Cushman, M.; Das, S.R.; Deo, R.; De Ferranti, S.D.; Floyd, J.; Fornage, M.; Gillespie, C.; et al. Heart Disease and Stroke Statistics' 2017 Update. *Circulation* **2017**, *135*, e146–e603. [CrossRef]
2. Virani, S.S.; Alonso, A.; Aparicio, H.J.; Benjamin, E.J.; Bittencourt, M.S.; Callaway, C.W.; Carson, A.P.; Chamberlain, A.M.; Cheng, S.; Delling, F.N.; et al. Heart Disease and Stroke Statistics—2021. *Circulation* **2021**, *143*, 254–743. [CrossRef] [PubMed]
3. Prabhakaran, D.; Jeemon, P.; Roy, A. Cardiovascular Diseases in India: Current Epidemiology and Future Directions. *Circulation* **2016**, *133*, 1605–1620. [CrossRef] [PubMed]
4. Timmis, A.; Townsend, N.; Gale, C.P.; Torbica, A.; Lettino, M.; Petersen, S.E.; Mossialos, E.A.; Maggioni, A.P.; Kazakiewicz, D.; May, H.T.; et al. European society of cardiology: Cardiovascular disease statistics 2019. *Eur. Heart J.* **2020**, *41*, 12–85. [CrossRef] [PubMed]
5. Pfuntner, A.; Wier, L.M.; Stocks, C. *Most Frequent Conditions in U.S. Hospitals, 2011: Statistical Brief#162*; Agency for Healthcare Research and Quality: Rockville, MD, USA, 2006.
6. Frangogiannis, N.G. Pathophysiology of myocardial infarction. *Compr. Physiol.* **2015**, *5*, 1841–1875. [PubMed]
7. Tonsho, M.; Michel, S.; Ahmed, Z.; Alessandrini, A.; Madsen, J.C. Heart transplantation: Challenges facing the field. *Cold Spring Harb. Perspect. Med.* **2014**, *4*, a015636. [CrossRef]
8. Feric, N.T.; Radisic, M. Maturing human pluripotent stem cell-derived cardiomyocytes in human engineered cardiac tissues. *Adv. Drug Deliv. Rev.* **2016**, *96*, 110–134. [CrossRef]
9. Vunjak-Novakovic, G.; Tandon, N.; Godier, A.; Maidhof, R.; Marsano, A.; Martens, T.P.; Radisic, M. Challenges in cardiac tissue engineering. *Tissue Eng. Part B Rev.* **2020**, *16*, 169–187. [CrossRef]
10. Cahill, T.J.; Choudhury, R.P.; Riley, P.R. Heart regeneration and repair after myocardial infarction: Translational opportunities for novel therapeutics. *Nat. Rev. Drug Discov.* **2017**, *16*, 699–717. [CrossRef]
11. Bagno, L.; Hatzistergos, K.E.; Balkan, W.; Hare, J.M. Mesenchymal Stem Cell-Based Therapy for Cardiovascular Disease: Progress and Challenges. *Mol. Ther.* **2018**, *26*, 1610–1623. [CrossRef]
12. Ban, K.; Park, H.-J.; Kim, S.; Andukuri, A.; Cho, K.-W.; Hwang, J.W.; Cha, H.J.; Kim, S.Y.; Kim, W.-S.; Jun, H.-W.; et al. Cell Therapy with Embryonic Stem Cell-Derived Cardiomyocytes Encapsulated in Injectable Nanomatrix Gel Enhances Cell Engraftment and Promotes Cardiac Repair. *ACS Nano* **2014**, *8*, 10815–10825. [CrossRef]
13. Cambria, E.; Pasqualini, F.S.; Wolint, P.; Günter, J.; Steiger, J.; Bopp, A.; Hoerstrup, S.P.; Emmert, M.Y. Translational cardiac stem cell therapy: Advancing from first-generation to next-generation cell types. *Npj Regen. Med.* **2017**, *2*, 1–9. [CrossRef] [PubMed]
14. Chetty, S.S.; Praneetha, S.; Vadivel Murugan, A.; Varthana, K.; Verma, R.S. Human Umbilical Cord Wharton's Jelly-Derived Mesenchymal Stem Cells Labeled with Mn^{2+} and Gd^{3+} Co-Doped CuInS2-ZnS Nanocrystals for Multimodality Imaging in a Tumor Mice Model. *ACS Appl. Mater. Interfaces* **2020**, *12*, 3415–3429. [CrossRef]
15. Behfar, A.; Perez-Terzic, C.; Faustino, R.S.; Arrell, D.K.; Hodgson, D.M.; Yamada, S.; Puceat, M.; Niederländer, N.; Alekseev, A.E.; Zingman, L.V.; et al. Cardiopoietic programming of embryonic stem cells for tumor-free heart repair. *J. Exp. Med.* **2007**, *19*, 405–420. [CrossRef] [PubMed]
16. Kim, J.; Shapiro, L.; Flynn, A. The clinical application of mesenchymal stem cells and cardiac stem cells as a therapy for cardiovascular disease. *Pharmacology* **2015**, *151*, 8–15. [CrossRef]
17. Ji, L.L.; Long, X.F.; Tian, H.; Liu, Y.F. Effect of transplantation of bone marrow stem cells on myocardial infarction size in a rabbit model. *World J. Emerg. Med.* **2013**, *4*, 304–310. [CrossRef]

18. Roura, S.; Gálvez-Montón, C.; Mirabel, C.; Vives, J.; Bayes-Genis, A. Mesenchymal stem cells for cardiac repair: Are the actors ready for the clinical scenario? *Stem Cell Res. Ther.* **2017**, *8*, 1–11. [CrossRef]
19. Shen, X.; Pan, B.; Zhou, H.; Liu, L.; Lv, T.; Zhu, J.; Huang, X.; Tian, J. Differentiation of mesenchymal stem cells into cardiomyocytes is regulated by miRNA-1-2 via WNT signaling pathway. *J. Biomed. Sci.* **2017**, *24*, 1–8. [CrossRef]
20. Tang, J.N.; Cores, J.; Huang, K.; Cui, X.L.; Luo, L.; Zhang, J.Y.; Li, T.S.; Qian, L.; Cheng, K. Concise Review: Is Cardiac Cell Therapy Dead? Embarrassing Trial Outcomes and New Directions for the Future. *Stem Cells Transl. Med.* **2018**, *7*, 354–359. [CrossRef] [PubMed]
21. Tompkins, B.A.; Balkan, W.; Winkler, J.; Gyöngyösi, M.; Goliasch, G.; Fernández-Avilés, F.; Hare, J.M. Preclinical Studies of Stem Cell Therapy for Heart Disease. *Circ. Res.* **2018**, *122*, 1006–1020. [CrossRef] [PubMed]
22. Chen, Z.; Chen, L.; Zeng, C.; Wang, W.E. Functionally improved mesenchymal stem cells to better treat myocardial infarction. *Stem Cells Int.* **2018**, *2018*, 7045245. [CrossRef]
23. Verma, R.S. Recent Advances in Induced Pluripotent Stem Cell (iPSC) based Therapeutics. *J. Stem Cell Res. Ther.* **2017**, *16*, 115–130. [CrossRef]
24. Xu, J.-Y.; Cai, W.-Y.; Tian, M.; Liu, D.; Huang, R.-C. Stem cell transplantation dose in patients with acute myocardial infarction: A meta-analysis. *Chronic Dis. Transl. Med.* **2016**, *2*, 92–101. [CrossRef]
25. Hartman, M.E.; Librande, J.R.; Medvedev, I.O.; Ahmad, R.N.; Moussavi-Harami, F.; Gupta, P.P.; Chien, W.M.; Chin, M.T. An optimized and simplified system of mouse embryonic stem cell cardiac differentiation for the assessment of differentiation modifiers. *PLoS ONE* **2014**, *9*, e93033. [CrossRef]
26. Hodgson, D.M.; Behfar, A.; Zingman, L.V.; Kane, G.C.; Alekseev, A.E.; Pucéat, M.; Terzic, A. Cellular Plasticity in the Cardiovascular System Stable benefit of embryonic stem cell therapy in myocardial infarction. *Am. J. Physiol.* **2009**, *55905*, 471–479.
27. Laflamme, M.A.; Chen, K.Y.; Naumova, A.V.; Muskheli, V.; Fugate, J.A.; Dupras, S.K.; Reinecke, H.; Xu, C.; Hassanipour, M.; Police, S.; et al. Cardiomyocytes derived from human embryonic stem cells in pro-survival factors enhance function of infarcted rat hearts. *Nat. Biotechnol.* **2007**, *25*, 1015–1024. [CrossRef] [PubMed]
28. Dementyeva, E.V.; Medvedev, S.P.; Kovalenko, V.R.; Vyatkin, Y.V.; Kretov, E.I.; Slotvitsky, M.M.; Shtokalo, D.N.; Pokushalov, E.A.; Zakian, S.M. Applying Patient-Specific Induced Pluripotent Stem Cells to Create a Model of Hypertrophic Cardiomyopathy. *Biochem. Biokhimiia* **2019**, *84*, 291–298. [CrossRef] [PubMed]
29. Mauritz, C.; Martens, A.; Rojas, S.V.; Schnick, T.; Rathert, C.; Schecker, N.; Menke, S.; Glage, S.; Zweigerdt, R.; Haverich, A.; et al. Induced pluripotent stem cell (iPSC)-derived Flk-1 progenitor cells engraft, differentiate, and improve heart function in a mouse model of acute myocardial infarction. *Eur. Heart J.* **2011**, *32*, 2634–2641. [CrossRef] [PubMed]
30. Pushp, P.; Nogueira, D.E.S.; Rodrigues, C.A.V.; Ferreira, F.C.; Cabral, J.M.S.; Gupta, M.K. A Concise Review on Induced Pluripotent Stem Cell-Derived Cardiomyocytes for Personalized Regenerative Medicine. *Stem Cell Rev. Rep.* **2020**, 748–776.
31. Maza, I.; Caspi, I.; Zviran, A.; Chomsky, E.; Rais, Y.; Viukov, S.; Geula, S.; Buenrostro, J.D.; Weinberger, L.; Krupalnik, V.; et al. Transient acquisition of pluripotency during somatic cell transdifferentiation with iPSC reprogramming factors. *Nat. Biotechnol.* **2015**, *33*, 769–774. [CrossRef]
32. Orlic, D.; Kajstura, J.; Chimenti, S.; Jakoniuk, I.; Anderson, S.M.; Li, B.; Pickel, J.; McKay, R.; Nadal-Ginard, B.; Bodine, D.M.; et al. Bone marrow cells regenerate infarcted myocardium. *Nature* **2001**, *410*, 701–705. [CrossRef] [PubMed]
33. Miao, C.; Lei, M.; Hu, W.; Han, S.; Wang, Q. A brief review: The therapeutic potential of bone marrow mesenchymal stem cells in myocardial infarction. *Stem Cell Res. Ther.* **2017**, *8*, 4–9. [CrossRef] [PubMed]
34. Esmaeili, R.; Darbandi-Azar, A.; Sadeghpour, A.; Majidzadeh, A.-K.; Eini, L.; Jafarbeik-Iravani, N.; Hoseinpour, P.; Vajhi, A.; Bakhshaiesh, T.O.; Masoudkabir, F.; et al. Mesenchymal stem cells Pretreatment with stromal-derived factor-1 alpha augments cardiac function and angiogenesis in infarcted myocardium. *Am. J. Med. Sci.* **2021**, *361*, 765–775. [CrossRef] [PubMed]
35. Feng, J.; Wu, Y.; Chen, W.; Li, J.; Wang, X.; Chen, Y.; Yu, Y.; Shen, Z.; Zhang, Y. Sustained release of bioactive IGF-1 from a silk fibroin microsphere-based injectable alginate hydrogel for the treatment of myocardial infarction. *J. Mater. Chem. B* **2020**, *8*, 308–315. [CrossRef] [PubMed]
36. Lakshmanan, R.; Kumaraswamy, P.; Krishnan, U.M.; Sethuraman, S. Engineering a growth factor embedded nanofiber matrix niche to promote vascularization for functional cardiac regeneration. *Biomaterials* **2016**, *97*, 176–195. [CrossRef]
37. Duelen, R.; Sampaolesi, M. Stem Cell Technology in Cardiac Regeneration: A Pluripotent Stem Cell Promise. *EBioMedicine* **2017**, *16*, 30–40. [CrossRef] [PubMed]
38. Tracy, E.P.; Gettler, B.C.; Zakhari, J.S.; Schwartz, R.J.; Williams, S.K.; Birla, R.K. 3D Bioprinting the Cardiac Purkinje System Using Human Adipogenic Mesenchymal Stem Cell Derived Purkinje Cells. *Cardiovasc. Eng. Technol.* **2020**, *11*, 587–604. [CrossRef]
39. Dawn, B.; Stein, A.B.; Urbanek, K.; Rota, M.; Whang, B.; Rastaldo, R.; Torella, D.; Tang, X.L.; Rezazadeh, A.; Kajstura, J.; et al. Cardiac stem cells delivered intravascularly traverse the vessel barrier, regenerate infarcted myocardium, and improve cardiac function. *Proc. Natl. Acad. Sci. USA* **2005**, *102*, 3766–3771. [CrossRef]
40. Li, N.; Huang, R.; Zhang, X.; Xin, Y.; Li, J.; Huang, Y.; Cui, W.; Stoltz, J.F.; Zhou, Y.; Kong, Q. Stem cells cardiac patch from decellularized umbilical artery improved heart function after myocardium infarction. *Bio-Med Mater. Eng.* **2017**, *28*, 87–94. [CrossRef] [PubMed]
41. Mauretti, A.; Spaans, S.; Bax, N.A.M.; Sahlgren, C.; Bouten, C.V.C. Cardiac Progenitor Cells and the Interplay with Their Microenvironment. *Stem Cells Int.* **2017**, *2017*, 7471582. [CrossRef]

42. Su, T.; Huang, K.; Daniele, M.A.; Hensley, M.T.; Young, A.T.; Tang, J.; Allen, T.A.; Vandergriff, A.C.; Erb, P.D.; Ligler, F.S.; et al. Cardiac Stem Cell Patch Integrated with Microengineered Blood Vessels Promotes Cardiomyocyte Proliferation and Neovascularization after Acute Myocardial Infarction. *ACS Appl. Mater. Interfaces* **2018**, *10*, 33088–33096. [CrossRef]
43. Toma, C.; Pittenger, M.F.; Cahill, K.S.; Byrne, B.J.; Kessler, P.D. Human mesenchymal stem cells differentiate to a cardiomyocyte phenotype in the adult murine heart. *Circulation* **2002**, *105*, 93–98. [CrossRef] [PubMed]
44. Makkar, R.R.; Smith, R.R.; Cheng, K.; Malliaras, K.; Thomson, L.E.J.; Berman, D.; Czer, L.S.C.; Marbán, L.; Mendizabal, A.; Johnston, P.V.; et al. Intracoronary cardiosphere-derived cells for heart regeneration after myocardial infarction (CADUCEUS): A prospective, randomised phase 1 trial. *Lancet* **2012**, *379*, 895–904. [CrossRef]
45. Malliaras, K.; Makkar, R.R.; Smith, R.R.; Cheng, K.; Wu, E.; Bonow, R.O.; Marbán, L.; Mendizabal, A.; Cingolani, E.; Johnston, P.V.; et al. Intracoronary cardiosphere-derived cells after myocardial infarction: Evidence of therapeutic regeneration in the final 1-year results of the CADUCEUS trial (CArdiosphere-derived aUtologous stem CElls to reverse ventricular dysfunction). *J. Am. Coll. Cardiol.* **2014**, *63*, 110–122. [CrossRef] [PubMed]
46. Reddy, K. Recent advances in the diagnosis and treatment of acute myocardial infarction. *World J. Cardiol.* **2015**, *7*, 243. [CrossRef]
47. Chow, A.; Stuckey, D.J.; Kidher, E.; Rocco, M.; Jabbour, R.J.; Mansfield, C.A.; Darzi, A.; Harding, S.E.; Stevens, M.M.; Athanasiou, T. Human Induced Pluripotent Stem Cell-Derived Cardiomyocyte Encapsulating Bioactive Hydrogels Improve Rat Heart Function Post Myocardial Infarction. *Stem Cell Rep.* **2017**, *9*, 1415–1422. [CrossRef] [PubMed]
48. Song, G.; Li, X.; Shen, Y. Qian, L.; Kong, X.; Chen, M.; Cao, K.; Zhang, F. Transplantation of iPSc Restores Cardiac Function by Promoting Angiogenesis and Ameliorating Cardiac Remodeling in a Post-infarcted Swine Model. *Cell Biochem. Biophys.* **2015**, *71*, 1463–1473. [CrossRef] [PubMed]
49. Ichihara, Y.; Kaneko, M.; Yamahara, K.; Koulouroudias, M.; Sato, N.; Uppal, R.; Yamazaki, K.; Saito, S.; Suzuki, K. Self-assembling peptide hydrogel enables instant epicardial coating of the heart with mesenchymal stromal cells for the treatment of heart failure. *Biomaterials* **2018**, *154*, 12–23. [CrossRef] [PubMed]
50. Mathieu, E.; Lamirault, G.; Toquet, C.; Lhommet, P.; Rederstorff, E.; Sourice, S.; Biteau, K.; Hulin, P.; Forest, V.; Weiss, P.; et al. Intramyocardial delivery of mesenchymal stem cell-seeded hydrogel preserves cardiac function and attenuates ventricular remodeling after myocardial infarction. *PLoS ONE* **2012**, *7*, e51991. [CrossRef]
51. Mayfield, A.E.; Tilokee, E.L.; Latham, N.; McNeill, B.; Lam, B.-K.; Ruel, M.; Suuronen, E.J.; Courtman, D.W.; Stewart, D.J.; Davis, D.R. The effect of encapsulation of cardiac stem cells within matrix-enriched hydrogel capsules on cell survival, post-ischemic cell retention and cardiac function. *Biomaterials* **2014**, *35*, 133–142. [CrossRef]
52. Durrani, S.; Konoplyannikov, M.; Ashraf, M.; Haider, K.H. Skeletal myoblasts for cardiac repair. *Regen. Med.* **2010**, *5*, 919–932. [CrossRef] [PubMed]
53. Suzuki, K.; Murtuza, B.; Beauchamp, J.R.; Smolenski, R.T.; Varela-Carver, A.; Fukushima, S.; Coppen, S.R.; Partridge, T.A.; Yacoub, M.H. Dynamics and mediators of acute graft attrition after myoblast transplantation to the heart. *FASEB J.* **2004**, *18*, 1153–1155. [CrossRef] [PubMed]
54. Menasché, P.; Alfieri, O.; Janssens, S.; McKenna, W.; Reichenspurner, H.; Trinquart, L.; Vilquin, J.-T.; Marolleau, J.-P.; Seymour, B.; Larghero, J.; et al. The Myoblast Autologous Grafting in Ischemic Cardiomyopathy (MAGIC) trial: First randomized placebo-controlled study of myoblast transplantation. *Circulation* **2008**, *117*, 1189–1200. [CrossRef] [PubMed]
55. Fernandes, S.; Rijen, H.V.M.; van Forest, V.; Evain, S.; Leblond, A.-L.; Mérot, J.; Charpentier, F.; Bakker, J.M.T.; de Lemarchand, P. Cardiac cell therapy: Overexpression of connexin43 in skeletal myoblasts and prevention of ventricular arrhythmias. *J. Cell. Mol. Med.* **2009**, *13*, 3703–3712. [CrossRef]
56. Hirata, Y.; Sata, M.; Motomura, N.; Takanashi, M.; Suematsu, Y.; Ono, M.; Takamoto, S. Human umbilical cord blood cells improve cardiac function after myocardial infarction. *Biochem. Biophys. Res. Commun.* **2005**, *327*, 609–614. [CrossRef] [PubMed]
57. Govarthanan, K.; Gupta, P.K.; Ramasamy, D.; Kumar, P.; Mahadevan, S.; Verma, R.S. DNA methylation microarray uncovers a permissive methylome for cardiomyocyte differentiation in human mesenchymal stem cells. *Genomics* **2020**, *112*, 1384–1395. [CrossRef]
58. Govarthanan, K.; Vidyasekar, P.; Gupta, P.K.; Lenka, N.; Verma, R.S. Glycogen synthase kinase 3β inhibitor-CHIR99021 augments the differentiation potential of mesenchymal stem cells. *Cytotherapy* **2020**, *22*, 91–105. [CrossRef] [PubMed]
59. Yeh, Y.-C.; Lee, W.-Y.; Yu, C.-L.; Hwang, S.-M.; Chung, M.-F.; Hsu, L.-W.; Chang, Y.; Lin, W.-W.; Tsai, M.-S.; Wei, H.-J.; et al. Cardiac repair with injectable cell sheet fragments of human amniotic fluid stem cells in an immune-suppressed rat model. *Biomaterials* **2010**, *31*, 6444–6453. [CrossRef]
60. Tang, J.; Cui, X.; Caranasos, T.G.; Hensley, M.T.; Vandergriff, A.C.; Hartanto, Y.; Shen, D.; Zhang, H.; Zhang, J.; Cheng, K. Heart Repair Using Nanogel-Encapsulated Human Cardiac Stem Cells in Mice and Pigs with Myocardial Infarction. *ACS Nano* **2017**, *11*, 9738–9749. [CrossRef]
61. Zhao, S.; Xu, Z.; Wang, H.; Reese, B.E.; Gushchina, L.V.; Jiang, M.; Agarwal, P.; Xu, j.; Zhang, M.; Shen, R.; et al. Bioengineering of injectable encapsulated aggregates of pluripotent stem cells for therapy of myocardial infarction. *Nat. Commun.* **2016**, *7*, 1–12. [CrossRef]
62. Bauer, M.; Kang, L.; Qiu, Y.; Wu, J.; Peng, M.; Chen, H.H.; Camci-Unal, G.; Bayomy, A.F.; Sosnovik, D.E.; Khademhosseini, A.; et al. Adult Cardiac Progenitor Cell Aggregates Exhibit Survival Benefit Both In Vitro and In Vivo. *PLoS ONE* **2012**, *7*, e50491.
63. Monsanto, M.M.; Wang, B.J.; Ehrenberg, Z.R.; Echeagaray, O.; White, K.S.; Alvarez, R.; Fisher, K.; Sengphanith, S.; Muliono, A.; Gude, N.A.; et al. Enhancing myocardial repair with CardioClusters. *Nat. Commun.* **2020**, *11*, 1–20. [CrossRef] [PubMed]

64. Zhang, B.; Montgomery, M.; Chamberlain, M.D.; Ogawa, S.; Korolj, A.; Pahnke, A.; Wells, L.A.; Massé, S.; Kim, J.; Reis, L.; et al. Biodegradable Scaffold with Built-in Vasculature for Organ-on-a-Chip Engineering and Direct Surgical Anastomosis. *Nat. Mater.* **2016**, *15*, 669–678. [CrossRef]
65. Dhandayuthapani, B.; Yoshida, Y.; Maekawa, T.; Kumar, D.S. Polymeric Scaffolds in Tissue Engineering Application: A Review. *Int. J. Polym. Sci.* **2011**, *2011*, 290602. [CrossRef]
66. Shahabipour, F.; Banach, M.; Johnston, T.P.; Pirro, M.; Sahebkar, A. Novel approaches toward the generation of bioscaffolds as a potential therapy in cardiovascular tissue engineering. *Int. J. Cardiol.* **2017**, *228*, 319–326. [CrossRef]
67. Fleischer, S.; Feiner, R.; Dvir, T. Cutting-edge platforms in cardiac tissue engineering. *Curr. Opin. Biotechnol.* **2017**, *47*, 23–29. [CrossRef]
68. Kim, S.; Kim, W.; Lim, S.; Jeon, J.S. Vasculature-on-a-chip for in vitro disease models. *Bioengineering* **2017**, *4*, 8. [CrossRef] [PubMed]
69. Navaei, A.; Moore, N.; Sullivan, R.T.; Truong, D.; Migrino, R.Q.; Nikkhah, M. Electrically conductive hydrogel-based micro-topographies for the development of organized cardiac tissues. *RSC Adv.* **2017**, *7*, 3302–3312. [CrossRef]
70. Monteiro, L.M.; Vasques-Nóvoa, F.; Ferreira, L.; Pinto-Do-ó, P.; Nascimento, D.S. Restoring heart function and electrical integrity: Closing the circuit. *Npj Regen. Med.* **2017**, *2*, 1–13. [CrossRef] [PubMed]
71. Gu, X.; Matsumura, Y.; Tang, Y.; Roy, S.; Hoff, R.; Wang, B.; Wagner, W.R. Sustained viral gene delivery from a micro-fibrous, elastomeric cardiac patch to the ischemic rat heart. *Biomaterials* **2017**, *133*, 132–143. [CrossRef]
72. Dong, R.; Ma, P.X.; Guo, B. Conductive biomaterials for muscle tissue engineering. *Biomaterials* **2020**, *229*, 119584. [CrossRef]
73. Sherrell, P.C.; Cieślar-Pobuda, A.; Ejneby, M.S.; Sammalisto, L.; Gelmi, A.; de Muinck, E.; Brask, J.; Łos, M.J.; Rafat, M. Rational Design of a Conductive Collagen Heart Patch. *Macromol. Biosci.* **2017**, *17*, 1600446. [CrossRef]
74. Fleischer, S.; Feiner, R.; Dvir, T. Cardiac tissue engineering: From matrix design to the engineering of bionic hearts. *Regen. Med.* **2017**, *12*, 275–284. [CrossRef]
75. Hussey, G.S.; Dziki, J.L.; Badylak, S.F. Extracellular matrix-based materials for regenerative medicine. *Nat. Rev. Mater.* **2018**, *3*, 159–173. [CrossRef]
76. KC, P.; Hong, Y.; Zhang, G. Cardiac tissue-derived extracellular matrix scaffolds for myocardial repair: Advantages and challenges. *Regen. Biomater.* **2019**, *6*, 185–199. [CrossRef]
77. Kuraitis, D.; Giordano, C.; Ruel, M.; Musarò, A.; Suuronen, E.J. Exploiting extracellular matrix-stem cell interactions: A review of natural materials for therapeutic muscle regeneration. *Biomaterials* **2012**, *33*, 428–443. [CrossRef]
78. Li, H.; Bao, M.; Nie, Y. Extracellular matrix–based biomaterials for cardiac regeneration and repair. *Heart Fail. Rev.* **2020**, *3*, 1–18. [CrossRef] [PubMed]
79. Lockhart, M.; Wirrig, E.; Phelps, A.; Wessels, A. Extracellular matrix and heart development. *Birth Defects Res. Part A Clin. Mol. Teratol.* **2011**, *91*, 535–550. [CrossRef] [PubMed]
80. Mackiewicz, Z.; Konttinen, Y.T.; Kaivosoja, E.; Stegajev, V.; Wagner, H.D.; Levón, J.; Tiainen, V.M. Extracellular matrix and tissue regeneration. In *Regenerative Medicine 2016—from Protocol to Patient: 1. Biology of Tissue Regeneration*; Springer: Cham, Switzerland, 2016; pp. 1–55.
81. Pattar, S.S.; Fatehi Hassanabad, A.; Fedak, P.W.M. Acellular Extracellular Matrix Bioscaffolds for Cardiac Repair and Regeneration. *Front. Cell Dev. Biol.* **2019**, *7*, 63. [CrossRef] [PubMed]
82. Batalov, I. Engineering 2D Cardiac Tissues Using Biomimetic Protein Micropatterns Based on the Extracellular Matrix in the Embryonic Heart. ProQuest Diss. Ph.D. Thesis, Carnegie Mellon University, Gloucester, VA, USA, 2017; p. 183.
83. Ciuffreda, M.C.; Malpasso, G.; Chokoza, C.; Bezuidenhout, D.; Goetsch, K.P.; Mura, M.; Pisano, F.; Davies, N.H.; Gnecchi, M. Synthetic extracellular matrix mimic hydrogel improves efficacy of mesenchymal stromal cell therapy for ischemic cardiomyopathy. *Acta Biomater.* **2018**, *70*, 71–83. [CrossRef] [PubMed]
84. Davis, M.E.; Hsieh, P.C.H.; Grodzinsky, A.J.; Lee, R.T. Custom design of the cardiac microenvironment with biomaterials. *Circ. Res.* **2005**, *97*, 8–15. [CrossRef]
85. Seif-Naraghi, S.B.; Singelyn, J.M.; Salvatore, M.A.; Osborn, K.G.; Wang, J.J.; Sampat, U.; Kwan, O.L.; Strachan, G.M.; Wong, J.; Schup-Magoffin, P.J.; et al. Safety and efficacy of an injectable extracellular matrix hydrogel for treating myocardial infarction. *Sci. Transl. Med.* **2013**, *5*, 173ra25. [CrossRef] [PubMed]
86. Radhakrishnan, J.; Krishnan, U.M.; Sethuraman, S. Hydrogel based injectable scaffolds for cardiac tissue regeneration. *Biotechnol. Adv.* **2014**, *32*, 449–461. [CrossRef]
87. Silvestri, A.; Boffito, M.; Sartori, S.; Ciardelli, G. Biomimetic materials and scaffolds for myocardial tissue regeneration. *Macromol. Biosci.* **2013**, *13*, 984–1019. [CrossRef] [PubMed]
88. Tamimi, M.; Rajabi, S.; Pezeshki-Modaress, M. Cardiac ECM/chitosan/alginate ternary scaffolds for cardiac tissue engineering application. *Int. J. Biol. Macromol.* **2020**, *164*, 389–402. [CrossRef] [PubMed]
89. Ruvinov, E.; Dvir, T.; Leor, J.; Cohen, S. Myocardial repair: From salvage to tissue reconstruction. *Expert Rev. Cardiovasc. Ther.* **2018**, *6*, 669–686. [CrossRef]
90. Asadi, N.; Del Bakhshayesh, A.R.; Davaran, S.; Akbarzadeh, A. Common biocompatible polymeric materials for tissue engineering and regenerative medicine. *Mater. Chem. Phys.* **2020**, *242*, 122528. [CrossRef]
91. Asti, A.; Gioglio, L. Natural and Synthetic Biodegradable Polymers: Different Scaffolds for Cell Expansion and Tissue Formation. *Int. J. Artif. Organs* **2014**, *37*, 187–205. [CrossRef]

92. Hinderer, S.; Brauchle, E.; Schenke-Layland, K. Generation and Assessment of Functional Biomaterial Scaffolds for Applications in Cardiovascular Tissue Engineering and Regenerative Medicine. *Adv. Healthc. Mater.* **2015**, *4*, 2326–2341. [CrossRef] [PubMed]
93. Reis, L.A.; Chiu, L.L.Y.; Feric, N.; Fu, L.; Radisic, M. Biomaterials in myocardial tissue engineering. *J. Tissue Eng. Regen. Med.* **2016**, *10*, 11–28. [CrossRef] [PubMed]
94. Joch, C. The safety of fibrin sealants. *Cardiovasc. Surg.* **2003**, *11*, 23–28. [CrossRef]
95. Kaiser, N.J.; Kant, R.J.; Minor, A.J.; Coulombe, K.L.K. Optimizing Blended Collagen-Fibrin Hydrogels for Cardiac Tissue Engineering with Human iPSC-derived Cardiomyocytes. *ACS Biomater. Sci. Eng.* **2019**, *5*, 887–899. [CrossRef] [PubMed]
96. Kobayashi, K.; Ichihara, Y.; Sato, N.; Umeda, N.; Fields, L.; Fukumitsu, M.; Tago, Y.; Ito, T.; Kainuma, S.; Podaru, M.; et al. On-site fabrication of Bi-layered adhesive mesenchymal stromal cell-dressings for the treatment of heart failure. *Biomaterials* **2019**, *209*, 41–53. [CrossRef] [PubMed]
97. Radosevich, M.; Goubran, H.A.; Burnouf, T. Fibrin sealant: Scientific rationale, production methods, properties, and current clinical use. *Vox Sang.* **1997**, *72*, 133–143. [CrossRef] [PubMed]
98. Domenech, M.; Polo-Corrales, L.; Ramirez-Vick, J.E.; Freytes, D.O. Tissue Engineering Strategies for Myocardial Regeneration: Acellular Versus Cellular Scaffolds? *Tissue Eng. Part B Rev.* **2016**, *22*, 438–458. [CrossRef]
99. Li, Y.; Meng, H.; Liu, Y.; Lee, B.P. Fibrin Gel as an Injectable Biodegradable Scaffold and Cell Carrier for Tissue Engineering. *Sci. World J.* **2015**, *2015*, 685690. [CrossRef]
100. Pieters, M.; Wolberg, A.S. Fibrinogen and fibrin: An illustrated review. *Res. Pract. Thromb. Haemost.* **2019**, *3*, 161–172. [CrossRef]
101. Croisier, F.; Jérôme, C. Chitosan-based biomaterials for tissue engineering. *Eur. Polym. J.* **2013**, *49*, 780–792. [CrossRef]
102. Hoeeg, C.; Dolatshahi-Pirouz, A.; Follin, B. Injectable hydrogels for improving cardiac cell therapy—in vivo evidence and translational challenges. *Gels* **2021**, *7*, 7. [CrossRef] [PubMed]
103. Del Valle, L.J.; Díaz, A.; Puiggalí, J. Hydrogels for Biomedical Applications: Cellulose, Chitosan, and Protein/Peptide Derivatives. *Gels* **2017**, *3*, 27. [CrossRef]
104. Tormos, C.J.; Madihally, S.V. Chitosan for cardiac tissue engineering and regeneration. In *Chitosan Based Biomaterials*; Woodhead Publishing: Shaston, UK, 2017; Volume 2.
105. Chen, J.; Zhan, Y.; Wang, Y.; Han, D.; Tao, B.; Luo, Z.; Ma, S.; Wang, Q.; Li, X.; Fan, L.; et al. Chitosan/silk fibroin modified nanofibrous patches with mesenchymal stem cells prevent heart remodeling post-myocardial infarction in rats. *Acta Biomater.* **2018**, *80*, 154–168. [CrossRef]
106. Hasan, A.; Khattab, A.; Islam, M.A.; Hweij, K.A.; Zeitouny, J.; Waters, R.; Sayegh, M.; Hossain, M.M.; Paul, A. Injectable Hydrogels for Cardiac Tissue Repair after Myocardial Infarction. *Adv. Sci.* **2015**, *2*, 1–18. [CrossRef] [PubMed]
107. Yu, L.; Ding, J. Injectable hydrogels as unique biomedical materials. *Chem. Soc. Rev.* **2008**, *37*, 1473–1481. [CrossRef]
108. Anker, S.D.; Coats, A.J.S.; Cristian, G.; Dragomir, D.; Pusineri, E.; Piredda, M.; Bettari, L.; Dowling, R.; Volterrani, M.; Kirwan, B.-A.; et al. A prospective comparison of alginate-hydrogel with standard medical therapy to determine impact on functional capacity and clinical outcomes in patients with advanced heart failure (AUGMENT-HF trial). *Eur. Heart J.* **2015**, *36*, 2297–2309. [CrossRef] [PubMed]
109. Burdick, J.A.; Prestwich, G.D. Hyaluronic Acid Hydrogels for Biomedical Applications. *Adv. Mater.* **2011**, *23*, H41–H56. [CrossRef] [PubMed]
110. Dahlmann, J.; Krause, A.; Möller, L.; Kensah, G.; Möwes, M.; Diekmann, A.; Martin, U.; Kirschning, A.; Gruh, I.; Dräger, G. Biomaterials Fully de fi ned in situ cross-linkable alginate and hyaluronic acid hydrogels for myocardial tissue engineering. *Biomaterials* **2013**, *34*, 940–951. [CrossRef] [PubMed]
111. Johnson, T.D.; Christman, K.L. Injectable hydrogel therapies and their delivery strategies for treating myocardial infarction. *Expert Opin. Drug Deliv.* **2013**, *10*, 59–72. [CrossRef]
112. Shen, Y.; Li, Q.; Tu, J.; Zhu, J. Synthesis and characterization of low molecular weight hyaluronic acid-based cationic micelles for efficient siRNA delivery. *Carbohydr. Polym.* **2009**, *77*, 95–104. [CrossRef]
113. Chi, N.H.; Yang, M.C.; Chung, T.W.; Chou, N.K.; Wang, S.S. Cardiac repair using chitosan-hyaluronan/silk fibroin patches in a rat heart model with myocardial infarction. *Carbohydr. Polym.* **2013**, *92*, 591–597. [CrossRef] [PubMed]
114. Kodavaty, J.; Deshpande, A.P. Regimes of microstructural evolution as observed from rheology and surface morphology of crosslinked poly(vinyl alcohol) and hyaluronic acid blends during gelation. *J. Appl. Polym. Sci.* **2014**, *131*. [CrossRef]
115. Chattopadhyay, S.; Raines, R.T. Review collagen-based biomaterials for wound healing. *Biopolymers* **2014**, *101*, 821–833. [CrossRef]
116. Wu, W.; Peng, S.; Song, Z.; Lin, S. Collagen biomaterial for the treatment of myocardial infarction: An update on cardiac tissue engineering and myocardial regeneration. *Drug Deliv. Transl. Res.* **2019**, *9*, 920–934. [CrossRef]
117. Camci-Unal, G.; Annabi, N.; Dokmeci, M.R.; Liao, R.; Khademhosseini, A. Hydrogels for cardiac tissue engineering. *NPG Asia Mater.* **2014**, *6*, 99. [CrossRef]
118. Sun, H.; Zhou, J.; Huang, Z.; Qu, L.; Lin, N.; Liang, C.; Dai, R.; Tang, L.; Tian, F. Carbon nanotube-incorporated collagen hydrogels improve cell alignment and the performance of cardiac constructs. *Int. J. Nanomed.* **2017**, *12*, 3109–3120. [CrossRef]
119. Sarig, U.; Sarig, H.; de-Berardinis, E.; Chaw, S.-Y.; Nguyen, E.B.V.; Ramanujam, V.S.; Thang, V.D.; Al-Haddawi, M.; Liao, S.; Seliktar, D.; et al. Natural myocardial ECM patch drives cardiac progenitor based restoration even after scarring. *Acta Biomater.* **2016**, *44*, 209–220. [CrossRef]
120. Gilpin, A.; Yang, Y. Decellularization Strategies for Regenerative Medicine: From Processing Techniques to Applications. *BioMed Res. Int.* **2017**, *2017*, 9831534. [CrossRef] [PubMed]

121. Ma, J.; Huang, C. Composition and Mechanism of Three-Dimensional Hydrogel System in Regulating Stem Cell Fate. *Tissue Engine. Part B Reviews.* **2020**, *26*, 498–518. [CrossRef] [PubMed]
122. Naahidi, S.; Jafari, M.; Logan, M.; Wang, Y.; Yuan, Y.; Bae, H.; Dixon, B.; Chen, P. Biocompatibility of hydrogel-based scaffolds for tissue engineering applications. *Biotechnol. Adv.* **2017**, *35*, 530–544. [CrossRef]
123. Jeffords, M.E.; Wu, J.; Shah, M.; Hong, Y.; Zhang, G. Tailoring material properties of cardiac matrix hydrogels to induce endothelial differentiation of human mesenchymal stem cells. *ACS Appl. Mater. Interfaces* **2015**, *7*, 11053–11061. [CrossRef]
124. Efraim, Y.; Sarig, H.; Cohen Anavy, N.; Sarig, U.; de Berardinis, E.; Chaw, S.-Y.; Krishnamoorthi, M.; Kalifa, J.; Bogireddi, H.; Duc, T.V.; et al. Biohybrid cardiac ECM-based hydrogels improve long term cardiac function post myocardial infarction. *Acta Biomater.* **2017**, *50*, 220–233. [CrossRef] [PubMed]
125. Bai, R.; Tian, L.; Li, Y.; Zhang, J.; Wei, Y.; Jin, Z.; Liu, Z.; Liu, H. Combining ECM Hydrogels of Cardiac Bioactivity with Stem Cells of High Cardiomyogenic Potential for Myocardial Repair. *Stem Cells Int.* **2019**, *2019*, 6708435. [CrossRef]
126. Mewhort, H.E.M.; Turnbull, J.D.; Meijndert, H.C.; Ngu, J.M.C.; Fedak, P.W.M. Epicardial infarct repair with basic fibroblast growth factor-enhanced CorMatrix-ECM biomaterial attenuates postischemic cardiac remodeling. *J. Thorac. Cardiovasc. Surg.* **2014**, *147*, 1650–1659. [CrossRef]
127. Sreejit, P.; Verma, R.S. Natural ECM as Biomaterial for Scaffold Based Cardiac Regeneration Using Adult Bone Marrow Derived Stem Cells. *Stem Cell Rev. Rep.* **2013**, *9*, 158–171. [CrossRef] [PubMed]
128. Heras-Bautista, C.O.; Mikhael, N.; Lam, J.; Shinde, V.; Katsen-Globa, A.; Dieluweit, S.; Molcanyi, M.; Uvarov, V.; Jütten, P.; Sahito, R.G.A.; et al. Cardiomyocytes facing fibrotic conditions re-express extracellular matrix transcripts. *Acta Biomater.* **2019**, *89*, 180–192. [CrossRef]
129. Carballo-Pedrares, N.; Fuentes-Boquete, I.; Díaz-Prado, S.; Rey-Rico, A. Hydrogel-based localized nonviral gene delivery in regenerative medicine approaches—An overview. *Pharmaceutics* **2020**, *12*, 752. [CrossRef]
130. Pascual-Gil, S.; Garbayo, E.; Díaz-Herráez, P.; Prosper, F.; Blanco-Prieto, M.J. Heart regeneration after myocardial infarction using synthetic biomaterials. *J. Control. Release* **2015**, *203*, 23–38. [CrossRef] [PubMed]
131. Ren, S.; Jiang, X.; Li, Z.; Wen, Y.; Chen, D.; Li, X.; Zhang, X.; Zhuo, R.; Chu, H. Physical properties of poly(N-isopropylacrylamide) hydrogel promote its effects on cardiac protection after myocardial infarction. *J. Int. Med. Res.* **2012**, *40*, 2167–2182. [CrossRef]
132. Nelson, D.M.; Ma, Z.; Fujimoto, K.L.; Hashizume, R.; Wagner, W.R. Intra-myocardial biomaterial injection therapy in the treatment of heart failure: Materials, outcomes and challenges. *Acta Biomater.* **2011**, *7*, 1–15. [CrossRef]
133. Perea-Gil, I.; Prat-Vidal, C.; Bayes-Genis, A. In vivo experience with natural scaffolds for myocardial infarction: The times they are a-changin'. *Stem Cell Res. Ther.* **2015**, *6*, 248. [CrossRef]
134. Mukherjee, R.; Zavadzkas, J.A.; Saunders, S.M.; McLean, J.E.; Jeffords, L.B.; Beck, C.; Stroud, R.E.; Leone, A.M.; Koval, C.N.; Rivers, W.T.; et al. Targeted myocardial microinjections of a biocomposite material reduces infarct expansion in pigs. *Ann. Thorac. Surg.* **2008**, *86*, 1268–1276. [CrossRef] [PubMed]
135. Pascual-Gil, S.; Simón-Yarza, T.; Garbayo, E.; Prósper, F.; Blanco-Prieto, M.J. Cytokine-loaded PLGA and PEG-PLGA microparticles showed similar heart regeneration in a rat myocardial infarction model. *Int. J. Pharm.* **2017**, *523*, 531–533. [CrossRef]
136. Wang, T.; Jiang, X.-J.; Tang, Q.-Z.; Li, X.-Y.; Lin, T.; Wu, D.-Q.; Zhang, X.-Z.; Okello, E. Bone marrow stem cells implantation with alpha-cyclodextrin/MPEG-PCL-MPEG hydrogel improves cardiac function after myocardial infarction. *Acta Biomater.* **2009**, *5*, 2939–2944. [CrossRef] [PubMed]
137. Diao, J.; Wang, H.; Chang, N.; Zhou, X.-H.; Zhu, X.; Wang, J.; Xiong, J.-W. PEG–PLA nanoparticles facilitate siRNA knockdown in adult zebrafish heart. *Dev. Biol.* **2015**, *406*, 196–202. [CrossRef] [PubMed]
138. Alcon, A.; Cagavi Bozkulak, E.; Qyang, Y. Regenerating functional heart tissue for myocardial repair. *Cell. Mol. Life Sci. CMLS* **2012**, *69*, 2635–2656. [CrossRef] [PubMed]
139. Venugopal, J.R.; Prabhakaran, M.P.; Mukherjee, S.; Ravichandran, R.; Dan, K.; Ramakrishna, S. Biomaterial strategies for alleviation of myocardial infarction. *J. R. Soc. Interface* **2012**, *9*, 1–19. [CrossRef]
140. Huang, C.-C.; Wei, H.-J.; Yeh, Y.-C.; Wang, J.-J.; Lin, W.-W.; Lee, T.-Y.; Hwang, S.-M.; Choi, S.-W.; Xia, Y.; Chang, Y.; et al. Injectable PLGA porous beads cellularized by hAFSCs for cellular cardiomyoplasty. *Biomaterials* **2012**, *33*, 4069–4077. [CrossRef]
141. McDevitt, T.C.; Angello, J.C.; Whitney, M.L.; Reinecke, H.; Hauschka, S.D.; Murry, C.E.; Stayton, P.S. In vitro generation of differentiated cardiac myofibers on micropatterned laminin surfaces. *J. Biomed. Mater. Res.* **2002**, *60*, 472–479. [CrossRef]
142. Stout, D.A.; Basu, B.; Webster, T.J. Poly(lactic–co-glycolic acid): Carbon nanofiber composites for myocardial tissue engineering applications. *Acta Biomater.* **2011**, *7*, 3101–3112. [CrossRef] [PubMed]
143. Park, H.; Radisic, M.; Lim, J.O.; Chang, B.H.; Vunjak-Novakovic, G. A novel composite scaffold for cardiac tissue engineering. *Vitr. Cell. Dev. Biol. Anim.* **2005**, *41*, 188–196. [CrossRef]
144. Ishii, O.; Shin, M.; Sueda, T.; Vacanti, J.P. In vitro tissue engineering of a cardiac graft using a degradable scaffold with an extracellular matrix–like topography. *J. Thorac. Cardiovasc. Surg.* **2005**, *130*, 1358–1363. [CrossRef]
145. Piao, H.; Kwon, J.-S.; Piao, S.; Sohn, J.-H.; Lee, Y.-S.; Bae, J.-W.; Hwang, K.-K.; Kim, D.-W.; Jeon, O.; Kim, B.-S.; et al. Effects of cardiac patches engineered with bone marrow-derived mononuclear cells and PGCL scaffolds in a rat myocardial infarction model. *Biomaterials* **2007**, *28*, 641–649. [CrossRef]
146. Kim, E.H.; Joo, M.K.; Bahk, K.H.; Park, M.H.; Chi, B.; Lee, Y.M.; Jeong, B. Reverse Thermal Gelation of PAF-PLX-PAF Block Copolymer Aqueous Solution. *Biomacromolecules* **2009**, *10*, 2476–2481. [CrossRef] [PubMed]

147. Wu, J.; Zeng, F.; Huang, X.-P.; Chung, J.; Konecny, C.-Y.F.; Weisel, R.D.; Li, R.-K. Infarct stabilization and cardiac repair with a VEGF-conjugated, injectable hydrogel. *Biomaterials* **2011**, *32*, 579–586. [CrossRef]
148. Pinto, A.R.; Ilinykh, A.; Ivey, M.J.; Kuwabara, J.T.; D'Antoni, M.L.; Debuque, R.; Chandran, A.; Wang, L.; Arora, K.; Rosenthal, N.A.; et al. Revisiting Cardiac Cellular Composition. *Circ. Res.* **2016**, *118*, 400–409. [CrossRef]
149. Navaei, A.; Truong, D.; Heffernan, J.; Cutts, J.; Brafman, D.; Sirianni, R.W.; Vernon, B.; Nikkhah, M. PNIPAAm-based biohybrid injectable hydrogel for cardiac tissue engineering. *Acta Biomater.* **2016**, *32*, 10–23. [CrossRef]
150. Klouda, L. Thermoresponsive hydrogels in biomedical applications: A seven-year update. *Eur. J. Pharm. Biopharm.* **2015**, *97*, 338–349. [CrossRef] [PubMed]
151. Fan, Z.; Fu, M.; Xu, Z.; Zhang, B.; Li, Z.; Li, H.; Zhou, X.; Liu, X.; Duan, Y.; Lin, P.-H.; et al. Sustained Release of a Peptide-Based Matrix Metalloproteinase-2 Inhibitor to Attenuate Adverse Cardiac Remodeling and Improve Cardiac Function Following Myocardial Infarction. *Biomacromolecules* **2017**, *18*, 2820–2829. [CrossRef] [PubMed]
152. Deep, N.; Mishra, P. Fabrication and characterization of thermally conductive PMMA/MWCNT nanocomposites. *Mater. Today Proc.* **2018**, *5*, 28328–28336. [CrossRef]
153. Koerner, H.; Price, G.; Pearce, N.A.; Alexander, M.; Vaia, R.A. Remotely actuated polymer nanocomposites—Stress-recovery of carbon-nanotube-filled thermoplastic elastomers. *Nat. Mater.* **2014**, *3*, 115–120. [CrossRef] [PubMed]
154. Li, X.; Zhou, J.; Liu, Z.; Chen, J.; Lü, S.; Sun, H.; Li, J.; Lin, Q.; Yang, B.; Duan, C.; et al. A PNIPAAm-based thermosensitive hydrogel containing SWCNTs for stem cell transplantation in myocardial repair. *Biomaterials* **2014**, *35*, 5679–5688. [CrossRef]
155. Balint, R.; Cassidy, N.J.; Cartmell, S.H. Conductive polymers: Towards a smart biomaterial for tissue engineering. *Acta Biomater.* **2014**, *10*, 2341–2353. [CrossRef] [PubMed]
156. Ravichandran, R.; Sundarrajan, S.; Venugopal, J.R.; Mukherjee, S.; Ramakrishna, S. Applications of conducting polymers and their issues in biomedical engineering. *J. R. Soc. Interface* **2010**, *7*, S559–S579. [CrossRef]
157. Dong, R.; Zhao, X.; Guo, B.; Ma, P.X. Self-Healing Conductive Injectable Hydrogels with Antibacterial Activity as Cell Delivery Carrier for Cardiac Cell Therapy. *ACS Appl. Mater. Interfaces* **2016**, *8*, 17138–17150. [CrossRef] [PubMed]
158. Klotz, B.J.; Gawlitta, D.; Rosenberg, A.J.W.P.; Malda, J.; Melchels, F.P.W. Gelatin-Methacryloyl Hydrogels: Towards Biofabrication-Based Tissue Repair. *Trends Biotechnol.* **2016**, *34*, 394–407. [CrossRef] [PubMed]
159. Yue, K.; Trujillo-de Santiago, G.; Alvarez, M.M.; Tamayol, A.; Annabi, N.; Khademhosseini, A. Synthesis, properties, and biomedical applications of gelatin methacryloyl (GelMA) hydrogels. *Biomaterials* **2015**, *73*, 254–271. [CrossRef]
160. Li, Y.; Poon, C.T.; Li, M.; Lu, T.J.; Pingguan-Murphy, B.; Xu, F. Hydrogel Fibers: Chinese-Noodle-Inspired Muscle Myofiber Fabrication (Adv. Funct. Mater. 37/2015). *Adv. Funct. Mater.* **2015**, *25*, 6020. [CrossRef]
161. Noshadi, I.; Hong, S.; Sullivan, K.E.; Shirzaei Sani, E.; Portillo-Lara, R.; Tamayol, A.; Shin, S.R.; Gao, A.E.; Stoppel, W.L.; Black III, L.D.; et al. In vitro and in vivo analysis of visible light crosslinkable gelatin methacryloyl (GelMA) hydrogels. *Biomater. Sci.* **2017**, *5*, 2093–2105. [CrossRef]
162. Liu, G.; Li, L.; Huo, D.; Li, Y.; Wu, Y.; Zeng, L.; Cheng, P.; Xing, M.; Zeng, W.; Zhu, C. A VEGF delivery system targeting MI improves angiogenesis and cardiac function based on the tropism of MSCs and layer-by-layer self-assembly. *Biomaterials* **2017**, *127*, 117–131. [CrossRef]
163. Jang, J.; Park, H.J.; Kim, S.W.; Kim, H.; Park, J.Y.; Na, S.J.; Kim, H.J.; Park, M.N.; Choi, S.H.; Park, S.H.; et al. 3D printed complex tissue construct using stem cell-laden decellularized extracellular matrix bioinks for cardiac repair. *Biomaterials* **2017**, *112*, 264–274. [CrossRef]
164. Shadrin, I.Y.; Allen, B.W.; Qian, Y.; Jackman, C.P.; Carlson, A.L.; Juhas, M.E.; Bursac, N. Cardiopatch platform enables maturation and scale-up of human pluripotent stem cell-derived engineered heart tissues. *Nat. Commun.* **2017**, *8*, 1825. [CrossRef]
165. Hou, D.; Youssef, E.A.-S.; Brinton, T.J.; Zhang, P.; Rogers, P.; Price, E.T.; Yeung, A.C.; Johnstone, B.H.; Yock, P.G.; March, K.L. Radiolabeled cell distribution after intramyocardial, intracoronary, and interstitial retrograde coronary venous delivery: Implications for current clinical trials. *Circulation* **2005**, *112* (Suppl. S9), I-150–I-156. [CrossRef] [PubMed]
166. Singh, A.; Singh, A.; Sen, D. Mesenchymal stem cells in cardiac regeneration: A detailed progress report of the last 6 years (2010–2015). *Stem Cell Res. Ther.* **2016**, *7*, 82. [CrossRef]
167. Wang, Q.L.; Wang, H.J.; Li, Z.H.; Wang, Y.L.; Wu, X.P.; Tan, Y.Z. Mesenchymal stem cell-loaded cardiac patch promotes epicardial activation and repair of the infarcted myocardium. *J. Cell. Mol. Med.* **2017**, *21*, 1751–1766. [CrossRef] [PubMed]
168. Jamaiyar, A.; Wan, W.; Ohanyan, V.; Enrick, M.; Janota, D.; Cumpston, D.; Song, H.; Stevanov, K.; Kolz, C.L.; Hakobyan, T.; et al. Alignment of inducible vascular progenitor cells on a micro-bundle scaffold improves cardiac repair following myocardial infarction. *Basic Res. Cardiol.* **2017**, *112*, 41. [CrossRef] [PubMed]
169. Tang, J.; Vandergriff, A.; Wang, Z.; Hensley, M.T.; Cores, J.; Allen, T.A.; Dinh, P.U.; Zhang, J.; Caranasos, T.G.; Cheng, K. A Regenerative Cardiac Patch Formed by Spray Painting of Biomaterials onto the Heart. *Tissue Eng. Part C Methods* **2017**, *23*, 146–155. [CrossRef]
170. Montgomery, M.; Ahadian, S.; Davenport Huyer, L.; Lo Rito, M.; Civitarese, R.A.; Vanderlaan, R.D.; Wu, J.; Reis, L.A.; Momen, A.; Akbari, S.; et al. Flexible shape-memory scaffold for minimally invasive delivery of functional tissues. *Nat. Mater.* **2017**, *16*, 1038–1046. [CrossRef]
171. Peña, B.; Bosi, S.; Aguado, B.A.; Borin, D.; Farnsworth, N.L.; Dobrinskikh, E.; Rowland, T.J.; Martinelli, V.; Jeong, M.; Taylor, M.R.G.; et al. Injectable Carbon Nanotube-Functionalized Reverse Thermal Gel Promotes Cardiomyocytes Survival and Maturation. *ACS Appl. Mater. Interfaces* **2017**, *9*, 31645–31656. [CrossRef]

172. Merimi, M.; Lewalle, P.; Meuleman, N.; Agha, D.M.; El-kehdy, H.; Bouhtit, F.; Ayoub, S.; Burny, A.; Fahmi, H.; Lagneaux, L.; et al. Mesenchymal stem/stromal cell therapeutic features: The Bridge between the Bench and the Clinic. *J. Clin. Med.* **2021**, *10*, 905. [CrossRef]
173. Inamdar, N.K.; Borenstein, J.T. Microfluidic cell culture models for tissue engineering. *Curr. Opin. Biotechnol.* **2011**, *22*, 681–689. [CrossRef]
174. Ni, M.; Tong, W.H.; Choudhury, D.; Rahim, N.A.A.; Iliescu, C.; Yu, H. Cell culture on MEMS platforms: A review. *Int. J. Mol. Sci.* **2009**, *10*, 5411–5441. [CrossRef]
175. Whitesides, G.M. The origins and the future of microfluidics. *Nature* **2006**, *442*, 368–373. [CrossRef]
176. Kobuszewska, A.; Tomecka, E.; Zukowski, K.; Jastrzebska, E.; Chudy, M.; Dybko, A.; Renaud, P.; Brzozka, Z. Heart-on-a-Chip: An Investigation of the Influence of Static and Perfusion Conditions on Cardiac (H9C2) Cell Proliferation, Morphology, and Alignment. *SLAS Technol. Transl. Life Sci. Innov.* **2017**, *22*, 536–546. [CrossRef] [PubMed]
177. Qiao, Y.; Dong, Q.; Li, B.; Obaid, S.; Miccile, C.; Yin, R.T.; Talapatra, T.; Lin, Z.; Li, S.; Li, Z.; et al. Multiparametric slice culture platform for the investigation of human cardiac tissue physiology. *Prog. Biophys. Mol. Biol.* **2019**, *144*, 139–150. [CrossRef]
178. Visone, R.; Talò, G.; Occhetta, P.; Cruz-moreira, D.; Lopa, S.; Pappalardo, O.A.; Redaelli, A.; Moretti, M.; Rasponi, M. A microscale biomimetic platform for generation and electro-mechanical stimulation of 3D cardiac microtissues. *APL Bioeng.* **2018**, *2*, 046102. [CrossRef] [PubMed]
179. Visone, R.; Ugolini, G.S.; Vinarsky, V.; Penati, M.; Redaelli, A.; Forte, G.; Rasponi, M. A Simple Vacuum-Based Microfluidic Technique to Establish High-Throughput Organs-On-Chip and 3D Cell Cultures at the Microscale. *Adv. Mater. Technol.* **2019**, *4*, 1–8. [CrossRef]
180. Ong, S.-G.; Huber, B.C.; Hee Lee, W.; Kodo, K.; Ebert, A.D.; Ma, Y.; Nguyen, P.K.; Diecke, S.; Chen, W.-Y.; Wu, J.C. Microfluidic Single-Cell Analysis of Transplanted Human Induced Pluripotent Stem Cell–Derived Cardiomyocytes After Acute Myocardial. *Infarction Circ.* **2015**, *132*, 762–771. [CrossRef] [PubMed]
181. Foster, G.A.; Gower, R.M.; Stanhope, K.L.; Havel, P.J.; Simon, S.I.; Armstrong, E.J. On-chip phenotypic analysis of inflammatory monocytes in atherogenesis and myocardial infarction. *Proc. Natl. Acad. Sci. USA* **2013**, *110*, 13944–13949. [CrossRef] [PubMed]
182. Kanda, P.; Benavente-Babace, A.; Parent, S.; Connor, M.; Soucy, N.; Steeves, A.; Lu, A.; Cober, N.D.; Courtman, D.; Variola, F.; et al. Deterministic paracrine repair of injured myocardium using microfluidic-based cocooning of heart explant-derived cells. *Biomaterials* **2020**, *247*, 120010. [CrossRef] [PubMed]
183. Doherty, E.L.; Aw, W.Y.; Hickey, A.J.; Polacheck, W.J. Microfluidic and Organ-on-a-Chip Approaches to Investigate Cellular and Microenvironmental Contributions to Cardiovascular Function and Pathology. *Front. Bioeng. Biotechnol.* **2021**, *9*, 1–14. [CrossRef] [PubMed]
184. Dixon, A.J.; Li, J.; Rickel, J.M.R.; Klibanov, A.L.; Zuo, Z.; Hossack, J.A. Efficacy of Sonothrombolysis Using Microbubbles Produced by a Catheter-Based Microfluidic Device in a Rat Model of Ischemic Stroke. *Ann. Biomed. Eng.* **2019**, *47*, 1012–1022. [CrossRef]
185. Ng, E.X.; Wang, M.; Neo, S.H.; Tee, C.A.; Chen5, C.-H.; Van Vliet, K.J. Dissolvable gelatin-based microcarriers generated through droplet microfluidics for expansion and culture of mesenchymal stromal cells. *Biotechnol. J.* **2020**, *16*, 2000048. [CrossRef]
186. Zhao, Y.; Rafatian, N.; Wang, E.Y.; Feric, N.T.; Lai, B.F.L.; Knee-Walden, E.J.; Backx, P.H.; Radisic, M. Engineering microenvironment for human cardiac tissue assembly in heart-on-a-chip platform. *Matrix Biol.* **2019**, *85*, 189–204. [CrossRef] [PubMed]
187. Huebsch, N.; Loskill, P.; Deveshwar, N.; Spencer, C.I.; Judge, L.M.; Mandegar, M.A.; Fox, B.C.; Mohamed, T.M.A.A.; Ma, Z.; Mathur, A.; et al. Miniaturized iPS-Cell-Derived Cardiac Muscles for Physiologically Relevant Drug Response Analyses. *Sci. Rep.* **2016**, *6*, 24726. [CrossRef]
188. Ma, Z.; Liu, Q.; Liu, H.; Yang, H.; Yun, J.X.; Eisenberg, C.; Borg, T.K.; Xu, M.; Gao, B.Z. Laser-patterned stem-cell bridges in a cardiac muscle model for on-chip electrical conductivity analyses. *Lab Chip* **2012**, *12*, 566–573. [CrossRef] [PubMed]
189. Ma, Z.; Wang, J.; Loskill, P.; Huebsch, N.; Koo, S.; Svedlund, F.L.; Marks, N.C.; Hua, E.W.; Grigoropoulos, C.P.; Conklin, B.R.; et al. Self-organizing human cardiac microchambers mediated by geometric confinement. *Nat. Commun.* **2015**, *6*, 7413. [CrossRef]
190. Richards, D.J.; Li, Y.; Kerr, C.M.; Yao, J.; Beeson, G.C.; Coyle, R.C.; Chen, X.; Jia, J.; Damon, B.; Wilson, R.; et al. Human Cardiac Organoids for the Modelling of Myocardial Infarction and Drug Cardiotoxicity. *Nat. Biomed. Eng.* **2020**, *4*, 446–462. [CrossRef]
191. Mathur, A.; Loskill, P.; Shao, K.; Huebsch, N.; Hong, S.G.; Marcus, S.G.; Marks, N.; Mandegar, M.; Conklin, B.R.; Lee, L.P.; et al. Human iPSC-based cardiac microphysiological system for drug screening applications. *Sci. Rep.* **2015**, *5*, 1–7. [CrossRef] [PubMed]
192. Tomecka, E.; Zukowski, K.; Jastrzebska, E.; Chudy Michałand Brzozka, Z.; Chudy, M.; Brzozka, Z. Microsystem with micropillar array for three- (gel-embaded) and two-dimensional cardiac cell culture. *Sens. Actuators B Chem.* **2018**, *254*, 973–983. [CrossRef]
193. Bergström, G.; Christoffersson, J.; Schwanke, K.; Zweigerdt, R.; Mandenius, C.F. Stem cell derived in vivo-like human cardiac bodies in a microfluidic device for toxicity testing by beating frequency imaging. *Lab Chip* **2015**, *15*, 3242–3249. [CrossRef]
194. Colosi, C.; Costantini, M.; Barbetta, A.; Dentini, M. Microfluidic bioprinting of heterogeneous 3d tissue constructs. *Methods Mol. Biol.* **2017**, *1612*, 369–380.
195. Aung, A.; Bhullar, I.S.; Theprungsirikul, J.; Davey, S.K.; Lim, H.L.; Chiu, Y.-J.J.; Ma, X.; Dewan, S.; Lo, Y.-H.H.; McCulloch, A.; et al. 3D cardiac μtissues within a microfluidic device with real-time contractile stress readout. *Lab Chip* **2016**, *16*, 153–162. [CrossRef] [PubMed]

196. Lind, J.U.; Busbee, T.A.; Valentine, A.D.; Pasqualini, F.S.; Yuan, H.; Yadid, M.; Park, S.J.; Kotikian, A.; Nesmith, A.P.; Campbell, P.H.; et al. Instrumented cardiac microphysiological devices via multimaterial three-dimensional printing. *Nat. Mater.* **2017**, *16*, 303–308. [CrossRef] [PubMed]
197. Wan, C.R.; Chung, S.; Kamm, R.D. Differentiation of embryonic stem cells into cardiomyocytes in a compliant microfluidic system. *Ann. Biomed. Eng.* **2011**, *39*, 1840–1847. [CrossRef]
198. Giridharan, G.A.; Nguyen, M.D.; Estrada, R.; Parichehreh, V.; Hamid, T.; Ismahil, M.A.; Prabhu, S.D.; Sethu, P. Microfluidic Cardiac Cell Culture Model. *Anal. Chem.* **2010**, *82*, 7581–7587. [CrossRef]
199. Sebastião, M.J.; Gomes-Alves, P.; Reis, I.; Sanchez, B.; Palacios, I.; Serra, M.; Alves, P.M. Bioreactor-based 3D human myocardial ischemia/reperfusion in vitro model: A novel tool to unveil key paracrine factors upon acute myocardial infarction. *Transl. Res.* **2020**, *215*, 57–74. [CrossRef] [PubMed]
200. Chen, T.; Vunjak-Novakovic, G. Human Tissue-Engineered Model of Myocardial Ischemia-Reperfusion Injury. *Tissue Eng. Part A* **2019**, *25*, 711–724. [CrossRef] [PubMed]
201. Amar, D.N.; Epshtein, M.; Korin, N. Endothelial cell activation in an embolic ischemia-reperfusion injury microfluidic model. *Micromachines* **2019**, *10*, 857. [CrossRef] [PubMed]
202. Marsano, A.; Conficconi, C.; Lemme, M.; Occhetta, P.; Gaudiello, E.; Votta, E.; Cerino, G.; Redaelli, A.; Rasponi, M. Beating heart on a chip: A novel microfluidic platform to generate functional 3D cardiac microtissues. *Lab Chip* **2016**, *16*, 599–610. [CrossRef] [PubMed]
203. Ugolini, G.S.; Rasponi, M.; Pavesi, A.; Santoro, R.; Kamm, R.; Fiore, G.B.; Pesce, M.; Soncini, M. On-chip assessment of human primary cardiac fibroblasts proliferative responses to uniaxial cyclic mechanical strain. *Biotechnol. Bioeng.* **2016**, *113*, 859–869. [CrossRef]
204. Castaño, A.G.; Hortigüela, V.; Lagunas, A.; Cortina, C.; Montserrat, N.; Samitier, J.; Martínez, E. Protein patterning on hydrogels by direct microcontact printing: Application to cardiac differentiation. *RSC Adv.* **2014**, *4*, 29120. [CrossRef]
205. Sakamiya, M.; Fang, Y.; Mo, X.; Shen, J.; Zhang, T. A heart-on-a-chip platform for online monitoring of contractile behavior via digital image processing and piezoelectric sensing technique. *Med. Eng. Phys.* **2020**, *75*, 36–44. [CrossRef]
206. Liu, J.; Miller, K.; Ma, X.; Dewan, S.; Lawrence, N.; Whang, G.; Chung, P.; McCulloch, A.D.; Chen, S. Direct 3D bioprinting of cardiac micro-tissues mimicking native myocardium. *Biomaterials* **2020**, *256*, 120204. [CrossRef]
207. Tandon, N.; Marsano, A.; Maidhof, R.; Numata, K.; Montouri-Sorrentino, C.; Cannizzaro, C.; Voldman, J.; Vunjak-Novakovic, G. Surface-patterned electrode bioreactor for electrical stimulation. *Lab Chip* **2010**, *10*, 692. [CrossRef]
208. Alassaf, A.; Tansik, G.; Mayo, V.; Wubker, L.; Carbonero, D.; Agarwal, A. Engineering anisotropic cardiac monolayers on microelectrode arrays for non-invasive analyses of electrophysiological properties. *Analyst* **2020**, *145*, 139–149. [CrossRef]
209. Li, L.; Chen, Z.; Shao, C.; Sun, L.; Sun, L.; Zhao, Y. Graphene Hybrid Anisotropic Structural Color Film for Cardiomyocytes' Monitoring. *Adv. Funct. Mater.* **2020**, *30*, 1–8. [CrossRef]
210. Fang, J.; Koh, J.; Fang, Q.; Qiu, H.; Archang, M.M.; Hasani-Sadrabadi, M.M.; Miwa, H.; Zhong, X.; Sievers, R.; Gao, D.W.; et al. Injectable Drug-Releasing Microporous Annealed Particle Scaffolds for Treating Myocardial Infarction. *Adv. Funct. Mater.* **2020**, *30*, 1–14. [CrossRef] [PubMed]
211. Huang, F.; Sun, L.; Zhao, C.; Qiu, Z.; Zhao, Y.; Jin, W. Protein microcapsules integrated hierarchical scaffolds for local treatment of acute myocardial infarction model. *Appl. Mater. Today* **2021**, *22*, 100901. [CrossRef]
212. Dinh, N.D.; Kukumberg, M.; Nguyen, A.T.; Keramati, H.; Guo, S.; Phan, D.T.; Ja'afar, N.B.; Birgersson, E.; Leo, H.L.; Huang, R.Y.J.; et al. Functional reservoir microcapsules generated: Via microfluidic fabrication for long-term cardiovascular therapeutics. *Lab Chip* **2020**, *20*, 2756–2764. [CrossRef]
213. Kamei, K.; Kato, Y.; Hirai, Y.; Ito, S.; Satoh, J.; Oka, A.; Tsuchiya, T.; Chen, Y.; Tabata, O. Integrated heart/cancer on a chip to reproduce the side effects of anti-cancer drugs in vitro. *RSC Adv.* **2017**, *7*, 36777–36786. [CrossRef]
214. Liu, H.; Bolonduro, O.A.; Hu, N.; Ju, J.; Rao, A.A.; Duffy, B.M.; Huang, Z.; Black, L.D.; Timko, B.P. Heart-on-a-Chip Model with Integrated Extra- And Intracellular Bioelectronics for Monitoring Cardiac Electrophysiology under Acute Hypoxia. *Nano Lett.* **2020**, *20*, 2585–2593. [CrossRef] [PubMed]
215. Huo, W.; Gao, Y.Y.; Zhang, L.; Shi, S.; Gao, Y.Y. A Novel High-Sensitivity Cardiac Multibiomarker Detection System Based on Microfluidic Chip and GMR Sensors. *IEEE Trans. Magn.* **2015**, *51*, 18–21. [CrossRef]
216. Murata, K.; Glaser, L.; Nardiello, M.; Ramanathan, L.V.; Carlow, D.C. Data from the analytical performance of the Abaxis Piccolo Xpress point of care analyzer in whole blood, serum, and plasma. *Data Brief* **2018**, *16*, 81–89. [CrossRef]
217. Li, F.; Guo, L.; Hu, Y.; Li, Z.; Liu, J.; He, J.; Cui, H. Multiplexed chemiluminescence determination of three acute myocardial infarction biomarkers based on microfluidic paper-based immunodevice dual amplified by multifunctionalized gold nanoparticles. *Talanta* **2020**, *207*, 120346. [CrossRef] [PubMed]
218. Lim, W.Y.; Thevarajah, T.M.; Goh, B.T.; Khor, S.M. Paper microfluidic device for early diagnosis and prognosis of acute myocardial infarction via quantitative multiplex cardiac biomarker detection. *Biosens. Bioelectron.* **2019**, *128*, 176–185. [CrossRef] [PubMed]
219. Wu, J.; Dong, M.; Santos, S.; Rigatto, C.; Liu, Y.; Lin, F. Lab-on-a-chip platforms for detection of cardiovascular disease and cancer biomarkers. *Sensors* **2017**, *17*, 2934. [CrossRef] [PubMed]
220. Mejía-Salazar, J.R.; Cruz, K.R.; Vásques, E.M.M.; de Oliveira, O.N. Microfluidic point-of-care devices: New trends and future prospects for ehealth diagnostics. *Sensors* **2020**, *20*, 1951. [CrossRef] [PubMed]
221. Sachdeva, S.; Davis, R.W.; Saha, A.K. Microfluidic Point-of-Care Testing: Commercial Landscape and Future Directions. *Front. Bioeng. Biotechnol.* **2021**, *8*, 1–14. [CrossRef]

cells MDPI

Review

Insight into Hypoxia Stemness Control

Miriam Di Mattia [1,†], Annunziata Mauro [1,*,†], Maria Rita Citeroni [1], Beatrice Dufrusine [2,3], Alessia Peserico [1], Valentina Russo [1], Paolo Berardinelli [1], Enrico Dainese [1], Annamaria Cimini [4,5] and Barbara Barboni [1]

1 Unit of Basic and Applied Biosciences, Faculty of Bioscience and Agro-Food and Environmental Technology, University of Teramo, 64100 Teramo, Italy; mdimattia@unite.it (M.D.M.); mrciteroni@unite.it (M.R.C.); apeserico@unite.it (A.P.); vrusso@unite.it (V.R.); pberardinelli@unite.it (P.B.); edainese@unite.it (E.D.); bbarboni@unite.it (B.B.)
2 Department of Innovative Technologies in Medicine & Dentistry, University of Chieti-Pescara, 66100 Chieti, Italy; bdufrusine@unite.it
3 Center of Advanced Studies and Technology (CAST), 66100 Chieti, Italy
4 Department of Life, Health and Environmental Sciences, University of L'Aquila, 67100 L'Aquila, Italy; annamaria.cimini@univaq.it
5 Sbarro Institute for Cancer Research and Molecular Medicine and Center for Biotechnology, Temple University, Philadelphia, PA 19122, USA
* Correspondence: amauro@unite.it; Tel.: +39-086-1426-6888; Fax: +39-08-6126-6860
† These authors contributed equally to this review.

Abstract: Recently, the research on stemness and multilineage differentiation mechanisms has greatly increased its value due to the potential therapeutic impact of stem cell-based approaches. Stem cells modulate their self-renewing and differentiation capacities in response to endogenous and/or extrinsic factors that can control stem cell fate. One key factor controlling stem cell phenotype is oxygen (O_2). Several pieces of evidence demonstrated that the complexity of reproducing O_2 physiological tensions and gradients in culture is responsible for defective stem cell behavior in vitro and after transplantation. This evidence is still worsened by considering that stem cells are conventionally incubated under non-physiological air O_2 tension (21%). Therefore, the study of mechanisms and signaling activated at lower O_2 tension, such as those existing under native microenvironments (referred to as hypoxia), represent an effective strategy to define if O_2 is essential in preserving naïve stemness potential as well as in modulating their differentiation. Starting from this premise, the goal of the present review is to report the status of the art about the link existing between hypoxia and stemness providing insight into the factors/molecules involved, to design targeted strategies that, recapitulating naïve O_2 signals, enable towards the therapeutic use of stem cell for tissue engineering and regenerative medicine.

Keywords: hypoxia; O_2 tension; hypoxia inducible factors; intracellular signaling; metabolism; stemness; hypoxia in vitro models

Citation: Di Mattia, M.; Mauro, A.; Citeroni, M.R.; Dufrusine, B.; Peserico, A.; Russo, V.; Berardinelli, P.; Dainese, E.; Cimini, A.; Barboni, B. Insight into Hypoxia Stemness Control. *Cells* **2021**, *10*, 2161. https://doi.org/10.3390/cells10082161

Academic Editor: Mehdi Najar

Received: 26 July 2021
Accepted: 19 August 2021
Published: 22 August 2021

1. Introduction

1.1. The Role of O_2 in Cell Biology

Molecular oxygen (O_2) is necessary for animal life and is essential for a variety of biological processes involved in the survival of prokaryotic and eukaryotic cells. The rate of O_2 usage by cells is various, depending on the cell type and function. In eukaryotic cells, O_2 uptake occurs by direct transport across the cell membrane and 90% of O_2 is consumed by mitochondria during respiration and oxidative phosphorylation processes [1]. Furthermore, the citric acid cycle and β-oxidation of fatty acids are tightly associated with the process of ATP production. Thus, O_2 availability is essential for cell functions, and decreased O_2 concentration represents a major stress factor for cells. In a homeostatic state, cells require a level of O_2 between 2–9% (14.4–64.8 mmHg), lower levels of O_2 in cells are related to a state of hypoxia 0.5–2% (<10 mmHg) [2]. Cells modulate gene expression in response to O_2 availability and these changes affect cell metabolism, immunity, and tissue reorganization [3]. Cellular adaptive responses to hypoxia are mainly mediated by the transcription factor hypoxia-inducible factor-1 (HIF-1) which induces transcriptional activation of various genes promoting angiogenesis, cell proliferation, and survival in hypoxic conditions. Cells activate multiple adaptative responses for O_2 supply: (I) reduce the rate of oxidative phosphorylation, (II) arrest cell cycle, (III) stimulate the formation of new blood vessels by releasing the major angiogenic factors (vascular endothelial growth factor (VEGF), angiopoietin1 (Ang-1), transforming growth factor β1 (TGF-β1), and fibroblast growth factor (FGF-2)), and (IV) switch to anaerobic glycolysis [4]. Furthermore, morphological cytoskeletal cellular changes occurring in hypoxia, such as alteration in protein polarization and aggregation, lead to an increase in membrane permeability [5]. O_2 dyshomeostasis, such as high O_2 levels, can also induce cytotoxicity due to the production of reactive O_2 species (ROS) during its utilization. ROS include peroxides, singlet O_2, hydroxy radical, and superoxide which are responsible for lipid, protein, and nucleic acids oxidation causing cellular dysfunction. Cells have different levels of antioxidants and redox enzymes to contrast the ROS accumulation. Unfortunately, these defense systems are not always adequate to contrast ROS production resulting in different levels of ROS tolerance.

1.2. The Role of O_2 in Tissue

Within an organism O_2 is up taken in the lungs, it passes into the alveoli and by simple diffusion across endothelial cells of the alveolar capillaries. Once in the circulation, O_2 is transported into the blood in two forms; mainly bound with hemoglobin or dissolved in plasma [6]. The O_2 tension of inspired air is 160 mmHg, in alveolar blood is 104 mmHg while in most tissues is around 40–50 mmHg [7,8] (Figure 1). However, in several tissues O_2 level is lower such as in the spleen, thymus, retina, and regions of the brain where it has been measured around 16, 10, 25, and 8 mmHg [8]. More in detail, low O_2 levels have been associated to various stem cell niches, such as mesenchymal stem cells (MSCs), neural stem cells (NSCs), and hematopoietic stem cells (HSCs), suggesting a pivotal role of O_2 in maintaining stem cells pluripotency as discussed along this review. All tissues have their own characteristic "tissue-normoxia" and oxygen dyshomeostasis induces damage depending on the tissue affected. Furthermore, O_2 levels vary in tissues during normal physiological states such as skeletal muscle exertion or embryo development. However, the hypoxic state in tissues is characteristic of pathological conditions that occur in infection, ischemic cardiovascular disease, chronic obstructive pulmonary disease, or cancer [2,9–12]. For example, oxygenation is very low in various areas of many solid tumors due to the uncontrolled proliferation of cells and abnormal blood vessels spreading. Moreover, hypoxic tissues are induced by impaired vascular function characterizing tissue wound. In the wound healing process, macrophages accumulate preferentially in hypoxic niches where respond rapidly by activating an array of adaptive genes [13]. Adaption of macrophages alters the expression of receptors and protein adhesion to further enhance their migration towards hypoxic sites. Furthermore, hypoxia-induced macrophages release growth factors and cytokines to recruit mesenchymal cells involved even in early wound healing events [13]. Hypoxia occurs in the later phases of reepithelization and restoration of tissue integrity and vasculature [14]. Furthermore, considering the heterogenicity of the cell populations characterizing each tissue niche it would be necessary to understand interactions between multiple cell types in hypoxic microenvironments and to investigate the response mechanisms to increased O_2 levels.

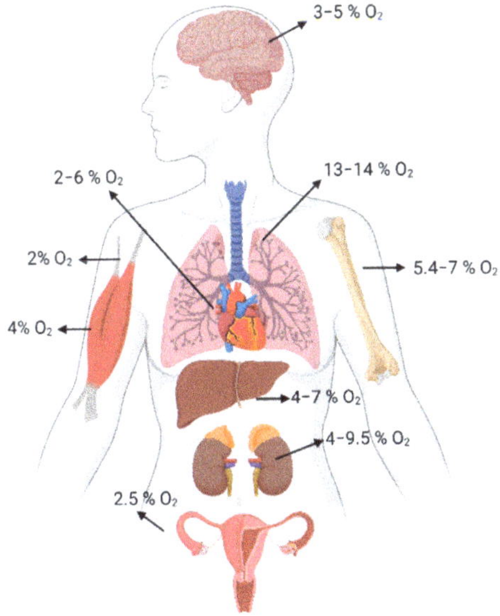

Figure 1. Different O_2 partial pressure in body districts.

1.3. Hypoxia-Inducible Factor (HIF)

Hypoxia-inducible factor (HIF) is the master regulator of O_2 homeostasis with hundreds of hypoxia-inducible target genes. HIF is a heterodimeric transcription factor consisting of two subunits: HIF-α and HIF-β [15]. HIF-β is considered a constitutively expressed gene while HIF-1α is the predominant regulator of hypoxia and it is mainly regulated at post-translational level [16]. Nowadays, three HIF-α subunits (HIF-1α, HIF-2α, and HIF-3α) have been recognized [17], with different functions [18], exhibiting high conservation of the protein domain structures and regulation of the hypoxia-dependent mechanisms. The three HIF isoforms differing in the oxygen-sensitive α subunit exist in vertebrates [19]. All the three isoforms form the heterodimer with HIF-1β binding to the same cis-element HIF-binding sites (HBS) [20].

HIF-1α and HIF-2α (EPAS1) are structurally similar and best characterized. HIF-3α (IPAS) exists as multiple splice variants, able to inhibit HIF-1α and HIF-2α activity [21]. HIF-1α is expressed in all cells, while HIF-2α and HIF-3α are selectively expressed in vascular endothelial cells, type II pneumocytes, renal interstitial cells, liver parenchymal cells, and cells of the myeloid lineage [22]. Even if they display similar biochemical properties, distinct physiological roles of HIF-1α and HIF-2α can be supposed [20]. It was demonstrated that embryos HIF-1$\alpha^{-/-}$ nor HIF-2$\alpha^{-/-}$ did not survive, suggesting that HIF-1α and HIF-2α have different and not complementary functions [20].

HIF is normally expressed in cells at the basal level, but in the presence of high an unphysiological air levels of O_2, 21% O_2 concentration generally defined as "normoxia", it is ubiquitinated and degraded. In detail, the DNA-binding domain of HIF-1α lies within the N-terminal region of the protein, whereas the C-terminal region holds the two transactivation domains. The central region of HIF-1α contains an O_2-dependent degradation (ODD) domain, located between the amino acids 401 and 603, which confers O_2-sensitiviy to the HIF-1α subunit [23]. Its conserved proline residues are hydroxylated by prolyl hydroxylase domain enzymes (PHDs) creating a binding site for the von Hippel–Lindau (VHL) protein, a component of the E3 ubiquitin ligase complex, which leads HIF-1α subunit to the proteasomal destruction [24]. As a result, HIF-1α is rapidly degraded in normoxic conditions. When low O_2 concentration, as hypoxia occurs, PHDs are inactive, HIF-1α is not transcriptionally upregulated but the protein was stabilized.

In a low O_2 environment HIF-α and HIF-β subunits form a heterodimer creating the aryl hydrocarbon receptor nuclear translocator complex (ARNT), which translocates into the nucleus [23] (Figure 2). The ARNT complex formation is O_2 concentration-dependent [18]. When the ARNT heterodimer is assembled in the nucleus, it could be recognized by the co-activator and could bind to the conserved consensus sequence 5′-(A/G)CGTG-3′ within the hypoxia-response elements (HRE) of O_2-regulated target genes modulating transcription [15,25]. HIFs can induce the transcription of more than 70 genes correlated with control O_2 homeostasis, angiogenesis, mitochondrial metabolism [25], and adaptive functions [26] including epigenetic DNA modification, mRNA, microRNA, and protein synthesis [4] related to different biological responses (Figure 2).

It has been indicated that HIF-1α and HIF-2α differ in their ability to transactivate hypoxia-inducible genes. Indeed, it was proved that some genes were transactivated exclusively by HIF-1α, such as genes coding for glycolytic enzymes, while others were transactivated by both [27]. By using siRNA interference, it was shown that a small group of genes having binding sites for the E-twenty-six (ETS) family of transcription factors in common, were regulated by HIF-2α. Knock-down of ELK-1, the most abundant member of ETS family, significantly reduced hypoxic induction of the HIF-2α—dependent genes [28]. HIF-2α is supposed to have a relevant role in angiogenesis since it specifically regulates VEGF receptor Flk-1 expression, even if the mechanism was not well elucidated [29].

HIF-1α and HIF-2α regulate, also, angiogenic *VEGF* genes [27,30,31]. In this context, it has been reported that, under a hypoxic environment, mutant mice with HIF-1α deletion in the endothelial stem cells (ECs) showed defective blood vessel growth and activation of VEGF and its receptor VEGFR2, accompanied by impaired cell proliferation and migration. The results obtained in the study lead the authors to hypothesize that HIF-1α induces an autocrine VEGF/VEGFR2 regulation in ECs promoting their functions in tissue angiogenesis [32].

Many studies have reported that HIF is involved in many pathways influencing, in particular, cells cycle, proliferation, metabolism, stem cells plasticity, angiogenesis, and immunomodulation [33] (Figure 2). HIF-1α can directly reprograms the metabolic state in cells. Both HIF-1α and HIF-2α can modulate the expression of cytochrome c oxidase isoforms and maximize the efficiency of the electron transport chain [34,35]. The deficit of this mechanism negatively affects the production of ATP and leads to a major ROS production in hypoxia. Moreover, HIF-1α encodes for the pyruvate dehydrogenase kinase 1 acting through the target gene PDK, which represses the flux of pyruvate into acetyl-CoA, suppressing O_2 consumption [35,36]. In knock-out HIF-1α cells, hypoxia contributes to reduced ATP levels, elevated ROS, and apoptosis [34,36].

HIF also has an active role in inflammatory conditions as it promotes nuclear factor-κB (NF-κB) activity, a family of inducible transcription factors regulating a large array of genes involved in different processes of the immune and inflammatory responses [37] in macrophages, neutrophils, and nonimmune cells [38]. The NF-κB proteins are normally sequestered in the cytoplasm by a family of inhibitory proteins, including IκB family members. Hypoxia inhibits PHD1 activity resulting in IKK activation and phosphorylation of IκB followed by its degradation with the consequent liberation of NF-kB from the cytoplasm inducing the transcription of inflammatory cytokines [39].

Moreover, HIF can promote the expression of several miRNAs [40]. Out of all miRNAs influenced by HIF, miR-210 is the most significantly induced by hypoxia in all cell lines [41]. Its expression is regulated by both HIF-1α [42] and HIF-2α [43]. Overexpression of miR-210 in HUVECs leads to enhanced VEGFA and VEGFR2 expression promoting angiogenesis [44]. Interestingly, it was proposed that miR-210 could contribute to the HIF switch between HIF-1/HIF-2 and HIF-3 in human chondrocytes [45] and hepatocellular carcinoma cells [46] as miR-210 directly targets HIF-3α and suppresses HIF-1α protein expression.

Even if PHD is recognized as the main regulator of HIF-1α [47], different factors influence the ultimate result of the HIF activity, such as the presence or absence of HRE in gene promoter; the structure of variable sequences in HRE element of gene influencing the selective co-operation of other transcription factors, co-activators, or co-inhibitors with HIF; the cell type that present specific expression, compartmentalization, and degradation location of HIF-α isomers [48].

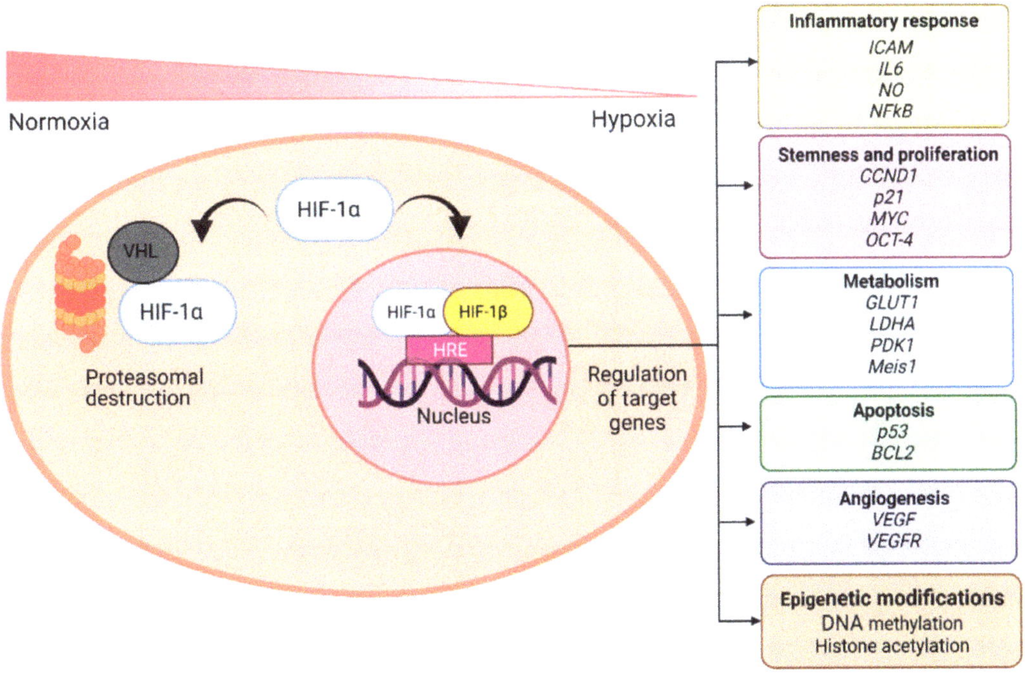

Figure 2. Scheme of HIF-1α activation during different O_2 level exposure. Under normoxia condition (≅21% O_2) HIF is rapidly degraded by proteasomal machinery. During hypoxia (≤10% O_2) HIF is stabilized and translocated into the nucleus where heterodimerized with HIF-1β. Heterodimer HIF-1α/β, regulates HRE target genes (some of which are indicated) involved in different biological responses.

2. Hypoxia and Stemness

O_2 concentration has been closely linked to the maintenance of stemness in stem cells that in vivo reside in specific "tissue niches", the anatomic locations that regulate their participation in tissue generation, maintenance, and repair [49]. Stem cell niche is a complex, heterotypic, and dynamic structure which includes supporting extracellular matrix, neighboring niche cells, secreted soluble signaling factors, physical, and environmental signals [50,51]. Comprehensive studies to clarify their critical components have been performed and stem cell niche's structures have been identified in many germlines and adult tissues [50,51]. It is known that hypoxia is recorded inside them and, even if the exact O_2 inside the niches in vivo cannot be recorded with the currently techniques, based on the closest approximations performed in human, an average O_2 of 3–13% O_2 exist in stem cell niche [7] (Figure 1). By residing in these dynamic tightly controlled in vivo environments that experience relatively low O_2 tensions, stem cells maintain a selective advantage suitable for their biological roles. Hence, reproducing the O_2 tension existing in native microenvironments represents a major challenge for researchers that might exploit it as a good strategy to preserve or enhance stem cells features with the advantage of their therapeutic value in regenerative medicine.

This review aims to clarify the state of art concerning the link between hypoxia and stemness to investigate the heterogeneity and complexity of the biological cues influencing the native local signal of stem cells, as well as to compare the hypoxia strategies and related aspects for preservation and improvement of stem cells properties.

Bibliographic papers dealing with this topic present in Scopus Database have been analyzed by using specific keywords, among these "hypoxia", "stemness", and "stem cells". To only select papers strictly related to the topic of the research, all articles associated with cancer stem cells and cancer progression were excluded (Figure 3).

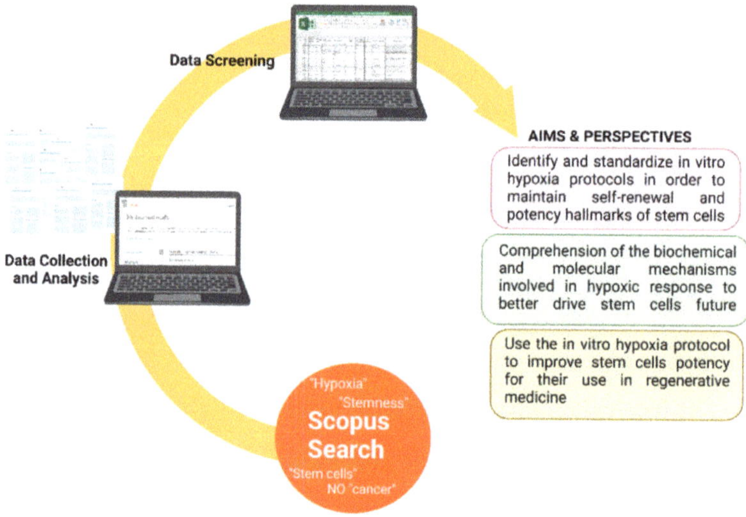

Figure 3. Schematic representation of methodology for data research and analysis performed with specific keywords in Scopus Database related to the link between hypoxia and stemness preservation elaborated in this review.

Many in vivo and in vitro approaches aiming to mimic the naive hypoxia cellular microenvironment experienced to stem cells, have been reported. Animal models are usually employed to study the effect of in vivo hypoxia, especially to characterize a wide variety of diseases, including reoxygenation injury, pre-eclampsia, diabetic retinopathy, and hypoxic insult of the brain [52]. However, animal study findings are often characterized by a great biological variability such as that recorded for values related to minute ventilation, tidal volume, peripheral O_2 saturation, arterial CO_2 pressure, and exhaled NO levels [52–54]. In addition, hypoxia animal models in vivo fail to recapitulate some of the key hallmarks of stem cell physiology, leading researchers to approach hypoxia studies by using in vitro cell cultures. More efforts have been made to exploit the possibility to grow stem cells in in vitro hypoxic conditions to mimic the niche microenvironment. However, outside their hypoxic natural environment, stem cells undergo physiological changes inducing high variabilities in their therapeutic efficacy remaining an open challenge for researchers and clinicians. In this context, for example, the O_2 gradient in culture conditions proposed for mesenchymal stem cells (MSCs) are different and, often, characterized by controversial results causing ambiguity in the interpretation of hypoxia effects. Main differences in methods, physical or chemical induction, O_2 percentages or chemical compound concentrations, stem cells models, time of hypoxic exposure, and different modality of HIF activation were reported, making it difficult for the optimization of hypoxic induction protocols to be used. In addition, stem cell metabolism has recently emerged as a critical determinant of cellular processes and is uniquely adapted to support proliferation, stemness, and commitment. Metabolic activation is also linked to HIF factor, which transcriptionally activates genes involved in O_2 homeostasis and metabolism [25,55]. Moreover, HIF-dependent mechanisms can influence many other processes as an epigenetic response including DNA methylation and histone acetylation, which in turn modulate hypoxia-responsive gene expression in cells. Nevertheless, HIF-1 pathway could be activated from stimuli different from hypoxic ones, as "hypoxic mimetic compounds", showing the same final effect on stemness preservation [56] but increasing the variability of hypoxic methods that can be applied in vitro.

Data collected by our scientometric analysis revealed a complexity of factors involved in the strategies mostly used for hypoxic induction and in the correlation between hypoxia and stemness. The purposes of this review are:

- Define and standardize in vitro hypoxia protocols to maintain self-renewal and potency hallmarks of stem cells.
- Comprehend the biochemical and molecular mechanisms involved in the hypoxic response to better drive stem cell future.
- Use in vitro hypoxia protocols to improve stem cell potency for their use in regenerative medicine.

We will consider the in vitro stimuli for hypoxia (physical, chemical, and biological) describing hypoxia exploitation for regenerative medicine (O_2 and HIF stemness preservation, stem metabolic state, and ROS), taking into account the most used stem cells sources (Figure 4).

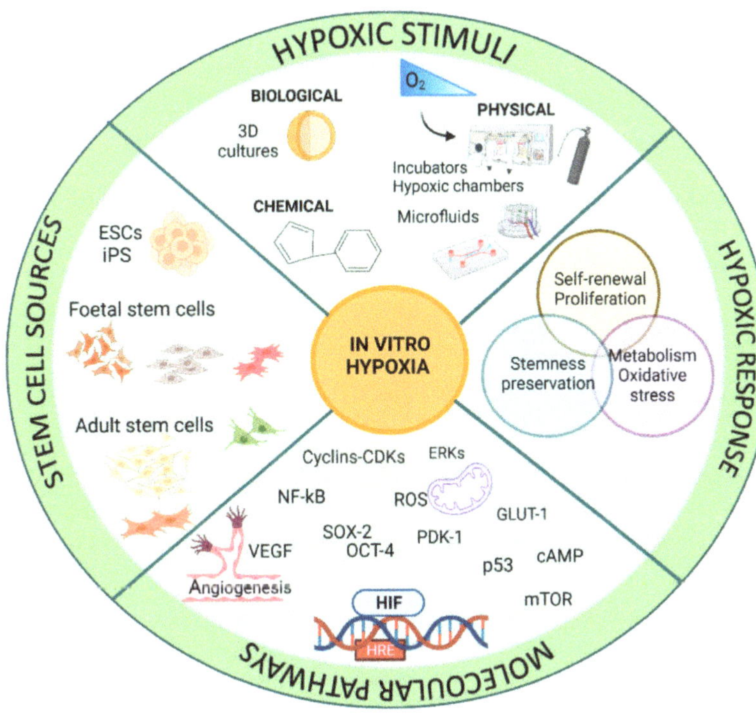

Figure 4. Parameters and outcomes for in vitro hypoxia exploitations. Different hypoxic stimuli, cells sources, biological responses, and pathways principally discussed in the review.

3. In Vitro Models of Hypoxia

Hypoxia can be differently induced in vitro, and several elements must be considered, as physical or chemical induction, O_2 percentages or chemical compound concentrations, and the time of exposure [57]. Cells can sense changes in the O_2 tension, defined as the activity of dissolved molecules in their microenvironment, influencing growth and differentiation processes. Routinely, in vitro cell cultures are performed in liquid culture media incubated at the atmospheric O_2 concentration of 21%, and it is difficult to know the exact O_2 tension that cells experience because it is strictly correlated to the rate of O_2 consumption by the cells [58]. Recently, it has been reported that in laboratory practice exists a considerable number of parameters, often non cited in literature, that generate a wide variability in O_2 delivery compromising results reproducibility. Culture dishes and their geometries, cell types, seeding density, media volume and its composition, culture temperature, and opening the doors of the incubator are frequently considered as principal factors influencing O_2 delivery in cell cultures [59] (Table 1).

Frequent handlings of cell cultures imply exposing cells at the atmospheric O_2 air tension where re-oxygenation rapidly occurs thus generating fluctuations in O_2 concentration. This factor can affect, for example, the lysis of those proteins that are susceptible to rapid oxygenation-dependent modifications [60]. Furthermore, medium changes interrupt the concentration gradient that is established over time in cell cultures since culture medium requires significant amounts of time to equilibrate to new O_2 concentrations [59]. This is a crucial aspect, and the improvement of the in vitro oxygenation control would be very advantageous to reach a functional in vivo resemblance [61].

Given the difficulties in controlling O_2 levels, appropriate O_2 sensors should be applied in cell cultures to measure O_2 concentration thus monitoring its fluctuations, however, they are not routinely used [62].

Table 1. Relevant parameters to be consider for hypoxia induction in in vitro cell cultures.

Parameter	Effects on O_2 Delivery	References
Cell type	Cells from different tissues have different O_2 consumption rate (OCR) that influences O_2 delivery.	[63]
Culture geometry dishes	O_2 diffusion through polystyrene changes between culture dishes geometries and it is responsible for up to 30% of O_2 delivery.	[58]
Seeding density	Seeding density influences the OCR. Experiments should be performed using always same cell densities, avoiding confluent cell cultures.	[59]
Medium volume	Culture medium holds less O_2 than air per unit volume and limits the movement of O_2 molecules. Medium depth determines the diffusive barrier to O_2 delivery.	[59] [58]
Medium composition	Protein and glucose, normally added to culture medium, reduce the capacity for dissolved O_2	[59]
Temperature	Cold medium holds significantly more O_2 than warm one. Increases in temperature cause conflicting effects of increasing the diffusion coefficient while decreasing O_2 solubility. As medium is cooled, the O_2 solubility increases. The concentration of dissolved O_2 depends on the temperature and partial pressure of O_2 in the gaseous phase.	[58] [58] [59]
Humidity	Humidity with carbon dioxide dilutes other atmospheric components–for dry air moving to saturation (~6% water vapor) and 5% CO_2 reduces the partial pressure of O_2 by 11% (or 8% for an initial atmosphere at 50% relative humidity).	[58,59]
Altitude	The decrease of atmospheric pressure with an increasing altitude, influences the amount of O_2 in the cell culture medium.	[52]
Handling cell cultures	Moving cells from hypoxia to ambient air generates O_2 fluctuations that influences O_2 concentration.	[60]

The use of hypoxic chambers or incubators with a specific mixture of nitrogen gas (N_2), carbon dioxide (CO_2), and O_2 represent a common approach to regulate O_2 tension in in vitro culture. To achieve low O_2 levels, more N_2 is introduced in the gas mixture, thus reducing the partial pressure of O_2. Alternatively, incubators can be connected to an external high-pressure liquid nitrogen tank that infuses N_2 displacing O_2. Among different methods adopted to this aim, the most innovative accredited system is the hypoxia workstation [60] that can offer precise control of O_2 and CO_2, as well as, control of temperature and relative humidity, providing to maintain a hypoxic environment in long-term cell culture. More details on O_2 different induction in cells culture are reported in Section 3.1 of this review.

Another approach to reproduce hypoxia in cell culture is represented by pharmacological treatment with chemical agents defined as "hypoxic mimetic compounds" (detailed in Section 3.3), among which cobalt chloride ($CoCl_2$) is one of the most used [64]. The chemical induction is cheap, easy to perform in cell cultures, allowing operators to open culture dishes or flasks without affecting hypoxic conditioning. However, these compounds may possess, in addition to HIF-1α induction [65], other unknown effects that limit the use depending on cells type and density and it is necessary to test preliminary their potential toxic effects to define the best concentration for inducing hypoxia without affecting cell viability. The time of exposure to hypoxia is another parameter to be considered as cells differently tune gene expression depending on short-or long-term hypoxia exposure [57,66]. Hypoxic induction with chemical compounds is kept just for a short period in cell cultures (maximum of 72 h usually), while performing physical hypoxia allows to maintain cultures in low O_2 condition for long term [64]. Moreover, keeping low O_2 conditions for a long time of exposure, allows cells to adapt to hypoxic environments justifying the variety of cellular outcomes. On the other hand, different cellular responses to hypoxia can be also dependent on different HIFs isoforms activated since they have a specific temporal role within cells [67]. Even if both HIF-1 and HIF-2 isoforms mediate the hypoxic response overlapping and target genes, it has been demonstrated that HIF-1 drives the early response to hypoxia within 24 h, while HIF-2 seems to manage the chronic response after 24 h, creating the so-called "HIF switch" mechanism in cells [67].

Literature data suggest that O_2 levels in standard cell culture experiments significantly deviate from a physiological range, as well as show that O_2 levels vary dramatically under different experimental settings, cell types investigated, cell confluency, volume and timing of media exchange, etc. (Tables 1 and 2).

In the following paragraphs some and most used approaches for in vitro hypoxia induction in cells culture are detailed.

3.1. Methods Providing Physical Hypoxia Conditions: Hypoxic Chambers, Tri-Gas Incubator, and Hypoxic Workstation

Different systems to reproduce low O_2 levels for in vitro cell culture has been proposed in the literature (Figure 5). Incubators or hypoxic chambers are the most used systems (Figure 5A,B). The easier way is represented by the use of modular gas chambers inside a standard CO_2 incubator. For investigators who want to test hypoxia effects for their own cells and projects, these small chambers could be a good solution. The chamber is made of solid materials in a fixed shape and size holding up to twelve 10-cm dishes and require additional equipment, as regulators, tubing, and pumps for gassing the chamber with a pre-mixed gas mixture. They must be recharged after each entry, and currently cannot be monitored for internal conditions. The inclusion of an extra dish with sterile water maintains the humidity within the chamber [68]. Hypoxic chambers were mostly used in the past decades. However, one of the common defects of this chamber is leakage, although it is not frequent, and the generation of an inner pressure if the operation is inappropriate. For this reason, most hypoxic cultures today are performed in a "Tri-gas" incubator (Figure 5C), a not properly adequate definition because only two gases CO_2 and N_2 are supplied causing the reduction of O_2 that can be lowered to 0.5–1%. Some manufacturers may claim O_2 levels as low as 0.1%, but this is hard to achieve due to the sensor detection limits. The first commercially available "Tri-gas" incubator was released in 1979 [68]. Inside the incubator can be placed separated compartments with their own glass doors reducing the O_2 fluctuations but also contaminations. The use of this incubator proved that cells in hypoxia grew better, healthier, and with longer lifespans. Other incubators reproduce hypoxic conditions using a gas mixture from a single tank without separate sensors, but this prevents cells from receiving proper amounts of premixed gases [68].

Even if "Tri-gas" incubators can provide a hypoxic environment, they do not protect cells from ambient conditions during any procedures that must take place outside the incubator, such as medium changes. This extra exposure to higher levels of O_2 concentration negatively impacts cell growth. It has been reported that few minutes of exposure to ambient O_2 conditions accumulated over several months adversely affect the results of culturing experiments [58]. Like the hypoxic chambers are the anaerobic bags (AnaeroPack) (Figure 5D), an innovative system that we found in two articles of bibliometric research [69,70]. These bags are very easy to handle without requiring water or catalyst, simply putting AnaeroPack in jars. This system can reproduce a suitable atmosphere specially to grow microorganisms allowing three types of cultivation (i.e., anaerobic, microaerophilic, and CO_2 cultivation with selected concentrations). The major advantages of this system are their low cost and easy handling, without preparing large equipment and generating high temperature. Furthermore, the use of a hypoxia workstation [60], which can offer precise control of O_2, and CO_2, seems to be an appropriate method (Figure 5E). The hypoxic workstation keeps cells to constant O_2 levels because cells can be passaged, and culture media can be changed without altering O_2 levels within it; cells can be handled for experiments or lysates preparation preventing all those O_2 dependent modifications that may alter some cellular constituents [60]. These operative functions are possible because the workstation has a chamber equipped with O_2 sensors to monitor the O_2 concentration and two gloves access ports for sample handling [60]. The workstation is especially useful for those researchers that need almost an "anoxic" environment or with very low O_2 concentrations difficult to manage with the use of a traditional incubator. However, "Tri-gas" incubators and hypoxic workstations are expensive, and this makes them not convenient for small laboratories that do not perform hypoxia experiments routinely.

The CulturePal (Trial Products; Mitsubishi Gas Chemical Company Inc.) with built-in deoxidizing reagent [71], represents a novel system for the induction of a hypoxic atmosphere. The principal constituent is sodium ascorbate, which absorbs O_2 and generates CO_2 by oxidative degradation [71]. In their experiment, Ito and colleagues [71] adopted two series of this system: the CulturePal-Zero, for the modulation of O_2 levels <0.1%, and CulturePal-Five (3–7% O_2 levels). According to the authors, CulturePal systems would be a suitable system for the induction of short-term hypoxia and the regulation of gas concentrations during cell transportation [71].

Recently, a sophisticated Microfluidic Devices have been proposed to reproduce a hypoxic environment with a precise control of O_2 tension over temporal dimension and spatial one, in order of microns (Figure 5F). The small dimension of this device allows to minimize the distance of O_2 diffusion creating a microvascular system of small volumes (in the range of microliters) [72]. An example of these microfluidic systems is represented by laser-cut polycarbonate foils, produced with a layer-by-layer manufacturing technology, and an elastomeric membrane joined together using thermal diffusion bonding. Mechanical strength, chemical resistance, and biocompatibility characterized the fluidic layers. Several O_2 sensing spots are integrated into the device and monitored O_2 content helping to adjust its levels and thus producing stable and defined hypoxic conditions for cells [73]. Another chip, described by Barmaki et al., utilizes two separate, but interdigitated microfluidic channels. The hypoxic microenvironment was created by sodium sulfite as an O_2 scavenger in one of the channels and started to increase after 100 min of pumping in the single channel [74]. Mathematical simulations contribute to support O_2 diffusion measurements rendering this kind of system very accurate [75].

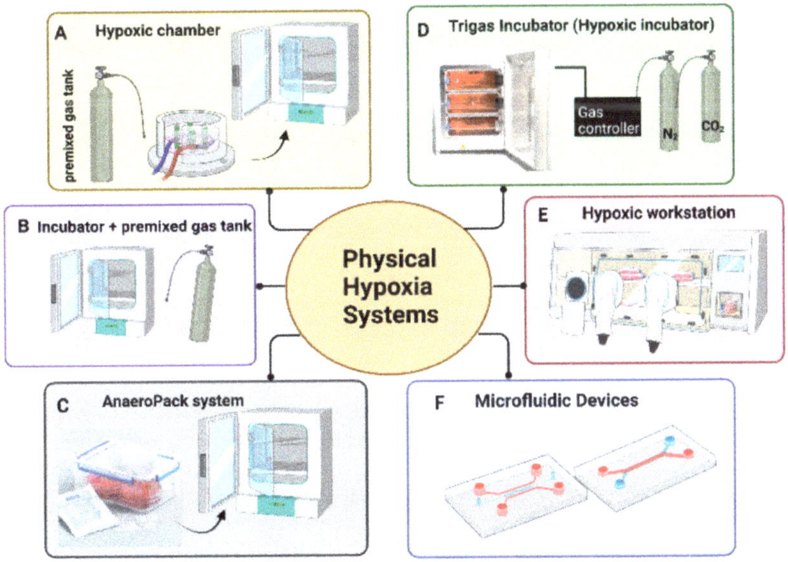

Figure 5. Most relevant Systems used to induce in vitro physical hypoxia. Gas mixture from a single tank can be connected directly to (**A**) hypoxic chambers before their incubation in standard CO_2 incubator or (**B**) incubator. A similar mechanism is used for (**C**) AnaeroPack system. (**D**) "Tri-Gas" incubator: only two gases are supplied, CO_2, as usual, and N_2, that can be modulated to reduce O_2. (**E**) hypoxic workstation. (**F**) example of microfluidic devices.

3.2. Biological-Mediated Approaches of Hypoxia: Spheroids 3D Cultures

Tissues are characterized by the presence of a gradient of O_2, nutrients, and paracrine factors but this state is not easily reproducible in typical 2D cultures [76]. For this reason, 3D culture model has received increased scientific interest as a favorable condition because it is closer to the physiological native environment [77] (Table 2).

The 3D culture is a common term used to refer to spheroid cultures, a method largely adopted for cancer cells cultures, but recently also to reproduce the physiological hypoxia cell niches [78]. Indeed, 3D in vitro model approaches may reproduce hypoxic conditions (Table 2) because of the lower O_2 concentration in the inner part of spheroids [79]. Although this is not properly a method to induce in vitro hypoxia, it has been hypothesized that spheroid formation potentiates cell function by the generation of a hypoxic core within spheroids with sufficient large diameters [79]. The spheroid size is a considerable variant as the difficulty in controlling spheroid diameters has an impact on the diffusion of nutrients, signaling molecules and O_2 concentration which decreases near the spheroid core [80].

Different developed methods, such as hanging drop, chitosan film cultures, or the use of bioreactors and rotating cultures [81] provide a suspension culture condition in which cell–cell adhesion and cell–matrix interactions improve the self-assembly of cells leading to the spheroid formation or 3D tissue-like aggregates [76].

Under these culture conditions, cells are stimulated to grow with the formation of numerous 3D proliferation centers, hypoxic regions, and specific microenvironments that allow them to express a tissue-like phenotype. The tendency of cells to form spheroids could be interpreted as a marker of the undifferentiated state of cells as stem cells. Indeed, literature data evidenced that adult MSCs possess a remarkable ability to coalesce and assemble in tri-dimensional (3D) structure which closely recapitulates the in vivo MSCs niche. MSCs cultured in 3D spheroid cultures showed a stable immuno-phenotypic profile, with a significant enhancement in survival, homing, stemness features differentiation potential, angiogenic effect, and anti-inflammatory properties [82]. The most relevant effects on 3D spheroids cultures are the high expansion and colony formation, the differentiation potential, and epigenetic changes in pluripotent genes such as *Oct-4*, *Sox-2*, and *Nanog* [83].

Zhang et colleagues observed that gingiva mesenchymal stem cells (GMSCs) spheroid expressed higher levels of HIF-1α and HIF-2α, against the adherent counterparts and increased production of ROS thus recapitulating features of low O_2 conditions [82]. Even if the correlation between hypoxic core within spheroids and HIF activation is well described in oncological research (solid tumors are characterized by regions permanently or transiently hypoxic due to the poor blood supply and lack of vascularization), the hypoxic adaptations in spheroids rely on the activation of the transcription factor HIF [84] and its influence on the maintenance of multipotency and self-renewal [55]. Consistent with this hypothesis, hypoxia-regulated genes, such as *VEGF*, are upregulated in MSCS spheroids [85].

An additional aspect to be considered in managing 3D cell cultures is the O_2 gradient as the three-dimensionality and the variable thickness of cellular structures introduce additional irregularities that hamper oxygen diffusion and lead to the formation of O_2 concentration gradients. In that context, several novel approaches and techniques have emerged tackling the challenges of O_2 gradient concentration as the use of oxygen-sensing microelectrodes [61]. However, the invasive nature of the approach represents a disadvantage as since time and technical challenges require repetitive calibrations and measurements in different spots inside the tissue construct motivating the search for alternative approaches.

For this aspect, the use of 3D cell culture systems for in vitro hypoxia induction is represented by the possibility of actively inducing a controlled O_2 gradient across the model, based on the experimental needs. These gradients can be induced with different methods, by perfusion with an oxygen scavenger in the medium, by positioning the culture between two micro-channel circuits perfused with gas, or by incorporation of an O_2-consuming reaction of specific hydrogel and thus regulating the O_2 levels [52].

3.3. Hypoxia Mimicking Agents

Some in vitro models utilize "hypoxia mimetic agents" biological or chemical molecules which simulate hypoxic conditions predominantly by increasing the availability of intrinsic HIF-1α in standard cell culture settings. This methodology can be used for both sustained and intermittent hypoxia models, the latter of which can be achieved by cyclic exposure to the agent. The precise mechanism of action of hypoxia mimetic agents may vary depending on the particular agent used (Table 2). Transcriptionally active HIF levels rise at sub-physiological concentrations of O_2 inducing upregulation of a range of genes with activities ranging from cell protection to apoptosis according to the specific context and cell type. As reported in Section 1.3, the regulation of HIF degradation requires hydroxylation by PHDs [56]. The PHDs are 2-oxoglutarate (2OG) and non-heme-Fe (II)-dependent dioxygenase family members, all requiring ferrous iron (Fe^{2+}), 2OG, O_2, and ascorbate for the enzymatic activity. Indeed, the reduction of substrate hydroxylation results in HIF-1α stabilization [86]. However, since HIF-1α is not the exclusive substrate of PHDs, it must be taken into account that its stabilization via PHDs inhibition could affect also other pathways [56].

The most used HIF stabilizers "hypoxia mimetic agent" are $CoCl_2$, dimethyloxalylglycine (DMOG), and deferoxamine (DFO) which hamper HIF degradation by the inhibition of PHDs although with different mechanisms [87].

More in detail, DMOG is a competitive inhibitor of the three PHD isoforms and of factor inhibiting HIF (FIH). DMOG acts as an analog of 2OG (co-substrate of PHDs), placed at the catalytic site-blocking enzymatic activity [87].

DFO is a Fe^{2+} chelator, another essential cofactor in PHD activity. A lack of Fe^{2+} availability causes inhibition of the activity of PHDs and FIH, provoking HIF-1α accumulation and an increase in activity [88].

The hypoxic $CoCl_2$ model is based on the inhibition of PHDs by substitution of the Fe^{2+}, thus increasing HIF-1α protein levels and inducing its transcriptional activity. However, experimental results have suggested different speculations regarding the mechanism of HIF-1α stabilization by Co^{2+} [64]. It has been shown that cobalt can prevent the binding of HIF-1α to von-Hippel–Lindau protein (pVHL), block HIF-1α degradation, or deplete ascorbate which is essential for maintaining the PHDs in the active state. An increase in HIF-1α levels after $CoCl_2$ treatment could be linked also to ROS generation [89].

However, even if the effects of hypoxia-mimetic agents are comparable to those resulting from reduced atmospheric O_2 levels [90], it should be noted that one of the most common downsides is their potential cellular cytotoxicity [52]. So, a hypoxic environment is induced by stabilization and accumulation of HIF-1α that occur because all of these chemical agents block the activity of the PHDs, disrupting the hydroxylation of HIF-1α and inhibiting the ubiquitin-dependent 26S proteasomal degradation pathway. In addition, they also have inhibitory effects on FIH. In this context, a newly developed PHD inhibitor, namely JNJ-42041935 1-(5-Chloro-6-(trifluoromethoxy)-1*H*-benzimidazol-2)-1*H*-pyrazole-4-carboxylic Acid (JNJ) has been identified through structure-based drug design methods and it seems to be highly selective for all the isoforms of PHDs relative to FIH [91]. Moreover, JNJ showed a high efficiency in stabilizing both HIF-1α and HIF-2α isoforms [92].

Other Hypoxic Mimetic Agents

Other stimuli and unusual chemical compounds are often used to activate HIF-1α and to mimic hypoxia (Table 2). Among these, there is ferulic acid (FA), a phytochemical found in the walls of plant cells with potential therapeutic effects in wound healing and ischemic diseases. It has been shown that FA can upregulate HIF-1α, VEGF, and platelet-derived growth factor (PDGF), which subsequently activate mitogen-activated protein kinase (MAPK) and phosphatidyl inositol 3-kinase (PI3K) pathways, improving angiogenesis [93].

Qiu et colleagues investigated the beneficial effect of FA on stemness of human tendon-derived stem cells (hTSCs) and demonstrated that FA treatment promoted proliferation, self-renewal, and multi-differentiation potential (adipogenesis, chondrogenesis, and osteogenesis) in hTSCs cultures, in a dose-dependent manner. However, the authors suggest that, in in vitro experiments, high FA concentration might present slightly adverse effects on cells suggesting that FA beneficial action falls within an optimal range [94]. Then, the critical involvement of HIF-1α in mediating the FA-elicited pro-stemness effect on hTSCs was further disclosed by a specific knockdown assay, which readily abolished this beneficial influence of FA [94].

Another natural product is celastrol, extracted from the Tripterygium wilfordii Hook, which, together with a variety of biological effects [95], appears able to stimulate hypoxia by HIF-1α stabilization protein synthesis [96]. It has been demonstrated that in hTSCs the celastrol treatment induced in vitro hypoxia via HIF-1α accumulation and significantly enhanced stemness of hTSCs in a HIF-1α dependent manner. Specific knockdown assay confirmed not only the function of HIF-1α in mediating celastrol pro-stemness effect on hTSCs, but also identified the mechanism of celastrol action in the HIF-1α-Smad7 axis pathway [96].

Additionally, an environmental factor such as ultraviolet A irradiation (UVA) is considered as hypoxia inductors on cells since they modulate HIF-1 expression. Indeed, the HIF-1 pathway seems to be susceptible to UVA which exerts an adverse effect on cells by promoting senescence [97], reducing the expression of stemness genes by activation of Prostaglandin E2 (PGE2)-cAMP-HIF-1α signaling [98]. UVA reduces the expression of stemness genes such as *Oct-4*, *Nanog*, and *Sox-2* through the downregulation of HIF-1α. Using an HRE-luciferase reporter assay it has been shown that the UVA irradiation reduced mRNA level of HIF-1α, negatively modulating stemness genes [99]. With a screening tool, several molecules have been selected as ideal candidates for reverting the negative UVA effect on stemness, by recovering HIF-1α stabilization via the inhibition of the PGE2-cAMP signaling in hMSCs from adipose tissue among which there are sinapic acid [100], aspartic acid [99], arctigenin [101], and ethylcystein [98].

Most interestingly, some other factors that are commonly used in cell cultures, such as glucose, normally added to the culture medium, can modulate HIF-1α protein expression by reducing the capacity to dissolved O_2 [59].

Several studies demonstrated that high-glucose may influence HIF-1α expression in various mammalian cells. As reported in literature, high glucose concentrations can increase intracellular superoxide levels leading to reduction of HIF-1α expression [102].

In nucleus pulpous-derived mesenchymal stem cell (NPMSC), cultured in high glucose condition, a lower expression of HIF-1α has been detected compared with NPMSC cultured with low glucose. At the same time, high glucose concentrations induce senescence and significantly decrease proliferation and stemness maintenance as indicated by the reduction of stemness genes expression (*Sox-2*, *Nanog*, and *Oct-4*), and this effect could be linked to HIF-1α reduction [103].

3.4. In Vitro Induced Hypoxia Using Multifactorial Approaches

Considering the difficulty of keeping a constant level of O_2, especially working within hypoxic chambers, it could be useful to add chemical compounds in a short range of time to preserve the stabilization of HIF protein. In this context, Večeřa and colleagues, induced hypoxia in NSCs cultivating them in an anaerobic chamber with 1% of O_2 and contextually treating cells with 300 μM of $CoCl_2$ for 6 h. Under this condition the NSCs were able to display their stemness features [104]. It has been shown that $CoCl_2$ can induce stemness maintenance in MSCs promoting the expression of stemness markers as Oct-4, Sox-2, or Nanog [105]. On the other hand, $CoCl_2$ can limit MSCs expansion inducing significant apoptosis due to the loss of the downstream nuclear factor of erythroid-derived 2-like 2 NRF2 [106]. In this context, the combinatorial overexpression of NRF2 and treatment with $CoCl_2$ could restore the maintenance of MSCs characteristics, promoted by $CoCl_2$ treatment, preventing apoptosis [106]. Similarly, in BMSCs, IGF-1 overexpression could restore Oct-4 and Nanog expression that decreased under 1% O_2 condition, performed with AnaeroPack system [69].

The possibility of combinate low O_2 culture with other culture conditions represents a system for the improvement of cellular response. For example, beneficial effects of hypoxic exposure combined with treatment with 1.8 mM of calcium ions (Ca^{2+}) have been reported on proliferation and self-renewal ability of small MSCs that showed also higher resistance response to passage-dependent senescence [107]. In the same way, in hUCB-MSCs treatment with hypoxia and Ca^{2+} exposure increased proliferation without losing Oct-4 and Nanog stem cell markers expression that resulted significantly higher in comparison to same cells treated with Ca^{2+} or 3% O_2 concentration alone. This combinatory approach was able to also enhance the hUCB-MSCs differentiation potential suggesting that the synergistic effect of Ca^{2+} and hypoxia in stem cells was dependent on HIF-1α expression and its downstream extracellular signal-regulated kinase (ERK) pathway [108].

Hypoxic culture can be easily coupled with seeding on chitosan films, as reported for cultured equine umbilical cord mesenchymal stem cells (UCM-MSCs), which showed an increase of *Oct-4*, *Sox-2*, and *Nanog* genes expression, after 7 days of culture [109]. These synergistic effects strongly support that the hypoxia and factor combinatory approach could be considered a good strategy to enhance the stemness potential of these cells to improve their positive response in healing tissues [110].

In a recent study performed in primary human Wharton's jelly MSCs, hypoxia has been induced by a combination of 5% O_2 levels with a pressure stimulus 2.0 PSI by using an AVATAR system increase of proliferation rate of cultured cells [111].

However, it is considered that not always the combination of more factors results in enhanced positive effects for cells. Indeed, even if hMSCs 3D spheroids and low O_2 culture were able to enhance stemness gene expression when used in e separate manner compared to flat substrate culture, their combination was not able to increase hMSCs stemness markers expression while maintaining Oct-4, Rex-1, and Sox-2 at constant levels also in spheroids exposed to different O_2 concentration [112].

Table 2. Summarized scientometric research articles for hypoxia induction and its effect on stemness modulation, in in vitro cell cultures.

Cell Source	Chemical Induction	Time of Exposure	HIF Analysis	Hypoxic Effect on Stemness	Reference
rat BM-MSCs (bone marrow derived MSCs)	BMC-CM (bone marrow cells conditioned medium)	1,3,5 passages	RNA expression (*HIF-1α*)	BMC-CM increases HIF-1α which suppresses OXPHOS activity and activates the anaerobic glycolytic metabolic pathway.	[113]
rat NPMSCs (Nucleus Pulposus mesenchymal stem cells)	High Glucose 4, 5 g/L vs. Low Glucose 1 g/L	3 passages	Protein expression (HIF-1α)	A significantly decreased expression of HIF-1α, Oct-4, Sox-2, Nanog, and GLUT-1 were found in High glucose NPMSCs in comparison to low glucose NPMSCs.	[103]
human TSCs (tendon stem cells)	Ferulic Acid (2, 10, 15 μM)	48 h	RNA, protein expression ChIP assay (HIF-1α)	Increase of self-renewal ability: colony number and average colony size were markedly increased in response to FA treatment.	[94]
human TSCs (tendon stem cells)	Celastrol (1, 2, 4 μM)	24 h	RNA, protein expression, ChIP assay (HIF-1α)	Improved self-renewal capacity evaluated through the proliferation rate and colony formation assays.	[96]
human SHED (Stem cells from human exfoliated deciduous teeth)	$CoCl_2$ (50, 100 μM)	3 days	Protein expression (HIF-1α)	Increase of stemness markers expression (Oct-4, Sox-2, Nanog, and c-Myc).	[114]
human DPSCs (Dental pulp stem cells)	$CoCl_2$ (10 μM)	48 h	RNA expression (*HIF-1α*)	Increase of stemness markers expression. (Oct-4 and Sox-2)	[115]
human UC-MSC (Umbilical cord derived mesenchymal stem cells)	$CoCl_2$ (5, 10 μM)	12, 24, 72 h	Protein expression and immuno-detection	Increase Nanog and NRF2 (nuclear factor erythroid-derived 2-like 2) expression. However, $CoCl_2$ limited MSCs expansion as it induced significant apoptosis that can be recovered with NRF2 overexpression.	[106]
human DPSCs (Dental Pulp Stem cells)	$CoCl_2$ (25 μM, 50 μM)	3 d	n.d. (Treatment with Apigenin, an HIF inhibitor, reverts $CoCl_2$ effects)	Increase of stemness markers expression, significant with 50 μM (Rex-1, Oct-4, Sox-2, and Nanog).	[105]

Table 2. *Cont.*

Cell Source	Chemical Induction	Time of Exposure	HIF Analysis	Hypoxic Effect on Stemness	Reference
human AAA-MSCs (Mesenchymal stem cells from abdominal aortic aneurism)	$CoCl_2$	24 h, 48 h, 72 h	n.d.	No differences in stemness gene expression. Stemness profile is unaffected by hypoxic treatment.	[116]
human PDLSCs (Periodontal ligament stem cells)	$CoCl_2$ (50 µM, 100 µM)	1, 3, 7 d	Protein expression (HIF-1α and HIF-2α)	Increase of stemness markers expression (7dRex-1 and Oct-4).	[117]
human ADSC (Adipose derived mesenchymal stem cells)	Arctigenin (1, 10, 50 µM)	3 d	RNA and protein expression	Increase stemness markers expression by antagonizing UVA irradiation effect. The effects of arctigenin are mediated by PGE2-cAMP signaling-dependent upregulation of HIF-1α.	[101]
human AMSCs (Adipose tissue derived mesenchymal stem cells)	Sinapic Acid (20, 200, 400 µM)	3 d	RNA and protein expression	Increase stemness markers expression by antagonizing UVA irradiation effect. The effects of sinapic acid are mediated by PGE2-cAMP signaling-dependent upregulation of HIF-1α.	[100]
human AMSCs (Adipose derived mesenchymal stem cells)	Aspartic Acid (1, 10, 100 µM)	3 d	RNA and protein expression	UVA irradiation decreases stemness via HIF-1α downregulation. Aspartic Acid increases stemness marker via upregulating HIF (antagonizing UVA irradiation).	[99]
human ASCs (Adipose derived mesenchymal stem cells)	Ethylcystein (1, 10, 200 µM)	3 d	RNA and protein expression	UVA irradiation decreases stemness via HIF-1α downregulation. Ethylcysteine recovers stemness by increasing HIF-1α levels.	[98]
Cell Source	**3D Cultures**	**Time of Exposure**	**HIF Analysis**	**Hypoxic Effect on Stemness**	**Reference**
Human MSCs, Human TMSC (Turbinate Mesenchymal Stem Cells) human ADSC (adipose derived stem cells)	3D cultures	7 d	n.d.	stemness is related to spheroids size. ADSCs expressed stemness markers Oct-4 and Nanog.	[118]
human DPCs (Dental pulp cells)	3D cultures	1, 4, 15 d	n.d.	compared to monolayer DPCs, spheroids showed higher expression levels of stem cell markers, Nanog, CD44, and TP63.	[119]
mouse GMSCs (Gingiva derived mesenchymal stem cells)	3D cultures	Up 3 d	Immunodetection (HIF-1α and HIF-2α)	Increase of stemness markers expression Oct-4 and Nanog.	[82]

Table 2. *Cont.*

Cell Source	Chemical Induction	Time of Exposure	HIF Analysis	Hypoxic Effect on Stemness	Reference
human UCMSCs (umbilical cord derived mesenchymal stem cells)	Hypoxic chamber (n.d.)/3D culture	2 days	Protein expression	Maintenance of stemness is related to the 3D cultures. Hypoxia is used to test resistance to hypoxic stress.	[77]
Cell Source	**Combinatorial Methods**	**Time of Exposure**	**HIF Analysis**	**Hypoxic Effect on Stemness**	**Reference**
mouse NECs (neuroepithelial cells)	Hypoxic chamber 1% O_2 + $CoCl_2$ 300μM	13 d + 6 h $CoCl_2$	Protein expression + co-IP and ChiP (HIF-1α)	Preservation of neural stemness via *Hes1* (hairy enhancer of split 1). HIF deficient-neurospheres showed reduced self-renewal properties and decreased expression of Tbr2, a marker of proliferating basal progenitors.	[104]
human UCB-MSC (umbilical cord blood MSCs)	3% O_2 + 1.8 mM Calcium	5 days	n.d.	Increase of stemness markers expression related to primitive stem cells including Oct-4, Nanog, STELLA, SALL-4, and BMI-1.	[107]
horse UCM-MSCs (umbilical cord matrix derived mesenchymal stem cells)	Incubator 5% O_2 + seeding on chitosan films	3 d and 7 d	n.d.	Increase of spheroids formation and size; increase of stemness markers expression, Oct-4, Sox-2, and Nanog.	[110]
horse UCMSCs (Umbilical cord derived mesenchymal stem cells)	Incubator 5% O_2 + seeding on chitosan films	3 d and 7 d	n.d.	7 d Hypoxic cultures of + seeding on chitosan films increases stemness	[109]
human PMSCs (Placenta derived stem cells)	Incubator 5% O_2 + HIF2α over-expression	4 h/24 h	Protein expression (HIF-2α)	Increase of stemness markers expression, CCND1 (CyclinD1), c-Myc, and POU5F1 (Oct4).	[120]
human MSCs (Mesenchymal stem cells)	Incubator 2% O_2 /3D cultures	From 24 h to 96 h	n.d.	Oct-4, Rex-1, Sox-2, and Notch-1 levels did not change significantly in spheroids, between different O_2 culture conditions. However, compared to flat substrate culture, *Sox-2* and *Notch-1* gene expression was significantly increased in low O_2 spheroids.	[112]
human ATSCs (Adipose tissue stromal cells)	Hypoxic chamber 1% O_2 + 10 ng/mL DHP (4-3,4-dihydroxyl phenyl)	2–6 h	Protein expression (HIF-1α)	Hypoxia + DHP increases stemness markers by inducing de-differentiation on ATSCs. De-differentiated ATSCs overexpress Oct-4, Sox-2, Nanog, Rex-1, and c-Myc.	[121]
human WJMSCs (Wharton's jelly MSCs)	Incubator 5% O_2 + pressure stimulus 2.0 PSI	24, 48, 72 h	n.d.	Increase of proliferation rate under hypoxic condition + pressure stimulus.	[111]

Table 2. *Cont.*

Cell Source	Chemical Induction	Time of Exposure	HIF Analysis	Hypoxic Effect on Stemness	Reference
Cell Source	**Physical Hypoxia**	**Time of Exposure**	**HIF Analysis**	**Hypoxic Effect on Stemness**	**Reference**
human MSCs (Mesenchymal stem cells)	Culture Pal System <0.1% and 5% O_2	24 h, 72 h	n.d.	No stemness markers but low O_2 suppresses cell senescence through down-regulation of $p16^{INK4A}$ mRNA expression.	[71]
mouse TSCs (Trophoblast stem cells)	Anaerobic bags 0% O_2; Incubator 0.5% and 2% O_2	12 h	n.d.	0.5% $O_2 \times$ 12 h causes loss of ERRB2 and ID2 (specific stemness markers); while 2% $O_2 \times$ 12 h maintains potency.	[70]
mouse TSCs (Trophoblast stem cells)	Hypoxic chamber 0.5% O_2	from 1 to 6 d	n.d.	0.5% O_2 reduces stemness in favor of differentiation.	[122]
rat BMSCs (Bone marrow stem cells)	AnaeroPack system in anaerobic jar 1% O_2	48 h	n.d.	Hypoxia decreases stemness. Overexpression of IGF1 maintains stemness under hypoxia.	[69]
human HSPCs (hematopoietic stem/progenitor cells) coculture with WJMSC	Incubator 1% O_2	7 days	RNA expression (*HIF-1α—HIF-2α*)	Hypoxia activates the Notch/Wnt/Hedgehog signaling pathway which plays an important role in preserving stemness.	[123]
human ESCs (Embryonic stem cells)	Incubator 1% O_2	-	HIF-1α is one of the top HUB gene	Hypoxia activates pathway involved in stemness maintenance.	[124]
rat MMSCs (Metanephric Mesenchymal stem cells)	Incubator 1% O_2	3 d	Protein expression (HIF-1α)	Decrease of stemness markers (Six-2/Cited-1 are specific markers of MMSCs).	[125]
human BMMSCs (Bone Marrow Mesenchymal stem cells)	Hypoxic chamber 1% O_2	1–2 weeks	n.d.	Increase of Oct-4 expression.	[126]
mouse BMMSCs (Bone Marrow Mesenchymal stem cells)	Incubator 1% O_2	6 weeks	RNA expression (*HIF-1α—HIF-2α*)	Increase of c-myc expression and colony numbers.	[127]
human ob/nV-ASCs (Adipose stem cells from Visceral fat of obese individuals/non obese)	Hypoxic chamber 1% O_2	2, 4, 8 h	RNA and protein expression (HIF-1α)	obV-ASCc obV-ASC, which showed a less stem-like phenotype, recovered stemness features after hypoxia. Increase of KLF4 and Oct-4 expression (after 8 h of hypoxia).	[128]
human MSCs (Mesenchymal stem cells)	Incubator 1% O_2	7 d	n.d.	Increase of stemness markers expression, Oct-4, Klf4, and c-myc.	[129]

Table 2. *Cont.*

Cell Source	Chemical Induction	Time of Exposure	HIF Analysis	Hypoxic Effect on Stemness	Reference
human BMMSCs (Bone Marrow Mesenchymal stem cells)	Incubator 1% O_2	1, 3, 5, 7 d	n.d.	Increase of stemness markers expression (7 d) Oct-4, Nanog, Klf4, and Sall4. Hypoxia increases proliferation and cyclin D1 (CCD1) expression.	[130]
human ASCs (Adipose derived mesenchymal stem cells)	Incubator + hypoxic workstation 1% O_2	21 d	n.d.	Expansion of ASCs in 20% O_2 led to a significant decrease in T/S ratio (relative length of the telomeres) compared to 1% O_2. ASCs in 1% O_2 proliferates faster, shows reduced aging, and preserves stemness.	[131]
mouse SPCs (Cochlear stem progenitor cells)	Incubator 1% O_2	24 h	RNA and protein expression (HIF-1α)	Increase of stemness markers expression, Nanog, Oct-4, and Musashi1.	[132]
human BMMSCs (Bone Marrow Mesenchymal stem cells)	Hypoxic chamber 1% O_2	2 passages /Until senescence	n.d.	Increase of stemness markers expression, Nanog, Oct-3/4, and Sox-2.	[133]
mouse ESCs (Embryonic stem cells)	Hypoxic chamber 1% O_2	24 h	RNA expression (*HIF-1α*)	Hypoxia favors differentiation through *H2afz* gene downregulation. While *H2afz* overexpression maintains stem markers (Nanog, Rex-1, andFgf4)	[134]
human ADMSCs (Adipose derived mesenchymal stem cells)	Hypoxia Incubator chamber 1–3% O_2	7 d	n.d.	Nanog and Sox-2 increased under low O_2 tension, although the differences were not statistically significant.	[135]
Mouse ESCs (Embryonic stem cells)	Incubator 1% O_2 or 5% O_2	24 h, 48 h	RNA and Protein expression (HIF1-α and HIF-2α)	Reduction of stemness.	[92]
human MSC in co-culture with HUVEC (human umbilical vein endothelial cells)	Incubator 2% O_2	7 d	RNA expression (*HIF-1α*)	Expression of stemness genes was lowered due to adipogenic differentiation of MSCs.	[136]
human ASCs (Adipose derived stem cells)—fresh vs. cryopreserved cultures	Incubator 2% O_2	3 passages	RNA expression (*HIF-1α*)	Increase stemness markers (Nanog, Sox-2, Oct-4, and Rex-1) especially in fresh hypoxic cultures vs. fresh normoxic ones.	[137]
human ASCs (Adipose derived mesenchymal stem cells)	Incubator 2% O_2	3 passages?	RNA expression (*HIF-1α*)	Increase of proliferation and stemness markers expression, Rex-1, Oct, Sox-2, and Nanog).	[138]

Table 2. *Cont.*

Cell Source	Chemical Induction	Time of Exposure	HIF Analysis	Hypoxic Effect on Stemness	Reference
human PDLSCs (Periodontal ligament stem cells), DPCs (Dental pulp cells)	Hypoxic chamber 2% O_2	24 h–1 w	Protein expression (HIF-1α)	Increase of stemness markers expression (Oct-4, Sox-2, and c-Myc) and 3D niche-like structures.	[139]
human HUCPVCs (Umbilical cord perivascular cells)	Incubator 2% O_2	2 w	Protein expression (HIF-1α and HIF-2α)	Increase of Oct-4 expression and colonies number.	[140]
mouse ASCs (Adipose derived mesenchymal stem cells)	Incubator 2% O_2 (100 mM $CoCl_2$ used aspositive control for HIF)	Passages 1, 5, 8over 6 w	Protein expression (HIF-1α)	Increase of stemness markers expression, Oct-3/4 and Nanog, at passage 5.	[141]
human MSCs (Mesenchymal stem cells)	Incubator 2% O_2	52–64 d	n.d.	Hypoxia prevents senescence since hypoxic hMSCs maintained their homogenous rapidly self-renewing morphology for up to 52 days.	[142]
rat CDCs (Cardiosphere derived cells)	Incubator 2% O_2or DMOG 1 mM or BIC 30 µM	P2 until 80% confluency	RNA and protein expression (HIF-1α)	Increase of stemness markers expression: Oct-4, Sox-2, Klf-4, Nanog, and c-Kit. Increase of GLUT-1 expression, lactate production, and glucose uptake	[143]
human UC-MSC (Umbilical cord derived mesenchymal stem cells)	Incubator 2.5% O_2	15′ 2,5% O_2 + 30′ 21% O_2 + 72 h 2.5% O_2	n.d.	Absence of stemness markers both in hypoxic and normoxic group: no expression of *Oct-4*, *Nanog*, and *Nt-3* genes were detected.	[144]
human ASCs (Adipose derived mesenchymal stem cells)	Incubator chamber 2–3% O_2	6 d	n.d.	Increase of stemness markers expression, Nanog and Sox-2.	[145]
human WJ-MSCs (Wharton's jelly mesenchymal stem cells)	Hypoxic chamber 3% O_2	3–6 days	RNA expression (*HIF-1α*)	Increase of stemness markers expression, of Nanog, Oct-4A, Oct-4B, and Sox-2. Enhanced WJ-MSCs clonogenicity and expansion capacity.	[146]
human DPCs (Dental Pulp Stem cells)	Hypoxic chamber 3% O_2	2 days	Protein expression (HIF-1α)	The expression of the cell surface markers, CD133, CD34 is increased, while CD105 and Oct-4 do not change significantly. Increase of colony forming units. ROS reduction under hypoxia.	[147]
human CDCs (Cardiac stem cells)	Incubator 3% O_2	48 h	n.d.	Increase of c-Kyt positive cells (Most primitive and undifferentiated population of cardiac stem cells).	[148]

Table 2. *Cont.*

Cell Source	Chemical Induction	Time of Exposure	HIF Analysis	Hypoxic Effect on Stemness	Reference
Mouse EG (Embrionic Germ cells) PGC (Primordial germ cells)	Incubator 3% O_2	1 d, 3 d, 7 d	Protein detection (immunofluorescent Staining for HIF-1α)	Hypoxia exposure does not induce Klf4 or c-Myc upregulation in PGCs. Hypoxia promotes a metabolic switch from OxPhos toward glycolysis in PGCs and hypoxic PHCs showed reduced ROS levels vs normoxic ones.	[149]
human RPCs (Retinal Progenitor cells)	Incubator 3% O_2	1, 5, 10 passages	Protein expression (HIF-1α and HIF-2α)	Increase of stemness markers expression, Oct-4, Sox-2, c-Myc, and Klf-4	[150]
mouse BMMSCs (Bone Marrow Mesenchymal stem cells)	Incubator 3% O_2	7 d	n.d.	Increase of stemness markers expression, Oct-4 and Rex-1. 3% O_2 augmented the amount of colony-forming cells by 1.6-fold vs. 21% O_2.	[151]
human MSCs (Mesenchymal stem cells)	Incubator 3% O_2	p0–p4 passages?	RNA expression (*HIF-1α*)	Hypoxia decreases differentiation potential and increases colony formations number. Under 20% the expression of Nanog and Rex-1 decreases and MSCs show a more senescent phenotype (evaluated with senescence associated markers expression).	[152]
human MIAMI cells (marrow-isolated adult multilineage inducible cells)	Incubator 3% O_2	3 w	RNA expression (*HIF-1α*)	Increase of stemness markers expression, Oct-4 and Rex-1.	[153]
buffalo ASCs (Adipose derived stem cells)	Hypoxic chamber 5% O_2	3 or 6 passages	RNA expression (*HIF-1α*)	Increase of stemness markers expression, Oct-4, Nanog, and c-Myc.	[154]
human RPCs (Retinal progenitor cells)	Incubator 5% O_2 with or without KOSR (Knock-out serum replacement)	Until confluency	n.d.	Increase of stemness markers expression, c-Myc and Oct-4, in both conditions(With or without KOSR) vs. normoxia.	[155]
Urine stem cells (USC), Dental pulp stem cells (DPSC), Amniuotic fluid stem cells (AFSC), Bone Marrow stem cells (BMSCs)	Hypoxic chamber 5% O_2	5 d	RNA expression (*HIF-1α*)	Increase of stemness markers expression (Oct-4, c-Myc, Nanog, and Nestin). Hypoxia showed also increased proliferation rate, inhibition of senescence, and increased differentiation ability.	[156]
human PMSCs (Placenta derived stem cells)	hypoxic chamber 5% O_2	3 d	n.d.	Increase of stemness markers expression, Oct-4, Sox-2, and Nanog.	[157]

Table 2. *Cont.*

Cell Source	Chemical Induction	Time of Exposure	HIF Analysis	Hypoxic Effect on Stemness	Reference
dog ADMSCs (Adipose derived Mesenchymal stem cells)	hypoxic chamber 5% O_2	3rd passage	RNA expression (*HIF-1α*)	Increase of stemness markers expression, Oct-4, Sox-2, and Nanog,	[158]
human ESCs (Embryonic stem cells)	Incubator 5% O_2	3 passages (followed by -reoxigenation)	RNA expression and ChIP analysis (*HIF-2α*)	HIF-2α enhances stemness by binding to an oct-sox cis-regulatory element in the promoter region of Nanog.	[159]
human WJMSCs (Wharton Jelly's mesenchymal stem cells)	Incubator 5% O_2	2–4 w	RNA expression (*HIF-1α* and *HIF-2α*) + immunodetection (HIF-1α)	Increase of stemness markers expression, Oct-4A, Sox-2, Rex-1, and Nanog.	[160]
mouse GSCs (Germline stem cells)	incubator 5% O_2	7 d	Protein expression (HIF-1α and HIF-2α)	Increase of stemness markers expression, Oct-4, Sox-2, Klf-4, and Nanog.	[161]
human TSCs (Tendon stem cells)	Incubator 5% O_2	3–5 d	n.d.	In cultures under 5% O_2, more hTSCs expressed the stem cell markers nucleostemin, Oct-4, Nanog and SSEA-4.	[162]
human ESCs (Embryonic stem cells)	Incubator 5% O_2	7–14 d	n.d.	No major difference in representative stemness genes (*Oct-3/4*, *Nanog*, and *Cripto*),	[163]
human TSCs (Tendon stem cells)	Incubator 0.5–5–10% O_2	2 passages	n.d.	10% and 5% O_2 increases stemness markers expression (5% has emerged as the optimal concentration) and both increases colony formation and number. 0.5% O_2 decreases stemness.	[164]
mouse iHepSCs (Induced-hepatic stem cells)	Incubator 10% O_2	6 h, 24 h, 72 h, 120 h	RNA and protein expression (HIF-1α and HIF-2α)	Enhanced stemness through faster progenitor proliferation rate; hypoxia accelerate G1/S transition through HIFs activation. Moreover, p53 and p21 in hypoxia-cultured iHepSCs were inhibited and CDKs (ciclin D kinase complexes) were upregulated.	[165]
mouse iSCs (ischemia induced stem cells) from brain ischemic regions	Ischemic areas/ cerebral infarction	-	n.d.	Cells from ischemic region, as well as control ES cells, exhibited pluripotency markers: c-Myc, Klf-4, Sox-2, and Nanog.	[53]

Table 2. *Cont.*

Cell Source	Chemical Induction	Time of Exposure	HIF Analysis	Hypoxic Effect on Stemness	Reference
Cell Source	**Other**	**Time of Exposure**	**HIF Analysis**	**Hypoxic Effect on Stemness**	**Reference**
Human WJ-MSCs (Wharton's Jelly MSCs)	None	-	RNA expression (*HIF-1α* and *HIF-2α*)	Gene expression related to the stemness properties showed differences according to ALDH activity. *HIFs*, *Glut-1* and stemness gene (*Oct-4*, *Nanog*, *and Rex-1*) are more expressed in ALDH+ positive cells compared to ALDH- cells.	[166]
Human AT-MSCs (Adipose Tissue MSCs)	None	-	RNA expression (*HIF-1α* and *HIF-2α*)	*HIFs*, *Glut-1*, and stemness genes (*Oct-4*, *Nanog*, and *Rex-1*) are more expressed in ALDH+ positive cells compared to ALDH- cells.	[167]
Human FSK-MSCs (Foreskin derived MSCs)	None	-	RNA expression (*HIF-1α* and *HIF-2α*)	*HIFs*, *Glut-1* and stemness gene (*Oct-4*, *Nanog* and *Rex-1*) are more expressed in ALDH+ positive cells compared to ALDH- cells.	[168]
Human BMMSCs (Bone marrow MSCs)	None	-	RNA expression (*HIF-1α* and *HIF-2α*)	*HIF* and stemness gene (especially *Oct-4* and *Rex-1*), *HIFs* and *Glut-1* are more expressed in ALDH+ positive cells compared to ALDH- cells.	[169]

4. Exploitation of Hypoxia for Regenerative Medicine Purposes

The potential application of stem cells for tissue regeneration represents an important challenge but it is necessary to optimize cell culture conditions to preserve the desired stem cell properties. As reported, some in vitro parameters can influence stem cell characteristics and reduce cellular proliferation supporting senescence [170]. One prominent characteristic of stem cells is their natural aptitude to reside in low O_2 condition niches. Of note, efforts have been made in the last years to exploit the possibility to grow these cells in hypoxic conditions to mimic the naive microenvironment looking for valid in vitro culture protocols improving the stemness phenotype [171]. The regulation of self-renewal properties by O_2 can also indirectly depend on HIF stabilization (in Section 1.3) as it represents a key determinant of the activation of stemness genes (such as *Oct-4*, *Sox-2*, or *Nanog*) and metabolic-related factors [172].

O_2 can even directly regulate the stem cell fate since it is a cellular metabolic substrate and pluripotency is also characterized by specific metabolic and mitochondrial responses. Although most studies explain the hypoxia induction mechanisms related to stemness preservation especially in pluripotent cells, as ESCs or pluripotent cells (PSCs), several types of cells benefit from low O_2 gradient for stemness retention as we further describe in Chapter 5. However, a briefly introduction on the role of stemness expression markers and metabolism in regulating stem cell fate and how hypoxia fits in stemness control are reported in the following sections.

4.1. Pluripotency-Related Markers of Stem Cells

Pluripotency requires the expression of important transcription factors such as *Oct-4*, *Sox-2*, and *Nanog* which are pivotal to orchestrate a complex interdependent transcriptional network in SCs [173] that might direct cell identity, as shown with genome-wide studies [174]. In addition, several proteins were identified as implicated in the control of cell self-renewal (Esrrb and Zfx), in the regulation of cell cycle progression (E2F1, c-Myc, and Klf4), and the maintenance of cell wellness (BMP-induced Smad1 and LIF) [124]. All these factors are also strictly correlated in their expression; for example, Nanog is an upstream regulator of the signal transducer and activator of transcription factor 3 (STAT3) and Oct-4, and its regulation of pluripotency and cell differentiation occurs through the interaction with a transcription factors complexes containing *Oct-4*, *Sox-2*, and *Klf4* [175]. Furthermore, Nanog and Oct-4 also can interact and co-occupy target genes of Nanog suggesting a cooperation of these transcription factors in the control of gene expression [175].

Although Sox-2, Oct-4, and Nanog are probably the most investigated stemness markers, Rex-1 is another factor which, in combination with the above-mentioned ones, plays a functional role in promoting cell cycle progression and stemness maintenance [176].

Together with genetic factors, epigenetic modifications contribute to create stem cells heterogeneity through different mechanisms such as DNA methylation, nucleosome remodeling, and post-translational modifications on histone tails generating a high variability of enhancers, conferring specific gene expression to cellular types [177].

Given that pluripotency is a dynamic state, the development and maintenance of stem cells is strictly dependent on a synergic regulation of these principal transcription factors expression, their epigenetic modifications and cellular localization [178].

4.2. Metabolic State of Stem Cells

The metabolic state of stem cells is characterized by a specific profile represented with a dynamic mitochondrial morphology shift from glycolysis to mitochondrial oxidative phosphorylation (OxPhos) when cells pass to a more differentiated phenotype [179]. Therefore, the undifferentiated stem cells are characterized by glycolysis instead of oxidative phosphorylation [180] (Figure 6).

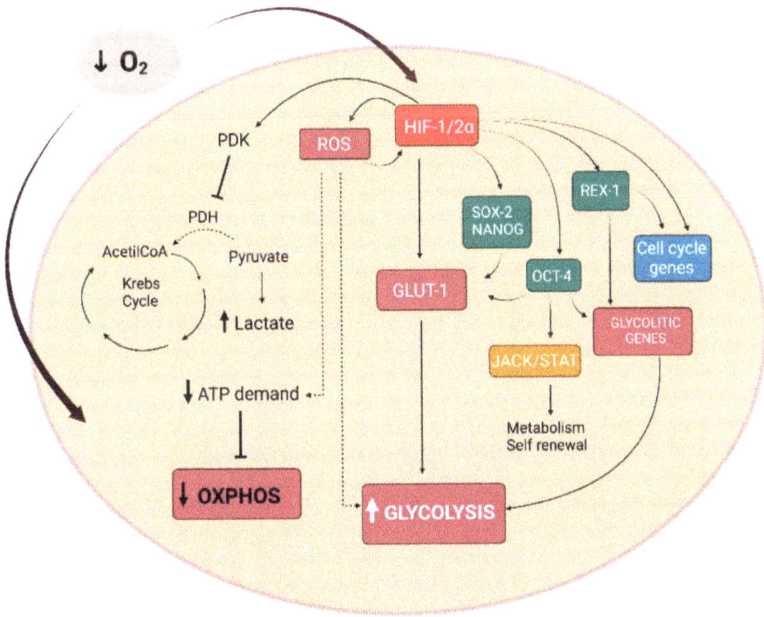

Figure 6. Summary of stem cell metabolic state during hypoxia condition and molecular pathways linking HIF-stemness genes-metabolism switch to glycolysis.

An elevated glycolytic flux is common to highly proliferating cell types, suggesting that this process has a central role in the acquisition and maintenance of pluripotency [181]. Self-renewing stem cells have significantly lower levels of mitochondrial activity, antioxidant enzymes, oxidative proteins, ROS levels, and lipid hydroperoxides [182]. So, manipulating metabolic pathways, with either genetic approaches or drugs or culture conditions, can directly affect whether stem cells remain quiescent, self-renew, or differentiate [183].

It has been described how ESCs and iPSCs, as totipotent stem cells, are characterized by high plasticity and can potentially be directed to any cell type. Certainly, the metabolic characteristics of iPSCs are not completely the same as ESCs, but it is not a detail that both primarily rely on glycolysis to meet energy requirements, in contrast to their somatic counterparts [184]. The set of evidence that correlate "stemness" and "metabolism" has driven the development of new systems that can be adopted to generate iPSC relying on small molecules that enhances cellular reprogramming through the promotion of glycolytic metabolism [185]. Coherently with this data, it has been reported that fibroblasts preconditioning with culture medium containing lactate resulted in a switch from OxPhos to glycolysis, in part, through ROS-mediated stabilization of HIF-1α [186]. Interestingly, reverting cells to an immature state, required the expression of factors involved in mitochondrial biogenesis, morphology and distribution, intracellular ATP production, and lactate generation, supporting the role of the metabolic state in stem cell fate [184]. Consistent with this aspect, the transcription factors involved in stemness preservation, such as Oct-4, also take part in the metabolic control as the key reprogramming factor Oct-4 has been shown to target multiple metabolic genes [187]. It has been demonstrated for example that in developing mice embryos, Oct-4 activate the JAK/STAT signaling pathway thus regulating cellular metabolic properties via energy metabolism, cell morphology, and chromatin accessibility [187]. Moreover, in mouse embryonic fibroblasts (MEFs), Rex-1 stimulates the expression of glycolytic genes, through the cyclin B activation, promoting glycolysis instead of OxPhos [176]. Glycolysis can be maintained by a constant glucose uptake by the glucose transporter 1 (Glut1) whose levels are finely regulated in hESCs [188].

The optimal levels of Glut1 expression are supported by the enhancer element of Glut1 (GE) which is accessible for the pluripotency factors Sox-2, Oct-4, and Nanog which can each bind to GE thus inducing the expression of Glut1 [188].

As mentioned above, the correlation between self-renewal properties and cellular metabolism is evidenced also in different type of multipotent stem cells. For example, no completely differentiated HSCs exhibit fewer mitochondria and higher glycolytic capacity in whole bone marrow [184]. Indeed, the levels of antioxidant enzymes, such as superoxide dismutase (SOD), are higher in circulating progenitor cells than in long-term quiescent HSCs which exhibits enhanced glycolytic status within mitochondrial activity and ROS balance cooperating and finely regulating self-renewal of these stem cells [184].

Moreover, in undifferentiated MSCs too, mitochondrial activities are maintained at low level while, at the same time, glycolytic activities are consistently maintained at high levels. Here, the glycolytic process contributes to greater than 97% of ATP production, in the energy metabolism of undifferentiated bone marrow MSCs [189]. The rapid uptake of glucose in MSC cultures confirms their dependence on glycolysis [190], and, moreover, undifferentiated MSCs produce high levels of lactate, suggesting a reliance on non-aerobic glycolysis to cover the bioenergetic needs [191].

The set of evidence reported in the literature supports the concept that cellular metabolism does not just represent only an energetic state but plays a central role in the determination of stem cell fate.

Finally, the preferential utilization of glycolysis over mitochondrial oxidative metabolism may also represent a mechanism to preserve the genomic integrity through the reduction of ROS production by OxPhos mechanism. Due to the ROS reduction, cells can limit possible damages within nuclear and mitochondrial DNA and even reduce the oxidation of proteins and lipids [192].

Although ROS have been considered just a metabolic waste product, in the past decades accumulating scientific evidence demonstrated their key role in cell fate signaling (Figure 6).

The acronym ROS refers to O_2 reactive species, but it may also include several nitrogen-containing compounds reactive nitrogen species (RNS), such as nitroxyl anion (NO^-), peroxynitrite ($ONOO^-$), and nitric oxide (NO) which are produced by the activity of inducible nitric oxide synthase (iNOS) and reacts with superoxide to give rise to the other RNS [193].

Emerging evidence has demonstrated how modulation of ROS level and metabolic flux has a key role in dedifferentiation processes, as reported for the generation of iPSCs from differentiated cells that benefits from careful regulation of ROS levels [194].

Metabolism can affect signaling pathways through the modulation of ROS levels which can react with various proteins, such as kinases, phosphatases, or transcription factors, to alter processes linked to cell cycle progression, quiescence, or differentiation [195]. In turn, ROS can also regulate metabolic processes such as glycolysis, OxPhos, pentose phosphate pathway activity, and autophagy [196]. This complex metabolic regulation can also occur through the triggering of HIF accumulation which, in a positive feedback loop, stabilizes ROS and enhances Glut1 expression, promoting the metabolic switch in favor of glycolysis [196].

However, ROS levels must be tightly regulated to preserve cellular senescence and proliferation while avoiding a dysregulated ROS production that occurs in pathological conditions [197]. The importance of ROS homeostasis is evidenced in ESCs that present mechanisms for enhanced ROS removal capacity as well as limited ROS production, despite this cell type possess a limited number of mitochondria. Accordingly, a recent study reveals that the human iPSC generation process can effectively reduce the mitochondrial genome copy number and cells have similar ROS levels and antioxidant defenses to those seen in ESCs [198].

Consistently, higher mitochondrial activity and oxidative stress were found as one of the major causes of functional decline in stemness features [184]. Overall, ROS should be considered as signaling molecules orchestrating the crosstalk between metabolism and stem cell fate decisions.

4.3. O_2 for Stemness Preservation

Self-renewal and potency hallmarks of stem cells are influenced by several intrinsic and extrinsic cell factors. As previously reported low O_2 concentration, hypoxia, has been closely linked to the maintenance of stemness. For most cell types, hypoxia acts as a modulator of cell proliferation decreasing the levels of respiratory enzymes meanwhile increasing the production of glycolytic enzymes and lactate, thus enhancing the glycolysis process [179]. In details, hypoxic conditions reduce the availability of molecular O_2 suppressing the activity of the mitochondrial electron transport chain. Cells switch to glycolysis also through HIFs activity, reducing the expression of mitochondrial enzymes and further enhancing the shift to glycolysis by upregulating glucose transporters and glycolytic enzymes [179].

The limited availability of molecular O_2 under hypoxic conditions results in the reduction of the mitochondrial electron transport chain (ETC) activity and cells switch from OxPhos to glycolysis to reach their energetic needs, since it does not require O_2 [199]. On the other hand, in presence of abundant O_2 levels, pyruvate dehydrogenase (PDH) converts pyruvate produced from glycolysis to acetyl coenzyme A (Acetyl-CoA) giving start to the tricarboxylic acid cycle (TCA cycle). This process is regulated by the pyruvate dehydrogenase kinase (PDK) that phosphorylates and inactivates PDH, thus playing a crucial role in metabolic adaptation of cell in response to hypoxia and it is transcriptionally regulated by HIF-α [200].

In addition, PDK attenuates mitochondrial ROS production, which is critical as increases in glycolytic flux can be associated with leakage of electrons from the respiratory chain resulting in unexpectedly elevated ROS levels [201].

Because high O_2 levels contribute to a decline in stem cell properties, low O_2 pressure should reflect the better physiological conditions for the cells and this aspect must be considered when culturing them. Stem cells cultured under hypoxic conditions can enhance their self-renewal ability and retain their pluripotent capacity, as it has been demonstrated in MSCs or HSCs [7]. In a low O_2 culture, MSCs improve the maintenance of their undifferentiated state through the suppression of mitochondrial activity and promote genetic stability [202]. Even more, adult HSCs, residing in low O_2 niches, have a metabolism that is mainly based on glycolysis for the energy demand, and an increase in mitochondrial activity is linked to a decline in stemness [203].

In literature, there are evidence supporting complex link connecting hypoxia, metabolism, HIFs, and several molecules with crucial roles in the regulation of stemness or differentiation (Figure 6). Low O_2 tension can upregulate proliferation and the expression of pluripotency-related genes, probably by mimicking the conditions that stem cells experience in vivo [170]. In turn, these stemness factors can regulate other subordinate genes involved in the metabolic control of stemness [204] allowing the preservation of a more undifferentiated state and genetic stability (Figure 6).

4.4. HIF Role in Stemness

It has been described that stem cells exist in physiological low O_2 environments with a metabolism that relies on glycolysis instead of oxidative phosphorylation [180], and that hypoxia can regulate stem cell plasticity through the action of HIFs. In the HSC stem cell model, it was demonstrated that HIF-1α stabilization is correlated to the maintenance of an anaerobic metabolism through the transcriptional activation of genes regulating glycolysis, such as *GLUT1* and *PDK1* [205,206], and transcriptional inactivation of genes encoding for key mitochondrial electron transport chain enzymes e.g., phosphoglycerate kinase-1 (*PGK1*), or lactate dehydrogenase-A (*LDHA*), and glucose transporters (such as *GLUT1* and *GLUT3*) [200] (Figures 2 and 6).

HIF-1α can regulate the HSCs metabolism after the transcriptional activation by the homeobox transcription factor myeloid ectotrophic viral integration site 1 (*Meis1*) [207], which is important also for transcriptional activation of HIF-2α. The loss of Meis1 in HSCs results in disruption of stem cells quiescence through increased ROS production, increased apoptosis and down-regulation of both HIF-1α and HIF-2α [208]. Thus, HIF-1α and ROS closely work together, along with O_2 homeostasis and energy metabolism, to maintain HSCs function [203].

HIF role in the metabolic status of stem cells can also affect stem cells immunoregulatory properties. HIF-1α silencing in MSCs resulted in affected capability to reduce inflammation and inhibit the generation of pro-inflammatory T cells. This impaired immunosuppressive potential was correlated to the metabolic switch from glycolysis to oxidative phosphorylation and a reduced ability to produce immunosuppressive mediators such as intercellular adhesion molecule (ICAM), interleukin-6 (IL-6), and nitric oxide (NO) [209].

HIF can preserve stemness also preventing apoptosis through the downregulation of p53 involved in cell survival [210]. It was demonstrated that HIF-1α overexpression, induced by transfection, suppressed p53, the downstream factors p21, and increased B-cell lymphoma 2 (BCL2) [211], an anti-apoptotic factor that can be suppressed by p53 [212]. On the other hand, p53 can induce the transcriptional activation of p21 that participates in apoptotic regulation modulating the cell cycle [213].

HIF capability to preserve stemness could be even correlated with aldehyde dehydrogenase (ALDH) activity as observed in Wharton's Jelly stem cells [166]. Adipose-derived stem cells [167], foreskin-derived mesenchymal stromal cells [168], and for bone marrow mesenchymal stromal cell [169]. ALDHs are enzymes responsible for the oxidation of aldehydes to their corresponding carboxylic acids. The main role of ALDHs is the catalysis of aldehydes [214] that can be toxic for the body. ALDHs are also involved in cell proliferation, embryogenesis, development, cell signaling, neurotransmission, protection from UV radiations, drug metabolism, osmoregulation, gene regulation, and redox balance [215,216]. It was shown that $ALDH^+$ sorted stem cells displayed a major hypoxia response compared to $ALDH^-$ stem cells, increasing of HIF-1α and HIF-2α. Moreover, $ALDH^+$ stem cells exhibited an increased mRNA expression of stemness correlated genes *Oct-4*, *Nanog*, *Sox-2*, and *Rex-1* compared to $ALDH^-$ cells [166].

HIF-1α downregulation by UVA irradiation was responsible for the decrease of MSC stemness properties. This effect was due to the upregulation of prostaglandin E2 (PGE2) and its downstream molecule, cyclic adenosine monophosphate (cAMP), through the activations of activator protein 1 (AP-1) and nuclear factor kappa-light-chain-enhancer of activated B cells (NF-κB) [98–101]. The cAMP molecule can reduce the expression of HIF-1α gene through the cAMP response element-binding protein (CREB) activation, downregulating the expression of the stemness genes *Nanog*, *Sox-2*, and *Oct-4*. However, some chemical compounds can attenuate the UVA-induced effects on the expression of the stemness genes by inhibiting p38 MAPK and NF-κB, the upstream factors in the PGE2 production, such as the arctigenin [101], sinapic acid [100], aspartic acid [99], or ethylcysteine [98].

In this regard, it has been demonstrated that HIF-1α is involved in a pathway that influences MSCs proliferation and migration [217]. In particular, hypoxia stimulates UCB-hMSC proliferation through the expression of the fatty acid synthase (FASN) and stearoyl-CoA desaturase-1 (SCD1), two lipogenic enzymes, whose expression was regulated by the HIF-1α/SCAP/SREBP1 pathway. This pathway was able to also induce the phosphorylation of the mammalian target of rapamycin (mTOR) [217] as CDK2, CDK4, cyclin D1, cyclin E, and F-actin expression as well as c-Myc, p-cofilin, profilin, and Rho GTPase, involved in cells cycle [217]. Moreover, stem cell proliferation was also related to HIF-1 phosphorylation, and other different substrate, by phosphatidylinositol 3-kinase (PI3K)/protein kinase B (Akt) signaling pathway activation [147]. Interestingly stemness can be influenced by HIF-2α as it targets specifically Oct-4 [218] that synergically cooperate with Nanog and Sox-2 to maintain stem cells properties and avoid differentiation [219]. Indeed, it was also demonstrated that HIF-2α expression preserved human placenta-derived mesenchymal stem cells (hPMSC) stemness and promoted their proliferation by regulating CyclinD1 (CCND1), c-Myc, and Oct-4 through the MAPK/ERK signaling pathway [120].

However, despite the increasing evidence reported in the literature about the link between stem cell plasticity and HIFs, the exact molecular mechanisms through which HIF influences stem cells preservation are not completely elucidated yet, due to the complexity and multiple crosstalk of signal pathways involved.

5. Hypoxia Cells Models

Hypoxia maintains a slow-cycling proliferation property, reduced oxidative stress, and undifferentiated status in several stem cell populations [220]. Different cellular models have a peculiar way to adapt to O_2 availability probably due to the different naïve niches. Indeed, cells derived from different tissues have different O_2 consumption rate and O_2 concentration can influence their behavior in vitro. Due to the importance to optimize protocols for preserving stemness properties of cultured cells and preventing senescent phenotype that occurs after a long period in culture, in this chapter it has been considered the wide stem cells models (embryonal, fetal, and adult stem cells) studied with different in vitro hypoxia approaches using different O_2 concentration. Table 3 summarized the principal results data specifically focusing the attention of hypoxic effects on growth and maintenance of pluripotency and differentiation.

Embryonal stem cells (ESCs). A growing number of studies confirm that hypoxia plays a role in the human ESCs niche through metabolic shifts and HIF regulatory elements [221]. Accordingly, the protein interaction (PPI) analysis performed by Murugesan and Premkumar [124] indicates specific genes with distinct roles in the regulation of metabolic shift contributing to hypoxic mediated stem cell niche. Moreover, low O_2 tension conditions could reduce the amount of spontaneous cell differentiation that normally occurs in hESCs in vitro, appearing as an important element that can help to maintain cells in a fully pluripotent state [222]. Therefore, it is not completely evident if reduced O_2 tensions are mandatory for the maintenance of full pluripotency. In vivo, inner cell mass normally undergoes differentiation, and therefore, it could be supposed that the maintenance of hESCs in an undifferentiated state does not represent a physiologic condition. This suggests that low O_2 culture may be optimal if the aim is to differentiate the hESCs thus reproducing the physiologic condition of embryo growth in vivo, whereas normoxic cultures may be better for the maintenance of hESCs in an undifferentiated state since hESCs naturally tend to differentiate [223] (Table 3).

Fetal stem cells (FSCs). A promising category of stem cell is now represented by FSCs cells that can be isolated from placenta and extraembryonic tissues. Their intermediate state between adult and embryonic stem cells makes them an ideal candidate to be used for regenerative medicine. Many multipotent stem cells have been isolated from different parts of the placenta, placenta-derived MSCs (PMSCs), and, even, from the amnion, chorion, umbilical cord, and fetal blood [224]. Amnion-derived stem cells can include amniotic epithelial stem cells (AECs), amniotic fluid stem cells (AFCs), and amniotic mesenchymal stem cells (AMCs).

Up to date few studies have focused on the hypoxic PMSCs although the first trimester of human placental occurs in a low O_2 environment, 2–3% O_2 percentage that can even promote stemness and proliferation of the trophoblast lineage stem cells (TSCs), the progenitors of placenta stem cells. These cells appear to be very vulnerable to high O_2 concentration indeed the choice of the proper O_2 physiological levels for TSCs cultures is a crucial aspect [70].

Amniotic stem cells (AECs) have received great interest because of their availability and their multilineage differentiation potential in vitro [225,226] and innate low immunogenicity [227,228] that make cells ideal candidates for tissue repair [229]. AECs, as well as umbilical cord-derived MSCs isolated from the perivascular region of the umbilical cord (HUCPVCs) or Wharton's jelly (WJ-MSCs) positively respond to hypoxia in favor of both stemness retention and differentiation (Table 3).

Table 3. Summarized hypoxia responses of mostly used stem cell sources in in vitro hypoxia cultures.

Stem Cell Source	Hypoxia Effects	Reference
Embryonic stem cells (ESCs)	5% O_2 tension did not change hESCs morphology" on day 7 of the first and fourth passages. After 10–14 days, hESC colonies were thinner and looked better morphologically in 5% O_2, but cells' proliferation was slower, and their sizes were larger. No significant differences in representative stemness genes (*Oct-3/4*, *Nanog* and *Cripto*), differentiation genes (*Desmin*, *α-fetoprotein* and *GDF-9*), and hypoxia-related genes *HIF-1α* and *VEGF*. The short-term stabilization of HIF, mediated by 1% or 5% hypoxia and PHD inhibitors, do not prevent the spontaneous loss of pluripotency of mouse ESCs naturally differentiating. In 1% O_2 condition, mESCs started to differentiate losing their self-renewal ability through the downregulation of H2afz gene that has been identified as new potential target gene involved in the maintenance of pluripotency in ESCs	[92,134,163]
Placental stem cells (PMSCs)	5% O_2 culture not altered morphology of PMSCS, while the expression of pluripotency markers Oct-4, Nanog, and Sox-2 is increased. In 5% O_2 exhibited a more naïve morphology and had a higher proliferative capability and higher HIF-2α expression than hPMSCs cultured in 21% O_2. PMSCs over-expressing HIF-2α showed higher proliferative potential and higher expression of CCND1 (CyclinD1), c-Myc, Oct-4 and components of the MAPK/ERK pathway. In contrast, these genes were down-regulated in the HIF-2α-silenced hPMSCs.	[120,157]
Amniotic Epithelial stem cells (AECs), Amniotic Fluid stem cells (AFCs), and Wharton's Jelly MSCs (WJ-MSC)	AECs positively respond to in vitro hypoxia (1% O_2) as combined islet-cell (IC) and hAECs organoids, cultured under hypoxic conditions, showed considerable protection from cell death, under ischemic stress. This protective effect of hAECs on islet cells can be linked to HIF-1α that orchestrate compensatory responses to hypoxia. 2% O_2 can be a suitable culture condition to induce tenogenesis in AECs as demonstrated by upregulation of tenomodulin in AECs. This suggest that hypoxic culture could be beneficial even for tendon structure formation. After low O_2 conditioning (1% O_2), the secretome of AFCs augments cardiomyocyte proliferation enhancing cardioprotective effects under hypoxic-ischemic conditions proving that O_2 modulation can be an efficient physical cue to produce secretomes enriched in soluble bioactive factors helpful for cardiac applications. 5% O_2 rejuvenates WJ-MSCs which appear less-differentiated, more primitive and faster-growing involving HIF-1α and HIF-2α. In correlation with the up-regulation of HIF-1α and HIF-2α there is an increase expression of stemness marker Oct-4, a direct down-stream target of HIF. Even 3% O_2 is a favorable concentration, enhancing WJ-MSCs clonogenicity and expansion capacity and stemness markers expression.	[146,160,230–232]
Trophoblast stem cells (TSCs)	0.5% O_2 exposure caused a reduction of ERB2 and ID2 expression, specific trophoblast stemness markers, in mouse TSCs, and favored differentiation while, under 2% O_2, the same cells maintained their potency.	[70,122]
iHepSCs	iHepSCs lost their stemness features and presented a diminished cellular expansion in 1% O_2 culture while 10% O_2 enhanced stemness through cell progenitor proliferation in iHepSCs appearing as the optimal concentration.	[165]
Hematopoietic stem cells (HSCs)	Low O_2 tension plays a crucial role in maintaining normal HSC function protecting cells from differentiation or senescence through the hypoxia-Fbxw7 pathway.	[233]

Table 3. *Cont.*

Stem Cell Source	Hypoxia Effects	Reference
Bone Marrow derived MSCs (BM-MSCs).	1% O_2 tension culture determined a delayed progression of cellular senescence through the activation of the serine/threonine kinase AKT pathway; improved the proliferation rate and increased the expression of stemness genes such as Oct-4, Klf4, and Nanog. 1% O_2, through HIF-1α activation, activated the Notch-2-c-myc pathway that is required for the inhibition of senescence and proliferation promotion in mouse BM-MSCs.Under 3% O_2, mouse BMMSCs cultures, exhibited a significative increase of proliferation rate and augmented colony formation number.	[127,129,130,133,151]
Adipose derived MSCs (AMSCs or ASCs).	A range from 1 to 5% O_2 seems to preserve stemness of AMSCs. These cells clearly show a differentiation potential towards the cell type of the mesoderm lineage (adipocytes, chondrocytes, and osteoblasts) right from the early passages in culture. The greater ability to preserve the stemness of hASCs is indicated by the increased expression of stemness genes *Nanog*, *Sox-2*, *Oct-4*, and *Rex-1*. It has been demonstrated that the upregulation of HIF-1α in hASCs in turn activates the downstream target genes resulting in a significantly favored cell proliferation and in the preservation of stemness genes expression.	[131,135,138,145]
Dental pulp stem cells (DPCs), Periodontal ligament cells (PDLCs), and stem cells from human exfoliated deciduous teeth (SHEDs).	2% O_2 favored stemness maintenance with enhanced expression of Oct-4, Sox-2, and c-Myc in PDLCs and DPCs while they inevitably undergo to replicative senescence under current culture conditions (21% O_2), resulting in cellular phenotypic changes. 3% O_2 showed PI3K/Akt pathway activation and inhibition of oxidative stress in a ROS-dependent manner suggesting that regulation of self-renewal in DPCs may involve ROS control. Cobalt Chloride treatment promoted stemness retention, in a dose dependent manner, in DPCs, PDLSC, and SHEDs.	[105,114,115,117,139,147]
Tendon stem cells (TSCs)	O_2 culture at 5% O_2 of human TSCs enhanced cell proliferation. The expression of stem cell marker genes, Nanog and Oct-4, was upregulated. After in vivo implantation, more tendon-like structures were formed in the 5% O_2. TSCs although deriving from a considerable hypoxic niche, showed a reduction of stemness markers expression when cultured under 0.5% O_2 while 10% O_2 and 5% O_2 preserved their stemness; in particular, 5% O_2 appeared to be the best concentration for the increase of stemness markers expression.	[162,164]
Cardiac progenitor cells (CPCs) and Cardiosphere derived cells (CDCs).	O_2 CPCs expanded in 5% O_2 increased cell yield, showed lower senescence and higher resistance to oxidative stress than those grown in 20% O_2. In vivo implantation of cells grown in 5% O_2 into mice infarcted hearts resulted in greater cell engraftment and better functional recovery. 3% O_2 or chemical compounds- preconditioning of cardio-spheres derived cells cultures improved stemness properties of CDCs which presented a higher expression of HIF, pluripotency gene markers, and a significantly higher number of c-kit-positive cells which are the most primitive undifferentiated population of cardiac stem cells. Hypoxia-preconditioned CDCs exhibited decreased O_2 consumption and improved glycolytic metabolism.	[143,148,234]

Table 3. *Cont.*

Stem Cell Source	Hypoxia Effects	Reference
Neural stem cells (NSCs)	In 2D (neuroepithelial cell line (NECs) and embryonic neural tissue) and 3D (ESCs-derived neuro-spheres (NSCs) cellular model, it was demonstrated that HIF-1α endogenous stabilization preserve stemness in a hypoxic environment prevented the premature neuronal differentiation through the direct activation of the neural repressor hairy and enhancer of split 1 (Hes1) pathway.	[104]
Retinal progenitor cells (RPCs)	3% O_2 to increases the proliferation rate of hRPCs shifting their proliferation limit. This increased proliferation was correlated with an upregulation of Ki67, CyclinD1, and telomerase activity and a decrease in p53 expression and apoptosis. Moreover, the increased c-Myc, Klf4, Oct-4, and Sox-2 expressions correlated with stabilization of both HIF-1α and HIF-2α was also detected in cells exposed to this hypoxic condition.	[155]

Adult stem cells. In HSCs or MSCs, hypoxia appears to prolong the lifespan of the stem cells, increases their proliferative capacity, and reduces differentiation in culture. A high level of HIF-1α expression was found in the primitive HSCs, which decreases as differentiation progresses and regulates several glycolytic enzymes that are under the control of HIF. Moreover, the primitive HSCs were low O_2 consuming and show high glycolytic flux and lower mitochondrial mass. Quiescent HSCs show low mitochondrial potential to avoid oxidative stress and ROS accumulation leading to decreased stemness and spontaneous differentiation [203]. Certainly, MSCs are the best-characterized stem cell type in hypoxic culture conditions. MSCs derived from several tissues, such as chondrocytes, adipocytes, osteocytes, bone marrow, tooth, and amniotic fluid, in hypoxic culture conditions expressed higher levels of stemness markers as *Oct-4*, *c-Myc*, *Nanog*, and *Nestin*. Moreover, hypoxia was able to inhibit senescence and increase the proliferation rate and differentiation ability [156]. Nevertheless, from literature data emerged that the different outcomes of MSC were strictly dependent on the O_2 concentration and the intrinsic properties of cell types. Indeed, in the study of Wagner et al., it was reported a list of MSC types with their respective O_2 consumption rates (OCRs) indicating that, in this category of cells, the redox environment depends on the levels of antioxidant and redox enzymes which influence cellular outcomes [63]. Therefore, MSCs take great advantage of hypoxic cultures, resulting in an advantageous condition to preserve stemness and differentiation potential for a long period in vitro.

6. Hypoxia, Aging, and Stem Cell Transplantation

In the last part of this review, a short section is dedicated to the link between hypoxia, aging, and effects of stem cell transplantation for the treatment of tissue injury or disease, as O_2 is a relevant factor even in the regulation of cell senescence.

When damage accumulates, mitotic cells from renewable tissues have two mechanisms to avoid replication. They can stop cell cycle progression and enter senescence or trigger cell death programs such as apoptosis. It is still unclear what determines if a cell undergoes senescence or apoptosis. Although most cells are capable of both, these processes seem to be exclusively yet linked to each other [7]. Inside tissues, O_2 gradients exist with stem cells residing in their hypoxic niches. These cells, more resistant to oxidative stress as a mechanism of self-preservation, benefit from their hypoxic environments by avoiding senescence, which would be detrimental to the tissue regenerative capacity. However, the tissues of a multicellular organism inevitably go toward a decline in organ function with aging. Although adult stem cells can self-renew and differentiate into multiple cell types within a tissue, they are not immune to damage accumulation over time [235]. Beyond the accumulation in DNA damage, mutations, and epigenetic alterations, the aging of the stem cell environment can also seriously alter stem cell functions even resulting in the niche deterioration [235]. Another hallmark of cellular aging is given by the shortening of telomerase with age [236]. Indeed, cellular senescence naturally occurs because of the gradual shortening of telomeres deriving from continuous replication [237]. Interestingly, it has been demonstrated that hypoxia can extend cell lifespan through the upregulation of the telomerase activity, reducing the senescent response. Of note, the telomerase reverse transcriptase (TERT) is another target of HIF-1α [238]. The telomere length is directly correlated to the age of the donor, to the time of culture before the senescence, and cell type. Slight differences exist in MSCs derived from Wharton jelly and those from the bone marrow regarding their phenotype, the telomerase activity, and the clonogenicity abilities after hypoxia or normoxia culture conditions, respectively [239]. These observations support the current view that MSCs properties are impacted by the tissue origin, especially if they derive from extraembryonic tissue or adult sources [240]. Importantly, as reported in the paper of Merini et al., [240] is strongly recommend that all biological issues related to the use of MSCs, such as the impact of tissue hypoxia, inflammation, and infection, should be well monitored to accelerate the transition from the bench to clinic [240]. Highlighting these features may improve the quality, safety, and efficiency of the future therapy.

Moreover, induction of senescence can be accelerated and prematurely induced by other environmental cues, including excessive oxidative stress [241]. Metabolism damage related to ROS increase is also a fundamental component in the aging process. In this context, as described above, hypoxia preconditioning prior to transplant, can effectively reduce ROS production in adult stem cells and improve their therapeutic efficiency in several in vivo ischemia or other disease models [242]. When cells are transplanted in the body, they face hypoxic in vivo environmental conditions, and a significant number of grafted cells die because of the severe in vivo environmental conditions at transplanted site [242]. The cell death due to hypoxic microenvironment is particularly considerable for those tissues that are not vascularized or already injured [243]. For example, in heart transplantation, donor hearts inevitably suffer from ischemia/reperfusion (I/R) injury, which leads to primary graft dysfunctions and affects patient survival rate. Remarkably, hypoxic conditioned medium derived from BMSCs enhances post-transplant graft functions, via paracrine effects that are improved by the hypoxic culture conditions [244].

Although these are encouraging reports, a consideration that deserves to be deeper investigated is whether replicative senescence limits the therapeutic potential of stem cells.

Nonetheless, the transplantation success of young or rejuvenated stem cells in aged patients is still problematic, since stem cell function is greatly influenced by extrinsic factors that become unsupportive with aging [245]. Confirming this, MSCs from aged donors did not perform as well as cells from younger donors in a transplantation following myocardial infarction [246] and similarly, MSCs obtained from young individuals have been induced to neuroectodermal differentiation in vitro, but this effect cannot be replicated in MSCs from elderly individuals [247].

Furthermore, hypoxia preconditioned BMSCs with up-regulated HIF-1α can enhance the bone healing process in geriatric individuals [248]. More in detail, the combination of hypoxia and DMOG preconditioning significantly increased the survival rate in bone defect site of transplanted BMSCs and may have great potential in regenerative cell therapy for bone defects in aged individuals [248].

Together, these data suggest that the use of stem cells from young donors or the rejuvenation of aged patient-derived stem cells may represent a promising system to improve the efficiency of transplantation. The preconditioning of MSCs in hypoxia triggers, even via the stabilization of HIF-1α, the upregulation of different functions, helping MSCs to survive after implantation, and increasing their curative potential [249]. O_2, in a range between 1% and 5%, is a low concentration, adequate to trigger adaptation, but not excessively low to cause apoptosis [250]. However, exact details of hypoxic treatment protocols (O_2 concentration, time of preconditioning, isolation under hypoxia, and reoxygenation) are still under examination to achieve a successful protocols optimization.

7. Conclusions

The present review aims to highlight the correlation existing between hypoxia and stemness focusing on cell culture models as invaluable research instruments for the comprehension of physiological hypoxia-induced mechanisms enabling the development of novel approaches to improve stem cell-based therapeutic strategies.

The O_2 tension, lowered to mimic niche microenvironment, has been successfully proposed to preserve cells phenotype during expansion for stem cell populations limited in supply. Alternatively, hypoxia has been adopted in vitro as a valuable stimulus to promote cell commitment into different tissue lineages. Beside this encouraging evidence, most of the protocols validated to date to expand stem cells recognize 21% O_2 tension (air O_2 concentration), which is about from 4- to 10-fold greater than gas levels in the natural niches by exposing the cells to cultural conditions that enhance oxidative stress and change cell metabolism with unpredictable and deleterious effects on stem cells phenotype and fate. Evidence emerged from available literature, mainly demonstrated a large spectrum of low O_2 in vitro effects on stemness maintenance, cell proliferation, senescence inhibition, and cell plasticity [137,156]. The hypoxic conditions reproduced in culture are mostly obtained by lowering O_2 tension in a range from 1% to 5% while levels lower than 1% are perceived from several cells as anoxic ones.

However, accurate control of O_2 levels has been suggested as a prerequisite to improving the reproducibility of the results as well as to compare them amongst laboratories by considering that small changes in pericellular O_2 levels can elicit profound molecular and signaling intracellular responses. To this aim, the use of O_2 sensors is considered a value approach to have a real-time O_2 monitoring under in vitro culture. Recently, several accessible, cost-effective, and high-throughput tools able to emulate controlled hypoxic conditions reproducing a steady or intermittent exposure closely mimicking the in vivo conditions have made available. Alternatively, hypoxia is induced by adopting chemical approaches that have been proposed taking advantage of their great flexibility and reproducibility in studying the acute effect of hypoxia.

In addition, higher levels of standardization of the in vitro conditions would be beneficial to interpret and compare the hypoxia-mediated HIF activation and stem cell response in term of hypoxic inductive physical or chemical methods (i.e., O_2 tension or drug concentrations), stem cell source, cultural parameters (cell concentration, degree of confluence, medium, and supplements), and time of hypoxic exposure (acute and chronic).

Of note, O_2 has been successfully suggested to provide a precommitment of stem cells before their therapeutic use. Indeed, the hypoxic imprinting of stem cells is an emerging in vitro strategy to improve tissue regeneration [251,252], strongly suggesting that decoding the mechanisms, by which cells sense O_2, could be useful for the development of new target molecules and stem cell-based treatment for several diseases, including cancer, stroke, and inflammation. Not by chance, in 2019, a trio of researchers, Gregg Semenza, William Kaelin, and Peter Ratcliffe, received the Nobel Prize in Physiology or Medicine for their discoveries on how cells sense and adapt to O_2 variations thus opening new cell biology paradigms recognizing the central role that O_2 may have in controlling cell response and adaptation.

Author Contributions: M.D.M. and A.M. are review coordinators who performed scientometric data analysis, wrote the draft paragraphs—Hypoxia and Stemness, In Vitro Models of Hypoxia, Hypoxia Cell Models, and Conclusions—produced figures and tables, and edited the manuscript. B.D. and E.D. wrote the draft paragraphs—The Role of O_2 in Cell Biology and The Role of O_2 in Tissue. M.R.C. wrote the draft paragraph about hypoxia-inducible factor (HIF) and edited the manuscript. A.P., V.R., and P.B. wrote the draft paragraphs about exploitation of hypoxia for regenerative medicine purposes. A.C. conducted a proofreading review. B.B. was involved in the conceptualization, scientometric data analysis, supervision of the review, and funding acquisition. All authors validated the data and reviewed the manuscript. All authors have read and agreed to the published version of the manuscript.

Funding: This research was funded by "Multiphase discoverY platform for EYE disease-MYEYE. Project granted from POR- Regione Lombardia 2014–2020_ID 1166606".

Conflicts of Interest: The authors declare no conflict of interest.

References

1. Bartz, R.R.; Piantadosi, C.A. Clinical review: Oxygen as a signaling molecule. *Crit. Care* **2010**, *14*, 234. [CrossRef]
2. Simon, M.C.; Keith, B. The role of oxygen availability in embryonic development and stem cell function. *Nat. Rev. Mol. Cell Biol.* **2008**, *9*, 285–296. [CrossRef]
3. Samanta, D.; Semenza, G.L. Metabolic adaptation of cancer and immune cells mediated by hypoxia-inducible factors. *Biochim. Biophys. Acta BBA Rev. Cancer* **2018**, *1870*, 15–22. [CrossRef]
4. Prabhakar, N.R.; Semenza, G.L. Oxygen Sensing and Homeostasis. *Physiology* **2015**, *30*, 340–348. [CrossRef]
5. Glass, J.J.; Phillips, P.A.; Gunning, P.W.; Stehn, J.R. Hypoxia alters the recruitment of tropomyosins into the actin stress fibres of neuroblastoma cells. *BMC Cancer* **2015**, *15*, 712. [CrossRef]
6. Habler, O.P.; Messmer, K.F.W. The physiology of oxygen transport. *Transfus. Sci.* **1997**, *18*, 425–435. [CrossRef]
7. Mas-Bargues, C.; Sanz-Ros, J.; Román-Domínguez, A.; Inglés, M.; Gimeno-Mallench, L.; El Alami, M.; Viña-Almunia, J.; Gambini, J.; Viña, J.; Borrás, C. Relevance of Oxygen Concentration in Stem Cell Culture for Regenerative Medicine. *Int. J. Mol. Sci.* **2019**, *20*, 1195. [CrossRef] [PubMed]
8. Trayhurn, P. Oxygen—A Critical, but Overlooked, Nutrient. *Front. Nutr.* **2019**, *6*, 10. [CrossRef] [PubMed]
9. Fang, F.C. Antimicrobial reactive oxygen and nitrogen species: Concepts and controversies. *Nat. Rev. Microbiol.* **2004**, *2*, 820–832. [CrossRef]

10. Hargreaves, M.; Spriet, L.L. Skeletal muscle energy metabolism during exercise. *Nat. Metab.* **2020**, *2*, 817–828. [CrossRef] [PubMed]
11. Nakazawa, M.S.; Keith, B.; Simon, M.C. Oxygen availability and metabolic adaptations. *Nat. Rev. Cancer* **2016**, *16*, 663–673. [CrossRef]
12. Wagner, P.D. The biology of oxygen. *Eur. Respir. J.* **2008**, *31*, 887–890. [CrossRef] [PubMed]
13. Cramer, T.; Yamanishi, Y. HIF-1 Is Essential for Myeloid Cell-Mediated Inflammation. *Cell* **2003**, *112*, 645–657. [CrossRef]
14. Santoro, M.; Gaudino, G. Cellular and molecular facets of keratinocyte reepithelization during wound healing. *Exp. Cell Res.* **2005**, *304*, 274–286. [CrossRef]
15. Muz, B.; Khan, M.N.; Kiriakidis, S.; Paleolog, E.M. The role of hypoxia and HIF-dependent signalling events in rheumatoid arthritis. *Arthritis Res. Ther.* **2009**, *11*, 201. [CrossRef] [PubMed]
16. Khan, W.S.; Adesida, A.B.; Hardingham, T.E. Hypoxic conditions increase hypoxia-inducible transcription factor 2α and enhance chondrogenesis in stem cells from the infrapatellar fat pad of osteoarthritis patients. *Arthritis Res. Ther.* **2007**, *9*, R55. [CrossRef] [PubMed]
17. Kaluz, S.; Kaluzová, M.; Stanbridge, E.J. Regulation of gene expression by hypoxia: Integration of the HIF-transduced hypoxic signal at the hypoxia-responsive element. *Clin. Chim. Acta* **2008**, *395*, 6–13. [CrossRef] [PubMed]
18. Berchner-Pfannschmidt, U.; Frede, S.; Wotzlaw, C.; Fandrey, J. Imaging of the hypoxia-inducible factor pathway: Insights into oxygen sensing. *Eur. Respir. J.* **2008**, *32*, 210–217. [CrossRef]
19. Tolonen, J.-P.; Heikkilä, M.; Malinen, M.; Lee, H.-M.; Palvimo, J.J.; Wei, G.-H.; Myllyharju, J. A long hypoxia-inducible factor 3 isoform 2 is a transcription activator that regulates erythropoietin. *Cell. Mol. Life Sci.* **2020**, *77*, 3627–3642. [CrossRef]
20. Huang, L.E.; Bunn, H.F. Hypoxia-inducible Factor and Its Biomedical Relevance. *J. Biol. Chem.* **2003**, *278*, 19575–19578. [CrossRef]
21. Kaelin, W.G.; Ratcliffe, P.J. Oxygen Sensing by Metazoans: The Central Role of the HIF Hydroxylase Pathway. *Mol. Cell* **2008**, *30*, 393–402. [CrossRef] [PubMed]
22. Bertout, J.A.; Patel, S.A.; Simon, M.C. The impact of O2 availability on human cancer. *Nat. Rev. Cancer* **2008**, *8*, 967–975. [CrossRef]
23. Jewell, U.R.; Kvietikova, I.; Scheid, A.; Bauer, C.; Wenger, R.H.; Gassmann, M. Induction of HIF–1α in response to hypoxia is instantaneous. *FASEB J.* **2001**, *15*, 1312–1314. [CrossRef] [PubMed]
24. Pollard, P.; Yang, M.; Su, H.; Soga, T.; Kranc, K. Prolyl hydroxylase domain enzymes: Important regulators of cancer metabolism. *Hypoxia* **2014**, 127. [CrossRef]
25. Choudhry, H.; Harris, A.L. Advances in Hypoxia-Inducible Factor Biology. *Cell Metab.* **2018**, *27*, 281–298. [CrossRef]
26. Majmundar, A.J.; Wong, W.J.; Simon, M.C. Hypoxia-Inducible Factors and the Response to Hypoxic Stress. *Mol. Cell* **2010**, *40*, 294–309. [CrossRef]
27. Hu, C.-J.; Wang, L.-Y.; Chodosh, L.A.; Keith, B.; Simon, M.C. Differential Roles of Hypoxia-Inducible Factor 1 (HIF-1) and HIF-2 in Hypoxic Gene Regulation. *Mol. Cell. Biol.* **2003**, *23*, 14. [CrossRef]
28. Aprelikova, O.; Wood, M.; Tackett, S.; Chandramouli, G.V.R.; Barrett, J.C. Role of ETS Transcription Factors in the Hypoxia-Inducible Factor-2 Target Gene Selection. *Cancer Res.* **2006**, *66*, 5641–5647. [CrossRef]
29. Elvert, G.; Kappel, A.; Heidenreich, R.; Englmeier, U.; Lanz, S.; Acker, T.; Rauter, M.; Plate, K.; Sieweke, M.; Breier, G.; et al. Cooperative Interaction of Hypoxia-inducible Factor-2α (HIF-2α) and Ets-1 in the Transcriptional Activation of Vascular Endothelial Growth Factor Receptor-2 (Flk-1). *J. Biol. Chem.* **2003**, *278*, 7520–7530. [CrossRef]
30. Kelly, B.D.; Hackett, S.F.; Hirota, K.; Oshima, Y.; Cai, Z.; Berg-Dixon, S.; Rowan, A.; Yan, Z.; Campochiaro, P.A.; Semenza, G.L. Cell Type–Specific Regulation of Angiogenic Growth Factor Gene Expression and Induction of Angiogenesis in Nonischemic Tissue by a Constitutively Active Form of Hypoxia-Inducible Factor 1. *Circ. Res.* **2003**, *93*, 1074–1081. [CrossRef] [PubMed]
31. Manalo, D.J.; Rowan, A.; Lavoie, T.; Natarajan, L.; Kelly, B.D.; Ye, S.Q.; Garcia, J.G.N.; Semenza, G.L. Transcriptional regulation of vascular endothelial cell responses to hypoxia by HIF-1. *Blood* **2005**, *105*, 659–669. [CrossRef]
32. Tang, C.; Liu, T.; Wang, K.; Wang, X.; Xu, S.; He, D.; Zeng, J. Transcriptional regulation of FoxM1 by HIF-1α mediates hypoxia-induced EMT in prostate cancer. *Oncol. Rep.* **2019**, *42*, 1307–1318. [CrossRef] [PubMed]
33. Widowati, W.; Rihibiha, D.D.; Khiong, K.; Widodo, M.A.; Sumitro, S.B.; Bachtiar, I. Hypoxia in Mesenchymal Stem Cell. In *Hypoxia and Human Diseases*; Zheng, J., Zhou, C., Eds.; InTech: London, UK, 2017; ISBN 978-953-51-2895-3.
34. Gordan, J.D.; Thompson, C.B.; Simon, M.C. HIF and c-Myc: Sibling Rivals for Control of Cancer Cell Metabolism and Proliferation. *Cancer Cell* **2007**, *12*, 108–113. [CrossRef] [PubMed]
35. Semenza, G.L. Evaluation of HIF-1 inhibitors as anticancer agents. *Drug Discov. Today* **2007**, *12*, 853–859. [CrossRef] [PubMed]
36. Simon, M.C. Coming up for air: HIF-1 and mitochondrial oxygen consumption. *Cell Metab.* **2006**, *3*, 150–151. [CrossRef] [PubMed]
37. Liu, T.; Zhang, L.; Joo, D.; Sun, S.-C. NF-κB signaling in inflammation. *Signal Transduct. Target. Ther.* **2017**, *2*, 17023. [CrossRef]
38. Nizet, V.; Johnson, R.S. Interdependence of hypoxic and innate immune responses. *Nat. Rev. Immunol.* **2009**, *9*, 609–617. [CrossRef]
39. Cummins, E.P.; Berra, E.; Comerford, K.M.; Ginouves, A.; Fitzgerald, K.T.; Seeballuck, F.; Godson, C.; Nielsen, J.E.; Moynagh, P.; Pouyssegur, J.; et al. Prolyl hydroxylase-1 negatively regulates IκB kinase-β, giving insight into hypoxia-induced NFκB activity. *Cell Biol.* **2006**, *103*, 18154–18159.
40. Serocki, M.; Bartoszewska, S.; Janaszak-Jasiecka, A.; Ochocka, R.J.; Collawn, J.F.; Bartoszewski, R. miRNAs regulate the HIF switch during hypoxia: A novel therapeutic target. *Angiogenesis* **2018**, *21*, 183–202. [CrossRef]
41. Chan, S.Y.; Loscalzo, J. MicroRNA-210: A unique and pleiotropic hypoxamir. *Cell Cycle* **2010**, *9*, 1072–1083. [CrossRef]

42. Camps, C.; Buffa, F.M.; Colella, S.; Moore, J.; Sotiriou, C.; Sheldon, H.; Harris, A.L.; Gleadle, J.M.; Ragoussis, J. hsa-miR-210 Is Induced by Hypoxia and Is an Independent Prognostic Factor in Breast Cancer. *Clin. Cancer Res.* **2008**, *14*, 1340–1348. [CrossRef] [PubMed]
43. Zhang, Z.; Sun, H.; Dai, H.; Walsh, R.; Imakura, M.; Schelter, J.; Burchard, J.; Dai, X.; Chang, A.N.; Diaz, R.L.; et al. MicroRNA miR-210 modulates cellular response to hypoxia through the MYC antagonist MNT. *Cell Cycle* **2009**, *8*, 2756–2768. [CrossRef]
44. Liu, F.; Lou, Y.-L.; Wu, J.; Ruan, Q.-F.; Xie, A.; Guo, F.; Cui, S.-P.; Deng, Z.-F.; Wang, Y. Upregulation of MicroRNA-210 Regulates Renal Angiogenesis Mediated by Activation of VEGF Signaling Pathway under Ischemia/Perfusion Injury in vivo and in vitro. *Kidney Blood Press. Res.* **2012**, *35*, 182–191. [CrossRef]
45. Li, Z.; Meng, D.; Li, G.; Xu, J.; Tian, K.; Li, Y. Overexpression of microRNA-210 promotes chondrocyte proliferation and extracellular matrix deposition by targeting HIF-3α in osteoarthritis. *Mol. Med. Rep.* **2016**, *13*, 2769–2776. [CrossRef]
46. Kai, A.K.-L.; Chan, L.K.; Lo, R.C.-L.; Lee, J.M.-F.; Wong, C.C.-L.; Wong, J.C.-M.; Ng, I.O.-L. Down-regulation of TIMP2 by HIF-1α/miR-210/HIF-3α regulatory feedback circuit enhances cancer metastasis in hepatocellular carcinoma. *Hepatology* **2016**, *64*, 473–487. [CrossRef]
47. Pientka, F.K.; Hu, J.; Schindler, S.G.; Brix, B.; Thiel, A.; Joehren, O.; Fandrey, J.; Berchner-Pfannschmidt, U.; Depping, R. Oxygen sensing by Prolyl-4-Hydroxylase PHD2 within the nuclear compartment and the influence of compartimentalisation on HIF-1 signalling. *J. Cell Sci.* **2012**, *125*, 5168–5176. [CrossRef] [PubMed]
48. Alijani, N.; Johari, B.; Moradi, M.; Kadivar, M. A review on transcriptional regulation responses to hypoxia in mesenchymal stem cells. *Cell Biol. Int.* **2020**, *44*, 14–26. [CrossRef]
49. Scadden, D.T. The stem-cell niche as an entity of action. *Nature* **2006**, *441*, 1075–1079. [CrossRef] [PubMed]
50. Discher, D.E.; Mooney, D.J.; Zandstra, P.W. Growth Factors, Matrices, and Forces Combine and Control Stem Cells. *Science* **2009**, *324*, 1673–1677. [CrossRef] [PubMed]
51. Lane, S.W.; Williams, D.A.; Watt, F.M. Modulating the stem cell niche for tissue regeneration. *Nat. Biotechnol.* **2014**, *32*, 795–803. [CrossRef] [PubMed]
52. Pavlacky, J.; Polak, J. Technical Feasibility and Physiological Relevance of Hypoxic Cell Culture Models. *Front. Endocrinol.* **2020**, *11*, 57. [CrossRef]
53. Nakagomi, T.; Nakano-Doi, A.; Narita, A.; Matsuyama, T. Concise Review: Are Stimulated Somatic Cells Truly Reprogrammed into an ES/iPS-Like Pluripotent State? Better Understanding by Ischemia-Induced Multipotent Stem Cells in a Mouse Model of Cerebral Infarction. *Stem Cells Int.* **2015**, *2015*, 1–6. [CrossRef]
54. Shimada, I.S.; Peterson, B.M.; Spees, J.L. Isolation of Locally Derived Stem/Progenitor Cells From the Peri-Infarct Area That Do Not Migrate From the Lateral Ventricle After Cortical Stroke. *Stroke* **2010**, *41*, e552–e560. [CrossRef] [PubMed]
55. Keith, B.; Simon, M.C. Hypoxia-Inducible Factors, Stem Cells, and Cancer. *Cell* **2007**, *129*, 465–472. [CrossRef] [PubMed]
56. Chen, R.; Forsyth, N. Editorial: The Development of New Classes of Hypoxia Mimetic Agents for Clinical Use. *Front. Cell Dev. Biol.* **2019**, *7*, 120. [CrossRef] [PubMed]
57. Buravkova, L.B.; Andreeva, E.R.; Gogvadze, V.; Zhivotovsky, B. Mesenchymal stem cells and hypoxia: Where are we? *Mitochondrion* **2014**, *19*, 105–112. [CrossRef] [PubMed]
58. Al-Ani, A.; Toms, D.; Kondro, D.; Thundathil, J.; Yu, Y.; Ungrin, M. Oxygenation in cell culture: Critical parameters for reproducibility are routinely not reported. *PLoS ONE* **2018**, *13*, e0204269. [CrossRef]
59. Place, T.L.; Domann, F.E.; Case, A.J. Limitations of oxygen delivery to cells in culture: An underappreciated problem in basic and translational research. *Free Radic. Biol. Med.* **2017**, *113*, 311–322. [CrossRef]
60. Esteban, M.A.; Maxwell, P.H. Manipulation of oxygen tensions for *in vitro* cell culture using a hypoxic workstation. *Expert Rev. Proteom.* **2005**, *2*, 307–314. [CrossRef]
61. Oomen, P.E.; Skolimowski, M.D.; Verpoorte, E. Implementing oxygen control in chip-based cell and tissue culture systems. *Lab. Chip* **2016**, *16*, 3394–3414. [CrossRef]
62. Rivera, K.R.; Yokus, M.A.; Erb, P.D.; Pozdin, V.A.; Daniele, M. Measuring and regulating oxygen levels in microphysiological systems: Design, material, and sensor considerations. *Analyst* **2019**, *144*, 3190–3215. [CrossRef]
63. Wagner, B.A.; Venkataraman, S.; Buettner, G.R. The rate of oxygen utilization by cells. *Free Radic. Biol. Med.* **2011**, *51*, 700–712. [CrossRef]
64. Muñoz-Sánchez, J.; Chánez-Cárdenas, M.E. The use of cobalt chloride as a chemical hypoxia model. *J. Appl. Toxicol.* **2019**, *39*, 556–570. [CrossRef]
65. Wu, D.; Yotnda, P. Induction and Testing of Hypoxia in Cell Culture. *J. Vis. Exp.* **2011**, 2899. [CrossRef] [PubMed]
66. Bahsoun, S.; Coopman, K.; Forsyth, N.R.; Akam, E.C. The Role of Dissolved Oxygen Levels on Human Mesenchymal Stem Cell Culture Success, Regulatory Compliance, and Therapeutic Potential. *Stem Cells Dev.* **2018**, *27*, 1303–1321. [CrossRef] [PubMed]
67. Koh, M.Y.; Powis, G. Passing the baton: The HIF switch. *Trends Biochem. Sci.* **2012**, *37*, 364–372. [CrossRef] [PubMed]
68. Bates, M.K. Culturing Cells Under Hypoxic Conditions for Biologically Relevant Results. *Am. Lab.* **2012**.
69. Lin, M.; Liu, X.; Zheng, H.; Huang, X.; Wu, Y.; Huang, A.; Zhu, H.; Hu, Y.; Mai, W.; Huang, Y. IGF-1 enhances BMSC viability, migration, and anti-apoptosis in myocardial infarction via secreted frizzled-related protein 2 pathway. *Stem Cell Res. Ther.* **2020**, *11*, 22. [CrossRef]
70. Yang, Y.; Jiang, Z.; Bolnick, A.; Dai, J.; Puscheck, E.E.; Rappolee, D.A. Departure from optimal O_2 level for mouse trophoblast stem cell proliferation and potency leads to most rapid AMPK activation. *J. Reprod. Dev.* **2017**, *63*, 87–94. [CrossRef]

71. Ito, A.; Aoyama, T.; Yoshizawa, M.; Nagai, M.; Tajino, J.; Yamaguchi, S.; Iijima, H.; Zhang, X.; Kuroki, H. The effects of short-term hypoxia on human mesenchymal stem cell proliferation, viability and p16INK4A mRNA expression: Investigation using a simple hypoxic culture system with a deoxidizing agent. *J. Stem Cells Regen. Med.* **2015**, *11*, 25–31. [CrossRef]
72. Lam, S.F.; Shirure, V.S.; Chu, Y.E.; Soetikno, A.G.; George, S.C. Microfluidic device to attain high spatial and temporal control of oxygen. *PLoS ONE* **2018**, *13*, e0209574. [CrossRef]
73. Busek, M.; Grünzner, S.; Steege, T.; Klotzbach, U.; Sonntag, F. Hypoxia-on-a-chip: Generating hypoxic conditions in microfluidic cell culture systems. *Curr. Dir. Biomed. Eng.* **2016**, *2*, 71–75. [CrossRef]
74. Barmaki, S.; Obermaier, D.; Kankuri, E.; Vuola, J.; Franssila, S.; Jokinen, V. A Microfluidic Chip Architecture Enabling a Hypoxic Microenvironment and Nitric Oxide Delivery in Cell Culture. *Micromachines* **2020**, *11*, 979. [CrossRef]
75. Uchida, H.; Sato, A.; Miyayama, A.; Tsukada, K. Generation of an Oxygen Gradient in a Microfluidic Device and Cellular Analysis in Hypoxia. *Adv. Biomed. Eng.* **2013**, *2*, 143–149. [CrossRef]
76. Antoni, D.; Burckel, H.; Josset, E.; Noel, G. Three-Dimensional Cell Culture: A Breakthrough in Vivo. *Int. J. Mol. Sci.* **2015**, *16*, 5517–5527. [CrossRef] [PubMed]
77. Qiao, Y.; Xu, Z.; Yu, Y.; Hou, S.; Geng, J.; Xiao, T.; Liang, Y.; Dong, Q.; Mei, Y.; Wang, B.; et al. Single cell derived spheres of umbilical cord mesenchymal stem cells enhance cell stemness properties, survival ability and therapeutic potential on liver failure. *Biomaterials* **2020**, *227*, 119573. [CrossRef] [PubMed]
78. Egger, D.; Tripisciano, C.; Weber, V.; Dominici, M.; Kasper, C. Dynamic Cultivation of Mesenchymal Stem Cell Aggregates. *Bioengineering* **2018**, *5*, 48. [CrossRef] [PubMed]
79. Murphy, K.C.; Hung, B.P.; Browne-Bourne, S.; Zhou, D.; Yeung, J.; Genetos, D.C.; Leach, J.K. Measurement of oxygen tension within mesenchymal stem cell spheroids. *J. R. Soc. Interface* **2017**, *14*, 20160851. [CrossRef] [PubMed]
80. Schmitz, C.; Potekhina, E.; Belousov, V.V.; Lavrentieva, A. Hypoxia Onset in Mesenchymal Stem Cell Spheroids: Monitoring With Hypoxia Reporter Cells. *Front. Bioeng. Biotechnol.* **2021**, *9*, 611837. [CrossRef]
81. Ryu, N.-E.; Lee, S.-H.; Park, H. Spheroid Culture System Methods and Applications for Mesenchymal Stem Cells. *Cells* **2019**, *8*, 1620. [CrossRef]
82. Zhang, Q.; Nguyen, A.L.; Shi, S.; Hill, C.; Wilder-Smith, P.; Krasieva, T.B.; Le, A.D. Three-Dimensional Spheroid Culture of Human Gingiva-Derived Mesenchymal Stem Cells Enhances Mitigation of Chemotherapy-Induced Oral Mucositis. *Stem Cells Dev.* **2012**, *21*, 937–947. [CrossRef]
83. Kouroupis, D.; Sanjurjo-Rodriguez, C.; Jones, E.; Correa, D. Mesenchymal Stem Cell Functionalization for Enhanced Therapeutic Applications. *Tissue Eng. Part. B Rev.* **2019**, *25*, 55–77. [CrossRef]
84. Petrova, V.; Annicchiarico-Petruzzelli, M.; Melino, G.; Amelio, I. The hypoxic tumour microenvironment. *Oncogenesis* **2018**, *7*, 10. [CrossRef] [PubMed]
85. Cesarz, Z.; Tamama, K. Spheroid Culture of Mesenchymal Stem Cells. *Stem Cells Int.* **2016**, *2016*, 1–11. [CrossRef] [PubMed]
86. Hanson, E.S.; Rawlins, M.L.; Leibold, E.A. Oxygen and Iron Regulation of Iron Regulatory Protein 2. *J. Biol. Chem.* **2003**, *278*, 40337–40342. [CrossRef] [PubMed]
87. Davis, C.K.; Jain, S.A.; Bae, O.-N.; Majid, A.; Rajanikant, G.K. Hypoxia Mimetic Agents for Ischemic Stroke. *Front. Cell Dev. Biol.* **2019**, *6*, 175. [CrossRef]
88. Triantafyllou, A.; Liakos, P.; Tsakalof, A.; Georgatsou, E.; Simos, G.; Bonanou, S. Cobalt induces hypoxia-inducible factor-1α (HIF-1α) in HeLa cells by an iron-independent, but ROS-, PI-3K- and MAPK-dependent mechanism. *Free Radic. Res.* **2006**, *40*, 847–856. [CrossRef]
89. Chandel, N.S.; McClintock, D.S.; Feliciano, C.E.; Wood, T.M.; Melendez, J.A.; Rodriguez, A.M.; Schumacker, P.T. Reactive Oxygen Species Generated at Mitochondrial Complex III Stabilize Hypoxia-inducible Factor-1α during Hypoxia. *J. Biol. Chem.* **2000**, *275*, 25130–25138. [CrossRef]
90. Zeng, H.-L.; Zhong, Q.; Qin, Y.-L.; Bu, Q.-Q.; Han, X.-A.; Jia, H.-T.; Liu, H.-W. Hypoxia-mimetic agents inhibit proliferation and alter the morphology of human umbilical cord-derived mesenchymal stem cells. *BMC Cell Biol.* **2011**, *12*, 32. [CrossRef]
91. Barrett, T.D.; Palomino, H.L.; Brondstetter, T.I.; Kanelakis, K.C.; Wu, X.; Haug, P.V.; Yan, W.; Young, A.; Hua, H.; Hart, J.C.; et al. Pharmacological Characterization of 1-(5-Chloro-6-(trifluoromethoxy)-1 *H* -benzoimidazol-2-yl)-1 *H* -pyrazole-4-carboxylic Acid (JNJ-42041935), a Potent and Selective Hypoxia-Inducible Factor Prolyl Hydroxylase Inhibitor. *Mol. Pharmacol.* **2011**, *79*, 910–920. [CrossRef]
92. Binó, L.; Kučera, J.; Štefková, K.; Šindlerová, L.Š.; Lánová, M.; Kudová, J.; Kubala, L.; Pacherník, J. The stabilization of hypoxia inducible factor modulates differentiation status and inhibits the proliferation of mouse embryonic stem cells. *Chem. Biol. Interact.* **2016**, *244*, 204–214. [CrossRef] [PubMed]
93. Kumar, N.; Pruthi, V. Potential applications of ferulic acid from natural sources. *Biotechnol. Rep.* **2014**, *4*, 86–93. [CrossRef]
94. Qiu, S.; Sun, Y.; Xu, J.; Wen, G.; Yu, Y.; Wu, T.; Chai, Y. Ferulic acid improves self-renewal and differentiation of human tendon-derived stem cells by upregulating early growth response 1 through hypoxia. *Genesis* **2019**, *57*. [CrossRef]
95. Gale, A.L.; Mammone, R.M.; Dodson, M.E.; Linardi, R.L.; Ortved, K.F. The effect of hypoxia on chondrogenesis of equine synovial membrane-derived and bone marrow-derived mesenchymal stem cells. *BMC Vet. Res.* **2019**, *15*, 201. [CrossRef]
96. Wu, T.; Liu, S.; Wen, G.; Xu, J.; Yu, Y.; Chai, Y. Celastrol improves self-renewal and differentiation of human tendon-derived stem cells by suppressing Smad7 through hypoxia. *Stem Cell Res. Ther.* **2017**, *8*, 274. [CrossRef]

97. Moon, K.-C.; Yang, J.-P.; Lee, J.-S.; Jeong, S.-H.; Dhong, E.-S.; Han, S.-K. Effects of Ultraviolet Irradiation on Cellular Senescence in Keratinocytes Versus Fibroblasts. *J. Craniofac. Surg.* **2019**, *30*, 270–275. [CrossRef]
98. Lee, J.; Jung, E.; Hyun, J.-W.; Park, D. Ultraviolet A regulates the stemness of human adipose tissue-derived mesenchymal stem cells through downregulation of the HIF-1α via activation of PGE_2 -cAMP signaling. *J. Cell. Biochem.* **2012**, *113*, 3681–3691. [CrossRef] [PubMed]
99. Jung, K.; Cho, J.Y.; Soh, Y.-J.; Lee, J.; Shin, S.W.; Jang, S.; Jung, E.; Kim, M.H.; Lee, J. Antagonizing Effects of Aspartic Acid against Ultraviolet A-Induced Downregulation of the Stemness of Human Adipose Tissue-Derived Mesenchymal Stem Cells. *PLoS ONE* **2015**, *10*, e0124417. [CrossRef] [PubMed]
100. Hwang, Y.S.; Park, S.-H.; Kang, M.; Oh, S.W.; Jung, K.; Park, Y.S.; Lee, J. Stemness and differentiation potential-recovery effects of sinapic acid against ultraviolet-A-induced damage through the regulation of p38 MAPK and NF-κB. *Sci. Rep.* **2017**, *7*, 909. [CrossRef]
101. Park, S.-H.; Cho, J.Y.; Oh, S.W.; Kang, M.; Lee, S.E.; Yoo, J.A.; Jung, K.; Lee, J.; Lee, S.Y.; Lee, J. Arctigenin protects against ultraviolet-A-induced damage to stemness through inhibition of the NF-κB/MAPK pathway. *Chem. Biol. Interact.* **2018**, *282*, 63–68. [CrossRef]
102. Ishizuka, T.; Hinata, T.; Watanabe, Y. Superoxide induced by a high-glucose concentration attenuates production of angiogenic growth factors in hypoxic mouse mesenchymal stem cells. *J. Endocrinol.* **2011**, *208*, 147–159. [CrossRef] [PubMed]
103. Liu, Y.; Li, Y.; Nan, L.; Wang, F.; Zhou, S.; Wang, J.; Feng, X.; Zhang, L. The effect of high glucose on the biological characteristics of nucleus pulposus-derived mesenchymal stem cells. *Cell Biochem. Funct.* **2020**, *38*, 130–140. [CrossRef]
104. Večeřa, J.; Procházková, J.; Šumberová, V.; Pánská, V.; Paculová, H.; Lánová, M.K.; Mašek, J.; Bohačiaková, D.; Andersson, E.R.; Pacherník, J. Hypoxia/Hif1α prevents premature neuronal differentiation of neural stem cells through the activation of Hes1. *Stem Cell Res.* **2020**, *45*, 101770. [CrossRef] [PubMed]
105. Laksana, K.; Sooampon, S.; Pavasant, P.; Sriarj, W. Cobalt Chloride Enhances the Stemness of Human Dental Pulp Cells. *J. Endod.* **2017**, *43*, 760–765. [CrossRef]
106. Yuan, Z.; Zhang, J.; Huang, Y.; Zhang, Y.; Liu, W.; Wang, G.; Zhang, Q.; Wang, G.; Yang, Y.; Li, H.; et al. NRF2 overexpression in mesenchymal stem cells induces stem-cell marker expression and enhances osteoblastic differentiation. *Biochem. Biophys. Res. Commun.* **2017**, *491*, 228–235. [CrossRef]
107. Kim, Y.; Jin, H.J.; Heo, J.; Ju, H.; Lee, H.-Y.; Kim, S.; Lee, S.; Lim, J.; Jeong, S.Y.; Kwon, J.; et al. Small hypoxia-primed mesenchymal stem cells attenuate graft-versus-host disease. *Leukemia* **2018**, *32*, 2672–2684. [CrossRef]
108. Choi, W.; Kwon, S.; Jin, H.J.; Jeong, S.Y.; Choi, S.J.; Oh, W.; Yang, Y.S.; Jeon, H.B.; Jeon, E.S. Optimization of culture conditions for rapid clinical-scale expansion of human umbilical cord blood-derived mesenchymal stem cells. *Clin. Transl. Med.* **2017**, *6*, 1–12. [CrossRef]
109. Griffon, D.J.; Cho, J.; Wagner, J.R.; Charavaryamath, C.; Wei, J.; Wagoner Johnson, A. Effects of Hypoxia and Chitosan on Equine Umbilical Cord-Derived Mesenchymal Stem Cells. *Stem Cells Int.* **2016**, *2016*, 1–11. [CrossRef] [PubMed]
110. Taguchi, T.; Cho, J.Y.; Hao, J.; Nout-Lomas, Y.S.; Kang, K.-S.; Griffon, D.J. Influence of hypoxia on the stemness of umbilical cord matrix-derived mesenchymal stem cells cultured on chitosan films: EFFECT OF CHITOSAN AND HYPOXIA ON UCM-MSCS. *J. Biomed. Mater. Res. B Appl. Biomater.* **2018**, *106*, 501–511. [CrossRef]
111. Park, S.E.; Kim, H.; Kwon, S.; Choi, S.; Oh, S.; Ryu, G.H.; Jeon, H.B.; Chang, J.W. Pressure Stimuli Improve the Proliferation of Wharton's Jelly-Derived Mesenchymal Stem Cells under Hypoxic Culture Conditions. *Int. J. Mol. Sci.* **2020**, *21*, 7092. [CrossRef] [PubMed]
112. Shearier, E.; Xing, Q.; Qian, Z.; Zhao, F. Physiologically Low Oxygen Enhances Biomolecule Production and Stemness of Mesenchymal Stem Cell Spheroids. *Tissue Eng. Part. C Methods* **2016**, *22*, 360–369. [CrossRef] [PubMed]
113. Miyaji, T.; Takami, T.; Fujisawa, K.; Matsumoto, T.; Yamamoto, N.; Sakaida, I. Bone marrow-derived humoral factors suppress oxidative phosphorylation, upregulate TSG-6, and improve therapeutic effects on liver injury of mesenchymal stem cells. *J. Clin. Biochem. Nutr.* **2020**, *66*, 213–223. [CrossRef]
114. Chen, Y.; Zhao, Q.; Yang, X.; Yu, X.; Yu, D.; Zhao, W. Effects of cobalt chloride on the stem cell marker expression and osteogenic differentiation of stem cells from human exfoliated deciduous teeth. *Cell Stress Chaperones* **2019**, *24*, 527–538. [CrossRef] [PubMed]
115. Bhandi, S.; Al Kahtani, A.; Mashyakhy, M.; Alsofi, L.; Maganur, P.C.; Vishwanathaiah, S.; Testarelli, L.; Del Giudice, A.; Mehta, D.; Vyas, N.; et al. Modulation of the Dental Pulp Stem Cell Secretory Profile by Hypoxia Induction Using Cobalt Chloride. *J. Pers. Med.* **2021**, *11*, 247. [CrossRef]
116. Ciavarella, C.; Fittipaldi, S.; Pedrini, S.; Vasuri, F.; Gallitto, E.; Freyrie, A.; Stella, A.; Gostjeva, E.; Pasquinelli, G. In vitro alteration of physiological parameters do not hamper the growth of human multipotent vascular wall-mesenchymal stem cells. *Front. Cell Dev. Biol.* **2015**, *3*, 36. [CrossRef] [PubMed]
117. Osathanon, T.; Vivatbutsiri, P.; Sukarawan, W.; Sriarj, W.; Pavasant, P.; Sooampon, S. Cobalt chloride supplementation induces stem-cell marker expression and inhibits osteoblastic differentiation in human periodontal ligament cells. *Arch. Oral Biol.* **2015**, *60*, 29–36. [CrossRef] [PubMed]
118. Byun, H.; Bin Lee, Y.; Kim, E.M.; Shin, H. Fabrication of size-controllable human mesenchymal stromal cell spheroids from micro-scaled cell sheets. *Biofabrication* **2019**, *11*, 035025. [CrossRef]
119. Xiao, L.; Tsutsui, T. Characterization of human dental pulp cells-derived spheroids in serum-free medium: Stem cells in the core. *J. Cell. Biochem.* **2013**, *114*, 2624–2636. [CrossRef]

120. Zhu, C.; Yu, J.; Pan, Q.; Yang, J.; Hao, G.; Wang, Y.; Li, L.; Cao, H. Hypoxia-inducible factor-2 alpha promotes the proliferation of human placenta-derived mesenchymal stem cells through the MAPK/ERK signaling pathway. *Sci. Rep.* **2016**, *6*, 35489. [CrossRef]
121. Jee, M.K.; Kim, J.H.; Han, Y.M.; Jung, S.J.; Kang, K.S.; Kim, D.W.; Kang, S.K. DHP-Derivative and Low Oxygen Tension Effectively Induces Human Adipose Stromal Cell Reprogramming. *PLoS ONE* **2010**, *5*, e9026. [CrossRef]
122. Yang, Y.; Arenas-Hernandez, M.; Gomez-Lopez, N.; Dai, J.; Parker, G.C.; Puscheck, E.E.; Rappolee, D.A. Hypoxic Stress Forces Irreversible Differentiation of a Majority of Mouse Trophoblast Stem Cells Despite FGF4. *Biol. Reprod.* **2016**, *95*, 1–10. [CrossRef] [PubMed]
123. Zhao, D.; Liu, L.; Chen, Q.; Wang, F.; Li, Q.; Zeng, Q.; Huang, J.; Luo, M.; Li, W.; Zheng, Y.; et al. Hypoxia with Wharton's jelly mesenchymal stem cell coculture maintains stemness of umbilical cord blood-derived CD34+ cells. *Stem Cell Res. Ther.* **2018**, *9*, 158. [CrossRef] [PubMed]
124. Murugesan, M.; Premkumar, K. Hypoxia stimulates microenvironment in human embryonic stem cell through inflammatory signalling: An integrative analysis. *Biochem. Biophys. Res. Commun.* **2018**, *498*, 437–444. [CrossRef] [PubMed]
125. Liu, S.; Song, N.; He, J.; Yu, X.; Guo, J.; Jiao, X.; Ding, X.; Teng, J. Effect of Hypoxia on the Differentiation and the Self-Renewal of Metanephrogenic Mesenchymal Stem Cells. *Stem Cells Int.* **2017**, *2017*, 1–16. [CrossRef] [PubMed]
126. Kim, D.S.; Ko, Y.J.; Lee, M.W.; Park, H.J.; Park, Y.J.; Kim, D.-I.; Sung, K.W.; Koo, H.H.; Yoo, K.H. Effect of low oxygen tension on the biological characteristics of human bone marrow mesenchymal stem cells. *Cell Stress Chaperones* **2016**, *21*, 1089–1099. [CrossRef] [PubMed]
127. Sato, Y.; Mabuchi, Y.; Miyamoto, K.; Araki, D.; Niibe, K.; Houlihan, D.D.; Morikawa, S.; Nakagawa, T.; Nakajima, T.; Akazawa, C.; et al. Notch2 Signaling Regulates the Proliferation of Murine Bone Marrow-Derived Mesenchymal Stem/Stromal Cells via c-Myc Expression. *PLoS ONE* **2016**, *11*, e0165946. [CrossRef]
128. Petrangeli, E.; Coroniti, G.; Brini, A.T.; de Girolamo, L.; Stanco, D.; Niada, S.; Silecchia, G.; Morgante, E.; Lubrano, C.; Russo, M.A.; et al. Hypoxia Promotes the Inflammatory Response and Stemness Features in Visceral Fat Stem Cells From Obese Subjects: Hypoxia effect on adipose-derived stem cells. *J. Cell. Physiol.* **2016**, *231*, 668–679. [CrossRef]
129. Kim, D.S.; Lee, M.W.; Ko, Y.J.; Park, H.J.; Park, Y.J.; Kim, D.-I.; Jung, H.L.; Sung, K.W.; Koo, H.H.; Yoo, K.H. Application of human mesenchymal stem cells cultured in different oxygen concentrations for treatment of graft-versus-host disease in mice. *Biomed. Res.* **2016**, *37*, 311–317. [CrossRef]
130. Hung, S.-P.; Ho, J.H.; Shih, Y.-R.V.; Lo, T.; Lee, O.K. Hypoxia promotes proliferation and osteogenic differentiation potentials of human mesenchymal stem cells: Hypoxia promotes MSCS proliferation. *J. Orthop. Res.* **2012**, *30*, 260–266. [CrossRef]
131. Weijers, E.M.; Van Den Broek, L.J.; Waaijman, T.; Van Hinsbergh, V.W.M.; Gibbs, S.; Koolwijk, P. The Influence of Hypoxia and Fibrinogen Variants on the Expansion and Differentiation of Adipose Tissue-Derived Mesenchymal Stem Cells. *Tissue Eng. Part A* **2011**, *17*, 2675–2685. [CrossRef]
132. Chen, H.-C.; Sytwu, H.-K.; Chang, J.-L.; Wang, H.-W.; Chen, H.-K.; Kang, B.-H.; Liu, D.-W.; Chen, C.-H.; Chao, T.-T.; Wang, C.-H. Hypoxia enhances the stemness markers of cochlear stem/progenitor cells and expands sphere formation through activation of hypoxia-inducible factor-1alpha. *Hear. Res.* **2011**, *275*, 43–52. [CrossRef]
133. Palumbo, S.; Tsai, T.-L.; Li, W.-J. Macrophage Migration Inhibitory Factor Regulates AKT Signaling in Hypoxic Culture to Modulate Senescence of Human Mesenchymal Stem Cells. *Stem Cells Dev.* **2014**, *23*, 852–865. [CrossRef]
134. Hwang, S.J.; Lee, H.-J. Identification of differentially expressed genes in mouse embryonic stem cell under hypoxia. *Genes Genom.* **2021**, *43*, 313–321. [CrossRef]
135. Fotia, C.; Massa, A.; Boriani, F.; Baldini, N.; Granchi, D. Hypoxia enhances proliferation and stemness of human adipose-derived mesenchymal stem cells. *Cytotechnology* **2015**, *67*, 1073–1084. [CrossRef]
136. Zhang, L.; Qian, Z.; Tahtinen, M.; Qi, S.; Zhao, F. Prevascularization of natural nanofibrous extracellular matrix for engineering completely biological three-dimensional prevascularized tissues for diverse applications. *J. Tissue Eng. Regen. Med.* **2018**, *12*. [CrossRef]
137. Wan Safwani, W.K.Z.; Choi, J.R.; Yong, K.W.; Ting, I.; Mat Adenan, N.A.; Pingguan-Murphy, B. Hypoxia enhances the viability, growth and chondrogenic potential of cryopreserved human adipose-derived stem cells. *Cryobiology* **2017**, *75*, 91–99. [CrossRef] [PubMed]
138. Choi, J.R.; Pingguan-Murphy, B.; Wan Abas, W.A.B.; Noor Azmi, M.A.; Omar, S.Z.; Chua, K.H.; Wan Safwani, W.K.Z. Impact of low oxygen tension on stemness, proliferation and differentiation potential of human adipose-derived stem cells. *Biochem. Biophys. Res. Commun.* **2014**, *448*, 218–224. [CrossRef]
139. Zhou, Y.; Fan, W.; Xiao, Y. The Effect of Hypoxia on the Stemness and Differentiation Capacity of PDLC and DPC. *BioMed Res. Int.* **2014**, *2014*, 1–7. [CrossRef] [PubMed]
140. Tsang, W.P.; Shu, Y.; Kwok, P.L.; Zhang, F.; Lee, K.K.H.; Tang, M.K.; Li, G.; Chan, K.M.; Chan, W.-Y.; Wan, C. CD146+ Human Umbilical Cord Perivascular Cells Maintain Stemness under Hypoxia and as a Cell Source for Skeletal Regeneration. *PLoS ONE* **2013**, *8*, e76153. [CrossRef] [PubMed]
141. Yamamoto, Y.; Fujita, M.; Tanaka, Y.; Kojima, I.; Kanatani, Y.; Ishihara, M.; Tachibana, S. Low Oxygen Tension Enhances Proliferation and Maintains Stemness of Adipose Tissue–Derived Stromal Cells. *BioResearch Open Access* **2013**, *2*, 199–205. [CrossRef]

142. Saller, M.M.; Prall, W.C.; Docheva, D.; Schönitzer, V.; Popov, T.; Anz, D.; Clausen-Schaumann, H.; Mutschler, W.; Volkmer, E.; Schieker, M.; et al. Increased stemness and migration of human mesenchymal stem cells in hypoxia is associated with altered integrin expression. *Biochem. Biophys. Res. Commun.* **2012**, *423*, 379–385. [CrossRef]
143. Tan, S.C.; Gomes, R.S.M.; Yeoh, K.K.; Perbellini, F.; Malandraki-Miller, S.; Ambrose, L.; Heather, L.C.; Faggian, G.; Schofield, C.J.; Davies, K.E.; et al. Preconditioning of Cardiosphere-Derived Cells with Hypoxia or Prolyl-4-Hydroxylase Inhibitors Increases Stemness and Decreases Reliance on Oxidative Metabolism. *Cell Transplant.* **2016**, *25*, 35–53. [CrossRef] [PubMed]
144. Kheirandish, M.; Gavgani, S.P.; Samiee, S. The effect of hypoxia preconditioning on the neural and stemness genes expression profiling in human umbilical cord blood mesenchymal stem cells. *Transfus. Apher. Sci.* **2017**, *56*, 392–399. [CrossRef] [PubMed]
145. Fotia, C.; Massa, A.; Boriani, F.; Baldini, N.; Granchi, D. Prolonged Exposure to Hypoxic Milieu Improves the Osteogenic Potential of Adipose Derived Stem Cells: Hypoxia Improves ASC Osteogenic Potential. *J. Cell. Biochem.* **2015**, *116*, 1442–1453. [CrossRef] [PubMed]
146. Obradovic, H.; Krstic, J.; Trivanovic, D.; Mojsilovic, S.; Okic, I.; Kukolj, T.; Ilic, V.; Jaukovic, A.; Terzic, M.; Bugarski, D. Improving stemness and functional features of mesenchymal stem cells from Wharton's jelly of a human umbilical cord by mimicking the native, low oxygen stem cell niche. *Placenta* **2019**, *82*, 25–34. [CrossRef] [PubMed]
147. Liu, F.; Huang, X.; Luo, Z.; He, J.; Haider, F.; Song, C.; Peng, L.; Chen, T.; Wu, B. Hypoxia-Activated PI3K/Akt Inhibits Oxidative Stress via the Regulation of Reactive Oxygen Species in Human Dental Pulp Cells. *Oxidative Med. Cell. Longev.* **2019**, *2019*, 1–10. [CrossRef]
148. RajendranNair, D.S.; Karunakaran, J.; Nair, R.R. Sub-physiological oxygen levels optimal for growth and survival of human atrial cardiac stem cells. *Mol. Cell. Biochem.* **2017**, *432*, 109–122. [CrossRef]
149. López-Iglesias, P.; Alcaina, Y.; Tapia, N.; Sabour, D.; Arauzo-Bravo, M.J.; Sainz de la Maza, D.; Berra, E.; O'Mara, A.N.; Nistal, M.; Ortega, S.; et al. Hypoxia Induces Pluripotency in Primordial Germ Cells by HIF1α Stabilization and Oct4 Deregulation. *Antioxid. Redox Signal.* **2015**, *22*, 205–223. [CrossRef] [PubMed]
150. Baranov, P.Y.; Tucker, B.A.; Young, M.J. Low-Oxygen Culture Conditions Extend the Multipotent Properties of Human Retinal Progenitor Cells. *Tissue Eng. Part A* **2014**, *20*, 1465–1475. [CrossRef]
151. Berniakovich, I.; Giorgio, M. Low Oxygen Tension Maintains Multipotency, Whereas Normoxia Increases Differentiation of Mouse Bone Marrow Stromal Cells. *Int. J. Mol. Sci.* **2013**, *14*, 2119–2134. [CrossRef]
152. Fehrer, C.; Brunauer, R.; Laschober, G.; Unterluggauer, H.; Reitinger, S.; Kloss, F.; Gülly, C.; Gaßner, R.; Lepperdinger, G. Reduced oxygen tension attenuates differentiation capacity of human mesenchymal stem cells and prolongs their lifespan: Mesenchymal stem cells and reduced oxygen tension. *Aging Cell* **2007**, *6*, 745–757. [CrossRef]
153. D'Ippolito, G.; Diabira, S.; Howard, G.A.; Roos, B.A.; Schiller, P.C. Low oxygen tension inhibits osteogenic differentiation and enhances stemness of human MIAMI cells. *Bone* **2006**, *39*, 513–522. [CrossRef]
154. Deng, Y.; Huang, G.; Chen, F.; Testroet, E.D.; Li, H.; Li, H.; Nong, T.; Yang, X.; Cui, J.; Shi, D.; et al. Hypoxia enhances buffalo adipose-derived mesenchymal stem cells proliferation, stemness, and reprogramming into induced pluripotent stem cells. *J. Cell. Physiol.* **2019**, *234*, 17254–17268. [CrossRef] [PubMed]
155. Singh, D.; Dromel, P.C.; Young, M. Low-oxygen and knock-out serum maintain stemness in human retinal progenitor cells. *Mol. Biol. Rep.* **2020**, *47*, 1613–1623. [CrossRef] [PubMed]
156. Kwon, S.Y.; Chun, S.Y.; Ha, Y.-S.; Kim, D.H.; Kim, J.; Song, P.H.; Kim, H.T.; Yoo, E.S.; Kim, B.S.; Kwon, T.G. Hypoxia Enhances Cell Properties of Human Mesenchymal Stem Cells. *Tissue Eng. Regen. Med.* **2017**, *14*, 595–604. [CrossRef]
157. Gu, Q.; Gu, Y.; Shi, Q.; Yang, H. Hypoxia Promotes Osteogenesis of Human Placental-Derived Mesenchymal Stem Cells. *Tohoku J. Exp. Med.* **2016**, *239*, 287–296. [CrossRef] [PubMed]
158. Lee, J.; Byeon, J.S.; Lee, K.S.; Gu, N.-Y.; Lee, G.B.; Kim, H.-R.; Cho, I.-S.; Cha, S.-H. Chondrogenic potential and anti-senescence effect of hypoxia on canine adipose mesenchymal stem cells. *Vet. Res. Commun.* **2016**, *40*, 1–10. [CrossRef]
159. Petruzzelli, R.; Christensen, D.R.; Parry, K.L.; Sanchez-Elsner, T.; Houghton, F.D. HIF-2a Regulates NANOG Expression in Human Embryonic Stem Cells following Hypoxia and Reoxygenation through the Interaction with an Oct-Sox Cis Regulatory Element. *PLoS ONE* **2014**, *9*, e108309. [CrossRef]
160. Drela, K.; Sarnowska, A.; Siedlecka, P.; Szablowska-Gadomska, I.; Wielgos, M.; Jurga, M.; Lukomska, B.; Domanska-Janik, K. Low oxygen atmosphere facilitates proliferation and maintains undifferentiated state of umbilical cord mesenchymal stem cells in an hypoxia inducible factor-dependent manner. *Cytotherapy* **2014**, *16*, 881–892. [CrossRef]
161. Huang, Y.-H.; Lin, M.-H.; Wang, P.-C.; Wu, Y.-C.; Chiang, H.-L.; Wang, Y.-L.; Chang, J.-H.; Huang, Y.-K.; Gu, S.-Y.; Ho, H.-N.; et al. Hypoxia inducible factor 2α/insulin-like growth factor receptor signal loop supports the proliferation and Oct-4 maintenance of mouse germline stem cells. *MHR Basic Sci. Reprod. Med.* **2014**, *20*, 526–537. [CrossRef]
162. Zhang, J.; Wang, J.H.-C. Human Tendon Stem Cells Better Maintain Their Stemness in Hypoxic Culture Conditions. *PLoS ONE* **2013**, *8*, e61424. [CrossRef]
163. Chen, H.-F.; Kuo, H.-C.; Chen, W.; Wu, F.-C.; Yang, Y.-S.; Ho, H.-N. A reduced oxygen tension (5%) is not beneficial for maintaining human embryonic stem cells in the undifferentiated state with short splitting intervals. *Hum. Reprod.* **2008**, *24*, 71–80. [CrossRef]
164. Yu, Y.; Lin, L.; Zhou, Y.; Lu, X.; Shao, X.; Lin, C.; Yu, K.; Zhang, X.; Hong, J.; Chen, Y. Effect of Hypoxia on Self-Renewal Capacity and Differentiation in Human Tendon-Derived Stem Cells. *Med. Sci. Monit.* **2017**, 23, 1334–1339. [CrossRef]

165. Zhi, X.; Xiong, J.; Wang, M.; Zhang, H.; Huang, G.; Zhao, J.; Zi, X.; Hu, Y.-P. Physiological Hypoxia Enhances Stemness Preservation, Proliferation, and Bidifferentiation of Induced Hepatic Stem Cells. *Oxidative Med. Cell. Longev.* **2018**, *2018*, 1–10. [CrossRef]
166. Najar, M.; Crompot, E.; van Grunsven, L.A.; Dollé, L.; Lagneaux, L. Aldehyde dehydrogenase activity of Wharton jelly mesenchymal stromal cells: Isolation and characterization. *Cytotechnology* **2019**, *71*, 427–441. [CrossRef]
167. Najar, M.; Crompot, E.; van Grunsven, L.A.; Dollé, L.; Lagneaux, L. Aldehyde Dehydrogenase Activity in Adipose Tissue: Isolation and Gene Expression Profile of Distinct Sub-population of Mesenchymal Stromal Cells. *Stem Cell Rev. Rep.* **2018**, *14*, 599–611. [CrossRef]
168. Najar, M.; Crompot, E.; van Grunsven, L.A.; Dollé, L.; Lagneaux, L. Foreskin-derived mesenchymal stromal cells with aldehyde dehydrogenase activity: Isolation and gene profiling. *BMC Cell Biol.* **2018**, *19*, 4. [CrossRef]
169. Najar, M.; Dollé, L.; Crompot, E.; Verhulst, S.; van Grunsven, L.A.; Busser, H.; Lagneaux, L. Isolation and Characterization of Bone Marrow Mesenchymal Stromal Cell Subsets in Culture Based on Aldehyde Dehydrogenase Activity. *Tissue Eng. Part C Methods* **2018**, *24*, 89–98. [CrossRef] [PubMed]
170. Samal, J.R.K.; Rangasami, V.K.; Samanta, S.; Varghese, O.P.; Oommen, O.P. Discrepancies on the Role of Oxygen Gradient and Culture Condition on Mesenchymal Stem Cell Fate. *Adv. Healthc. Mater.* **2021**, *10*, 2002058. [CrossRef] [PubMed]
171. Abdollahi, H.; Harris, L.J.; Zhang, P.; McIlhenny, S.; Srinivas, V.; Tulenko, T.; DiMuzio, P.J. The Role of Hypoxia in Stem Cell Differentiation and Therapeutics. *J. Surg. Res.* **2011**, *165*, 112–117. [CrossRef] [PubMed]
172. Dengler, V.L.; Galbraith, M.D.; Espinosa, J.M. Transcriptional regulation by hypoxia inducible factors. *Crit. Rev. Biochem. Mol. Biol.* **2014**, *49*, 1–15. [CrossRef]
173. Boyer, L.A.; Lee, T.I.; Cole, M.F.; Johnstone, S.E.; Levine, S.S.; Zucker, J.P.; Guenther, M.G.; Kumar, R.M.; Murray, H.L.; Jenner, R.G.; et al. Core Transcriptional Regulatory Circuitry in Human Embryonic Stem Cells. *Cell* **2005**, *122*, 947–956. [CrossRef]
174. Chambers, I.; Tomlinson, S.R. The transcriptional foundation of pluripotency. *Development* **2009**, *136*, 2311–2322. [CrossRef]
175. Zhang, W.; Sui, Y.; Ni, J.; Yang, T. Insights into the *Nanog* gene: A propeller for stemness in primitive stem cells. *Int. J. Biol. Sci.* **2016**, *12*, 1372–1381. [CrossRef]
176. Son, M.-Y.; Choi, H.; Han, Y.-M.; Sook Cho, Y. Unveiling the critical role of REX1 in the regulation of human stem cell pluripotency: REX1-Mediated Control of Pluripotency. *Stem Cells* **2013**, *31*, 2374–2387. [CrossRef]
177. Boland, M.J.; Nazor, K.L.; Loring, J.F. Epigenetic Regulation of Pluripotency and Differentiation. *Circ. Res.* **2014**, *115*, 311–324. [CrossRef]
178. van Schaijik, B.; Davis, P.F.; Wickremesekera, A.C.; Tan, S.T.; Itinteang, T. Subcellular localisation of the stem cell markers OCT4, SOX2, NANOG, KLF4 and c-MYC in cancer: A review. *J. Clin. Pathol.* **2018**, *71*, 88–91. [CrossRef] [PubMed]
179. Rehman, J. Empowering self-renewal and differentiation: The role of mitochondria in stem cells. *J. Mol. Med.* **2010**, *88*, 981–986. [CrossRef] [PubMed]
180. Shyh-Chang, N.; Ng, H.-H. The metabolic programming of stem cells. *Genes Dev.* **2017**, *31*, 336–346. [CrossRef]
181. Coller, H.A. The paradox of metabolism in quiescent stem cells. *FEBS Lett.* **2019**, *593*, 2817–2839. [CrossRef] [PubMed]
182. Prigione, A.; Fauler, B.; Lurz, R.; Lehrach, H.; Adjaye, J. The Senescence-Related Mitochondrial/Oxidative Stress Pathway is Repressed in Human Induced Pluripotent Stem Cells. *Stem Cells* **2010**, *28*, 721–733. [CrossRef] [PubMed]
183. Takubo, K.; Nagamatsu, G.; Kobayashi, C.I.; Nakamura-Ishizu, A.; Kobayashi, H.; Ikeda, E.; Goda, N.; Rahimi, Y.; Johnson, R.S.; Soga, T.; et al. Regulation of Glycolysis by Pdk Functions as a Metabolic Checkpoint for Cell Cycle Quiescence in Hematopoietic Stem Cells. *Cell Stem Cell* **2013**, *12*, 49–61. [CrossRef] [PubMed]
184. Hu, C.; Fan, L.; Cen, P.; Chen, E.; Jiang, Z.; Li, L. Energy Metabolism Plays a Critical Role in Stem Cell Maintenance and Differentiation. *Int. J. Mol. Sci.* **2016**, *17*, 253. [CrossRef]
185. Liu, K.; Yu, C.; Xie, M.; Li, K.; Ding, S. Chemical Modulation of Cell Fate in Stem Cell Therapeutics and Regenerative Medicine. *Cell Chem. Biol.* **2016**, *23*, 893–916. [CrossRef]
186. Kozlov, A.M.; Lone, A.; Betts, D.H.; Cumming, R.C. Lactate preconditioning promotes a HIF-1α-mediated metabolic shift from OXPHOS to glycolysis in normal human diploid fibroblasts. *Sci. Rep.* **2020**, *10*, 8388. [CrossRef]
187. Stirparo, G.G.; Kurowski, A.; Yanagida, A.; Bates, L.E.; Strawbridge, S.E.; Hladkou, S.; Stuart, H.T.; Boroviak, T.E.; Silva, J.C.R.; Nichols, J. OCT4 induces embryonic pluripotency via STAT3 signaling and metabolic mechanisms. *Proc. Natl. Acad. Sci. USA* **2021**, *118*, e2008890118. [CrossRef] [PubMed]
188. Yu, L.; Ji, K.; Zhang, J.; Xu, Y.; Ying, Y.; Mai, T.; Xu, S.; Zhang, Q.; Yao, K.; Xu, Y. Core pluripotency factors promote glycolysis of human embryonic stem cells by activating GLUT1 enhancer. *Protein Cell* **2019**, *10*, 668–680. [CrossRef]
189. Fillmore, N.; Huqi, A.; Jaswal, J.S.; Mori, J.; Paulin, R.; Haromy, A.; Onay-Besikci, A.; Ionescu, L.; Thébaud, B.; Michelakis, E.; et al. Effect of Fatty Acids on Human Bone Marrow Mesenchymal Stem Cell Energy Metabolism and Survival. *PLoS ONE* **2015**, *10*, e0120257. [CrossRef]
190. Nuschke, A.; Rodrigues, M.; Wells, A.W.; Sylakowski, K.; Wells, A. Mesenchymal stem cells/multipotent stromal cells (MSCs) are glycolytic and thus glucose is a limiting factor of in vitro models of MSC starvation. *Stem Cell Res. Ther.* **2016**, *7*, 179. [CrossRef]
191. Chen, C.-T.; Shih, Y.-R.V.; Kuo, T.K.; Lee, O.K.; Wei, Y.-H. Coordinated Changes of Mitochondrial Biogenesis and Antioxidant Enzymes During Osteogenic Differentiation of Human Mesenchymal Stem Cells. *Stem Cells* **2008**, *26*, 960–968. [CrossRef]
192. Perales-Clemente, E.; Folmes, C.D.L.; Terzic, A. Metabolic Regulation of Redox Status in Stem Cells. *Antioxid. Redox Signal.* **2014**, *21*, 1648–1659. [CrossRef] [PubMed]

193. Tafani, M.; Sansone, L.; Limana, F.; Arcangeli, T.; De Santis, E.; Polese, M.; Fini, M.; Russo, M.A. The Interplay of Reactive Oxygen Species, Hypoxia, Inflammation, and Sirtuins in Cancer Initiation and Progression. *Oxidative Med. Cell. Longev.* **2016**, *2016*, 1–18. [CrossRef]
194. Folmes, C.D.L.; Nelson, T.J.; Martinez-Fernandez, A.; Arrell, D.K.; Lindor, J.Z.; Dzeja, P.P.; Ikeda, Y.; Perez-Terzic, C.; Terzic, A. Somatic Oxidative Bioenergetics Transitions into Pluripotency-Dependent Glycolysis to Facilitate Nuclear Reprogramming. *Cell Metab.* **2011**, *14*, 264–271. [CrossRef]
195. Bigarella, C.L.; Liang, R.; Ghaffari, S. Stem cells and the impact of ROS signaling. *Development* **2014**, *141*, 4206–4218. [CrossRef]
196. Forrester, S.J.; Kikuchi, D.S.; Hernandes, M.S.; Xu, Q.; Griendling, K.K. Reactive Oxygen Species in Metabolic and Inflammatory Signaling. *Circ. Res.* **2018**, *122*, 877–902. [CrossRef]
197. Zhou, D.; Shao, L.; Spitz, D.R. Reactive Oxygen Species in Normal and Tumor Stem Cells. In *Advances in Cancer Research*; Elsevier: Amsterdam, The Netherlands, 2014; Volume 122, pp. 1–67. ISBN 978-0-12-420117-0.
198. Chaudhari, P.; Ye, Z.; Jang, Y.-Y. Roles of Reactive Oxygen Species in the Fate of Stem Cells. *Antioxid. Redox Signal.* **2014**, *20*, 1881–1890. [CrossRef] [PubMed]
199. Wheaton, W.W.; Chandel, N.S. Hypoxia. 2. Hypoxia regulates cellular metabolism. *Am. J. Physiol. Cell Physiol.* **2011**, *300*, C385–C393. [CrossRef]
200. Kierans, S.J.; Taylor, C.T.; Forsythe, I.; Greenhaff, P. Regulation of glycolysis by the hypoxia-inducible factor (HIF): Implications for cellular physiology. *J. Physiol* **2020**, *599*, 23–37. [CrossRef] [PubMed]
201. Kim, J.; Tchernyshyov, I.; Semenza, G.L.; Dang, C.V. HIF-1-mediated expression of pyruvate dehydrogenase kinase: A metabolic switch required for cellular adaptation to hypoxia. *Cell Metab.* **2006**, *3*, 177–185. [CrossRef] [PubMed]
202. Haque, N.; Rahman, M.T.; Abu Kasim, N.H.; Alabsi, A.M. Hypoxic Culture Conditions as a Solution for Mesenchymal Stem Cell Based Regenerative Therapy. *Sci. World J.* **2013**, *2013*, 1–12. [CrossRef] [PubMed]
203. Roy, I.M.; Biswas, A.; Verfaillie, C.; Khurana, S. Energy Producing Metabolic Pathways in Functional Regulation of the Hematopoietic Stem Cells: HSC proliferation, ageing. *IUBMB Life* **2018**, *70*, 612–624. [CrossRef]
204. Kim, H.; Jang, H.; Kim, T.W.; Kang, B.-H.; Lee, S.E.; Jeon, Y.K.; Chung, D.H.; Choi, J.; Shin, J.; Cho, E.-J.; et al. Core Pluripotency Factors Directly Regulate Metabolism in Embryonic Stem Cell to Maintain Pluripotency: ESC Metabolism by Core Pluripotency Factors. *Stem Cells* **2015**, *33*, 2699–2711. [CrossRef]
205. Suda, T.; Takubo, K.; Semenza, G.L. Metabolic Regulation of Hematopoietic Stem Cells in the Hypoxic Niche. *Cell Stem Cell* **2011**, *9*, 298–310. [CrossRef] [PubMed]
206. Zhang, C.C.; Sadek, H.A. Hypoxia and Metabolic Properties of Hematopoietic Stem Cells. *Antioxid. Redox Signal.* **2014**, *20*, 1891–1901. [CrossRef]
207. Simsek, T.; Kocabas, F.; Zheng, J.; DeBerardinis, R.J.; Mahmoud, A.I.; Olson, E.N.; Schneider, J.W.; Zhang, C.C.; Sadek, H.A. The Distinct Metabolic Profile of Hematopoietic Stem Cells Reflects Their Location in a Hypoxic Niche. *Cell Stem Cell* **2010**, *7*, 380–390. [CrossRef] [PubMed]
208. Kocabas, F.; Zheng, J.; Thet, S.; Copeland, N.G.; Jenkins, N.A.; DeBerardinis, R.J.; Zhang, C.; Sadek, H.A. Meis1 regulates the metabolic phenotype and oxidant defense of hematopoietic stem cells. *Blood* **2012**, *120*, 4963–4972. [CrossRef]
209. Contreras-Lopez, R.; Elizondo-Vega, R.; Paredes, M.J.; Luque-Campos, N.; Torres, M.J.; Tejedor, G.; Vega-Letter, A.M.; Figueroa-Valdés, A.; Pradenas, C.; Oyarce, K.; et al. HIF1α-dependent metabolic reprogramming governs mesenchymal stem/stromal cell immunoregulatory functions. *FASEB J.* **2020**, *34*, 8250–8264. [CrossRef] [PubMed]
210. Gillory, L.A.; Stewart, J.E.; Megison, M.L.; Waters, A.M.; Beierle, E.A. Focal adhesion kinase and p53 synergistically decrease neuroblastoma cell survival. *J. Surg. Res.* **2015**, *196*, 339–349. [CrossRef]
211. Lv, B.; Li, F.; Fang, J.; Xu, L.; Sun, C.; Han, J.; Hua, T.; Zhang, Z.; Feng, Z.; Jiang, X. Hypoxia inducible factor 1α promotes survival of mesenchymal stem cells under hypoxia. *Am. J. Transl. Res.* **2017**, *9*, 1521–1529.
212. Iervolino, A.; Trisciuoglio, D.; Ribatti, D.; Candiloro, A.; Biroccio, A.; Zupi, G.; Del Bufalo, D. Bcl-2 overexpression in human melanoma cells increases angiogenesis through VEGF mRNA stabilization and HIF-1mediated transcriptional activity. *FASEB J.* **2002**, *16*, 1453–1455. [CrossRef]
213. Pavlová, S.; Vašíček, D.; Kotwica, J.V.; Sirotkin, A. Involvement of Cell Cycle and Apoptosis-Related Protein p21 in Control of Secretory Activity of Porcine Ovarian Cells. *J. Steroids Horm. Sci.* **2010**, *1*, 102. [CrossRef]
214. Marchitti, S.A.; Brocker, C.; Stagos, D.; Vasiliou, V. Non-P450 aldehyde oxidizing enzymes: The aldehyde dehydrogenase superfamily. *Expert Opin. Drug Metab. Toxicol.* **2008**, *4*, 697–720. [CrossRef]
215. Dollé, L.; Gao, B. Pharmacological chaperone therapies: Can aldehyde dehydrogenase activator make us healthier? *J. Hepatol.* **2015**, *62*, 1228–1230. [CrossRef] [PubMed]
216. Croker, A.; Rodriguez-Torres, M.; Xia, Y.; Pardhan, S.; Leong, H.; Lewis, J.; Allan, A. Differential Functional Roles of ALDH1A1 and ALDH1A3 in Mediating Metastatic Behavior and Therapy Resistance of Human Breast Cancer Cells. *Int. J. Mol. Sci.* **2017**, *18*, 2039. [CrossRef] [PubMed]
217. Lee, H.J.; Ryu, J.M.; Jung, Y.H.; Oh, S.Y.; Lee, S.-J.; Han, H.J. Novel Pathway for Hypoxia-Induced Proliferation and Migration in Human Mesenchymal Stem Cells: Involvement of HIF-1α, FASN, and mTORC1. *Stem Cells Dayt. Ohio* **2015**, *33*, 2182–2195. [CrossRef] [PubMed]
218. Covello, K.L. HIF-2 regulates Oct-4: Effects of hypoxia on stem cell function, embryonic development, and tumor growth. *Genes Dev.* **2006**, *20*, 557–570. [CrossRef]

219. Boiani, M.; Eckardt, S.; Schöler, H.R.; McLaughlin, K.J. Oct4 distribution and level in mouse clones: Consequences for pluripotency. *Genes Dev.* **2002**, *16*, 1209–1219. [CrossRef]
220. Huang, X.; Trinh, T.; Aljoufi, A.; Broxmeyer, H.E. Hypoxia Signaling Pathway in Stem Cell Regulation: Good and Evil. *Curr. Stem Cell Rep.* **2018**, *4*, 149–157. [CrossRef]
221. Arthur, S.A.; Blaydes, J.P.; Houghton, F.D. Glycolysis Regulates Human Embryonic Stem Cell Self-Renewal under Hypoxia through HIF-2α and the Glycolytic Sensors CTBPs. *Stem Cell Rep.* **2019**, *12*, 728–742. [CrossRef]
222. Ezashi, T.; Das, P.; Roberts, R.M. Low O_2 tensions and the prevention of differentiation of hES cells. *Proc. Natl. Acad. Sci. USA* **2005**, *102*, 4783–4788. [CrossRef] [PubMed]
223. Chen, X.; Xu, H.; Yuan, P.; Fang, F.; Huss, M.; Vega, V.B.; Wong, E.; Orlov, Y.L.; Zhang, W.; Jiang, J.; et al. Integration of External Signaling Pathways with the Core Transcriptional Network in Embryonic Stem Cells. *Cell* **2008**, *133*, 1106–1117. [CrossRef] [PubMed]
224. Miki, T. Amnion-derived stem cells: In quest of clinical applications. *Stem Cell Res. Ther.* **2011**, *2*, 25. [CrossRef] [PubMed]
225. Koike, C.; Zhou, K.; Takeda, Y.; Fathy, M.; Okabe, M.; Yoshida, T.; Nakamura, Y.; Kato, Y.; Nikaido, T. Characterization of Amniotic Stem Cells. *Cell. Reprogramming* **2014**, *16*, 298–305. [CrossRef] [PubMed]
226. Miki, T.; Lehmann, T.; Cai, H.; Stolz, D.B.; Strom, S.C. Stem Cell Characteristics of Amniotic Epithelial Cells. *Stem Cells* **2005**, *23*, 1549–1559. [CrossRef]
227. Barboni, B.; Russo, V.; Curini, V.; Martelli, A.; Berardinelli, P.; Mauro, A.; Mattioli, M.; Marchisio, M.; Bonassi Signoroni, P.; Parolini, O.; et al. Gestational stage affects amniotic epithelial cells phenotype, methylation status, immunomodulatory and stemness properties. *Stem Cell Rev. Rep.* **2014**, *10*, 725–741. [CrossRef]
228. Magatti, M.; Vertua, E.; Cargnoni, A.; Silini, A.; Parolini, O. The Immunomodulatory Properties of Amniotic Cells: The Two Sides of the Coin. *Cell Transplant.* **2018**, *27*, 31–44. [CrossRef]
229. Barboni, B.; Russo, V.; Berardinelli, P.; Mauro, A.; Valbonetti, L.; Sanyal, H.; Canciello, A.; Greco, L.; Muttini, A.; Gatta, V.; et al. Placental Stem Cells from Domestic Animals: Translational Potential and Clinical Relevance. *Cell Transplant.* **2018**, *27*, 93–116. [CrossRef]
230. Lebreton, F.; Lavallard, V.; Bellofatto, K.; Bonnet, R.; Wassmer, C.H.; Perez, L.; Kalandadze, V.; Follenzi, A.; Boulvain, M.; Kerr-Conte, J.; et al. Insulin-producing organoids engineered from islet and amniotic epithelial cells to treat diabetes. *Nat. Commun.* **2019**, *10*, 4491. [CrossRef]
231. Citeroni, M.R.; Mauro, A.; Ciardulli, M.C.; Di Mattia, M.; El Khatib, M.; Russo, V.; Turriani, M.; Santer, M.; Della Porta, G.; Maffulli, N.; et al. Amnion-Derived Teno-Inductive Secretomes: A Novel Approach to Foster Tendon Differentiation and Regeneration in an Ovine Model. *Front. Bioeng. Biotechnol.* **2021**, *9*, 649288. [CrossRef]
232. Kukumberg, M.; Phermthai, T.; Wichitwiengrat, S.; Wang, X.; Arjunan, S.; Chong, S.Y.; Fong, C.-Y.; Wang, J.-W.; Rufaihah, A.J.; Mattar, C.N.Z. Hypoxia-induced amniotic fluid stem cell secretome augments cardiomyocyte proliferation and enhances cardioprotective effects under hypoxic-ischemic conditions. *Sci. Rep.* **2021**, *11*, 163. [CrossRef]
233. Iriuchishima, H.; Takubo, K.; Matsuoka, S.; Onoyama, I.; Nakayama, K.I.; Nojima, Y.; Suda, T. Ex vivo maintenance of hematopoietic stem cells by quiescence induction through Fbxw7α overexpression. *Blood* **2011**, *117*, 2373–2377. [CrossRef]
234. Li, T.-S.; Cheng, K.; Lee, S.-T.; Matsushita, S.; Davis, D.; Malliaras, K.; Zhang, Y.; Matsushita, N.; Smith, R.R.; Marbán, E. Cardiospheres Recapitulate a Niche-Like Microenvironment Rich in Stemness and Cell-Matrix Interactions, Rationalizing Their Enhanced Functional Potency for Myocardial Repair. *Stem Cells* **2010**, *28*, 2088–2098. [CrossRef]
235. Schultz, M.B.; Sinclair, D.A. When stem cells grow old: Phenotypes and mechanisms of stem cell aging. *Development* **2016**, *143*, 3–14. [CrossRef]
236. Ferrón, S.R.; Marqués-Torrejón, M.Á.; Mira, H.; Flores, I.; Taylor, K.; Blasco, M.A.; Fariñas, I. Telomere Shortening in Neural Stem Cells Disrupts Neuronal Differentiation and Neuritogenesis. *J. Neurosci.* **2009**, *29*, 14394–14407. [CrossRef]
237. Hernandez-Segura, A.; Nehme, J.; Demaria, M. Hallmarks of Cellular Senescence. *Trends Cell Biol.* **2018**, *28*, 436–453. [CrossRef] [PubMed]
238. Welford, S.M.; Giaccia, A.J. Hypoxia and Senescence: The Impact of Oxygenation on Tumor Suppression. *Mol. Cancer Res.* **2011**, *9*, 538–544. [CrossRef]
239. Laroye, C.; Gauthier, M.; Antonot, H.; Decot, V.; Reppel, L.; Bensoussan, D. Mesenchymal Stem/Stromal Cell Production Compliant with Good Manufacturing Practice: Comparison between Bone Marrow, the Gold Standard Adult Source, and Wharton's Jelly, an Extraembryonic Source. *J. Clin. Med.* **2019**, *8*, E2207. [CrossRef] [PubMed]
240. Merimi, M.; Lagneaux, L.; Moussa Agha, D.; Lewalle, P.; Meuleman, N.; Burny, A.; Fahmi, H.; Najar, M. Mesenchymal Stem/Stromal Cells in Immunity and Disease: A Better Understanding for an Improved Use. *J. Clin. Med.* **2020**, *9*, E1516. [CrossRef] [PubMed]
241. Davalli, P.; Mitic, T.; Caporali, A.; Lauriola, A.; D'Arca, D. ROS, Cell Senescence, and Novel Molecular Mechanisms in Aging and Age-Related Diseases. *Oxidative Med. Cell. Longev.* **2016**, *2016*, 3565127. [CrossRef]
242. Bahir, B.; Choudhery, M.S.; Hussain, I. *Hypoxic Preconditioning as a Strategy to Maintain the Regenerative Potential of Mesenchymal Stem Cells*; IntechOpen: London, UK, 2020; ISBN 978-1-83881-928-6.
243. Ezquer, F.E.; Ezquer, M.E.; Vicencio, J.M.; Calligaris, S.D. Two complementary strategies to improve cell engraftment in mesenchymal stem cell-based therapy: Increasing transplanted cell resistance and increasing tissue receptivity. *Cell Adhes. Migr.* **2017**, *11*, 110–119. [CrossRef]

244. Zhou, P.; Liu, H.; Liu, X.; Ling, X.; Xiao, Z.; Zhu, P.; Zhu, Y.; Lu, J.; Zheng, S. Donor heart preservation with hypoxic-conditioned medium-derived from bone marrow mesenchymal stem cells improves cardiac function in a heart transplantation model. *Stem Cell Res. Ther.* **2021**, *12*, 56. [CrossRef] [PubMed]
245. Narbonne, P. The effect of age on stem cell function and utility for therapy. *Cell Med.* **2018**, *10*, 215517901877375. [CrossRef] [PubMed]
246. Fan, M.; Chen, W.; Liu, W.; Du, G.-Q.; Jiang, S.-L.; Tian, W.-C.; Sun, L.; Li, R.-K.; Tian, H. The Effect of Age on the Efficacy of Human Mesenchymal Stem Cell Transplantation after a Myocardial Infarction. *Rejuvenation Res.* **2010**, *13*, 429–438. [CrossRef] [PubMed]
247. Hermann, A.; List, C.; Habisch, H.-J.; Vukicevic, V.; Ehrhart-Bornstein, M.; Brenner, R.; Bernstein, P.; Fickert, S.; Storch, A. Age-dependent neuroectodermal differentiation capacity of human mesenchymal stromal cells: Limitations for autologous cell replacement strategies. *Cytotherapy* **2010**, *12*, 17–30. [CrossRef]
248. Zhang, J.; Feng, Z.; Wei, J.; Yu, Y.; Luo, J.; Zhou, J.; Li, Y.; Zheng, X.; Tang, W.; Liu, L.; et al. Repair of Critical-Sized Mandible Defects in Aged Rat Using Hypoxia Preconditioned BMSCs with Up-regulation of Hif-1α. *Int. J. Biol. Sci.* **2018**, *14*, 449–460. [CrossRef]
249. Lavrentieva, A.; Hatlapatka, T.; Neumann, A.; Weyand, B.; Kasper, C. Potential for Osteogenic and Chondrogenic Differentiation of MSC. In *Mesenchymal Stem Cells-Basics and Clinical Application I*; Weyand, B., Dominici, M., Hass, R., Jacobs, R., Kasper, C., Eds.; Springer: Berlin/Heidelberg, Germany, 2012; Volume 129, pp. 73–88. ISBN 978-3-642-35670-4.
250. El-Badawy, A.; Amer, M.; Abdelbaset, R.; Sherif, S.N.; Abo-Elela, M.; Ghallab, Y.H.; Abdelhamid, H.; Ismail, Y.; El-Badri, N. Adipose Stem Cells Display Higher Regenerative Capacities and More Adaptable Electro-Kinetic Properties Compared to Bone Marrow-Derived Mesenchymal Stromal Cells. *Sci. Rep.* **2016**, *6*, 37801. [CrossRef]
251. Beegle, J.; Lakatos, K.; Kalomoiris, S.; Stewart, H.; Isseroff, R.R.; Nolta, J.A.; Fierro, F.A. Hypoxic Preconditioning of Mesenchymal Stromal Cells Induces Metabolic Changes, Enhances Survival, and Promotes Cell Retention In Vivo: Hypoxia Affects Metabolism and Survival of MSC. *Stem Cells* **2015**, *33*, 1818–1828. [CrossRef] [PubMed]
252. Wang, W.; Huang, X.; Lin, W.; Qiu, Y.; He, Y.; Yu, J.; Xi, Y.; Ye, X. Hypoxic preconditioned bone mesenchymal stem cells ameliorate spinal cord injury in rats via improved survival and migration. *Int. J. Mol. Med.* **2018**, *42*, 2538–2550. [CrossRef]

cells

MDPI

Review

Human Granulosa Cells—Stemness Properties, Molecular Cross-Talk and Follicular Angiogenesis

Claudia Dompe [1], Magdalena Kulus [2], Katarzyna Stefańska [3], Wiesława Kranc [4], Błażej Chermuła [5], Rut Bryl [4], Wojciech Pieńkowski [6], Mariusz J. Nawrocki [4], James N. Petitte [7], Bogusława Stelmach [5], Paul Mozdziak [7,8], Michal Jeseta [9,10], Leszek Pawelczyk [5], Jędrzej M. Jaśkowski [11], Hanna Piotrowska-Kempisty [12,13], Robert Z. Spaczyński [5], Michał Nowicki [3] and Bartosz Kempisty [2,3,4,7,*]

1 The School of Medicine, Medical Sciences and Nutrition, University of Aberdeen, Aberdeen AB25 2ZD, UK; u16cd16@abdn.ac.uk
2 Department of Veterinary Surgery, Institute of Veterinary Medicine, Nicolaus Copernicus University in Torun, 87-100 Torun, Poland; magdalena.kulus@umk.pl
3 Department of Histology and Embryology, Poznan University of Medical Sciences, 60-781 Poznan, Poland; k.stefanska94@o2.pl (K.S.); mnowicki@ump.edu.pl (M.N.)
4 Department of Anatomy, Poznan University of Medical Sciences, 60-781 Poznan, Poland; wkranc@ump.edu.pl (W.K.); rutbryl@gmail.com (R.B.); mjnawrocki@ump.edu.pl (M.J.N.)
5 Division of Infertility and Reproductive Endocrinology, Department of Gynecology, Obstetrics and Gynecological Oncology, Poznan University of Medical Sciences, 60-535 Poznan, Poland; blazej.chermula@wp.pl (B.C.); b.stelmach@wp.pl (B.S.); pawelczyk.leszek@ump.edu.pl (L.P.); rspaczynski@yahoo.com (R.Z.S.)
6 Division of Perinatology and Women's Diseases, Poznan University of Medical Sciences, 60-535 Poznan, Poland; wpienkowski@poczta.onet.pl
7 Prestage Department of Poultry Science, North Carolina State University, Raleigh, NC 27607, USA; jnppo@ncsu.edu (J.N.P.); pemozdzi@ncsu.edu (P.M.)
8 Physiology Graduate Program, North Carolina State University, Raleigh, NC 27695, USA
9 Department of Obstetrics and Gynecology, Faculty of Medicine, Masaryk University and University Hospital Brno, 602 00 Brno, Czech Republic; jeseta.michal@fnbrno.cz
10 Department of Veterinary Sciences, Czech University of Life Sciences in Prague, 165 00 Prague, Czech Republic
11 Department of Diagnostics and Clinical Sciences, Institute of Veterinary Medicine, Nicolaus Copernicus University in Torun, 87-100 Torun, Poland; jmjaskowski@umk.pl
12 Department of Toxicology, Poznan University of Medical Sciences, 60-631 Poznan, Poland; hpiotrow@ump.edu.pl
13 Department of Basic and Preclinical Sciences, Institute of Veterinary Medicine, Nicolaus Copernicus University in Torun, 87-100 Torun, Poland
* Correspondence: bkempisty@ump.edu.pl; Tel.: +48-61-854-6567; Fax: +48-61-854-6568

Citation: Dompe, C.; Kulus, M.; Stefańska, K.; Kranc, W.; Chermuła, B.; Bryl, R.; Pieńkowski, W.; Nawrocki, M.J.; Petitte, J.N.; Stelmach, B.; et al. Human Granulosa Cells—Stemness Properties, Molecular Cross-Talk and Follicular Angiogenesis. *Cells* **2021**, *10*, 1396. https://doi.org/10.3390/cells10061396

Academic Editor: Mehdi Najar

Received: 23 April 2021
Accepted: 2 June 2021
Published: 5 June 2021

Abstract: The ovarian follicle is the basic functional unit of the ovary, comprising theca cells and granulosa cells (GCs). Two different types of GCs, mural GCs and cumulus cells (CCs), serve different functions during folliculogenesis. Mural GCs produce oestrogen during the follicular phase and progesterone after ovulation, while CCs surround the oocyte tightly and form the cumulus oophurus and corona radiata inner cell layer. CCs are also engaged in bi-directional metabolite exchange with the oocyte, as they form gap-junctions, which are crucial for both the oocyte's proper maturation and GC proliferation. However, the function of both GCs and CCs is dependent on proper follicular angiogenesis. Aside from participating in complex molecular interplay with the oocyte, the ovarian follicular cells exhibit stem-like properties, characteristic of mesenchymal stem cells (MSCs). Both GCs and CCs remain under the influence of various miRNAs, and some of them may contribute to polycystic ovary syndrome (PCOS) or premature ovarian insufficiency (POI) occurrence. Considering increasing female fertility problems worldwide, it is of interest to develop new strategies enhancing assisted reproductive techniques. Therefore, it is important to carefully consider GCs as ovarian stem cells in terms of the cellular features and molecular pathways involved in their development and interactions as well as outline their possible application in translational medicine.

Keywords: stem cells; granulosa cells; cumulus cells; translational medicine; miRNA; follicular angiogenesis

1. Introduction

Female infertility is a worldwide problem, affecting millions of women. It can be caused by various factors and reproductive system disorders, especially those affecting the ovaries. For example, polycystic ovary syndrome (PCOS) affects 5% to 10% of females of reproductive age [1]. The cause of PCOS is multifactorial, including both genetic and environmental factors. PCOS signs and symptoms include enlarged ovaries with numerous small cysts, irregular menstrual cycles, hirsutism, increased androgen levels and problems with ovulation [1]. Another relatively common disease affecting female fertility is premature ovarian insufficiency (POI), occurring in about 1% of all women worldwide. POI is characterised by amenorrhea, hypergonadotropism and hypoestrogenism before the age of 40, effectively causing female infertility [2]. Furthermore, ovarian cysts also constitute a serious problem, affecting both human and animal reproduction [3]. As therapies for reproductive diseases are often characterised by low efficacy, there is a need to develop new treatment strategies for possibly enhancing or restoring fertility.

Stem cells have gained a lot of interest in recent years, especially in the context of regenerative medicine and cellular therapies. As embryonic stem cell (ESC) use is associated with significant ethical concerns, stem cells isolated from adult tissues such as bone marrow [4], adipose tissue [5] or the umbilical cord [6] are an excellent alternative for cell-based therapies. This group includes "mesenchymal stem cells" (MSCs), characterised by a specific set of criteria established by the International Society for Cellular Therapy [7]. MSCs were primarily isolated from the bone marrow by Friedenstein et al. [8], but their populations have since been found in other adult tissues.

The cells building the follicle, a primary functional unit of the ovary, such as granulosa cells (GCs), were demonstrated to possess certain stem-like properties. Furthermore, GCs express markers specific for MSCs such as CD105, CD90 and CD44 and differentiate into other cell types such as osteoblasts, neurons and chondrocytes [9]. Therefore, it seems that there is a potential for their application in translational medicine. Hence, it is important to consider ovarian stem cell cellular features and the molecular pathways involved in their development and interactions as well as outline their possible applications in translational medicine. For this reasons, this article reviews the cellular and molecular aspects of human folliculogenesis as well as follicular angiogenesis, a crucial event for achieving dominance by the follicle; the stemness properties of GCs and their molecular cross-talk; and the influence of miRNAs on function of GCs and cumulus cells (CCs). Finally, we review studies concerning the potential clinical utilisation of ovarian stem cells or their derivatives.

2. Cellular and Molecular Aspects of Folliculogenesis

Ovaries are covered with a single layer of flattened or cuboid cells known as the coelomic epithelium, whereas the ovarian stroma consists mostly of fibroblast-like cells and can be subdivided into cortex and medulla, with ovarian follicles located in the cortex [10].

The ovarian follicle, the ovary's basic functional unit, consists of an oocyte surrounded by one or several layers of somatic cells, including GCs and theca cells. Depending on the folliculogenesis stage, follicles' exact appearance varies in terms of the number of somatic cell layers and their cellular structure.

Folliculogenesis begins during the fourth month of foetal life, resulting in the formation of the stock of resting follicles (RF), comprising mostly primordial follicles [11]. The primordial follicle contains an oocyte arrested in prophase I, surrounded by a flat layer of GCs enveloped by a basement membrane [12]. Subsequently, due to an unknown selection mechanism, primordial follicles undergo growth and differentiation, becoming primary follicles characterised by cuboidalisation of GCs [13]. Simultaneously, a glycoprotein layer (zona pellucida), connecting GCs with the oocyte, is formed [14].

Primary follicles develop into secondary follicles when GCs proliferate and form three to six layers of cells. Simultaneously, theca cells are recruited from connective tissue to surround the basal lamina and form an outer layer of the follicle [15]. Later, theca cells differentiate into theca externa and theca interna, the former composed of cells similar to undifferentiated theca cells and the latter resembling epithelioid cells [11]. Subsequently, these small, fluid-filled cavities develop within the follicle, resulting in antral follicle formation. The antrum enlarges in response to FSH stimulation, and the GCs surrounding the oocyte differentiate into CCs, building the cumulus oophorus, which is characteristic for the antral (preovulatory) follicle [12]. Therefore, the preovulatory follicle contain four different layers of GCs, namely the granulosa membrane on the furthest layer, which becomes vascularised with capillaries sprouting from the theca interna [16]; the periantral granulosa layer; the cumulus oophorus in the intermediate layer; and the corona radiata in the layer closest to the oocyte. GCs exert different functions in each layer, secreting various molecules and expressing a range of receptors [17,18].

Due to the LH surge at the final stage of follicular development, ovulation occurs, and GCs differentiate into luteal cells. The follicle ruptures, forming a gap filled by GCs, which transform into lutein cells with the primary purpose of maintaining the pregnancy [19]. Simultaneously, the oocyte resumes meiotic division. Furthermore, dynamic angiogenetic processes in the corpus luteum ensure a proper supply of blood containing nutrients, oxygen and hormones for the cells showing hormonal activity [20]. However, corpus luteum may regress, and production of oestradiol by lutein cells may be inhibited due to TNFα [21], which was shown to be expressed by GC during short-term in vitro culture [22].

While both CCs and mural GCs differentiate from the same common progenitor during folliculogenesis, they exert different functions [23]. CCs remain in close contact with the oocyte via gap junctions formed by trans-zonal cytoplasmic projections transverse the zona pellucida matrix, which results in cumulus–oocyte complex (COC) development [24]. Mural GCs produce oestrogen during the follicular phase and progesterone after ovulation, while CCs surround the oocyte tightly and form the cumulus oophorus and corona radiata inner cell layer. Gap junctions connecting CCs and the oocyte allow bi-directional cellular communication, exchanging nutrients and metabolites, which results in the stimulation of oocyte maturation [25,26]. Finally, both types of GCs work together toward the oocyte's full maturation, maintaining its quality as well as conducting steroidogenesis [27]. The co-occurrence of folliculogenesis and oogenesis allows the mature Graafian follicle to protect the matured ovum [28].

Although the GCs and epithelial cells share similar characteristics, the follicular epithelium is more dynamic than other epithelia in the body, as follicular development requires significant changes. As the follicle grows, the epithelium expands and the number of layers of GCs forming the follicular epithelium grows around the oocyte, followed by their lateral expansion [29]. Furthermore, gonadotropins regulate the changes characteristic for the ovarian gland that occur from embryogenesis until menopause, while transcription factors influence gene expression during oogenesis and the development of the antral follicles. Hence, any mutations in these genes may lead to ovarian insufficiency and infertility in mammals [30].

There are several important factors involved in folliculogenesis and development of GCs. Follistatin (FST) and bone morphogenetic protein pathways are engaged in folliculogenesis, whereas the forkhead transcription factor (FOXL2) is a master regulator of GC formation [31]. The addition of growth factors such as bFGF, activin A, BMP4, wingless-type mouse mammary tumour virus integration site family member 3A (WNT3A) and follistatin to an in vitro culture of human embryonic stem cells resulted in their differentiation into granulosa-like cells [31].

3. Ovarian Cell Stem-Like Plasticity

Two different populations of ovarian stem cells resembling MSCs have been discovered to reside in ovarian epithelial layers. Ovarian surface epithelial cells were reported to be a possible source of germ cells, while the epithelium-derived epithelial nests might represent primitive GCs; both germ cells and epithelial nests differentiate de novo from mesenchymal progenitor cells located in the ovarian tunica albuginea. It was suggested that germ cells derived from ovarian surface epithelial cells could assemble within primitive GC nests, forming primary follicles [16]. Similarly, Bowen et al. [32] revealed that human ovarian surface epithelium was multipotential because it expressed genes associated with adult stem cell maintenance. In addition, ovarian epithelial cells in patients with severe ovarian infertility were shown to express markers of pluripotency such as SOX-2 or SSEA-4 [33].

Bhartiya et al. [34] established that the ovary contains two stem cell populations: very small, embryonic-like stem cells (VSELs) and ovarian stem cells (OSCs), both located in the ovarian surface epithelium (OSE). Pluripotent VSELs are small (2–4 μm), quiescent stem cells equivalent to primordial germ cells and are able to self-renew and give rise to OSCs, which subsequently form germ cell nests by clonal expansion. While OSCs express OCT-4A, VSELs are characterised by the expression of OCT-4B [34]. The variety in the expression of OCT-4 occurs due to asymmetric cell division of VSELs embedded in the OSE, causing relocation of OCT-4 from the nuclear membrane to the cytoplasm in MSCs [35]. In addition, VSELs present in all adult organs are thought to migrate to different tissues to replace potentially damaged cells [36]. Furthermore, both VSELs and OSCs respond to FSH, participating in the formation of oocytes and primordial follicles, which is also preceded by an epithelial–mesenchymal transition [35].

Recently, a single-cell RNA sequencing of the ovarian cortex revealed six populations of cells: oocytes, GCs, immune cells, endothelial cells, perivascular cells and stromal cells [37]. However, cortical GCs clustered separately from GCs found in antral follicles. The aim of this study was to obtain OSCs by DDX4 (DEAD-Box Helicase 4) antibody, a common marker for germ cells, isolation; moreover, OSCs were suggested to express it on their cellular membrane. The OSCs markers, such as DDX4, were found only in oocytes and perivascular cells, indicating that germline stem cells were absent in adult ovaries [37]. This would contradict the findings of Bhartiya et al. [34]; however, it is important to note that the study by Wanger et al. [37] focused on the ovarian cortex, whereas Bhartiya et al. [34] reported the presence of OSCs in OSE. Moreover, the latter study addressed the controversy regarding the utilisation of the DDX4 marker to obtain OSCs and concluded that additional markers should be used, but technical confusion should not be the reason to doubt the existence of OSCs [34]. Wagner et al. [37], on the other hand, concluded that the DDX4 antibody used in their experiment recognised an epitope specifically expressed on perivascular cells, even though they did not express the *DDX4* transcript. Because DDX4 has been found in perivascular cells, it is interesting to consider that these cells may give rise to multipotent progenitors. It would be consistent with the results obtained by Crisan et al. [38], who obtained perivascular cells from various tissues such as pancreas, muscle, adipose tissue and others. When cultured in vitro, these cells were positive for markers characteristic for MSCs such as CD44, CD73, CD90 and CD105 and exhibited trilineage differentiation potential, suggesting that perivascular cells might have given rise to stem cells, especially MSCs [38].

Another single-cell RNA sequencing study of the inner cortex of adult ovaries was recently performed [39]. As a result, different populations of GCs were distinguished based on specific gene expression. The GCs of small antral follicles exhibited high expression of WT1 (Wilms tumour 1) and EGR4 (early growth response 4) and low expression of VCAN (versican) and FST (follistatin). GCs from selectable follicles were categorised as CCs and mural GCs. CCs showed high expression of VCAN, FST, IGFBP2 (insulin-like growth factor binding protein 2), HTRA1 (high-temperature requirement A serine peptidase 1), INHBB (inhibin subunit beta B) and IHH (Indian hedgehog signalling molecule). Mural GCs expressed high levels of KRT18 (keratin 18), CITED2 (CBP/p300-interacting transactivator

2) and AKIRIN1 and low levels of WT1 and EGR4. In contrast, GCs from atretic follicles did not express VCAN, FST or KRT18 and expressed lower levels of GJA1 (gap junction protein alpha 1) and CDH2 (cadherin 2) compared to other clusters of GCs [39]. However, other cells with stem-like properties were found to reside in the ovary.

As GCs proliferate rapidly during folliculogenesis and exert a variety of specialised functions, it has been suggested that their population contains cells at various stages of differentiation and therefore is not uniform. Kossowska-Tomaszczuk et al. [9] demonstrated, for the first time, that GCs had high proliferation capability and differentiation potential toward various cell lineages and were not terminally differentiated as it was previously assumed. The authors aspirated follicular fluid during oocyte collection for assisted reproduction and isolated a subpopulation of cells expressing FSHR, which were subsequently pooled from different patients. Since these cells expressed both FSHR and aromatase, they were recognised as luteinising GCs. When cultured with LIF (leukaemia inhibitory factor), commonly used to support stem cell growth, they remained viable up to 4 months and retained their morphology. Therefore, they could be cultivated in vitro for a prolonged period. Importantly, LIF is present in follicular fluid, and it has been suggested that it promotes the primordial-to-primary follicle transition [40]. These GCs were positive for OCT-4, a transcription factor considered one of the primary regulators of differentiation and self-renewal. However, other markers of pluripotency such as Nanog, Stellar and or Vasa were negative. Additionally, GCs expressed markers characteristic for MSCs such as CD166, CD90, CD105, CD29 or CD44, but not CD73. GCs were also subjected to neurogenic, osteogenic and chondrogenic differentiation, which resulted in the expression of specific markers such as nestin, neurofilament, BSP (bone sialoprotein), OC (osteocalcin) or COLL1 (collagen 1) [9]. In addition, a subpopulation of GCs was reported to express CD117 (c-kit), a mesenchymal lineage marker and stem cell factor receptor, which is assumed to be involved in the survival of human ovarian follicles [16]. Markers characteristic for stem cells, expressed by GCs, are presented in Figure 1.

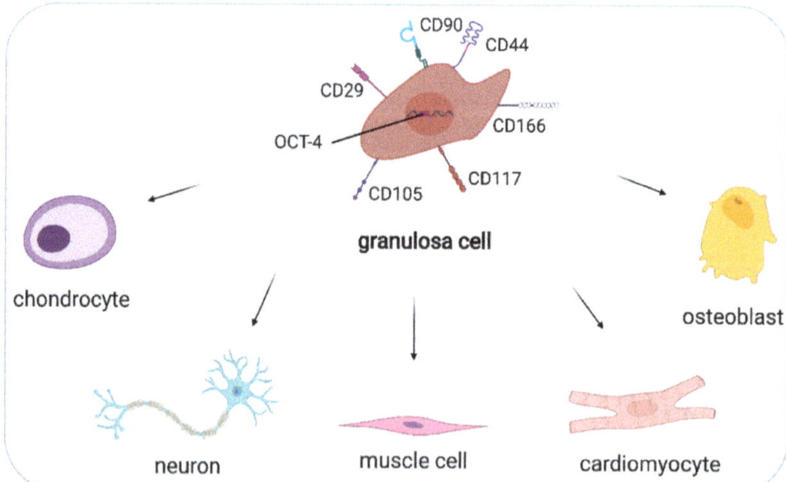

Figure 1. Expression of MSC- and stemness-specific markers, namely CD90, CD44, CD166, CD117, CD105, CD29 and OCT-4 as well as the multipotency of granulosa cells. (Created with BioRender.com, accession date: 29 May 2021).

Further studies showed the ability of GCs to differentiate into muscle cells and cardiac cells [41,42]. However, to maintain stemness properties, these cells may express FSHR but not LHR. It was observed that an increase in LHR expression promoted GC phenotype, and cells expressing both FSHR and LHR would enter apoptosis as in vitro culture progressed [16]. Furthermore, GCs undergo cell division in contact with neighbouring cells in the absence of the substratum that is characteristic for stem cells as well as form basal lamina in vitro, proving that this structure is usually composed of GCs [43].

Dzafic et al. [44] isolated follicular cells aspirated during in vitro fertilisation procedures and reported that these cells, consisting mostly of GCs and theca cells, exhibited expression of MSC-related genes that varied as compared to those expressed by bone marrow MSCs. *IL10, CD45, RUNX2, CD106, OCT4* and 11 other genes were significantly upregulated in follicular cells, suggesting MSC phenotypes. However, *CD73, CD90* and *CD105* were downregulated. Importantly, follicular cells were shown to express a degree of plasticity because they were successfully differentiated toward osteogenic, adipogenic and pancreatic-like cells [44].

Furthermore, GCs share some characteristics with epithelial origin cells such as the secretion of the basal lamina and the presence of adherens junctions, believed to be fundamental for the initiation of follicular growth. However, they lack desmosomes and do not express epithelial cell markers [45]. Therefore, although GCs exhibit some epithelial characteristics, they also resemble the MSC phenotype, expressing vimentin, while their luteinised counterparts express OCT-4, an essential protein involved in cell differentiation and self-renewal.

Further stemness of GCs becomes apparent at the start of ovulation, when the basal lamina is ruptured, and the inner layer of GCs becomes more loosely packed, less polarised and increasingly proliferative [45]. Another example of epithelial–mesenchymal transition in adult tissue, a rare event happening in physiology, occurs when mural GCs increase their size, forming luteinised cells [46]. Furthermore, GCs shift from steroid production of oestradiol to progesterone before the follicle ruptures and after ovulation, if they stay in contact with endothelial cells and produce an extracellular matrix, GCs undergo hypertrophy, differentiating into large luteal cells. Simultaneously, the process of centripetal angiogenesis starts from the vascular network around the follicle, the follicular basement membrane is destroyed and endothelial cells migrate toward the inner GC layer.

A blood platelet lysate was reported to stimulate luteinisation of porcine GCs by converting oestradiol synthesis to progesterone. Furthermore, platelets containing hemoattractive substances were observed to induce angiogenesis [47]. To support the idea of using platelets in ovarian regenerative medicine, it was also suggested that blood platelets regulated endothelial cell migration and GCs' luteinisation during corpus luteum formation in human ovaries [48].

Apart from the characteristics above suggesting the stem-like potential of GCs, they also exhibit higher telomerase activity than other adult cells without stem-like properties [49]. However, telomerase activity decreases with age and is suggested to be related to primordial follicle depletion [50].

Ovarian follicular cells exhibit stem-like potential and present phenotype changes and multipotency during long-term in vitro cultures. Under specific culture conditions eventually leading to successful differentiation into multiple cell types, these stem cells can lose GC functional markers including FSHR and aromatase, and express markers characteristic of mesenchymal stem cell phenotypes [51]. Although the expression of stem cell markers in GCs including OCT-4, Nanog and SOX-2 have been described, their expression varies between different species and maturational stages of GCs [52].

Understanding the processes underlying the differentiation of GCs toward different cell lineages and the related molecular pathways of this mechanism is fundamental to uncovering other possible stemness markers of GCs, allowing for taking full advantage of their MSC-like characteristics. Improved cell cultures to prolong the lifespan of GCs are being developed. Utilising 3D cultures, employing MEF (mouse embryonic fibroblast) medium and the addition of follicular fluid and LIF were observed to prolong the lifespan, encourage proliferation and maintain cells in an undifferentiated state in vitro [9,53,54].

4. Regulation of Angiogenesis in Ovarian Follicles

The proper function of ovaries and ovarian follicles is maintained by continual angiogenesis, a process of new blood vessel formation from those already existing. In response to an angiogenic stimulus such as hypoxia or wounding, the endothelial and mural cells become destabilised. Subsequently, they migrate towards the angiogenic stimuli and proliferate, forming a new vessel [53].

Follicular vasculature begins to develop at the secondary follicle stage, as previously mentioned, within the theca cell layer, with the GC layer remaining avascular, separated by the basement membrane. However, GCs seem to play an important role in this process, producing angiogenic factors and influencing events occurring in the theca layer [45]. Yang et al. [54] suggested that vascular endothelial growth factor (VEGF) was a factor responsible for initial vasculature recruitment, probably due to oocyte-secreted factors influencing its expression in theca and GCs [55]. Follicular vasculature seems to be crucial in achieving dominance by the follicle, as indicated by another study conducted in bovine follicles [56], because the vascularity was shown to be positively correlated with the expression of VEGF in oestrogen-active follicles.

The gonadotropin surge occurring in the middle of the menstrual cycle results in the breakdown of the basement membrane, invasion of the blood vessels to the granulosa layer and subsequent corpus luteum formation. Therefore, luteal angiogenesis originates from the developing follicle, especially because accumulation of pro-angiogenic growth factors occurs during preovulatory follicle development [57]. Angiogenesis occurring during corpus luteum development is more intense than follicular angiogenesis because up to 85% of the cells proliferating in the corpus luteum are of vascular origin [58]. In a previous study, it was suggested that granulosa cells might participate in the vascularisation of the corpus luteum, as they were observed to express phenotypic (Tie, Tek, cKit, Flt-1, CD-31, vWF proteins) and functional (rapid AcLDL uptake and tube-forming ability in vitro) markers associated with endothelial or endothelial-like cells [59]. Furthermore, Merkwitz et al. identified the presence of somatic, KIT-positive progenitor cells in GC cultures as well as expression of CD14, CD45, CD133 or VEGF-R2. In this study, from KIT-positive cell cultures, they obtained cultures of granulosa cells or endothelial cells, showing heterogeneity of microvascular sources. These results suggested that progenitor cells could be obtained from harvested GCs and could be important for further research on the angiogenesis of the corpus luteum [60].

Amongst pro-angiogenic factors, VEGF, basic fibroblast growth factor (FGF2), platelet-derived growth factor (PDGF) and angiopoietin (ANGPT) can be distinguished, which, in general, promote the proliferation and migration of endothelial cells, whereas thrombospondin and angiostatin are factors acting in an anti-angiogenic manner, inhibiting endothelial cell proliferation and migration or stimulating their apoptosis. Although thrombospondins (both THBS1 and THBS2) were associated with decreased vascularity and proliferation of GCs in growing follicles, a study conducted in macaques revealed that the levels of THBS1 mRNA and protein were increased in the GCs of preovulatory follicles after the gonadotropin surge. In addition, the treatment of monkey ovarian microvascular endothelial cells with THBS1 resulted in increased migration, proliferation and capillary sprout formation, suggesting that THBS1 acted also in a pro-angiogenic manner during ovulation and corpus luteum formation [61]. In contrast, Garside et al. [62] demonstrated that THBS1 both inhibited angiogenesis and promoted follicular atresia via direct induction of apoptosis of GCs.

FGF2 is expressed in mature follicles and the corpus luteum and was the first angiogenic factor identified in the ovary [63]. As indicated by Grasselli et al. [64], FGF2 exerted an inhibitory effect on nitric oxide production in porcine GCs, which might be associated with its pro-angiogenic properties. Moreover, FGF2 seemed to be a critical factor regulating early luteinisation, possibly due to its response to the LH surge because LH was demonstrated to stimulate FGF2 production in dispersed luteal cells. Furthermore, the concentration of FGF2 in bovine preovulatory follicular fluid was higher in animals experiencing an LH surge [65]. The assumption of FGF2 being vital for luteal angiogenesis is consistent with the findings of Yamashita et al. [66] in cows. Intraluteal injections of FGF2 antibody resulted in decreased progesterone secretion and corpus luteum volume. Similar results were obtained for the VEGF antibody, suggesting a prominent role of both factors in corpus luteum formation and function [66].

The induction of angiogenesis is also dependent on the occurrence of hypoxia, with hypoxia-inducible factor 1 (HIF-1) being responsible for the response to oxygen deprivation. A study conducted by Calvani et al. [67] revealed that human umbilical endothelial vein cells (HUVECs) cultured under hypoxic conditions formed tube-like structures. Neutralisation of FGF2 resulted in blocked survival and sprouting of HUVECs under hypoxic conditions, whereas HIF-1A activity was required for hypoxic induction of FGF2 mRNA and protein expression [67]. Hypoxia was also reported to influence the level of VEGF, as indicated by Nishimura et al. [68], because oxygen deprivation increased the amount of VEGF mRNA and protein in cultured bovine luteal cells.

There is no doubt that VEGF is a master regulator of both follicular and luteal angiogenesis as well as new capillary formation within the ovulatory follicle because its inhibition results in a marked decrease in endothelial and granulosa cell proliferation in developing antral follicles, alongside inhibition of follicular growth and ovulation [69]. VEGFA (vascular endothelial growth factor A) exists in several isoforms as a result of an alternative splicing, with $VEGFA_{165}$ being the most predominant protein isoform in humans [57]. It has been suggested that various isoforms of VEGFA may be differentially expressed during folliculogenesis, as indicated by a study conducted in cows. The authors investigated the effects of progesterone on the gene expression of hypoxia-inducible factor 1 alpha (HIF-1A), $VEGFA_{120}$ and $VEGFA_{164}$. Progesterone might have stimulated the expression of $VEGFA_{120}$ via HIF-1A, whereas the expression of $VEGFA_{164}$ expression was inhibited by this hormone [70]. VEGFA binds to the receptors VEGFR1 and VEGFR2, which are located on the surfaces of endothelial cells [57]. However, several co-receptors (such as neuropilin 1 and neuropilin 2) were reported to influence the effects exerted by VEGFA. Shimizu et al. [71] revealed that the neuropilin 1 (NRP-1) gene was expressed in the GCs and theca cells of pre- and post-selection bovine follicles, whereas the NRP-2 gene was expressed only in theca cells of these follicles. Moreover, the NRP-1 gene was shown to be regulated by sex hormones; namely, oestradiol increased its expression in cultured GCs, while progesterone decreased NRP-1 expression [71].

However, other factors were also reported to play important roles in follicular angiogenesis. VEGF combined with PDGF was shown to suppress angiogenesis, suggesting an antagonistic relationship between both aforementioned factors [72]. PDGF is responsible for the recruitment of pericytes, which are important constituents of microvessels, to the blood vessel wall as well as their interactions with endothelial cells [57]. PDGF acts through PDGF receptor β (PDGFRβ). Inhibition of this signalling resulted in severe haemorrhage in mice due to the blockade of pericyte recruitment in the angiogenic corpus luteum, highlighting the importance of PDGF signalling during ovarian angiogenesis [73]. In addition, intraovarian injection of an inhibitor of PDGF receptor activity caused a significant decrease in corpora lutea in rats [74].

Angiopoietins (ANGPT1 and ANGPT2) act via endothelial, cell-specific tyrosine kinase receptor Tie2. ANGPT1 stimulates vessel maturation and is essential for normal vasculature development, whereas ANGPT2 is a naturally occurring antagonist for ANGPT1 and Tie2, destabilising endothelial–pericyte contacts [75]. While ANGPT1 is expressed in all tissues, ANGPT2 is present mostly in the ovaries, uterus and placenta [63]. During the pre-ovulatory stage, the ANGPT1/ANGPT2 ratio increases, suggesting that the process of vascular maturation is more evident than vascular expansion. After the LH surge, the ANGPT2/ANGPT1 ratio is increased, probably inducing destabilisation of existing vessels and pericyte detachment. In the corpus luteum, new vessels are formed due to the activity of ANGPT1 and the recruitment of pericytes, resulting in fully vascularised luteal cells [53].

Therefore, the ratio of ANGPT1/ANGPT2 changes during follicular–luteal transition and plays an important role in vessel fate, which is schematically presented in Figure 2.

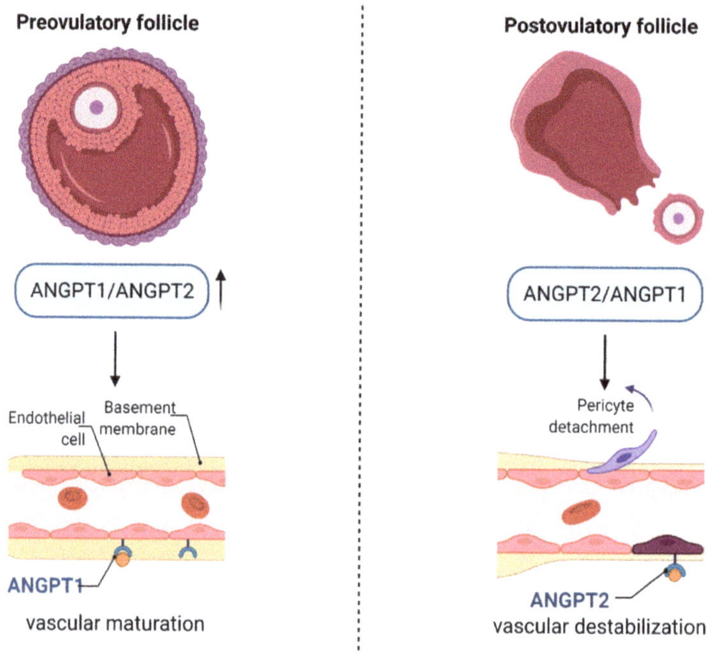

Figure 2. The role of angiopoietins in pre- and post-ovulatory follicles. Before ovulation, the ANGPT1/ANGPT2 ratio increases, leading to vascular maturation, whereas after the LH surge, the level of ANGPT2 increases compared to ANGPT1, leading to the destabilisation of vessels and pericyte detachment. Abbreviations—ANGPT1: angiopoietin 1; ANGPT2: angiopoietin 2. (Created with BioRender.com, accession date: 29 May 2021).

5. The Role of miRNA in the Regulation of Granulosa and Cumulus Cells' Function

MiRNAs are small, non-coding molecules involved in post-transcriptional gene expression regulation by base-pairing to mRNAs. They regulate various cellular processes such as proliferation, differentiation and apoptosis, targeting multiple functionally related genes constituting gene expression networks [76]. MiRNAs are expressed in ovarian tissue, GCs, oocytes and follicular fluid, among others, affecting mammalian fertility. It seems that miRNAs exert vital roles in folliculogenesis because differential expression of miRNAs involved in follicular cell proliferation, steroidogenesis, luteinisation and oocyte maturation between small and large bovine follicles, or between healthy and atretic ones, has been observed [77].

MiRNAs present in GCs or CCs may affect oocyte maturation, as indicated in several studies. Sinha et al. [78] increased or inhibited the expression of miR-130b in bovine GCs and CCs cultured in vitro, revealing that the SMAD5 and MSK1 genes were direct targets of miR-130b. Furthermore, overexpression of miR-130b resulted in increased GC and CC proliferation, whereas its inhibition during oocyte in vitro maturation caused reduction in the first polar body extrusion, mitochondrial activity and number of oocytes reaching metaphase II. Therefore, miR-130b was demonstrated to be involved in both GCs' and CCs' proliferation and survival and in oocyte maturation [78]. Similarly, miR-375 was reported to be involved in the regulation of bovine oocyte in vitro maturation via targeting ADAMTS1 and PGR in CCs, leading to the suppression of cumulus–oocyte complex maturation [79]. Moreover, miR-375 was shown to regulate the expression of BMPR2, thereby affecting the expression of BMP15 and GDF9 receptors, influencing the proliferation and apoptosis of bovine CCs [80]. MiR-21-3p, on the other hand, influences bovine GCs autophagy, as demonstrated by Ma et al. [81], by targeting VEGFA and attenuating PI3K/AKT signalling, resulting in autophagy inhibition.

In the case of humans, miRNAs were reported to regulate the cumulus–oocyte complex as indicated by Assou et al. [82], who compared the expression of miRNA in human metaphase II oocytes and in CCs. As a result, the most abundant miRNAs in CCs were let-7b, let-7c and miR-21, whereas miR-184 and miR-10a were the most abundant in oocytes. Further analyses revealed differentially expressed genes in CCs and oocytes that were predicted to be targeted by the aforementioned miRNAs and were associated with regulation of the cell cycle or apoptosis [82].

GCs and CCs may differ in terms of various miRNA content, as indicated by Andrei et al. [83]. Human GCs and CCs were obtained from healthy women undergoing in vitro fertilisation and subjected to small RNA sequencing. As a result, 53 miRNAs were reported to be significantly differentially expressed between GCs and CCs. Most of the highly abundant miRNAs such as miR-21-5p, let-7a-5p and let-7f-5p were present both in GCs and CCs; however, miR-30a-5p was uniquely expressed in the top 10 miRNAs of GCs, whereas miR-320a was present only in the top 10 miRNAs of CCs. Differentially expressed miRNAs were implicated to be involved in the regulation of steroidogenesis, as well as in the apoptosis and proliferation of GCs; for example, miR-146a promoted apoptosis by targeting interleukin-1 receptor-associated kinase 1 (IRAK1) and tumour necrosis factor receptor-associated factor 6 (TRAF6) [83].

MiRNAs are important regulators of GC proliferation and apoptosis. Sirotkin et al. [84] demonstrated that miR-15a inhibited the proliferation of human GCs by decreasing the level of proliferating cell nuclear antigen (PCNA). In addition, miR-15a was reported to promote the release of progesterone and testosterone, but not oestradiol, when transfected in GCs in vitro [85]. MiR-143 also participates in the regulation of hormonal production, as indicated by Zhang et al. [86]. After the transfection of miR-143 inhibitor into primary cultured GCs, the production of oestradiol was significantly increased, as well as steroidogenesis-related gene expression. Further studies revealed that miR-143 negatively regulated the signalling pathway of FSH, whereas FSH decreased miR-143 expression [86].

MiRNAs could also be implicated in various pathological conditions such as premature ovarian insufficiency or polycystic ovary syndrome. Indeed, Yang et al. [87] discovered differentially expressed miRNAs in the plasmas of POI patients and normal cycling women, revealing that miR-23a and miR-27a were highly expressed in POI patients' plasma. After the transfection of human ovarian GCs with pre-miR-23a, there was an increase in the occurrence of apoptosis, probably due to a decrease in the X-linked inhibitor of apoptosis protein (XIAP) and caspase-3 levels [87]. Further studies aimed to clarify the exact anti-apoptotic mechanism. Using luciferase reporter assay, RT-PCR and Western blotting, Nie et al. [88] discovered that SMAD5 was a target gene for both miR-23a and miR-27, and that the regulation of apoptosis in GCs occurred via the FasL-Fas pathway. Several other miRNAs may be involved in the occurrence of POI, namely the already-mentioned miR-146a. Cho et al. [89] aimed to identify new target genes for polymorphism of miR-146a (miR-146aC > G) in POI. Altered miRNA was transfected into human GCs, and further analysis revealed that miR-146aC > G led to significantly altered regulation of cyclin D2 (CCND2) and forkhead box O3 (FOXO3), genes associated with POI. Therefore, genetic variants of miR-146a may contribute to the occurrence of POI [89]. Similarly, miR938G > A polymorphisms were also identified as factors associated with POI due to altered binding to the gonadotropin-releasing hormone receptor (GnRHR) mRNA [90]. Mir-139-5p was also reported to be involved in POI occurrence, as indicated by Zhao et al. [91]. Progesterone receptor membrane component 1 (PGRMC1), crucial for GC survival, was found to be upregulated in GCs due to hyaluronic acid-mediated suppression of miR-139-5p. In POI patients, however, levels of miR-139-5p were significantly increased, presenting inverse correlation with PGRMC1 level [91].

In the case of PCOS, miR-93 and miR-21 have been highlighted as androgen-responsive factors, positively correlated with free testosterone and the free androgen index because both aforementioned miRNAs were increased in GC-form hyperandrogenic PCOS patients as compared to normoandrogenic patients [92]. Oestrogen deficiency is a hallmark of PCOS, and downregulation of miR-320a expression in CCs from PCOS patients has been found to influence steroidogenesis due to modulation of CYP11A1 and CYP19A1 by directly targeting osteogenic transcription factor RUNX2 [93]. MiR-483 was also found to be downregulated in PCOS, and its overexpression resulted in decreased cell viability and proliferation as well as cell cycle arrest induction in human granulosa-like tumour cell line KGN. Such inhibition of proliferation was a result of miR-483-targeting IGF1 [94]. MiR-145 negatively regulates the proliferation of GCs from PCOS patients. However, miR-145 inhibited the expression of insulin receptor substrate 1 (IRS1), which resulted in inhibition of the activation of p38 mitogen-activated protein kinase (p38 MAPK) and extracellular signal-regulated kinase (ERK). Furthermore, high concentrations of insulin decreased the expression of miR-145, upregulated IRS1 and promoted cell proliferation, which was consistent with the downregulation of miR-145 in PCOS patients [95]. The important miRNAs and their functions in PCOS are presented in Figure 3.

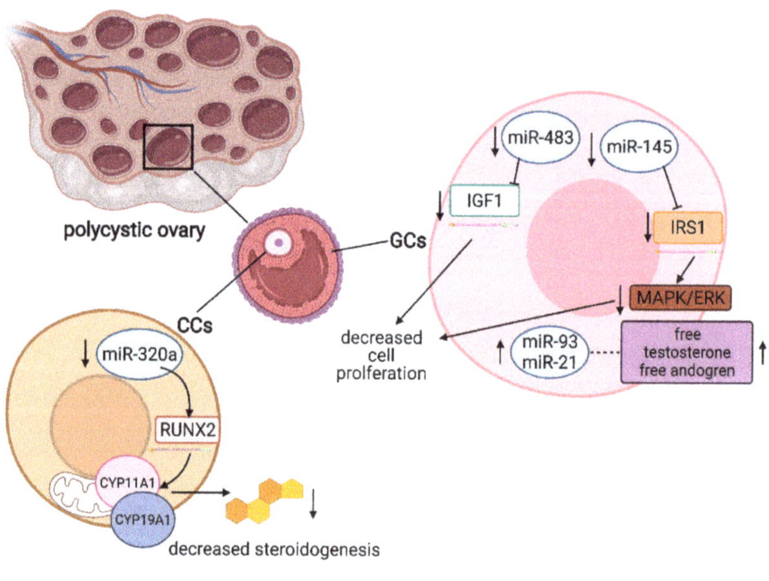

Figure 3. Schematic representation of miRNAs influencing the occurrence of PCOS. MiR-320a is significantly downregulated in CCs, resulting in decreased steroidogenesis due to dysregulation of RUNX2 and CYP11A1/CYP19A1. In GCs, both miR-93 and miR-21 are upregulated and positively correlated with free testosterone and the androgen index, whereas miR-483 and miR-145 are downregulated. MiR-483 directly targets the IGF1 mRNA, leading to decreased cell proliferation. IRS1 mRNA is a direct target of miR-145, leading to inhibition of the MAPK/ERK signalling pathway, resulting in decreased cell proliferation. Abbreviations—RUNX2: runt-related transcription factor 2; CYP11A1: cytochrome P450 family 11 subfamily A member 1; CYP19A1: cytochrome P450 family 19 subfamily A member 1; IGF1: insulin-like growth factor 1; IRS1: insulin receptor substrate 1; MAPK: mitogen-activated protein kinase; ERK: extracellular signal-regulated kinase; GCs: granulosa cells; CCs: cumulus cells. (Created with BioRender.com, accession date: 29 May 2021).

6. Ovarian Follicular Cells' Molecular "Cross-Talk" and Interaction

The oocyte's developmental ability to undergo meiosis, be fertilised and form a healthy embryo is a factor determining female fertility. The growth of the mammalian oocyte is simultaneous with follicular growth, both of which are regulated via signals from molecular cross-talk between the germline and somatic cells. The oocyte grows within the follicle, which also comprises the somatic granulosa or theca cells, extending through an inward division of the outermost layer that results in additional layer formation [14]. When the antrum develops, GCs differentiate into CCs and mural GCs [23]. Meanwhile, the oocyte grows and develops the ability to undergo meiosis and be fertilised before ovulation, supported by GCs [96,97]. Folliculogenesis, oogenesis and ovulation are regulated by pituitary gonadotropins (reviewed by Amsterdam A et al. [98], whose action is mediated by intra-ovarian signals from paracrine factors and cell–cell communication through gap junctions [6]. Specifically, the bi-directional communication between the oocyte and CCs is crucial for proper folliculogenesis and oogenesis. CCs strongly support the growth and maturation of the oocyte, which, in turn, influences proliferation and differentiation of CCs as well as their production of extracellular matrix and steroid hormones. Moreover, both GCs and CCs protect the oocyte from oxidative stress through the antioxidant system [99,100].

Considerable evidence suggests the involvement of CCs in the development of the oocyte. However, only a few paracrine factors have been recognised. KIT ligand (KITL) and the KIT tyrosine kinase receptor mediate the CC–oocyte interaction and are both expressed in developing and preovulatory follicles of the postnatal ovary. KITL is expressed in rat, mouse and human CCs and was reported to stimulate the oocyte's growth through the KIT receptor located on the oolemma. Moreover, it was identified as necessary and sufficient to induce primordial follicle development and might trigger oocyte's growth initiation [101–105]. Mice with mutations in KIT or KITL were reported to be infertile, and KITL-dependant activation of KIT is essential for ovarian follicle growth when FSH is not yet expressed [106,107]. KITL binding to KIT activates diverse signalling pathways involved in cell survival and apoptosis. Furthermore, phosphatidylinositol (PI) 3-kinase (PI3K) is a KIT effector, leading to changes in the expression of key players of the apoptotic pathway [108]. PI3K inhibitors were shown to block the anti-apoptotic effect of KITL in germ cells during foetal oogenesis. Simultaneously, PTEN negatively interacted with the PI3K signalling pathway, and oocyte-specific PTEN knockout led to premature ovarian insufficiency [109]. Furthermore, KIT activation induced the phosphorylation of AKT and FKHRL1, activating the former and functionally suppressing the latter [110].

In CCs, KITL is expressed as either a soluble or membrane-bound protein, which, respectively, can be cleaved or not (remaining stably on the membrane). Furthermore, the latter (KITL2) is thought to be the principal isoform necessary for the oocyte's growth, whereas the former (KITL1) is associated with fully grown oocytes [111]. FSH is the main endocrine factor regulating ovarian functions and stimulating follicle growth. It influences the expression of KITL—its low concentration increases KITL2 level and stimulates oocyte growth, while high expression increases the KITL1/KITL2 ratio, failing to promote development of the oocyte. Therefore, the correct concentration of FSH is crucial for proper development of the oocyte [105]. The role of KIT signalling in CC–oocyte interaction is presented in Figure 4.

Furthermore, the understanding of molecular mechanisms and regulation of gene expression in CCs may shed light on oocyte quality and their ability to acquire developmental competence. Long-term in vitro cultured CCs were demonstrated to exhibit significant changes in the expression of genes such as *DKK1*, *ANXA3*, *KIAA1199*, *VCAM1* and *HTR2B*, all of which were upregulated. *DKK1* is an antagonist of the WNT signalling pathway responsible for pluripotency regulation. Therefore, it was hypothesised that elevated expression of DKK1 in CCs might influence oocyte developmental competence acquisition and sustain its pluripotency [100].

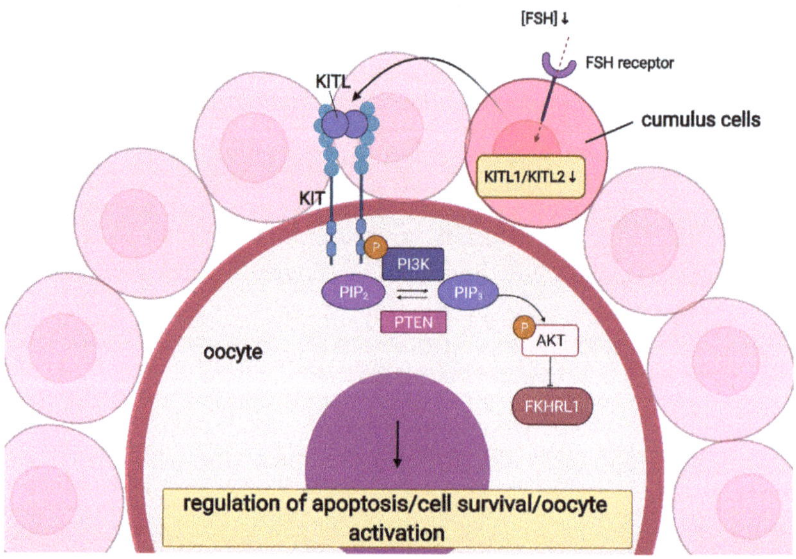

Figure 4. KIT signalling in CC–oocyte interaction: regulation of oocyte growth and development. Abbreviations—AKT: protein kinase B; FKHRL1: forkhead1; FSH: follicle-stimulating hormone; KITL: KIT ligand; PI3K: phosphatidylinositol 3-kinase; PIP_2: phosphatidylinositol 4,5-bisphosphate; PIP_3: phosphatidylinositol (3,4,5)-trisphosphate; PTEN: phosphatase and tensin homolog. (Created with BioRender.com, accession date: 29 May 2021).

Oocyte or oocyte-secreted factors contribute to the suppression of luteinisation and are required for successful cumulus expansion and extrusion of the oocyte–cumulus cell mass from the follicle at ovulation [112]. This process depends on gonadotropins, epidermal growth factor (EGF) and paracrine factors secreted from the oocyte. CCs respond to these signals thanks to oocyte-secreted factors, leading to the expression of the transcripts necessary to form the extracellular matrix [113,114]. Oocytes stimulate proliferation of CCs and secrete paracrine factors, which inhibit LH receptors and progesterone production in CCs [115–117].

Oocyte-secreted factors also act in an anti-apoptotic manner, resulting in a low incidence of apoptosis in CCs. A study conducted on bovine COCs revealed that the removal of the oocyte from this complex resulted in increased CC apoptosis. This effect was reversed by the addition of denuded oocytes to CC in vitro culture, which, on the molecular level, promoted the expression of anti-apoptotic BCL-2 and inhibited pro-apoptotic BAX expression. Bone morphogenetic protein 15 (BMP-15) and bone morphogenetic protein 6 (BMP-6), both secreted by the oocyte, maintained the anti-apoptotic effect via establishment of a localised gradient of bone morphogenetic proteins [118]. During long-term in vitro culture of human CCs, the expression of the apoptosis-regulatory genes BAX, CASP9 and TP53 decreased, as indicated by the results of RT-qPCR [119]. However, the study conducted with GCs cultured in vitro for seven days revealed that the ratio of BCL-2/BAX was improved after that period, indicating these cells' survival [120].

Furthermore, BMP-15 and growth differentiation factor 9 (GDF-9) are fundamental for follicular development [121,122]. Oocytes communicate via these factors, and their secretion is regulated by bi-directional communication between oocytes and CCs [123]. These oocyte-secreted factors prevent spontaneous luteinisation of CCs and control their endocrine function [124,125]. Oocytes produce GDF-9, deficiency in which causes an arrest of follicular development, leading to infertility. The absence of BMP-15, on the other hand, results in decreased ovulation and fertilisation rates. These two factors belong to the TGF-β superfamily and are essential for regulating the differentiation and proper functioning of CCs [126]. The superfamily of TGF-β consists of factors crucial in mammalian reproductive functions, participating in folliculogenesis, oogenesis and embryo development [127].

BMP-15 upregulates KITL1 and KITL2. However, the ratio of KITL1/KITL2 remains unchanged, which is crucial for the stimulation of oocyte development. The soluble isoform of KITL inhibits BMP-15 expression in oocytes. Therefore, a negative feedback loop between KITL and BMP-15 may be a primary regulatory pathway [105]. Moreover, because FSH influences KITL expression, and its high concentration decreases BMP-15 expression, FSH may regulate BMP-15 expression via KIT signalling [105].

GDF-9 may inhibit the expression of KITL. Similarly, KITL expression was inhibited by GC co-culture with fully grown, but not with partly grown, oocytes. Therefore, it was suggested that GDF-9 mediated the effect of fully grown oocytes on KITL expression in GCs [128].

Additionally, other components of the TGF-β superfamily were reported to play an essential part in follicular cell cross-talk. The expression of TGF-β1 after 7 days in vitro culture was unchanged, probably due to the regulatory functions of GC proliferation and differentiation as well as stimulation of preantral follicle growth [129].

CCs maintain contact with the oocyte via cellular processes known as transzonal projections (TZPs), which penetrate the zona pellucida, an extracellular coat formed around the oocyte after it enters the growth phase [130]. TZPs originate from CCs and are composed of a strong backbone of actin filaments or tubulin [131,132]. Microtubule-containing TZPs are assumed to be involved in paracrine communication, whereas actin-containing TZPs appear to mediate gap junctional communication and CC–oocyte adhesion. FSH was proven to be an essential modulator of microtubule-containing TZP organisation [24]. The amount and shape of TZPs change during follicular development, with a high number contributing to growth and a lower number during the maturation of fully grown oocytes upon gonadotropic surges [133]. Gap junctions form at the tip of these TZPs and are composed of connexins, a family of 20 proteins [134]. Six connexins form a connexon that, together with another connexon from an adjacent cell, forms a channel between cells, connecting the oocyte and CCs or two CCs in the follicle [135,136]. Amino acids are transported through these gap junctions, enhancing uptake of the glycine, alanine, lysine and taurine necessary for oocyte growth [137]. Other molecules passing via gap junctions include ions, metabolites and cAMP. Finally, mammalian oocytes rely on CCs containing additional GLUT that have a high affinity to glucose as well as phosphofructokinase activity to convert glucose into the substrates necessary for energy metabolism during the oocyte's growth [138,139].

Apart from bi-directional communication between the oocyte and CCs or GCs, ovarian follicular cells also interact with each other. Both GCs and theca cells produce steroidal and nonsteroidal factors, influencing each other's proliferation and differentiation during folliculogenesis. Specifically, growth factors secreted by theca cells such as epidermal growth factor (EGF), transforming growth factor α (TGFα), keratinocyte growth factor (KGF), hepatocyte growth factor (HGF) and bone morphogenetic protein 7 (BMP-7) are thought to promote the proliferation of GCs and suppress their apoptosis in early antral follicles [140]. Moreover, theca cells produce androgens, with androgen receptors localised on GCs. In mice, these theca cell-derived androgens were shown to stimulate GC mitosis and preantral follicle growth, as indicated by in vitro studies [141].

The molecular interaction between the oocyte and theca cells has been described, as well. As previously mentioned, GDF-9 belongs to the oocyte-secreted factors. A study conducted in rats revealed that GDF-9 enhanced follicular development, while its antagonist suppressed follicular growth and androgen production. Moreover, when the androgen receptor antagonist flutamide was used, it suppressed the preantral follicle growth mediated by GDF-9 in vitro. Therefore, GDF-9 was proposed to play a role in the promotion of preantral follicle growth via androgen synthesis upregulation by theca cells [142].

7. Application of Ovarian Follicular Stem Cells in Translational Medicine

Although ovarian follicular cells exhibit promising properties such as stem-like potential and high proliferation capability, the data on their possible application in translational medicine remain minimal. Because the ability of GCs to differentiate into osteogenic cells had already been suggested, Mattioli et al. [143] aimed to understand the osteo-regenerative potential of GCs. Mattioli et al. isolated GCs from growing and luteinising porcine follicles and subjected them to osteogenic differentiation, which resulted in marked extracellular matrix mineralisation, alkaline phosphatase activation, upregulation of osteocalcin and Runx2 expression. GCs were also differentiated after incorporating polylactic-co-glycolic acid (PLGA) scaffolds and being subsequently transplanted subcutaneously in the dorsal region of SCID mice. After the implant retrieval, the viable GCs surrounding nodules of calcification were revealed. The above results indicated that GCs possessed osteogenic potential both in vitro and in vivo and therefore could be used for bone regeneration [143]. Chandramohan et al. [144] proposed a scaffold based on chitosan and polycaprolactone (PCL) for ovarian stem cell engineering. The cells were collected from human follicular fluid and exhibited properties of MSCs while seeded on the chitosan/PCL scaffold, coated with zinc divalent ions to impart osteogenic properties. As a result, calcium deposition and alkaline phosphatase activity were increased, together with the expression of Runx2, osteonectin or osteocalcin, indicating that such a scaffold was compatible with MSCs and could be used in bone tissue engineering [144]. GCs were also demonstrated as a useful tool for obtaining functional oocytes. Tian et al. [145] isolated GCs from adult mouse ovaries and induced them to generate germline-competent, pluripotent stem cells (gPSCs) by a chemical approach, using crotonic sodium or acid. These gPSCs could then be differentiated into primordial germ cell-like cells and form functional oocytes [145].

Apart from the possibility to use the cells themselves, the supernatant of cultured granulosa–cumulus cells was proven to exhibit therapeutic properties [146]. Madkour et al. isolated immature denuded oocytes from patients with polycystic ovary syndrome, which were subsequently subjected to in vitro maturation (IVM) with the addition of autologous or heterologous follicular fluid and supernatant of cumulus–granulosa cells. The latter resulted in an improved yield of the developed blastocyst and an IVM rate higher than that of an in vivo maturation rate. Therefore, the supernatant of cumulus–granulosa cells could be a useful IVF-enhancing tool [146]. Atrabi et al. [147] investigated the influence of the conditioned medium of GCs and CCs on the activation of primordial follicles in mice. One-day-old mice ovaries were cultured for six days with conditioned media of GCs or CCs or co-cultured with these cells. Obtained results indicated that the conditioned medium of GCs could contribute to primordial follicle activation, probably due to downregulation of PTEN [147].

Ovarian stem cells could serve as an in vitro model to examine germ cell development because they are able to differentiate into oocyte-like cells. As indicated in a study by Taheri et al. [148], such differentiation could be induced with BMP-15. Lee et al. [149], on the other hand, overexpressed OCT-4 in ovarian stem cells, which resulted in higher oogenesis potential compared to the controls. In addition, CC and GC biomarkers may be useful in predicting oocyte and embryo quality in assisted reproductive techniques [12,150]. For example, matrix metalloproteinases (MMPs) and their tissue inhibitors (TIMPs) in GCs and CCs were assessed as potential biomarkers of oocyte quality by Luddi et al. [151]. As a result, the expression of MMP2 in GCs was shown to be increased in infertile patients as compared to fertile patients. Increased expression of MMP2 in GCs and CCs was also associated with the production of fewer oocytes, while the expression of MMP9 in GCs was positively correlated with the number of retrieved oocytes [151]. The potential application of ovarian follicular cells or their derivatives in translational medicine is presented in Figure 5.

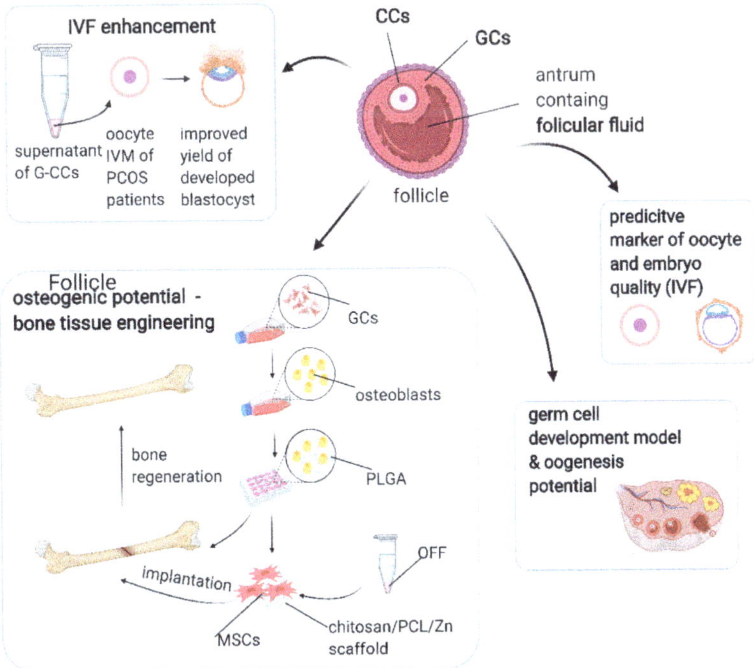

Figure 5. Possible clinical applications of ovarian follicular cells. Abbreviations—CCs: cumulus cells; G–CCs: granulosa–cumulus cells; GCs: granulosa cells; IVF: in vitro fertilisation; IVM: in vitro maturation; MSCs: mesenchymal stem cells; OFF: ovarian follicular fluid; PCL: polycaprolactone; PCOS: polycystic ovarian syndrome; PLGA: polylactic-co-glycolic acid. (Created with BioRender.com, accession date: 29 May 2021).

8. Conclusions

Follicular cells contained within the ovary exhibit several important functions during folliculogenesis and oogenesis. They comprise cells like GCs, CCs or theca cells, involved in the synthesis, expression and metabolism of various hormones essential for gamete maturation, regulation of ovulation and pregnancy sustenance such as progesterone production upon ovulation and oestradiol production during the growth of the follicle. Moreover, they form gap junctions with the oocyte, which enable a bidirectional exchange of nutrients and metabolites, essential for gamete maturation stimulation. The developing oocyte influences follicular cell proliferation and expansion, and regulation of follicular angiogenesis is crucial for achieving dominance by the follicle. The proper function of GCs and CCs is dependent on miRNAs, and alterations in these interactions may lead to PCOS or POI. Since GCs were suggested to exhibit properties characteristic of mesenchymal stem cells such as multilineage differentiation potential, specific antigen pattern expression and high proliferation rate, their potential utilisation in translational medicine has been suggested. Although the available data are still limited, ovarian stem cells have been successfully used for bone tissue engineering, in vitro maturation and predicting the quality of oocytes and embryos. However, further studies are required to fully exploit these cells' potential in regenerative and experimental medicine.

Following the discovery by Kossowska-Tomaszczuk et al. of the multipotency of GCs to differentiate in other cell lineages, their long-life span may be applied in a broad range of therapies and medical research specifically related to the treatment of ovarian pathologies such as cancer, endometriosis and polycystic ovary syndrome [9].

Furthermore, as prolonged cultures of GCs with LIF have shown the ability to differentiate toward different lineages both in vitro and in vivo, they exhibit characteristics of mesenchymal cells. These, in turn, have been widely studied for their therapeutic potential, with further developments potentially achieved through the study of differentiation-promoting agents. Furthermore, CGs could be employed as starting materials to obtain tissue made up of cells of other lineages, with these advances potentially representing a breakthrough for modern clinical medicine.

Author Contributions: Conceptualization, H.P.-K., R.Z.S., J.M.J. and L.P.; methodology, M.J.N., P.M. and J.N.P.; investigation, C.D., M.K., K.S. and B.S.; resources, W.P., M.J., W.K. and B.C.; writing—original draft preparation, C.D., M.K., K.S. and B.S.; writing—review and editing, W.K., B.C. and P.M.; visualization, R.B.; supervision, B.K., M.N., W.P., H.P.-K. and L.P.; project administration, M.J., M.J.N., J.N.P. and J.M.J.; funding acquisition, B.K., R.Z.S and M.N. All authors have read and agreed to the published version of the manuscript.

Funding: This research was funded by the National Science Centre, grant number 2018/31/B/NZ5/02475. The APC was funded by the National Science Centre.

Conflicts of Interest: The authors declare no conflict of interest.

References

1. Ndefo, U.A.; Eaton, A.; Green, M.R. Polycystic ovary syndrome: A review of treatment options with a focus on pharmacological approaches. *P T* **2013**, *38*, 336–355.
2. Zhang, C. The Roles of Different Stem Cells in Premature Ovarian Failure. *Curr. Stem Cell Res. Ther.* **2019**, *15*, 473–481. [CrossRef] [PubMed]
3. Rybska, M.; Knap, S.; Jankowski, M.; Borowiec, B.; Jeseta, M.; Bukowska, D.; Antosik, P.; Nowicki, M.; Zabel, M.; Kempisty, B.; et al. Pathogenesis and pathophysiology of ovarian follicular cysts in mammals. *Med. J. Cell Biol.* **2018**, *6*, 120–124. [CrossRef]
4. Zhang, T.; Lee, W.Y.W.; Rui, Y.F.; Cheng, T.Y.; Jiang, X.H.; Li, G. Bone marrow-derived mesenchymal stem cells promote growth and angiogenesis of breast and prostate tumors. *Stem Cell Res. Ther.* **2013**, *4*. [CrossRef]
5. Song, Y.; Du, H.; Dai, C.; Zhang, L.; Li, S.; Hunter, D.J.; Lu, L.; Bao, C. Human adipose-derived mesenchymal stem cells for osteoarthritis: A pilot study with long-term follow-up and repeated injections. *Regen. Med.* **2018**, *13*, 295–307. [CrossRef] [PubMed]
6. Can, A.; Celikkan, F.T.; Cinar, O. Umbilical cord mesenchymal stromal cell transplantations: A systemic analysis of clinical trials. *Cytotherapy* **2017**, *19*, 1351–1382. [CrossRef]
7. Dominici, M.; Le Blanc, K.; Mueller, I.; Slaper-Cortenbach, I.; Marini, F.C.; Krause, D.S.; Deans, R.J.; Keating, A.; Prockop, D.J.; Horwitz, E.M. Minimal criteria for defining multipotent mesenchymal stromal cells. The International Society for Cellular Therapy position statement. *Cytotherapy* **2006**, *8*, 315–317. [CrossRef]
8. Friedenstein, A.J.; Chailakhjan, R.K.; Lalykina, K.S. The Development of Fibroblast Colonies in Monolayer Cultures of Guinea-Pig Bone Marrow and Spleen Cells. *Cell Prolif.* **1970**, *3*, 393–403. [CrossRef]
9. Kossowska-Tomaszczuk, K.; De Geyter, C.; De Geyter, M.; Martin, I.; Holzgreve, W.; Scherberich, A.; Zhang, H. The Multipotency of Luteinizing Granulosa Cells Collected from Mature Ovarian Follicles. *Stem Cells* **2009**, *27*, 210–219. [CrossRef] [PubMed]
10. Weidner, N.; Cote, R.J.; Suster, S.; Weiss, L.M. *Modern Surgical Pathology*; Saunders: Philadelphia, PA, USA, 2009; ISBN 9781416039662.
11. Gougeon, A. Regulation of ovarian follicular development in primates: Facts and hypotheses. *Endocr. Rev.* **1996**, *17*, 121–155. [CrossRef]
12. Uyar, A.; Torrealday, S.; Seli, E. Cumulus and granulosa cell markers of oocyte and embryo quality. *Fertil. Steril.* **2013**, *99*, 979–997. [CrossRef]
13. McGee, E.A.; Hsueh, A.J.W. Initial and Cyclic Recruitment of Ovarian Follicles*. *Endocr. Rev.* **2000**, *21*, 200–214. [CrossRef]
14. Da Silva-Buttkus, P.; Jayasooriya, G.S.; Mora, J.M.; Mobberley, M.; Ryder, T.A.; Baithun, M.; Stark, J.; Franks, S.; Hardy, K. Effect of cell shape and packing density on granulosa cell proliferation and formation of multiple layers during early follicle development in the ovary. *J. Cell Sci.* **2008**, *121*, 3890–3900. [CrossRef]
15. Shah, J.S.; Sabouni, R.; Cayton Vaught, K.C.; Owen, C.M.; Albertini, D.F.; Segars, J.H. Biomechanics and mechanical signaling in the ovary: A systematic review. *J. Assist. Reprod. Genet.* **2018**, *35*, 1135–1148. [CrossRef]
16. Kossowska-Tomaszczuk, K.; De Geyter, C. Cells with stem cell characteristics in somatic compartments of the ovary. *Biomed Res. Int.* **2013**, *2013*, 310859. [CrossRef] [PubMed]
17. Findlay, J.K.; Kerr, J.B.; Britt, K.; Liew, S.H.; Simpson, E.R.; Rosairo, D.; Drummond, A. Ovarian physiology: Follicle development, oocyte and hormone relationships. *Anim. Reprod.* **2009**, *6*, 16–19.

18. Nguyen, T.; Lee, S.; Hatzirodos, N.; Hummitzsch, K.; Sullivan, T.R.; Rodgers, R.J.; Irving-Rodgers, H.F. Spatial differences within the membrana granulosa in the expression of focimatrix and steroidogenic capacity. *Mol. Cell. Endocrinol.* **2012**, *363*, 62–73. [CrossRef]
19. Kranc, W.; Chachula, A.; Wojtanowicz-Markiewicz, K.; Ciesiólka, S.; Ociepa, E.; Bukowska, D.; Borys, S.; Piotrowska, H.; Bryja, A.; Antosik, P.; et al. The Insight into Developmental Capacity of Mammalian Cocs and Cumulus-Granulosa Cells-Recent Studies and Perspectives. *Austin J. Invit. Fertilzation* **2015**, *2*, 1023.
20. Fraser, H.M.; Wulff, C. Angiogenesis in the corpus luteum. *Reprod. Biol. Endocrinol.* **2003**, *1*, 88. [CrossRef] [PubMed]
21. Rybska, M.; Knap, S.; Jankowski, M.; Jeseta, M.; Bukowska, D.; Antosik, P.; Nowicki, M.; Zabel, M.; Kempisty, B.; Jaśkowski, J.M. Characteristic of factors influencing the proper course of folliculogenesis in mammals. *Med. J. Cell Biol.* **2018**, *6*, 33–38. [CrossRef]
22. Kałuzna, S.; Bryl, R.; Chermuła, B.; Sibiak, R.; Stefańska, K.; Pieńkowski, W.; Kranc, W.; Jeseta, M.; Ventruba, P.; Zakova, J.; et al. Expression of genes involved in the inflammatory response in human granulosa cells in short-term in vitro culture. *Med. J. Cell Biol.* **2020**, *8*, 190–195. [CrossRef]
23. Erickson, G.F.; Hofeditz, C.; Unger, M.; Allen, W.R.; Dulbecco, R. A monoclonal antibody to a mammary cell line recognizes two distinct subtypes of ovarian granulosa cells. *Endocrinology* **1985**, *117*, 1490–1499. [CrossRef]
24. Plancha, C.E.; Sanfins, A.; Rodrigues, P.; Albertini, D. Cell polarity during folliculogenesis and oogenesis. *Reprod. Biomed. Online* **2005**, *10*, 478–484. [CrossRef]
25. Saeed-Zidane, M.; Linden, L.; Salilew-Wondim, D.; Held, E.; Neuhoff, C.; Tholen, E.; Hoelker, M.; Schellander, K.; Tesfaye, D. Cellular and exosome mediated molecular defense mechanism in bovine granulosa cells exposed to oxidative stress. *PLoS ONE* **2017**, *12*. [CrossRef]
26. Mora, J.M.; Fenwick, M.A.; Castle, L.; Baithun, M.; Ryder, T.A.; Mobberley, M.; Carzaniga, R.; Franks, S.; Hardy, K. Characterization and significance of adhesion and junction-related proteins in mouse ovarian follicles. *Biol. Reprod.* **2012**, *86*, 1–14. [CrossRef]
27. Vanderhyden, B.C.; Tonary, A.M. Differential regulation of progesterone and estradiol production by mouse cumulus and mural granulosa cells by a factor(s) secreted by the oocyte. *Biol. Reprod.* **1995**, *53*, 1243–1250. [CrossRef]
28. Kranc, W.; Budna, J.; Kahan, R.; Chachuła, A.; Bryja, A.; Ciesiółka, S.; Borys, S.; Antosik, M.P.; Bukowska, D.; Brussow, K.P.; et al. Molecular basis of growth, proliferation, and differentiation of mammalian follicular granulosa cells. *J. Biol. Regul. Homeost. Agents* **2017**, *31*, 1–8. [PubMed]
29. Rodgers, R.J.; Lavranos, T.C.; Van Wezel, I.L.; Irving-Rodgers, H.F. Development of the ovarian follicular epithelium. *Mol. Cell. Endocrinol.* **1999**, *151*, 171–179. [CrossRef]
30. Jagarlamudi, K.; Rajkovic, A. Oogenesis: Transcriptional regulators and mouse models. *Mol. Cell. Endocrinol.* **2012**, *356*, 31–39. [CrossRef] [PubMed]
31. Lan, C.-W.; Chen, M.-J.; Jan, P.-S.; Chen, H.-F.; Ho, H.-N. Differentiation of Human Embryonic Stem Cells Into Functional Ovarian Granulosa-like Cells. *J. Clin. Endocrinol. Metab.* **2013**, *98*, 3713–3723. [CrossRef] [PubMed]
32. Bowen, N.J.; Walker, L.D.; Matyunina, L.V.; Logani, S.; Totten, K.A.; Benigno, B.B.; McDonald, J.F. Gene expression profiling supports the hypothesis that human ovarian surface epithelia are multipotent and capable of serving as ovarian cancer initiating cells. *BMC Med. Genomics* **2009**, *2*, 71. [CrossRef] [PubMed]
33. Virant-Klun, I.; Skutella, T.; Stimpfel, M.; Sinkovec, J. Ovarian surface epithelium in patients with severe ovarian infertility: A potential source of cells expressing markers of pluripotent/multipotent stem cells. *J. Biomed. Biotechnol.* **2011**, *2011*, 381928. [CrossRef]
34. Bhartiya, D.; Patel, H. Ovarian stem cells—resolving controversies. *J. Assist. Reprod. Genet.* **2018**, *35*, 393–398. [CrossRef]
35. Stefańska, K.; Sibiak, R.; Hutchings, G.; Dompe, C.; Moncrieff, L.; Janowicz, K.; Jeseta, M.; Kempisty, B.; Machatkova, M.; Mozdziak, P. Evidence for existence of molecular stemness markers in porcine ovarian follicular granulosa cells. *Med. J. Cell Biol.* **2019**, *7*, 183–188. [CrossRef]
36. Moncrieff, L.; Mozdziak, P.; Jeseta, M.; Machatkova, M.; Kranc, W.; Kempisty, B. Ovarian follicular cells—Living in the shadow of stemness cellular competence. *Med. J. Cell Biol.* **2019**, *7*, 134–140. [CrossRef]
37. Wagner, M.; Yoshihara, M.; Douagi, I.; Damdimopoulos, A.; Panula, S.; Petropoulos, S.; Lu, H.; Pettersson, K.; Palm, K.; Katayama, S.; et al. Single-cell analysis of human ovarian cortex identifies distinct cell populations but no oogonial stem cells. *Nat. Commun.* **2020**, *11*, 1147. [CrossRef] [PubMed]
38. Crisan, M.; Yap, S.; Casteilla, L.; Chen, C.W.; Corselli, M.; Park, T.S.; Andriolo, G.; Sun, B.; Zheng, B.; Zhang, L.; et al. A Perivascular Origin for Mesenchymal Stem Cells in Multiple Human Organs. *Cell Stem Cell* **2008**, *3*, 301–313. [CrossRef]
39. Fan, X.; Bialecka, M.; Moustakas, I.; Lam, E.; Torrens-Juaneda, V.; Borggreven, N.V.; Trouw, L.; Louwe, L.A.; Pilgram, G.S.K.; Mei, H.; et al. Single-cell reconstruction of follicular remodeling in the human adult ovary. *Nat. Commun.* **2019**, *10*, 3164. [CrossRef]
40. Nilsson, E.E.; Kezele, P.; Skinner, M.K. Leukemia inhibitory factor (LIF) promotes the primordial to primary follicle transition in rat ovaries. *Mol. Cell. Endocrinol.* **2002**, *188*, 65–73. [CrossRef]
41. Kranc, W.; Brązert, M.; Celichowski, P.; Bryja, A.; Nawrocki, M.J.; Ożegowska, K.; Jankowski, M.; Jeseta, M.; Pawelczyk, L.; Bręborowicz, A.; et al. 'Heart development and morphogenesis' is a novel pathway for human ovarian granulosa cell differentiation during long-term in vitro cultivation-a microarray approach. *Mol. Med. Rep.* **2019**, *19*, 1705–1715. [CrossRef]
42. Brevini, T.A.L.; Pennarossa, G.; Rahman, M.M.; Paffoni, A.; Antonini, S.; Ragni, G.; deEguileor, M.; Tettamanti, G.; Gandolfi, F. Morphological and Molecular Changes of Human Granulosa Cells Exposed to 5-Azacytidine and Addressed Toward Muscular Differentiation. *Stem Cell Rev. Reports* **2014**, *10*, 633–642. [CrossRef] [PubMed]

43. Rodgers, R.J.; Lavranos, T.C.; Rodgers, H.F.; Young, F.M.; Vella, C.A. The physiology of the ovary: Maturation of ovarian granulosa cells and a novel role for antioxidants in the corpus luteum. *J. Steroid Biochem. Mol. Biol.* **1995**, *53*, 241–246. [CrossRef]
44. Dzafic, E.; Stimpfel, M.; Novakovic, S.; Cerkovnik, P.; Virant-Klun, I. Expression of Mesenchymal Stem Cells-Related Genes and Plasticity of Aspirated Follicular Cells Obtained from Infertile Women. *Biomed Res. Int.* **2014**, *2014*. [CrossRef] [PubMed]
45. Chermuła, B.; Brazert, M.; Izycki, D.; Ciesiółka, S.; Kranc, W.; Celichowski, P.; Ozegowska, K.; Nawrocki, M.J.; Jankowski, M.; Jeseta, M.; et al. New Gene Markers of Angiogenesis and Blood Vessels Development in Porcine Ovarian Granulosa Cells during Short-Term Primary Culture in Vitro. *Biomed Res. Int.* **2019**, *2019*. [CrossRef]
46. Bowdridge, E.C.; Vernon, M.W.; Flores, J.A.; Clemmer, M.J. In vitro progesterone production by luteinized human mural granulosa cells is modulated by activation of AMPK and cause of infertility. *Reprod. Biol. Endocrinol.* **2017**, *15*, 76. [CrossRef]
47. Furukawa, K.; Fujiwara, H.; Sato, Y.; Zeng, B.X.; Fujii, H.; Yoshioka, S.; Nishi, E.; Nishio, T. Platelets are novel regulators of neovascularization and luteinization during human corpus luteum formation. *Endocrinology* **2007**, *148*, 3056–3064. [CrossRef]
48. Basini, G.; Bussolati, S.; Grolli, S.; Ramoni, R.; Conti, V.; Quintavalla, F.; Grasselli, F. Platelets are involved in in vitro swine granulosa cell luteinization and angiogenesis. *Anim. Reprod. Sci.* **2018**, *188*, 51–56. [CrossRef]
49. Dzafic, E.; Stimpfel, M.; Virant-Klun, I. Plasticity of granulosa cells: On the crossroad of stemness and transdifferentiation potential. *J. Assist. Reprod. Genet.* **2013**, *30*, 1255–1261. [CrossRef]
50. Kinugawa, C.; Murakami, T.; Okamura, K.; Yajima, A. Telomerase activity in normal ovaries and premature ovarian failure. *Tohoku J. Exp. Med.* **2000**, *190*, 231–238. [CrossRef]
51. Hoang, S.N.; Ho, C.N.Q.; Nguyen, T.T.P.; Doan, C.C.; Tran, D.H.; Le, L.T. Evaluation of stemness marker expression in bovine ovarian granulosa cells. *Anim. Reprod.* **2019**, *16*, 277–281. [CrossRef] [PubMed]
52. Bezerra, M.É.S.; Gouveia, B.B.; Barberino, R.S.; Menezes, V.G.; Macedo, T.J.S.; Cavalcante, A.Y.P.; Monte, A.P.O.; Santos, J.M.S.; Matos, M.H.T. Resveratrol promotes in vitro activation of ovine primordial follicles by reducing DNA damage and enhancing granulosa cell proliferation via phosphatidylinositol 3-kinase pathway. *Reprod. Domest. Anim.* **2018**, *53*, 1298–1305. [CrossRef] [PubMed]
53. Robinson, R.S.; Woad, K.J.; Hammond, A.J.; Laird, M.; Hunter, M.G.; Mann, G.E. Angiogenesis and vascular function in the ovary. *Reproduction* **2009**, *138*, 869–881. [CrossRef] [PubMed]
54. Yang, M.Y.; Fortune, J.E. Vascular endothelial growth factor stimulates the primary to secondary follicle transition in bovine follicles in vitro. *Mol. Reprod. Dev.* **2007**, *74*, 1095–1104. [CrossRef]
55. Nilsson, E.E.; Detzel, C.; Skinner, M.K. Platelet-derived growth factor modulates the primordial to primary follicle transition. *Reproduction* **2006**, *131*, 1007–1015. [CrossRef]
56. Grazul-Bilska, A.T.; Navanukraw, C.; Johnson, M.L.; Vonnahme, K.A.; Ford, S.P.; Reynolds, L.P.; Redmer, D.A. Vascularity and expression of angiogenic factors in bovine dominant follicles of the first follicular wave. *J. Anim. Sci.* **2007**, *85*, 1914–1922. [CrossRef] [PubMed]
57. Woad, K.J.; Robinson, R.S. Luteal angiogenesis and its control. *Theriogenology* **2016**, *86*, 221–228. [CrossRef]
58. Reynolds, L.P.; Redmer, D.A. Growth and development of the corpus luteum. *J. Reprod. Fertil. Suppl.* **1999**, *54*, 181–191. [CrossRef]
59. Antczak, M.; Van Blerkom, J. The vascular character of ovarian follicular granulosa cells: Phenotypic and functional evidence for an endothelial-like cell population. *Hum. Reprod.* **2000**, *15*, 2306–2318. [CrossRef]
60. Merkwitz, C.; Ricken, A.M.; Lösche, A.; Sakurai, M.; Spanel-Borowski, K. Progenitor cells harvested from bovine follicles become endothelial cells. *Differentiation* **2010**, *79*, 203–210. [CrossRef]
61. Bender, H.R.; Campbell, G.E.; Aytoda, P.; Mathiesen, A.H.; Duffy, D.M. Thrombospondin 1 (THBS1) Promotes Follicular Angiogenesis, Luteinization, and Ovulation in Primates. *Front. Endocrinol.* **2019**, *10*, 727. [CrossRef]
62. Garside, S.A.; Harlow, C.R.; Hillier, S.G.; Fraser, H.M.; Thomas, F.H. Thrombospondin-1 Inhibits Angiogenesis and Promotes Follicular Atresia in a Novel in Vitro Angiogenesis Assay. *Endocrinology* **2010**, *151*, 1280–1289. [CrossRef] [PubMed]
63. Tamanini, C.; De Ambrogi, M. Angiogenesis in Developing Follicle and Corpus Luteum. *Reprod. Domest. Anim.* **2004**, *39*, 206–216. [CrossRef]
64. Grasselli, F.; Basini, G.; Bussolati, S.; Tamanini, C. Effects of VEGF and bFGF on proliferation and production of steroids and nitric oxide in porcine granulosa cells. *Reprod. Domest. Anim.* **2002**, *37*, 362–368. [CrossRef] [PubMed]
65. Robinson, R.S.; Nicklin, L.T.; Hammond, A.J.; Schams, D.; Hunter, M.G.; Mann, G.E. Fibroblast Growth Factor 2 Is More Dynamic than Vascular Endothelial Growth Factor A During the Follicle-Luteal Transition in the Cow1. *Biol. Reprod.* **2007**, *77*, 28–36. [CrossRef] [PubMed]
66. Yamashita, H.; Kamada, D.; Shirasuna, K.; Matsui, M.; Shimizu, T.; Kida, K.; Berisha, B.; Schams, D.; Miyamoto, A. Effect of local neutralization of basic fibroblast growth factor or vascular endothelial growth factor by a specific antibody on the development of the corpus luteum in the cow. *Mol. Reprod. Dev.* **2008**, *75*, 1449–1456. [CrossRef]
67. Calvani, M.; Rapisarda, A.; Uranchimeg, B.; Shoemaker, R.H.; Melillo, G. Hypoxic induction of an HIF-1α-dependent bFGF autocrine loop drives angiogenesis in human endothelial cells. *Blood* **2006**, *107*, 2705–2712. [CrossRef] [PubMed]
68. Nishimura, R.; Okuda, K. Hypoxia is important for establishing vascularization during corpus luteum formation in cattle. *J. Reprod. Dev.* **2010**, *56*, 110–116. [CrossRef]
69. Fraser, H.M.; Duncan, W.C. Vascular morphogenesis in the primate ovary. *Angiogenesis* **2005**, *8*, 101–116. [CrossRef]
70. Shimizu, T.; Miyamoto, A. Progesterone induces the expression of vascular endothelial growth factor (VEGF) 120 and Flk-1, its receptor, in bovine granulosa cells. *Anim. Reprod. Sci.* **2007**, *102*, 228–237. [CrossRef]

71. Shimizu, T.; Jayawardana, B.C.; Nishimoto, H.; Kaneko, E.; Tetsuka, M.; Miyamoto, A. Hormonal regulation and differential expression of neuropilin (NRP)-1 and NRP-2 genes in bovine granulosa cells. *Reproduction* **2006**, *131*, 555–559. [CrossRef]
72. Greenberg, J.I.; Shields, D.J.; Barillas, S.G.; Acevedo, L.M.; Murphy, E.; Huang, J.; Scheppke, L.; Stockmann, C.; Johnson, R.S.; Angle, N.; et al. A role for VEGF as a negative regulator of pericyte function and vessel maturation. *Nature* **2008**, *456*, 809–814. [CrossRef]
73. Kuhnert, F.; Tam, B.Y.Y.; Sennino, B.; Gray, J.T.; Yuan, J.; Jocson, A.; Nayak, N.R.; Mulligan, R.C.; McDonald, D.M.; Kuo, C.J. Soluble receptor-mediated selective inhibition of VEGFR and PDGFRβ signaling during physiologic and tumor angiogenesis. *Proc. Natl. Acad. Sci. USA* **2008**, *105*, 10185–10190. [CrossRef] [PubMed]
74. Sleer, L.S.; Taylor, C.C. Platelet-derived growth factors and receptors in the rat corpus luteum: Localization and identification of an effect on luteogenesis. *Biol. Reprod.* **2007**, *76*, 391–400. [CrossRef] [PubMed]
75. Maisonpierre, P.C.; Suri, C.; Jones, P.F.; Bartunkova, S.; Wiegand, S.J.; Radziejewski, C.; Compton, D.; McClain, J.; Aldrich, T.H.; Papadopoulos, N.; et al. Angiopoietin-2, a natural antagonist for Tie2 that disrupts in vivo angiogenesis. *Science.* **1997**, *277*, 55–60. [CrossRef] [PubMed]
76. Gurtan, A.M.; Sharp, P.A. The role of miRNAs in regulating gene expression networks. *J. Mol. Biol.* **2013**, *425*, 3582–3600. [CrossRef] [PubMed]
77. Salilew-Wondim, D.; Gebremedhn, S.; Hoelker, M.; Tholen, E.; Hailay, T.; Tesfaye, D. The role of micrornas in mammalian fertility: From gametogenesis to embryo implantation. *Int. J. Mol. Sci.* **2020**, *21*, 585. [CrossRef] [PubMed]
78. Sinha, P.B.; Tesfaye, D.; Rings, F.; Hossien, M.; Hoelker, M.; Held, E.; Neuhoff, C.; Tholen, E.; Schellander, K.; Salilew-Wondim, D. MicroRNA-130b is involved in bovine granulosa and cumulus cells function, oocyte maturation and blastocyst formation. *J. Ovarian Res.* **2017**, *10*, 37. [CrossRef]
79. Zhang, J.; Guan, Y.; Shen, C.; Zhang, L.; Wang, X. MicroRNA-375 regulates oocyte in vitro maturation by targeting ADAMTS1 and PGR in bovine cumulus cells. *Biomed. Pharmacother.* **2019**, *118*, 109350. [CrossRef]
80. Chen, H.; Liu, C.; Jiang, H.; Gao, Y.; Xu, M.; Wang, J.; Liu, S.; Fu, Y.; Sun, X.; Xu, J.; et al. Regulatory Role of miRNA-375 in Expression of BMP15/GDF9 Receptors and its Effect on Proliferation and Apoptosis of Bovine Cumulus Cells. *Cell. Physiol. Biochem.* **2017**, *41*, 439–450. [CrossRef]
81. Ma, L.; Zheng, Y.; Tang, X.; Gao, H.; Liu, N.; Gao, Y.; Hao, L.; Liu, S.; Jiang, Z. MiR-21-3p inhibits autophagy of bovine granulosa cells by targeting VEGFA via PI3K/AKT signaling. *Reproduction* **2019**, *158*, 441–452. [CrossRef]
82. Assou, S.; Al-Edani, T.; Haouzi, D.; Philippe, N.; Lecellier, C.H.; Piquemal, D.; Commes, T.; Aït-Ahmed, O.; Dechaud, H.; Hamamah, S. MicroRNAs: New candidates for the regulation of the human cumulus-oocyte complex. *Hum. Reprod.* **2013**, *28*, 3038–3049. [CrossRef]
83. Andrei, D.; Nagy, R.A.; van Montfoort, A.; Tietge, U.; Terpstra, M.; Kok, K.; van den Berg, A.; Hoek, A.; Kluiver, J.; Donker, R. Differential miRNA Expression Profiles in Cumulus and Mural Granulosa Cells from Human Pre-ovulatory Follicles. *MicroRNA* **2018**, *8*, 61–67. [CrossRef] [PubMed]
84. Sirotkin, A.V.; Lauková, M.; Ovcharenko, D.; Brenaut, P.; Mlynček, M. Identification of microRNAs controlling human ovarian cell proliferation and apoptosis. *J. Cell. Physiol.* **2010**, *223*, 49–56. [CrossRef]
85. Sirotkin, A.V.; Kisova, G.; Brenaut, P.; Ovcharenko, D.; Grossmann, R.; Mlyncek, M. Involvement of MicroRNA Mir15a in Control of Human Ovarian Granulosa Cell Proliferation, Apoptosis, Steroidogenesis, and Response to FSH. *MicroRNA* **2014**, *3*, 29–36. [CrossRef] [PubMed]
86. Zhang, L.; Zhang, X.X.; Zhang, X.; Lu, Y.; Li, L.; Cui, S. MiRNA-143 mediates the proliferative signaling pathway of FSH and regulates estradiol production. *J. Endocrinol.* **2017**, *234*, 1–14. [CrossRef]
87. Yang, X.; Zhou, Y.; Peng, S.; Wu, L.; Lin, H.Y.; Wang, S.; Wang, H. Differentially expressed plasma microRNAs in premature ovarian failure patients and the potential regulatory function of mir-23a in granulosa cell apoptosis. *Reproduction* **2012**, *144*, 234–244. [CrossRef]
88. Nie, M.; Yu, S.; Peng, S.; Fang, Y.; Wang, H.; Yang, X. miR-23a and miR-27a promote human granulosa cell apoptosis by targeting SMAD51. *Biol. Reprod.* **2015**, *98*, 1–10. [CrossRef] [PubMed]
89. Cho, S.H.; An, H.J.; Kim, K.A.; Ko, J.J.; Kim, J.H.; Kim, Y.R.; Ahn, E.H.; Rah, H.C.; Lee, W.S.; Kim, N.K. Single nucleotide polymorphisms at miR-146a/196a2 and their primary ovarian insufficiency-related target gene regulation in granulosa cells. *PLoS ONE* **2017**, *12*, e0183479. [CrossRef]
90. Cho, S.H.; Ahn, E.H.; An, H.J.; Kim, J.H.; Ko, J.J.; Kim, Y.R.; Lee, W.S.; Kim, N.K. Association of mir-938G>A polymorphisms with primary ovarian insufficiency (POI)-related gene expression. *Int. J. Mol. Sci.* **2017**, *18*, 1255. [CrossRef]
91. Zhao, G.; Zhou, X.; Fang, T.; Hou, Y.; Hu, Y. Hyaluronic acid promotes the expression of progesterone receptor membrane component 1 via epigenetic silencing of miR-139-5p in human and rat granulosa cells. *Biol. Reprod.* **2014**, *91*, 116. [CrossRef]
92. Naji, M.; Aleyasin, A.; Nekoonam, S.; Arefian, E.; Mahdian, R.; Amidi, F. Differential Expression of miR-93 and miR-21 in Granulosa Cells and Follicular Fluid of Polycystic Ovary Syndrome Associating with Different Phenotypes. *Sci. Rep.* **2017**, *7*. [CrossRef]
93. Zhang, C.L.; Wang, H.; Yan, C.Y.; Gao, X.F.; Ling, X.J. Deregulation of RUNX2 by miR-320a deficiency impairs steroidogenesis in cumulus granulosa cells from polycystic ovary syndrome (PCOS) patients. *Biochem. Biophys. Res. Commun.* **2017**, *482*, 1469–1476. [CrossRef] [PubMed]

94. Xiang, Y.; Song, Y.; Li, Y.; Zhao, D.; Ma, L.; Tan, L. miR-483 is down-regulated in polycystic ovarian syndrome and inhibits KGN cell proliferation via targeting insulin-like growth factor 1 (IGF1). *Med. Sci. Monit.* **2016**, *22*, 3383–3393. [CrossRef] [PubMed]
95. Cai, G.; Ma, X.; Chen, B.; Huang, Y.; Liu, S.; Yang, H.; Zou, W. MicroRNA-145 Negatively Regulates Cell Proliferation Through Targeting IRS1 in Isolated Ovarian Granulosa Cells from Patients with Polycystic Ovary Syndrome. *Reprod. Sci.* **2017**, *24*, 902–910. [CrossRef] [PubMed]
96. De La Fuente, R.; Eppig, J.J. Transcriptional activity of the mouse oocyte genome: Companion granulosa cells modulate transcription and chromatin remodeling. *Dev. Biol.* **2001**, *229*, 224–236. [CrossRef] [PubMed]
97. Brower, P.T.; Schultz, R.M. Intercellular communication between granulosa cells and mouse oocytes: Existence and possible nutritional role during oocyte growth. *Dev. Biol.* **1982**, *90*, 144–153. [CrossRef]
98. Amsterdam, A.; Tajima, K.; Frajese, V.; Seger, R. Analysis of signal transduction stimulated by gonadotropins in granulosa cells. *Mol. Cell. Endocrinol.* **2003**, *202*, 77–80.
99. Kocherova, I.; Bryl, R.; Crha, I.; Ventruba, P.; Zakova, J.; Ješeta, M. The extracellular reactive oxygen species levels in primary in vitro culture of human ovarian granulosa and cumulus cells. *Med. J. Cell Biol.* **2020**, *8*, 112–117. [CrossRef]
100. Brazert, M.; Kranc, W.; Jopek, K.; Kempisty, B.; Pawelczyk, L. New markers of human cumulus oophorus cells cultured in vitro-transcriptomic profile. *Med. J. Cell Biol.* **2020**, *8*, 60–72. [CrossRef]
101. Ismail, R.S.; Okawara, Y.; Fryer, J.N.; Vanderhyden, B.C. Hormonal regulation of the ligand for c-kit in the rat ovary and its effects on spontaneous oocyte meiotic maturation. *Mol. Reprod. Dev.* **1996**, *43*, 458–469. [CrossRef]
102. Manova, K.; Huang, E.J.; Angeles, M.; De Leon, V.; Sanchez, S.; Pronovost, S.M.; Besmer, P.; Bachvarova, R.F. The expression pattern of the c-kit ligand in gonads of mice supports a role for the c-kit receptor in oocyte growth and in proliferation of spermatogonia. *Dev. Biol.* **1993**, *157*, 85–99. [CrossRef]
103. Laitinen, M.; Rutanen, E.M.; Ritvos, O. Expression of c-kit ligand messenger ribonucleic acids in human ovaries and regulation of their steady state levels by gonadotropins in cultured granulosa-luteal cells. *Endocrinology* **1995**, *136*, 4407–4414. [CrossRef]
104. Parrott, J.A.; Skinner, M.K. Kit-ligand/stem cell factor induces primordial follicle development and initiates folliculogenesis. *Endocrinology* **1999**, *140*, 4262–4271. [CrossRef]
105. Thomas, F.H.; Ethier, J.F.; Shimasaki, S.; Vanderhyden, B.C. Follicle-stimulating hormone regulates oocyte growth by modulation of expression of oocyte and granulosa cell factors. *Endocrinology* **2005**, *146*, 941–949. [CrossRef]
106. Huang, E.J.; Manova, K.; Packer, A.I.; Sanchez, S.; Bachvarova, R.F.; Besmer, P. The murine Steel Panda mutation affects kit ligand expression and growth of early ovarian follicles. *Dev. Biol.* **1993**, *157*, 100–109. [CrossRef] [PubMed]
107. Yoshida, H.; Takakura, N.; Kataoka, H.; Kunisada, T.; Okamura, H.; Nishikawa, S.I. Stepwise requirement of c-kit tyrosine kinase in mouse ovarian follicle development. *Dev. Biol.* **1997**, *184*, 122–137. [CrossRef]
108. Jin, X.; Han, C.S.; Zhang, X.S.; Yuan, J.X.; Hu, Z.Y.; Liu, Y.X. Signal transduction of stem cell factor in promoting early follicle development. *Mol. Cell. Endocrinol.* **2005**, *229*, 3–10. [CrossRef] [PubMed]
109. Reddy, P.; Liu, L.; Adhikari, D.; Jagarlamudi, K.; Rajareddy, S.; Shen, Y.; Du, C.; Tang, W.; Hämäläinen, T.; Peng, S.L.; et al. Oocyte-specific deletion of pten causes premature activation of the primordial follicle pool. *Science.* **2008**, *319*, 611–613. [CrossRef] [PubMed]
110. Reddy, P.; Shen, L.; Ren, C.; Boman, K.; Lundin, E.; Ottander, U.; Lindgren, P.; Liu, Y.X.; Sun, Q.Y.; Liu, K. Activation of Akt (PKB) and suppression of FKHRL1 in mouse and rat oocytes by stem cell factor during follicular activation and development. *Dev. Biol.* **2005**, *281*, 160–170. [CrossRef]
111. Honda, A.; Hirose, M.; Inoue, K.; Hiura, H.; Miki, H.; Ogonuki, N.; Sugimoto, M.; Abe, K.; Kanatsu-Shinohara, M.; Kono, T.; et al. Large-scale production of growing oocytes in vitro from neonatal mouse ovaries. *Int. J. Dev. Biol.* **2009**, *53*, 605–613. [CrossRef] [PubMed]
112. Salustri, A.; Ulisse, S.; Yanagishita, M.; Hascall, V.C. Hyaluronic acid synthesis by mural granulosa cells and cumulus cells in vitro is selectively stimulated by a factor produced by oocytes and by transforming growth factor-β. *J. Biol. Chem.* **1990**, *265*, 19517–19523. [CrossRef]
113. Diaz, F.J.; O'Brien, M.J.; Wigglesworth, K.; Eppig, J.J. The preantral granulosa cell to cumulus cell transition in the mouse ovary: Development of competence to undergo expansion. *Dev. Biol.* **2006**, *299*, 91–104. [CrossRef]
114. Dragovic, R.A.; Ritter, L.J.; Schulz, S.J.; Amato, F.; Armstrong, D.T.; Gilchrist, R.B. Role of oocyte-secreted growth differentiation factor 9 in the regulation of mouse cumulus expansion. *Endocrinology* **2005**, *146*, 2798–2806. [CrossRef]
115. Gilchrist, R.B.; Lane, M.; Thompson, J.G. Oocyte-secreted factors: Regulators of cumulus cell function and oocyte quality. *Hum. Reprod. Update* **2008**, *14*, 159–177. [CrossRef]
116. Eppig, J.J.; Wigglesworth, K.; Pendola, F.; Hirao, Y. Murine oocytes suppress expression of luteinizing hormone receptor messenger ribonucleic acid by granulosa cells. *Biol. Reprod.* **1997**, *56*, 976–984. [CrossRef] [PubMed]
117. Li, R.; Norman, R.J.; Armstrong, D.T.; Gilchrist, R.B. Oocyte-secreted factor(s) determine functional differences between bovine mural granulosa cells and cumulus cells. *Biol. Reprod.* **2000**, *63*, 839–845. [CrossRef]
118. Hussein, T.S.; Froiland, D.A.; Amato, F.; Thompson, J.G.; Gilchrist, R.B. Oocytes prevent cumulus cell apoptosis by maintaining a morphogenic paracrine gradient of bone morphogenetic proteins. *J. Cell Sci.* **2005**, *118*, 5257–5268. [CrossRef] [PubMed]
119. Sibiak, R.; Bryl, R.; Stefańska, K.; Chermuła, B.; Pieńkowski, W.; Jeseta, M.; Pawelczyk, L.; Mozdziak, P.; Spaczyński, R.Z.; Kempisty, B. Expression of the apoptosis regulatory gene family in the long-term in vitro cultured human cumulus cells. *Med. J. Cell Biol.* **2021**, *9*, 8–13. [CrossRef]

120. Kocherova, I.; Stefańska, K.; Bryl, R.; Perek, J.; Pieńkowski, W.; Zakova, J.; Crha, I.; Ventruba, P.; Mozdziak, P.; Ješeta, M. Apoptosis-related genes expression in primary in vitro culture of human ovarian granulosa cells. *Med. J. Cell Biol.* **2020**, *8*, 176–182. [CrossRef]
121. Dong, J.; Albertini, D.F.; Nishimori, K.; Kumar, T.R.; Lu, N.; Matzuk, M.M. Growth differentiation factor-9 is required during early ovarian folliculogenesis. *Nature* **1996**, *383*, 531–535. [CrossRef] [PubMed]
122. Yan, C.; Wang, P.; Demayo, J.; Demayo, F.J.; Elvin, J.A.; Carino, C.; Prasad, S.V.; Skinner, S.S.; Dunbar, B.S.; Dube, J.L.; et al. Synergistic roles of bone morphogenetic protein 15 and growth differentiation factor 9 in ovarian function. *Mol. Endocrinol.* **2001**, *15*, 854–866. [CrossRef]
123. Matzuk, M.M.; Burns, K.H.; Viveiros, M.M.; Eppig, J.J. Intercellular communication in the mammalian ovary: Oocytes carry the conversation. *Science (80-.).* **2002**, *296*, 2178–2180. [CrossRef]
124. Hanrahan, J.P.; Gregan, S.M.; Mulsant, P.; Mullen, M.; Davis, G.H.; Powell, R.; Galloway, S.M. Mutations in the Genes for Oocyte-Derived Growth Factors GDF9 and BMP15 Are Associated with Both Increased Ovulation Rate and Sterility in Cambridge and Belclare Sheep (Ovis aries). *Biol. Reprod.* **2004**, *70*, 900–909. [CrossRef]
125. Sugiura, K.; Su, Y.Q.; Diaz, F.J.; Pangas, S.A.; Sharma, S.; Wigglesworth, K.; O'Brien, M.J.; Matzuk, M.M.; Shimasaki, S.; Eppig, J.J. Oocyte-derived BMP15 and FGFs cooperate to promote glycolysis in cumulus cells. *Development* **2007**, *134*, 2593–2603. [CrossRef] [PubMed]
126. Mottershead, D.G.; Ritter, L.J.; Gilchrist, R.B. Signalling pathways mediating specific synergistic interactions between GDF9 and BMP15. *Mol. Hum. Reprod.* **2012**, *18*, 121–128. [CrossRef]
127. Rybska, M.; Knap, S.; Stefańska, K.; Jankowski, M.; Chamier-Gliszczyńska, A.; Popis, M.; Jeseta, M.; Bukowska, D.; Antosik, P.; Kempisty, B.; et al. Transforming growth factor (TGF)—Is it a key protein in mammalian reproductive biology? *Med. J. Cell Biol.* **2018**, *6*, 127–130. [CrossRef]
128. Joyce, I.M.; Clark, A.T.; Pendola, F.L.; Eppig, J.J. Comparison of recombinant growth differentiation factor-9 and oocyte regulation of KIT ligand messenger ribonucleic acid expression in mouse ovarian follicles. *Biol. Reprod.* **2000**, *63*, 1669–1675. [CrossRef] [PubMed]
129. Bryja, A.; Pieńkowski, W.; Stefańska, K.; Chermuła, B.; Bryl, R.; Wieczorkiewicz, M.; Kulus, J.; Wąsiatycz, G.; Bukowska, D.; Ratajczak, K.; et al. Analysis of TGFB1, CD105 and FSP1 expression in human granulosa cells during a 7-day primary in vitro culture. *Med. J. Cell Biol.* **2020**, *8*, 152–157. [CrossRef]
130. Baena, V.; Terasaki, M. Three-dimensional organization of transzonal projections and other cytoplasmic extensions in the mouse ovarian follicle. *Sci. Rep.* **2019**, *9*, 1262. [CrossRef]
131. Macaulay, A.D.; Gilbert, I.; Caballero, J.; Barreto, R.; Fournier, E.; Tossou, P.; Sirard, M.A.; Clarke, H.J.; Khandjian, É.W.; Richard, F.J.; et al. The gametic synapse: RNA transfer to the bovine oocyte. *Biol. Reprod.* **2014**, *91*, 90. [CrossRef]
132. Albertini, D.F.; Rider, V. Patterns of intercellular connectivity in the mammalian cumulus-oocyte complex. *Microsc. Res. Tech.* **1994**, *27*, 125–133. [CrossRef]
133. Motta, P.M.; Correr, S.; Makabe, S.; Naguro, T. Oocyte Follicle Cells Association during Development of Human Ovarian Follicle. A Study by High Resolution Scanning and Transmission Electron Microscopy. *Arch. Histol. Cytol.* **1994**, *57*, 369–394. [CrossRef]
134. Simon, A.M.; Goodenough, D.A. Diverse functions of vertebrate gap junctions. *Trends Cell Biol.* **1998**, *8*, 477–483. [CrossRef]
135. Willecke, K.; Eiberger, J.; Degen, J.; Eckardt, D.; Romualdi, A.; Güldenagel, M.; Deutsch, U.; Söhl, G. Structural and functional diversity of connexin genes in the mouse and human genome. *Biol. Chem.* **2002**, *383*, 725–737. [CrossRef] [PubMed]
136. Kidder, G.M.; Vanderhyden, B.C. Bidirectional communication between oocytes and follicle cells: Ensuring oocyte developmental competence. *Can. J. Physiol. Pharmacol.* **2010**, *88*, 399–413. [CrossRef]
137. Pelland, A.M.D.; Corbett, H.E.; Baltz, J.M. Amino acid transport mechanisms in mouse oocytes during growth and meiotic maturation. *Biol. Reprod.* **2009**, *81*, 1041–1054. [CrossRef]
138. Purcell, S.H.; Moley, K.H. Glucose transporters in gametes and preimplantation embryos. *Trends Endocrinol. Metab.* **2009**, *20*, 483–489. [CrossRef] [PubMed]
139. Augustin, R.; Pocar, P.; Navarrete-Santos, A.; Wrenzycki, C.; Gandolfi, F.; Niemann, H.; Fischer, B. Glucose transporter expression is developmentally regulated in in vitro derived bovine preimplantation embryos. *Mol. Reprod. Dev.* **2001**, *60*, 370–376. [CrossRef] [PubMed]
140. Orisaka, M.; Tajima, K.; Tsang, B.K.; Kotsuji, F. Oocyte-granulosa-theca cell interactions during preantral follicular development. *J. Ovarian Res.* **2009**, *2*, 9. [CrossRef] [PubMed]
141. Murray, A.A.; Gosden, R.G.; Allison, V.; Spears, N. Effect of androgens on the development of mouse follicles growing in vitro. *J. Reprod. Fertil.* **1998**, *113*, 27–33. [CrossRef]
142. Orisaka, M.; Jiang, J.Y.; Orisaka, S.; Kotsuji, F.; Tsang, B.K. Growth differentiation factor 9 promotes rat preantral follicle growth by up-regulating follicular androgen biosynthesis. *Endocrinology* **2009**, *150*, 2740–2748. [CrossRef]
143. Mattioli, M.; Gloria, A.; Turriani, M.; Berardinelli, P.; Russo, V.; Nardinocchi, D.; Curini, V.; Baratta, M.; Martignani, E.; Barboni, B. Osteo-regenerative potential of ovarian granulosa cells: An in vitro and in vivo study. *Theriogenology* **2012**, *77*, 1425–1437. [CrossRef] [PubMed]
144. Chandramohan, Y.; Jeganathan, K.; Sivanesan, S.; Koka, P.; Amritha, T.M.S.; Vimalraj, S.; Dhanasekaran, A. Assessment of human ovarian follicular fluid derived mesenchymal stem cells in chitosan/PCL/Zn scaffold for bone tissue regeneration. *Life Sci.* **2021**, *264*, 118502. [CrossRef] [PubMed]

145. Tian, C.; Liu, L.; Ye, X.; Fu, H.; Sheng, X.; Wang, L.; Wang, H.; Heng, D.; Liu, L. Functional Oocytes Derived from Granulosa Cells. *Cell Rep.* **2019**, *29*, 4256–4267.e9. [CrossRef]
146. Madkour, A.; Bouamoud, N.; Kaarouch, I.; Louanjli, N.; Saadani, B.; Assou, S.; Aboulmaouahib, S.; Sefrioui, O.; Amzazi, S.; Copin, H.; et al. Follicular fluid and supernatant from cultured cumulus-granulosa cells improve in vitro maturation in patients with polycystic ovarian syndrome. *Fertil. Steril.* **2018**, *110*, 710–719. [CrossRef]
147. Atrabi, M.J.; Akbarinejad, V.; Khanbabaee, R.; Dalman, A.; Amorim, C.A.; Najar-Asl, M.; Valojerdi, M.R.; Fathi, R. Formation and activation induction of primordial follicles using granulosa and cumulus cells conditioned media. *J. Cell. Physiol.* **2019**, *234*, 10148–10156. [CrossRef]
148. Taheri, M.; Saki, G.; Nikbakht, R.; Eftekhari, A.R. Bone morphogenetic protein 15 induces differentiation of mesenchymal stem cells derived from human follicular fluid to oocyte-like cell. *Cell Biol. Int.* **2021**, *45*, 127–139. [CrossRef]
149. Lee, Y.M.; Kim, T.H.; Lee, J.H.; Lee, W.J.; Jeon, R.H.; Jang, S.J.; Ock, S.A.; Lee, S.L.; Park, B.W.; Rho, G.J. Overexpression of Oct4 in porcine ovarian stem/stromal cells enhances differentiation of oocyte-like cells in vitro and ovarian follicular formation in vivo. *J. Ovarian Res.* **2016**, *9*, 24. [CrossRef]
150. Kordus, R.J.; LaVoie, H.A. Granulosa cell biomarkers to predict pregnancy in ART: Pieces to solve the puzzle. *Reproduction* **2017**, *153*, R69–R83. [CrossRef] [PubMed]
151. Luddi, A.; Gori, M.; Marrocco, C.; Capaldo, A.; Pavone, V.; Bianchi, L.; Boschi, L.; Morgante, G.; Piomboni, P.; de Leo, V. Matrix metalloproteinases and their inhibitors in human cumulus and granulosa cells as biomarkers for oocyte quality estimation. *Fertil. Steril.* **2018**, *109*, 930–939.e3. [CrossRef]

cells

MDPI

Review

Towards Physiologic Culture Approaches to Improve Standard Cultivation of Mesenchymal Stem Cells

Ilias Nikolits †, Sabrina Nebel †, Dominik Egger *, Sebastian Kreß and Cornelia Kasper

Institute for Cell and Tissue Culture Technology, Department of Biotechnology, University of Natural Resources and Life Sciences, Muthgasse 18, 1190 Vienna, Austria; ilias.nikolits@boku.ac.at (I.N.); sabrina.nebel@boku.ac.at (S.N.); sebastian.kress@boku.ac.at (S.K.); cornelia.kasper@boku.ac.at (C.K.)
* Correspondence: dominik.egger@boku.ac.at
† These authors contributed equally to this work.

Abstract: Mesenchymal stem cells (MSCs) are of great interest for their use in cell-based therapies due to their multipotent differentiation and immunomodulatory capacities. In consequence of limited numbers following their isolation from the donor tissue, MSCs require extensive expansion performed in traditional 2D cell culture setups to reach adequate amounts for therapeutic use. However, prolonged culture of MSCs in vitro has been shown to decrease their differentiation potential and alter their immunomodulatory properties. For that reason, preservation of these physiological characteristics of MSCs throughout their in vitro culture is essential for improving the efficiency of therapeutic and in vitro modeling applications. With this objective in mind, many studies already investigated certain parameters for enhancing current standard MSC culture protocols with regard to the effects of specific culture media components or culture conditions. Although there is a lot of diversity in the final therapeutic uses of the cells, the primary stage of standard isolation and expansion is imperative. Therefore, we want to review on approaches for optimizing standard MSC culture protocols during this essential primary step of in vitro expansion. The reviewed studies investigate and suggest improvements focused on culture media components (amino acids, ascorbic acid, glucose level, growth factors, lipids, platelet lysate, trace elements, serum, and xenogeneic components) as well as culture conditions and processes (hypoxia, cell seeding, and dissociation during passaging), in order to preserve the MSC phenotype and functionality during the primary phase of in vitro culture.

Keywords: mesenchymal stem cells; physiologic cultivation; optimization cultivation

Citation: Nikolits, I.; Nebel, S.; Egger, D.; Kreß, S.; Kasper, C. Towards Physiologic Culture Approaches to Improve Standard Cultivation of Mesenchymal Stem Cells. *Cells* **2021**, *10*, 886. https://doi.org/10.3390/cells10040886

Academic Editor: Mehdi Najar

Received: 23 March 2021
Accepted: 12 April 2021
Published: 13 April 2021

1. Introduction

Cell-based therapies aim to repair or regenerate defective tissues or organs due to physical or congenital damage, ageing, and degenerative diseases. From a regulatory viewpoint cell-based therapies can be divided between minimally manipulated cells for homologous use like transplants or transfusions and somatic cell, gene therapy, and tissue engineered products, which are referred to as Advanced Therapy Medicinal Products (ATMPs) by EU authorities [1]. These are considered medicines and need to comply to quality, safety, and efficacy standards before getting a market authorization. Generally, when referring to cell-based therapies, not only cells as medicinal product, but also products derived from them are included [2].

One possible treatment approach in cell-based therapies is the engraftment of cells into the damaged site to replace or restore the defective cells or tissue regarded as classical cell therapy. Another mechanism is based on the stimulating effect of trophic factors, through targeted drug delivery or cell genetic manipulation, to promote endogenous self-regeneration of the injured tissue. Cell-based therapies are a very promising strategy for the treatment of many severe and until recently considered incurable vascular, neurological, autoimmune, ophthalmologic, and skeletal diseases [3].

Among the different cell types tested for their therapeutic potential, stem cells are the most prominent, considering their intrinsic self-renewal and differentiation capacities [4]. Particularly, mesenchymal stem cells (MSCs) are of great interest for their use in cell-based therapies. MSCs are highly proliferative multipotent stromal cells that can differentiate into the mesodermal cell lineage [5]. Due to the therapeutic potential of MSCs, many studies focus on their characterization and identification. According to the International Society for Cellular Therapy (ISCT), MSCs must be plastic-adherent and differentiate into osteoblasts, chondroblasts, and adipocytes when cultured in vitro. In addition, they should express CD90, CD73, and CD105 surface markers and should not express CD34, CD45, CD14 or CD11b, CD19 or CD79A and HLA-DR surface molecules [6]. MSCs can be derived from various tissue sources such as the bone marrow [7], adipose tissue [8], umbilical cord [9], peripheral blood [10], and dental pulp [11], but the number of primary MSCs isolated from a donor tissue is rather limited [5] and source-dependent [12].

With the introduction of cell culture techniques, traditional two-dimensional (2D) cell culture has been for decades the backbone method for medical and biological research and drug development due to its simplicity and robustness. However, the static and planar conditions of 2D culture fail to simulate the physiological environment of the cells. For that reason, during the last years, concerns over the biological relevance and reproducibility of in vitro culture techniques have led to the development of advanced three-dimensional (3D) and dynamic cell culture methods. These methods serve as better models for the representation of in vivo physiological conditions [13,14], as well as platforms for up-scaling production of cells for clinical applications [15,16]. As a result, several 3D cell culture techniques now exist, from multicellular spheroids/organoids [17,18] and cell-laden biomaterials [19–22], to dynamic bioreactor systems [23–26] and organ/body on a chip systems [27–29]. It has been shown that such 3D culture systems preserve many characteristics and properties that resembled MSC behavior and differentiation in vivo [30–32]. Therefore, the gained results are more relevant for the translation into clinical applicability. However, these 3D advanced cell cultures require high numbers of cells [16], are more complex to establish and maintain, and are therefore more laborious.

Due to the fact that MSCs are a very limited adult tissue population [33,34], following their isolation from donor tissue, harvested primary MSCs should be expanded first in 2D culture setups, to reach adequate numbers for use in various in vitro modelling and therapeutic applications [5]. However, prolonged culture of MSCs in vitro in 2D systems has shown to decrease their differentiation potential [35,36] and alter their phenotype [37] and immunomodulatory properties [38]. In order to improve the relevance of in vitro modelling and efficiency of therapeutic applications, it is essential to preserve the physiological properties and characteristics of MSCs throughout their in vitro culture.

In view of this, there is a considerable research focus on the optimization of culture conditions and protocols for MSCs. Since there is yet no medium that is as close to natural in vivo fluids to support physiological MSC culture, specialized culture media are composed based on the cell type, type of culture, and focus of analysis. Additional focus is put on the effects of other cell culture variables, including physiochemical conditions and subculture protocols. As a result, between different cell culture laboratories, there is divergence on these culture variables. Therefore, it is necessary to define what constitutes an "optimal practice" for maintaining the physiological status of MSCs, during their primary step of in vitro 2D expansion. Further, during optimization of cultivation procedures considerations on other markers of success next to population growth should be considered to circumvent potentially highly proliferative, but severely altered cell populations. In a recent position statement, the ISCT underlines that MSCs often lose their functional properties during in vitro expansion, and it is suggested to verify relevant functionalities of MSCs by applying a matrix of functional assays based on the later therapeutic use [39]. Therefore, we want to review on optimizing approaches for standard 2D in vitro expansion of human MSCs, focusing on culture media components (amino acids, ascorbic acid, glucose level, growth factors, lipids and fatty acids, platelet lysate, trace elements, serum, and xenogeneic components) as well as culture conditions and processes (hypoxia, seeding density and cell dissociation during passaging).

2. Basal Media for Isolation and Expansion

In vitro culture of MSCs is performed using a defined basal medium. Several basal media with different formulations are commercially available for culture of human MSCs, and the choice of basal media used varies between different laboratories. These media primarily contain glucose, amino acids, and trace elements which are necessary for cellular growth and survival. Regarding the selection of basal medium for optimal expansion MSCs, Sotiropoulou et al., cultured primary bone marrow mononuclear cells and passaged bone marrow MSCs (BM-MSCs) in the presence of different media (see Table 1). According to their results, cultures of primary and passaged MSCs in aMEM/GL and aMEM/L-G media generated the highest number of cells, with minor differences in their phenotype over subsequent passages [40]. In another study comparing 4 different basal media, adipose tissue MSCs showed increased proliferation when aMEM, but also DMEM/LG, were used as basal media, without changes in their morphology and viability during subsequent passages [41]. In a similar study comparing different culture media, a variation of "Verfaillie" medium with DMEM/HG/L-G yielded higher number of adherent MSCs from primary mononuclear bone marrow cell populations and higher proliferation rate of BM-MSCs at early passages, compared to other media formulations, including aMEM/L-G and DMEM/LG/L-G based media [42]. In addition, studies comparing different types of media with BM-MSCs reported significant changes in their phenotypic expression profile [42,43], in contrast with a study using adipose-derived MSCs where different basal media did not alter the MSC specific marker expression after multiple passages [41]. As indicated by the literature, the optimal choice of basal media can vary, depending on the type and of MSCs cultured, although in general, aMEM based media have been indicated as the most suitable for isolation and expansion of MSCs. Furthermore, evaluation of the optimal choice of basal media should not only depend on its effect on the proliferative capacity of MSCs, but also on the preservation of their intrinsic characteristics. Researches should test and consider both aspects regarding their choice of basal media for their MSC cultures.

Table 1. Different types of basal media.

Media Name	Description
DMEM	Dulbecco's modified Eagle's medium (MEM)
DMEM/LG/L-G	Dulbecco's MEM (DMEM) with 1000 mg/mL glucose and L-glutamine
DMEM/HG/L-G	DMEM with 4500 mg/mL glucose and L-glutamine
DMEM/HG/GL	DMEM with 4500 mg/mL glucose and Glutamax
IMDM	Iscove's modified Dulbecco's medium with L-glutamine
aMEM	MEM alpha
aMEM/L-G	MEM alpha with L-glutamine
aMEM/GL	MEM alpha with Glutamax

3. Glucose

Glucose is a basic source of energy for the cells and is involved in the synthesis of proteins and lipids. In vivo, plasma glucose concentration for healthy people ranges between 700–1000 mg/L throughout the day, and it does not exceed 1600 mg/L after meals. Glucose plays an essential role in survival, metabolism, and physiological function of MSCs. In cell culture media formulations glucose concentration ranges from 1000 mg/L to resemble physiological in vivo conditions, up to 10,000 mg/L. Culture media formulations with glucose concentrations higher than 1000 mg/L simulate in vivo diabetic conditions [44].

Due to the highly glycolytic phenotype of MSCs, glucose is more rapidly consumed and its depletion has stronger effects on the cells compared to other nutrients and serum [45]. According to the literature however, the effects of the amount glucose on MSC cultures are diverse, depending on the metabolic activity and the type of MSCs. Most studies under normoxic culture conditions support that high glucose concentration (5000 mg/L) in the culture media can suppress proliferation of bone marrow [46–48], Bichat's buccal fat pad [49] and nucleus pulposus MSCs [50], reduce their colony forming ability [50], induce cellular senescence [47,50,51], alter MSC spindle-shape morphology [47], and upregulate autophagy [50,51], compared with low glucose (1000 mg/L) media concentration. In addition, MSCs cultured in low glucose media have been shown to maintain their differentiation capacities and stemness [46,47,50], and increase antioxidant enzyme expression and mitochondrial respiration [47]. In contrast with these results, a few studies have reported that high glucose does not have an effect on the apoptosis and proliferation rates of adipose-derived [52] and primary BM-MSCs [53], while it can significantly enhance proliferation of telomerase-immortalized MSCs. While it is known that under hypoxic culture conditions the metabolic activity of MSCs is enhanced [54], Deschepper et al., observed that under continuous hypoxic culture, excess of glucose provided by high glucose concentration media prevents glucose shortage during culture, thus preserving the proliferative capacity, morphology, and viability of BM-MSCs [55].

4. Amino Acids

Already in the first days of cell culture, it became clear that for cultivation ex vivo essential nutrients have to be supplied to cells, one of the most prominent categories being amino acids. Well known in the field of cell culture are Dulbecco and Eagle, who pioneered basal medium for cell culture that is still in use today, known as DMEM (Dulbecco's modified eagle's medium), usually supplemented with fetal bovine serum (FBS) [56]. With the aspiration to get rid of ill-defined xenogeneic supplements and to transition to chemically defined media, understanding and optimizing amino acid formulations is becoming even more important. In the past amino acid concentrations in culture medium have been chosen according to the consumption by the cells. However, degree of availability of an amino acid can have influence on the cellular metabolism therefore refined studies on their cellular metabolic properties are called for. Other parameters like solubility, stability, and transport into the cells also determine availability and further consumption of amino acids [57]. In the last years, it also became obvious that different mammalian cell types have completely different amino acid requirements and that medium optimization in terms of amino acid metabolism have to be fitted to the individual culture [58,59].

Of the 20 naturally occurring amino acids l-Arginine, l-Cysteine, l-Glutamine, l-Histidine, l-Isoleucine, l-Leucine, l-Lysine, l-Methionine, l-Phenylalanine, l-Threonine, l-Tryptophan, l-Tyrosine, and l-Valine are regarded as essential and Glycine, l-Alanine l-Asparagine, l-Aspartic, l-Glutamic, l-Proline, and l-Serine as non-essential, thus generally supplied with the basal media are the essential ones, as non-essential ones can be synthesized by the cells [60]. In a 2007 paper by Choi et al. [61], six different medium compositions with essential amino acids (EAAs) and non-essential amino-acids (NEAAs) were tested for their efficacy to promote proliferation of BM-MSCs. Their basal media were DMEM (basically containing EAAs) and IMDM (Iscove's modified Dulbecco's Medium) (basically containing EAAs and NEAAs), where either essential or non-essential amino acids or both were additionally added. They could see that oversupply with essential AAs as well as supplementation with non-essential AAs improved proliferation, whereas oversupply of NEAAs resulted in lowered proliferation. In none of the medium conditions did a change in stemness markers occur. Although this study did not perform design of experiment to tune each amino acid concentration individually it is one step in the direction of establishing improved culture conditions. Another approach to determine the nutritional requirements of MSCs in particular is the analysis of metabolic patterns, in detail which and in what concentration are amino acids consumed or secreted. They also tested whether this was consistent in 2D monolayer culture compared to dynamic culture on microcarriers. EAAs were consumed in both set ups as expected, but NEAAs were differently metabolized. MSCs consumed cysteine and proline in dynamic culture, but secreted them in static culture. However, results of static and dynamic agree on alanine, glutamate, glycine, and ornithine being produced, and arginine, asparagine, aspartate, glutamine, serine, and tyrosine being consumed by the cells. This consumption was not seen in other cell types before and might be unique to stem cells [62]. The difference of amino acid metabolism from static to dynamic culture indicates that the mechanisms are even more complex than thought, and optimizing medium composition is also influenced by cultivation techniques. Regarding one specific amino acid, already Eagle, et al. [63] describe the importance of glutamine for mammalian cell proliferation. Specifically for BM-MSCs, both dos Santos, et al. [64] and Zhou, et al. [65] report improved proliferation and very importantly maintenance of stemness. Additionally, oversupply of branched chain AAs (BCAAs), valine, leucine, and isoleucine increased S/G2/M cell cycle phases of a murine MSC lineage and could also improve the immunomodulatory capacity of the cells [66]. Although the minimal supply of amino acids is provided by the basal medium, these results show oversupply of certain amino acids can improve MSC proliferation, stemness, or immunomodulatory effects. The choice of AAs to be supplemented has to be made according to the intended application of the cells as well as the culture method.

5. Lipids

Lipids are next to amino acids, sugars, and nucleic acids as one of the major classes of biomolecules that play important roles in the cell's membrane structure, metabolism, and signaling. During the last decades, it has become obvious that disturbances in the lipid homeostasis are involved in the development of diseases like diabetes, cardiovascular disorders, autoimmune diseases, and even cancer development [67]. Not only understanding the lipidomics in vivo is of utter importance, but also the mechanisms of lipid metabolism in vitro are important to improve cell-based treatment strategies of those disorders. The most common forms of lipids in the cell are phosphatidylcholine (PC), phosphatidylserine (PS), phosphatidylinositol (PI), phosphatidylethanolamine (PE) [68], and cholesterol, which make up the outer cell membrane and the membrane of different organelles [60]. Next to this, triglycerides serve as energy storage within reservoirs (lipid droplets). The triglycerides can be broken down to fatty acids and used to produce ATP if needed. Lipids can also bind to proteins, making them useable as signaling molecules for extra- or intracellular messaging [60,69]. Almost all lipids necessary for basic cellular functions in mammalian cells can be produced by themselves, only two are regarded as "essential": Linoleic and linolenic acid. However, in a lot of cases, even media without them allowed cells to proliferate, whereas addition drastically improved the results, suggesting differentiation between minimal requirement and optimal performance [70]. Requirements needed for substantial cellular proliferation, were traditionally met by the excess of lipids in the supplemented FBS. With the increasing interest to omit animal products (discussed in detail in "human platelet lysate" and "Serum-, xeno-free- and chemically defined media"), other delivery strategies have to be found. The hydrophobic nature of lipids needs different strategies to disperse lipids, allow cellular uptake, and remain stable in the medium. Three approaches have been described: (1) Adsorption to a soluble carrier molecule, (2) self-assembly to the required size, and (3) dispersion [71]. The mechanism in FBS is the adsorption to the serum proteins, especially to the bovine serum albumin (BSA), a strategy that can be used also in chemically defined media that are supplemented with some form of serum albumin (human or bovine, natural or recombinant origin). Dispersion techniques include liposomes, emulsion, and microemulsions and have the advantage that they can be used in animal component free and even protein free medium (see "Serum-, xeno-free- and chemically defined media"). These requirements are true for mammalian cell lines in general, yet better understanding of the exact lipidomics especially within stem cells is necessary to understand possible alterations during long term ex vivo maintenance and lead to establishment of improved cultivation environments [72]. For example, in a 2017 paper by Chatgilialoglu et al., they showed that the fatty acid composition of the membrane of primary human fetal membrane-derived MSCs changed in 2D culture over time [73]. The amount of omega-6 fatty acids decreased, whereas monounsaturated fatty acids (MUFA) and omega-3 fatty acids increased. Additionally, Kilpinen et al., showed changes in membrane composition of BM-MSCs [74]. In order to decrease the effect of prolonged culture on the cell membranes, Chatgilialoglu et al., developed a cell medium supplement. Refeed® supplement is a completely defined combination of lipids and lipophilic antioxidants in ethanol. The supplementation was able to increase cellular proliferation compared to the control, without changes in surface marker composition [73]. With the advances in the field of omics, more complex interactions can be monitored. For example, differences in whole lipidomics of early and late passage cells were investigated using ultraperformance liquid chromatography coupled to mass spectrometry (UPLC–MS) and multivariate analysis. Global analysis revealed that changes in lipid significance and alterations during BM-MSC aging could be mostly contributed to chain length alterations and level of unsaturation [75]. An important aspect to consider was taken up by Fillmore et al. [76], who cultured BM-MSCs in physiological levels of fatty acids and serum albumin, which are higher than in standard cultivation conditions, to have a look on energy metabolism and survival. They could show that the addition of physiological levels of palmitate, a saturated fatty acid, significantly decreased BM-MSC viability in a time and concentration dependent manner. They could rescue this decrease in survival with the addition of equal amounts of oleate, an unsaturated fatty acid. By investigation of the metabolic activities, they conclude that decreasing fatty acid oxidation may be the source of cell death. These results are important to understand limited efficacy of cell-based therapies at the moment. Decreasing the amount of circulating saturated fatty acids in the patient during treatment might overcome current limitations in cell homing and survival after implantation [76]. With novel lipidomic tools, like, e.g., tandem MS-based, high mass accuracy-based, or multi-dimensional MS-based shotgun lipidomics [77] and improved analysis software [78,79], more insight is gained into the impact of lipids and fatty acids on MSCs ex vivo, which is the first step towards finding solutions to prevent unwanted alterations. As a general rule, media supplemented with FBS grants the cells a sufficient fatty acid supply, whereas human platelet lysate supplemented media might (depending on the manufacturing method) and serum/xeno-free medium need ancillary supply with lipids. Available are enriched forms or extracts from natural sources like palm oil, wool, serum, or fish [80], however for xeno-free culture synthetic alternatives to animal-derived supplements need to be used.

6. Growth Factors

Over the last decades, different growth factors have been widely studied for their role of improving ex vivo expansion of human MSCs. Due to the pleiotropic behavior of these factors, they can affect multiple biological behaviors like proliferation, morphology, immunophenotype, survival, and differentiation capacity of MSCs. Selection of which growth factor or combination of factors can promote physiological culture of MSCs in vitro, is still under investigation. Many different growth factors have been investigated for their potential effects on expanding MSCs. From them, fibroblast growth factors-2 and -4 (FGF-2, FGF-4), platelet derived growth factor-BB (PDGF-BB), and epidermal growth factor (EGF) have been reported to influence human MSC proliferation and survival in vitro [81].

6.1. Fibroblast Growth Factor-2 and -4

FGFs is involved in tissue repair and cell growth. FGF-2 used in many culture studies can enhance the proliferative [40,82–87] and colony-forming capacity [88] of MSCs, suppress cellular senescence [87,89], reduce trypsinization time [84], and preserve the thin spindle-shape morphology of MSCs [84,90]. In a study by Battula et al., primary bone marrow and placental MSCs cultured in gelatin-coated flasks with serum-free medium supplemented with FGF-2 demonstrated significantly higher proliferation rates than control cultures with serum containing media in uncoated flasks. Combination of FGF-2 with platelet derived growth factor-BB (PDGF) and/or transforming growth factor-β (TGF-β) in the culture media was also reported to increase the proliferation of human MSCs [83,88,90] and preserve the spindle-shaped morphology of MSCs in serum-free media conditions [90]. However, some studies have reported that FGF-2 supplement in human MSC cultures can affect the phenotype and intrinsic properties of the cells [40,83]. FGF-4, similarly to FGF-2, has been reported to increase proliferation, reduce autophagy, and delay the decrease of cellular renewal capacity of human MSCs [87].

6.2. Platelet Derived Growth Factor-BB

PDGF-BB is a known regulatory factor of cellular growth and division. It has been noted that when PDGF-BB is supplement alone [88,91] or in combination with other factors [84,88,90] in the culture media, it can highly increase human MSC proliferation, while inhibition of its receptor can significantly impair their proliferative capacity [92]. Regardless, reports have indicated that it can also decrease colony forming efficiency [88] and osteogenic differentiation potential of cultured MSCs [91].

6.3. Epidermal Growth Factor

EGF and Heparin-Binding Epidermal Growth Factor (HB-EGF) have been shown to play a role in promoting wound healing and cell growth. Human BM-MSC cultures supplemented with EGF have demonstrated increased proliferation similar to the effect of PDGF stimulated proliferation [88,93], but without diminishing their osteogenic potential [91,93]. Krampera et al., have demonstrated that BM-MSCs cultured with HB-EGF can proliferate more rapidly without undergoing spontaneous differentiation, thus maintaining their multilineage differentiation potential during culture [94].

In the study by Fraz et al., vascular endothelial growth factor (VEGF), and insulin-like growth factor (IGF) in combination with FGF-2 were reported to improve MSC proliferation. More interestingly, from the same study it was reported that high initial growth factors contents in the MSC population, or pre-incubation of the cells with growth factors, can significantly diminish their impact on the proliferative capacity of MSCs when added as supplements in the culture medium [82].

Overall, the growth factors discussed above have been shown to be beneficial for MSC expansion when added to the medium and consist considerable components for standard culture media formulations. However, most studies have examined the effects of these factors when added solely to the media. That gives room to further investigation and analysis regarding the exerted effect from the combination of these growth factors during MSC culture.

7. Trace Elements

Culture media apart from the organic substances, also contain inorganic elements such as Cu, Mg, Mn, Zn, Se, Ca, and Fe in trace quantities. Serum used in cell cultures usually contains many of these elements, and therefore many basal media do not include them in the formulations. These trace elements can influence important biochemical functions of the cells, but their effect and exact biological role on MSCs have not been thoroughly investigated.

Zinc is involved with nucleic acid metabolizing enzymes and stabilization of cell membranes. Serum used usually in cell cultures contain the necessary amounts of zinc for cell growth and survival. For that reason, many basal media do not contain zinc. Addition of $ZnCl_2$ in adipose tissue derived MSCs culture has been shown to highly enhance their proliferation, while zinc chelation reversed this effect. The exact mechanism of this effect has not been fully elucidated yet [95].

Magnesium acts as a cofactor for many enzymatic processes and as a counter ion for nucleic acids and ATP [96]. Human MSCs cultured with magnesium-conditioned medium have demonstrated increased viability and proliferation. Furthermore, magnesium-conditioned medium with Zinc and/or Manganese additions were shown to further promote human MSC proliferation and viability in the same study [97].

Calcium is an important element for a variety of biological functions including cell attachment, signaling, and morphology. Studies have shown that increased concentrations of extracellular calcium ions can significantly promote proliferation of BM-MSCs cultured in serum [98] and serum-free [99] media conditions, compared to controls. Increased calcium supplementation in the culture medium can also induce morphological changes on MSCs, resulting in more spread out cytoskeleton arrangements, which could be an indication towards their differentiation commitment [98].

Iron plays an essential role in cell metabolism and respiration. It protects cells from oxidative damage acting as an enzyme cofactor, but it can also cause toxic effects. Its delivery to the cells is controlled by transferrin, included in serum. That is why serum-free media formulations have a higher risk to cause iron toxic effects to cells. Borriello et al. have demonstrated that iron supplementation to the culture medium of primary human BM-MSCs, can greatly enhance their proliferation and initiate their S phase. Iron addition also hindered the innate osteogenic commitment of BM-MSCs, by blocking matrix calcification and preserving their undifferentiated status [100].

Selenium is included in enzymes that prevent oxidative damage to the cells by reducing peroxides and other oxidative free radicals to non-harmful forms. In cell culture media is usually added as selenium, sodium selenite, or selenium dioxide. The literature results on the effects of selenium supplementation on human MSCs are quite diverse. Selenium can restore the low antioxidative capacity of primary human BM-MSCs, and prevent cellular damage [101]. Park et al., found out that selenium can enhance the proliferation of human amniotic fluid MSCs, while combination of selenium with FGF-2 supplementation exerts an additive effect on the proliferative capacity of the cells [102]. Conversely, other studies have reported that selenium reduces the colony-forming ability of human MSCs [88] and does not have any significant effect on their doubling time [101].

Copper is important for cell growth and development, notably of the skeleton by affecting bone turnover and metabolism. Addition of copper in different concentrations in human BM-MSC cultures has been reported to decrease their proliferation in a dose dependent manner [103]. Since increased copper concentrations did not affect cell viability or cell size to justify contact inhibition of proliferating cells, it was hypothesized that copper could interfere with multiple cell cycle stages to hamper MSC proliferation [103].

Although studies have shown that trace elements included in the media can affect the biological behavior and characteristics of MSCs, further investigation is required on their exact roles and mechanisms. In addition, while performing cultures with serum-free media, studies should pay attention to the source or amounts of trace elements added to the MSC media.

8. Ascorbic Acid

Ascorbic acid acts as an antioxidant factor in cell cultures. Although it is not so commonly added in MSC media, studies have shown that it is involved in promoting MSC growth and proliferation when supplied to the culture medium. The study of Choi et al., revealed that the addition of L-ascorbate-2-phosphate in the culture medium enhanced the proliferation rate of human BM-MSCs in a dose-dependent manner [104]. Addition of 250 μM of L-ascorbate-2-phosphate induces the highest proliferation rates without any effect on cell morphology and MSC phenotype, while addition of 500 μM hinders the proliferation [104]. Same results using the same concentration of ascorbic acid supplementation were observed by other studies using human MSCs [83] from adipose tissue [105]. Another study using also human BM-MSCs showed similar results with the addition of L-ascorbate-2-phosphate up to 3 mM, but without differences between the concentrations tested [106]. In the same study, L-ascorbate-2-phosphate promoted MSC expansion from mononuclear cell population. These results support that ascorbic acid is a beneficial additive for MSC cultures.

9. Human Platelet Lysate

As mentioned earlier, culture medium enrichment by the addition of fetal bovine (fetal calf) serum (FBS/FCS) is still a common practice in many laboratories worldwide. It is rich in hormones, vitamins, transport proteins, trace elements, spreading, and growth factors [107,108]. In vivo, these are released upon blood coagulation and destined to help repair the site of injury, while in vitro, they promote cellular growth. The production raises both scientific as well as ethical concerns, in terms of xenogeneic compounds, reproducibility, and animal welfare, respectively. The whole blood needed is collected from unborn calves, which are a "by-product" of the meat producing unit. The use of fetal animals instead of adults is regarded to the low number of immunoglobulins in the calf serum. Collection is performed by harvesting the blood from the umbilical vein or more invasively directly from the heart by cardiac puncture. These procedures are performed right after the mother dams' neck has been cut to ensure fetal death by anoxia and blood loss. To prevent suffering of the calf during the whole procedure, it has to be unconscious and must not breathe air at any stage, however, such safeguards are only recommendations and not legally defined [109]. After coagulation the liquid serum part is separated from the solid blood clot by centrifugation and can be further processed (filtration, quality control, packaging) for sale. When considering that in vitro cell culture is supposed to reduce the need for animals and the associated painful procedures, using animal derived compounds is contradistinctive. An estimation of the amount of FBS needed per anno leads to a calculated number of 1 million fetal calves necessary to meet the global requirements [109]. The fact that the main production countries are situated in regions with less legal restrictions concerning animal welfare, gives rise to even more ethical concerns regarding the use of FBS. Furthermore, the dependence from global meat markets and price variations, adds even more concerns [110,111].

On another aspect, addition of such ill-defined growth factor and protein cocktails and especially xenogeneic compounds is also not in the spirit of good manufacturing practice (GMP) guidelines and hinders transition of cell-based therapies to the clinics. In a desperate search for alternatives, various derivatives of human blood have been investigated [112]. Due to lower levels of growth factors and nutrients, plasma and serum prepared from anti-coagulated or coagulated whole human blood, respectively, are not widely used. However, human platelet lysate (hPL) has been shown to be an adequate substitute for FBS. This blood derivative can be manufactured from platelet concentrates by induction of bioactive molecule-release from the platelets. As platelet concentrates (PCs) are of human origin, already routinely collected by blood banks for the treatment of thrombocytopenia, and can be repurposed after a short shelf life of 5–7 days for research purposes [113–118], and its use is not accompanied by ethical or animal welfare concerns. Human platelet lysate has been used in cell culture already in the 1980s and by now has even proven to be superior to FBS at the same concentration in regard to promoting cell proliferation and maintaining MSC characteristics in several studies [113–117,119–125]. A summary from selected studies from 2005 to 2020 is listed in Table 2. Only studies that compared the performance to FBS as supplement were included. Used concentrations ranged from as low as 0.1% hPL supplementation to 10%, where in a majority of the trials, 5% hPL performed similar to 10% FBS. Stemness characteristics could be maintained or even improved, while one difference observed in some studies was a reversible change in morphology in media with one or the other supplement, with generally smaller cell sizes in hPL cultures. However, the mode of preparation may possibly influence the properties of the hPL. PCs can be prepared in 3 different ways: By the platelet rich plasma (PRP) method, from buffy coats, or by direct single donor platelet apheresis. For the PRP method whole blood is centrifuged to separate red blood cells from platelet rich plasma which is further centrifuged to reach the required concentration and material from 4–5 donors is pooled to reach the desired volume [126]. For the second method 4 buffy coat units from centrifuged whole blood are mixed with one plasma unit and centrifuged again, followed by separation and leukocyte depletion. The only method producing single donor

PCs is by platelet apheresis. Optionally a pathogen inactivation or irradiation step is then performed. In order to produce hPL from the PCs there are again multiple options. A very common, efficient, and economical method is using single or multiple freeze/thaw cycles to lyse the platelets [113,116–118,120,123–125,127–129]. The concentrates are usually shock frozen at either −30 °C or −80 °C and reheated to 37 °C. Next to that, platelets can be activated by inducing thrombin generation, thus fibrinogen polymerization to fibrin and platelet degranulation. This can be achieved by the addition of calcium salts or one step further down the cascade, recombinant thrombin. Further methods are mechanical rupture by sonication, sometimes in combination with freeze thaw cycles or solvent/detergent treatment. With this in parallel to platelet activation, lipid enveloped viruses can be inactivated. Generally, to allow for more consistency multiple hPL units are again pooled (up to 109 units [130]). Lastly, any remaining fragments are removed, the pool is aliquoted and controlled before release. In recent years multiple commercial options have become available and some of them have already been under investigation in comparison to FBS (see Table 2) [131–134]. Suppliers include Sigma-Aldrich, Merck, PL Bioscience, Compass Biomedical and Stemcell Technologies. All of them offer their hPL in research grade, PL Bioscience additionally offers GMP grade and even pharmaceutical grade quality. For all hPL products in which fibrinogen is not depleted, heparin is needed to prevent medium gelation (coagulation) during culture, which can hamper cell proliferation and due to often porcine origin prevents the system from becoming fully humanized. To achieve this either recombinant human heparin or fibrinogen depleted hPL should be used in the future. In instances were only a small number of cells needs to be provided for a cell-based therapy (CBT), thus lower culture medium amounts are required autologous hPL can be used. Though in general, larger allogenic pooled hPL is regarded as the more promising alternative, due to higher availability. The downside of pooling is the increased risk of contamination with different pathogens.

All in all, hPL has proven to be equally suitable, if not an improvement over the classically supplemented FBS, represents an alternative, and its usage is one important step towards the establishment of xeno-free, humanized culture models.

Table 2. Human platelet lysate (hPL) as alternative to bovine serum; selected studies from 2005 to 2020 including commercially available products. Information is provided on starting material, preparation technique and culture conditions.

Starting Material	Solution	Pooled (PC)	Platelet Counts ($\times 10^9$/mL)	Platelet Lysis (Freeze/Thaw)	Platelet Lysis (Others)	Expanded Cells	Supplementation	Outperformed FBS?	Ref.
Aph-PC	Plasma	10	1	−80 °C	-	BM-MSC	5%	yes	[120]
BC-PC	Plasma	10–13	0.95	−30 °C	-	UCB-MSC	10%	yes	[124]
Exp Aph-PC	Plasma	yes	-	−80 °C	-	BM-MSC	10%	yes	[113]
Exp Aph-PC	Plasma	-	-	−80 °C	-	BM-MSC	5%	no	[118]
Aph-PC	Plasma	-	-	−80 °C	-	BM-MSC	8%	no	[129]
Aph-PC	Plasma	-	1.0–1.3	-	S/D	AD-MSC	10%	no	[135]
Aph-PC	Plasma	10	-	−80 °C	-	BM-MSC, UCB-MSC	5–10%	yes	[128]
BC-PC, leuko depleted	Plasma	-	-	−80 °C	-	AD-SC	5%	yes	[123]
Aph-PC	Plasma	4–6 don	-	−80 °C	Sonication	BM-MSC	10%	yes	[119]
BC-PC	Plasma	2	-	−30 °C	-	BM-MSC, AD-MSC	2.5–10%	yes	[121]
BC-PC	Plasma	2	-	−30 °C	Thrombin	BM-MSC, AD-MSC	2.5–10%	yes	[121]
BC-PC	Saline	6	3.34	−80 °C	-	AD-MSC	5%	yes	[125]
BC-PC	Plasma	6	3.58	−80 °C	-	AD-MSC	5%	yes	[125]
Exp Aph-PC	Plasma	5 don	1	−80 °C	$CaCl_2$	BM-MSC	10%	yes	[114]
Exp BC-PC	Plasma	yes	-	−80 °C	-	BM-MSC	10%	yes	[117]
BC-PC, path. Inactivated	Add Sol	12 don	1	−80 °C	-	BM-MSC	10%	yes	[127]
BC-PRP	Plasma	10–20	10	−196 °C	Lyophilization, Irradiation	BM-MSC	5%	yes	[122]
PL-Serum	-	49–109	-	-	$CaCl_2$ 20% w/v	commercial BM-MSC	10%	yes	[130]
BC-PC	Plasma	2	2.03	−25 °C	-	AD-MSC	1–10%	yes	[116]

Table 2. *Cont.*

Starting Material	Solution	Pooled (PC)	Platelet Counts ($\times 10^9$/mL)	Platelet Lysis		Expanded Cells	Supplementation	Outperformed FBS?	Ref.
				(Freeze/Thaw)	(Others)				
BC-PC	TSOL	2	0.91	−25 °C	-	AD-MSC	1–10%	yes	[116]
Exp BC-PC	Plasma	2	0.41	−25 °C	-	AD-MSC	1–10%	yes	[116]
Exp BC-PC	TSOL	2	0.19	−25 °C	-	AD-MSC	1–10%	yes	[116]
Exp Aph-PC	-	3–4 don	-	−80 °C	-	AD-MSC	0.1–1%	yes	[115]
Brand Name	**Supplier**	**Pooled (PC)**	**Platelet Counts ($\times 10^9$/mL)**	**Platelet Lysis (freeze/thaw)**	**(others)**	**Expanded Cells**	**Supplementation**	**Outperformed FBS?**	**REF**
PLTMAX	Sigma Aldrich	yes	-	-	-	AD-MSC	5%	yes	[134]
MesenCult hPL media	Stemcell Technologies	yes	-	-	-	AD-MSC	10%	yes	[134]
PLUS™ hPL	Compass Biomedical	yes	-	-	-	BM-MSC	5%	yes	[131]
PLTmax	MERCK	yes	-	-	-	AD-MSC	1–10%	yes	[132]
phPL	PL BioScience	yes	-		-	BM-MSC	10%	yes	[133]

(Abbreviations: AD—adipose tissue derived, add sol—additive solution, Aph—apheresis, BC—buffy coat, BM—bone marrow derived, don—donor, Exp—expired, FBS—fetal bovine serum, MSC -mesenchymal stem cell, PC—platelet concentrate, PL—platelet lysate, PRP—platelet rich plasma, S/D—solvent/detergent, UC—umbilical cord derived UCB—umbilical cord blood derived).

10. Serum-, Xeno-Free-, and Chemically Defined Media

Next to the switch from bovine or other animal-origin sera to human platelet lysate, other alternatives to fully humanize or potentially even develop chemically defined media are investigated for various reasons. Improving cell growth as well as phenotype is of great interest and specially to reduce the high variability encountered with non- or ill-defined medium components. To understand all the different approaches, firstly the various definitions used have to be examined. Chemically defined medium means that there is no supplementation with unknown composition, this includes blood derived supplements like serum, platelet lysates, but also plant hydrolysates. Highly purified components of different origin or recombinant products but with exact concentrations, can be added. As the origin can be from animals as well, chemically defined is not the same as animal-derived component free. Animal-derived component free (ADCF) stands for medium without supplements originating from animals, this usually includes humans as well. When no supplementation of serum is necessary the medium is called serum-free (SF), but may contain other protein fractions with plant or animal origin. As there could in both cases still be hydrolysates from bacteria or yeast for the ADCF media or other animal derived protein cocktails for the SF medium added, these terms not necessarily mean chemically defined. The term xeno-free is especially tricky as it depends on the origin of the to be cultured cell line. Xeno is of Greek origin and means "strange" and denotes compounds derived from a different species. Therefore, for human MSCs medium with human serum or platelet lysate is xeno-free, but the same media would not be if one would culture mouse MSCs in it [107]. Lastly, protein-free media are media that do not contain large proteins, but only small peptides. In this case, one can already see the limitations associated to these definitions. Where is the border to be set between protein and peptide? Further, recombinant human proteins can be produced in either bacterial or other mammal cell lines. This again raises question if they can be considered xeno-free and needs to be addressed in future media formulations [136].

Considering human cells for cell-based therapies, omitting xenogeneic compounds will rule out further immune reactions of the transplanted material and especially for bovine serum, and also the risk of transmitting prion diseases such as spongiform encephalopathy. Additionally, FBS production inevitably has high lot-to-lot variations, which requires pretesting before a new lot can be used for culture of MSCs [137,138]. With the transition to completely serum free chemically defined medium, further advantages arise. Blood-derived components, whether they are of animal or human origin, are heterogenous and, as already said, will inevitably have differences from batch to batch. This leads to less reproducible results and also causes a more heterogenous cell population [139]. With the use of serum-free medium (SFM), it can be specifically tailored for the needs of the MSC fraction, enhancing selection already during isolation. Enhanced consistency is also in the spirit of GMP regulation and thus the logical next step. There are already ready-to-use media specifically for MSCs available from different suppliers that are either serum-free, xeno-free, or both. To give a few examples, StemPro™ MSC SFM and StemPro™ MSC SFM XenoFree™ (Thermo Fisher Scientific, Waltham, MA, USA), StemXVivo (R&D Systems, Minneapolis, MN, USA), PowerStem MSC1 (Pan Biotech, Aidenbach, Germany), MSC NutriStem® XF (Biological Industries, Kibbutz Beit-Haemek, Israel), PRIME XV MSC Expansion SFM (IrvineScientific, Santa Ana, CA, USA), MesenCult™-ACF Plus (Stem Cell Technologies, Vancouver, Canada), and MSCGM-CD (Lonza, Basel, Switzerland). Other options are SFM designed, intended for culture of embryonic or induced pluripotent stem cells (iPSCs) that are adapted to suit MSC as well, e.g., mTeSR (Stem Cell Technologies). The advantage of these commercial options is that they comply to GMP regulations and undergo strict testing by the supplier, and as they are premixed, handling times and material requirements are minimized on the user side. Further, a lot-to-lot variation of the individual components can also be averted. On the downside, however, the exact formulations are not disclosed by the companies, and thus adaption and optimization to the specific cell source or intended application is not possible, and in comparison to other media options they are costlier. A summary of tested media including the origin of the cells they were tested on is listed in Table 4. Interestingly, in a study by Al-Saqi from 2015, bone marrow derived cells did not perform well in Mesencult XF medium compared to MSCs derived from adipose tissues. This further indicates unique needs of different MSC populations. For the generation of chemically defined medium in-house all aspects regarding medium ingredients and concentrations above are of great importance to support MSC growth; optimization of serum albumin, fatty acids, trace minerals, glucose content, and combination of growth factors.

Common supplements, next to what has been discussed before, include insulin, transferrin, and ethanolamine in combination with the trace element selenium (e.g., ITS, SITE). The role of insulin is to help cells use glucose and amino acids, transferrin serves as iron transport protein as free iron has toxic effects on cells [140] and ethanolamine is necessary for phospholipid synthesis [141]. Exemplary compositions from different disclosed serum free media include simpler compositions of only a few ingredients: IMDM supplemented with bFGF, human albumin (most abundant protein in vertebrate plasma, important for colloidal osmotic pressure and binding of ions, fatty acids, steroids and drugs [142]), hydrocortisone (steroid hormone, promotes cell growth [143]), and SITE [144]. A more complex chemically defined medium for hMSCs was patented already in 1995 by Marshak [145]: IMDM supplemented with serum albumin, lipoprotein, transferrin, insulin, MEM vitamins, MEM essential amino acids, MEM nonessential amino acids, sodium pyruvate, GlutaMAX-I supplement (substitution for L-glutamine, which is more stable in culture conditions [146]), folic acid (central role in one-carbon metabolism, important for growth, differentiation, and survival [147]), Ascorbic acid 2-phosphate, Biotin (essential vitamin, relevant for fatty acid synthesis, amino acid, and energy metabolism [148]), vitamin B12 mix, trace element mix, $FeSO_4$, nucleoside mix, antibiotic/antimycotic, and either PDGF ββ homodimer or M 5-hydroxytryptamine (serotonin, micromolar amounts can increase cell growth [149]). Both chemically defined and xeno-free is the formulation of Wu et al. [150] (IMDM, l-glutamine, sodium bicarbonate, insulin, transferrin, serum albumin, β-mercaptoethanol (potent reducing agent, prevents toxic levels of oxygen radicals [151]), chemically defined lipid concentrate, MEM essential amino acids solution, MEM non-essential amino acid solution, Vitamins solution, trace elements solution, hydrocortisone, l-ascorbic acid-2-phosphate, fibronectin, progesterone (hormone, enhances immunomodulatory function of MSCs [152]), putrescine (non-protein nitrogen base, associated with proliferation in mammalian cells [153]), serotonin, epidermal growth factor, basic fibroblast growth factor, platelet-derived growth factor, insulin-like growth factor) [150]. The exact compositions of each of the three examples can be found in Table 3. Another option for labs is to get a custom-made medium from a commercial vendor. Such tailormade media can be procured from Cell Culture Technologies, a Swiss biotech company [154], or also larger biotech suppliers like Thermo Fisher Scientific, R&D Systems, and Biological Industries. Improvement on current animal component and undefined media compositions is the first step to improved cell culture concepts, further improvements could be made in investigation of culture media that are designed to specifically meet the needs of the cells in different "phases" of their life cycle, as current approaches do not consider such differences during culture.

Although the classical serum supplemented medium is still widely used and produces satisfying cell yields, transition away from animal-derived as well as poorly defined medium components is an essential step towards clinical relevance and further product development. Clearly, increased costs and challenges in the transition phase hinder a lot of research groups to switch to chemically defined medium. However, even small changes, one at a time, starting by omitting animal-derived products over to disposal of undefined sera and ultimately even proteins can improve the cultivation system and ensure more consistency.

Table 3. Composition of three disclosed chemically defined cell culture media for MSCs.

Ref.	[144]	[145]	[150]
Basal Media	IMDM	IMDM	17.7 g/L IDMD
Supplemented With	17.91 ng bovine FGF/mL	5 mg/mL human serum albumin	5 mM l-glutamine
	2.80 mg/mL human albumin	100 µg/mL human Ex-Cyte lipoprotein	3.024 g/L sodium bicarbonate
	27. 65 µM hydrocortisone	2 µg/mL saturated human transferrin	10 mg/L rh insulin
	1.18% SITE (S4920; containing 0.5 µg/mL sodium selenite, 1.0 mg/mL bovine insulin, 0.55 mg/mL human transferrin, 0.2 mg/mL ethanolamine; 100-fold concentrate)	10 µg/mL rh insulin	10 mg/L rh transferrin
		1.0% 100 × MEM vitamins	4 g/L rh serum albumin
		0.89% MEM essential amino acids	55 µM β-mercaptoethanol
		0.4% MEM nonessential amino acids	0.1% chemically defined lipid concentrate
		1 mM sodium pyruvate	2% MEM essential amino acids solution
		1 mM GlutaMAX-I supplement	1% MEM non-essential amino acid solution
		10 µg/mL folic acid	1% Vitamins solution
		10 µM ascorbic acid 2-phosphate	0.1% trace elements solution
		1.0 µg/mL Biotin	50 µg/L hydrocortisone
		1.36 µg/mL vitamin B12 mix	50 mg/L l-ascorbic acid-2-phosphate
		500fold diluted trace element mix	5 mg/L rh fibronectin
		4×10^{-8} M $FeSO_4$	5 µg/L progesterone
		10 µg/mL nucleoside mix (ribonucleosides, 2′-deoxyribo-nucleosides, uridine, and thymidine)	10 mg/L putrescine
			2 mg/L serotonin
			10 ng/mL rh EGF
		1.0% antibiotic/antimycotic	10 ng/mL rh basic FGF
		10–20 ng/mL rh PDGF ββ homodimer or 10^{-5} to 10^{-6} M 5-hydroxytryptamine	10 ng/mL rh PDGF
			10 ng/mL rh IGF

(Abbreviations: EGF—epidermal growth factor, FGF—fibroblast growth factor, rh—recombinant human, IGF—insulin-like growth factor, IMDM—Iscove's modified Dulbecco's medium, MEM—minimum essential media, PDGF—platelet-derived growth factor).

Table 4. Commercially available, chemically defined/serum-free/xeno-free media tested for expansion of hMSCs; including selected studies from 2010–2021. Information is given on brand name, supplier, any added supplements, the cell source, cultivation parameters, and stem cell characteristics.

Basal Medium	Supplier	Add. Supplements	Cells Expanded	Cultivation Time	Outperformed CTL Medium?	Diff Capacity	Cell Surface Markers	Ref.
MSCGM-CD	Lonza	-	UC-MSC	5–7 passages	n.t.	unaltered	unaltered	[38]
MesenCult	Stemcell Technologies	-	hESC-derived MSCs	-	yes	n.t.	unaltered	[155]
Mesencult-XF™	Stemcell Technologies	-	AD-MSC, BM-MSC	-	yes	improved (AD-MSC), decreased (BM-MSC)	altered (BM-MSC)	[156]
StemPro MSC SFM XenoFree™	Life Technologies	-	BM-MSC	up to p4	no	n.t.	n.t.	[139]
Mesencult-XF™	Stemcell Technologies	-	BM-MSC	up to p4	no	unaltered	unaltered	[139]
BD Mosaic™ Mesenchymal Stem Cell Serum-Free media	BD Biosciences	-	BM-MSC	up to p4	no	unaltered	unaltered	[139]
StemPro® MSC SFM XenoFree, Invitrogen	Life Technologies	-	AD-MSC, BM-MSC	-	yes	altered	unaltered	[157]
StemPro MSC SFM XenoFree™	Life Technologies	PDGF-BB, bFGF, TGF-β1	ASC line, BM-MSC	up to p9	yes	unaltered	unaltered	[158]
StemPro® MSC SFM	Life Technologies	PDGF-BB, bFGF, TGF-β1	BM-MSC	8 passages	no	unaltered	unaltered	[159]
StemPro MSC SFM Xenofree	Life Technologies	-	BM-MSC, UC-MSC, AD-MSC	7 days	yes	n.t.	n.t.	[160]
MSC Nutristem XF	Biological Industries	-	BM-MSC, UC-MSC, AD-MSC	7 days	yes	n.t.	n.t.	[160]
MesenCult-XF	Stemcell Technologies	-	BM-MSC, UC-MSC, AD-MSC	7 days	yes	n.t.	n.t.	[160]

Table 4. *Cont.*

Basal Medium	Supplier	Add. Supplements	Cells Expanded	Cultivation Time	Outperformed CTL Medium?	Diff Capacity	Cell Surface Markers	Ref.
StemXVivo SFM Human MSC Expansion Medium	R&D Systems	-	BM-MSC, UC-MSC, AD-MSC	7 days	yes	n.t.	n.t.	[160]
RoosterNourish-MSC XF	RoosterBio, Inc.		BM-MSC	up to p5	no	unaltered	unaltered	[161]
StemMACS-MSC Expansion Media Kit XF	Miltenyi Biotec		BM-MSC	up to p5	no	unaltered	unaltered	[161]
MSC NutriStem XF	Biological Industries		BM-MSC	up to p5	no	unaltered	unaltered	[161]
StemXVivo SFM Human MSC Expansion Medium	R&D Systems		BM-MSC	up to p5	no	unaltered	unaltered	[161]

(Abbreviations: AD—adipose tissue derived. Add.—additional, BM—bone marrow derived, CD—chemically defined, CTL—control, diff—differentiation, hPL - human platelet lysate, MSC—mesenchymal stem cell, n.t.—not tested, p—passage, SFM—serum free medium, UC—umbilical cord derived, XF—xeno-free).

11. Hypoxia

Thinking of parameters outside of the culture medium, itself long-overlooked, is the so called "normoxic" environment. When laboratories report culture of cells under normoxic conditions, they mean ambient oxygen concentrations of 21%, yet the question arises if this is really physiologically "normal" for the cultured cells. Some cell types in the human body are adapted to these levels, like cells of the respiratory system, skin, and the oral cavity. However, if we follow the dissolved oxygen within the blood stream to its destination in organs and tissues, a lot of lower concentrations are the norm than what is termed "normoxic". Depending on the source tissue in vivo, oxygen levels between 18% (lungs) and as low as 1% (cartilage, bone marrow) are found [162]. Therefore, the actual physiological normoxic conditions for MSCs are actually rather low and are regarded as hypoxia. One must also take into account that atmospheric oxygen levels are not the same as the amount of oxygen dissolved in the medium. In case of a standard culture flask in a typical incubator, 20% oxygen in the surroundings means an estimated dissolved oxygen content of 100% in the medium. The amount of diffusion of oxygen into the culture medium is determined by physical laws influenced by a plethora of parameters. Ambient O_2 concentration [137], O_2 partial pressure determined by altitude, temperature, culture volume, and surface area have to be taken into account and monitored closely for studies on effect of oxygen levels on cells. With higher levels of dissolved oxygen other problem can emerge, the formation of so-called reactive oxygen species (ROS), for example peroxides, super-oxides, and hydroxyl radicals. Although they are a by-product in normal cell metabolism, dramatic increase of them can harm cellular structures, this is termed oxidative stress. Decrease of environmental O_2 levels also reduces the formation of these ROS [163,164]. In the last years more and more studies have indicated, that mimicking the native niche conditions where the cells to be expanded reside in can improve their viability, promote proliferation, and maintain their in vivo properties. An indication of how important hypoxic conditions are in a cellular context is the large group of gene promoter regions that are responsive to hypoxia (hypoxia responsive elements HRE). The key player for activation is hypoxia inducible factor I (HIF1), which is a dimeric transcription factor. Expression of the two subunits is not oxygen dependent, but stability of the α-subunit is. Under normoxic conditions it is rapidly targeted for degradation, only in low oxygen concentration it is able to accumulate intracellularly and associate with the β-subunit and become functional [165]. In order to benefit from the positive effects of hypoxia in in vitro settings, such an environment has to be established first. Practical considerations of pros and cons of different methods shall be discussed.

One of the easiest and simplest forms is the acquisition of a hypoxic sub-chamber (for example from BioSpherix [166] or Stemcell Technologies [167]), which is cost-saving and easy to handle. On the downside is of course the inevitable reoxygenation events every time a media change, microscopic evaluation, etc. is performed. Most models do not allow for tight oxygen control; thus, leakage and consistent low levels cannot be secured. For very low levels repeated re-flushing with nitrogen might be necessary to keep them. A bit more controlled is the use of a standard incubator with nitrogen gas supply, which is able to establish hypoxia by controlling oxygen levels next to humidity and temperature. However, handling with the culture vessels still requires removal from the hypoxic environment and especially for laboratories with multiple persons using the same hypoxia incubator, constant open and closing can lead to unwanted reoxygenation. The increase of size can be regarded as a two-sided coin; on one side, increased space allows for larger experiments and multi-person use, but on the other side, more gas is needed to establish the desired atmosphere as a sub-chamber does. Complete hypoxia workstations, optionally with gloves, circumvent any disruptions in oxygen levels, but are very costly and require a lot of attention for maintenance. They can be large enough to fit microscopes or other imaging systems within and allow for routine maintenance of cells like media exchanges or passaging in the desired environmental conditions [168]. Downsides to these workstations are that, on the other hand, access to the inside, for example for cleaning, is rather difficult.

A different approach is to only mimic hypoxia by addition of Cobalt Chloride ($CoCl_2$) or Deferoxamine Mesylate (DFO). These chelating agents can, after short hypoxic treatment, stabilize HIF at non-hypoxic conditions. This treatment has many limitations, cannot emulate real hypoxia, and has known and maybe even not yet discovered side effects on the cells, thus we will not go further into detail on this method.

Although a general improvement of cell yield under hypoxic conditions can be concluded (see Table 5) [61,113,169–181], some research groups could not confirm the beneficial effects of hypoxic culture conditions [182–185]. This might also be caused by high variations within the applied parameters. First and foremost is the inconsistent definition of hypoxia, tested O_2 concentrations range from 1% to 5%, and secondly, fluctuations due to the aforementioned different methods of establishing hypoxic conditions in a laboratory setting are not taken into account. An important step towards harmonizing future results is the employment of non-invasive monitoring of the dissolved oxygen. Another point is the distinction between short term hypoxic treatments, or continuous culture under physiological O_2 levels. Cells destined for transplantation are sometimes pre-conditioned for a few hours before administration, whereas other approaches try to keep cells already starting from the isolation procedure [186] until the transplantation at lowered oxygen concentrations.

Nevertheless, a switch from the prevailing atmospheric oxygen levels during culture is an important advance in providing the cells a more physiological environment.

Table 5. Hypoxic conditions for expansion of hMSC; selected studies from 2007–2020. Information is given on the cell source, exposure to hypoxic conditions and on stem cell characteristics.

Cells	O_2 (%)	Exposure Time	Outperformed 21% O_2?	Diff Capacity	Cell Surface Markers	Other Cultivation Parameters Analyzed	Ref.
BM-MSC	2	6 weeks	yes	unaltered	n.t.		[169]
BM-MSC	1–3	16 h	no	n.t.	n.t.	HGF stimulation	[184]
BM-MSC	5	up to p4	yes	unaltered	unaltered	hPL	[113]
BM-MSC	2	14 d	yes	n.t.	unaltered		[113]
UC-MSC	1.5–5	3 d	yes	n.t.	n.t.		[171]
BM-MSC	1–5	14 d	no	decreased	unaltered		[182]
BM-MSC	1	84 days	yes	improved	unaltered		[172]
BM-MSC	5	up to p3	yes	improved	unaltered		[173]
AD-MSC	2	7 d	yes	improved	unaltered		[174]
BM-MSC	1	up to 90 d	yes	improved	upregulated		[175]
BM-MSC, AD-MSC, AF-MSC, UCB-MSC	1	7 d	depending on cell source (prenatal yes, postnatal no)	n.t.	n.t.	prenatal + postnatal material	[185]
AD-MSC	2	up to 21 d	yes	unaltered	unaltered		[176]
UCB-MSC	5	5 d	yes	n.t.	unaltered		[179]
AD-MSC	5	up to 14 d	yes	n.t.	slightly altered	lean + obese donors	[178]
BM-MSC	5	up to p15	yes	unaltered	n.t.	donor age	[177]
BM-MSC	1–4	up to p2	no	unaltered	unaltered		[183]
UCB-MSC	3	5 d	yes	unaltered	unaltered	Ca^{2+}	[61]
AD-MSC	5	up to p28	yes (until passage 23)	n.t.	n.t.		[180]
AD-MSC	1	48 h	yes	decreased (osteogenic) increased (chondrogenic)	unaltered		[181]

(Abbreviations: AD—adipose tissue derived, AF—amniotic fluid BM—bone marrow derived, diff—differentiation, HGF—human growth factors, hPL—human platelet lysate, MSC -mesenchymal stem cell, n.t.—not tested, p—passage, UC—umbilical cord derived, UCB—umbilical cord blood derived).

12. Isolation and Passage Cell Seeding Densities

Studies on cell seeding density, although limited, indicate that it does have an effect on the proliferation rate and morphology of the cells. For primary MSCs, Sotiropoulou et al., found that seeding density of 1000 cells/cm^2 of primary bone marrow mononuclear cells can result in significantly increased number of adherent MSCs after 1 week of culture compared to higher seeding densities tested, without any alterations on their phenotype, colony-forming, and immunosuppressive capacities [40]. Regarding passaged cells, studies have shown that seeding BM-MSCs in low densities (50–100 cells/cm^2) increases their proliferation rates compared to higher densities (500–5000 cells/cm^2) [40,187–189]. In addition, studies using extremely low seeding densities (1.5–10 cells/cm^2) have showed that proliferation rates are inversely proportional to the seeding density of BM-MSCs, and that the cells maintain their immunophenotype, and morphological and self-renewal characteristics during culture [187,189–191]. Furthermore, differences in the seeding density of MSCs can also alter their gene expression patterns. In particular, studies have shown that human MSCs cultured in low seeding density (200 cells/cm^2) can maintain their stem and immunomodulatory gene expression profiles better than when seeded in higher density (5000 cells/cm^2) [192,193]. In general, seeding MSCs in low densities results in higher proliferation rates and better preservation of their physiological characteristics. This is also advantageous with regard to the limited number of primary MSCs that must be expanded after isolation from the tissue.

13. Cell Detaching Methods

During in vitro expansion, MSCs are usually cultured for multiple passages. It has been reported that human BM-MSCs are genetically stable and maintain their immunomodulatory and differentiation capacities up to passage 4 [194]. Therefore, in clinical applications, MSCs can be successfully used up to 4 passages. During passaging, efficient cell detachment is imperative in cultures of adherent cells like MSCs, without causing any damage. The most common and effective methods involve the use of enzymatic solutions to release the cells from the plastic culture surface by degrading their surface attachment proteins. Trypsin is a serine protease and is the most common enzymatic mean for dissociation of adherent cells. An alternative to Trypsin is Accutase, a proteolytic and collagenolytic enzymatic solution, which in comparison to Trypsin does not contain mammalian or bacterially derived proteins. Trypsin and Accutase have been reported to be the most suitable cell detaching reagents for human primary [195] and cultured MSCs [196–198] compared to other enzymatic (Collagenase, Prolyl-specific Peptidase) and non-enzymatic cell dissociation solutions (cell dissociation buffers, ethylenediamine tetra-acetic acid (EDTA)) and cell scraping, with significantly lower incubation times, increased viability, and higher yields of detached cells. Although different types of culture surface coatings do not significantly influence the effectiveness of enzymatic dissociation solutions, the results from the study of Salzig et al., suggest that the efficiency of different enzymatic detachment methods and corresponding cell viability following detachment are significantly affected by the culture media used [198]. However, due to its proteolytic activity, it has been reported that trypsin can significantly reduce the expression of MSC surface markers [195,198] and change the expression of several non-coding RNAs, resulting in several transcriptomic changes [199]. Although some non-enzymatic buffers seem to minimize the adverse effects of other methods on the detached MSCs [195,198,200], most studies suggest that due to the long incubation times required for non-enzymatic detachment means, Trypsin is comparatively a more beneficial method for quick preparation of MSC suspensions [195,196,200].

Although trypsinization is the most common laboratory method used for detaching MSCs, in order to avoid the negative impact of proteolytic degradation on cell membrane during enzymatic dissociation, new techniques are developed, for example thermoresponsive detachment of adherent cells. Adherent MSCs can be detached from the culture surface by changing the temperature. However, without appropriate culture substrate, MSCs must be exposed to very low/non-physiologic temperatures to facilitate detachment [201]. For this reason, a thermoresponsive polymeric coating, usually poly(N-isopropylacrylamide), is applied on the cell culture surface. These thermoresponsive surfaces have been reported to promote human MSC adhesion, colony formation and proliferation [202,203], and enable their detachment with temperature reduction below a critical solution temperature [202–204], without any changes on the MSC specific marker profile and morphology [205,206]. A comparative study by Yang et al., suggests that thermoresponsive detachment of MSC from poly(N-isopropylacrylamide) films is as effective a method as trypsinization [206]. These cell detachment methods are promising in avoiding the proteolytic adverse effects of trypsin, although they are still not so well optimized as routine methods such as trypsinization.

Other cell dissociation methods have been recently developed, using for example acoustic pressure [207], light-induced plasmonic substrates [208], and non-enzymatic arginine-based solutions [209] to detach adherent cells. These techniques show very promising results, but their application on MSCs has not been investigated yet.

14. Conclusions

Human MSCs are considered one of the most prominent types of cells for cellular therapies, due to their multipotent and regenerative capacities. However, due to their limited number after isolation, in vitro expansion is required to achieve a sufficient number of cells for clinical and research applications. This expansion step is usually performed in a standard 2D culture setup that requires a lower number of seeded cells due to its robustness and simplicity compared to more recently developed advanced 3D culture methods. Therefore, in order to improve the relevance of in vitro applications, it is imperative to preserve the physiological characteristics of MSCs during this necessary in vitro expansion step.

Many optimization approaches for in vitro expansion of human MSCs focus on well-formulated and chemically defined culture media. An overview of the major components of culture media and additional optimization approaches and their biological effects mentioned in this review are summarized in Figure 1. However, apart from the components mentioned there, there are many more included in their formulations that can potentially affect cell behavior and were not given enough attention so far, and therefore require focused research. Furthermore, apart from their chemical properties, alterations in the physical properties of the media like osmolarity are often neglected. Similarly, the influences of exposure to light and temperature on the characteristics and behavior of the cells during culture are widely ignored and present a promising aspect for further investigation [210–213].

In order to maximize the benefits of these aspects on MSC culture, it is important that each of these elements is thoroughly investigated, so their exact underlining mechanisms and concomitant effects on the cells are identified. This can also expand the range and validity of analytics for assessing MSC expansion cultures. Many of the studies in this review evaluated the effect of different aspects on expanding MSCs by mainly measuring total number of cells, doubling times or time to reach confluency. Although these methods may give representative indications of the effects on cultured MSCs, due to the different metric approaches, comparison and relevance between the results are not always accurate. In addition, these methods only cover cell attributes like proliferation and growth rates. Thereby, possible differences in a more thorough biological level, like gene expression, or surface marker and proteomic alterations, are often disregarded. Attempts to identify alterations on the cells immunomodulatory profile by prolonged ex vivo cultivation have been made in "standard" cultivation settings [214], as understanding the cellular response to expansion can influence the therapeutic potential of a CBT. Optimization approaches can also greatly influence these cell properties, which can be even more relevant for treatment success than solely greater cell numbers. For that reason, the introduction of specific criteria and analytics for the evaluation of the effects of different aspects during in vitro expansion of human MSCs, similar for example to the ones the ISCT established for defining MSCs, would be beneficial for the significance and relevance of future studies in this research topic.

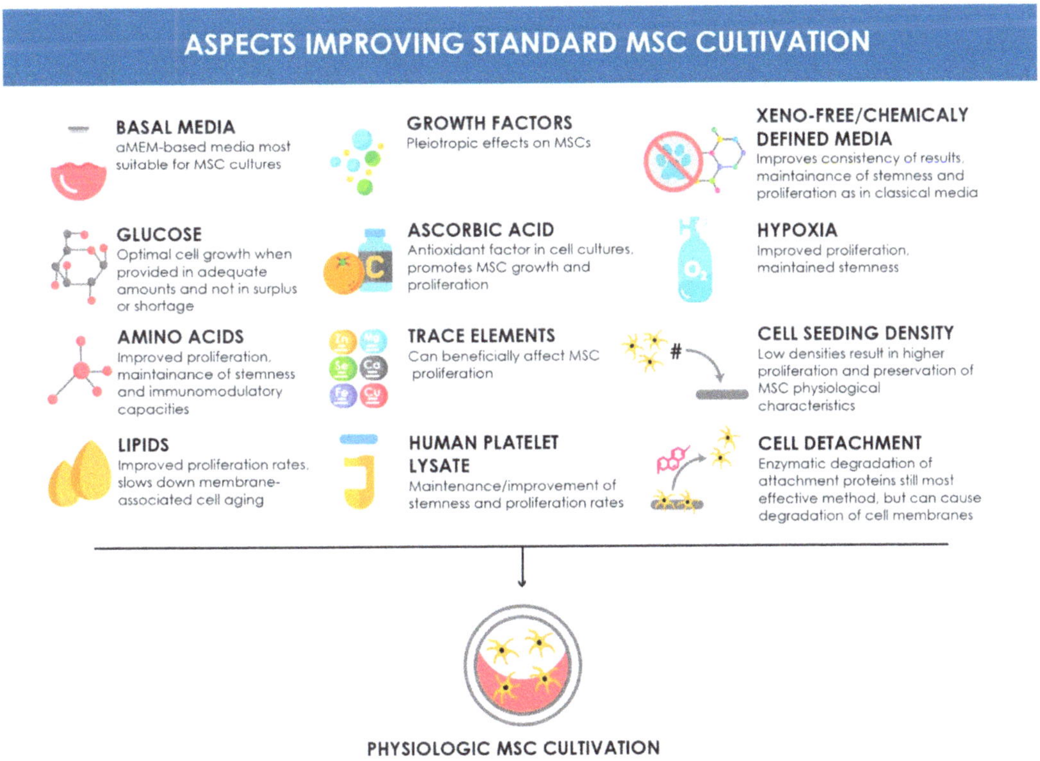

Figure 1. Schematic overview of the aspects improving the standard MSC cultivation setup discussed in this review. The beneficial biological effects are summarized for each topic.

While most of the elements mentioned in this review have been proven to be beneficial for human MSC in vitro cultures when examined solely, the impact of the combination of different elements should also be investigated. Most papers studied only one parameter, and only rarely were two investigated. For future media composition optimization, multiparametric analysis with carefully considered design of experiments should reveal more information on the interplay of different components. Furthermore, MSCs inherently present high donor and source variability in their characteristics and properties. For that reason, some of the studies included in this review have reported different effects of the same investigated in vitro culture parameter on different types of MSCs. Therefore, further research should be directed towards formulation of MSC type-specific culture media and protocols to avoid variability in preserving the physiological status of cultured MSCs.

In conclusion, the approaches discussed in this review constitute the basic principles for MSC culture and seem to be advantageous for the preservation of physiological MSC characteristics during in vitro expansion. To fully recapitulate the in vivo conditions, further strategies such as 3D passage free culture will have to be developed further. The purpose of this review is to provide the readers with a broad variety of up-to-date approaches on standard expansion of human MSCs, that can help them optimize their current culture protocols and improve the efficacy and physiological relevance of their MSC cultures, as well as serve as a baseline for additional investigation and further optimization of standard cultivation of MSCs in vitro.

Author Contributions: D.E., S.K., and C.K. designed the study, I.N. and S.N., contributed to the conception of the study. I.N. and S.N. performed the literature research and writing of the review. D.E., S.K., and C.K. are responsible for the design and structure of the manuscript. All authors contributed to manuscript revision, read and approved the submitted form. All authors have read and agreed to the published version of the manuscript.

Funding: Open access funding provided by BOKU Vienna Open Access Publishing Fund.

Institutional Review Board Statement: Not applicable.

Informed Consent Statement: Not applicable.

Data Availability Statement: Not applicable.

Conflicts of Interest: The authors declare no conflict of interest. The funders had no role in the design of the study; in the collection, analyses, or interpretation of data; in the writing of the manuscript, or in the decision to publish the results.

Abbreviations

2D	two-dimensional
3D	three-dimensional
ADCF	animal-derived component free
BM-MSCs	bone marrow-mesenchymal stem cells
EAAs	essential amino acids
EGF	epidermal growth factor
FBS	fetal bovine serum
FGF-2 and -4	fibroblast growth factor-2 and -4
GMP	good manufacturing practice
HB-EGF	heparin-Binding Epidermal Growth Factor
hPL	human platelet lysate
HIF1	hypoxia inducible factor I
IGF	insulin-like growth factor
ISCT	International Society for Cellular Therapy
MSC(s)	mesenchymal stem cell(s)
NEAAs	non-essential amino-acids
PDGF-BB	platelet derived growth factor-BB
PCs	platelet concentrates
ROS	reactive oxygen species
PRP	platelet rich plasma
SF	serum-free
VEGF	endothelial growth factor

References

1. Hanna, E.; Rmuzat, C.; Auquier, P.; Toumi, M. Advanced therapy medicinal products: Current and future perspectives. *J. Mark. Access Health Policy* **2016**, *4*, 31036. [CrossRef] [PubMed]
2. Mount, N.M.; Ward, S.J.; Kefalas, P.; Hyllner, J. Cell-based therapy technology classifications and translational challenges. *Philos. Trans. R. Soc. B Biol. Sci.* **2015**, *370*, 20150017. [CrossRef]
3. Buzhor, E.; Leshansky, L.; Blumenthal, J.; Barash, H.; Warshawsky, D.; Mazor, Y.; Shtrichman, R. Cell-based therapy approaches: The hope for incurable diseases. *Regen. Med.* **2014**, *9*, 649–672. [CrossRef] [PubMed]
4. Chagastelles, P.C.; Nardi, N.B. Biology of stem cells: An overview. *Kidney Int. Suppl.* **2011**, *1*, 63–67. [CrossRef] [PubMed]
5. Pittenger, M.F.; Mackay, A.M.; Beck, S.C.; Jaiswal, R.K.; Douglas, R.; Mosca, J.D.; Moorman, M.A.; Simonetti, D.W.; Craig, S.; Marshak, D.R. Multilineage potential of adult human mesenchymal stem cells. *Science* **1999**, *284*, 143–147. [CrossRef]
6. Dominici, M.; Le Blanc, K.; Mueller, I.; Slaper-Cortenbach, I.; Marini, F.; Krause, D.; Deans, R.; Keating, A.; Prockop, D.; Horwitz, E. Minimal criteria for defining multipotent mesenchymal stromal cells. The International Society for Cellular Therapy position statement. *Cytotherapy* **2006**, *8*, 315–317. [CrossRef]
7. Gardner, O.F.; Alini, M.; Stoddart, M.J. Mesenchymal Stem Cells Derived from Human Bone Marrow. *Methods Mol. Biol.* **2015**, *1340*, 41–52. [CrossRef]
8. Mahmoudifar, N.; Doran, P.M. Mesenchymal Stem Cells Derived from Human Adipose Tissue. *Methods Mol. Biol.* **2015**, *1340*, 53–64. [CrossRef]

9. Van Pham, P.; Truong, N.C.; Le, P.T.; Tran, T.D.; Vu, N.B.; Bui, K.H.; Phan, N.K. Isolation and proliferation of umbilical cord tissue derived mesenchymal stem cells for clinical applications. *Cell Tissue Bank.* **2016**, *17*, 289–302. [CrossRef]
10. Ouryazdanpanah, N.; Dabiri, S.; Derakhshani, A.; Vahidi, R.; Farsinejad, A. Peripheral Blood-Derived Mesenchymal Stem Cells: Growth Factor-Free Isolation, Molecular Characterization and Differentiation. *Iran. J. Pathol.* **2018**, *13*, 461–466.
11. Ledesma-Martínez, E.; Mendoza-Núñez, V.M.; Santiago-Osorio, E. Mesenchymal Stem Cells Derived from Dental Pulp: A Review. *Stem Cells Int.* **2016**, *2016*, 4709572. [CrossRef] [PubMed]
12. Hass, R.; Kasper, C.; Böhm, S.; Jacobs, R. Different populations and sources of human mesenchymal stem cells (MSC): A comparison of adult and neonatal tissue-derived MSC. *Cell Commun. Signal.* **2011**, *9*, 12. [CrossRef] [PubMed]
13. Duval, K.; Grover, H.; Han, L.-H.; Mou, Y.; Pegoraro, A.F.; Fredberg, J.; Chen, Z. Modeling Physiological Events in 2D vs. 3D Cell Culture. *Physiology* **2017**, *32*, 266–277. [CrossRef] [PubMed]
14. Chaicharoenaudomrung, N.; Kunhorm, P.; Noisa, P. Three-dimensional cell culture systems as an in vitro platform for cancer and stem cell modeling. *World J. Stem Cells* **2019**, *11*, 1065–1083. [CrossRef] [PubMed]
15. Kropp, C.; Massai, D.; Zweigerdt, R. Progress and challenges in large-scale expansion of human pluripotent stem cells. *Process. Biochem.* **2017**, *59*, 244–254. [CrossRef]
16. Rafiq, Q.A.; Coopman, K.; Hewitt, C.J. Scale-up of human mesenchymal stem cell culture: Current technologies and future challenges. *Curr. Opin. Chem. Eng.* **2013**, *2*, 8–16. [CrossRef]
17. Cesarz, Z.; Tamama, K. Spheroid Culture of Mesenchymal Stem Cells. *Stem Cells Int.* **2016**, *2016*, 9176357. [CrossRef]
18. Yin, X.; Mead, B.E.; Safaee, H.; Langer, R.; Karp, J.M.; Levy, O. Engineering Stem Cell Organoids. *Cell Stem Cell* **2016**, *18*, 25–38. [CrossRef]
19. Chen, Y.; Shu, Z.; Qian, K.; Wang, J.; Zhu, H. Harnessing the Properties of Biomaterial to Enhance the Immunomodulation of Mesenchymal Stem Cells. *Tissue Eng. Part. B Rev.* **2019**, *25*, 492–499. [CrossRef]
20. Hanson, S.; D'Souza, R.N.; Hematti, P. Biomaterial-mesenchymal stem cell constructs for immunomodulation in composite tissue engineering. *Tissue Eng. Part A* **2014**, *20*, 2162–2168. [CrossRef]
21. Schneider, R.K.; Knüchel, R.; Neuss, S. Mesenchymal stem cells and their interaction with biomaterials: Potential applications in tissue engineering. *Pathologe* **2011**, *32* (Suppl. S2), 296–303. [CrossRef]
22. Azandeh, S.; Mohammad Gharravi, A.; Orazizadeh, M.; Khodadi, A.; Hashemi Tabar, M. Improvement of mesenchymal stem cell differentiation into the endoderm lineage by four step sequential method in biocompatible biomaterial. *Bioimpacts* **2016**, *6*, 9–13. [CrossRef]
23. Das, R.; Roosloot, R.; van Pel, M.; Schepers, K.; Driessen, M.; Fibbe, W.E.; de Bruijn, J.D.; Roelofs, H. Preparing for cell culture scale-out: Establishing parity of bioreactor- and flask-expanded mesenchymal stromal cell cultures. *J. Transl. Med.* **2019**, *17*, 241. [CrossRef]
24. Jossen, V.; van den Bos, C.; Eibl, R.; Eibl, D. Manufacturing human mesenchymal stem cells at clinical scale: Process and regulatory challenges. *Appl. Microbiol. Biotechnol.* **2018**, *102*, 3981–3994. [CrossRef]
25. Lawson, T.; Kehoe, D.E.; Schnitzler, A.C.; Rapiejko, P.J.; Der, K.A.; Philbrick, K.; Punreddy, S.; Rigby, S.; Smith, R.; Feng, Q.; et al. Process development for expansion of human mesenchymal stromal cells in a 50L single-use stirred tank bioreactor. *Biochem. Eng. J.* **2017**, *120*, 49–62. [CrossRef]
26. Hoch, A.I.; Duhr, R.; Di Maggio, N.; Mehrkens, A.; Jakob, M.; Wendt, D. Expansion of Bone Marrow Mesenchymal Stromal Cells in Perfused 3D Ceramic Scaffolds Enhances In Vivo Bone Formation. *Biotechnol. J.* **2017**, *12*. [CrossRef] [PubMed]
27. Lee, S.H.; Sung, J.H. Organ-on-a-chip technology for reproducing multiorgan physiology. *Adv. Healthc. Mater.* **2018**, *7*, 1700419. [CrossRef] [PubMed]
28. Kimura, H.; Sakai, Y.; Fujii, T. Organ/body-on-a-chip based on microfluidic technology for drug discovery. *Drug Metab. Pharmacokinet.* **2018**, *33*, 43–48. [CrossRef]
29. Wu, Q.; Liu, J.; Wang, X.; Feng, L.; Wu, J.; Zhu, X.; Wen, W.; Gong, X. Organ-on-a-chip: Recent breakthroughs and future prospects. *Biomed. Eng. Online* **2020**, *19*, 9. [CrossRef]
30. Frith, J.E.; Thomson, B.; Genever, P.G. Dynamic three-dimensional culture methods enhance mesenchymal stem cell properties and increase therapeutic potential. *Tissue Eng. Part C Methods* **2010**, *16*, 735–749. [CrossRef] [PubMed]
31. Stiehler, M.; Bünger, C.; Baatrup, A.; Lind, M.; Kassem, M.; Mygind, T. Effect of dynamic 3-D culture on proliferation, distribution, and osteogenic differentiation of human mesenchymal stem cells. *J. Biomed. Mater. Res. A* **2009**, *89*, 96–107. [CrossRef] [PubMed]
32. Wang, W.; Itaka, K.; Ohba, S.; Nishiyama, N.; Chung, U.I.; Yamasaki, Y.; Kataoka, K. 3D spheroid culture system on micropatterned substrates for improved differentiation efficiency of multipotent mesenchymal stem cells. *Biomaterials* **2009**, *30*, 2705–2715. [CrossRef]
33. Friedenstein, A.J.; Latzinik, N.W.; Grosheva, A.G.; Gorskaya, U.F. Marrow microenvironment transfer by heterotopic transplantation of freshly isolated and cultured cells in porous sponges. *Exp. Hematol.* **1982**, *10*, 217–227.
34. Wexler, S.A.; Donaldson, C.; Denning-Kendall, P.; Rice, C.; Bradley, B.; Hows, J.M. Adult bone marrow is a rich source of human mesenchymal 'stem'cells but umbilical cord and mobilized adult blood are not. *Br. J. Haematol.* **2003**, *121*, 368–374. [CrossRef] [PubMed]
35. Bonab, M.M.; Alimoghaddam, K.; Talebian, F.; Ghaffari, S.H.; Ghavamzadeh, A.; Nikbin, B. Aging of mesenchymal stem cell in vitro. *BMC Cell Biol.* **2006**, *7*, 14. [CrossRef] [PubMed]

36. Chen, J.; Sotome, S.; Wang, J.; Orii, H.; Uemura, T.; Shinomiya, K. Correlation of in vivo bone formation capability and in vitro differentiation of human bone marrow stromal cells. *J. Med. Dent. Sci.* **2005**, *52*, 27–34.
37. Yang, Y.-H.K.; Ogando, C.R.; Wang See, C.; Chang, T.-Y.; Barabino, G.A. Changes in phenotype and differentiation potential of human mesenchymal stem cells aging in vitro. *Stem Cell Res. Ther.* **2018**, *9*, 131. [CrossRef]
38. Wu, M.; Han, Z.B.; Liu, J.F.; Wang, Y.W.; Zhang, J.Z.; Li, C.T.; Xin, P.L.; Han, Z.C.; Zhu, X.P. Serum-free media and the immunoregulatory properties of mesenchymal stem cells in vivo and in vitro. *Cell. Physiol. Biochem.* **2014**, *33*, 569–580. [CrossRef]
39. Viswanathan, S.; Shi, Y.; Galipeau, J.; Krampera, M.; Leblanc, K.; Martin, I.; Nolta, J.; Phinney, D.G.; Sensebe, L. Mesenchymal stem versus stromal cells: International Society for Cell & Gene Therapy (ISCT (R)) Mesenchymal Stromal Cell committee position statement on nomenclature. *Cytotherapy* **2019**, *21*, 1019–1024. [CrossRef]
40. Sotiropoulou, P.A.; Perez, S.A.; Salagianni, M.; Baxevanis, C.N.; Papamichail, M. Characterization of the optimal culture conditions for clinical scale production of human mesenchymal stem cells. *Stem Cells* **2006**, *24*, 462–471. [CrossRef]
41. Dhanasekaran, M.; Indumathi, S.; Rashmi, M.; Rajkumar, J.S.; Sudarsanam, D. Unravelling the retention of proliferation and differentiation potency in extensive culture of human subcutaneous fat-derived mesenchymal stem cells in different media. *Cell Prolif.* **2012**, *45*, 516–526. [CrossRef]
42. Hagmann, S.; Moradi, B.; Frank, S.; Dreher, T.; Kämmerer, P.W.; Richter, W.; Gotterbarm, T. Different culture media affect growth characteristics, surface marker distribution and chondrogenic differentiation of human bone marrow-derived mesenchymal stromal cells. *BMC Musculoskelet. Disord.* **2013**, *14*, 223. [CrossRef] [PubMed]
43. Haack-Sorensen, M.; Friis, T.; Bindslev, L.; Mortensen, S.; Johnsen, H.E.; Kastrup, J. Comparison of different culture conditions for human mesenchymal stromal cells for clinical stem cell therapy. *Scand. J. Clin. Lab. Investig.* **2008**, *68*, 192–203. [CrossRef]
44. Woerle, H.J.; Gerich, J.E. Glucose Physiology, Normal. In *Encyclopedia of Endocrine Diseases*; Martini, L., Ed.; Elsevier: New York, NY, USA, 2004; pp. 263–270.
45. Nuschke, A.; Rodrigues, M.; Wells, A.W.; Sylakowski, K.; Wells, A. Mesenchymal stem cells/multipotent stromal cells (MSCs) are glycolytic and thus glucose is a limiting factor of in vitro models of MSC starvation. *Stem Cell Res. Ther.* **2016**, *7*, 179. [CrossRef]
46. Keats, E.; Khan, Z.A. Unique responses of stem cell-derived vascular endothelial and mesenchymal cells to high levels of glucose. *PLoS ONE* **2012**, *7*, e38752. [CrossRef] [PubMed]
47. Lo, T.; Ho, J.H.; Yang, M.-H.; Lee, O.K. Glucose Reduction Prevents Replicative Senescence and Increases Mitochondrial Respiration in Human Mesenchymal Stem Cells. *Cell Transplant.* **2011**, *20*, 813–826. [CrossRef] [PubMed]
48. Weil, B.R.; Abarbanell, A.M.; Herrmann, J.L.; Wang, Y.; Meldrum, D.R. High glucose concentration in cell culture medium does not acutely affect human mesenchymal stem cell growth factor production or proliferation. *Am. J. Physiol. Regul. Integr. Comp. Physiol.* **2009**, *296*, R1735–R1743. [CrossRef]
49. D'Esposito, V.; Lecce, M.; Marenzi, G.; Cabaro, S.; Ambrosio, M.R.; Sammartino, G.; Misso, S.; Migliaccio, T.; Liguoro, P.; Oriente, F.; et al. Platelet-rich plasma counteracts detrimental effect of high-glucose concentrations on mesenchymal stem cells from Bichat fat pad. *J. Tissue Eng. Regen. Med.* **2020**. [CrossRef]
50. Liu, Y.; Li, Y.; Nan, L.-P.; Wang, F.; Zhou, S.-F.; Wang, J.-C.; Feng, X.-M.; Zhang, L. The effect of high glucose on the biological characteristics of nucleus pulposus-derived mesenchymal stem cells. *Cell Biochem. Funct.* **2020**. [CrossRef]
51. Chang, T.-C.; Hsu, M.-F.; Wu, K.K. High Glucose Induces Bone Marrow-Derived Mesenchymal Stem Cell Senescence by Upregulating Autophagy. *PLoS ONE* **2015**, *10*, e0126537. [CrossRef]
52. Liang, C.; Li, H.; Tao, Y.; Zhou, X.; Li, F.; Chen, G.; Chen, Q. Responses of human adipose-derived mesenchymal stem cells to chemical microenvironment of the intervertebral disc. *J. Transl. Med.* **2012**, *10*, 49. [CrossRef]
53. Li, Y.-M.; Schilling, T.; Benisch, P.; Zeck, S.; Meissner-Weigl, J.; Schneider, D.; Limbert, C.; Seufert, J.; Kassem, M.; Schütze, N.; et al. Effects of high glucose on mesenchymal stem cell proliferation and differentiation. *Biochem. Biophys. Res. Commun.* **2007**, *363*, 209–215. [CrossRef] [PubMed]
54. Ejtehadifar, M.; Shamsasenjan, K.; Movassaghpour, A.; Akbarzadehlaleh, P.; Dehdilani, N.; Abbasi, P.; Molaeipour, Z.; Saleh, M. The effect of hypoxia on mesenchymal stem cell biology. *Adv. Pharm. Bull.* **2015**, *5*, 141–149. [CrossRef] [PubMed]
55. Deschepper, M.; Oudina, K.; David, B.; Myrtil, V.; Collet, C.; Bensidhoum, M.; Logeart-Avramoglou, D.; Petite, H. Survival and function of mesenchymal stem cells (MSCs) depend on glucose to overcome exposure to long-term, severe and continuous hypoxia. *J. Cell. Mol. Med.* **2011**, *15*, 1505–1514. [CrossRef]
56. Jedrzejczak-Silicka, M. History of Cell Culture. In *New Insights into Cell Culture Technology*; IntechOpen: Rijeka, Croatia, 2017.
57. Salazar, A.; Keusgen, M.; von Hagen, J. Amino acids in the cultivation of mammalian cells. *Amino Acids* **2016**, *48*, 1161–1171. [CrossRef] [PubMed]
58. Doverskog, M.; Ljunggren, J.; Öhman, L.; Häggström, L. Physiology of cultured animal cells. *J. Biotechnol.* **1997**, *59*, 103–115. [CrossRef]
59. Kontoravdi, C.; Wong, D.; Lam, C.; Lee, Y.Y.; Yap, M.G.S.; Pistikopoulos, E.N.; Mantalaris, A. Modeling Amino Acid Metabolism in Mammalian Cells-Toward the Development of a Model Library. *Biotechnol. Prog.* **2007**, *23*, 1261–1269. [CrossRef]
60. Alberts, B.; Johnson, A.; Lewis, J.; Raff, M.; Roberts, K.; Walter, P. *Molecular Biology of The Cell*; Norton & Company: New York, NY, USA, 2007; Volume 230.
61. Choi, W.; Kwon, S.-J.; Jin, H.J.; Jeong, S.Y.; Choi, S.J.; Oh, W.; Yang, Y.S.; Jeon, H.B.; Jeon, E.S. Optimization of culture conditions for rapid clinical-scale expansion of human umbilical cord blood-derived mesenchymal stem cells. *Clin. Transl. Med.* **2017**, *6*, 38. [CrossRef]

62. Higuera, G.A.; Schop, D.; Spitters, T.W.G.M.; van Dijkhuizen-Radersma, R.; Bracke, M.; de Bruijn, J.D.; Martens, D.; Karperien, M.; van Boxtel, A.; van Blitterswijk, C.A. Patterns of amino acid metabolism by proliferating human mesenchymal stem cells. *Tissue Eng. Part A* **2012**, *18*, 654–664. [CrossRef]
63. Eagle, H.; Oyama, V.I.; Levy, M.A. The growth response of mammalian cells in tissue culture to L-glutamine and L-glutamic acid. *J. Biol. Chem.* **1956**, *218*, 607–616. [CrossRef]
64. Dos Santos, G.G.; Hastreiter, A.A.; Sartori, T.; Borelli, P.; Fock, R.A. L-Glutamine in vitro Modulates some Immunomodulatory Properties of Bone Marrow Mesenchymal Stem Cells. *Stem Cell Rev. Rep.* **2017**, *13*, 482–490. [CrossRef]
65. Zhou, T.; Yang, Y.; Chen, Q.; Xie, L.; Li, J. Glutamine Metabolism Is Essential for Stemness of Bone Marrow Mesenchymal Stem Cells and Bone Homeostasis. *Stem Cells Int.* **2019**, *2019*, 8928934. [CrossRef] [PubMed]
66. Sartori, T.; Santos, A.C.A.A. Branched chain amino acids improve mesenchymal stem cell proliferation, reducing nuclear factor kappa B expression and modulating some inflammatory properties. *Nutrition* **2020**, *78*, 110935. [CrossRef] [PubMed]
67. Wymann, M.P.; Schneiter, R. Lipid signalling in disease. *Nat. Rev. Mol. Cell Biol.* **2008**, *9*, 162–176. [CrossRef]
68. Van Meer, G.; de Kroon, A.I.P.M. Lipid map of the mammalian cell. *J. Cell Sci.* **2011**, *124*, 5. [CrossRef] [PubMed]
69. Clémot, M.; Sênos Demarco, R.; Jones, D.L. Lipid Mediated Regulation of Adult Stem Cell Behavior. *Front. Cell Dev. Biol.* **2020**, *8*, 115. [CrossRef]
70. Grammatikos, S.I.; Subbaiah, P.V.; Victor, T.A.; Miller, W.M. Diverse effects of essential (n-6 and n-3) fatty acids on cultured cells. In *Cell Culture Engineering I.*; Springer: Berlin/Heidelberg, Germany, 1994; pp. 31–50.
71. Whitford, W.; Manwaring, J. Lipids in Cell Culture Media. *Fish. Appl. Notes* **2004**, 152–154.
72. Rampler, E.; Egger, D.; Schoeny, H.; Rusz, M.; Pacheco, M.P.; Marino, G.; Kasper, C.; Naegele, T.; Koellensperger, G. The power of LC-MS based multiomics: Exploring adipogenic differentiation of human mesenchymal stem/stromal cells. *Molecules* **2019**, *24*, 3615. [CrossRef]
73. Chatgilialoglu, A.; Rossi, M.; Alviano, F.; Poggi, P.; Zannini, C.; Marchionni, C.; Ricci, F.; Tazzari, P.L.; Taglioli, V.; Calder, P.C.; et al. Restored in vivo-like membrane lipidomics positively influence in vitro features of cultured mesenchymal stromal/stem cells derived from human placenta. *Stem Cell Res. Ther.* **2017**, *8*, 31. [CrossRef]
74. Kilpinen, L.; Tigistu-Sahle, F.; Oja, S.; Greco, D.; Parmar, A.; Saavalainen, P.; Nikkilä, J.; Korhonen, M.; Lehenkari, P.; Käkelä, R.; et al. Aging bone marrow mesenchymal stromal cells have altered membrane glycerophospholipid composition and functionality. *J. Lipid Res.* **2013**, *54*, 622–635. [CrossRef] [PubMed]
75. Lu, X.; Chen, Y.; Wang, H.; Bai, Y.; Zhao, J.; Zhang, X.; Liang, L.; Chen, Y.; Ye, C.; Li, Y.; et al. Integrated Lipidomics and Transcriptomics Characterization upon Aging-Related Changes of Lipid Species and Pathways in Human Bone Marrow Mesenchymal Stem Cells. *J. Proteome Res.* **2019**, *18*, 2065–2077. [CrossRef]
76. Fillmore, N.; Huqi, A.; Jaswal, J.S.; Mori, J.; Paulin, R.; Haromy, A.; Onay-Besikci, A.; Ionescu, L.; Thbaud, B.; Michelakis, E.; et al. Effect of fatty acids on human bone marrow mesenchymal stem cell energy metabolism and survival. *PLoS ONE* **2015**, *10*, e0120257. [CrossRef] [PubMed]
77. Hu, C.; Wang, C.; He, L.; Han, X. Novel strategies for enhancing shotgun lipidomics for comprehensive analysis of cellular lipidomes. *TrAC Trends Anal. Chem.* **2019**, *120*, 115330. [CrossRef] [PubMed]
78. Goracci, L.; Tortorella, S.; Tiberi, P.; Pellegrino, R.M.A. Lipostar, a Comprehensive Platform-Neutral Cheminformatics Tool for Lipidomics. *Anal. Chem.* **2017**, *89*, 6257–6264. [CrossRef] [PubMed]
79. Mohamed, A.; Molendijk, J.; Hill, M.M. Lipidr: A Software Tool for Data Mining and Analysis of Lipidomics Datasets. *J. Proteome Res.* **2020**, *19*, 2890–2897. [CrossRef] [PubMed]
80. MERCK. Lipids and Lipid Carriers—Solubilizing Agents | Sigma-Aldrich. Available online: https://www.sigmaaldrich.com/life-science/cell-culture/cell-culture-products.html?TablePage=22608488 (accessed on 12 January 2021).
81. Rodrigues, M.; Griffith, L.G.; Wells, A. Growth factor regulation of proliferation and survival of multipotential stromal cells. *Stem Cell Res. Ther.* **2010**, *1*, 32. [CrossRef]
82. Franz, K.C.; Suschek, C.V.; Grotheer, V.; Akbas, M.; Pallua, N. Impact of growth factor content on proliferation of mesenchymal stromal cells derived from adipose tissue. *PLoS ONE* **2020**, *15*, e0230265. [CrossRef]
83. Gharibi, B.; Hughes, F.J. Effects of medium supplements on proliferation, differentiation potential, and in vitro expansion of mesenchymal stem cells. *Stem Cells Transl. Med.* **2012**, *1*, 771–782. [CrossRef]
84. Kabiri, A.; Esfandiari, E.; Hashemibeni, B.; Kazemi, M.; Mardani, M.; Esmaeili, A. Effects of FGF-2 on human adipose tissue derived adult stem cells morphology and chondrogenesis enhancement in Transwell culture. *Biochem. Biophys. Res. Commun.* **2012**, *424*, 234–238. [CrossRef]
85. Solchaga, L.A.; Penick, K.; Porter, J.D.; Goldberg, V.M.; Caplan, A.I.; Welter, J.F. FGF-2 enhances the mitotic and chondrogenic potentials of human adult bone marrow-derived mesenchymal stem cells. *J. Cell. Physiol.* **2005**, *203*, 398–409. [CrossRef] [PubMed]
86. Tsutsumi, S.; Shimazu, A.; Miyazaki, K.; Pan, H.; Koike, C.; Yoshida, E.; Takagishi, K.; Kato, Y. Retention of Multilineage Differentiation Potential of Mesenchymal Cells during Proliferation in Response to FGF. *Biochem. Biophys. Res. Commun.* **2001**, *288*, 413–419. [CrossRef]
87. Eom, Y.W.; Oh, J.-E.; Lee, J.I.; Baik, S.K.; Rhee, K.-J.; Shin, H.C.; Kim, Y.M.; Ahn, C.M.; Kong, J.H.; Kim, H.S.; et al. The role of growth factors in maintenance of stemness in bone marrow-derived mesenchymal stem cells. *Biochem. Biophys. Res. Commun.* **2014**, *445*, 16–22. [CrossRef]

88. Jung, S.; Sen, A.; Rosenberg, L.; Behie, L.A. Identification of growth and attachment factors for the serum-free isolation and expansion of human mesenchymal stromal cells. *Cytotherapy* **2010**, *12*, 637–657. [CrossRef] [PubMed]
89. Ito, T.; Sawada, R.; Fujiwara, Y.; Seyama, Y.; Tsuchiya, T. FGF-2 suppresses cellular senescence of human mesenchymal stem cells by down-regulation of TGF-β2. *Biochem. Biophys. Res. Commun.* **2007**, *359*, 108–114. [CrossRef] [PubMed]
90. Ng, F.; Boucher, S.; Koh, S.; Sastry, K.S.R.; Chase, L.; Lakshmipathy, U.; Choong, C.; Yang, Z.; Vemuri, M.C.; Rao, M.S.; et al. PDGF, TGF-β, and FGF signaling is important for differentiation and growth of mesenchymal stem cells (MSCs): Transcriptional profiling can identify markers and signaling pathways important in differentiation of MSCs into adipogenic, chondrogenic, and osteogenic lineages. *Blood* **2008**, *112*, 295–307. [CrossRef]
91. Gruber, R.; Karreth, F.; Kandler, B.; Fuerst, G.; Rot, A.; Fischer, M.B.; Watzek, G. Platelet-released supernatants increase migration and proliferation, and decrease osteogenic differentiation of bone marrow-derived mesenchymal progenitor cells under in vitro conditions. *Platelets* **2004**, *15*, 29–35. [CrossRef]
92. Fierro, F.; Illmer, T.; Jing, D.; Schleyer, E.; Ehninger, G.; Boxberger, S.; Bornhäuser, M. Inhibition of platelet-derived growth factor receptorbeta by imatinib mesylate suppresses proliferation and alters differentiation of human mesenchymal stem cells in vitro. *Cell Prolif.* **2007**, *40*, 355–366. [CrossRef]
93. Tamama, K.; Fan, V.H.; Griffith, L.G.; Blair, H.C.; Wells, A. Epidermal Growth Factor as a Candidate for Ex Vivo Expansion of Bone Marrow–Derived Mesenchymal Stem Cells. *Stem Cells* **2006**, *24*, 686–695. [CrossRef]
94. Krampera, M.; Pasini, A.; Rigo, A.; Scupoli, M.T.; Tecchio, C.; Malpeli, G.; Scarpa, A.; Dazzi, F.; Pizzolo, G.; Vinante, F. HB-EGF/HER-1 signaling in bone marrow mesenchymal stem cells: Inducing cell expansion and reversibly preventing multilineage differentiation. *Blood* **2005**, *106*, 59–66. [CrossRef] [PubMed]
95. Moon, M.-Y.; Kim, H.J.; Choi, B.Y.; Sohn, M.; Chung, T.N.; Suh, S.W. Zinc Promotes Adipose-Derived Mesenchymal Stem Cell Proliferation and Differentiation towards a Neuronal Fate. *Stem Cells Int.* **2018**, *2018*, 5736535. [CrossRef]
96. Jahnen-Dechent, W.; Ketteler, M. Magnesium basics. *Clin. Kidney J.* **2012**, *5*, i3–i14. [CrossRef] [PubMed]
97. Siti Noor Fazliah, M.N.; Yusuf, M.M.; Abdullah, T.K.; Zuhailawati, H. Human Mesenchymal Stem Cells Response to Magnesium-based Biomaterials. *Proc. Chem.* **2016**, *19*, 75–82. [CrossRef]
98. Barradas, A.M.; Fernandes, H.A.; Groen, N.; Chai, Y.C.; Schrooten, J.; van de Peppel, J.; van Leeuwen, J.P.; van Blitterswijk, C.A.; de Boer, J. A calcium-induced signaling cascade leading to osteogenic differentiation of human bone marrow-derived mesenchymal stromal cells. *Biomaterials* **2012**, *33*, 3205–3215. [CrossRef] [PubMed]
99. Lei, Q.; Chen, J.; Huang, W.; Wu, D.; Lin, H.; Lai, Y. Proteomic analysis of the effect of extracellular calcium ions on human mesenchymal stem cells: Implications for bone tissue engineering. *Chem Biol. Interact.* **2015**, *233*, 139–146. [CrossRef] [PubMed]
100. Borriello, A.; Caldarelli, I.; Speranza, M.C.; Scianguetta, S.; Tramontano, A.; Bencivenga, D.; Stampone, E.; Negri, A.; Nobili, B.; Locatelli, F.; et al. Iron overload enhances human mesenchymal stromal cell growth and hampers matrix calcification. *Biochim. Biophys. Acta* **2016**, *1860*, 1211–1223. [CrossRef] [PubMed]
101. Ebert, R.; Ulmer, M.; Zeck, S.; Meissner-Weigl, J.; Schneider, D.; Stopper, H.; Schupp, N.; Kassem, M.; Jakob, F. Selenium supplementation restores the antioxidative capacity and prevents cell damage in bone marrow stromal cells in vitro. *Stem Cells* **2006**, *24*, 1226–1235. [CrossRef]
102. Park, J.; Lee, J.H.; Yoon, B.S.; Jun, E.K.; Lee, G.; Kim, I.Y.; You, S. Additive effect of bFGF and selenium on expansion and paracrine action of human amniotic fluid-derived mesenchymal stem cells. *Stem Cell Res.* **2018**, *9*, 293. [CrossRef]
103. Rodríguez, J.; Ríos, S.; González, M. Modulation of the proliferation and differentiation of human mesenchymal stem cells by copper. *J. Cell. Biochem.* **2002**, *85*, 92–100. [CrossRef]
104. Choi, K.-M.; Seo, Y.-K.; Yoon, H.-H.; Song, K.-Y.; Kwon, S.-Y.; Lee, H.-S.; Park, J.-K. Effect of ascorbic acid on bone marrow-derived mesenchymal stem cell proliferation and differentiation. *J. Biosci. Bioeng.* **2008**, *105*, 586–594. [CrossRef]
105. Potdar, P.D.; D'souza, S.B. Ascorbic acid induces in vitro proliferation of human subcutaneous adipose tissue derived mesenchymal stem cells with upregulation of embryonic stem cell pluripotency markers Oct4 and SOX 2. *Hum. Cell* **2010**, *23*, 152–155. [CrossRef] [PubMed]
106. Fujisawa, K.; Hara, K.; Takami, T.; Okada, S.; Matsumoto, T.; Yamamoto, N.; Sakaida, I. Evaluation of the effects of ascorbic acid on metabolism of human mesenchymal stem cells. *Stem Cell Res.* **2018**, *9*, 93. [CrossRef]
107. Brunner, D.; Frank, J.; Appl, H.; Schoffl, H.; Pfaller, W.; Gstraunthaler, G. Serum-free cell culture: The serum-free media interactive online database. *ALTEX* **2010**, *27*, 53–62. [CrossRef]
108. Van der Valk, J.; Brunner, D.; De Smet, K.; Fex Svenningsen, A.; Honegger, P.; Knudsen, L.E.; Lindl, T.; Noraberg, J.; Price, A.; Scarino, M.L.; et al. Optimization of chemically defined cell culture media–replacing fetal bovine serum in mammalian in vitro methods. *Toxicol. Vitr.* **2010**, *24*, 1053–1063. [CrossRef]
109. Van der Valk, J.; Mellor, D.; Brands, R.; Fischer, R.; Gruber, F.; Gstraunthaler, G.; Hellebrekers, L.; Hyllner, J.; Jonker, F.H.; Prieto, P.; et al. The humane collection of fetal bovine serum and possibilities for serum-free cell and tissue culture. *Toxicol. Vitr.* **2004**, *18*, 1–12. [CrossRef]
110. Gstraunthaler, G. Alternatives to the use of fetal bovine serum: Serum-free cell culture. *Altex Altern Tierexp* **2003**, *20*, 275–281.
111. Van der Valk, J.; Bieback, K.; Buta, C.; Cochrane, B.; Dirks, W.G.; Fu, J.; Hickman, J.J.; Hohensee, C.; Kolar, R.; Liebsch, M.; et al. Fetal Bovine Serum (FBS): Past—Present—Future. *Altex* **2018**, *35*, 99–118. [CrossRef] [PubMed]
112. Bieback, K.; Hecker, A.; Kocaömer, A.; Lannert, H.; Schallmoser, K.; Strunk, D.; Klüter, H. Human Alternatives to Fetal Bovine Serum for the Expansion of Mesenchymal Stromal Cells from Bone Marrow. *Stem Cells* **2009**, *27*, 2331–2341. [CrossRef]

113. Carrancio, S.; Lopez-Holgado, N.; Sanchez-Guijo, F.M.; Villarn, E.; Barbado, V.; Tabera, S.; Dez-Campelo, M.; Blanco, J.A. Optimization of mesenchymal stem cell expansion procedures by cell separation and culture conditions modification. *Exp. Hematol.* **2008**, *36*, 1014–1021. [CrossRef] [PubMed]
114. Copland, I.B.; Garcia, M.A.; Waller, E.K.; Roback, J.D.; Galipeau, J. The effect of platelet lysate fibrinogen on the functionality of MSCs in immunotherapy. *Biomaterials* **2013**, *34*, 7840–7850. [CrossRef] [PubMed]
115. Cowper, M.; Frazier, T.; Wu, X.; Curley, J.; Ma, M.; Mohiuddin, O.; Dietrich, M.; McCarthy, M.; Bukowska, J.; Gimble, J. Human Platelet Lysate as a Functional Substitute for Fetal Bovine Serum in the Culture of Human Adipose Derived Stromal/Stem Cells. *Cells* **2019**, *8*, 724. [CrossRef] [PubMed]
116. Dessels, C.; Durandt, C.; Pepper, M.S. Comparison of human platelet lysate alternatives using expired and freshly isolated platelet concentrates for adipose-derived stromal cell expansion. *Platelets* **2019**, *30*, 356–367. [CrossRef] [PubMed]
117. Jonsdottir-Buch, S.M.; Lieder, R.; Sigurjonsson, O.E. Platelet Lysates Produced from Expired Platelet Concentrates Support Growth and Osteogenic Differentiation of Mesenchymal Stem Cells. *PLoS ONE* **2013**, *8*, e68984. [CrossRef] [PubMed]
118. Pérez-Ilzarbe, M.; Dez-Campelo, M.; Aranda, P.; Tabera, S.; Lopez, T.; del Caizo, C.; Merino, J.; Moreno, C.; Andreu, E.J.; Prsper, F.; et al. Comparison of ex vivo expansion culture conditions of mesenchymal stem cells for human cell therapy. *Transfusion* **2009**, *49*, 1901–1910. [CrossRef] [PubMed]
119. Bernardi, M.; Albiero, E.; Alghisi, A.; Chieregato, K.; Lievore, C.; Madeo, D.; Rodeghiero, F.; Astori, G. Production of human platelet lysate by use of ultrasound for ex vivo expansion of human bone marrow–derived mesenchymal stromal cells. *Cytotherapy* **2013**, *15*. [CrossRef] [PubMed]
120. Doucet, C.; Ernou, I.; Zhang, Y.; Llense, J.R.; Begot, L.; Holy, X.; Lataillade, J.J. Platelet lysates promote mesenchymal stem cell expansion: A safety substitute for animal serum in cell-based therapy applications. *J. Cell. Physiol.* **2005**, *205*, 228–236. [CrossRef]
121. Kinzebach, S.; Dietz, L.; Klter, H.; Thierse, H.J.; Bieback, K. Functional and differential proteomic analyses to identify platelet derived factors affecting ex vivo expansion of mesenchymal stromal cells. *BMC Cell Biol.* **2013**, *14*. [CrossRef]
122. Muraglia, A.; Ottonello, C. Span. Biological activity of a standardized freeze-dried platelet derivative to be used as cell culture medium supplement. *Platelets* **2014**, *25*, 211–220. [CrossRef]
123. Naaijkens, B.A.; Niessen, H.W.M.; Prins, H.J.; Krijnen, P.A.J.; Kokhuis, T.J.A. Human platelet lysate as a fetal bovine serum substitute improves human adipose-derived stromal cell culture for future cardiac repair applications. *Cell Tissue Res.* **2012**, *348*, 119–130. [CrossRef]
124. Reinisch, A.; Bartmann, C.; Rohde, E.; Schallmoser, K.; Bjelic-Radisic, V.; Lanzer, G.; Linkesch, W.; Strunk, D. Humanized system to propagate cord blood-derived multipotent mesenchymal stromal cells for clinical application. *Regen. Med.* **2007**, *2*, 371–382. [CrossRef]
125. Witzeneder, K.; Lindenmair, A.; Gabriel, C.; Höller, K.; Theiß, D.; Redl, H.; Hennerbichler, S. Human-Derived Alternatives to Fetal Bovine Serum in Cell Culture. *Transfus. Med. Hemother.* **2013**, *40*, 417–423. [CrossRef]
126. Burnouf, T.; Strunk, D.; Koh, M.B.; Schallmoser, K. Human platelet lysate: Replacing fetal bovine serum as a gold standard for human cell propagation? *Biomaterials* **2016**, *76*, 371–387. [CrossRef]
127. Iudicone, P.; Fioravanti, D.; Bonanno, G.; Miceli, M.; Lavorino, C.; Totta, P.; Frati, L.; Nuti, M.; Pierelli, L. Pathogen-free, plasma-poor platelet lysate and expansion of human mesenchymal stem cells. *J. Transl. Med.* **2014**, *12*. [CrossRef] [PubMed]
128. Jenhani, F.; Durand, V.; Ben Azouna, N.; Thallet, S.; Ben Othmen, T.; Bejaoui, M.; Domenech, J. Human cytokine expression profile in various conditioned media for in vitro expansion bone marrow and umbilical cord blood immunophenotyped mesenchymal stem cells. *Transpl. Proc.* **2011**, *43*, 639–643. [CrossRef]
129. Xia, W.; Li, H.; Wang, Z.; Xu, R.; Fu, Y.; Zhang, X.; Ye, X.; Huang, Y.; Xiang, A.P.; Yu, W. Human platelet lysate supports ex vivo expansion and enhances osteogenic differentiation of human bone marrow-derived mesenchymal stem cells. *Cell Biol. Int.* **2011**, *35*, 639–643. [CrossRef] [PubMed]
130. Pierce, J.; Benedetti, E.; Preslar, A.; Jacobson, P.; Jin, P.; Stroncek, D.F.; Reems, J.-A. Comparative analyses of industrial-scale human platelet lysate preparations. *Transfusion* **2017**, *57*, 2858–2869. [CrossRef] [PubMed]
131. Bandeiras, C.; Koc, J.R.; Ma, Y.; Samberg, M.; Finkelstein, S.; Ferreira, F. Cost effectiveness analysis of allogeneic, just-in-time expansion of mesenchymal stem cells with PLUS™ human platelet lysate for a clinical trial. *Cytotherapy* **2018**, *20*, S60. [CrossRef]
132. Kakudo, N.; Morimoto, N.; Ma, Y.; Kusumoto, K. Differences between the Proliferative Effects of Human Platelet Lysate and Fetal Bovine Serum on Human Adipose-Derived Stem Cells. *Cells* **2019**, *8*, 1218. [CrossRef] [PubMed]
133. Karadjian, M.; Senger, A.-S.; Essers, C.; Wilkesmann, S.; Heller, R.; Fellenberg, J.; Simon, R.; Westhauser, F. Human Platelet Lysate Can Replace Fetal Calf Serum as a Protein Source to Promote Expansion and Osteogenic Differentiation of Human Bone-Marrow-Derived Mesenchymal Stromal Cells. *Cells* **2020**, *9*, 918. [CrossRef]
134. Lensch, M.; Muise, A.; White, L.; Badowski, M.; Harris, D. Comparison of Synthetic Media Designed for Expansion of Adipose-Derived Mesenchymal Stromal Cells. *Biomedicines* **2018**, *6*, 54. [CrossRef]
135. Shih, D.T.-B.; Chen, J.-C.; Chen, W.-Y.; Kuo, Y.-P.; Su, C.-Y.; Burnouf, T. Expansion of adipose tissue mesenchymal stromal progenitors in serum-free medium supplemented with virally inactivated allogeneic human platelet lysate. *Transfusion* **2011**, *51*, 770–778. [CrossRef]
136. Sargent, B. What Does Xeno-Free Really Mean, and Why Does It Matter to Cell Culture Scientists Today? Available online: https://cellculturedish.com/what-does-xeno-free-really-mean-and-why-does-it-matter-to-cell-culture-scientists-today/ (accessed on 27 February 2021).

137. Bahsoun, S.; Coopman, K.; Forsyth, N.R.; Akam, E.C. The Role of Dissolved Oxygen Levels on Human Mesenchymal Stem Cell Culture Success, Regulatory Compliance, and Therapeutic Potential. *Stem Cells Dev.* **2018**, *27*, 1303–1321. [CrossRef]
138. Gottipamula, S.; Muttigi, M.S.; Kolkundkar, U.; Seetharam, R.N. Serum-free media for the production of human mesenchymal stromal cells: A review. *Cell Prolif.* **2013**, *46*, 608–627. [CrossRef]
139. Gottipamula, S.; Ashwin, K.M.; Muttigi, M.S.; Kannan, S.; Kolkundkar, U.; Seetharam, R.N. Isolation, expansion and characterization of bone marrow-derived mesenchymal stromal cells in serum-free conditions. *Cell Tissue Res.* **2014**, *356*, 123–135. [CrossRef] [PubMed]
140. Guilbert, L.; Iscove, N. Partial replacement of serum by selenite, transferrin, albumin and lecithin in haemopoitec cell cultures. *Nature* **1976**, *263*, 594–595. [CrossRef] [PubMed]
141. Murakami, H.; Masui, H.; Sato, G.H.; Sueoka, N.; Chow, T.P.; Kano-Sueoka, T. Growth of hybridoma cells in serum-free medium: Ethanolamine is an essential component. *Proc. Natl. Acad. Sci. USA* **1982**, *79*, 1158–1162. [CrossRef]
142. Sargent, B. Albumin in Cell Culture Media—An Examination of Quality and Function. Available online: https://cellculturedish.com/albumin-in-cell-culture-media-an-examination-of-quality-and-function/ (accessed on 30 November 2020).
143. Jung, S.; Panchalingam, K.M.; Rosenberg, L.; Behie, L.A. Ex vivo expansion of human mesenchymal stem cells in defined serum-free media. *Stem Cells Int.* **2012**, *2012*, 123030. [CrossRef]
144. Liu, C.-H.; Wu, M.-L.; Hwang, S.-M. Optimization of serum free medium for cord blood mesenchymal stem cells. *Biochem. Eng. J.* **2007**, *33*, 1–9. [CrossRef]
145. Marshak, D.R. *US5908782A—Chemically Defined Medium for Human Mesenchymal Stem Cells—Google Patents*; Patent and Trademark Office: Washington, DC, USA, 1999.
146. ThermoFisher Scientific. L-Glutamine & GlutaMAX Cell Culture Supplements. Available online: https://www.thermofisher.com/at/en/home/life-science/cell-culture/mammalian-cell-culture/media-supplements/glutamax-media.html?gclid=EAIaIQobChMIuYj0ya7F6wIV05TVCh0CCQFFEAAYASAAEgJXG_D_BwE&s_kwcid=AL!3652!3!345664482627!p!!g!!glutamax&ef_id=EAIaIQobChMIuYj0ya7F6wIV05TVCh0CCQFFEAAYASAAEgJXG_D_BwE:G:s&s_kwcid=AL!3652!3!345664482627!p!!g!!glutamax&cid=bid_clb_cce_r01_co_cp0000_pjt0000_bid00000_0se_gaw_bt_pur_con (accessed on 27 February 2021).
147. Sigma Aldrich. Folic Acid and Tetrahydrofolates in Cell Culture. Available online: https://www.sigmaaldrich.com/life-science/cell-culture/learning-center/media-expert/folic-acid.html (accessed on 27 February 2021).
148. Sigma Aldrich. Biotin in Cell Culture. Available online: https://www.sigmaaldrich.com/life-science/cell-culture/learning-center/media-expert/biotin.html (accessed on 25 February 2021).
149. Boucek, R.J.; Alvarez, T.R. 5-Hydroxytryptamine: A cytospecific growth stimulator of cultured fibroblasts. *Science* **1970**, *167*, 898–899. [CrossRef]
150. Wu, X.; Kang, H.; Liu, X.; Gao, J.; Zhao, K.; Ma, Z. Serum and xeno-free, chemically defined, no-plate-coating-based culture system for mesenchymal stromal cells from the umbilical cord. *Cell Prolif.* **2016**, *49*, 579–588. [CrossRef] [PubMed]
151. ThermoFisher Scientific. Gibco™ 2-Mercaptoethanol. Available online: https://www.thermofisher.com/order/catalog/product/21985023#/21985023 (accessed on 27 February 2021).
152. Zhao, X.; Liu, L.; Liu, D.; Fan, H.; Wang, Y.; Hu, Y.; Hou, Y. Progesterone Enhances Immunoregulatory Activity of Human Mesenchymal Stem Cells Via PGE 2 and IL-6. *Am. J. Reprod. Immunol.* **2012**, *68*, 290–300. [CrossRef]
153. Devireddy, L.R.; Myers, M.; Screven, R.; Liu, Z.; Boxer, L. A serum-free medium formulation efficiently supports isolation and propagation of canine adipose-derived mesenchymal stem/stromal cells. *PLoS ONE* **2019**, *14*, e0210250. [CrossRef]
154. Cell Culture Technologies LLC. Cell Culture Technologies. Available online: https://www.cellculture.com/ (accessed on 2 March 2021).
155. Li, E.; Zhang, Z.; Jiang, B.; Yan, L.; Park, J.W.; Xu, R.H. Generation of mesenchymal stem cells from human embryonic stem cells in a complete serum-free condition. *Int. J. Biol. Sci.* **2018**, *14*, 1901–1909. [CrossRef] [PubMed]
156. Al-Saqi, S.H.; Saliem, M.; Quezada, H.C.; Ekblad, A.; Jonasson, A.F.; Hovatta, O.; Gotherstrom, C. Defined serum- and xeno-free cryopreservation of mesenchymal stem cells. *Cell Tissue Bank.* **2015**, *16*, 181–193. [CrossRef] [PubMed]
157. Oikonomopoulos, A.; Van Deen, W.K.; Manansala, A.R.; Lacey, P.N.; Tomakili, T.A.; Ziman, A.; Hommes, D.W. Optimization of human mesenchymal stem cell manufacturing: The effects of animal/xeno-free media. *Sci. Rep.* **2015**, *5*, 1–11. [CrossRef] [PubMed]
158. Chase, L.G.; Yang, S.; Zachar, V.; Yang, Z.; Lakshmipathy, U.; Bradford, J.; Boucher, S.E.; Vemuri, M.C. Development and Characterization of a Clinically Compliant Xeno-Free Culture Medium in Good Manufacturing Practice for Human Multipotent Mesenchymal Stem Cells. *Stem Cells Transl. Med.* **2012**, *1*, 750–758. [CrossRef] [PubMed]
159. Chase, L.G.; Lakshmipathy, U.; Solchaga, L.A.; Rao, M.S.; Vemuri, M.C. A novel serum-free medium for the expansion of human mesenchymal stem cells. *Stem Cell Res. Ther.* **2010**, *1*, 1–11. [CrossRef]
160. Tan, K.Y.; Teo, K.L.; Lim, J.F.Y.; Chen, A.K.L.; Reuveny, S.; Oh, S.K.W. Serum-free media formulations are cell line-specific and require optimization for microcarrier culture. *Cytotherapy* **2015**, *17*, 1152–1165. [CrossRef]
161. Bhat, S.; Viswanathan, P.; Chandanala, S.; Prasanna, S.J.; Seetharam, R.N. Expansion and characterization of bone marrow derived human mesenchymal stromal cells in serum-free conditions. *Sci. Rep.* **2021**, *11*, 3403. [CrossRef]
162. McKeown, S.R. Defining normoxia, physoxia and hypoxia in tumours-implications for treatment response. *Br. J. Radiol.* **2014**, *87*, 20130676. [CrossRef]
163. Jackson, S.P.; Bartek, J. The DNA-damage response in human biology and disease. *Nature* **2009**, *461*, 1071–1078. [CrossRef]

164. Kim, J.H.; Park, S.-H.; Park, S.G.; Choi, J.-S.; Xia, Y.; Sung, J.-H. The Pivotal Role of Reactive Oxygen Species Generation in the Hypoxia-Induced Stimulation of Adipose-Derived Stem Cells. *Stem Cells Dev.* **2011**, *20*, 1753–1761. [CrossRef]
165. Wenger, R.H. Cellular adaptation to hypoxia: O2-sensing protein hydroxylases, hypoxia-inducible transcription factors, and O2-regulated gene expression. *FASEB J.* **2002**, *16*, 1151–1162. [CrossRef]
166. BioSpherix. C-Chamber Incubator Subchamber. Available online: https://www.biospherix.com/products/c-chamber (accessed on 20 February 2021).
167. Stemcell Technologies. Hypoxia Incubator Chamber. Available online: https://www.stemcell.com/hypoxia-incubator-chamber.html (accessed on 20 February 2021).
168. Kay, A.G.; Dale, T.P.; Akram, K.M.; Mohan, P.; Hampson, K.; Maffulli, N.; Spiteri, M.A.; El Haj, A.J.; Forsyth, N.R. BMP2 repression and optimized culture conditions promote human bone marrow-derived mesenchymal stem cell isolation. *Regen. Med.* **2015**, *10*, 109–125. [CrossRef] [PubMed]
169. Grayson, W.L.; Zhao, F.; Bunnell, B.; Ma, T. Hypoxia enhances proliferation and tissue formation of human mesenchymal stem cells. *Biochem. Biophys. Res. Commun.* **2007**, *358*, 948–953. [CrossRef]
170. Dos Santos, F.; Andrade, P.Z.; Boura, J.S.; Abecasis, M.M.; da Silva, C.L.; Cabral, J.M. Ex vivo expansion of human mesenchymal stem cells: A more effective cell proliferation kinetics and metabolism under hypoxia. *J. Cell. Physiol.* **2010**, *223*, 27–35. [CrossRef] [PubMed]
171. Lavrentieva, A.; Majore, I.; Kasper, C.; Hass, R. Effects of hypoxic culture conditions on umbilical cord-derived human mesenchymal stem cells. *Cell Commun. Signal.* **2010**, *8*, 1–9. [CrossRef]
172. Tsai, C.C.; Chen, Y.J.; Yew, T.L.; Chen, L.L.; Wang, J.Y.; Chiu, C.H.; Hung, S.C. Hypoxia inhibits senescence and maintains mesenchymal stem cell properties through down-regulation of E2A-p21 by HIF-TWIST. *Blood* **2011**, *117*, 459–469. [CrossRef]
173. Basciano, L.; Nemos, C.; Foliguet, B.; de Isla, N.; de Carvalho, M.; Tran, N.; Dalloul, A. Long term culture of mesenchymal stem cells in hypoxia promotes a genetic program maintaining their undifferentiated and multipotent status. *BMC Cell Biol.* **2011**, *12*, 12. [CrossRef] [PubMed]
174. Valorani, M.G.; Montelatici, E.; Germani, A.; Biddle, A.; D'Alessandro, D.; Strollo, R.; Patrizi, M.P.; Lazzari, L.; Nye, E.; Otto, W.R.; et al. Pre-culturing human adipose tissue mesenchymal stem cells under hypoxia increases their adipogenic and osteogenic differentiation potentials. *Cell Prolif.* **2012**, *45*, 225–238. [CrossRef] [PubMed]
175. Palumbo, S.M.; Tsai, T.L.; Li, W.J. Macrophage migration inhibitory factor regulates AKT signaling in hypoxic culture to modulate senescence of human mesenchymal stem cells. *Stem Cells Dev.* **2014**, *23*, 852–865. [CrossRef]
176. Choi, J.R.; Pingguan-Murphy, B.A. In Situ Normoxia Enhances Survival and Proliferation Rate of Human Adipose Tissue-Derived Stromal Cells without Increasing the Risk of Tumourigenesis. *PLoS ONE* **2015**, *10*, e0115034. [CrossRef]
177. Ali, N.M.; Boo, L.; Yeap, S.K.; Ky, H.; Satharasinghe, D.A.; Liew, W.C.; Ong, H.K.; Cheong, S.K.; Kamarul, T. Probable impact of age and hypoxia on proliferation and microRNA expression profile of bone marrow-derived human mesenchymal stem cells. *PeerJ* **2016**, *2016*, e1536. [CrossRef]
178. Pachón-Peña, G.; Serena, C.; Ejarque, M.; Petriz, J.; Duran, X.; Oliva-Olivera, W.; Simó, R.; Tinahones, F.J. Obesity Determines the Immunophenotypic Profile and Functional Characteristics of Human Mesenchymal Stem Cells from Adipose Tissue. *Stem Cells Transl. Med.* **2016**, *5*, 464–475. [CrossRef]
179. Peng, L.; Shu, X.; Lang, C.; Yu, X. Effects of hypoxia on proliferation of human cord blood-derived mesenchymal stem cells. *Cytotechnology* **2016**, *68*, 1615–1622. [CrossRef] [PubMed]
180. Ratushnyy, A.Y.; Rudimova, Y.V.; Buravkova, L.B. Alteration of Hypoxia-Associated Gene Expression in Replicatively Senescent Mesenchymal Stromal Cells under Physiological Oxygen Level. *Biochemistry* **2019**, *84*, 263–271. [CrossRef] [PubMed]
181. Hwang, O.K.; Noh, Y.W.; Hong, J.T.; Lee, J.W. Hypoxia Pretreatment Promotes Chondrocyte Differentiation of Human Adipose-Derived Stem Cells via Vascular Endothelial Growth Factor. *Tissue Eng. Regen. Med.* **2020**, *17*, 335–350. [CrossRef]
182. Holzwarth, C.; Vaegler, M.; Gieseke, F.; Pfister, S.M.; Handgretinger, R.; Kerst, G.; Mller, I. Low physiologic oxygen tensions reduce proliferation and differentiation of human multipotent mesenchymal stromal cells. *BMC Cell Biol.* **2010**, *11*, 11. [CrossRef] [PubMed]
183. Pezzi, A.; Amorin, B.; Laureano, A.; Valim, V.; Dahmer, A.; Zambonato, B.; Sehn, F.; Wilke, I.; Bruschi, L.; Silva, M.; et al. Effects Of Hypoxia in Long-Term In Vitro Expansion of Human Bone Marrow Derived Mesenchymal Stem Cells. *J Cell Biochem* **2017**, *118*, 3072–3079. [CrossRef]
184. Rosová, I.; Dao, M.; Capoccia, B.; Link, D.; Nolta, J.A. Hypoxic Preconditioning Results in Increased Motility and Improved Therapeutic Potential of Human Mesenchymal Stem Cells. *Stem Cells* **2008**, *26*, 2173–2182. [CrossRef]
185. Dionigi, B.; Ahmed, A.; Pennington, E.C.; Zurakowski, D.; Fauza, D.O. A comparative analysis of human mesenchymal stem cell response to hypoxia in vitro: Implications to translational strategies. *J. Pediatr. Surg.* **2014**, *49*, 915–918. [CrossRef]
186. Egger, D.; Oliveira, A.C.; Mallinger, B.; Hemeda, H.; Charwat, V.; Kasper, C. From 3D to 3D: Isolation of mesenchymal stem/stromal cells into a three-dimensional human platelet lysate matrix. *Stem Cell Res. Ther.* **2019**, *10*, 248. [CrossRef]
187. Bartmann, C.; Rohde, E.; Schallmoser, K.; Pürstner, P.; Lanzer, G.; Linkesch, W.; Strunk, D. Two steps to functional mesenchymal stromal cells for clinical application. *Transfusion* **2007**, *47*, 1426–1435. [CrossRef]
188. Both, S.K.; Muijsenberg, A.J.C.V.D.; Blitterswijk, C.A.V.; Boer, J.D.; Bruijn, J.D.D. A Rapid and Efficient Method for Expansion of Human Mesenchymal Stem Cells. *Tissue Eng.* **2007**, *13*, 3–9. [CrossRef]

189. Sekiya, I.; Larson, B.L.; Smith, J.R.; Pochampally, R.; Cui, J.G.; Prockop, D.J. Expansion of human adult stem cells from bone marrow stroma: Conditions that maximize the yields of early progenitors and evaluate their quality. *Stem Cells* **2002**, *20*, 530–541. [CrossRef] [PubMed]
190. Colter, D.C.; Class, R.; DiGirolamo, C.M.; Prockop, D.J. Rapid expansion of recycling stem cells in cultures of plastic-adherent cells from human bone marrow. *Proc. Natl. Acad. Sci. USA* **2000**, *97*, 3213. [CrossRef]
191. Prockop, D.J.; Sekiya, I.; Colter, D.C. Isolation and characterization of rapidly self-renewing stem cells from cultures of human marrow stromal cells. *Cytotherapy* **2001**, *3*, 393–396. [CrossRef] [PubMed]
192. Kim, D.S.; Lee, M.W.; Lee, T.H.; Sung, K.W.; Koo, H.H.; Yoo, K.H. Cell culture density affects the stemness gene expression of adipose tissue-derived mesenchymal stem cells. *Biomed. Rep.* **2017**, *6*, 300–306. [CrossRef] [PubMed]
193. Lee, M.W.; Kim, D.S.; Yoo, K.H.; Kim, H.R.; Jang, I.K.; Lee, J.H.; Kim, S.Y.; Son, M.H.; Lee, S.H.; Jung, H.L.; et al. Human bone marrow-derived mesenchymal stem cell gene expression patterns vary with culture conditions. *Blood Res.* **2013**, *48*, 107–114. [CrossRef] [PubMed]
194. Binato, R.; de Souza Fernandez, T.; Lazzarotto-Silva, C.; Du Rocher, B.; Mencalha, A.; Pizzatti, L.; Bouzas, L.F.; Abdelhay, E. Stability of human mesenchymal stem cells during in vitro culture: Considerations for cell therapy. *Cell Prolif.* **2013**, *46*, 10–22. [CrossRef]
195. Tsuji, K.; Ojima, M.; Otabe, K.; Horie, M.; Koga, H.; Sekiya, I.; Muneta, T. Effects of Different Cell-Detaching Methods on the Viability and Cell Surface Antigen Expression of Synovial Mesenchymal Stem Cells. *Cell Transplant.* **2017**, *26*, 1089–1102. [CrossRef]
196. Heng, B.C.; Cowan, C.M.; Basu, S. Comparison of Enzymatic and Non-Enzymatic Means of Dissociating Adherent Monolayers of Mesenchymal Stem Cells. *Biol. Proc. Online* **2009**, *11*, 161. [CrossRef]
197. Salzig, D.; Schmiermund, A.; Grace, P.P.; Elseberg, C.; Weber, C.; Czermak, P. Enzymatic detachment of therapeutic mesenchymal stromal cells grown on glass carriers in a bioreactor. *Open Biomed. Eng. J.* **2013**, *7*, 147–158. [CrossRef]
198. Salzig, D.; Leber, J.; Merkewitz, K.; Lange, M.C.; Köster, N.; Czermak, P. Attachment, Growth, and Detachment of Human Mesenchymal Stem Cells in a Chemically Defined Medium. *Stem Cells Int.* **2016**, *2016*, 5246584. [CrossRef] [PubMed]
199. Della Bella, E.; Stoddart, M.J. Cell detachment rapidly induces changes in noncoding RNA expression in human mesenchymal stromal cells. *BioTechniques* **2019**, *67*, 286–293. [CrossRef] [PubMed]
200. Ojima, M.; Tsuji, K.; Otabe, K.; Horie, M.; Koga, H.; Sekiya, I.; Muneta, T. Different methods of detaching adherent cells significantly affect the detection of stem cell antigens in synovial mesenchymal stem cells. *Osteoarthr. Cartil.* **2016**, *24*, S509–S510. [CrossRef]
201. Kozanoglu, I.; Boga, C.; Ozdogu, H.; Maytalman, E.; Ovali, E.; Sozer, O. A detachment technique based on the thermophysiologic responses of cultured mesenchymal cells exposed to cold. *Cytotherapy* **2008**, *10*, 686–689. [CrossRef]
202. Liao, T.; Moussallem, M.D.; Kim, J.; Schlenoff, J.B.; Ma, T. N-isopropylacrylamide-based thermoresponsive polyelectrolyte multilayer films for human mesenchymal stem cell expansion. *Biotechnol. Prog.* **2010**, *26*, 1705–1713. [CrossRef]
203. Patel, N.G.; Cavicchia, J.P.; Zhang, G.; Zhang Newby, B.-M. Rapid cell sheet detachment using spin-coated pNIPAAm films retained on surfaces by an aminopropyltriethoxysilane network. *Acta Biomater.* **2012**, *8*, 2559–2567. [CrossRef]
204. Nagase, K.; Hatakeyama, Y.; Shimizu, T.; Matsuura, K.; Yamato, M.; Takeda, N.; Okano, T. Thermoresponsive Cationic Copolymer Brushes for Mesenchymal Stem Cell Separation. *Biomacromolecules* **2015**, *16*, 532–540. [CrossRef]
205. Nash, M.E.; Fan, X.; Carroll, W.M.; Gorelov, A.V.; Barry, F.P.; Shaw, G.; Rochev, Y.A. Thermoresponsive substrates used for the expansion of human mesenchymal stem cells and the preservation of immunophenotype. *Stem Cell Rev. Rep.* **2013**, *9*, 148–157. [CrossRef]
206. Yang, L.; Cheng, F.; Liu, T.; Lu, J.R.; Song, K.; Jiang, L.; Wu, S.; Guo, W. Comparison of mesenchymal stem cells released from poly(N -isopropylacrylamide) copolymer film and by trypsinization. *Biomed. Mater.* **2012**, *7*, 035003. [CrossRef] [PubMed]
207. Kurashina, Y.; Imashiro, C.; Hirano, M.; Kuribara, T.; Totani, K.; Ohnuma, K.; Friend, J.; Takemura, K. Enzyme-free release of adhered cells from standard culture dishes using intermittent ultrasonic traveling waves. *Commun. Biol.* **2019**, *2*, 393. [CrossRef] [PubMed]
208. Giner-Casares, J.J.; Henriksen-Lacey, M.; García, I.; Liz-Marzán, L.M. Plasmonic Surfaces for Cell Growth and Retrieval Triggered by Near-Infrared Light. *Angew. Chem. Int. Ed.* **2016**, *55*, 974–978. [CrossRef]
209. Ikeda, T.; Ichikawa, K.; Shigeto, H.; Ishida, T.; Hirota, R.; Funabashi, H.; Kuroda, A. Arginine-mediated dissociation of single cells and cell sheets from a polystyrene culture dish. *Biosci. Biotechnol. Biochem.* **2019**, *83*, 2272–2275. [CrossRef] [PubMed]
210. Chaudhry, M.A.; Bowen, B.D.; Piret, J.M. Culture pH and osmolality influence proliferation and embryoid body yields of murine embryonic stem cells. *Biochem. Eng. J.* **2009**, *45*, 126–135. [CrossRef]
211. Fekrazad, R.; Asefi, S.; Allahdadi, M.; Kalhori, K.A.M. Effect of Photobiomodulation on Mesenchymal Stem Cells. *Photomed. Laser Surg.* **2016**, *34*, 533–542. [CrossRef]
212. Stolzing, A.; Scutt, A. Effect of reduced culture temperature on antioxidant defences of mesenchymal stem cells. *Free Radic. Biol. Med.* **2006**, *41*, 326–338. [CrossRef]

213. Ahmadyan, S.; Kabiri, M.; Hanaee-Ahvaz, H.; Farazmand, A. Osmolyte Type and the Osmolarity Level Affect Chondrogenesis of Mesenchymal Stem Cells. *Appl. Biochem. Biotechnol.* **2018**, *185*, 507–523. [CrossRef]
214. Buyl, K.; Merimi, M.; Rodrigues, R.M.; Moussa Agha, D.; Melki, R.; Vanhaecke, T.; Bron, D.; Lewalle, P.; Meuleman, N.; Fahmi, H.; et al. The Impact of Cell-Expansion and Inflammation on The Immune-Biology of Human Adipose Tissue-Derived Mesenchymal Stromal Cells. *J. Clin. Med.* **2020**, *9*, 696. [CrossRef]

cells

MDPI

Review

Trends in Articular Cartilage Tissue Engineering: 3D Mesenchymal Stem Cell Sheets as Candidates for Engineered Hyaline-Like Cartilage

Hallie Thorp [1,2], Kyungsook Kim [1,*], Makoto Kondo [1], Travis Maak [3], David W. Grainger [1,2] and Teruo Okano [1,4,*]

[1] Cell Sheet Tissue Engineering Center (CSTEC), Department of Pharmaceutics and Pharmaceutical Chemistry, University of Utah, 30 South 2000 East, Salt Lake City, UT 84112, USA; hallie.thorp@utah.edu (H.T.); makoto.kondo@utah.edu (M.K.); David.Grainger@hsc.utah.edu (D.W.G.)
[2] Department of Biomedical Engineering, University of Utah, 36 S Wasatch Dr, Salt Lake City, UT 84112, USA
[3] Department of Orthopaedic Surgery, University of Utah, 590 Wakara Way, Salt Lake City, UT 84108, USA; Travis.Maak@hsc.utah.edu
[4] Institute of Advanced Biomedical Engineering and Science, Tokyo Women's Medical University, Wakamatsucho, 2−2, Shinjuku-ku, Tokyo 162-8480, Japan
* Correspondence: kyungsook.kim@utah.edu (K.K.); teruo.okano@utah.edu (T.O.); Tel.: +1-801-585-0070 (K.K. & T.O.); Fax: +1-801-581-3674 (K.K. & T.O.)

Citation: Thorp, H.; Kim, K.; Kondo, M.; Maak, T.; Grainger, D.W.; Okano, T. Trends in Articular Cartilage Tissue Engineering: 3D Mesenchymal Stem Cell Sheets as Candidates for Engineered Hyaline-Like Cartilage. *Cells* **2021**, *10*, 643. https://doi.org/10.3390/cells10030643

Academic Editors: Mehdi Najar and Hélène Boeuf

Received: 16 February 2021
Accepted: 10 March 2021
Published: 13 March 2021

Abstract: Articular cartilage defects represent an inciting factor for future osteoarthritis (OA) and degenerative joint disease progression. Despite multiple clinically available therapies that succeed in providing short term pain reduction and restoration of limited mobility, current treatments do not reliably regenerate native hyaline cartilage or halt cartilage degeneration at these defect sites. Novel therapeutics aimed at addressing limitations of current clinical cartilage regeneration therapies increasingly focus on allogeneic cells, specifically mesenchymal stem cells (MSCs), as potent, banked, and available cell sources that express chondrogenic lineage commitment capabilities. Innovative tissue engineering approaches employing allogeneic MSCs aim to develop three-dimensional (3D), chondrogenically differentiated constructs for direct and immediate replacement of hyaline cartilage, improve local site tissue integration, and optimize treatment outcomes. Among emerging tissue engineering technologies, advancements in cell sheet tissue engineering offer promising capabilities for achieving both in vitro hyaline-like differentiation and effective transplantation, based on controlled 3D cellular interactions and retained cellular adhesion molecules. This review focuses on 3D MSC-based tissue engineering approaches for fabricating "ready-to-use" hyaline-like cartilage constructs for future rapid in vivo regenerative cartilage therapies. We highlight current approaches and future directions regarding development of MSC-derived cartilage therapies, emphasizing cell sheet tissue engineering, with specific focus on regulating 3D cellular interactions for controlled chondrogenic differentiation and post-differentiation transplantation capabilities.

Keywords: chondrogenesis; chondral defects; differentiation; cellular interactions; adhesion; transplantation

1. Introduction

A plethora of therapies are clinically available for treating articular cartilage defects, all seeking to improve outcomes and mitigate osteoarthritis (OA) in the global population [1–4]. Advanced approaches employ cells prepared in vitro to increase control of cell populations, phenotypes, and dosing, with the goal of achieving more reliable hyaline cartilage regeneration [5,6]. Mesenchymal stem cells (MSCs) have been thoroughly researched as cell sources for cartilage tissue engineering due to accessibility, extended in vitro expansion capabilities, and chondrogenic lineage capacity [7–10]. However, MSC therapies are often limited by poor survival, engraftment, and control of MSC chondrogenic differentiation fate in vivo [7,11]. Therefore, one unique method of advanced cartilage regeneration aims to prepare MSC-derived cartilage constructs that express hyaline-like characteristics at the time of transplantation with the goal of more rapidly and reliably replacing damaged hyaline articular cartilage [12].

To prepare these MSC-derived pre-differentiated cartilage therapies, design considerations must include both the extent and stability of in vitro chondrogenesis and in vivo transplantation capabilities to ensure robust and lasting hyaline regeneration. MSC chondrogenic potential is known to be increased in three-dimensional (3D) structures [13–16]; therefore, development of tailored 3D constructs that promote transition of cells toward stable hyaline-like cartilage in vitro is crucial for success. Three-dimensional structures influence chondrogenesis in part by increasing 3D cellular interactions compared to two dimensional (2D) constructs [17,18]. As a result, developing a 3D platform that optimizes and controls these cellular interactions should subsequently improve the final construct's hyaline-chondral characteristics.

Even when cells are successfully differentiated, delivery and retention in the joint represent two major translational hurdles. Traditional suspended cell injections for cartilage regeneration demonstrate no homing ability if injected intravenously and poor engraftment and cellular retention at injured or diseased sites even when administered directly to the synovial space, offering only transient pain reduction [8,19,20]. Recent data show only ~3% cellular retention in the knee joint a few days post-injection with very few cells attached to the cartilage surface [8]. Obvious limitations in cell delivery result in inconsistent and suboptimal regeneration in vivo. Therefore, many current cell therapies utilize support materials to maintain cellular localization at the injury or defect sites [21,22]. Unfortunately, these additional support materials present added biocompatibility concerns [23]. As a result, MSC cartilage tissue engineering research has increasingly trended toward developing scaffold-free platforms that not only offer superior in vitro chondrogenic differentiation and optimized control 3D cellular interactions, but also support direct, unassisted delivery for robust engraftment with improved surgical versatility. Of these approaches, cell sheet tissue engineering specifically presents a unique scaffold-free platform that retains endogenous 3D cellular interactions and tissue-like organization for promoting stable in vitro hyaline-like chondrogenesis, while preserving intact adhesion molecules along the transplantation surface for direct in vivo transplantation [12,24,25]. The goal of this review is to discuss current and future directions in the development of tissue-engineered 3D MSC-derived hyaline cartilage, emphasizing cell sheet tissue engineering, with specific focus on controlled chondrogenic differentiation through 3D cellular interactions and post-differentiation engraftment capabilities.

2. Hyaline Cartilage Structure and Function

Hyaline articular cartilage is an avascular and aneural tissue that covers articulating surfaces, such as the knee, and has minimal intrinsic ability to regenerate without intervention. Hyaline cartilage structure and function (Figure 1) have been thoroughly reviewed in recent literature [9,26–30]. Briefly, it has a unique architecture and biochemical composition, comprising a sole cell type, chondrocytes, and their deposited extracellular matrix (ECM). Hyaline cartilage is characterized by predominantly rounded chondrocytes, organized in lacunae, at low cellular density, and the ECM deposited by these chondrocytes is rich in collagens type II, type IX, and type XI in addition to aggrecan, hyaluronic acid, glycosaminoglycans (GAGs), and other proteoglycans. The structure and relationship between the type II collagen and proteoglycans play a crucial role in providing hyaline cartilage's shock absorbing functionality through releasing and absorbing water in response to joint loading. Distinct from hyaline cartilage, fibrocartilage, a common clinical outcome from chondral defect therapies, is characterized by dense-packed, aligned collagen fibrils (rich in type I relative to type II collagen) lacking robust dynamic compression capabilities of hyaline cartilage [27,31,32]. To successfully develop hyaline cartilage replacement therapies, tissue-engineered cartilage constructs must satisfy key design specifications relative to native hyaline cartilage: be biocompatible, comprise viable rounded chondrocytes in lacunae structures, contain ECM rich in type II collagen, aggrecan, and sulfated proteoglycans and lacking type I and X collagens and MMP13, be able to integrate with the native cartilage, and be able to survive repeated loading within the knee joint.

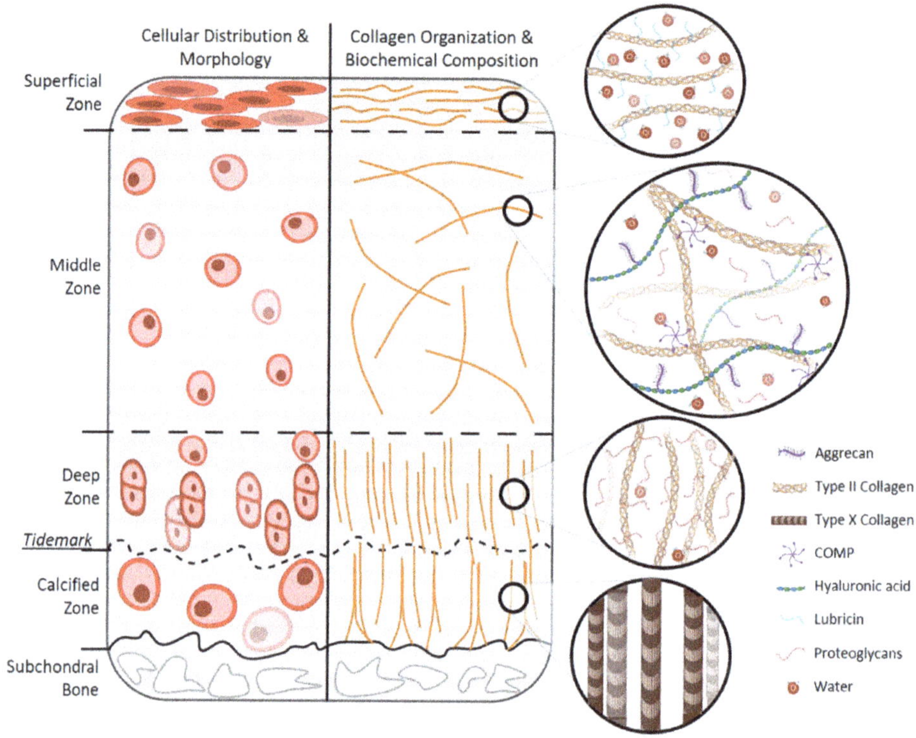

Figure 1. Hyaline cartilage structure and biochemical composition. Schematic representation of hyaline cartilage zonal structure and variable cellular distribution, morphology, collagen organization, and biochemical composition. Created with BioRender.com (accessed on 1 March 2021).

3. Current Clinical Cartilage Regeneration Therapies

Articular cartilage defects are increasingly responsible for morbidity and compromised quality of life in the global population and remain a significant precursor to osteoarthritis (OA) [28,33–35]. Based on a compelling need to regenerate durable cartilage in these defects, the past several decades witnessed numerous new therapeutic strategies designed to restore functional hyaline cartilage, increase patient quality of life, and reduce degenerative joint disease progression [21,28,36,37]. A multitude of clinical therapeutic options are currently available for treating chondral and osteochondral articular cartilage defects, thoroughly summarized in recent reviews [1–4]. These therapies include arthroscopic debridement, osteochondral allograft transplant (OCA), osteochondral autograft transplantation (OAT), mosaicplasty, and marrow stimulation techniques, among others [1–4]. Optimal therapy selection depends on numerous factors such as grade and location of the defect, patient age, and desired activity level.

For most smaller focal chondral defects, marrow stimulation, such as microfracture, is often the first-line treatment option [1,2,38]. Microfracture involves mechanical stimulation of the subchondral bone to repopulate the defect with autologous bone marrow that contains populations of regenerative stem cells [39]. Microfracture has shown clinical success in filling small focal chondral defects of the knee (<3.6 cm^2) and reducing pain short-term [40,41]. However, long-term follow-up data show that regenerated cartilage tissue is predominantly fibrocartilage with subsequent higher failure after two to five years [40,42,43]. Limitations of microfracture are often attributed to the low relative population of endogenous multipotent stem cells recruited to blood clots that fill the defect post-surgery, hindering the therapy's regenerative capacity [44].

Advanced approaches to regenerate native cartilage in chondral defects aim to specifically prepare the patients' own chondral cells (autologous chondrocytes from cartilage biopsy) ex vivo to support greater control of cell culture population, phenotype, and dosing upon re-implantation, with the goal of more reliable hyaline cartilage regeneration and enduring function in vivo. Autologous chondrocytes are the primary cell source used in these clinical cell-based cartilage regeneration therapies because chondrocytes are the primary cell source in articular cartilage [28,29]. Significantly, autologous cell sourcing presents few immunological hurdles based on the patient being both the donor and recipient of the ex vivo-processed cells. The first cell-based approach to treat articular cartilage defects—autologous chondrocyte implantation (ACI)—was FDA-approved in 1997 [45] with several new "generations" of ACI reported recently [1,28,44,46,47]. ACI harvests autologous chondrocytes from a healthy, low load-bearing area of the patient's cartilage, followed by cell expansion ex vivo, and then staged reimplantation of the expanded cells back to the defect as suspended cell injections under a sutured periosteal flap [22]. Unlike microfracture, ACI provides more reliable and improved pain reduction and mobility outcomes at 5-year follow-ups [48,49]. Further development of this therapy led to the use of porcine collagen support membranes for matrix-supported autologous cultured chondrocyte therapy (MACI) [22], FDA-approved in 2016 [50]. The collagen support membrane is intended to preserve chondrocyte characteristics during culture and retain cells in the defect site during transplantation. MACI has shown some in vivo therapeutic benefit in treating chondral defects [22,49,51,52]. Short-term 2-year clinical follow-ups reported 75% of tissue filling the defects was hyaline-like [53], and long-term 15-year follow-ups showed increases in Lysholm [54], International Knee Documentation Committee (IKDC) [55], and Tegner activity [56] scores compared to preoperative baselines [57]. However, superiority of MACI relative to ACI remains controversial. In randomized trials with 2-year follow-ups, no significant improvements (IKDC and Tegner activity scores) were noted for MACI compared to ACI, with ACI reporting slightly better International Cartilage Repair Society (ICRS) [58] and Lysholm functionality scores [1,3,22,59,60]. Few additional cell-based therapies have gained clinical approval in recent decades around the world, but ACI and MACI remain the only cell and tissue engineering cartilage therapies approved in the U.S. (Table 1). Hundreds more are currently in the clinical trial pipeline [1,21,61] (www.clinicaltrials.gov; accessed on 1 March 2021).

Table 1. Clinically approved cell and tissue engineered cartilage regeneration therapies.

Product Name	Company	Cell Type *	Support Material(s)	Country of Approval—Approval Body	Year Approved	Refs.
Carticel (1st gen. ACI)	Vericel	Autologous chondrocytes	Surgical application of periosteal flap	U.S.—FDA	1997 (2017 phased out)	[62,63]
Chondron™	Sewon Cellontech	Autologous chondrocytes	Fibrin gel	Korea—MFDS	2001	[64]
ChondroCelect®	TiGenix	Autologous chondrocytes	Surgical application of periosteal flap or commercially available collagen membrane (not included)	E.U.—EMA	2009 (2016 withdrawn)	[65,66]
Cartistem®	Medipost	Allogeneic umbilical cord blood-derived mesenchymal stem cells	N/A (injection into synovial space)	Korea—MFDS	2012	[67]
JACC®	J-Tec	Autologous cultured chondrocytes	Collagen gel	Japan—MHLW	2012	[68]
Novocart® 3D	Aesculap Biologics	Autologous chondrocytes	Three-dimensional collagen-chondroitin sulphate scaffolds	Germany/Switzerland	2014	[69]
MACI®	Vericel	Autologous cultured chondrocytes	Porcine type I/III collagen membrane	E.U.—EMA	2013 (2018 withdrawn)	[70]
				U.S.—FDA	2016	[71]
Ortho-ACI® (3rd gen. MACI)	Orthocell	Autologous chondrocytes	Porcin type I/III collagen scaffold	Australia	2017	[72]
Spherox (chondrosphere®)	co.don	Autologous matrix-associated chondrocytes	N/A (self-adhering)	E.U.—EMA	2017	[73]
Invossa™ (TissueGene-C)	Kolon Life Sciences	Allogeneic chondrocytes (retrovirally transduced to be TGF-β-expressing)	N/A (injection into synovial space)	Korea—MFDS	2017 (2019 revoked)	[74,75]

* All cell types are human unless otherwise noted. ACI: Articular Chondrocyte Implantation. MACI: matrix-supported autologous cultured chondrocyte therapy. U.S.: United States of America. E.U.: European Union. FDA: Federal Drug Admnistration. MFDS: Ministry of Food and Drug Safety. EMA: European Medicines Agency. MHLW: Ministry of Health, Labour and Welfare. Not Applicable (N/A) refers to products that are not prepared, or indicated to be used, with any biomaterials for supporting adhesion or localization.

4. Limitations of Current Autologous Cell-Based Cartilage Regeneration Therapies

Despite clinical availability of several generations of these autologous cell-based cartilage regeneration therapies, clinical outcomes remain heterogeneous and unconvincing, and difficulties persist in enabling broader patient population applications [1,4,43,76]. One primary limitation of these therapies is reliance on autologous chondrocyte cell sourcing. Chondrocytes are known to dedifferentiate during in vitro culture and expansion, transitioning during preparation from their mature phenotype to fibroblast-like phenotypes, and also exhibit limited capacity for in vitro expansion before becoming senescent [1,2]. Autologous sourcing of these chondrocytes also introduces patient burden through multiple surgeries, donor site morbidity, and extended time between donation and treatment. Additionally, cell quality and quantity from autologous sources are donor-dependent, increasing procedural cost and complexity [6,25,45,77,78], and making it difficult, if not impossible, to predict, control, and standardize therapeutic potency [19,79]. Due to these limitations, further efforts focus on selecting improved, appropriate cell sources for cartilage tissue engineering and regenerative purposes. Greater consistency and control over cellular characteristics are needed to ensure reliable chondrogenic construct production and understand implant performance. Moreover, these sources should ideally be broadly applicable and efficacious for treating a wide range of patient populations [4,19,48,80,81].

5. Allogeneic Mesenchymal Stem Cells as Promising Cell Sources for Cartilage Applications

Developing tissue-engineered constructs for articular cartilage focal defect therapies increasingly focuses on transitioning from non-standard, heterogeneous autologous to standardized allogeneic cell sourcing [21,79,82]. In contrast to autologous cell sourcing issues, allogeneic cells offer greater control over cell quality and characteristics, improved accessibility, and potentially broader use [5,6,83]. Allogeneic sourcing also permits greater in vitro expansion capacity, and cells with various profiles and characteristics can be profiled, selected, validated, and banked, enabling "off-the-shelf" products [5,6,83]. Concerns regarding allogeneic cell immune rejection remain. However, with a long history of osteochondral allografting [38,84] and new insights into immune-matching [85,86], paired with reported immunomodulatory characteristics of certain allogeneic cell sources [87,88], translational prospects for human allogeneic cells are seemingly more feasible.

Advanced cell-based therapies also seek to replace chondrocytes with MSCs as the chondrogenic cell source. Chondrocyte sourcing is tissue-specific, whereas MSCs are adult progenitor cells isolated from a variety of tissues (e.g., bone marrow, adipose, dental pulp, umbilical cord, etc.), offering a widely accessible cell source [14,15,87,89]. Additionally, chondrocytes are limited by de-differentiation during culture and passaging, while MSCs exhibit strong capacity for in vitro expansion while maintaining their identity and unique capacity for in vitro self-renewal [11,16,82,88,90]. Although not standardized, MSC identity is generally confirmed via several accepted surface markers: $CD90^+$, $CD44^+$, $CD73^+$, $CD105^+$, $CD11^-$, $CD34^-$, $CD45^-$ [91,92]. When selecting appropriate MSC sources it is important to account and test for reduced in vitro self-renewal and differentiation capacities induced by extensive passaging, occurring at different rates for different MSCs [11,83,93–95]. Specific to chondral regeneration, MSCs have utility for fabricating cartilage in vitro based on their multilineage differentiation potential, including the capacity to transition to chondrocytes [15,16,87,88]. Many reports have described undifferentiated MSC therapies exhibiting some therapeutic efficacy in delaying cartilage degeneration and reducing pain [8,19,96,97]. However, in vitro and in vivo MSC differentiation fate and maintenance are still not easily controlled, limiting these therapies' capabilities to induce lasting cartilage regeneration [7,20,98]. Advanced approaches in MSC-based cartilage regeneration aim to employ allogeneic MSC sources and exploit innate MSC chondrogenic potential to better control their differentiation in vitro, preparing hyaline-like transplantable constructs for rapid structural cartilage regeneration through direct tissue replacement in vivo, applicable to a broader range of patients with a more consistent cell-based product.

6. Three-Dimensional Culture for MSC Chondrogenesis

MSC-derived hyaline-like cartilage constructs prepared in vitro actively exploit recent advances in 3D culture systems (Figure 2).

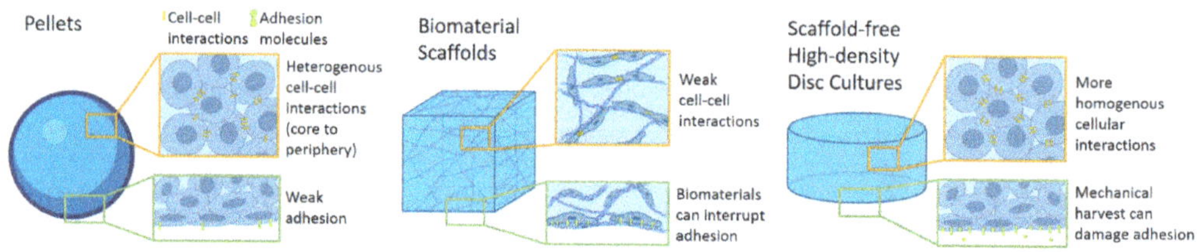

Figure 2. Categories of three-dimensional culture platforms for in vitro mesenchymal stem cell (MSC) differentiation. Cellular interactions (yellow linkers) and surface interface adhesion molecules (green markers) among the constructs varies in response to biomaterials and construct cellular organization. Created with BioRender.com (accessed on 1 March 2021).

MSC multipotency enables directed cell differentiation to hyaline-like chondrocyte phenotypes in vitro within 3D cultures both with and without supporting biomaterials [14–16,87]. Successful MSC chondrogenesis is generally verified by detecting positive expression of hyaline cartilage markers within the cells and their deposited ECM (e.g., Sox9, sulfated proteoglycans, type II collagen, and aggrecan) [7,16,99,100]. A persisting limitation in MSC chondrogenesis is the expression of transient hyaline-like cartilage phenotypes with the inevitable and undesired transition toward hypertrophic or fibrocartilage phenotypes [7–9,44]. Therefore, hyaline differentiation must also exhibit persistent negative marker expression of type X and type 1 collagens and MMP13 [7,16,99,100]. Researchers have long noted that 3D culture conditions and 3D cellular interactions are essential for inducing and maintaining this stable hyaline-like chondrogenesis [7,14,18,99,101–104]. Standard 2D culture conditions limit chondrogenesis because they are unable to promote requisite 3D cellular interactions and structures associated with chondrogenic condensation and further maturation [17,105,106]. Unlike traditional adherent 2D cell culture methods, 3D culture platforms allow cells to assume rounded morphologies associated with mature chondrocytes [13,107,108] and promote 3D cellular interactions, mimicking early condensation stages during cartilage development and playing an important role in stabilizing terminally differentiated cartilage [13,99,101].

In addition to three-dimensionality, appropriate culture conditions are critical for inducing MSC chondrogenesis. Cartilage tissue's innate avascularity results in a naturally hypoxic environment that directly impacts chondrogenic development and cellular functionality [27,109,110]. Likewise, experimentally recapitulating this low oxygen environment in vitro, via hypoxic culture conditions (1–7% O_2), is essential for eliciting hyaline-like ECM deposition [27,111–114]. In vitro hypoxic culture specifically upregulates type II collagen and aggrecan synthesis for both chondrocytes and MSCs [27,111,114]. As such, in vitro MSC chondrogenic differentiation generally utilizes 3D cultures, chondrogenic induction media, and humidified hypoxic culture conditions [7,115].

The most common method for assessing MSC chondrogenic potential in vitro employs spheroids [116], usually as pellet or micromass cultures [14–16]. Beyond their simplicity of fabrication, these cultures allow cells to self-aggregate and assume rounded morphologies while establishing 3D cellular interactions necessary for chondrogenesis [14–16,117,118]. Although pellet cultures allow cells to assume rounded morphologies, these cultures regularly produce heterogenous tissue in vitro that does not mimic native cartilage in structure, phenotype, or function. Such heterogeneity is often attributed to media and oxygen diffusion limitations influencing 3D cellular interactions, resulting in variable differentiation between the pellet's periphery and hypoxic core [119–121].

In an attempt to offer improved control over cell differentiation, many MSC differentiation platforms employ natural or synthetic biomaterial scaffolds, such as collagens, alginates, hyaluronic acid, agarose, chitosan, decellularized "native" ECM, and polyglycolic acid (PGA)/polylactic acid (PLA), to accommodate cells in 3D structures and promote MSC chondrogenic differentiation [107,122,123]. These biomaterial scaffolds permit a high degree of control over 3D construct architecture, a key component in controlling MSC chondrogenesis [124,125]. Extensive work is reported for further tailoring these biomaterial scaffolds, via fabrication techniques (e.g., bioprinting, electrospinning, molding, etc.) and combinations of cell ligands and binding motifs, macro- and micro- structure, stiffness, and other biomaterials properties [23,98,107,123,126–128] seeking to promote and maintain cellular interactions and functionality, supporting transitions toward hyaline-like phenotypes [23,122,129]. However, these approaches are often limited by poor cell–cell communication due to interruptive scaffold materials hindering requisite direct cell–cell and cell-ECM interactions, hyaline-like cell transitions, and reliable hyaline-like phenotypic preservation [23,124,129].

Scaffold-free approaches offer increasing benefits compared to scaffold-based methods, supporting MSC differentiation in 3D conditions, within their endogenous ECM and in continuous, direct 3D contact, promoting necessary cellular interactions without scaffold interference [23]. Scaffold-free cell-based constructs can also accommodate higher cell densities than scaffold-based approaches, and despite native cartilage's intrinsic low cell density [30,130], cell-dense constructs are recognized as necessary for promoting in vitro MSC chondrogenesis [15,118,131,132]. Recently proposed advanced scaffold-free methods employ high-density seeding cultures that create disc-like cartilage constructs in vitro by seeding MSCs into porous cell culture inserts at very high concentrations [100,133–136]. These high-density 3D cultures induce more homogenous chondrogenesis compared to pellet cultures, and produce more ergonomic implant forms to more completely fill cartilage defects [100,133–135]. However, these approaches are hindered by exorbitant cell seeding densities and limited control over cellular interactions in culture, based solely on cell aggregation forced by over-confluence [100,133–136]. Such high-density 3D constructs are sometimes referred to as "cell sheets" [134,135], but differ significantly from temperature-responsive culture dish (TRCD) derived cell sheets discussed in Sections 8 and 9 based on their (1) three-dimensionality achieved solely through over-confluent culture, and (2) harvest methods reliant on mechanical detachment that damage the cultured construct's adhesion interface. Despite extensive work focused on promoting in vitro hyaline-like chondrogenesis within a wide range of 3D culture constructs, these platforms are still broadly unable to sufficiently control both structure and 3D cellular interactions, hindering resulting chondrogenic stability and homogeneity in vitro.

7. Transplantation Capabilities of 3D MSC Chondrogenic Cultures

Even when 3D culture platforms achieve hyaline-like chondrogenesis in vitro, these resulting cellular constructs are still unable to directly adhere and interface with host tissues in vivo. Most constructs require additional transplantation support materials (e.g., suturing, fibrin glue, periosteal flap, etc.), increasing biocompatibility concerns and disrupting direct communication between the transplanted cells and host tissue [19,27,137,138]. Limited unassisted in vivo tissue engraftment is often attributed to chondrogenic constructs' inadequate endogenous expression of surface adhesion molecules [12,80,107,123,124,129,139]. Poor in vivo tissue site engraftment leads to construct delamination, loss of transplanted cell viability, mechanical instability, and decreased integration with host tissue, common precursors for fibrocartilage tissue formation [26] and suboptimal pre-clinical in vivo outcomes [8,27,31,140,141]. Discrepancies between in vitro and in vivo pre-clinical results may be partly due to the high variability among animal models employed [142–147], but inferior engraftment and retention remain driving factors of pre-clinical failure regardless of the model employed [26].

Cartilage tissue transplant failure is also attributed to insufficient interfacial properties [148]. Native hyaline cartilage exhibits a low coefficient of friction at the joint interface, allowing free sliding of adjacent cartilage surfaces under high pressure during joint articulation [149,150]. To successfully replace hyaline cartilage at focal defect sites, transplanted cartilage constructs must be able to not only adhere and engraft into the defect site, but also present a suitable articulating surface that mitigates excessive frictional forces during joint function. As superficial chondrocytes naturally produce lubricating agents, such as lubricin and hyaluronic acid [31,151], some approaches focus on functionalizing the cells within 3D structures to tailor their secretion abilities and recreate this lubricated articular surface [7,152]. Other approaches, specifically those employing cell-seeded hydrogels, focus on selecting scaffold biomaterials with low intrinsic coefficients of friction [31,153]. However, the inability of current constructs to both strongly adhere and recapitulate this lubrication interface increases associated friction during articulation, causing pain, abnormal stress and wear on the transplant, and increased risk of tissue delamination [149].

Despite 3D cell delivery platforms being designed to create hyaline-like chondrogenic constructs capable of engraftment and retention at the defect site, to date, no platform has yielded robust evidence of success, necessitating further investigation in controlled clinical trials to verify translational potential of these therapies [1,21,23,154]. A clear unmet need persists for improved 3D MSC platforms that not only control 3D cellular interactions in vitro to reliably yield more stable hyaline-like cartilage constructs, but also enhance their adhesion for mechanical and physiological integration in vivo to better address current translational limitations in MSC-based cartilage regeneration.

8. Cell Sheet Technology as a Transplantable 3D Tissue-Like Platform

Cell sheet technology supports fabrication of transplantable, scaffold-free, 3D, tissue-like cell constructs [155–159] (Figure 3).

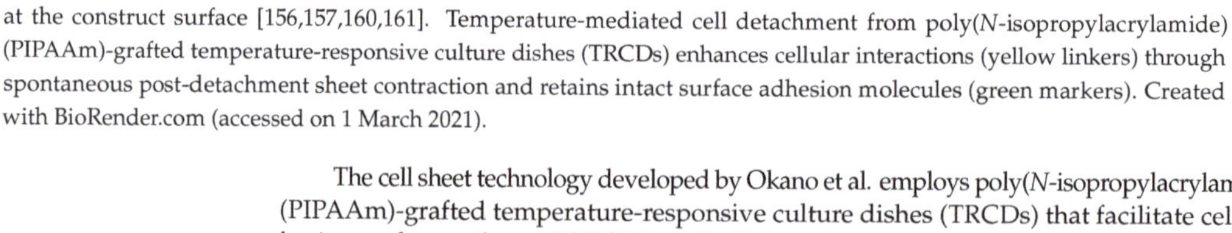

Figure 3. Cell sheet fabrication leads to increased three-dimensional (3D) cellular interactions and intact adhesion molecules at the construct surface [156,157,160,161]. Temperature-mediated cell detachment from poly(*N*-isopropylacrylamide) (PIPAAm)-grafted temperature-responsive culture dishes (TRCDs) enhances cellular interactions (yellow linkers) through spontaneous post-detachment sheet contraction and retains intact surface adhesion molecules (green markers). Created with BioRender.com (accessed on 1 March 2021).

The cell sheet technology developed by Okano et al. employs poly(*N*-isopropylacrylamide) (PIPAAm)-grafted temperature-responsive culture dishes (TRCDs) that facilitate cell adhesion and growth at 37 °C [158–160]. Below the PIPAAm lower critical solution temperature (32 °C), cells spontaneously detach from the culture surface, bypassing typical culture requirements for damaging enzymatic cell harvesting [160,162]. This temperature-mediated detachment retains endogenous cell–cell and cell-ECM interactions and preserves cellular environments, allowing cultured cells to be harvested as intact cell sheets [83,156,157,160,162–164]. As cells are seeded and grown under adherent 2D conditions, this abrupt temperature-mediated detachment prompts established cytoskeletal filaments and retained ECM to naturally contract when released from culture surfaces [165,166]. This post-detachment cell sheet contraction spontaneously yields 3D, multi-nuclei thick, scaffold-free cell sheet structures [12,161]. Cell sheet three-dimensionality can be further controlled by cell sheet layering to produce tissues of specified thicknesses and cellular densities, even combining cell sheets from different cell sources [157,167–170]. Cell sheet post-detachment contraction and layering both increase 3D cellular interactions, areas of hypoxia within the construct, and functionality relative to suspended cells and 2D conditions [167,171,172].

In addition to promoting 3D architecture with increased 3D cellular interactions, cell sheets naturally retain innate surface receptors, ECM, and tissue adhesion capabilities, allowing spontaneous engraftment to tissue sites and rapid initiation of direct cell–cell communication [156,157]. Cell sheets fabricated from a wide range of cell sources have been applied to a multiple tissue targets and show significant adhesion and localization capabilities [157,173,174]. Specifically, for cartilage regeneration therapies, significant translational work has focused on cell sheet technology approaches for repairing and replacing hyaline cartilage using various cell sources and preparation methods (Table 2).

Table 2. Cell sheet tissue engineering cartilage regeneration studies.

Cell Source	Study Type	In Vitro Chondrogenic Enhancement	Refs.
Human articular chondrocytes	In vitro	Layering	[172]
Articular chondrocytes (human, rabbit)	In vitro/in vivo (allogeneic rabbit)	Layering	[173]
Rat articular chondrocytes and synoviocytes	In vivo (allogeneic rat)	Layering	[175]
Rabbit articular chondrocytes and synoviocytes	In vivo (allogeneic rabbit)	Layering	[176]
Porcine articular chondrocytes	In vivo (allogeneic minipig)	Layering	[177]
Human articular chondrocytes	In vitro	Co-culture with synoviocytes + layering	[178]
Human articular chondrocytes	In vitro	Co-culture with synoviocytes + layering	[179]
Human articular chondrocytes	In vivo (xenogeneic immunosuppressed rabbit)	Co-culture with synoviocytes + layering	[180]
Human articular chondrocytes and synoviocytes	In vivo (athymic rat)	Co-culture with synoviocytes + layering	[181]
Autologous human articular chondrocytes (with microfracture)	In vivo (autologous human—small cohort clinical study)	Co-culture with synoviocytes + layering	[182]
Rat articular chondrocytes	In vitro/in vivo (allogeneic rat)	None	[183]
Human juvenile polydactyly chondrocytes	In vitro/in vivo(xenogeneic immunosuppressed rabbit)	None	[184]
Human juvenile polydactyly chondrocytes	In vivo (athymic rat)	None	[25]
Human endometrial gland-derived MSCs	In vitro	Layering	[171]
Human bone marrow-derived MSCs	In vitro	Chondrogenic induction medium + hypoxia (5% O_2)	[12]

Cell sheet technology employing chondrocyte sources has shown preliminary success in both pre-clinical models and small cohort clinical studies [24,25,173,175–177,180–184]. Chondrocyte sheets adhere directly and spontaneously to cartilage tissue via retained endogenous ECM and adhesion proteins. Notably, this strength of defect site adhesion for the undifferentiated chondrocyte sheets is sufficient to allow initial defect retention without suturing, and to withstand knee joint mechanical forces while maintaining long-term localization of transplanted cells [24,173,177,180,182,185]. This engraftment capability facilitates successful chondrocyte sheet induction of hyaline-like cartilage regeneration in articular cartilage focal chondral defects by 4 weeks post-transplantation [24,25,173,177,180–182] (Figure 4a–d).

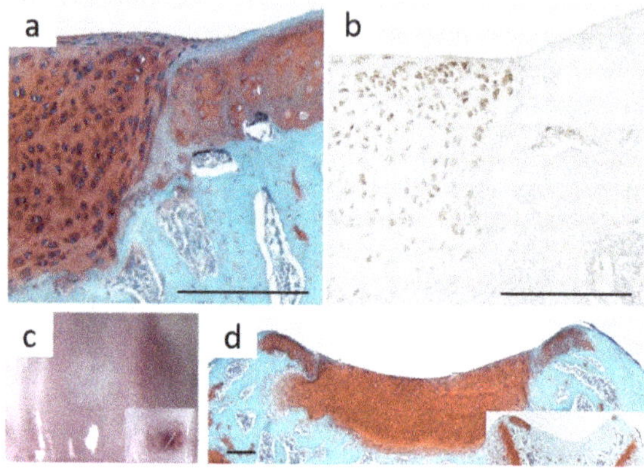

Figure 4. Cell sheets adhere, remain localized, and induce cartilage regeneration in vivo without any additional support materials. Histological and immunohistochemical staining of rat knee cross-sections 4-weeks post cartilage sheet transplantation show close interfacing with the native cartilage and (**a**) areas of positive hyaline-like regeneration (Safranin-O), correlating to (**b**) retention and viability of human cells (hVimentin—brown color) in trochlear groove chondral defects. Regenerated tissue filling the defect at 4 weeks post-transplantation (**c**) macroscopically and (**d**) histologically (Safranin-O) (right corner boxes shows defect only controls) resembles native cartilage. Scale bars = 200 μm. Adapted and reprinted from Kondo M., Kameishi S., Grainger D. W. & Okano T. Novel therapies using cell sheets engineered from allogeneic mesenchymal stem/stromal cells. Adapted with permission from *Emerg. Top. Life Sci.* **4** (6): 677–689 (2020). Copyright 2020 Portland Press.

9. Three-Dimensional MSC Sheets as In Vitro Platforms for Fabricating Transplantable Hyaline-Like Cartilage

Emerging cell sheet approaches prepare in vitro chondrogenically differentiated MSC sheets that are directly transplantable in vivo, which should support more rapid hyaline cartilage replacement at defect sites for future in vivo regenerative therapies. Reliable fabrication of 3D MSC sheets increases cell–cell interactions, promotes hyaline-like chondrogenesis, and retains construct adhesion capabilities [12], all of which are essential to support robust and direct replacement of damaged or missing hyaline cartilage. Sheet-enhanced 3D cellular interactions specifically benefit MSC chondrogenesis in vitro, resulting in stable hyaline-like phenotypes and delayed hypertrophic transitions compared to standard pellet cultures [12]. Cell sheet 3D manipulation affords greater control over the induction of pro-chondrogenic 3D cell–cell and cell-ECM interactions and increased control of the final chondrogenic cell sheet characteristics (Figure 5).

Cell sheet technology employs multiple manipulation techniques for promoting specific pro-chondrogenic interactions. Post-detachment cell sheet contraction, occurring spontaneously following temperature-mediated detachment from adherent culture, and sheet multilayering are primary strategies used to control and influence cellular interactions and MSC chondrogenic differentiation in scaffold-free cell sheet forms [25,157,167–172] (Figure 5a). Cell sheet contraction can be modified by changing cell seeding density, culture time, MSC source, or use of removable support membranes [155,166,167,186]. Cell sheet multilayering has also been utilized extensively in various cell sheet tissue engineering applications [167,169,170,187,188]. Specifically, multilayering chondrocyte sheets has been shown to directly increase 3D cellular interactions, promoting enhanced chondrogenic characteristics within those sheets [173,178,179]. Moreover, layering endometrial cell sheets increased glycosaminoglycan and collagen development within as little as 24 h [171] (Figure 5b). This multilayering manipulation should facilitate similar control of 3D cellular interactions within MSC-derived sheets, as well as construct thickness and density. These factors directly impact the oxygen tension and hypoxic conditions within the MSC construct, stimulating more controlled transitions to hyaline-like phenotypes in vitro. Multilayering may also prompt more rapid chondrogenesis, decreasing MSC-derived hypertrophic characteristics commonly associated with extended in vitro media induction [18,103].

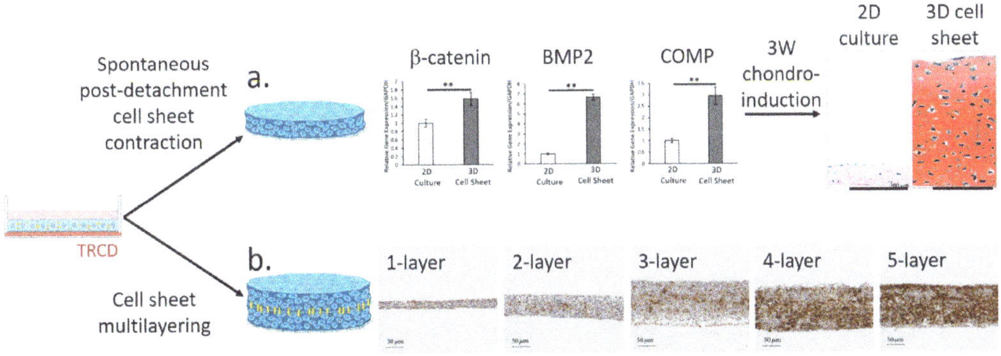

Figure 5. Cell sheet manipulation techniques using TRCDs increase cellular communication and ECM characteristics related to enhanced in vitro chondrogenesis potential. Cell sheet manipulation techniques include (**a**) spontaneous, post-detachment cell sheet contraction and (**b**) sheet multilayering utilizing either contracted or non-contracted cell sheets [157,167–171,173]. These manipulation techniques increase chondrogenic potential of the MSCs as shown with (**a**) Safranin O staining and (**b**) Type II collagen immunohistochemical (IHC) staining. For graphs in (**a**), error bars represent means ± standard deviations (** $p < 0.01$). (**a**) Adapted and reprinted from Thorp H., Kim K., Kondo M., Grainger D. W. & Okano T. Fabrication of hyaline-like cartilage constructs using mesenchymal stem cell sheets. Adapted with permission from *Sci. Rep.* **10**, (2020). Copyright 2020 Springer Nature. (**b**) Adapted and reprinted from Waki S., Yuji H., Tatsuya S., Masayuki Y., Akihiro U., Teruo O. Chondrocyte Differentiation of Human Endometrial Gland-Derived MSCs in Layered Cell Sheets. Adapted with permission from *Sci. World J.*, Article ID 359109, (2013). Copyright 2013 Hindawi. Created in part with BioRender.com (accessed on 1 March 2021).

In addition to promoting stable hyaline-like chondrogenesis in vitro, MSC sheets retain strong adhesion capabilities after chondrogenic differentiation [12]. Post-differentiation temperature-mediated harvest does not damage cell sheet characteristics, thereby allowing maintenance of critical adhesion molecule expression for cells along the basal side of the sheet. MSC-derived hyaline-like cell sheets can strongly adhere to fresh ex vivo cartilage tissue and rapidly initiate mechanical and biochemical signaling interactions between the cell sheet and adjacent native cartilage [12]. Based on previous adhesion studies conducted with chondrocyte sheets [173] and their successful integration and maintained adhesion in vivo [24,177,180,182], these adhesion capabilities of chondrogenically differentiated MSC sheets are expected to promote similar stable engraftment and enhanced cellular communication in this environment.

Cell sheet in vitro chondrogenesis studies support prior assertions that three-dimensional cell interactions play essential roles in fabrication and stability of in vitro hyaline-like cartilage. Furthermore, cell sheet manipulation techniques allow greater control over these 3D cellular interactions and related hypoxic culture conditions, while maintaining known cell sheet adhesion capabilities. Additional application of hypoxic culture conditions for chondrogenic induction not only significantly increases the MSC sheets' chondrogenic capacity, but should also condition them for the hypoxic in vivo environment, allowing greater retention of cellular functionality post-transplantation. These chondrogenic capacity and adhesion capabilities position MSC cell sheet technology as a prospective next-generation platform for fabricating future translational allogeneic MSC therapies offering direct, unassisted transplantation of hyaline-like cartilage constructs for improved future articular cartilage regeneration. To improve upon current cell-based approaches for cartilage regeneration in human defects, these implanted MSC-derived cartilage sheets will have to demonstrate key regenerative behaviors in vivo, notably: complete filling of the focal defect, lateral and basal integration with the host tissue, lasting retention of hyaline-like phenotypes within the defect, and mechanical properties similar to native cartilage once integrated.

10. Summary

Articular cartilage defects represent inciting events and a significant cause of degenerative joint disease with inevitable progression to generalized OA [28,33–35]. Although many clinical therapies exist for treating these defects, none achieve lasting, robust regeneration of hyaline cartilage [1,4,43,76]. Advanced cell therapy products are continually being developed to address the limitations of current clinical therapies, but few have shown much clinical promise to date in practically addressing diverse chondral defects [1,21,23,154]. Overall, tissue engineering cartilage therapies are still largely limited in their control over in vitro cellular interactions necessary for producing robust hyaline-like cartilage and inconsistent in vivo engraftment, hindering integration with the host tissue and lasting replacement of hyaline cartilage [23,124,129]. Some 3D MSC-based approaches, specifically those employing banked, standardized allogeneic MSCs within scaffold-free 3D constructs, offer very promising platforms for producing cartilage constructs in vitro via controlled 3D structures and key cellular interactions that are capable of inducing reliable, rapid regeneration of hyaline-like cartilage in vivo in articular cartilage focal defects.

Although in vitro chondrogenic differentiation is extensively published for pellet cultures, cell seeded scaffolds, and scaffold-free high-density seeding cultures [14–16,87], these 3D constructs are limited in their abilities to achieve both robust hyaline-like differentiation and direct, unassisted transplantation to defect sites [1,21,23,154]. To address these concerns, cell sheet tissue engineering constructs afford improved control of 3D cellular interactions, maintenance of chondrogenic characteristics via established manipulation techniques, and optimize endogenous adhesion abilities [83,156,157,160,162–164]. To date, autologous chondrocyte cell sheets have exhibited experimental and some clinical success in adhering, surviving, and inducing regeneration in articular cartilage defects [24,173,177,180,182,185,189]. These data provide an important precedent for further development of cell sheet therapies that support more rapid cartilage regeneration. The chondral regeneration field is currently transitioning toward the creation of single-stage, immediately available cell-based chondral restoration options [21]. In this vein, cell sheet tissue engineering employing allogeneic MSCs presents a unique platform capable of (1) producing stable in vitro hyaline-like cartilage from banked MSCs, (2) providing an off-the-shelf, pre-validated cartilage tissue construct without biomaterials support, and (3) maintaining and sustaining endogenous cellular adhesion and signaling for direct transplantation to cartilage tissues applicable to a broad patient population.

11. Future Perspectives

Despite decades of research on tissue engineering and MSC chondrogenesis, current chondrogenic approaches are largely unable to reliably create stable hyaline-like cartilage in vitro that is directly transplantable in vivo to a broad patient population. However, cell sheet tissue engineering offers a unique scaffold-free platform to facilitate enhanced in vitro hyaline-like differentiation, to support direct in vivo transplantation to defects without biomaterials support. Combining cell sheet technology with allogeneic MSC sourcing, specifically for MSCs that have been screened for cell potency and differentiation capacity, should facilitate more rapid and reliable cartilage regeneration for a broader patient population.

Although hundreds of MSC-derived cell therapy clinical trials are ongoing, no MSC-based regenerative medicine applications have clinical validation for cartilage regeneration. While causes for failure with these MSC therapies are not fully understood, the current inability to properly control cellular interactions and cellular phenotypes in vitro to reliably yield stable hyaline-like cartilage, combined with poor tissue site engraftment and retention in vivo necessary to restore normal cartilage functional properties through mechanical and biochemical signaling, are central hypotheses. To improve upon cell-based and MSC therapies, specific considerations and attention must be paid to (1) selecting and validating appropriate cell sources, essential to regulatory and manufacturing challenges during translation, (2) the importance of three dimensionality in tissue-like structures and its role in inducing and maintaining 3D cellular interactions required for stable in vitro hyaline like chondrogenesis, (3) robust engraftment and integration of the transplanted construct with host tissue, and (4) the long-term stability of hyaline features in vivo without reversion to fibrocartilage. Focusing on these essential performance specifications will support progress in developing MSC-derived therapies that are both transplantable and phenotypically stable as hyaline-like cartilage to robustly regenerating hyaline articular cartilage at the site of articular cartilage defects.

Furthermore, future approaches may additionally enhance MSC chondrogenic potential and robust tissue regeneration and integration through the use of CRISPR or other gene editing techniques [91,190–192] to bias MSCs using guided genetic instructions. Incorporating these modified allogeneic MSCs into established transplantable 3D cell sheets could yield even more robust hyaline-like tissues with greater regenerative potential, but will likely face greater regulatory scrutiny and manufacturing hurdles in their path to clinical approval [45,78,193,194].

Author Contributions: H.T. prepared review outline, conducted literature searches, wrote main manuscript text, and prepared all figures. K.K. prepared review outline, assisted with preparation of main manuscript text, and supervised, reviewed, and edited final manuscript. T.O. and D.W.G. advised on review outline and T.O., M.K., T.M., and D.W.G. supervised, reviewed, and edited final manuscript. All authors have read and agreed to the published version of the manuscript.

Funding: This work was supported in part by the University Technology Acceleration Grant (UTAG) from the Utah Science, Technology, and Research (USTAR) program and a Whitaker International Fellowship from the Institute of International Education (IIE).

Institutional Review Board Statement: Not Applicable.

Informed Consent Statement: Not Applicable.

Acknowledgments: Referenced cell sheet data were cited from research conducted at The Cell Sheet Tissue Engineering Center at the University of Utah (CSTEC@Utah), Tokyo Women's Medical University, and Tokai University. We thank S. Bou-Ghannam for continued technical advice. We acknowledge pioneering contributions of M. Sato and his team (Tokai University, Japan) in establishing cell sheet cartilage regeneration preclinical and clinical precedents, and T. Shimizu and T. Iwata (Tokyo Women's Medical University, Japan) for contributing to ongoing collaborative information exchange relevant to cell sheet technology.

Conflicts of Interest: T.O. holds equity in CellSeed, Inc. (Japan) and is an inventor/developer designated on the patent for Cell Seed's commercialized temperature-responsive cultureware. No other competing financial interests exist and all authors declare that they have no other competing interests.

References

1. Makris, E.A.; Gomoll, A.H.; Malizos, K.N.; Hu, J.C.; Athanasiou, K.A. Repair and Tissue Engineering Techniques for Articular Cartilage. *Nat. Rev. Rheumatol.* **2015**, *11*, 21–34. [CrossRef]
2. Memon, A.R.; Quinlan, J.F. Surgical Treatment of Articular Cartilage Defects in the Knee: Are We Winning? *Adv. Orthop.* **2012**, *2012*. [CrossRef]
3. Falah, M.; Nierenberg, G.; Soudry, M.; Hayden, M.; Volpin, G. Treatment of Articular Cartilage Lesions of the Knee. *Int. Orthop.* **2010**, 621–630. [CrossRef]
4. Rodeo, S.A. Cell Therapy in Orthopaedics: Where Are We in 2019? *Bone Jt. J.* **2019**, 361–364. [CrossRef]
5. Abou-El-Enein, M.; Elsanhoury, A.; Reinke, P. Overcoming Challenges Facing Advanced Therapies in the EU Market. *Cell Stem Cell* **2016**, 293–297. [CrossRef]

6. Haddock, R.; Lin-Gibson, S.; Lumelsky, N.; McFarland, R.; Roy, K.; Saha, K.; Zhang, J.; Zylberberg, C. Manufacturing Cell Therapies: The Paradigm Shift in Health Care of This Century. *NAM Perspect.* **2017**, *7*. [CrossRef]
7. Somoza, R.A.; Welter, J.F.; Correa, D.; Caplan, A.I. Chondrogenic Differentiation of Mesenchymal Stem Cells: Challenges and Unfulfilled Expectations. *Tissue Eng. Part B Rev.* **2014**, *20*, 596–608. [CrossRef] [PubMed]
8. Barry, F. MSC Therapy for Osteoarthritis: An Unfinished Story. *J. Orthop. Res.* **2019**, 1229–1235. [CrossRef]
9. Goldring, M.B. Chondrogenesis, Chondrocyte Differentiation, and Articular Cartilage Metabolism in Health and Osteoarthritis. *Ther. Adv. Musculoskelet. Dis.* **2012**, *4*, 269–285. [CrossRef] [PubMed]
10. Pittenger, M.F.; Discher, D.E.; Péault, B.M.; Phinney, D.G.; Hare, J.M.; Caplan, A.I. Mesenchymal Stem Cell Perspective: Cell Biology to Clinical Progress. *NPJ Regen. Med.* **2019**, 1–15. [CrossRef] [PubMed]
11. Kim, K.; Bou-Ghannam, S.; Kameishi, S.; Oka, M.; Grainger, D.W.; Okano, T. Allogeneic Mesenchymal Stem Cell Sheet Therapy: A New Frontier in Drug Delivery Systems. *J. Control. Release* **2021**, *330*, 696–704. [CrossRef] [PubMed]
12. Thorp, H.; Kim, K.; Kondo, M.; Grainger, D.W.; Okano, T. Fabrication of Hyaline-like Cartilage Constructs Using Mesenchymal Stem Cell Sheets. *Sci. Rep.* **2020**, *10*. [CrossRef]
13. Blain, E.J. Involvement of the Cytoskeletal Elements in Articular Cartilage Homeostasis and Pathology. *Int. J. Exp. Pathol.* **2009**, 1–15. [CrossRef]
14. Johnstone, B.; Hering, T.M.; Caplan, A.I.; Goldberg, V.M.; Yoo, J.U. In Vitro Chondrogenesis of Bone Marrow-Derived Mesenchymal Progenitor Cells. *Exp. Cell Res.* **1998**, *238*, 265–272. [CrossRef] [PubMed]
15. Pittenger, M.F.; Mackay, A.M.; Beck, S.C.; Jaiswal, R.K.; Douglas, R.; Mosca, J.D.; Moorman, M.A.; Simonetti, D.W.; Craig, S.; Marshak, D.R. Multilineage Potential of Adult Human Mesenchymal Stem Cells. *Science* **1999**, *284*, 143–147. [CrossRef]
16. Mackay, A.M.; Beck, S.C.; Murphy, J.M.; Barry, F.P.; Chichester, C.O.; Pittenger, M.F. Chondrogenic Differentiation of Cultured Human Mesenchymal Stem Cells from Marrow. *Tissue Eng.* **1998**, *4*, 415–428. [CrossRef]
17. Merceron, C.; Portron, S.; Masson, M.; Lesoeur, J.; Fellah, B.H.; Gauthier, O.; Geffroy, O.; Weiss, P.; Guicheux, J.; Vinatier, C. The Effect of Two- and Three-Dimensional Cell Culture on the Chondrogenic Potential of Human Adipose-Derived Mesenchymal Stem Cells after Subcutaneous Transplantation with an Injectable Hydrogel. *Cell Transplant.* **2011**, *20*, 1575–1588. [CrossRef] [PubMed]
18. Zhang, T.; Wen, F.; Wu, Y.; Goh, G.S.H.; Ge, Z.; Tan, L.P.; Hui, J.H.P.; Yang, Z. Cross-Talk between TGF-Beta/SMAD and Integrin Signaling Pathways in Regulating Hypertrophy of Mesenchymal Stem Cell Chondrogenesis under Deferral Dynamic Compression. *Biomaterials* **2015**, *38*, 72–85. [CrossRef]
19. Goldberg, A.; Mitchell, K.; Soans, J.; Kim, L.; Zaidi, R. The Use of Mesenchymal Stem Cells for Cartilage Repair and Regeneration: A Systematic Review. *J. Orthop. Surg. Res.* **2017**, *12*, 1–30. [CrossRef]
20. Gomez-Salazar, M.; Gonzalez-Galofre, Z.N.; Casamitjana, J.; Crisan, M.; James, A.W.; Péault, B. Five Decades Later, Are Mesenchymal Stem Cells Still Relevant? *Front. Bioeng. Biotechnol.* **2020**, 148. [CrossRef] [PubMed]
21. Negoro, T.; Takagaki, Y.; Okura, H.; Matsuyama, A. Trends in Clinical Trials for Articular Cartilage Repair by Cell Therapy. *NPJ Regen. Med.* **2018**, *3*, 17. [CrossRef] [PubMed]
22. Kon, E.; Filardo, G.; Di Martino, A.; Marcacci, M. ACI and MACI. *J. Knee Surg.* **2012**, 17–22. [CrossRef]
23. Zhang, L.; Hu, J.; Athanasiou, K.A. The Role of Tissue Engineering in Articular Cartilage Repair and Regeneration. *Crit. Rev. Biomed. Eng.* **2009**, *37*, 1–57. [CrossRef] [PubMed]
24. Sato, M.; Yamato, M.; Hamahashi, K.; Okano, T.; Mochida, J. Articular Cartilage Regeneration Using Cell Sheet Technology. *Anat. Rec.* **2014**, *297*, 36–43. [CrossRef]
25. Kondo, M.; Kameishi, S.; Grainger, D.W.; Okano, T. Novel Therapies Using Cell Sheets Engineered from Allogeneic Mesenchymal Stem/Stromal Cells. *Emerg. Top. Life Sci.* **2020**. [CrossRef]
26. Benjamin, M.; Ralphs, J. Biology of Fibrocartilage Cells. *Int. Rev. Cytol.* **2004**, *233*, 1–45. [CrossRef]
27. Armiento, A.R.; Alini, M.; Stoddart, M.J. Articular Fibrocartilage–Why Does Hyaline Cartilage Fail to Repair? *Adv. Drug Deliv. Rev.* **2019**, *146*, 289–305. [CrossRef] [PubMed]
28. Minas, T. A Primer in Cartilage Repair. *J. Bone Jt. Surg. Br.* **2012**, *94*, 1482–1486. [CrossRef] [PubMed]
29. Sophia Fox, A.J.; Bedi, A.; Rodeo, S.A. The Basic Science of Articular Cartilage: Structure, Composition, and Function. *Sports Health* **2009**, *1*, 461–468. [CrossRef] [PubMed]
30. Kamisan, N.; Naveen, S.V.; Ahmad, R.E.; Tunku, K. Chondrocyte Density, Proteoglycan Content and Gene Expressions from Native Cartilage Are Species Specific and Not Dependent on Cartilage Thickness: A Comparative Analysis between Rat, Rabbit and Goat. *BMC Vet. Res.* **2013**, *9*. [CrossRef]
31. Armiento, A.R.; Stoddart, M.J.; Alini, M.; Eglin, D. Biomaterials for Articular Cartilage Tissue Engineering: Learning from Biology. *Acta Biomater.* **2018**, *65*, 1–20. [CrossRef]
32. Gharpuray, V.M. Fibrocartilage. In *Handbook of Biomaterial Properties*, 2nd ed.; Springer: New York, NY, USA, 2016; Volume 171, pp. 45–54. [CrossRef]
33. Ferket, B.S.; Feldman, Z.; Zhou, J.; Oei, E.H.; Bierma-Zeinstra, S.M.A.; Mazumdar, M. Impact of Total Knee Replacement Practice: Cost Effectiveness Analysis of Data from the Osteoarthritis Initiative. *BMJ* **2017**, *356*, 1–12. [CrossRef] [PubMed]
34. Schinhan, M.; Gruber, M.; Vavken, P.; Dorotka, R.; Samouh, L.; Chiari, C.; Gruebl-Barabas, R.; Nehrer, S. Critical-Size Defect Induces Unicompartmental Osteoarthritis in a Stable Ovine Knee. *J. Orthop. Res.* **2012**, *30*, 214–220. [CrossRef] [PubMed]
35. Houck, D.A.; Kraeutler, M.J.; Belk, J.W.; Frank, R.M.; McCarty, E.C.; Bravman, J.T. Do Focal Chondral Defects of the Knee Increase the Risk for Progression to Osteoarthritis? A Review of the Literature. *Orthop. J. Sports Med.* **2018**. [CrossRef]

36. Gomoll, A.H.; Farr, J.; Gillogly, S.D.; Kercher, J.; Minas, T. Surgical Management of Articular Cartilage Defects of the Knee. *J. Bone Jt. Surg. Am.* **2010**, *92*, 2470–2490.
37. Akkiraju, H.; Nohe, A. Role of Chondrocytes in Cartilage Formation, Progression of Osteoarthritis and Cartilage Regeneration. *J. Dev. Biol.* **2016**, *3*, 177. [CrossRef] [PubMed]
38. Marcacci, M.; Filardo, G.; Kon, E. Treatment of Cartilage Lesions: What Works and Why? *Injury* **2013**, *44* (Suppl. 1), S11–S15. [CrossRef]
39. Steadman, J.R.; Rodkey, W.G.; Singleton, S.B.; Briggs, K.K. Microfracture Technique for Full-Thickness Chondral Defects: Technique and Clinical Results. *Oper. Tech. Orthop.* **1997**, *7*, 300–304. [CrossRef]
40. Gao, L.; Orth, P.; Cucchiarini, M.; Madry, H. Autologous Matrix-Induced Chondrogenesis: A Systematic Review of the Clinical Evidence. *Am. J. Sports Med.* **2019**, 222–231. [CrossRef]
41. Mithoefer, K.; McAdams, T.; Williams, R.J.; Kreuz, P.C.; Mandelbaum, B.R. Clinical Efficacy of the Microfracture Technique for Articular Cartilage Repair in the Knee: An Evidence-Based Systematic Analysis. *Database Abstr. Rev. Eff. Qual. Rev.* **2009**, *38*. [CrossRef]
42. Solheim, E.; Hegna, J.; Inderhaug, E.; Øyen, J.; Harlem, T.; Strand, T. Results at 10–14 Years after Microfracture Treatment of Articular Cartilage Defects in the Knee. *Knee Surg. Sports Traumatol. Arthrosc.* **2016**, *24*, 1587–1593. [CrossRef]
43. Gobbi, A.; Karnatzikos, G.; Kumar, A. Long-Term Results after Microfracture Treatment for Full-Thickness Knee Chondral Lesions in Athletes. *Knee Surg. Sports Traumatol. Arthrosc.* **2014**, *22*, 1986–1996. [CrossRef]
44. Dewan, A.K.; Gibson, M.A.; Elisseeff, J.H.; Trice, M.E. Evolution of Autologous Chondrocyte Repair and Comparison to Other Cartilage Repair Techniques. *Biomed. Res. Int.* **2014**, *2014*, 1–11. [CrossRef] [PubMed]
45. McGowan, K.B.; Stiegman, G. Regulatory Challenges for Cartilage Repair Technologies. *Cartilage* **2013**, *4*, 4–11. [CrossRef]
46. Ahmed, T.A.E.; Hincke, M.T. Strategies for Articular Cartilage Lesion Repair and Functional Restoration. *Tissue Eng. Part B Rev.* **2010**, *16*, 305–329. [CrossRef] [PubMed]
47. Hunziker, E.B.; Lippuner, K.; Keel, M.J.B.; Shintani, N. An Educational Review of Cartilage Repair: Precepts & Practice-Myths & Misconceptions-Progress & Prospects. *Osteoarthr. Cartil.* **2015**, 334–350. [CrossRef]
48. Kraeutler, M.J.; Belk, J.W.; Purcell, J.M.; McCarty, E.C. Microfracture versus Autologous Chondrocyte Implantation for Articular Cartilage Lesions in the Knee: A Systematic Review of 5-Year Outcomes. *Am. J. Sports Med.* **2017**, *46*, 995. [CrossRef] [PubMed]
49. DiBartola, A.C.; Wright, B.M.; Magnussen, R.A.; Flanigan, D.C. Clinical Outcomes after Autologous Chondrocyte Implantation in Adolescents' Knees: A Systematic Review. *Arthrosc. J. Arthrosc. Relat. Surg.* **2016**, *32*, 1905–1916. [CrossRef]
50. FDA Approves First Autologous Cellularized Scaffold for the Repair of Cartilage Defects of the Knee. 2016. Available online: https://cartilage.org/news/fda-approves-first-autologous-cellularized-scaffold-for-the-repair-of-cartilage-defects-of-the-knee/ (accessed on 1 March 2021).
51. Jones, C.W.; Willers, C.; Keogh, A.; Smolinski, D.; Fick, D.; Yates, P.J.; Kirk, T.B.; Zheng, M.H. Matrix-Induced Autologous Chondrocyte Implantation in Sheep: Objective Assessments Including Confocal Arthroscopy. *J. Orthop. Res.* **2008**, *26*, 292–303. [CrossRef]
52. Ebert, J.R.; Robertson, W.B.; Woodhouse, J.; Fallon, M.; Zheng, M.H.; Ackland, T.; Wood, D.J. Clinical and Magnetic Resonance Imaging-Based Outcomes to 5 Years after Matrix-Induced Autologous Chondrocyte Implantation to Address Articular Cartilage Defects in the Knee. *Am. J. Sports Med.* **2011**, *39*, 753–763. [CrossRef]
53. Zheng, M.H.; Willers, C.; Kirilak, L.; Yates, P.; Xu, J.; Wood, D.; Shimmin, A. Matrix-Induced Autologous Chondrocyte Implantation (MACI®): Biological and Histological Assessment. *Tissue Eng.* **2007**, *13*, 737–746. [CrossRef]
54. Smith, H.J.; Richardson, J.B.; Tennant, A. Modification and Validation of the Lysholm Knee Scale to Assess Articular Cartilage Damage. *Osteoarthr. Cartil.* **2009**, *17*, 53–58. [CrossRef]
55. Barber-Westin, S.D.; Noyes, F.R. The International Knee Documentations Committee Rating System. In *Noyes' Knee Disorders: Surgery, Rehabilitation, Clinical Outcomes*; Elsevier: Amsterdam, The Netherlands, 2010; pp. e28–e47. [CrossRef]
56. Hambly, K. The Use of the Tegner Activity Scale for Articular Cartilage Repair of the Knee: A Systematic Review. *Knee Surg. Sports Traumatol. Arthrosc.* **2011**, *19*, 604–614. [CrossRef] [PubMed]
57. Gille, J.; Behrens, P.; Schulz, A.P.; Oheim, R.; Kienast, B. Matrix-Associated Autologous Chondrocyte Implantation: A Clinical Follow-Up at 15 Years. *Cartilage* **2016**, *7*, 309–315. [CrossRef] [PubMed]
58. Van den Borne, M.P.J.; Raijmakers, N.J.H.; Vanlauwe, J.; Victor, J.; de Jong, S.N.; Bellemans, J.; Saris, D.B.F. International Cartilage Repair Society (ICRS) and Oswestry Macroscopic Cartilage Evaluation Scores Validated for Use in Autologous Chondrocyte Implantation (ACI) and Microfracture. *Osteoarthr. Cartil.* **2007**, *15*, 1397–1402. [CrossRef]
59. Bartlett, W.; Skinner, J.A.; Gooding, C.R.; Carrington, R.W.J.; Flanagan, A.M.; Briggs, T.W.R.; Bentley, G. Autologous Chondrocyte Implantation versus Matrix-Induced Autologous Chondrocyte Implantation for Osteochondral Defects of the Knee. A Prospective, Randomised Study. *J. Bone Jt. Surg. Ser. B* **2005**, *87*, 640–645. [CrossRef]
60. Zeifang, F.; Oberle, D.; Nierhoff, C.; Richter, W.; Moradi, B.; Schmitt, H. Autologous Chondrocyte Implantation Using the Original Periosteum-Cover Technique versus Matrix-Associated Autologous Chondrocyte Implantation: A Randomized Clinical Trial. *Am. J. Sports Med.* **2010**, *38*, 924–933. [CrossRef]
61. Kim, Y.S.; Smoak, M.M.; Melchiorri, A.J.; Mikos, A.G. An Overview of the Tissue Engineering Market in the United States from 2011 to 2018. *Tissue Eng. Part A* **2019**, *25*, 1–8. [CrossRef]

62. BlueCross BlueShield. *ACI (Carticel) Phased Out*; BlueCross BlueShield: Chicago, IL, USA, 2018; Available online: https://www.bluecrossnc.com/sites/default/files/document/attachment/services/public/pdfs/medicalpolicy/autologous_chondrocyte_implantation.pdf (accessed on 1 March 2021).
63. Rheinstein, P.H.; Adbari, B. Significant FDA Approvals in 1997. *Am. Fam. Physician* **1998**, *57*, 2865.
64. Ministry of Food and Drug Safety. Chondron | Ministry of Food and Drug Safety. Available online: https://www.mfds.go.kr/eng/brd/m_30/view.do?seq=70954 (accessed on 1 March 2021).
65. European Medicines Agency. *ChondroCelect | European Medicines Agency (Withdrawal)*; European Medicines Agency: Amsterdam, The Netherlands, 2016; Available online: https://www.ema.europa.eu/en/documents/public-statement/public-statement-chondrocelect-withdrawal-marketing-authorisation-european-unionen.pdf (accessed on 1 March 2021).
66. European Medicines Agency. ChondroCelect | European Medicines Agency. Available online: https://www.ema.europa.eu/en/medicines/human/EPAR/chondrocelect (accessed on 1 March 2021).
67. Medipost. MEDIPOST Clears Product Approval Renewal of CARTISTEM by the Ministry of Food and Drug Safety (MFDS). Available online: http://www.medi-post.com/stem-cell-therapeutic/2019/03/36443/ (accessed on 1 March 2021).
68. Sakakibara, N. J-TEC Received Government Approval to Manufacture and Sell Autologous Cultured Cartilage JACC®in Japan. 2012. Available online: http://www.jpte.co.jp/english/ir/library/JACC_20120730E.pdf (accessed on 1 March 2021).
69. NIHR. *NOVOCART 3D for Articular Cartilage Defects of the Knee*; NIHR: London, UK, 2019. Available online: http://www.io.nihr.ac.uk/wp-content/uploads/2019/09/13181-Autologous-Chondrocyte-Implant-for-Articular-Cartilage-Defects-V1.0-SEP2019-NON-CONF.pdf (accessed on 1 March 2021).
70. European Medicines Agency. Maci | European Medicines Agency. Available online: https://www.ema.europa.eu/en/medicines/human/EPAR/maci (accessed on 1 March 2021).
71. U.S. Food and Drug Administration. MACI (Autologous Cultured Chondrocytes on a Porcine Collagen Membrane) | FDA. Available online: https://www.fda.gov/vaccines-blood-biologics/cellular-gene-therapy-products/maci-autologous-cultured-chondrocytes-porcine-collagen-membrane (accessed on 1 March 2021).
72. Australian Department of Health-Therapeutic Goods Administration. Ortho-ACI | Australian Department of Health. Available online: https://www.ebs.tga.gov.au/servlet/xmlmillr6?dbid=ebs/PublicHTML/pdfStore.nsf&docid=289402&agid=(PrintDetailsPublic)&actionid=1 (accessed on 1 March 2021).
73. European Medicines Agency. Spherox | European Medicines Agency. Available online: https://www.ema.europa.eu/en/medicines/human/EPAR/spherox (accessed on 1 March 2021).
74. prnewswire.com. Korea Approves the World's First Cell and Gene Therapy for Knee Osteoarthritis. Available online: https://www.prnewswire.com/news-releases/korea-approves-the-worlds-first-cell-and-gene-therapy-for-knee-osteoarthritis-300486969.html (accessed on 1 March 2021).
75. Won Shin, J. Invossa Approval Revoked As Korea Confirms False Data Submission. 2019. Available online: https://pink.pharmaintelligence.informa.com/PS125369 (accessed on 1 March 2021).
76. Crawford, D.; Deberardino, T.; Williams, R. NeoCart, an Autologous Cartilage Tissue Implant, Compared with Microfracture for Treatment of Distal Femoral Cartilage Lesions. *J. Bone Jt. Surg.* **2012**, *94*, 979–989. [CrossRef] [PubMed]
77. Simaria, A.S.; Hassan, S.; Varadaraju, H.; Rowley, J.; Warren, K.; Vanek, P.; Farid, S.S. Allogeneic Cell Therapy Bioprocess Economics and Optimization: Single-Use Cell Expansion Technologies. *Biotechnol. Bioeng.* **2014**, *111*, 69–83. [CrossRef]
78. Heathman, T.R.; Nienow, A.W.; McCall, M.J.; Coopman, K.; Kara, B.; Hewitt, C.J. The Translation of Cell-Based Therapies: Clinical Landscape and Manufacturing Challenges. *Regen. Med.* **2015**, 49–64. [CrossRef]
79. Trounson, A.; McDonald, C. Stem Cell Therapies in Clinical Trials: Progress and Challenges. *Cell Stem Cell* **2015**, *17*. [CrossRef]
80. Mekhileri, N.V.; Lim, K.S.; Brown, G.C.J.; Mutreja, I.; Schon, B.S.; Hooper, G.J.; Woodfield, T.B.F. Automated 3D Bioassembly of Micro-Tissues for Biofabrication of Hybrid Tissue Engineered Constructs. *Biofabrication* **2018**, *10*. [CrossRef]
81. Musumeci, G.; Loreto, C.; Castorina, S.; Imbesi, R.; Leonardi, R.; Castrogiovanni, P. Current Concepts in the Treatment of Cartilage Damage. A Review. *Ital. J. Anat. Embryol.* **2013**, *118*, 189–203.
82. Park, Y.-B.; Ha, C.-W.; Rhim, J.H.; Lee, H.-J. Stem Cell Therapy for Articular Cartilage Repair: Review of the Entity of Cell Populations Used and the Result of the Clinical Application of Each Entity. *Am. J. Sports Med.* **2018**, *46*, 2540–2552. [CrossRef] [PubMed]
83. Kim, K.; Bou-Ghannam, S.; Thorp, H.; Grainger, D.W.; Okano, T. Human Mesenchymal Stem Cell Sheets in Xeno-Free Media for Possible Allogenic Applications. *Sci. Rep.* **2019**, *9*, 1–12. [CrossRef] [PubMed]
84. León, S.A.; Mei, X.Y.; Safir, O.A.; Gross, A.E.; Kuzyk, P.R. Long-Term Results of Fresh Osteochondral Allografts and Realignment Osteotomy for Cartilage Repair in the Knee. *Bone Jt. J.* **2019**, *101-B*, 46–52. [CrossRef] [PubMed]
85. Kot, M.; Baj-Krzyworzeka, M.; Szatanek, R.; Musiał-Wysocka, A.; Suda-Szczurek, M.; Majka, M. The Importance of HLA Assessment in "off-the-Shelf" Allogeneic Mesenchymal Stem Cells Based-Therapies. *Int. J. Mol. Sci.* **2019**, *20*, 5680. [CrossRef]
86. García-Sancho, J.; Sánchez, A.; Vega, A.; Noriega, D.C.; Nocito, M. Influence of HLA Matching on the Efficacy of Allogeneic Mesenchymal Stromal Cell Therapies for Osteoarthritis and Degenerative Disc Disease. *Transplant. Direct* **2017**, *3*, e205. [CrossRef] [PubMed]
87. Caplan, A.I. Mesenchymal Stem Cells. *J. Orthop. Res.* **1991**, *9*, 641–650. [CrossRef] [PubMed]
88. Caplan, A.I.; Sorrell, J.M. The MSC Curtain That Stops the Immune System. *Immunol. Lett.* **2015**. [CrossRef]

89. Dominici, M.; Le Blanc, K.; Mueller, I.; Slaper-Cortenbach, I.; Marini, F.; Krause, D.S.; Deans, R.J.; Keating, A.; Prockop, D.J.; Horwitz, E.M. Minimal Criteria for Defining Multipotent Mesenchymal Stromal Cells. The International Society for Cellular Therapy Position Statement. *Cytotherapy* **2006**, *8*, 315–317. [CrossRef]
90. Harrell, C.R.; Markovic, B.S.; Fellabaum, C.; Arsenijevic, A.; Volarevic, V. Mesenchymal Stem Cell-Based Therapy of Osteoarthritis: Current Knowledge and Future Perspectives. *Biomed. Pharmacother.* **2019**, *109*, 2318–2326. [CrossRef] [PubMed]
91. Lee, W.Y.; Wang, B. Cartilage Repair by Mesenchymal Stem Cells: Clinical Trial Update and Perspectives. *J. Orthop. Transl.* **2017**, 76–88. [CrossRef] [PubMed]
92. Schachtele, S.; Clouser, C.; Aho, J. Markers and Methods to Validate Mesenchymal Stem Cell Identity, Potency, and Quality. *R&D Syst. White Pap.* **2018**. Available online: https://www.rndsystems.com/resources/articles/markers-and-methods-verify-mesenchymal-stem-cell-identity-potency-and-quality (accessed on 4 February 2021).
93. Martin, C.; Olmos, É.; Collignon, M.L.; De Isla, N.; Blanchard, F.; Chevalot, I.; Marc, A.; Guedon, E. Revisiting MSC Expansion from Critical Quality Attributes to Critical Culture Process Parameters. *Process Biochem.* **2017**, *59*, 231–243. [CrossRef]
94. Bravery, C.A.; Carmen, J.; Fong, T.; Oprea, W.; Hoogendoorn, K.H.; Woda, J.; Burger, S.R.; Rowley, J.A.; Bonyhadi, M.L.; Van'T Hof, W. Potency Assay Development for Cellular Therapy Products: An ISCT* Review of the Requirements and Experiences in the Industry. *Cytotherapy* **2013**, 9–19.e9. [CrossRef] [PubMed]
95. 95. FDA.; CBER. Draft Guidance for Industry: Potency Tests for Cellular and Gene Therapy Products. *Biotechnol. Law Rep.* **2011**. [CrossRef]
96. Squillaro, T.; Peluso, G.; Galderisi, U. Clinical Trials with Mesenchymal Stem Cells: An Update. *Cell Transplant.* **2016**, *25*, 829–848. [CrossRef] [PubMed]
97. Merckx, G.; Bronckaers, A.; Hilkens, P.; Ratajczak, J.; Lo Monaco, M.; Clegg, P.; Vandeweerd, J.-M.; Gervois, P.; Lambrichts, I. Stem Cells for Cartilage Repair: Preclinical Studies and Insights in Translational Animal Models and Outcome Measures. *Stem Cells Int.* **2018**, *2018*, 1–22. [CrossRef]
98. Saidova, A.A.; Vorobjev, I.A. Lineage Commitment, Signaling Pathways, and the Cytoskeleton Systems in Mesenchymal Stem Cells. *Tissue Eng. Part B Rev.* **2019**, *26*, 13–25. [CrossRef]
99. Yamashita, A.; Nishikawa, S.; Rancourt, D.E. Identification of Five Developmental Processes during Chondrogenic Differentiation of Embryonic Stem Cells. *PLoS ONE* **2010**, *5*, e10998. [CrossRef] [PubMed]
100. Murdoch, A.D.; Grady, L.M.; Ablett, M.P.; Katopodi, T.; Meadows, R.S.; Hardingham, T.E. Chondrogenic Differentiation of Human Bone Marrow Stem Cells in Transwell Cultures: Generation of Scaffold-Free Cartilage. *Stem Cells* **2007**, *25*, 2786–2796. [CrossRef]
101. Xu, J.; Wang, W.; Ludeman, M.; Cheng, K.; Hayami, T.; Lotz, J.C.; Kapila, S. Chondrogenic Differentiation of Human Mesenchymal Stem Cells in Three-Dimensional Alginate Gels. *Tissue Eng. Part A* **2008**, *14*, 667–680. [CrossRef] [PubMed]
102. Liu, Y.; Jiang, X.; Zhang, X.; Chen, R.; Sun, T.; Fok, K.L.; Dong, J.; Tsang, L.L.; Yi, S.; Ruan, Y.; et al. Dedifferentiation-Reprogrammed Mesenchymal Stem Cells with Improved Therapeutic Potential. *Stem Cells* **2011**, *29*, 2077–2089. [CrossRef]
103. Mueller, M.B.; Fischer, M.; Zellner, J.; Berner, A.; Dienstknecht, T.; Prantl, L.; Kujat, R.; Nerlich, M.; Tuan, R.S.; Angele, P. Hypertrophy in Mesenchymal Stem Cell Chondrogenesis: Effect of TGF-β Isoforms and Chondrogenic Conditioning. *Cells Tissues Organs* **2010**, *192*, 158–166. [CrossRef] [PubMed]
104. Deng, Y.; Lei, G.; Lin, Z.; Yang, Y.; Lin, H.; Tuan, R.S. Engineering Hyaline Cartilage from Mesenchymal Stem Cells with Low Hypertrophy Potential via Modulation of Culture Conditions and Wnt/β-Catenin Pathway. *Biomaterials* **2019**, *192*, 569–578. [CrossRef]
105. DeLise, A.M.; Fischer, L.; Tuan, R.S. Cellular Interactions and Signaling in Cartilage Development. *Osteoarthr. Cartil.* **2000**, *8*, 309–334. [CrossRef]
106. Watts, A.E.; Ackerman-Yost, J.C.; Nixon, A.J. A Comparison of Three-Dimensional Culture Systems to Evaluate in Vitro Chondrogenesis of Equine Bone Marrow-Derived Mesenchymal Stem Cells. *Tissue Eng. Part A* **2013**, *19*, 2275–2283. [CrossRef]
107. Li, Y.Y.; Choy, T.H.; Ho, F.C.; Chan, P.B. Scaffold Composition Affects Cytoskeleton Organization, Cell-Matrix Interaction and the Cellular Fate of Human Mesenchymal Stem Cells upon Chondrogenic Differentiation. *Biomaterials* **2015**, *52*, 208–220. [CrossRef] [PubMed]
108. Sasazaki, Y.; Seedhom, B.B.; Shore, R. Morphology of the Bovine Chondrocyte and of Its Cytoskeleton in Isolation and in Situ: Are Chondrocytes Ubiquitously Paired through the Entire Layer of Articular Cartilage? *Rheumatology* **2008**, *47*, 1641–1646. [CrossRef]
109. McMurtrey, R.J. Analytic Models of Oxygen and Nutrient Diffusion, Metabolism Dynamics, and Architecture Optimization in Three-Dimensional Tissue Constructs with Applications and Insights in Cerebral Organoids. *Tissue Eng. Part C Methods* **2016**, *22*, 221–249. [CrossRef]
110. Samal, J.R.K.; Rangasami, V.K.; Samanta, S.; Varghese, O.P.; Oommen, O.P. Discrepancies on the Role of Oxygen Gradient and Culture Condition on Mesenchymal Stem Cell Fate. *Adv. Healthc. Mater.* **2021**, 2002058. [CrossRef] [PubMed]
111. Shang, J.; Liu, H.; Li, J.; Zhou, Y. Roles of Hypoxia During the Chondrogenic Differentiation of Mesenchymal Stem Cells. *Curr. Stem Cell Res. Ther.* **2014**, *9*, 141–147. [CrossRef]
112. Bae, H.C.; Park, H.J.; Wang, S.Y.; Yang, H.R.; Lee, M.C.; Han, H.S. Hypoxic Condition Enhances Chondrogenesis in Synovium-Derived Mesenchymal Stem Cells. *Biomater. Res.* **2018**, *22*, 28. [CrossRef]
113. Malladi, P.; Xu, Y.; Chiou, M.; Giaccia, A.J.; Longaker, M.T. Effect of Reduced Oxygen Tension on Chondrogenesis and Osteogenesis in Adipose-Derived Mesenchymal Cells. *Am. J. Physiol. Cell Physiol.* **2006**, *290*. [CrossRef] [PubMed]

114. Lafont, J.E.; Talma, S.; Hopfgarten, C.; Murphy, C.L. Hypoxia Promotes the Differentiated Human Articular Chondrocyte Phenotype through SOX9-Dependent and -Independent Pathways. *J. Biol. Chem.* **2008**, *283*, 4778–4786. [CrossRef]
115. Vater, C.; Kasten, P.; Stiehler, M. Culture Media for the Differentiation of Mesenchymal Stromal Cells. *Acta Biomater.* **2011**, 463–477. [CrossRef] [PubMed]
116. Achilli, T.M.; Meyer, J.; Morgan, J.R. Advances in the Formation, Use and Understanding of Multi-Cellular Spheroids. *Expert Opin. Biol. Ther.* **2012**, 1347–1360. [CrossRef]
117. Sekiya, I.; Vuoristo, J.T.; Larson, B.L.; Prockop, D.J. In Vitro Cartilage Formation by Human Adult Stem Cells from Bone Marrow Stroma Defines the Sequence of Cellular and Molecular Events during Chondrogenesis. *Proc. Natl. Acad. Sci. USA* **2002**, *99*, 4397–4402. [CrossRef]
118. Zhang, L.; Su, P.; Xu, C.; Yang, J.; Yu, W.; Huang, D. Chondrogenic Differentiation of Human Mesenchymal Stem Cells: A Comparison between Micromass and Pellet Culture Systems. *Biotechnol. Lett.* **2010**, *32*, 1339–1346. [CrossRef]
119. Tamamura, Y.; Otani, T.; Kanatani, N.; Koyama, E.; Kitagaki, J.; Komori, T.; Yamada, Y.; Costantini, F.; Wakisaka, S.; Pacifici, M.; et al. Developmental Regulation of Wnt/β-Catenin Signals Is Required for Growth Plate Assembly, Cartilage Integrity, and Endochondral Ossification. *J. Biol. Chem.* **2005**, *280*, 19185–19195. [CrossRef] [PubMed]
120. Li, J.; Dong, S. The Signaling Pathways Involved in Chondrocyte Differentiation and Hypertrophic Differentiation. *Stem Cells Int.* **2016**, *2016*, 1–12. [CrossRef] [PubMed]
121. Dong, Y.F.; Soung, D.Y.; Schwarz, E.M.; O'Keefe, R.J.; Drissi, H. Wnt Induction of Chondrocyte Hypertrophy through the Runx2 Transcription Factor. *J. Cell. Physiol.* **2006**, *208*, 77–86. [CrossRef]
122. Madeira, C.; Santhagunam, A.; Salgueiro, J.B.; Cabral, J.M.S. Advanced Cell Therapies for Articular Cartilage Regeneration. *Trends Biotechnol.* **2015**, 35–42. [CrossRef]
123. Vinatier, C.; Mrugala, D.; Jorgensen, C.; Guicheux, J.; Noël, D. Cartilage Engineering: A Crucial Combination of Cells, Biomaterials and Biofactors. *Trends Biotechnol.* **2009**, 307–314. [CrossRef]
124. Tatman, P.D.; Gerull, W.; Sweeney-Easter, S.; Davis, J.I.; Gee, A.O.; Kim, D.-H. Multiscale Biofabrication of Articular Cartilage: Bioinspired and Biomimetic Approaches. *Tissue Eng. Part B Rev.* **2015**, *21*, 543–559. [CrossRef] [PubMed]
125. Du, Y.; Liu, H.; Yang, Q.; Wang, S.; Wang, J.; Ma, J.; Noh, I.; Mikos, A.G.; Zhang, S. Selective Laser Sintering Scaffold with Hierarchical Architecture and Gradient Composition for Osteochondral Repair in Rabbits. *Biomaterials* **2017**, *137*, 37–48. [CrossRef] [PubMed]
126. Di Bella, C.; Duchi, S.; O'Connell, C.D.; Blanchard, R.; Augustine, C.; Yue, Z.; Thompson, F.; Richards, C.; Beirne, S.; Onofrillo, C.; et al. In Situ Handheld Three-Dimensional Bioprinting for Cartilage Regeneration. *J. Tissue Eng. Regen. Med.* **2018**, *12*, 611–621. [CrossRef]
127. Ansari, M.; Eshghanmalek, M. Biomaterials for Repair and Regeneration of the Cartilage Tissue. *Bio-Des. Manuf.* **2019**, *2*, 41–49. [CrossRef]
128. Bernhard, J.C.; Vunjak-Novakovic, G. Should We Use Cells, Biomaterials, or Tissue Engineering for Cartilage Regeneration? *Stem Cell Res. Ther.* **2016**, 56. [CrossRef]
129. Campos, Y.; Almirall, A.; Fuentes, G.; Bloem, H.L.; Kaijzel, E.L.; Cruz, L.J. Tissue Engineering: An Alternative to Repair Cartilage. *Tissue Eng. Part B Rev.* **2019**, 357–373. [CrossRef] [PubMed]
130. Bobacz, K.; Erlacher, L.; Smolen, J.; Soleiman, A.; Graninger, W.B. Chondrocyte Number and Proteoglycan Synthesis in the Aging and Osteoarthritic Human Articular Cartilage. *Ann. Rheum. Dis.* **2004**, *63*, 1618–1622. [CrossRef] [PubMed]
131. Mahboudi, H.; Kazemi, B.; Soleimani, M. Comparison between High Cell-Density Culture Systems for Chondrogenic Differentiation and Articular Cartilage Reconstruction of Human Mesenchymal Stem Cells: A Literature Review. *Regen. Reconstr. Restor.* **2017**, *2*, 7–15. [CrossRef]
132. Mcbeath, R.; Pirone, D.M.; Nelson, C.M.; Bhadriraju, K.; Chen, C.S. Cell Shape, Cytoskeletal Tension, and RhoA Regulate Stem Cell Lineage Commitment. *Dev. Cell* **2004**, *6*, 483–495. [CrossRef]
133. Jubel, A.; Andermahr, J.; Schiffer, G.; Fischer, J.; Rehm, K.E.; Stoddart, M.J.; Häuselmann, H.J. Transplantation of de Novo Scaffold-Free Cartilage Implants into Sheep Knee Chondral Defects. *Am. J. Sports Med.* **2008**, *36*, 1555–1564. [CrossRef]
134. Jang, J.; Lee, J.; Lee, E.; Lee, E.A.; Son, Y. Disc-Type Hyaline Cartilage Reconstruction Using 3D-Cell Sheet Culture of Human Bone Marrow Stromal Cells and Human Costal Chondrocytes and Maintenance of Its Shape and Phenotype after Transplantation. *Tissue Eng. Regen. Med.* **2016**, *13*, 352–363. [CrossRef] [PubMed]
135. Itokazu, M.; Wakitani, S.; Mera, H.; Tamamura, Y.; Sato, Y.; Takagi, M.; Nakamura, H. Transplantation of Scaffold-Free Cartilage-Like Cell-Sheets Made from Human Bone Marrow Mesenchymal Stem Cells for Cartilage Repair: A Preclinical Study. *Cartilage* **2016**, *7*, 361–372. [CrossRef] [PubMed]
136. Elder, S.H.; Cooley, A.J.; Borazjani, A.; Sowell, B.L.; To, H.; Tran, S.C. Production of Hyaline-like Cartilage by Bone Marrow Mesenchymal Stem Cells in a Self-Assembly Model. *Tissue Eng. Part A* **2009**, *15*, 3025–3036. [CrossRef]
137. Wood, J.J.; Malek, M.A.; Frassica, F.J.; Polder, J.A.; Mohan, A.K.; Bloom, E.T.; Braun, M.M.; Coté, T.R. Autologous Cultured Chondrocytes: Adverse Events Reported to the United States Food and Drug Administration. *J. Bone Jt. Surg. Am.* **2006**, *88*, 503–507. [CrossRef]
138. Middendorf, J.M.; Griffin, D.J.; Shortkroff, S.; Dugopolski, C.; Kennedy, S.; Siemiatkoski, J.; Cohen, I.; Bonassar, L.J. Mechanical Properties and Structure-Function Relationships of Human Chondrocyte-Seeded Cartilage Constructs after in Vitro Culture. *J. Orthop. Res.* **2017**, *35*, 2298–2306. [CrossRef] [PubMed]

139. Meyer, U.; Wiesmann, H.P.; Libera, J.; Depprich, R.; Naujoks, C.; Handschel, J. Cartilage Defect Regeneration by Ex Vivo Engineered Autologous Microtissue-Preliminary Results. *In Vivo* **2012**, *26*, 251–257.
140. Iwasa, J.; Engebretsen, L.; Shima, Y.; Ochi, M. Clinical Application of Scaffolds for Cartilage Tissue Engineering. *Knee Surg. Sports Traumatol. Arthrosc.* **2009**, *17*, 561–577. [CrossRef]
141. Malinin, T.; Ouellette, E.A. Articular Cartilage Nutrition Is Mediated by Subchondral Bone: A Long-Term Autograft Study in Baboons. *Osteoarthr. Cartil.* **2000**, *8*, 483–491. [CrossRef]
142. Chu, C.R.; Szczodry, M.; Bruno, S. Animal Models for Cartilage Regeneration and Repair. *Tissue Eng. Part B Rev.* **2010**, *16*. [CrossRef] [PubMed]
143. Pfeifer, C.G.; Fisher, M.B.; Carey, J.L.; Mauck, R.L. Impact of Guidance Documents on Translational Large Animal Studies of Cartilage Repair. *Sci. Transl. Med.* **2015**, *7*, 310re9. [CrossRef] [PubMed]
144. Gregory, M.H.; Capito, N.; Kuroki, K.; Stoker, A.M.; Cook, J.L.; Sherman, S.L. A Review of Translational Animal Models for Knee Osteoarthritis. *Arthritis* **2012**. [CrossRef]
145. Goetz, J.E.; Fredericks, D.; Petersen, E.; Rudert, M.J.; Baer, T.; Swanson, E.; Roberts, N.; Martin, J.; Tochigi, Y. A Clinically Realistic Large Animal Model of Intra-Articular Fracture That Progresses to Post-Traumatic Osteoarthritis. *Osteoarthr. Cartil.* **2015**. [CrossRef]
146. Ahern, B.J.; Parvizi, J.; Boston, R.; Schaer, T.P. Preclinical Animal Models in Single Site Cartilage Defect Testing: A Systematic Review. *Osteoarthr. Cartil.* **2009**, *17*. [CrossRef] [PubMed]
147. Hoemann, C.; Kandel, R.; Roberts, S.; Saris, D.B.F.; Creemers, L.; Mainil-Varlet, P.; Méthot, S.; Hollander, A.P.; Buschmann, M.D. International Cartilage Repair Society (ICRS) Recommended Guidelines for Histological Endpoints for Cartilage Repair Studies in Animal Models and Clinical Trials. *Cartilage* **2011**, *2*, 153–172. [CrossRef]
148. Klein, J. Chemistry: Repair or Replacement–A Joint Perspective. *Science* **2009**, *323*, 47–48. [CrossRef]
149. Wong, B.L.; Bae, W.C.; Chun, J.; Gratz, K.R.; Lotz, M.; Sah, R.L. Biomechanics of Cartilage Articulation: Effects of Lubrication and Degeneration on Shear Deformation. *Arthritis Rheum.* **2008**, *58*, 2065–2074. [CrossRef] [PubMed]
150. Unsworth, A.; Dowson, D.; Wright, V. The Frictional Behavior of Human Synovial Joints–Part 1: Natural Joints. *J. Lubr. Technol.* **1975**, *97*, 369–376. [CrossRef]
151. Patel, J.M.; Wise, B.C.; Bonnevie, E.D.; Mauck, R.L. A Systematic Review and Guide to Mechanical Testing for Articular Cartilage Tissue Engineering. *Tissue Eng. Part C Methods* **2019**, 593–608. [CrossRef]
152. Nakagawa, Y.; Muneta, T.; Otabe, K.; Ozeki, N.; Mizuno, M.; Udo, M.; Saito, R.; Yanagisawa, K.; Ichinose, S.; Koga, H.; et al. Cartilage Derived from Bone Marrow Mesenchymal Stem Cells Expresses Lubricin in Vitro and in Vivo. *PLoS ONE* **2016**, *11*, e0148777. [CrossRef] [PubMed]
153. Lin, W.; Kluzek, M.; Iuster, N.; Shimoni, E.; Kampf, N.; Goldberg, R.; Klein, J. Cartilage-Inspired, Lipid-Based Boundary-Lubricated Hydrogels. *Science* **2020**, *370*, 335–338. [CrossRef] [PubMed]
154. Davatchi, F.; Abdollahi, B.S.; Mohyeddin, M.; Shahram, F.; Nikbin, B. Mesenchymal Stem Cell Therapy for Knee Osteoarthritis. Preliminary Report of Four Patients. *Int. J. Rheum. Dis.* **2011**, *14*, 211–215. [CrossRef]
155. Haraguchi, Y.; Shimizu, T.; Yamato, M.; Okano, T. Concise Review: Cell Therapy and Tissue Engineering for Cardiovascular Disease. *Stem Cells Transl. Med.* **2012**, *1*, 136–141. [CrossRef]
156. Yang, J.; Yamato, M.; Shimizu, T.; Sekine, H.; Ohashi, K.; Kanzaki, M.; Ohki, T.; Nishida, K.; Okano, T. Reconstruction of Functional Tissues with Cell Sheet Engineering. *Biomaterials* **2007**, 5033–5043. [CrossRef]
157. Yang, J.; Yamato, M.; Kohno, C.; Nishimoto, A.; Sekine, H.; Fukai, F.; Okano, T. Cell Sheet Engineering: Recreating Tissues without Biodegradable Scaffolds. *Biomaterials* **2005**, *26*. [CrossRef]
158. Li, M.; Ma, J.; Gao, Y.; Yang, L. Cell Sheet Technology: A Promising Strategy in Regenerative Medicine. *Cytotherapy* **2019**, *21*, 3–16. [CrossRef] [PubMed]
159. Owaki, T.; Shimizu, T.; Yamato, M.; Okano, T. Cell Sheet Engineering for Regenerative Medicine: Current Challenges and Strategies. *Biotechnol. J.* **2014**, 904–914. [CrossRef]
160. Yamada, N.; Okano, T.; Sakai, H.; Karikusa, F.; Sawasaki, Y.; Sakurai, Y. Thermo-responsive Polymeric Surfaces; Control of Attachment and Detachment of Cultured Cells. *Die Makromol. Chem. Rapid Commun.* **1990**, *11*, 571–576. [CrossRef]
161. Kobayashi, J.; Kikuchi, A.; Aoyagi, T.; Okano, T. Cell Sheet Tissue Engineering: Cell Sheet Preparation, Harvesting/Manipulation, and Transplantation. *J. Biomed. Mater. Res. Part A* **2019**, *107*, 955–967. [CrossRef]
162. Kushida, A.; Yamato, M.; Konno, C.; Kikuchi, A.; Sakurai, Y.; Okano, T. Decrease in Culture Temperature Releases Monolayer Endothelial Cell Sheets Together with Deposited Fibronectin Matrix from Temperature-Responsive Culture Surfaces. *J. Biomed. Mater. Res.* **1999**, *45*, 355–362. [CrossRef]
163. Matsuda, N.; Shimizu, T.; Yamato, M.; Okano, T. Tissue Engineering Based on Cell Sheet Technology. *Adv. Mater.* **2007**, *19*, 3089–3099. [CrossRef]
164. Nakao, M.; Kim, K.; Nagase, K.; Grainger, D.W.; Kanazawa, H.; Okano, T. Phenotypic Traits of Mesenchymal Stem Cell Sheets Fabricated by Temperature-Responsive Cell Culture Plate: Structural Characteristics of MSC Sheets. *Stem Cell Res. Ther.* **2019**, *10*, 353. [CrossRef]
165. Ingber, D.E.; Dike, L.; Hansen, L.; Karp, S.; Liley, H.; Maniotis, A.; McNamee, H.; Mooney, D.; Plopper, G.; Sims, J.; et al. Cellular Tensegrity: Exploring How Mechanical Changes in the Cytoskeleton Regulate Cell Growth, Migration, and Tissue Pattern during Morphogenesis. *Int. Rev. Cytol.* **1994**, *150*, 173–224. [CrossRef] [PubMed]

166. Wei, Q.; Reidler, D.; Shen, M.Y.; Huang, H. Keratinocyte Cytoskeletal Roles in Cell Sheet Engineering. *BMC Biotechnol.* **2013**, *13*, 17. [CrossRef] [PubMed]
167. Tsuda, Y.; Shimizu, T.; Yamato, M.; Kikuchi, A.; Sasagawa, T.; Sekiya, S.; Kobayashi, J.; Chen, G.; Okano, T. Cellular Control of Tissue Architectures Using a Three-Dimensional Tissue Fabrication Technique. *Biomaterials* **2007**, *28*. [CrossRef]
168. Kikuchi, T.; Shimizu, T.; Wada, M.; Yamato, M.; Okano, T. Automatic Fabrication of 3-Dimensional Tissues Using Cell Sheet Manipulator Technique. *Biomaterials* **2014**, *35*. [CrossRef]
169. Williams, C.; Xie, A.W.; Yamato, M.; Okano, T.; Wong, J.Y. Stacking of Aligned Cell Sheets for Layer-by-Layer Control of Complex Tissue Structure. *Biomaterials* **2011**, *32*, 5625–5632. [CrossRef]
170. Haraguchi, Y.; Shimizu, T.; Sasagawa, T.; Sekine, H.; Sakaguchi, K.; Kikuchi, T.; Sekine, W.; Sekiya, S.; Yamato, M.; Umezu, M.; et al. Fabrication of Functional Three-Dimensional Tissues by Stacking Cell Sheets in Vitro. *Nat. Protoc.* **2012**, *7*, 850–858. [CrossRef] [PubMed]
171. Sekine, W.; Haraguchi, Y.; Shimizu, T.; Yamato, M.; Umezawa, A.; Okano, T. Chondrocyte Differentiation of Human Endometrial Gland-Derived MSCs in Layered Cell Sheets. *Sci. World J.* **2013**, *2013*. [CrossRef] [PubMed]
172. Mitani, G.; Sato, M.; Lee, J.I.; Kaneshiro, N.; Ishihara, M.; Ota, N.; Kokubo, M.; Sakai, H.; Kikuchi, T.; Mochida, J. The Properties of Bioengineered Chondrocyte Sheets for Cartilage Regeneration. *BMC Biotechnol.* **2009**, *9*, 17. [CrossRef] [PubMed]
173. Kaneshiro, N.; Sato, M.; Ishihara, M.; Mitani, G.; Sakai, H.; Mochida, J. Bioengineered Chondrocyte Sheets May Be Potentially Useful for the Treatment of Partial Thickness Defects of Articular Cartilage. *Biochem. Biophys. Res. Commun.* **2006**, *349*, 723–731. [CrossRef]
174. Sekine, H.; Shimizu, T.; Dobashi, I.; Matsuura, K.; Hagiwara, N.; Takahashi, M.; Kobayashi, E.; Yamato, M.; Okano, T. Cardiac Cell Sheet Transplantation Improves Damaged Heart Function via Superior Cell Survival in Comparison with Dissociated Cell Injection. *Tissue Eng. Part A* **2011**, *17*, 2973–2980. [CrossRef]
175. Shimizu, R.; Kamei, N.; Adachi, N.; Hamanishi, M.; Kamei, G.; Mahmoud, E.E.; Nakano, T.; Iwata, T.; Yamato, M.; Okano, T.; et al. Repair Mechanism of Osteochondral Defect Promoted by Bioengineered Chondrocyte Sheet. *Tissue Eng. Part A* **2014**, *21*, 1131–1141. [CrossRef]
176. Ito, S.; Sato, M.; Yamato, M.; Mitani, G.; Kutsuna, T.; Nagai, T.; Ukai, T.; Kobayashi, M.; Kokubo, M.; Okano, T.; et al. Repair of Articular Cartilage Defect with Layered Chondrocyte Sheets and Cultured Synovial Cells. *Biomaterials* **2012**, *33*, 5278–5286. [CrossRef]
177. Ebihara, G.; Sato, M.; Yamato, M.; Mitani, G.; Kutsuna, T.; Nagai, T.; Ito, S.; Ukai, T.; Kobayashi, M.; Kokubo, M.; et al. Cartilage Repair in Transplanted Scaffold-Free Chondrocyte Sheets Using a Minipig Model. *Biomaterials* **2012**, *33*, 3846–3851. [CrossRef]
178. Hamahashi, K.; Sato, M.; Yamato, M.; Kokubo, M.; Mitani, G.; Ito, S.; Nagai, T.; Ebihara, G.; Kutsuna, T.; Okano, T.; et al. Studies of the Humoral Factors Produced by Layered Chondrocyte Sheets. *J. Tissue Eng. Regen. Med.* **2015**, *9*, 24–30. [CrossRef]
179. Kokubo, M.; Sato, M.; Yamato, M.; Mitani, G.; Kutsuna, T.; Ebihara, G.; Okano, T.; Mochida, J. Characterization of Chondrocyte Sheets Prepared Using a Co-Culture Method with Temperature-Responsive Culture Inserts. *J. Tissue Eng. Regen. Med.* **2016**, *10*, 486–495. [CrossRef]
180. Takahashi, T.; Sato, M.; Toyoda, E.; Maehara, M.; Takizawa, D.; Maruki, H.; Tominaga, A.; Okada, E.; Okazaki, K.; Watanabe, M. Rabbit Xenogeneic Transplantation Model for Evaluating Human Chondrocyte Sheets Used in Articular Cartilage Repair. *J. Tissue Eng. Regen. Med.* **2018**, *12*, 2067–2076. [CrossRef]
181. Takizawa, D.; Sato, M.; Okada, E.; Takahashi, T.; Maehara, M.; Tominaga, A.; Sogo, Y.; Toyoda, E.; Watanabe, M. Regenerative Effects of Human Chondrocyte Sheets in a Xenogeneic Transplantation Model Using Immune-Deficient Rats. *J. Tissue Eng. Regen. Med.* **2020**, *14*, 1296–1306. [CrossRef] [PubMed]
182. Sato, M.; Yamato, M.; Mitani, G.; Takagaki, T.; Hamahashi, K.; Nakamura, Y.; Ishihara, M.; Matoba, R.; Kobayashi, H.; Okano, T.; et al. Combined Surgery and Chondrocyte Cell-Sheet Transplantation Improves Clinical and Structural Outcomes in Knee Osteoarthritis. *NPJ Regen. Med.* **2019**, *4*. [CrossRef] [PubMed]
183. Takatori, N.; Sato, M.; Toyoda, E.; Takahashi, T.; Okada, E.; Maehara, M.; Watanabe, M. Cartilage Repair and Inhibition of the Progression of Cartilage Degeneration after Transplantation of Allogeneic Chondrocyte Sheets in a Nontraumatic Early Arthritis Model. *Regen. Ther.* **2018**, *9*, 24–31. [CrossRef] [PubMed]
184. Toyoda, E.; Sato, M.; Takahashi, T.; Maehara, M.; Okada, E.; Wasai, S.; Watanabe, M.; Iijima, H.; Nonaka, K.; Kawaguchi, Y. Transcriptomic and Proteomic Analyses Reveal the Potential Mode of Action of Chondrocyte Sheets in Hyaline Cartilage Regeneration. *Int. J. Mol. Sci.* **2020**, *21*, 149. [CrossRef] [PubMed]
185. Yokoyama, M.; Sato, M.; Umezawa, A.; Mitani, G.; Takagaki, T.; Yokoyama, M.; Kawake, T.; Okada, E.; Kokubo, M.; Ito, N.; et al. Assessment of the Safety of Chondrocyte Sheet Implantation for Cartilage Regeneration. *Tissue Eng. Part C Methods* **2016**, *22*, 59–68. [CrossRef]
186. Kim, K.; Bou-Ghannam, S.; Okano, T. Cell Sheet Tissue Engineering for Scaffold-Free Three-Dimensional (3D) Tissue Reconstruction. In *Methods in Cell Biology*; Academic Press Inc.: Cambridge, MA, USA; Elsevier: Cambridge, MA, USA, 2020; Volume 157, pp. 143–167. [CrossRef]
187. Patel, N.G.; Zhang, G. Stacked Stem Cell Sheets Enhance Cell-Matrix Interactions. *Organogenesis* **2014**, *10*, 170–176. [CrossRef]
188. Kim, K.; Utoh, R.; Ohashi, K.; Kikuchi, T.; Okano, T. Fabrication of Functional 3D Hepatic Tissues with Polarized Hepatocytes by Stacking Endothelial Cell Sheets in Vitro. *J. Tissue Eng. Regen. Med.* **2017**, *11*, 2071–2080. [CrossRef] [PubMed]

189. Maehara, M.; Sato, M.; Toyoda, E.; Takahashi, T.; Okada, E.; Kotoku, T.; Watanabe, M. Characterization of Polydactyly-Derived Chondrocyte Sheets versus Adult Chondrocyte Sheets for Articular Cartilage Repair. *Inflamm. Regen.* **2017**, *37*, 1–10. [CrossRef] [PubMed]
190. Madry, H.; Kaul, G.; Cucchiarini, M.; Stein, U.; Zurakowski, D.; Remberger, K.; Menger, M.D.; Kohn, D.; Trippel, S.B. Enhanced Repair of Articular Cartilage Defects in Vivo by Transplanted Chondrocytes Overexpressing Insulin-like Growth Factor I (IGF-I). *Gene Ther.* **2005**, *12*, 1171–1179. [CrossRef] [PubMed]
191. Alkaya, D.; Gurcan, C.; Kilic, P.; Yilmazer, A.; Gurman, G. Where Is Human-Based Cellular Pharmaceutical R&D Taking Us in Cartilage Regeneration? *3 Biotech* **2020**, 1–13. [CrossRef]
192. Farhang, N.; Davis, B.; Weston, J.; Ginley-Hidinger, M.; Gertz, J.; Bowles, R.D. Synergistic CRISPRa-Regulated Chondrogenic Extracellular Matrix Deposition without Exogenous Growth Factors. *Tissue Eng. Part A* **2020**, *26*, 1169–1179. [CrossRef]
193. Sridharan, B.; Sharma, B.; Detamore, M.S. A Road Map to Commercialization of Cartilage Therapy in the United States of America. *Tissue Eng. Part B Rev.* **2015**, *22*, 15–33. [CrossRef]
194. Schacker, M.; Seimetz, D. From Fiction to Science: Clinical Potentials and Regulatory Considerations of Gene Editing. *Clin. Transl. Med.* **2019**, *8*, 27. [CrossRef]

cells MDPI

Review

The Future of Regenerative Medicine: Cell Therapy Using Pluripotent Stem Cells and Acellular Therapies Based on Extracellular Vesicles

Margot Jarrige [1,2,3,†], Elie Frank [1,2,3,†], Elise Herardot [1,2,†], Sabrina Martineau [1,2], Annabelle Darle [1,2,3], Manon Benabides [1,2,3], Sophie Domingues [1,2,3], Olivier Chose [1,2,3], Walter Habeler [1,2,3], Judith Lorant [1,2], Christine Baldeschi [1,2], Cécile Martinat [1,2], Christelle Monville [1,2], Lise Morizur [1,2,3,*,‡] and Karim Ben M'Barek [1,2,3,*,‡]

[1] INSERM U861, I-Stem, AFM, Institute for Stem Cell Therapy and Exploration of Monogenic Diseases, 91100 Corbeil-Essonnes, France; mjarrige@istem.fr (M.J.); efrank@istem.fr (E.F.); eherardot@istem.fr (E.H.); smartineau@istem.fr (S.M.); adarle@istem.fr (A.D.); mbenabides@istem.fr (M.B.); sdomingues@istem.fr (S.D.); ochose@istem.fr (O.C.); whabeler@istem.fr (W.H.); jlorant@istem.fr (J.L.); cbaldeschi@istem.fr (C.B.); cmartinat@istem.fr (C.M.); cmonville@istem.fr (C.M.)
[2] Université Paris-Saclay, Université d'Evry, U861, 91100 Corbeil-Essonnes, France
[3] Centre d'Etude des Cellules Souches, 91100 Corbeil-Essonnes, France
* Correspondence: lmorizur@istem.fr (L.M.); kbenmbarek@istem.fr (K.B.M.)
† Equal contribution.
‡ These authors share equal senior authorship.

Citation: Jarrige, M.; Frank, E.; Herardot, E.; Martineau, S.; Darle, A.; Benabides, M.; Domingues, S.; Chose, O.; Habeler, W.; Lorant, J.; et al. The Future of Regenerative Medicine: Cell Therapy Using Pluripotent Stem Cells and Acellular Therapies Based on Extracellular Vesicles. *Cells* **2021**, *10*, 240. https://doi.org/10.3390/cells10020240

Academic Editor: Hélène Boeuf
Received: 30 November 2020
Accepted: 23 January 2021
Published: 27 January 2021

Abstract: The rapid progress in the field of stem cell research has laid strong foundations for their use in regenerative medicine applications of injured or diseased tissues. Growing evidences indicate that some observed therapeutic outcomes of stem cell-based therapy are due to paracrine effects rather than long-term engraftment and survival of transplanted cells. Given their ability to cross biological barriers and mediate intercellular information transfer of bioactive molecules, extracellular vesicles are being explored as potential cell-free therapeutic agents. In this review, we first discuss the state of the art of regenerative medicine and its current limitations and challenges, with particular attention on pluripotent stem cell-derived products to repair organs like the eye, heart, skeletal muscle and skin. We then focus on emerging beneficial roles of extracellular vesicles to alleviate these pathological conditions and address hurdles and operational issues of this acellular strategy. Finally, we discuss future directions and examine how careful integration of different approaches presented in this review could help to potentiate therapeutic results in preclinical models and their good manufacturing practice (GMP) implementation for future clinical trials.

Keywords: pluripotent stem cells; cell therapy; extracellular vesicles; exosome; acellular therapy

1. Introduction

In the background of many diseases, the degeneration and/or dysfunction of a particular cell type affects tissues and organs, which will deteriorate over time and lose their functions. In this context, regenerative medicine offers new perspectives for pathologies that otherwise remain untreated [1,2]. Hundreds of clinical trials have been initiated in the last few years for a large panel of indications. Strategies are more and more refined to comply with large-scale productions and identify the most efficient formulation to maximize therapeutic effects [3]. Tissue engineering has recently emerged and constitutes the second generation of cell-based regenerative therapy by incorporating three-dimensional (3D) biodegradable compounds mimicking the extracellular matrix and/or multiple cell types [4,5].

Besides the therapeutic effects of cells elicited by their direct presence, some of the observed benefits are mediated by their indirect actions and, as such, may not require them [6,7]. Extracellular vesicles (EVs) have emerged as important mediators of paracrine signaling and could exert such functions [8–10]. Intensive research on EVs over the last half-century has led to an in-depth understanding of their origin and biological functions and has placed them at the forefront of various disease treatments.

Herein, we first review the state of the art of cell therapy and its current limitations and challenges, with a particular focus on pluripotent stem cell-derived therapeutic products. In a second part, we elaborate on functions of EVs and discuss their interest in the field of "acellular therapies" for pathological conditions that affect the eye, skin, heart and skeletal muscle. Finally, we discuss the future prospects of these therapies and their combination to potentiate therapeutic outcomes.

2. Regenerative Medicine: State of the Art

2.1. Introduction

The source material for cell and tissue-based products is a major point to consider for regenerative medicine. A prerequisite for the broad application of cell products is the production of large quantities of cells, with consistency between different batches or manufacturing processes to ensure the same quality and therapeutic efficacy to each patient.

This section will describe the numerous cell sources that may be used in cell therapy and their use in selected pathological conditions. In addition, we will discuss the transition from cell therapies formulated as a cell suspension to more complex tissue-engineered products.

2.2. Cell Sources

Various cell sources have been considered for cell therapy and regenerative medicine, including adult material from living donors or cadavers, fetal materials and pluripotent stem cell lines.

Cell material from adult origin can be obtained directly from patients and purified/amplified in vitro (like mesenchymal stem cells (MSCs) or epithelial cells from the skin). Such autologous approach prevents the risk of rejection. However, the manufacturing process and supply logistics of autologous cell-based therapy products are highly complicated and hinder their scale out. These limitations are particularly important when addressing widespread diseases that affect millions of patients like Age-related Macular Degeneration (AMD) or when the number of cells needed for therapy is important, as in the case of large burns. Alternatively, cells can be obtained from cadavers or, when possible, living donors. For all adult sources, donor variability—histocompatibility, age, genotype, etc.—drives cell product variability. In addition, adult cell sources have a reduced potential for expansion. This is especially true for terminally differentiated adult cell types (i.e., skeletal muscle cells) that hardly proliferate, or for organs for which there is limited access to endogenous stem cells.

Another source of materials for cell therapy is aborted fetuses. Harvested cells retain a proliferation potential that is superior to that of adult cells. Although fetus stem/progenitor cells are restricted in their potentiality, giving rise to limited cell types, some of these cells are no longer present in adult tissues and could thus be used to generate specific cell types. Nevertheless, their access is highly variable around the world, depending on the abortion law and policy in place in each country. Indeed, in some countries, it is strictly prohibited; in others, abortion is prohibited after different gestational ages, which could be restrictive. The use of human fetal tissue also raises a number of ethical considerations. Besides these restrictions, the access to fetuses at a defined gestational age is complex, as highlighted for Huntington's disease (HD) cell therapy as an example [11]. Clinical trials have aimed to restore degenerated striatal cells in the brain of HD patients through the transplantation of fetal ganglionic eminences (GE) containing future striatal cells. In the MIG-HD clinical study, 86 fetal GE tissues were grafted into HD patients but a total of 163 surgeries were cancelled due to inadequate/insufficient donor fetal material as fetal cells should be grafted within 2 days following abortion [12].

Since their first derivation, human pluripotent stem cells (hPSCs) have been considered a promising cell source for regenerative medicine. hPSCs are self-renewable and give rise to any cell type of the human body. They can be obtained from supernumerary in vitro fertilized embryos (human embryonic stem cells or hESCs) or after the conversion of adult primary cells to pluripotency by the overexpression of a cocktail of factors (human induced pluripotent stem cells or hiPSCs) [13–15]. hPSCs are compatible with large-scale industrial productions in good manufacturing practices (GMP) facilities and quality controlled as any other more conventional pharmaceutical products. Whereas the majority of hPSC-based clinical trials have used hESCs up to now, the field is moving toward hiPSCs as they do not require the destruction of embryos and consequently can be used worldwide without restriction [1].

In the following section, we highlight some recent therapeutic applications of hPSC-based cell therapy products.

2.3. *hPSC-Based Cell Therapy: Selected Examples*

2.3.1. Retinal Degeneration

Different causes of visual impairment, such as AMD, Retinitis Pigmentosa (RP) and diabetic retinopathy [16], are characterized by the degeneration of the light-sensitive part of the eye: the retina [17]. Depending on the nature and/or stage of the disease, different cell types are susceptible to degeneration. In particular, the retinal pigment epithelium (RPE), an essential supportive tissue of the retina affected in AMD and some forms of RP, triggers the degeneration of photoreceptors (PRs). Consequently, a specific focus has emerged to replace dead or dysfunctional RPE cells in order to prevent PR degeneration and subsequent vision loss.

In the last decade, several protocols based on spontaneous differentiation have been developed to convert hPSCs into RPE cells [7,18–20]. The differentiation is initiated by FGF2 withdrawal (used to maintain the pluripotency state of hPSCs). After several weeks, pigmented clusters start to appear among cultured cells. These clusters, corresponding to RPE cells, can be manually isolated and then expanded. At the end of the process, hPSC-RPE cells are banked in cryovials and can be thawed on demand. More recently, defined protocols were also described to improve the yield and size of the final cell bank through the addition of cytokines and/or small molecules [21–24]. These strategies allowed to bypass manual enrichment and critical steps of the manufacturing process could be programmed in automated cell culture systems [25–27].

The therapeutic potential of these hPSC-RPE cells has been evaluated after injection as a cell suspension into the subretinal space between endogenous PRs and RPE/Bruch's membrane of a rat model of RP. All these studies demonstrated the benefit of hPSC-RPE cells with a preservation of PRs and a restoration of the visual function [18,23,28–30]. Recent optimizations of the final formulation with the reconstruction of an organized RPE epithelium on top of different scaffolds improved graft survival and therapeutic outcomes [18,31,32]. This formulation as an organized tissue required the development of specific devices to prepare and deliver the therapeutic product to the eye [33–36].

In the context of PR degeneration, hPSCs are able to differentiate into PRs [17] through the formation of retinal organoids [37–40]. hPSC-PRs are then enriched following dissociation and cell sorting [17,41]. First transplantations of PR cell suspensions in rodent models of retinal degeneration suggested a therapeutic effect mediated by cell integration and maturation [42]. However, further studies contradicted these first results and revealed a phenomenon of cytoplasmic material transfer with remaining PRs [43–45]. Retinal organoids dissected into small sections were also transplanted into the subretinal space of rodents and primates. With this grafting approach, PRs create connections and elicit some signs of visual recovery [46–48]. However, these sheets tend to form rosettes and PRs are transplanted with other cell types present in organoids, which limits the number of connections with the other neurons present in the retina. An improvement of the PR sheet formulation is ongoing to produce an organized PR sheet with or without a RPE epithelium [3]. In particular, polymeric 3D micro-structured scaffolds could be used to organize and structure PRs [49–52]. To overcome limitations of PR functionality caused by maturation defects, another strategy is to genetically modify hPSC-PRs with an optogene that brings light sensitivity to immature PRs [53].

2.3.2. Cutaneous Wounds

Skin wounds, principally caused by traumas or thermal burns, can be self-repaired by the body [5,54]. The different stages of self-wound healing include: (1) hemostasis to stop blood loss and provide a scaffold for cell migration, (2) inflammation to eliminate pathogens and tissue debris, (3) proliferation in particular of keratinocytes to achieve wound coverage and (4) remodeling of collagens, which are secreted by dermal fibroblasts [5,55,56]. However, the healing process can be impaired by the size of the wound, environmental factors (stress, smoking, medications or recreational drugs) or genetic disorders affecting the skin or wound healing capacity [5,57].

Occlusive dressings may be applied to the wound in order to improve healing by providing a mechanical protection and a moist and clean environment that favors re-epithelialization [5,56,58]. When this is not sufficient, skin substitutes composed of allogenic or autologous cells can be obtained from the in vitro expansion of epidermal progenitors [59]. These skin substitutes may be composed of epidermal, dermal, or dermo-epidermal (composite) layers [56,57,59]. For example, sheet of autologous keratinocytes from donors are expanded in vitro to generate large epidermal sheets. However, this approach is time consuming [5,56] and, in the case of severe burns, strategies to generate ready-to-use banks of keratinocytes will improve patient care. Such objectives are achieved by the preparation of skin substitutes derived from hPSCs, which facilitate the production of large banks suitable for large-scale use and clinically compatible industrial productions.

hPSCs have been successfully differentiated into basal keratinocytes capable to form a pluristratified epidermis in vitro and in vivo in rodents [60–62]. However, depending on the disease context, inflammation should be controlled and angiogenesis should be promoted in order to maintain the graft. Therefore, complex tissue engineering is needed to improve graft integration and survival. Endothelial cells, fibroblast and collagen may be part of the reconstructed skin substitute to mimic the dermal layer and favor revascularization [61,63,64]. An inherent difficulty in working with several cell types is their correct spatial organization to recreate the complex skin structure to ensure functional integration of the graft. Promising results were obtained using micro-patterned 3D vascular networks with iPSC-derived endothelial cells in skin substitutes, with a better engraftment and long-term survival as well as an improved functionality when grafted in immunodeficient mice [65]. The choice of the bioink to create the 3D vasculature has also an impact on the physical properties of the reconstructed skin. Indeed, skin-derived extracellular matrix bioinks reduce the shrinkage and contraction observed with the use of collagen [66]. In addition, it favors epithelial organization and barrier function.

Nevertheless, other characteristics of the skin are still missing (i.e., melanocytes, sebaceous glands and hairiness). Different protocols based on hPSCs have recently been developed to obtain melanocytes [67–69] and hair follicles [70,71] but the development of clinically compatible complex skin substitutes with all appendages require further developments.

2.3.3. Heart Infarction

Despite advances in interventional and medication therapies, heart failure is the first cause of human death worldwide [72]. The human cardiac muscle, or myocardium, has almost no regenerative potential compared to that of lower vertebrates, the lost tissue being replaced by a fibrotic scar [73]. Existing treatments are not curative and do not trigger heart regeneration [74]. Cell therapy may address this bottleneck through the transplantation of cells sharing the same contractile properties as heart muscle cells [74,75].

Various protocols have been developed to produce cardiomyocytes from hPSCs (hPSC-CMs) both from 3D or 2D cultures [76–80], some of them with a high yield and purity using bioreactors [81]. For example, hPSCs are committed towards a cardiac mesodermal lineage through exposure to BMP-2 [82]. Cardiovascular progenitor cells are then maintained to a progenitor state by inhibiting the fibroblast growth factor receptor with SU-5402 [83]. Alternatively, a temporal regulation of WNT signaling pathway (i.e., initial activation followed by inhibition) is sufficient to trigger hPSC-CM differentiation [84]. hPSC-CMs display some main characteristics of cardiomyocytes such as contractility, ion channels, calcium handling and excitation propagation [85,86].

The preclinical evaluation of hPSC-CM functionality has been conducted in rodents [87] and larger animals (pigs or primates) [88–90]. Cells are delivered either as a cell suspension through an intramuscular route or after tissue engineering. Following injection of allogenic PSC-CMs or hESC-CMs as cell suspensions in myocardial infarcted primates, cardiac contractile functions improved [88–90]. In addition, injections of PSC-CMs were shown to be superior to skeletal myoblasts or MSCs for the restoration of cardiac functions and oxygen consumption in a porcine model of ischemic injury [91]. A recent phase I clinical trial delivered epicardially hESC-CM embedded in a fibrin patch into patients with severe left ventricular dysfunction (see Section 2.4. for a more detailed description; [86]).

A major concern with the use of hPSC-CMs is the apparition of a ventricular arrhythmia probably due to immaturity and automaticity of the graft [90,92]. Indeed, differentiation protocols also give rise to nodal subtypes with automaticity-associated genes like *HCN4* susceptible to trigger pacemaker currents [93,94]. Exclusion of these cells may reduce ventricular arrhythmia [75]. Overall, hPSC-CMs delivered as a cell suspension appear functionally coupled with the host myocardium, but this observation is still debated regarding tissue-engineered hearts [75,95]. In addition, the grafted cells have limited survival in vivo. Interestingly, hPSC-CMs co-transplanted with hPSC-derived epicardial cells or MSC-loaded patches improved both graft survival (i.e., size of the graft) and maturation in rodents [96,97]. Nevertheless, studies in larger animals should confirm these results to validate long-term integration and survival of transplanted hPSC-CMs.

2.3.4. Skeletal Muscle Regeneration

Muscle regeneration involves the activation of PAX7 positive quiescent satellite cells that respond to tissue injury by proliferation and differentiation to give rise to MyoD positive progenitors called myoblasts (MBs) [98]. MBs then differentiate and fuse with myofibers to regenerate the damaged muscle [99]. Despite an important regenerative potential, skeletal muscle atrophy is common following trauma or congenital muscle diseases, such as Duchenne Muscular Dystrophy (DMD), but remains an unmet medical need [100].

Upon transplantation, freshly isolated rodent satellite cells are able to regenerate chemically injured skeletal muscles that were depleted by irradiation of endogenous satellite cells [101]. The dystrophin-deficient mdx mouse model of DMD was also rescued through this strategy [102–104]. These results hint at the promising potential of cell therapy to tackle muscle atrophy. However, satellite cells amplified through cell culture loss their regenerative potential in vivo [103]. Thus, a renewable source of cells is required to treat patients.

Protocols have been described allowing the conversion of hPSCs into myoblasts through cytokines or small molecules exposition, recapitulating in vivo developmental cues [98,99]. Briefly, hPSCs are induced to presomitic mesoderm progenitors after activation of WNT and inhibition BMP signaling [99,105]. Then, myoblast progenitors are obtained after FGF, HGF and IGF stimulations [105]. Satellite-like cells (PAX7+ cells) represented 22% of the final cell population at 4 weeks [105,106]. These cells could then be subcultured without the loss of PAX7+ population [106]. PSC-derived satellite-like cells were able to repopulate endogenous satellite cell niche and regenerate skeletal muscles [105]. In addition, the presence of ERBB3 and NGFR surface markers allowed selective enrichment of a myogenic population with increased regenerative potential in vivo in mdx mice [107]. Therefore, cell sorting of ERBB3+ cells to enrich a myogenic cell population is suitable for cell therapy. Recently, a myogenic population was obtained after only 15 days of differentiation following CD10+/CD24- cell sorting [108]. These cells are suggested to be more homogenous compared to ERBB3+/NGFR+ and more myogenic in vivo in mdx mice [108]. Similar protocols were developed to produce large-scale banks of cryopreserved hPSC-derived myogenic progenitors (expanded for a maximum of 5×10^{11}-fold) [109].

To date, clinically compatible protocols are still missing [98]. For volumetric muscle loss, new muscle fibers should be regenerated to reconstruct the tissue cytoarchitecture. This will require specific scaffolds [110]. In addition, supportive cells (i.e., muscle resident cells), such as endothelial cells, are necessary to ensure proper tissue vascularization [98,110]. Finally, for de novo reconstruction of muscle fibers, strategies to promote innervation should be developed [111].

2.4. hPSC-Based Clinical Trials

Approved in 2009 by the FDA, the first clinical trial led by Geron Corporation paved the way for the use of hESC-derived cells into the clinic [112]. The company had to fill an investigational new drug application of 22,000 pages to demonstrate the safety, functionality and quality of their hESC-derived oligodendrocyte progenitors for the treatment of spinal cord injuries. Unfortunately, during this first phase I clinical trial, only half of the patient cohort had been treated before it was halted prematurely for economic reasons [113]. Asterias Biotherapeutics (acquired later by Lineage Cell Therapeutics) pursued the development of this cell therapy in a new phase I/IIa dose escalation clinical trial and announced in 2019 that signs of motor improvements without safety concerns at 12-month were observed in the 25 treated patients [114].

Since then, a number of indications were evaluated, including eye diseases, diabetes and ischemic hearts. A recent manually curated database identified 54 planned or initiated clinical trials based on hPSCs worldwide [1]. Starting from 2018, a switch was observed regarding the source material with the number of hiPSC-based clinical trials beginning to be more important compared to that of hESC-based trials.

Most of these clinical trials focus on hPSC-derived RPE (24 clinical trials) to target macular degeneration, some forms of RP and Stargardt's disease [1,3]. This could be explained by the easy access and available live imaging technologies, as well as low number of required cells and immune privilege of the eye. hESC-RPE cells initially injected as a cell suspension were demonstrated to be safe in a total of 38 patients [115]. The two major risks associated with hPSC therapy (i.e., teratoma formation due to residual hPSCs and cell dispersion to other organs) were not reported. hESC-RPE cells were also organized as an epithelial tissue before implantation on a synthetic [34,116] or biological scaffold [18,36]. Strikingly, one of two patients treated for a severe exudative AMD gained enough vision to be able to read again a year after the transplantation of a hESC-RPE sheet seeded on a polyester scaffold [34]. These results are encouraging but should be taken with caution due to the limited number of patients.

The first autologous hiPSC clinical trial was also conducted with hiPSC-RPE epithelial sheets to target the wet form of AMD [117]. The trial was halted after the first patient due to a Japanese regulation change leading to a switch to an allogeneic hiPSC-RPE graft strategy [3]. The only treated patient maintained her visual acuity for one year [117].

Ischemic heart disease is another critical indication with four clinical trials planned or already started [1]. A completed phase I clinical trial delivered epicardially hESC-derived cardiovascular progenitors embedded in a fibrin patch into six patients with severe left ventricular dysfunction [86]. The primary outcomes were safety through the evaluation of cardiac or off-target tumors, occurrence of arrhythmia and alloimmunization. Patients had a history of myocardial infarction at least 6 months before screening. hESC-derived cardiovascular progenitors were immunomagnetically sorted for the expression of SSEA-1, a marker of pluripotency loss and expressed the cardiac transcription factor ISL1. A nearly pure population of cardiovascular progenitor (median of 97.5%) was obtained. After a follow-up of 4-and-a-half years at maximum, no signs of tumors or arrhythmias were detected in patients [114]. An alloimmunization was detected through antibodies directed against donor cells in three patients without clinically relevant signs [86]. In fact, patients were under an immunosuppressive regimen only for one month as grafted cells are suggested to be short-lived and to act through a paracrine effect. More importantly, this trial was one of the first to suggest the safety of such cell therapy products. Since 2018, other clinical trials in China and Japan have been planned with autologous or allogenic hiPSC-derived cardiomyocytes [1].

To address type 1 diabetes, Viacyte launched four clinical trials based on hESC-derived beta-like cells delivered through subcutaneous implantation [1]. These cells, able to secrete insulin, are evaluated following subcutaneous injection in an immune-encapsulation device. No safety issues were reported in a phase I/II clinical trial of 19 implanted patients (June, 2018) [114]. In addition, the differentiation of implanted cells into insulin- and glucagon-producing cells has been reported. Concerns with the long-term cell engraftment led to an evolution of the implanted device. A new version currently under evaluation is composed of two membranes: the outer one is cell permeable to support neovascularization, contrary to the inner one, which prevents immune cell infiltration [118].

Up to now, all these clinical trials have demonstrated the safety of cell therapy products based on hPSC-derivatives. While the therapeutic application of these innovative cell products is still in its infancy, encouraging observations are also associated with some signs of efficacy. However, larger clinical trials are required to confirm the therapeutic efficacy of this strategy.

2.5. Limitations/Challenges

Preclinical and first clinical studies have shed light on different issues and challenges. First, immune reaction to donor cells is a critical drawback. Autologous transplantation does not need an immunosuppressive regimen and favors an optimal cell survival. However, the cost of autologous cell therapy is prohibitive [119]. The delay inherent to derivate hPSCs from a patient, to differentiate the cells to a particular cell type and finally quality control the cells to ensure safety is long. This strategy is not viable when the need for a treatment concerns millions of patients, as in AMD or in conditions that affect the heart. In contrast, allogenic cell banks allow an off-the-shelf product that could be distributed and used on demand. Cell banks could be designed large enough to treat an important number of patients at an industrial scale, reducing the cost of each graft unit. As the human leukocyte antigen (HLA) of donor is not matched to the patient, the risk of immune rejection is high and the so-called immune privilege of specific organs (i.e., central nervous system or eye) not always clearly demonstrated upon allogenic transplantation. To prevent graft rejection, an immunosuppressive regimen is required but associated side effects can be important and deleterious [120]. Current immunosuppressive protocols tested for hPSC-RPE transplantation, the most advanced hPSC-based therapy, ranged from local to systemic administration [2,34,116,121]. To bypass this, two strategies are proposed: matching the HLA between donor and recipient through multiple cell banks or to genetically engineer a hypoimmunogenic hPSC bank. Human PSC banks with known HLA could be facilitated by the identification of few selected homozygous HLA-typed donors that have the potential to match the HLA of a large part of the population in a particular country [122,123]. For example, 10 selected donors could cover nearly 40% of the UK population and 150 donors cover the majority (93%) of the population [124]. A clinical trial in Japan used one HLA-matched hiPSC-RPE cell line to treat five wet-AMD patients [125]. Transplanted cells survived at one year. The authors reported a mild immune rejection that resolved following local steroid administration. Interestingly, 18.8% of the 105 AMD patients followed in this hospital had a matched HLA with this hiPSC-RPE cell line.

Other approaches under investigation depend on genetic engineering to modulate immunogenicity of donor cells in order to generate a "universal" cell bank suitable to treat any patient, instead of multiple cell banks. For example, AAV (serotype 3B)-mediated knock-in of HLA-E at the beta-2-microglobulin (B2M) locus in B2M-/- PSCs leads to overexpression of HLA-E without surface expression of HLA-A, -B or -C. This prevents stimulation of allogenic T cells or natural killer (NK) cell mediated lysis [126]. Another approach is to knockout HLA-A and –B using CRISPR-Cas9 technology while retaining HLA-C to prevent NK mediated lysis. This approach improves donor compatibility and only 12 lines of HLA-C-retained hiPSC lines could cover 90% of the world population [127]. Alternatively, the knockout of HLA I and II was combined to the expression of immunomodulatory factors PD-L1, HLA-G, and CD47 inserted in AAVS1 locus to prevent immune responses [128]. Finally, the lentiviral-mediated overexpression of only CD47 combined with the CRISPR-Cas9 knockout of HLA I and II was found to be sufficient to generate hypoimmunogenic hiPSCs. Cardiomyocytes, endothelial or smooth muscle cells derived from these hiPSCs did not elicit immune rejection [129]. However, such genetic manipulations need a careful safety assessment of the cell bank to prevent unwanted genetic alterations and/or residual transducing vectors. In addition, the insertion of a suicide gene inducing the selective death of grafted cells upon drug supplementation may provide a safety switch against uncontrolled proliferation of hypoimmunogenic hPSCs [130]. Implementation of one of these strategies will greatly improve the benefice/risk ratio and may extend cell therapy to a larger number of patients while reducing the cost.

The formulation of cell therapy is the object of intense research for achieving optimal functionality and integration within host surrounding tissues. When cells are injected as a suspension, it could affect their survival, retention into target organ or their functionality [18,31,114]. In addition, depending on the target organ, donor hPSC-derived cells are not mature enough upon delivery and, therefore, need to complete maturation in vivo following transplantation. However, additional cues may be required, especially in a diseased tissue environment. For example, a supportive micro-structured scaffold could be associated with cells to provide a suitable environment for cell maturation or retention [3,114]. In some instances, multiple cell types may be combined to improve cell integration and functionality. This is the case when multiple cell types degenerate or when a microenvironment should be recreated [7,61,63,64,96,97,110]. Stimulation of the vascularization of the grafted hPSC-derived tissue (i.e., muscle, heart, skin) is also required for its long-term survival and integration [61,63,64,98,110,114]. First hPSC-derived clinical trials addressed the safety of "simple" systems. Additional studies are still needed to determine whether the use of more complex tissue reconstructions will go hand-in-hand with better efficiency.

Finally, cell survival is not always required to achieve functional recovery of a specific organ, suggesting that a temporary paracrine effect is sufficient to stimulate endogenous regeneration in organs that retained this potency. This is shown in the case of ischemic hearts or through the action of MSCs in acute myocardial infarction, burned skins, liver or traumatic brain injuries and many other diseases [114,131–134]. Such observations raise the question of whether cell-derived products may be sufficient for a therapeutic improvement in some specific disease conditions. Among materials released by cells, EVs hold characteristics that focus the attention of the scientific community [131,133,135].

3. Application of EVs in Regenerative Medicine

3.1. Extracellular Vesicles: Definition

Extracellular vesicles (EVs) form a heterogeneous group of double layered lipid membrane-enclosed vesicles, with distinct biophysical properties and functions both in physiology and under pathological conditions [136]. They have emerged as important mediators of intercellular communication due to their ability to shuttle a variety of nucleic acids (including mRNAs, miRNAs), proteins, and lipids between cells (Figure 1) [137–139]. EVs can transmit information to target cells through different mechanisms. First, the mere interaction of EVs with surface molecules on the cell membrane can trigger intracellular signaling cascades, without delivery of their content. EVs can also fuse with acceptor cells to release their cargo, either by direct fusion with the cell membrane, phagocytosis or through a variety of endocytic pathways, including clathrin- and caveolin-mediated endocytosis, macropinocytosis, micropinocytosis, and lipid raft-mediated internalization [140,141].

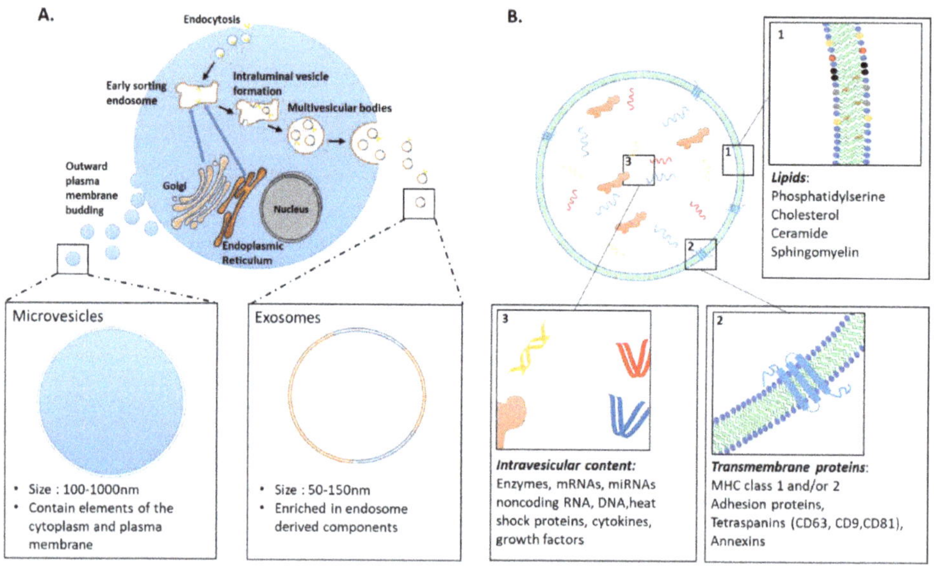

Figure 1. Biogenesis and general composition of EVs. (**A**) Scheme describing the biogenesis of EVs: Microvesicles are produced via the outward budding of the plasma membrane whereas exosomes arise from the fusion of multivesicular bodies with the plasma membrane. Early sorting endosomes receive materials from endoplasmic reticulum, golgi and the endocytic pathway. Multivesicular bodies are generated through the formation of intraluminal vesicles in the late sorting endosome. (**B**) Illustration of nucleic acids, proteins and lipids that can be present in EVs (this list is not exhaustive). Exact nature of EV cargo depends on the cell type, culture conditions (i.e., stress) and pathological state.

EVs are categorized into three major classes based on their biogenesis: (i) Exosomes are formed by the endocytic pathway through invagination of the endosomal membrane, which ultimately forms multivesicular bodies (MVBs; Figure 1). Upon the fusion of MVBs with the plasma membrane, exosomes are released into the extracellular environment. (ii) Microvesicles are shed directly by outward vesicle budding of the plasma membrane. Exosomes are typically 50–150 nm, whereas microvesicles are 100–1000 nm in size. (iii) Apoptotic bodies (size over 500 nm) are formed by blebbing of the plasma membrane of cells undergoing apoptosis. Recently updated guidelines of the International Society for Extracellular Vesicles (ISEV) recommend use of the term extracellular vesicle as "the generic term for particles naturally released from the cell that are delimited by a lipid bilayer and cannot replicate" [142]. Unless authors are able to assign an EV to a particular biogenesis pathway, with appropriate characterization of subcellular origin, the term EV should be used exclusively. Indeed, much overlap has been present in the literature with the term "exosome" used interchangeably with "extracellular vesicle" in many studies. For the sake of clarity and coherence, we chose here to refer to them as EVs.

3.2. Clinical Trials Using EVs

In recent years, a growing body of studies suggests that EVs might hold remarkable potential as therapeutics, either as active agents or delivery systems. EVs are primarily being developed as tools for anti-tumor/immunomodulatory therapies, drug delivery, and regenerative therapies [143]. In addition, EVs are extensively studied in the probing of pathophysiological states of the host as potential biomarkers in biological fluids for the diagnosis and monitoring of various diseases [144]. An increasing number of ongoing, planned or completed clinical trials have been undertaken in recent years. In order to evaluate the use of EVs in translational clinical trials, we searched for the keywords "exosomes", "extracellular vesicles" or "microvesicles" on the ClinicalTrials.gov website. Results are presented for each EV subpopulation in relation with the disease context and source of EVs (Figure 2).

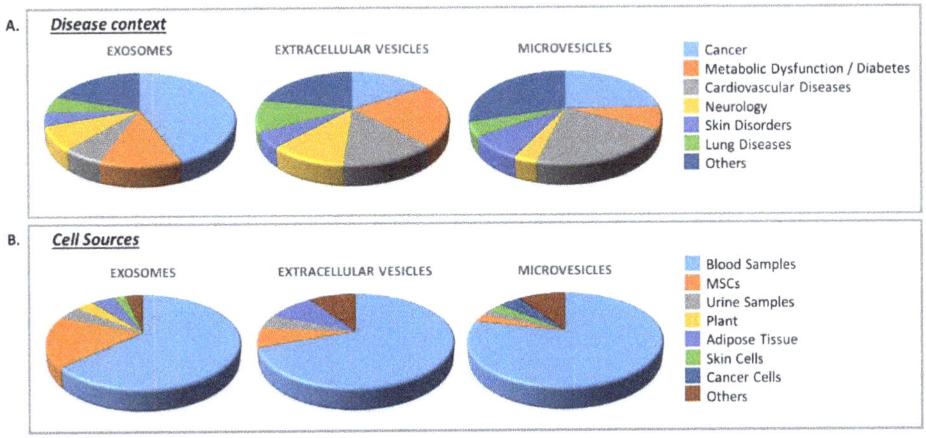

Figure 2. Planned or initiated clinical trials based on EVs. Diagrams representing the proportion of clinical trials targeting a group of diseases (**A**) or using a specific EV source (as biomarker or therapeutic); (**B**). These diagrams were obtained following a database search on the Clinicaltrial.gov website with indicated search terms (exosomes, extracellular vesicles, microvesicles). The majority of clinical trials use EVs as biomarkers from the blood or urine.

So far, most of the studies have investigated the use of EVs as reliable biomarkers in biological fluids, such as blood and urine, for diagnostic and prognostic. The immunomodulatory properties of EVs from dendritic and tumor cells are also being explored as cell-free anti-tumor agents. Promising results have been obtained in phase I and II clinical studies (reviewed in [145]). A phase II clinical trial (NCT01159288) evaluated the therapeutic potential of dendritic cell (DC)-derived EVs loaded with major histocompatibility complex (MHC) class I- and class II-restricted cancer antigens as maintenance immunotherapy after induction chemotherapy in patients bearing inoperable non-small cell lung cancer (NSCLC) [146]. This phase II trial showed that DC-derived EVs exerted NK cell effector functions in patients with advanced NSCLC, boosting the NK arm of antitumor immunity.

Recent attention has focused on the potential interest of EVs as therapeutic tools for acellular regenerative medicine. Thirty interventional clinical trials are based on the use of EVs for therapeutic purpose (Table S1). Only 16 of them are specifically evaluating the therapeutic efficacy and safety of stem-cell derived EVs in patients. Of note, none of these trials use EVs from either hPSCs or hPSC derivatives and the vast majority of them are based on MSCs. For example, a phase I clinical trial aims at studying the therapeutic potentials of condition medium from MSCs in wound healing on patients with chronic skin ulcer (NCT04134676). Similarly, a phase I/II trial uses MSC-derived EVs from normal donors to improve cutaneous wound healing of skin lesions in Epidermolysis Bullosa (EB) patients (NCT04173650). The safety and efficacy of MSC-derived EVs is also being investigated in a phase I clinical trial to promote functional recovery of large and refractory macular holes (NCT03437759). To date, none of these trials has published results yet.

3.3. EVs as Potential Therapeutic Tools: Selected Examples

Increasing evidence suggests that EVs could recapitulate the beneficial effect of their parental cells in a number of applications. To draw a parallel with our previous section, we will focus here on the recent developments obtained with EVs as a therapeutic tool for diseases affecting the eye, skin, heart and skeletal muscle. We also explore how to integrate cell-based and acellular therapies to take advantages of these two approaches and potentiate therapeutic results (Figure 3B).

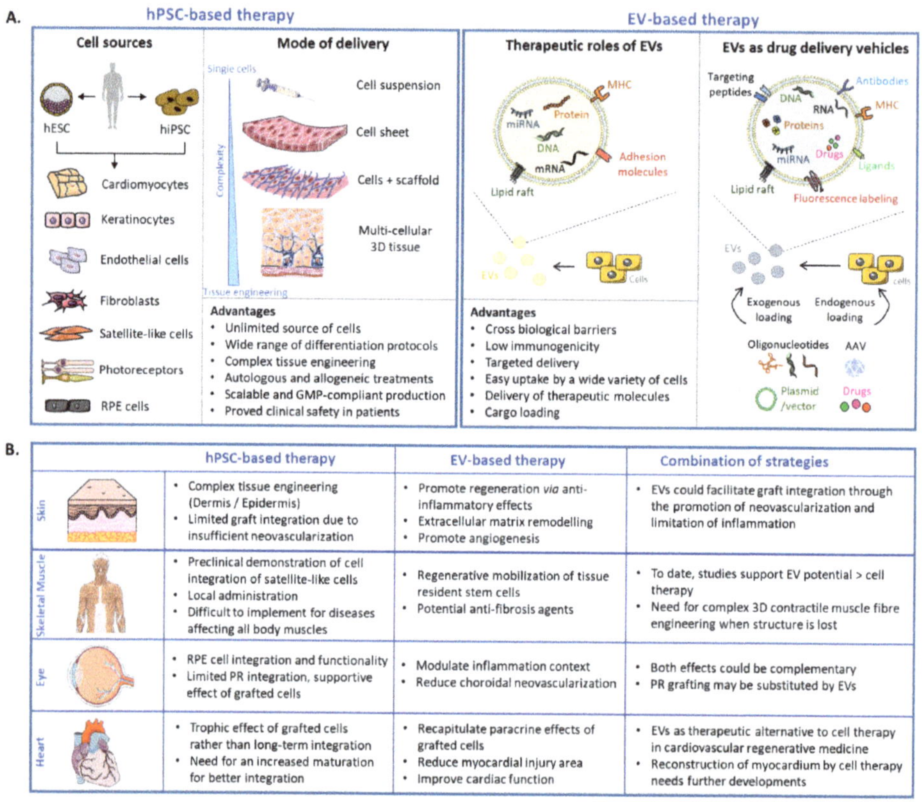

Figure 3. Disease-specific regenerative medicine strategy: toward tailored approaches. (**A**) Source material, mode of delivery and advantages of EV and hPSC-based therapies. (**B**) Therapeutic effects of EVs and cell therapies in preclinical animal models for selected organs (eye, skin, heart, skeletal muscle) and expected cumulative effects.

Of note, studies described herein include, but are not limited to, PSCs and derivatives. Indeed, EVs from cells with well-described paracrine actions, for example, MSCs, have been shown to exert beneficial effects in various applications. Depending on the disease and affected tissue, the therapeutic potential of EVs should therefore be examined from a variety of cell sources.

3.3.1. Tissue Restoration in the Eye

Therapeutic effects of EVs have been evaluated for RP, AMD, and following injuries to the eye. Most of these studies focused on the potential of EVs to limit inflammation that would otherwise trigger the death of retinal cells [147].

As previously mentioned, different studies have started to reveal that the presence of cells might not be required to generate a beneficial effect. As an example, subretinal implantation of human neural progenitor cells preserved the vision of a model of RP through a paracrine effect [148]. The same effect is elicited by human fetal retinal progenitor cells in rats modeling RP and human patients [149,150]. As these transplantations were performed before PR degeneration, a replacement of dead cells was clearly not expected. Building on this, EVs derived from neural progenitor cells were injected subretinally in rats modeling RP before vision loss [151]. Following a single injection, the visual function and PR survival was temporarily improved (up to 28 days post-surgery). EVs were mostly internalized by Iba1+ microglial cells that had migrated from the inner retina to the subretinal space. EVs induced the downregulation of pro-inflammatory cytokines and inhibited microglia, whose suppressed activation is involved in PR survival in RP [152].

A choroidal neovascularization (CNV) is characteristic in wet AMD and can be induced in vivo with laser injuries in rodent (laser-induced CNV model). Human umbilical cord blood MSC (hUCMSC)-derived EVs injected once intravitreally were sufficient to reduce vessel leakage and the development of CNV via downregulation of *VEGF-A* [153]. Preservation of retinal functions and suppression of inflammation is equivalent when MSCs or their EVs are injected intravitreally [154]. In the same vein, retinal astroglial cell-derived EVs inhibited laser-induced CNV in mice when injected daily through the subtenon route for 7 days [155]. Interestingly, injections of EVs derived from RPE cells did not recapitulate these results [155]. Retinal neovascularization was also observed following oxygen-induced retinopathy in mice. The injection of EVs derived from hMSCs cultured under hypoxic conditions also reduced neovascularization [156].

Therefore, EVs are able to modulate retinal degeneration via the inflammatory response. This function is similar to transplanted cells, suggesting that the action of transplanted cells is likely through a paracrine effect (i.e., MSCs, retinal astroglial cells, human neural progenitors). An important point raised by these studies is that EVs or their parental cells need to be delivered at an early stage before complete degeneration. As EVs may not need complex surgeries or immunosuppression, it could be envisioned as a first line of treatment to support retinal survival and delay the requirement for a cell-based intervention. Future preclinical studies need however to determine a delivery route that allows repeated injections for long-term efficiency.

In the context of an advanced RPE cell degeneration as in late AMD, EVs may not be sufficient to recapitulate all RPE functions in order to preserve surrounding retina. Thus, endogenous RPE may be replaced through cell therapy (Figure 3A). A patch of RPE cells could be proposed as an ideal therapeutic substrate, using supporting scaffolds made of polymers or of biological composition. As EVs modulate inflammation during retinal degeneration, it should be determined whether a combined approach of RPE transplantation and EV therapy may improve visual outcomes. This is particularly true as RPE-derived EVs are not able to reduce the CNV when compared to EVs derived from paracrine-acting cells [155]. Such combined approaches may be of interest for multi-factorial diseases like AMD and may also preserve grafted cells from degeneration (Figure 3B).

Another potential EV therapy is related to PR transplantation studies for RP. Early studies showed the feasibility of transplanting post-mitotic PR precursors that achieved some degree of integration into the host mouse retina, ultimately resulting in partial visual function recovery [42,157–159]. The vast majority of transplanted PR precursors were later found to remain in the subretinal space where they engaged in a process of material transfer of functional proteins with host PRs [43–45,160,161]. The precise mechanism of this material exchange between PRs remains elusive but preclinical studies clearly demonstrated that this exchange occurs only with donor PR precursors and not with other cell types [7,160]. It could involve direct donor/host plasma membrane fusion or other methods of intercellular trafficking including EVs [162]. Therefore, it should be determined whether EVs secreted from PR precursors could recapitulate partial visual recovery obtained with PR precursors. If similar results were obtained, it would be an attractive strategy to preserve endogenous PRs without the need of cell grafting and all associated constraints (i.e., surgery, immunosuppression). When PRs have already degenerated, replacement strategies with PRs susceptible to integrate into the host retina will still be required. As discussed earlier, micro-structured scaffolds and/or addition of RPE cells could improve integration and structuration.

Taking advantage of EV ability to carry nucleic acids, they could be used as vehicles in gene therapy strategies for RP [9]. As an example, EV-associated AAV2 vector was delivered to the retina and outperformed conventional AAV2 [163]. Interestingly, these EV-AAV2 served as a robust gene delivery tool in the inner nuclear/outer plexiform and the outer nuclear layer, targeting retinal ganglion cells, bipolar cells, Müller cells as well as PRs [163]. The natural ability of EVs to deliver bioactive nucleic acids to multiple layers in the inner retina suggests that cell-free EV therapies may also benefit other traumatic or neurodegenerative ocular diseases [164].

3.3.2. Cutaneous Wound Healing

Major skin injuries, resulting from extensive burns, infection or trauma, require medical interventions to heal properly [54]. Therapeutic strategies aim at facilitating the 4 phases of cutaneous wound healing–homeostasis, inflammation, proliferation, remodeling-to accelerate wound repair and regeneration [165]. Molecular and cellular events in these phases are tightly coordinated and many cell types interact with each other in a highly coordinated sequence to restore the damaged tissue [166].

EVs hold the potential to promote all phases of wound healing and facilitate skin regeneration (reviewed in [166,167]). Transition from inflammatory to proliferative phase is a key step for successful wound healing. During the early stages of inflammation, the vast majority of macrophages differentiate towards a pro-inflammatory (M1) phenotype. As the wound matures, the ratio switches to an M2 phenotype that promotes tissue remodeling and wound healing [168]. EVs obtained from lipopolysaccharide-preconditioned MSCs could convert M1 macrophage polarization to an M2 phenotype, which alleviated inflammation, and enhanced diabetic cutaneous wound healing in rats by shuttling let-7b miRNA [169]. Similarly, MSC-derived EVs promoted cutaneous wound healing in mice by regulating macrophage polarization through miR-223 [170]. In line with these results, hUCMSCs significantly decreased the number of inflammatory cells and pro-inflammatory cytokines TNF-a, IL-1, IL-6 levels while increasing the production of the anti-inflammatory cytokine IL-10 in wounds of severe burn rats [171]. The same team later found that miR-181c expression in hUCMSC-derived EVs reduced burn-induced excessive inflammation by downregulating the TLR4 signaling pathway [172].

During the proliferative phase, re-epithelization, wound contraction and angiogenesis are essential processes for the restoration of normal tissue architecture. Early recruitment of resident keratinocytes and fibroblasts is particularly important as abnormalities in the intercellular epidermal-dermal crosstalk impairs the skin repair efficiency [173]. In this context, EVs from both fetal and adult stem cell sources can improve migration and proliferation of both fibroblasts and keratinocytes [174–183]. EVs from hUCMSCs and MSCs activate signaling pathways important in wound healing, including RAC-alpha serine/threonine-protein kinase (AKT) pathway [174,177,184] and Notch signaling [182]. Increased phosphorylation of extracellular signal-regulated kinase (ERK)-1/2 [175,180,184] and inhibition of phosphatase and tensin homolog (PTEN) [184] have also been reported. Additionally, an increased extracellular matrix (ECM) deposition by fibroblasts has been observed, facilitating wound contraction. Indeed, fibroblasts increased collagen I and III production following systemic administration of MSC-derived EVs at wound sites in a mice full-thickness wound model [176]. Similar results were obtained with EVs derived from hiPSC-MSCs [185] and hUCMSCs [186]. Human adipose MSCs-derived EVs also prevented the differentiation of fibroblasts into myofibroblasts, increased the ratio of transforming growth factor-β3 (TGF-β3) to TGF-β1 and upregulated the matrix metalloproteinases-3 (*MMP3*) expression of skin dermal fibroblasts through the activation of ERK/MAPK pathway [187]. As such, EVs could be used to promote extracellular matrix remodeling and reduce scar formation.

Transplantation of cellular skin substitutes have shown considerable potential to treat both acute and chronic wounds [58]. Complex multicellular 3D models are being developed with the goal of making engineered tissues similar to their natural counterpart (Figure 3A;

reviewed in [59]). However, there is still an urgent need for improving the vascularization of these substitutes to prevent necrosis and provide better long-term function and integration in clinical practice. This is doubly important as patients with chronic skin wounds usually present defects in the angiogenesis process, which consequently leads to delayed wound healing. One possibility would be the use of pre-vascularized skin substitutes that combine dermal fibroblasts, endothelial cells, and epidermal keratinocytes [63,65,66,188]. An alternative strategy would be to supplement dermo-epidermal skin grafts with EVs conveying pro-angiogenic signals to activate tissue-resident endothelial progenitor cells (Figure 3B). Indeed, exogenous EVs were shown to promote local angiogenesis in murine models of wound healing [185,189,190]. For example, EVs derived from hUCMSC, hiPSC-MSC and human urine-derived stem cells (USC) enhanced in vitro endothelial cell proliferation, migration, and tube formation [178,185,191]. HUCMSC-derived EVs promoted angiogenesis in vivo to repair deep second-degree burn injury by delivering Wnt4 and activating Wnt/B-catenin signaling in endothelial cells [191]. EVs derived from human umbilical cord blood enriched in miR-21-3p promoted the proliferation and migration of fibroblasts as well as enhanced the angiogenic activities of endothelial cells in a full-thickness skin wound mice model, thus accelerating re-epithelialization and cutaneous wound healing [184]. Similarly, EVs derived from hiPSC-MSC applied to wound sites in a full-thickness skin defect rat model promoted not only the generation of newly formed vessels, but also accelerated their maturation [185]. In another study, human USC-derived EVs markedly enhanced the generation of newly formed blood vessels in diabetic mice, in part via the transfer of pro-angiogenic protein deleted in malignant brain tumors 1 (DMBT1) [178]. Alternatively, EVs derived from hESC facilitated pressure ulcer healing by reducing endothelial senescence and promoting local angiogenesis at wound site in aged mice [190].

At present, one of the main obstacles in the treatment of skin wounds is achieving healing over time, particularly in patients with underlying skin disorders. Biomaterial-based wound dressings could be loaded with EVs to achieve sustained release to the wound sites [192,193]. For instance, Tao et al. used the polymer chitosan to prolong delivery of EVs derived from miR-126-3p-overexpressing synovium MSCs to diabetic wounds [193]. They tested this system in a diabetic rat model and found that it increased formation of granulation tissue, which provides a scaffold for the assembly of neighboring cells at wound margins, along with angiogenesis [193].

Overall, all these proofs-of-concept experiments raised considerable interest of EVs for skin repair. Of interest, their delivery to skin wounds is relatively simple due to easy access and could be sustained over time by the use of biomaterial or repeated topical applications. However, additional preclinical studies are needed to evaluate the synergic effects of combined acellular and cellular strategies.

3.3.3. Heart

EVs have been investigated as promising therapeutic options for various cardiac diseases such as ischemic heart diseases and myocardial infarctions. One of the main objectives is to promote vascular repair mechanisms to reduce myocardial injury that would lead to cell death and therefore improve cardiac functions.

hESC-derived MSC conditioned medium (hESC-MSC-CoM), collected with clinically compatible processes, were shown to contain factors susceptible to modulate cardiovascular-related pathways [194]. Administration of hESC-MSC-CoM recapitulated the benefit of hESC-MSC injections in a context of post-myocardial infarction [195]. Indeed, in a porcine model of myocardial infarction, hESC-MSC-CoM intravenous treatment for 7 days enhanced capillary density, reduced the myocardial infarct size and preserved systolic and diastolic functions [195]. In addition, hESC-MSC-CoM reduced myocardial apoptosis and oxidative stress in another porcine model of ischemia and reperfusion injury [196]. This hESC-MSC-CoM contained large particles of 50–100 nm that were purified and characterized as EVs [197]. hESC-MSC-derived EVs similarly diminished the infarct size in an ex vivo mouse model of myocardial ischemia and reperfusion injury.

It was also proposed that ESC-derived EVs could stimulate endogenous myocardial regeneration [198]. Their delivery via an intramyorcardial route following mouse myocardial infarction stimulated endogenous repair (i.e., revascularization, cardiomyocyte proliferation/survival and reduced fibrosis). Interestingly, fibroblasts-derived EVs did not improve cardiac functions when compared to ESC-derived EVs in this model, highlighting differences between EV sources.

The functionality of cells from the cardiac lineage is superior to MSCs in the different heart disease models [199,200]. Therefore, EVs derived from these cells might achieve the most efficient heart recovery. Indeed, hPSC-CM-derived EVs recapitulated the therapeutic effects of their parental cells in the mouse model of chronic heart failure [201]. In this study, EVs or their parental cells were delivered once intramyocardially. Gene expression profiling identified 927 genes similarly upregulated in hearts treated with hPSC-CM-derived EVs and their parental cells as compared to control. The majority of enriched biological processes associated with these genes were predicted to improve heart regeneration and decrease fibrosis [201]. A recent study further highlighted the importance of determining which cellular source is the best candidate to produce therapeutic EVs. While both hPSC- and hPSC-CM-derived EVs protected CMs from hypoxia in vitro, only hPSC-CM-derived EVs completely improved the hypoxia-induced phenotype [202].

In order to maintain a sustained delivery, hPSC-CM-derived EVs were loaded into a collagen-based hydrogel patch [202]. Such system allows the release of EVs during 21 days in vitro. Patches loaded with EVs were implanted directly into the myocardium following an ischemic insult in a rat model of acute myocardial infarction. Rats recovered with this treatment, with improved heart contractile function and a reduction of the infarct size [202].

Overall, recent results indicate that EVs recapitulate the beneficial effects of their parental cells in the treatment of heart diseases (Figure 3B). Importantly, overall complexity associated with cell manufacturing, graft survival and patient immunosuppression are bypassed by this strategy. Future studies are nevertheless required to validate sustained EV release in large animal models as well as reproducibility across hPSC-CM-derived EV production protocols.

3.3.4. Skeletal Muscle

Severe muscle injuries and genetic defects like muscular dystrophies cause myofiber death. Spontaneous reparation to regenerate skeletal myofibers do occur but are insufficient [203]. A central goal of therapeutic approaches is to re-establish the muscle structural integrity and functionality by re-populating the satellite cell niche, promoting vascularization while inhibiting fibrosis formation, and stimulating the formation of contractile muscle fibers [100]. Several cellular candidates with myogenic or non-myogenic origins have been proposed for skeletal muscle regeneration, and their transplantation has been a widely investigated therapeutic strategy [100]. However, the massive donor cell death and cellular dispersion observed after delivery of cells via injection limit their therapeutic potential. Cell therapy products are still a long way from being able to reconstruct the muscle architecture, let alone to reconstruct it with nerve and sufficient vascularization. Overall, cell therapy could be considered for small muscles but is difficult to implement for diseases affecting all body muscles.

A mounting body of evidence suggests that EVs are actively produced by skeletal muscles cells and contribute to muscle repair and regeneration [203]. For example, EVs secreted during the differentiation of human skeletal myoblasts (HSkM) into myotubes contain specific biochemical cues that promote and regulate the myogenic differentiation of human adipose-derived stem cells (HASCs) [204]. Treatment of lacerated muscle sites with these differentiating HSkM-derived EVs led to an improved muscle regeneration with a large number of regenerative myofibers associated to minimal fibrosis compared to the control group [204]. In addition to the facilitation of myofiber repair, EVs also attenuate excessive ECM deposition for optimal muscle remodeling. In muscular dystrophies and severe muscle injuries, fibrogenic cells are overactivated and hyperproliferate, leading to the substitution of skeletal muscle with nonfunctional fibrotic tissue [205]. This excessive accumulation of extracellular matrix components not only alters muscle function but also reduces the amount of tissue available for therapy and repair. Establishing new anti-fibrotic therapeutic strategies is one of the major clinical options to improve muscle function in patients. In response to hypertrophic stimuli, satellite cells give rise to myogenic progenitor cells (MPCs) able to secrete EVs containing miR-206, which represses cell collagen expression through ribosomal binding protein 1 (Rrbp1) by neighboring fibroblasts, thus preventing excessive ECM deposition [206]. Similarly, fibroblasts derived from muscle biopsies of DMD patients secreted exosomes with increased levels of miR199a-5p, causing increased fibrosis in skeletal muscle and surrounding matrix [207]. These data indicate that EVs could be of interest as potential anti-fibrotic agents.

EVs are also evaluated as potential therapeutic agents to counteract muscle wasting and skeletal muscle dysfunction. Chronic kidney disease (CKD), which ultimately leads to end-stage renal failure, often leads to muscle wasting. Intramuscular injection of EV-encapsulated miR29, previously shown to have anti-fibrotic activity, could attenuate UUO-induced body weight loss and muscle atrophy [208]. Similar results were obtained after injection of EV-miR26a in CKD mice [209]. DMD is a heritable myodegenerative disease characterized by the absence of functional dystrophin leading to progressive muscle weakness and degeneration. Recent data suggest that a treatment with EVs from cardiosphere-derived cells (CDCs) originally targeted at DMD cardiomyopathy could potentially benefit both cardiac and skeletal muscle [210]. CDC-derived EVs injected into the soleus of mdx mouse model of DMD enhanced muscle regeneration, decreased inflammation and fibrosis, allowing complete restoration of contractile forces. More surprisingly, detectable levels of full-length dystrophin were evident in the diaphragm and soleus up to three weeks after systemic CDC-derived EV delivery [210]. Dystrophin protein and transcript were undetectable in CDC-derived EVs. Moreover, analysis of exon-intron junctions for dystrophins transcripts after CDC-derived EV treatment showed no exon skipping or alternative splicing. However, RNA-seq of CDC-EVs revealed a 144-fold increase in miR-148a. Intramyocardial injection of miR-148a restored expression of dystrophin in mdx hearts 3 weeks after administration, implicating this miRNA as a potential mediator of enhanced full-length dystrophin protein synthesis [210]. Targeting EV-derived miRNAs appears as a promising strategy to improve muscle function [211].

Finally, EVs could be used as efficient delivery tools of functional cargoes in vivo to restore expression of missing proteins in patients. For example, Gao et al. demonstrated that systemic administration of exosomes loaded with CP05-conjugated dystrophin splice increased dystrophin protein expression in dystrophin-deficient mdx mice with functional improvements [212].

It has become clear that EVs enable intercellular signaling that facilitate myofiber regeneration, limit excessive ECM deposits and improve muscle functions. More strikingly, their systemic delivery improves muscle function in diseases like DMD or CDK. Future studies are now required to characterize their biogenesis, compositions and biological activities on recipient cells in both physiological and pathophysiological conditions to determine whether they might be envisioned for therapy [213].

3.4. Challenges

As discussed, some of the benefits observed with cell therapy are likely due to paracrine effects that can be recapitulated by EVs derived from these cells, rather than long-term engraftment and survival of transplanted cells [214]. Overall, EVs and grafted cells can elicit different outcomes according to the target organ (Figure 3B). In some instances, therapeutic effects could be similar and therefore, EV therapy may have the highest benefice/risk ratio. In other instances, both therapies are complementary and a combination of both may achieve the best therapeutic effects.

Despite their promising roles, several challenges associated with EVs as acellular therapy products still need to be overcome.

3.4.1. Manufacturing EVs: Considerations

As products of cells, the identity and functions of EVs are directly correlated to their cell source. Consequently, it is essential to identify the most appropriate cell source for EV production [9]. As illustrated in Section 3.3, studies have identified a growing list of cell sources, some being more relevant than others, that could be suitable in acellular approaches for regenerative medicine. Our capacity to decipher in depth the various mechanisms by which EVs mediate biological and regenerative functions will be determinant to identify the best candidates in the future. It is also crucial to optimize the upstream processing conditions to increase the notably low EV yields [215]. Finding the optimal conditions for EV production by a specific cell type remains a challenge. Indeed, it is always a compromise between optimal conditions for growth and phenotype and those for EV production and isolation [216]. This is particularly true for the protocols that rely on the induction of a cell stress to obtain specific EVs (for example, hypoxia [156]).

Another challenge is the development of robust procedures to isolate and purify EVs with high yield and purity while preserving their structure and activity [215]. As with GMP-compliant cell manufacturing, large quantities of EVs should be produced with defined medium conditions, devoid as much as possible of xenogeneic substances and serum-derived vesicles, which otherwise have a high risk of contaminating the isolated EV sample [9]. Of note, while EVs secreted by cells cultured in serum-free conditions did not exhibit significant biophysical or size differences compared with cells cultured with serum, the expression levels of certain vesicular proteins (e.g., small GTPases, G-protein complexes, mRNA processing proteins and splicing factors) were found differentially expressed in EVs [217]. The use of commercial exosome-depleted fetal bovine serum (FBS) is another alternative to serum-free conditions. However, they are obtained by various exosome-reduction means, with each inducing their own effect upon the serum constituents and characteristics [218]. Alternatively, cells could be cultured during the EV release period with medium that has been pre-depleted of EVs [142]. However, FBS-derived EV elimination protocols have profound impacts on the cells themselves as they were shown to decrease cell growth and survival-promoting effects of FBS [219,220].

EVs are isolated from a variety of different sources, including body fluids with highly variable compositions (e.g., plasma, milk, urine, saliva) and cell culture media [221]. Importantly, the non-EV contaminants found in EV preparations differ substantially depending on their source. Specific EV isolation procedures should, therefore, be carefully determined accordingly to the starting material [221]. So far, there is no consensus on a "gold standard" technology to isolate EVs, for either therapeutic application or basic research [222,223]. Differential centrifugation/ultracentrifugation, polymer-based precipitation and density-gradient centrifugation are the most commonly used techniques but have very limited scalability and are associated with low EV recovery, disruption of EV integrity and risk of undesirable co-isolation of contaminants [9,136,223–225].

To purify EVs in a GMP-grade scalable manner, studies have employed ultrafiltration or tangential flow filtration (TFF) to concentrate EVs based on their size [226–229]. For example, GMP-EV-cardiac progenitor cell manufacturing was implemented for up to 8 L of conditioned medium, allowing high final product yield ($\geq$58%) with concomitant consistent reduction of contaminants (total protein removal 97–98%) [228]. Subsequent size exclusion chromatography (SEC) allows the separation of EVs from other media components without altering their integrity [227,229–231]. However, size-based isolation techniques do not purify a specific EV subpopulation but rather yield complex mixtures because of the overlap in size of EVs (described in Section 3.1; [232]). Additional purification steps may also still be necessary to remove contaminants with overlapping sizes, such as bovine serum-EVs. Recently, alternative isolation techniques based on EV-surface markers have been described and seem to be promising but still in infancy [233,234]. Overall, existing techniques to isolate EVs do not have a one-size-fits-all model. The optimal isolation method(s) and acceptable level of impurity need to be carefully considered to prevent loss of function due to damage to EV-intrinsic effectors or loss of associated factors that act with EVs to exert function [142,143,221].

As with cells and other cell-derived products, maintenance of EV biological activity during storage is critical for their therapeutic use [235]. Stability should rely on the monitoring of both changes in physicochemical parameters (size, particle concentration, and morphology) and EV bioactivity. A number of studies have evaluated the impact of different storage temperatures (4, −20, and −80 °C) and repetitive free-thaw cycles on the composition and functionality of EVs isolated from various body fluids (reviewed in [235,236]). Current evidences suggest that storage at −80 °C is best-suited. Nonetheless, our current understanding of storage-mediated effects is still limited and standard criterion of EV preservation should be established in the future.

3.4.2. EV-Based Therapeutics: Regulatory Aspects

Current legislation does not provide specific regulatory guidelines for EV-based therapies. In the European Union (EU) and United States of America, they enter in the framework of the biological medicinal products regulation under the definition of "biological medicine" (i.e., a medicine that contains one or more active substances made by or derived from a biological cell) [222]. The pharmaceutical classification of any biological product is determined by its active substance(s). In that regard, EV-based therapeutics are particularly challenging as it is not clear in many cases whether the biological effect depends on the vesicle internal content (EV-associated cargo), the vesicle membranes, or a combination of both [143]. As such, they share characteristics from both cell and gene therapies, which makes them difficult to classify in an existing pharmaceutical category.

Three subcategories of EV products are possible, each with specific regulatory guidance: 1/native EVs from unmodified cells, 2/EVs from genetically manipulated cells which do not contain transgene products, and 3/EVs from genetically modified cells containing transgene products [222]. In addition to the existing guidance on the manufacture of biological medicinal products, safety and quality standards for tissue- or cell-based products (DIRECTIVE 2004/23/EC and DIRECTIVE 2006/17/EC in the EU) may serve as roadmaps for EV-based therapeutics derived from human tissues and cells [222].

Stem cell therapy and tissue engineering are intimately intertwined in the development of regenerative medicine products. In recent years, research has focused on the use of biocompatible scaffolds functionalized with EVs, with or without cells, to improve regenerative capacity of the grafted products. Regulatory strategies for advanced therapy medicinal products (ATMPs)-encompassing gene therapy, somatic cell therapy, and tissue-engineered products–could be applied to such complex EV-based products (DIRECTIVE 2001/83/EC). EVs produced by genetically modified cells with biologically active transgene products could also fall within this subcategory [143]. Of note, EVs are considered as an excipient instead of an active substance when used as drug delivery systems loaded with molecules.

4. Conclusions

As outlined in this review, a number of key steps to develop regenerative therapies has been achieved. Preclinical evidences for many indications were the basis for first-in-man clinical trials. Main concerns related to the use of hPSCs were ruled out in clinical trials (i.e., risks of teratoma formation and cell dispersion). So far, simplified cell therapy systems have been used with some promising therapeutic results. However, preclinical evidences highlight the need for more complex cell therapy products containing various cells and/or biomaterials, structured as the native tissue. In parallel, the roles played by EVs as intercellular effectors of paracrine signaling has led to a strong interest in their use as cell-free therapeutic products to stimulate endogenous factors that would limit cell death or improve tissue regeneration. Strikingly, as these two strategies could act at different levels depending on the pathological context, their complementarity opens new perspectives to maximize therapeutic outcomes. EVs could be substituted to cell therapy in a number of indications including heart diseases, simplifying treatment complexity. However, much preclinical work is still required to characterize optimal "acellular" and cellular therapies and their potential synergistic effects. In addition, definition of appropriate conditions of GMP production need to be elaborated and customized for each application. All these future improvements may help to achieve the ultimate goal of regenerative medicine: the replacement of a degenerated tissue.

5. Patents

WH, CM and KB are inventors of a patent (FR3078712) related to medical devices for the preparation of retinal tissues for regenerative medicine. LM, CM and KB are inventors of a pending patent related to the automated differentiation of hPSC into RPE cells (EP3754014). LM and CB are inventors of a pending patent related to the automated differentiation of hPSC into keratinocytes (EP20305217.0). SD and CB are inventors of two pending patents related to the differentiation of hPSCs into keratinocytes (EP20305218.0) and fibroblasts (EP 20305214.7) and to the formation of a skin substitute composed of hPSCs derived cells (EP 20305213.9). AD and CB are inventors of a pending patent related to the differentiation of hPSCs into endothelial cells (EP 20305215.4).

Supplementary Materials: The following are available online at https://www.mdpi.com/2073-4409/10/2/240/s1, Table S1: List of planned or initiated clinical trials using EVs as therapeutics.

Author Contributions: All authors contributed to the manuscript writing and approved the final version. All authors have read and agreed to the published version of the manuscript.

Funding: This work was supported by a grant from the ANR (RebuildingRETINA: ANR-19-CE18-0004) to KB. EH and SM are recipient of PhD fellowships from the French Ministry of Higher Education and Research. I-Stem is part of the Biotherapies Institute for Rare Diseases supported by the Association Française contre les Myopathies-Téléthon.

Informed Consent Statement: Not applicable.

Data Availability Statement: Not applicable.

Acknowledgments: Some elements of figures were created using Servier Medical Art templates (https://smart.servier.com), which are licensed under a Creative Commons Attribution 3.0 Unported License.

Conflicts of Interest: CB collaborates with the company URGO to develop cell therapy products. The other authors declare no conflict of interest.

References

1. Kobold, S.; Guhr, A.; Mah, N.; Bultjer, N.; Seltmann, S.; Seiler Wulczyn, A.E.M.; Stacey, G.; Jie, H.; Liu, W.; Loser, P.; et al. A Manually Curated Database on Clinical Studies Involving Cell Products Derived from Human Pluripotent Stem Cells. *Stem Cell Rep.* **2020**, *15*, 546–555. [CrossRef]
2. Schwartz, S.D.; Hubschman, J.P.; Heilwell, G.; Franco-Cardenas, V.; Pan, C.K.; Ostrick, R.M.; Mickunas, E.; Gay, R.; Klimanskaya, I.; Lanza, R. Embryonic stem cell trials for macular degeneration: A preliminary report. *Lancet* **2012**, *379*, 713–720. [CrossRef]
3. Ben M'Barek, K.; Monville, C. Cell Therapy for Retinal Dystrophies: From Cell Suspension Formulation to Complex Retinal Tissue Bioengineering. *Stem Cells Int.* **2019**, *2019*, 4568979. [CrossRef]
4. Ghareeb, A.E.; Lako, M.; Steel, D.H. Coculture techniques for modeling retinal development and disease, and enabling regenerative medicine. *Stem Cells Transl. Med.* **2020**, *9*, 1531–1548. [CrossRef]
5. Sun, B.K.; Siprashvili, Z.; Khavari, P.A. Advances in skin grafting and treatment of cutaneous wounds. *Science* **2014**, *346*, 941–945. [CrossRef] [PubMed]
6. Sun, J.; Mandai, M.; Kamao, H.; Hashiguchi, T.; Shikamura, M.; Kawamata, S.; Sugita, S.; Takahashi, M. Protective Effects of Human iPS-Derived Retinal Pigmented Epithelial Cells in Comparison with Human Mesenchymal Stromal Cells and Human Neural Stem Cells on the Degenerating Retina in rd1 mice. *Stem Cells* **2015**, *33*, 1543–1553. [CrossRef] [PubMed]
7. Morizur, L.; Herardot, E.; Monville, C.; Ben M'Barek, K. Human pluripotent stem cells: A toolbox to understand and treat retinal degeneration. *Mol. Cell Neurosci.* **2020**, *107*, 103523. [CrossRef] [PubMed]
8. Hargett, L.A.; Bauer, N.N. On the origin of microparticles: From "platelet dust" to mediators of intercellular communication. *Pulm. Circ.* **2013**, *3*, 329–340. [CrossRef] [PubMed]

9. Wiklander, O.P.B.; Brennan, M.A.; Lotvall, J.; Breakefield, X.O.; El Andaloussi, S. Advances in therapeutic applications of extracellular vesicles. *Sci. Transl. Med.* **2019**, *11*. [CrossRef] [PubMed]
10. S, E.L.A.; Mager, I.; Breakefield, X.O.; Wood, M.J. Extracellular vesicles: Biology and emerging therapeutic opportunities. *Nat. Rev. Drug Discov.* **2013**, *12*, 347–357. [CrossRef]
11. Bachoud-Levi, A.C.; Perrier, A.L. Regenerative medicine in Huntington's disease: Current status on fetal grafts and prospects for the use of pluripotent stem cell. *Rev. Neurol. (Paris)* **2014**, *170*, 749–762. [CrossRef] [PubMed]
12. Bachoud-Levi, A.C.; Massart, R.; Rosser, A. Cell therapy in Huntington's disease: Taking stock of past studies to move the field forward. *Stem Cells* **2020**. [CrossRef]
13. Nakagawa, M.; Koyanagi, M.; Tanabe, K.; Takahashi, K.; Ichisaka, T.; Aoi, T.; Okita, K.; Mochiduki, Y.; Takizawa, N.; Yamanaka, S. Generation of induced pluripotent stem cells without Myc from mouse and human fibroblasts. *Nat. Biotechnol.* **2008**, *26*, 101–106. [CrossRef] [PubMed]
14. Takahashi, K.; Tanabe, K.; Ohnuki, M.; Narita, M.; Ichisaka, T.; Tomoda, K.; Yamanaka, S. Induction of pluripotent stem cells from adult human fibroblasts by defined factors. *Cell* **2007**, *131*, 861–872. [CrossRef]
15. Thomson, J.A.; Itskovitz-Eldor, J.; Shapiro, S.S.; Waknitz, M.A.; Swiergiel, J.J.; Marshall, V.S.; Jones, J.M. Embryonic stem cell lines derived from human blastocysts. *Science* **1998**, *282*, 1145–1147. [CrossRef]
16. Flaxman, S.R.; Bourne, R.R.A.; Resnikoff, S.; Ackland, P.; Braithwaite, T.; Cicinelli, M.V.; Das, A.; Jonas, J.B.; Keeffe, J.; Kempen, J.H.; et al. Global causes of blindness and distance vision impairment 1990-2020: A systematic review and meta-analysis. *Lancet Glob. Health* **2017**, *5*, e1221–e1234. [CrossRef]
17. Gagliardi, G.; Ben M'Barek, K.; Goureau, O. Photoreceptor cell replacement in macular degeneration and retinitis pigmentosa: A pluripotent stem cell-based approach. *Prog. Retin. Eye Res.* **2019**, *71*, 1–25. [CrossRef]
18. Ben M'Barek, K.; Habeler, W.; Plancheron, A.; Jarraya, M.; Regent, F.; Terray, A.; Yang, Y.; Chatrousse, L.; Domingues, S.; Masson, Y.; et al. Human ESC-derived retinal epithelial cell sheets potentiate rescue of photoreceptor cell loss in rats with retinal degeneration. *Sci. Transl. Med.* **2017**, *9*. [CrossRef]
19. Klimanskaya, I.; Hipp, J.; Rezai, K.A.; West, M.; Atala, A.; Lanza, R. Derivation and comparative assessment of retinal pigment epithelium from human embryonic stem cells using transcriptomics. *Cloning Stem Cells* **2004**, *6*, 217–245. [CrossRef]
20. Buchholz, D.E.; Hikita, S.T.; Rowland, T.J.; Friedrich, A.M.; Hinman, C.R.; Johnson, L.V.; Clegg, D.O. Derivation of functional retinal pigmented epithelium from induced pluripotent stem cells. *Stem Cells* **2009**, *27*, 2427–2434. [CrossRef]
21. Buchholz, D.E.; Pennington, B.O.; Croze, R.H.; Hinman, C.R.; Coffey, P.J.; Clegg, D.O. Rapid and efficient directed differentiation of human pluripotent stem cells into retinal pigmented epithelium. *Stem Cells Transl. Med.* **2013**, *2*, 384–393. [CrossRef] [PubMed]
22. Leach, L.L.; Buchholz, D.E.; Nadar, V.P.; Lowenstein, S.E.; Clegg, D.O. Canonical/beta-catenin Wnt pathway activation improves retinal pigmented epithelium derivation from human embryonic stem cells. *Investig. Ophthalmol. Vis. Sci.* **2015**, *56*, 1002–1013. [CrossRef] [PubMed]
23. Idelson, M.; Alper, R.; Obolensky, A.; Ben-Shushan, E.; Hemo, I.; Yachimovich-Cohen, N.; Khaner, H.; Smith, Y.; Wiser, O.; Gropp, M.; et al. Directed differentiation of human embryonic stem cells into functional retinal pigment epithelium cells. *Cell Stem Cell* **2009**, *5*, 396–408. [CrossRef]
24. Maruotti, J.; Sripathi, S.R.; Bharti, K.; Fuller, J.; Wahlin, K.J.; Ranganathan, V.; Sluch, V.M.; Berlinicke, C.A.; Davis, J.; Kim, C.; et al. Small-molecule-directed, efficient generation of retinal pigment epithelium from human pluripotent stem cells. *Proc. Natl. Acad. Sci. USA* **2015**, *112*, 10950–10955. [CrossRef] [PubMed]
25. Crombie, D.E.; Daniszewski, M.; Liang, H.H.; Kulkarni, T.; Li, F.; Lidgerwood, G.E.; Conquest, A.; Hernandez, D.; Hung, S.S.; Gill, K.P.; et al. Development of a Modular Automated System for Maintenance and Differentiation of Adherent Human Pluripotent Stem Cells. *SLAS Discov.* **2017**, *22*, 1016–1025. [CrossRef]
26. Regent, F.; Morizur, L.; Lesueur, L.; Habeler, W.; Plancheron, A.; Ben M'Barek, K.; Monville, C. Automation of human pluripotent stem cell differentiation toward retinal pigment epithelial cells for large-scale productions. *Sci. Rep.* **2019**, *9*, 10646. [CrossRef]
27. Matsumoto, E.; Koide, N.; Hanzawa, H.; Kiyama, M.; Ohta, M.; Kuwabara, J.; Takeda, S.; Takahashi, M. Fabricating retinal pigment epithelial cell sheets derived from human induced pluripotent stem cells in an automated closed culture system for regenerative medicine. *PLoS ONE* **2019**, *14*, e0212369. [CrossRef]
28. Carr, A.J.; Vugler, A.A.; Hikita, S.T.; Lawrence, J.M.; Gias, C.; Chen, L.L.; Buchholz, D.E.; Ahmado, A.; Semo, M.; Smart, M.J.; et al. Protective effects of human iPS-derived retinal pigment epithelium cell transplantation in the retinal dystrophic rat. *PLoS ONE* **2009**, *4*, e8152. [CrossRef]
29. Lu, B.; Malcuit, C.; Wang, S.; Girman, S.; Francis, P.; Lemieux, L.; Lanza, R.; Lund, R. Long-term safety and function of RPE from human embryonic stem cells in preclinical models of macular degeneration. *Stem Cells* **2009**, *27*, 2126–2135. [CrossRef]
30. Lund, R.D.; Wang, S.; Klimanskaya, I.; Holmes, T.; Ramos-Kelsey, R.; Lu, B.; Girman, S.; Bischoff, N.; Sauve, Y.; Lanza, R. Human embryonic stem cell-derived cells rescue visual function in dystrophic RCS rats. *Cloning Stem Cells* **2006**, *8*, 189–199. [CrossRef]
31. Diniz, B.; Thomas, P.; Thomas, B.; Ribeiro, R.; Hu, Y.; Brant, R.; Ahuja, A.; Zhu, D.; Liu, L.; Koss, M.; et al. Subretinal implantation of retinal pigment epithelial cells derived from human embryonic stem cells: Improved survival when implanted as a monolayer. *Investig. Ophthalmol. Vis. Sci.* **2013**, *54*, 5087–5096. [CrossRef]
32. Sharma, R.; Khristov, V.; Rising, A.; Jha, B.S.; Dejene, R.; Hotaling, N.; Li, Y.; Stoddard, J.; Stankewicz, C.; Wan, Q.; et al. Clinical-grade stem cell-derived retinal pigment epithelium patch rescues retinal degeneration in rodents and pigs. *Sci. Transl. Med.* **2019**, *11*. [CrossRef] [PubMed]

33. Stanzel, B.V.; Liu, Z.; Somboonthanakij, S.; Wongsawad, W.; Brinken, R.; Eter, N.; Corneo, B.; Holz, F.G.; Temple, S.; Stern, J.H.; et al. Human RPE stem cells grown into polarized RPE monolayers on a polyester matrix are maintained after grafting into rabbit subretinal space. *Stem Cell Rep.* **2014**, *2*, 64–77. [CrossRef] [PubMed]
34. da Cruz, L.; Fynes, K.; Georgiadis, O.; Kerby, J.; Luo, Y.H.; Ahmado, A.; Vernon, A.; Daniels, J.T.; Nommiste, B.; Hasan, S.M.; et al. Phase 1 clinical study of an embryonic stem cell-derived retinal pigment epithelium patch in age-related macular degeneration. *Nat. Biotechnol.* **2018**, *36*, 328–337. [CrossRef] [PubMed]
35. Fernandes, R.A.B.; Stefanini, F.R.; Falabella, P.; Koss, M.J.; Wells, T.; Diniz, B.; Ribeiro, R.; Schor, P.; Maia, M.; Penha, F.M.; et al. Development of a new tissue injector for subretinal transplantation of human embryonic stem cell derived retinal pigmented epithelium. *Int. J. Retina Vitr.* **2017**, *3*, 41. [CrossRef]
36. Ben M'Barek, K.; Bertin, S.; Brazhnikova, E.; Jaillard, C.; Habeler, W.; Plancheron, A.; Fovet, C.M.; Demilly, J.; Jarraya, M.; Bejanariu, A.; et al. Clinical-grade production and safe delivery of human ESC derived RPE sheets in primates and rodents. *Biomaterials* **2020**, *230*, 119603. [CrossRef]
37. Nakano, T.; Ando, S.; Takata, N.; Kawada, M.; Muguruma, K.; Sekiguchi, K.; Saito, K.; Yonemura, S.; Eiraku, M.; Sasai, Y. Self-formation of optic cups and storable stratified neural retina from human ESCs. *Cell Stem Cell* **2012**, *10*, 771–785. [CrossRef]
38. Zhong, X.; Gutierrez, C.; Xue, T.; Hampton, C.; Vergara, M.N.; Cao, L.H.; Peters, A.; Park, T.S.; Zambidis, E.T.; Meyer, J.S.; et al. Generation of three-dimensional retinal tissue with functional photoreceptors from human iPSCs. *Nat. Commun.* **2014**, *5*, 4047. [CrossRef]
39. Capowski, E.E.; Samimi, K.; Mayerl, S.J.; Phillips, M.J.; Pinilla, I.; Howden, S.E.; Saha, J.; Jansen, A.D.; Edwards, K.L.; Jager, L.D.; et al. Reproducibility and staging of 3D human retinal organoids across multiple pluripotent stem cell lines. *Development* **2019**, *146*. [CrossRef]
40. Reichman, S.; Slembrouck, A.; Gagliardi, G.; Chaffiol, A.; Terray, A.; Nanteau, C.; Potey, A.; Belle, M.; Rabesandratana, O.; Duebel, J.; et al. Generation of Storable Retinal Organoids and Retinal Pigmented Epithelium from Adherent Human iPS Cells in Xeno-Free and Feeder-Free Conditions. *Stem Cells* **2017**, *35*, 1176–1188. [CrossRef]
41. Welby, E.; Lakowski, J.; Di Foggia, V.; Budinger, D.; Gonzalez-Cordero, A.; Lun, A.T.L.; Epstein, M.; Patel, A.; Cuevas, E.; Kruczek, K.; et al. Isolation and Comparative Transcriptome Analysis of Human Fetal and iPSC-Derived Cone Photoreceptor Cells. *Stem Cell Rep.* **2017**, *9*, 1898–1915. [CrossRef] [PubMed]
42. MacLaren, R.E.; Pearson, R.A.; MacNeil, A.; Douglas, R.H.; Salt, T.E.; Akimoto, M.; Swaroop, A.; Sowden, J.C.; Ali, R.R. Retinal repair by transplantation of photoreceptor precursors. *Nature* **2006**, *444*, 203–207. [CrossRef]
43. Singh, M.S.; Balmer, J.; Barnard, A.R.; Aslam, S.A.; Moralli, D.; Green, C.M.; Barnea-Cramer, A.; Duncan, I.; MacLaren, R.E. Transplanted photoreceptor precursors transfer proteins to host photoreceptors by a mechanism of cytoplasmic fusion. *Nat. Commun.* **2016**, *7*, 13537. [CrossRef] [PubMed]
44. Santos-Ferreira, T.; Llonch, S.; Borsch, O.; Postel, K.; Haas, J.; Ader, M. Retinal transplantation of photoreceptors results in donor-host cytoplasmic exchange. *Nat. Commun.* **2016**, *7*, 13028. [CrossRef] [PubMed]
45. Decembrini, S.; Martin, C.; Sennlaub, F.; Chemtob, S.; Biel, M.; Samardzija, M.; Moulin, A.; Behar-Cohen, F.; Arsenijevic, Y. Cone Genesis Tracing by the Chrnb4-EGFP Mouse Line: Evidences of Cellular Material Fusion after Cone Precursor Transplantation. *Mol. Ther.* **2017**, *25*, 634–653. [CrossRef]
46. Tu, H.Y.; Watanabe, T.; Shirai, H.; Yamasaki, S.; Kinoshita, M.; Matsushita, K.; Hashiguchi, T.; Onoe, H.; Matsuyama, T.; Kuwahara, A.; et al. Medium- to long-term survival and functional examination of human iPSC-derived retinas in rat and primate models of retinal degeneration. *EBioMedicine* **2019**, *39*, 562–574. [CrossRef] [PubMed]
47. Iraha, S.; Tu, H.Y.; Yamasaki, S.; Kagawa, T.; Goto, M.; Takahashi, R.; Watanabe, T.; Sugita, S.; Yonemura, S.; Sunagawa, G.A.; et al. Establishment of Immunodeficient Retinal Degeneration Model Mice and Functional Maturation of Human ESC-Derived Retinal Sheets after Transplantation. *Stem Cell Rep.* **2018**, *10*, 1059–1074. [CrossRef] [PubMed]
48. McLelland, B.T.; Lin, B.; Mathur, A.; Aramant, R.B.; Thomas, B.B.; Nistor, G.; Keirstead, H.S.; Seiler, M.J. Transplanted hESC-Derived Retina Organoid Sheets Differentiate, Integrate, and Improve Visual Function in Retinal Degenerate Rats. *Investig. Ophthalmol. Vis. Sci.* **2018**, *59*, 2586–2603. [CrossRef]
49. Jung, Y.H.; Phillips, M.J.; Lee, J.; Xie, R.; Ludwig, A.L.; Chen, G.; Zheng, Q.; Kim, T.J.; Zhang, H.; Barney, P.; et al. 3D Microstructured Scaffolds to Support Photoreceptor Polarization and Maturation. *Adv. Mater.* **2018**, *30*, e1803550. [CrossRef]
50. Worthington, K.S.; Wiley, L.A.; Kaalberg, E.E.; Collins, M.M.; Mullins, R.F.; Stone, E.M.; Tucker, B.A. Two-photon polymerization for production of human iPSC-derived retinal cell grafts. *Acta Biomater.* **2017**, *55*, 385–395. [CrossRef]
51. Thompson, J.R.; Worthington, K.S.; Green, B.J.; Mullin, N.K.; Jiao, C.; Kaalberg, E.E.; Wiley, L.A.; Han, I.C.; Russell, S.R.; Sohn, E.H.; et al. Two-photon polymerized poly(caprolactone) retinal cell delivery scaffolds and their systemic and retinal biocompatibility. *Acta Biomater.* **2019**, *94*, 204–218. [CrossRef] [PubMed]
52. Singh, D.; Wang, S.B.; Xia, T.; Tainsh, L.; Ghiassi-Nejad, M.; Xu, T.; Peng, S.; Adelman, R.A.; Rizzolo, L.J. A biodegradable scaffold enhances differentiation of embryonic stem cells into a thick sheet of retinal cells. *Biomaterials* **2018**, *154*, 158–168. [CrossRef] [PubMed]
53. Garita-Hernandez, M.; Lampic, M.; Chaffiol, A.; Guibbal, L.; Routet, F.; Santos-Ferreira, T.; Gasparini, S.; Borsch, O.; Gagliardi, G.; Reichman, S.; et al. Restoration of visual function by transplantation of optogenetically engineered photoreceptors. *Nat. Commun.* **2019**, *10*, 4524. [CrossRef]

54. Chu, G.Y.; Chen, Y.F.; Chen, H.Y.; Chan, M.H.; Gau, C.S.; Weng, S.M. Stem cell therapy on skin: Mechanisms, recent advances and drug reviewing issues. *J. Food Drug Anal.* **2018**, *26*, 14–20. [CrossRef] [PubMed]
55. Gurtner, G.C.; Werner, S.; Barrandon, Y.; Longaker, M.T. Wound repair and regeneration. *Nature* **2008**, *453*, 314–321. [CrossRef]
56. Przekora, A. A Concise Review on Tissue Engineered Artificial Skin Grafts for Chronic Wound Treatment: Can We Reconstruct Functional Skin Tissue In Vitro? *Cells* **2020**, *9*, 1622. [CrossRef]
57. Dai, C.; Shih, S.; Khachemoune, A. Skin substitutes for acute and chronic wound healing: An updated review. *J. Dermatolog. Treat.* **2020**, *31*, 639–648. [CrossRef]
58. Nourian Dehkordi, A.; Mirahmadi Babaheydari, F.; Chehelgerdi, M.; Raeisi Dehkordi, S. Skin tissue engineering: Wound healing based on stem-cell-based therapeutic strategies. *Stem Cell Res. Ther.* **2019**, *10*, 111. [CrossRef]
59. Vig, K.; Chaudhari, A.; Tripathi, S.; Dixit, S.; Sahu, R.; Pillai, S.; Dennis, V.A.; Singh, S.R. Advances in Skin Regeneration Using Tissue Engineering. *Int. J. Mol. Sci.* **2017**, *18*, 789. [CrossRef]
60. Guenou, H.; Nissan, X.; Larcher, F.; Feteira, J.; Lemaitre, G.; Saidani, M.; Del Rio, M.; Barrault, C.C.; Bernard, F.X.; Peschanski, M.; et al. Human embryonic stem-cell derivatives for full reconstruction of the pluristratified epidermis: A preclinical study. *Lancet* **2009**, *374*, 1745–1753. [CrossRef]
61. Umegaki-Arao, N.; Pasmooij, A.M.; Itoh, M.; Cerise, J.E.; Guo, Z.; Levy, B.; Gostynski, A.; Rothman, L.R.; Jonkman, M.F.; Christiano, A.M. Induced pluripotent stem cells from human revertant keratinocytes for the treatment of epidermolysis bullosa. *Sci. Transl. Med.* **2014**, *6*, 264ra164. [CrossRef] [PubMed]
62. Sebastiano, V.; Zhen, H.H.; Haddad, B.; Bashkirova, E.; Melo, S.P.; Wang, P.; Leung, T.L.; Siprashvili, Z.; Tichy, A.; Li, J.; et al. Human COL7A1-corrected induced pluripotent stem cells for the treatment of recessive dystrophic epidermolysis bullosa. *Sci. Transl. Med.* **2014**, *6*, 264ra163. [CrossRef] [PubMed]
63. Miyazaki, H.; Tsunoi, Y.; Akagi, T.; Sato, S.; Akashi, M.; Saitoh, D. A novel strategy to engineer pre-vascularized 3-dimensional skin substitutes to achieve efficient, functional engraftment. *Sci. Rep.* **2019**, *9*, 7797. [CrossRef] [PubMed]
64. Itoh, M.; Umegaki-Arao, N.; Guo, Z.; Liu, L.; Higgins, C.A.; Christiano, A.M. Generation of 3D skin equivalents fully reconstituted from human induced pluripotent stem cells (iPSCs). *PLoS ONE* **2013**, *8*, e77673. [CrossRef]
65. Abaci, H.E.; Guo, Z.; Coffman, A.; Gillette, B.; Lee, W.H.; Sia, S.K.; Christiano, A.M. Human Skin Constructs with Spatially Controlled Vasculature Using Primary and iPSC-Derived Endothelial Cells. *Adv. Healthc. Mater.* **2016**, *5*, 1800–1807. [CrossRef]
66. Kim, B.S.; Kwon, Y.W.; Kong, J.S.; Park, G.T.; Gao, G.; Han, W.; Kim, M.B.; Lee, H.; Kim, J.H.; Cho, D.W. 3D cell printing of in vitro stabilized skin model and in vivo pre-vascularized skin patch using tissue-specific extracellular matrix bioink: A step towards advanced skin tissue engineering. *Biomaterials* **2018**, *168*, 38–53. [CrossRef]
67. Hosaka, C.; Kunisada, M.; Koyanagi-Aoi, M.; Masaki, T.; Takemori, C.; Taniguchi-Ikeda, M.; Aoi, T.; Nishigori, C. Induced pluripotent stem cell-derived melanocyte precursor cells undergoing differentiation into melanocytes. *Pigment Cell Melanoma Res.* **2019**, *32*, 623–633. [CrossRef]
68. Ohta, S.; Imaizumi, Y.; Okada, Y.; Akamatsu, W.; Kuwahara, R.; Ohyama, M.; Amagai, M.; Matsuzaki, Y.; Yamanaka, S.; Okano, H.; et al. Generation of human melanocytes from induced pluripotent stem cells. *PLoS ONE* **2011**, *6*, e16182. [CrossRef]
69. Nissan, X.; Larribere, L.; Saidani, M.; Hurbain, I.; Delevoye, C.; Feteira, J.; Lemaitre, G.; Peschanski, M.; Baldeschi, C. Functional melanocytes derived from human pluripotent stem cells engraft into pluristratified epidermis. *Proc. Natl. Acad. Sci. USA* **2011**, *108*, 14861–14866. [CrossRef]
70. Lee, J.; Rabbani, C.C.; Gao, H.; Steinhart, M.R.; Woodruff, B.M.; Pflum, Z.E.; Kim, A.; Heller, S.; Liu, Y.; Shipchandler, T.Z.; et al. Hair-bearing human skin generated entirely from pluripotent stem cells. *Nature* **2020**, *582*, 399–404. [CrossRef]
71. Yang, R.; Zheng, Y.; Burrows, M.; Liu, S.; Wei, Z.; Nace, A.; Guo, W.; Kumar, S.; Cotsarelis, G.; Xu, X. Generation of folliculogenic human epithelial stem cells from induced pluripotent stem cells. *Nat. Commun.* **2014**, *5*, 3071. [CrossRef] [PubMed]
72. GBD 2017 Causes of Death Collaborators. Global, regional, and national age-sex-specific mortality for 282 causes of death in 195 countries and territories, 1980-2017: A systematic analysis for the Global Burden of Disease Study 2017. *Lancet* **2018**, *392*, 1736–1788. [CrossRef]
73. Nakamura, K.; Murry, C.E. Function Follows Form- A Review of Cardiac Cell Therapy. *Circ. J.* **2019**, *83*, 2399–2412. [CrossRef] [PubMed]
74. Muller, P.; Lemcke, H.; David, R. Stem Cell Therapy in Heart Diseases—Cell Types, Mechanisms and Improvement Strategies. *Cell Physiol. Biochem.* **2018**, *48*, 2607–2655. [CrossRef]
75. Kadota, S.; Tanaka, Y.; Shiba, Y. Heart regeneration using pluripotent stem cells. *J. Cardiol.* **2020**, *76*, 459–463. [CrossRef]
76. Yang, L.; Soonpaa, M.H.; Adler, E.D.; Roepke, T.K.; Kattman, S.J.; Kennedy, M.; Henckaerts, E.; Bonham, K.; Abbott, G.W.; Linden, R.M.; et al. Human cardiovascular progenitor cells develop from a KDR+ embryonic-stem-cell-derived population. *Nature* **2008**, *453*, 524–528. [CrossRef]
77. Laflamme, M.A.; Chen, K.Y.; Naumova, A.V.; Muskheli, V.; Fugate, J.A.; Dupras, S.K.; Reinecke, H.; Xu, C.; Hassanipour, M.; Police, S.; et al. Cardiomyocytes derived from human embryonic stem cells in pro-survival factors enhance function of infarcted rat hearts. *Nat. Biotechnol.* **2007**, *25*, 1015–1024. [CrossRef]
78. Karakikes, I.; Ameen, M.; Termglinchan, V.; Wu, J.C. Human induced pluripotent stem cell-derived cardiomyocytes: Insights into molecular, cellular, and functional phenotypes. *Circ. Res.* **2015**, *117*, 80–88. [CrossRef]
79. Machiraju, P.; Greenway, S.C. Current methods for the maturation of induced pluripotent stem cell-derived cardiomyocytes. *World J. Stem Cells* **2019**, *11*, 33–43. [CrossRef]

80. Burridge, P.W.; Matsa, E.; Shukla, P.; Lin, Z.C.; Churko, J.M.; Ebert, A.D.; Lan, F.; Diecke, S.; Huber, B.; Mordwinkin, N.M.; et al. Chemically defined generation of human cardiomyocytes. *Nat. Methods* **2014**, *11*, 855–860. [CrossRef]
81. Kempf, H.; Kropp, C.; Olmer, R.; Martin, U.; Zweigerdt, R. Cardiac differentiation of human pluripotent stem cells in scalable suspension culture. *Nat. Protoc.* **2015**, *10*, 1345–1361. [CrossRef] [PubMed]
82. Verma, V.; Purnamawati, K.; Manasi; Shim, W. Steering signal transduction pathway towards cardiac lineage from human pluripotent stem cells: A review. *Cell Signal* **2013**, *25*, 1096–1107. [CrossRef] [PubMed]
83. Ellis, P.; Fagan, B.M.; Magness, S.T.; Hutton, S.; Taranova, O.; Hayashi, S.; McMahon, A.; Rao, M.; Pevny, L. SOX2, a persistent marker for multipotential neural stem cells derived from embryonic stem cells, the embryo or the adult. *Dev. Neurosci.* **2004**, *26*, 148–165. [CrossRef] [PubMed]
84. Lian, X.; Hsiao, C.; Wilson, G.; Zhu, K.; Hazeltine, L.B.; Azarin, S.M.; Raval, K.K.; Zhang, J.; Kamp, T.J.; Palecek, S.P. Robust cardiomyocyte differentiation from human pluripotent stem cells via temporal modulation of canonical Wnt signaling. *Proc. Natl. Acad. Sci. USA* **2012**, *109*, E1848–E1857. [CrossRef]
85. Kadota, S.; Minami, I.; Morone, N.; Heuser, J.E.; Agladze, K.; Nakatsuji, N. Development of a reentrant arrhythmia model in human pluripotent stem cell-derived cardiac cell sheets. *Eur. Heart J.* **2013**, *34*, 1147–1156. [CrossRef]
86. Menasche, P.; Vanneaux, V.; Hagege, A.; Bel, A.; Cholley, B.; Parouchev, A.; Cacciapuoti, I.; Al-Daccak, R.; Benhamouda, N.; Blons, H.; et al. Transplantation of Human Embryonic Stem Cell-Derived Cardiovascular Progenitors for Severe Ischemic Left Ventricular Dysfunction. *J. Am. Coll. Cardiol.* **2018**, *71*, 429–438. [CrossRef]
87. Guan, X.; Xu, W.; Zhang, H.; Wang, Q.; Yu, J.; Zhang, R.; Chen, Y.; Xia, Y.; Wang, J.; Wang, D. Transplantation of human induced pluripotent stem cell-derived cardiomyocytes improves myocardial function and reverses ventricular remodeling in infarcted rat hearts. *Stem Cell Res. Ther.* **2020**, *11*, 73. [CrossRef]
88. Shiba, Y.; Gomibuchi, T.; Seto, T.; Wada, Y.; Ichimura, H.; Tanaka, Y.; Ogasawara, T.; Okada, K.; Shiba, N.; Sakamoto, K.; et al. Allogeneic transplantation of iPS cell-derived cardiomyocytes regenerates primate hearts. *Nature* **2016**, *538*, 388–391. [CrossRef]
89. Liu, Y.W.; Chen, B.; Yang, X.; Fugate, J.A.; Kalucki, F.A.; Futakuchi-Tsuchida, A.; Couture, L.; Vogel, K.W.; Astley, C.A.; Baldessari, A.; et al. Erratum: Human embryonic stem cell-derived cardiomyocytes restore function in infarcted hearts of non-human primates. *Nat. Biotechnol.* **2018**, *36*, 899. [CrossRef]
90. Chong, J.J.; Yang, X.; Don, C.W.; Minami, E.; Liu, Y.W.; Weyers, J.J.; Mahoney, W.M.; Van Biber, B.; Cook, S.M.; Palpant, N.J.; et al. Human embryonic-stem-cell-derived cardiomyocytes regenerate non-human primate hearts. *Nature* **2014**, *510*, 273–277. [CrossRef]
91. Ishida, M.; Miyagawa, S.; Saito, A.; Fukushima, S.; Harada, A.; Ito, E.; Ohashi, F.; Watabe, T.; Hatazawa, J.; Matsuura, K.; et al. Transplantation of Human-induced Pluripotent Stem Cell-derived Cardiomyocytes Is Superior to Somatic Stem Cell Therapy for Restoring Cardiac Function and Oxygen Consumption in a Porcine Model of Myocardial Infarction. *Transplantation* **2019**, *103*, 291–298. [CrossRef] [PubMed]
92. Romagnuolo, R.; Masoudpour, H.; Porta-Sanchez, A.; Qiang, B.; Barry, J.; Laskary, A.; Qi, X.; Masse, S.; Magtibay, K.; Kawajiri, H.; et al. Human Embryonic Stem Cell-Derived Cardiomyocytes Regenerate the Infarcted Pig Heart but Induce Ventricular Tachyarrhythmias. *Stem Cell Rep.* **2019**, *12*, 967–981. [CrossRef] [PubMed]
93. Kehat, I.; Khimovich, L.; Caspi, O.; Gepstein, A.; Shofti, R.; Arbel, G.; Huber, I.; Satin, J.; Itskovitz-Eldor, J.; Gepstein, L. Electromechanical integration of cardiomyocytes derived from human embryonic stem cells. *Nat. Biotechnol.* **2004**, *22*, 1282–1289. [CrossRef]
94. Zhang, J.Z.; Termglinchan, V.; Shao, N.Y.; Itzhaki, I.; Liu, C.; Ma, N.; Tian, L.; Wang, V.Y.; Chang, A.C.Y.; Guo, H.; et al. A Human iPSC Double-Reporter System Enables Purification of Cardiac Lineage Subpopulations with Distinct Function and Drug Response Profiles. *Cell Stem Cell* **2019**, *24*, 802–811. [CrossRef]
95. Gerbin, K.A.; Yang, X.; Murry, C.E.; Coulombe, K.L. Enhanced Electrical Integration of Engineered Human Myocardium via Intramyocardial versus Epicardial Delivery in Infarcted Rat Hearts. *PLoS ONE* **2015**, *10*, e0131446. [CrossRef]
96. Bargehr, J.; Ong, L.P.; Colzani, M.; Davaapil, H.; Hofsteen, P.; Bhandari, S.; Gambardella, L.; Le Novere, N.; Iyer, D.; Sampaziotis, F.; et al. Epicardial cells derived from human embryonic stem cells augment cardiomyocyte-driven heart regeneration. *Nat. Biotechnol.* **2019**, *37*, 895–906. [CrossRef] [PubMed]
97. Park, S.J.; Kim, R.Y.; Park, B.W.; Lee, S.; Choi, S.W.; Park, J.H.; Choi, J.J.; Kim, S.W.; Jang, J.; Cho, D.W.; et al. Dual stem cell therapy synergistically improves cardiac function and vascular regeneration following myocardial infarction. *Nat. Commun.* **2019**, *10*, 3123. [CrossRef]
98. Judson, R.N.; Rossi, F.M.V. Towards stem cell therapies for skeletal muscle repair. *NPJ Regen. Med.* **2020**, *5*, 10. [CrossRef]
99. Chal, J.; Pourquie, O. Making muscle: Skeletal myogenesis in vivo and in vitro. *Development* **2017**, *144*, 2104–2122. [CrossRef]
100. Qazi, T.H.; Duda, G.N.; Ort, M.J.; Perka, C.; Geissler, S.; Winkler, T. Cell therapy to improve regeneration of skeletal muscle injuries. *J. Cachexia Sarcopenia Muscle* **2019**, *10*, 501–516. [CrossRef]
101. Sacco, A.; Doyonnas, R.; Kraft, P.; Vitorovic, S.; Blau, H.M. Self-renewal and expansion of single transplanted muscle stem cells. *Nature* **2008**, *456*, 502–506. [CrossRef] [PubMed]
102. Cerletti, M.; Jurga, S.; Witczak, C.A.; Hirshman, M.F.; Shadrach, J.L.; Goodyear, L.J.; Wagers, A.J. Highly efficient, functional engraftment of skeletal muscle stem cells in dystrophic muscles. *Cell* **2008**, *134*, 37–47. [CrossRef] [PubMed]
103. Montarras, D.; Morgan, J.; Collins, C.; Relaix, F.; Zaffran, S.; Cumano, A.; Partridge, T.; Buckingham, M. Direct isolation of satellite cells for skeletal muscle regeneration. *Science* **2005**, *309*, 2064–2067. [CrossRef] [PubMed]

104. Arpke, R.W.; Darabi, R.; Mader, T.L.; Zhang, Y.; Toyama, A.; Lonetree, C.L.; Nash, N.; Lowe, D.A.; Perlingeiro, R.C.; Kyba, M. A new immuno-, dystrophin-deficient model, the NSG-mdx(4Cv) mouse, provides evidence for functional improvement following allogeneic satellite cell transplantation. *Stem Cells* **2013**, *31*, 1611–1620. [CrossRef] [PubMed]
105. Chal, J.; Oginuma, M.; Al Tanoury, Z.; Gobert, B.; Sumara, O.; Hick, A.; Bousson, F.; Zidouni, Y.; Mursch, C.; Moncuquet, P.; et al. Differentiation of pluripotent stem cells to muscle fiber to model Duchenne muscular dystrophy. *Nat. Biotechnol.* **2015**, *33*, 962–969. [CrossRef]
106. Chal, J.; Al Tanoury, Z.; Hestin, M.; Gobert, B.; Aivio, S.; Hick, A.; Cherrier, T.; Nesmith, A.P.; Parker, K.K.; Pourquie, O. Generation of human muscle fibers and satellite-like cells from human pluripotent stem cells in vitro. *Nat. Protoc.* **2016**, *11*, 1833–1850. [CrossRef]
107. Hicks, M.R.; Hiserodt, J.; Paras, K.; Fujiwara, W.; Eskin, A.; Jan, M.; Xi, H.; Young, C.S.; Evseenko, D.; Nelson, S.F.; et al. ERBB3 and NGFR mark a distinct skeletal muscle progenitor cell in human development and hPSCs. *Nat. Cell. Biol.* **2018**, *20*, 46–57. [CrossRef]
108. Wu, J.; Matthias, N.; Lo, J.; Ortiz-Vitali, J.L.; Shieh, A.W.; Wang, S.H.; Darabi, R. A Myogenic Double-Reporter Human Pluripotent Stem Cell Line Allows Prospective Isolation of Skeletal Muscle Progenitors. *Cell Rep.* **2018**, *25*, 1966–1981.e4. [CrossRef]
109. van der Wal, E.; Herrero-Hernandez, P.; Wan, R.; Broeders, M.; In 't Groen, S.L.M.; van Gestel, T.J.M.; van, I.W.F.J.; Cheung, T.H.; van der Ploeg, A.T.; Schaaf, G.J.; et al. Large-Scale Expansion of Human iPSC-Derived Skeletal Muscle Cells for Disease Modeling and Cell-Based Therapeutic Strategies. *Stem Cell Rep.* **2018**, *10*, 1975–1990. [CrossRef]
110. Quarta, M.; Cromie, M.; Chacon, R.; Blonigan, J.; Garcia, V.; Akimenko, I.; Hamer, M.; Paine, P.; Stok, M.; Shrager, J.B.; et al. Bioengineered constructs combined with exercise enhance stem cell-mediated treatment of volumetric muscle loss. *Nat. Commun.* **2017**, *8*, 15613. [CrossRef]
111. Wolf, M.T.; Dearth, C.L.; Sonnenberg, S.B.; Loboa, E.G.; Badylak, S.F. Naturally derived and synthetic scaffolds for skeletal muscle reconstruction. *Adv. Drug Deliv. Rev.* **2015**, *84*, 208–221. [CrossRef] [PubMed]
112. Alper, J. Geron gets green light for human trial of ES cell-derived product. *Nat. Biotechnol.* **2009**, *27*, 213–214. [CrossRef]
113. Scott, C.T.; Magnus, D. Wrongful termination: Lessons from the Geron clinical trial. *Stem Cells Transl. Med.* **2014**, *3*, 1398–1401. [CrossRef] [PubMed]
114. Desgres, M.; Menasche, P. Clinical Translation of Pluripotent Stem Cell Therapies: Challenges and Considerations. *Cell Stem Cell* **2019**, *25*, 594–606. [CrossRef]
115. Qiu, T.G. Transplantation of human embryonic stem cell-derived retinal pigment epithelial cells (MA09-hRPE) in macular degeneration. *NPJ Regen. Med.* **2019**, *4*, 19. [CrossRef] [PubMed]
116. Kashani, A.H.; Lebkowski, J.S.; Rahhal, F.M.; Avery, R.L.; Salehi-Had, H.; Dang, W.; Lin, C.M.; Mitra, D.; Zhu, D.; Thomas, B.B.; et al. A bioengineered retinal pigment epithelial monolayer for advanced, dry age-related macular degeneration. *Sci. Transl. Med.* **2018**, *10*. [CrossRef]
117. Mandai, M.; Watanabe, A.; Kurimoto, Y.; Hirami, Y.; Morinaga, C.; Daimon, T.; Fujihara, M.; Akimaru, H.; Sakai, N.; Shibata, Y.; et al. Autologous Induced Stem-Cell-Derived Retinal Cells for Macular Degeneration. *N. Engl. J. Med.* **2017**, *376*, 1038–1046. [CrossRef]
118. Sneddon, J.B.; Tang, Q.; Stock, P.; Bluestone, J.A.; Roy, S.; Desai, T.; Hebrok, M. Stem Cell Therapies for Treating Diabetes: Progress and Remaining Challenges. *Cell Stem Cell* **2018**, *22*, 810–823. [CrossRef]
119. Bravery, C.A. Do human leukocyte antigen-typed cellular therapeutics based on induced pluripotent stem cells make commercial sense? *Stem Cells Dev.* **2015**, *24*, 1–10. [CrossRef]
120. Malat, G.; Culkin, C. The ABCs of Immunosuppression: A Primer for Primary Care Physicians. *Med. Clin. N. Am.* **2016**, *100*, 505–518. [CrossRef]
121. Schwartz, S.D.; Regillo, C.D.; Lam, B.L.; Eliott, D.; Rosenfeld, P.J.; Gregori, N.Z.; Hubschman, J.P.; Davis, J.L.; Heilwell, G.; Spirn, M.; et al. Human embryonic stem cell-derived retinal pigment epithelium in patients with age-related macular degeneration and Stargardt's macular dystrophy: Follow-up of two open-label phase 1/2 studies. *Lancet* **2015**, *385*, 509–516. [CrossRef]
122. Taylor, C.J.; Bolton, E.M.; Pocock, S.; Sharples, L.D.; Pedersen, R.A.; Bradley, J.A. Banking on human embryonic stem cells: Estimating the number of donor cell lines needed for HLA matching. *Lancet* **2005**, *366*, 2019–2025. [CrossRef]
123. Wilmut, I.; Leslie, S.; Martin, N.G.; Peschanski, M.; Rao, M.; Trounson, A.; Turner, D.; Turner, M.L.; Yamanaka, S.; Taylor, C.J. Development of a global network of induced pluripotent stem cell haplobanks. *Regen. Med.* **2015**, *10*, 235–238. [CrossRef] [PubMed]
124. Taylor, C.J.; Peacock, S.; Chaudhry, A.N.; Bradley, J.A.; Bolton, E.M. Generating an iPSC bank for HLA-matched tissue transplantation based on known donor and recipient HLA types. *Cell Stem Cell* **2012**, *11*, 147–152. [CrossRef]
125. Sugita, S.; Mandai, M.; Hirami, Y.; Takagi, S.; Maeda, T.; Fujihara, M.; Matsuzaki, M.; Yamamoto, M.; Iseki, K.; Hayashi, N.; et al. HLA-Matched Allogeneic iPS Cells-Derived RPE Transplantation for Macular Degeneration. *J. Clin. Med.* **2020**, *9*, 2217. [CrossRef]
126. Gornalusse, G.G.; Hirata, R.K.; Funk, S.E.; Riolobos, L.; Lopes, V.S.; Manske, G.; Prunkard, D.; Colunga, A.G.; Hanafi, L.A.; Clegg, D.O.; et al. HLA-E-expressing pluripotent stem cells escape allogeneic responses and lysis by NK cells. *Nat. Biotechnol.* **2017**, *35*, 765–772. [CrossRef]
127. Xu, H.; Wang, B.; Ono, M.; Kagita, A.; Fujii, K.; Sasakawa, N.; Ueda, T.; Gee, P.; Nishikawa, M.; Nomura, M.; et al. Targeted Disruption of HLA Genes via CRISPR-Cas9 Generates iPSCs with Enhanced Immune Compatibility. *Cell Stem Cell* **2019**, *24*, 566–578.e7. [CrossRef]

128. Han, X.; Wang, M.; Duan, S.; Franco, P.J.; Kenty, J.H.; Hedrick, P.; Xia, Y.; Allen, A.; Ferreira, L.M.R.; Strominger, J.L.; et al. Generation of hypoimmunogenic human pluripotent stem cells. *Proc. Natl. Acad. Sci. USA* **2019**, *116*, 10441–10446. [CrossRef]
129. Deuse, T.; Hu, X.; Gravina, A.; Wang, D.; Tediashvili, G.; De, C.; Thayer, W.O.; Wahl, A.; Garcia, J.V.; Reichenspurner, H.; et al. Hypoimmunogenic derivatives of induced pluripotent stem cells evade immune rejection in fully immunocompetent allogeneic recipients. *Nat. Biotechnol.* **2019**, *37*, 252–258. [CrossRef]
130. Shi, Z.D.; Tchao, J.; Wu, L.; Carman, A.J. Precision installation of a highly efficient suicide gene safety switch in human induced pluripotent stem cells. *Stem Cells Transl. Med.* **2020**, *9*, 1378–1388. [CrossRef]
131. Jing, H.; He, X.; Zheng, J. Exosomes and regenerative medicine: State of the art and perspectives. *Transl. Res.* **2018**, *196*, 1–16. [CrossRef] [PubMed]
132. Brini, A.T.; Amodeo, G.; Ferreira, L.M.; Milani, A.; Niada, S.; Moschetti, G.; Franchi, S.; Borsani, E.; Rodella, L.F.; Panerai, A.E.; et al. Therapeutic effect of human adipose-derived stem cells and their secretome in experimental diabetic pain. *Sci. Rep.* **2017**, *7*, 9904. [CrossRef] [PubMed]
133. Maumus, M.; Rozier, P.; Boulestreau, J.; Jorgensen, C.; Noel, D. Mesenchymal Stem Cell-Derived Extracellular Vesicles: Opportunities and Challenges for Clinical Translation. *Front. Bioeng. Biotechnol.* **2020**, *8*, 997. [CrossRef]
134. Cai, Y.; Liu, W.; Lian, L.; Xu, Y.; Bai, X.; Xu, S.; Zhang, J. Stroke treatment: Is exosome therapy superior to stem cell therapy? *Biochimie* **2020**, *179*, 190–204. [CrossRef] [PubMed]
135. Kennedy, T.L.; Russell, A.J.; Riley, P. Experimental limitations of extracellular vesicle-based therapies for the treatment of myocardial infarction. *Trends Cardiovasc. Med.* **2020**. [CrossRef]
136. Fuster-Matanzo, A.; Gessler, F.; Leonardi, T.; Iraci, N.; Pluchino, S. Acellular approaches for regenerative medicine: On the verge of clinical trials with extracellular membrane vesicles? *Stem Cell Res. Ther.* **2015**, *6*, 227. [CrossRef]
137. Thery, C.; Zitvogel, L.; Amigorena, S. Exosomes: Composition, biogenesis and function. *Nat. Rev. Immunol.* **2002**, *2*, 569–579. [CrossRef]
138. Valadi, H.; Ekstrom, K.; Bossios, A.; Sjostrand, M.; Lee, J.J.; Lotvall, J.O. Exosome-mediated transfer of mRNAs and microRNAs is a novel mechanism of genetic exchange between cells. *Nat. Cell Biol.* **2007**, *9*, 654–659. [CrossRef]
139. Skog, J.; Wurdinger, T.; van Rijn, S.; Meijer, D.H.; Gainche, L.; Sena-Esteves, M.; Curry, W.T., Jr.; Carter, B.S.; Krichevsky, A.M.; Breakefield, X.O. Glioblastoma microvesicles transport RNA and proteins that promote tumour growth and provide diagnostic biomarkers. *Nat. Cell Biol.* **2008**, *10*, 1470–1476. [CrossRef]
140. Mulcahy, L.A.; Pink, R.C.; Carter, D.R. Routes and mechanisms of extracellular vesicle uptake. *J. Extracell. Vesicles* **2014**, *3*. [CrossRef]
141. Mathieu, M.; Martin-Jaular, L.; Lavieu, G.; Thery, C. Specificities of secretion and uptake of exosomes and other extracellular vesicles for cell-to-cell communication. *Nat. Cell Biol.* **2019**, *21*, 9–17. [CrossRef] [PubMed]
142. Thery, C.; Witwer, K.W.; Aikawa, E.; Alcaraz, M.J.; Anderson, J.D.; Andriantsitohaina, R.; Antoniou, A.; Arab, T.; Archer, F.; Atkin-Smith, G.K.; et al. Minimal information for studies of extracellular vesicles 2018 (MISEV2018): A position statement of the International Society for Extracellular Vesicles and update of the MISEV2014 guidelines. *J. Extracell. Vesicles* **2018**, *7*, 1535750. [CrossRef] [PubMed]
143. Reiner, A.T.; Witwer, K.W.; van Balkom, B.W.M.; de Beer, J.; Brodie, C.; Corteling, R.L.; Gabrielsson, S.; Gimona, M.; Ibrahim, A.G.; de Kleijn, D.; et al. Concise Review: Developing Best-Practice Models for the Therapeutic Use of Extracellular Vesicles. *Stem Cells Transl. Med.* **2017**, *6*, 1730–1739. [CrossRef] [PubMed]
144. Gandham, S.; Su, X.; Wood, J.; Nocera, A.L.; Alli, S.C.; Milane, L.; Zimmerman, A.; Amiji, M.; Ivanov, A.R. Technologies and Standardization in Research on Extracellular Vesicles. *Trends Biotechnol.* **2020**, *38*, 1066–1098. [CrossRef] [PubMed]
145. Markov, O.; Oshchepkova, A.; Mironova, N. Immunotherapy Based on Dendritic Cell-Targeted/-Derived Extracellular Vesicles-A Novel Strategy for Enhancement of the Anti-tumor Immune Response. *Front. Pharmacol.* **2019**, *10*, 1152. [CrossRef]
146. Besse, B.; Charrier, M.; Lapierre, V.; Dansin, E.; Lantz, O.; Planchard, D.; Le Chevalier, T.; Livartoski, A.; Barlesi, F.; Laplanche, A.; et al. Dendritic cell-derived exosomes as maintenance immunotherapy after first line chemotherapy in NSCLC. *Oncoimmunology* **2016**, *5*, e1071008. [CrossRef]
147. Mead, B.; Tomarev, S. Extracellular vesicle therapy for retinal diseases. *Prog. Retin. Eye Res.* **2020**, 100849. [CrossRef]
148. Wang, S.; Girman, S.; Lu, B.; Bischoff, N.; Holmes, T.; Shearer, R.; Wright, L.S.; Svendsen, C.N.; Gamm, D.M.; Lund, R.D. Long-term vision rescue by human neural progenitors in a rat model of photoreceptor degeneration. *Investig. Ophthalmol. Vis. Sci.* **2008**, *49*, 3201–3206. [CrossRef]
149. Liu, Y.; Chen, S.J.; Li, S.Y.; Qu, L.H.; Meng, X.H.; Wang, Y.; Xu, H.W.; Liang, Z.Q.; Yin, Z.Q. Long-term safety of human retinal progenitor cell transplantation in retinitis pigmentosa patients. *Stem Cell Res. Ther.* **2017**, *8*, 209. [CrossRef]
150. Luo, J.; Baranov, P.; Patel, S.; Ouyang, H.; Quach, J.; Wu, F.; Qiu, A.; Luo, H.; Hicks, C.; Zeng, J.; et al. Human retinal progenitor cell transplantation preserves vision. *J. Biol. Chem.* **2014**, *289*, 6362–6371. [CrossRef]
151. Bian, B.; Zhao, C.; He, X.; Gong, Y.; Ren, C.; Ge, L.; Zeng, Y.; Li, Q.; Chen, M.; Weng, C.; et al. Exosomes derived from neural progenitor cells preserve photoreceptors during retinal degeneration by inactivating microglia. *J. Extracell. Vesicles* **2020**, *9*, 1748931. [CrossRef] [PubMed]
152. Peng, B.; Xiao, J.; Wang, K.; So, K.F.; Tipoe, G.L.; Lin, B. Suppression of microglial activation is neuroprotective in a mouse model of human retinitis pigmentosa. *J. Neurosci. Off. J. Soc. Neurosci.* **2014**, *34*, 8139–8150. [CrossRef] [PubMed]

153. He, G.H.; Zhang, W.; Ma, Y.X.; Yang, J.; Chen, L.; Song, J.; Chen, S. Mesenchymal stem cells-derived exosomes ameliorate blue light stimulation in retinal pigment epithelium cells and retinal laser injury by VEGF-dependent mechanism. *Int. J. Ophthalmol.* **2018**, *11*, 559–566. [CrossRef] [PubMed]
154. Yu, B.; Shao, H.; Su, C.; Jiang, Y.; Chen, X.; Bai, L.; Zhang, Y.; Li, Q.; Zhang, X.; Li, X. Exosomes derived from MSCs ameliorate retinal laser injury partially by inhibition of MCP-1. *Sci. Rep.* **2016**, *6*, 34562. [CrossRef]
155. Hajrasouliha, A.R.; Jiang, G.; Lu, Q.; Lu, H.; Kaplan, H.J.; Zhang, H.G.; Shao, H. Exosomes from retinal astrocytes contain antiangiogenic components that inhibit laser-induced choroidal neovascularization. *J. Biol. Chem.* **2013**, *288*, 28058–28067. [CrossRef]
156. Moisseiev, E.; Anderson, J.D.; Oltjen, S.; Goswami, M.; Zawadzki, R.J.; Nolta, J.A.; Park, S.S. Protective Effect of Intravitreal Administration of Exosomes Derived from Mesenchymal Stem Cells on Retinal Ischemia. *Curr. Eye Res.* **2017**, *42*, 1358–1367. [CrossRef]
157. Bartsch, U.; Oriyakhel, W.; Kenna, P.F.; Linke, S.; Richard, G.; Petrowitz, B.; Humphries, P.; Farrar, G.J.; Ader, M. Retinal cells integrate into the outer nuclear layer and differentiate into mature photoreceptors after subretinal transplantation into adult mice. *Exp. Eye Res.* **2008**, *86*, 691–700. [CrossRef]
158. Pearson, R.A.; Barber, A.C.; Rizzi, M.; Hippert, C.; Xue, T.; West, E.L.; Duran, Y.; Smith, A.J.; Chuang, J.Z.; Azam, S.A.; et al. Restoration of vision after transplantation of photoreceptors. *Nature* **2012**, *485*, 99–103. [CrossRef]
159. Barber, A.C.; Hippert, C.; Duran, Y.; West, E.L.; Bainbridge, J.W.; Warre-Cornish, K.; Luhmann, U.F.; Lakowski, J.; Sowden, J.C.; Ali, R.R.; et al. Repair of the degenerate retina by photoreceptor transplantation. *Proc. Natl. Acad. Sci. USA* **2013**, *110*, 354–359. [CrossRef]
160. Pearson, R.A.; Gonzalez-Cordero, A.; West, E.L.; Ribeiro, J.R.; Aghaizu, N.; Goh, D.; Sampson, R.D.; Georgiadis, A.; Waldron, P.V.; Duran, Y.; et al. Donor and host photoreceptors engage in material transfer following transplantation of post-mitotic photoreceptor precursors. *Nat. Commun.* **2016**, *7*, 13029. [CrossRef]
161. Ortin-Martinez, A.; Tsai, E.L.; Nickerson, P.E.; Bergeret, M.; Lu, Y.; Smiley, S.; Comanita, L.; Wallace, V.A. A Reinterpretation of Cell Transplantation: GFP Transfer From Donor to Host Photoreceptors. *Stem Cells* **2017**, *35*, 932–939. [CrossRef] [PubMed]
162. Singh, M.S.; Park, S.S.; Albini, T.A.; Canto-Soler, M.V.; Klassen, H.; MacLaren, R.E.; Takahashi, M.; Nagiel, A.; Schwartz, S.D.; Bharti, K. Retinal stem cell transplantation: Balancing safety and potential. *Prog. Retin. Eye Res.* **2020**, *75*, 100779. [CrossRef] [PubMed]
163. Wassmer, S.J.; Carvalho, L.S.; Gyorgy, B.; Vandenberghe, L.H.; Maguire, C.A. Exosome-associated AAV2 vector mediates robust gene delivery into the murine retina upon intravitreal injection. *Sci. Rep.* **2017**, *7*, 45329. [CrossRef] [PubMed]
164. van der Merwe, Y.; Steketee, M.B. Extracellular Vesicles: Biomarkers, Therapeutics, and Vehicles in the Visual System. *Curr. Ophthalmol. Rep.* **2017**, *5*, 276–282. [CrossRef]
165. Gonzalez, A.C.; Costa, T.F.; Andrade, Z.A.; Medrado, A.R. Wound healing—A literature review. *An. Bras. Dermatol.* **2016**, *91*, 614–620. [CrossRef]
166. Ha, D.H.; Kim, H.K.; Lee, J.; Kwon, H.H.; Park, G.H.; Yang, S.H.; Jung, J.Y.; Choi, H.; Lee, J.H.; Sung, S.; et al. Mesenchymal Stem/Stromal Cell-Derived Exosomes for Immunomodulatory Therapeutics and Skin Regeneration. *Cells* **2020**, *9*, 1157. [CrossRef]
167. Ferreira, A.D.F.; Gomes, D.A. Stem Cell Extracellular Vesicles in Skin Repair. *Bioengineering* **2018**, *6*, 4. [CrossRef]
168. Hesketh, M.; Sahin, K.B.; West, Z.E.; Murray, R.Z. Macrophage Phenotypes Regulate Scar Formation and Chronic Wound Healing. *Int. J. Mol. Sci.* **2017**, *18*, 1545. [CrossRef]
169. Ti, D.; Hao, H.; Tong, C.; Liu, J.; Dong, L.; Zheng, J.; Zhao, Y.; Liu, H.; Fu, X.; Han, W. LPS-preconditioned mesenchymal stromal cells modify macrophage polarization for resolution of chronic inflammation via exosome-shuttled let-7b. *J. Transl. Med.* **2015**, *13*, 308. [CrossRef]
170. He, X.; Dong, Z.; Cao, Y.; Wang, H.; Liu, S.; Liao, L.; Jin, Y.; Yuan, L.; Li, B. MSC-Derived Exosome Promotes M2 Polarization and Enhances Cutaneous Wound Healing. *Stem Cells Int.* **2019**, *2019*, 7132708. [CrossRef]
171. Liu, L.; Yu, Y.; Hou, Y.; Chai, J.; Duan, H.; Chu, W.; Zhang, H.; Hu, Q.; Du, J. Human umbilical cord mesenchymal stem cells transplantation promotes cutaneous wound healing of severe burned rats. *PLoS ONE* **2014**, *9*, e88348. [CrossRef] [PubMed]
172. Li, X.; Liu, L.; Yang, J.; Yu, Y.; Chai, J.; Wang, L.; Ma, L.; Yin, H. Exosome Derived From Human Umbilical Cord Mesenchymal Stem Cell Mediates MiR-181c Attenuating Burn-induced Excessive Inflammation. *EBioMedicine* **2016**, *8*, 72–82. [CrossRef] [PubMed]
173. Werner, S.; Krieg, T.; Smola, H. Keratinocyte-fibroblast interactions in wound healing. *J. Investig. Dermatol.* **2007**, *127*, 998–1008. [CrossRef] [PubMed]
174. Zhang, B.; Wang, M.; Gong, A.; Zhang, X.; Wu, X.; Zhu, Y.; Shi, H.; Wu, L.; Zhu, W.; Qian, H.; et al. HucMSC-Exosome Mediated-Wnt4 Signaling Is Required for Cutaneous Wound Healing. *Stem Cells* **2015**, *33*, 2158–2168. [CrossRef] [PubMed]
175. Shabbir, A.; Cox, A.; Rodriguez-Menocal, L.; Salgado, M.; Van Badiavas, E. Mesenchymal Stem Cell Exosomes Induce Proliferation and Migration of Normal and Chronic Wound Fibroblasts, and Enhance Angiogenesis In Vitro. *Stem Cells Dev.* **2015**, *24*, 1635–1647. [CrossRef]
176. Hu, L.; Wang, J.; Zhou, X.; Xiong, Z.; Zhao, J.; Yu, R.; Huang, F.; Zhang, H.; Chen, L. Exosomes derived from human adipose mensenchymal stem cells accelerates cutaneous wound healing via optimizing the characteristics of fibroblasts. *Sci. Rep.* **2016**, *6*, 32993. [CrossRef]

177. Ferreira, A.D.F.; Cunha, P.D.S.; Carregal, V.M.; da Silva, P.C.; de Miranda, M.C.; Kunrath-Lima, M.; de Melo, M.I.A.; Faraco, C.C.F.; Barbosa, J.L.; Frezard, F.; et al. Extracellular Vesicles from Adipose-Derived Mesenchymal Stem/Stromal Cells Accelerate Migration and Activate AKT Pathway in Human Keratinocytes and Fibroblasts Independently of miR-205 Activity. *Stem Cells Int.* **2017**, *2017*, 9841035. [CrossRef]
178. Chen, C.Y.; Rao, S.S.; Ren, L.; Hu, X.K.; Tan, Y.J.; Hu, Y.; Luo, J.; Liu, Y.W.; Yin, H.; Huang, J.; et al. Exosomal DMBT1 from human urine-derived stem cells facilitates diabetic wound repair by promoting angiogenesis. *Theranostics* **2018**, *8*, 1607–1623. [CrossRef]
179. Cooper, D.R.; Wang, C.; Patel, R.; Trujillo, A.; Patel, N.A.; Prather, J.; Gould, L.J.; Wu, M.H. Human Adipose-Derived Stem Cell Conditioned Media and Exosomes Containing MALAT1 Promote Human Dermal Fibroblast Migration and Ischemic Wound Healing. *Adv. Wound Care (New Rochelle)* **2018**, *7*, 299–308. [CrossRef]
180. Kim, S.; Lee, S.K.; Kim, H.; Kim, T.M. Exosomes Secreted from Induced Pluripotent Stem Cell-Derived Mesenchymal Stem Cells Accelerate Skin Cell Proliferation. *Int. J. Mol. Sci.* **2018**, *19*, 3119. [CrossRef]
181. Kobayashi, H.; Ebisawa, K.; Kambe, M.; Kasai, T.; Suga, H.; Nakamura, K.; Narita, Y.; Ogata, A.; Kamei, Y. Editors' Choice Effects of exosomes derived from the induced pluripotent stem cells on skin wound healing. *Nagoya J. Med. Sci.* **2018**, *80*, 141–153. [CrossRef] [PubMed]
182. Wang, X.; Jiao, Y.; Pan, Y.; Zhang, L.; Gong, H.; Qi, Y.; Wang, M.; Gong, H.; Shao, M.; Wang, X.; et al. Fetal Dermal Mesenchymal Stem Cell-Derived Exosomes Accelerate Cutaneous Wound Healing by Activating Notch Signaling. *Stem Cells Int.* **2019**, *2019*, 2402916. [CrossRef] [PubMed]
183. Yan, Y.; Wu, R.; Bo, Y.; Zhang, M.; Chen, Y.; Wang, X.; Huang, M.; Liu, B.; Zhang, L. Induced pluripotent stem cells-derived microvesicles accelerate deep second-degree burn wound healing in mice through miR-16-5p-mediated promotion of keratinocytes migration. *Theranostics* **2020**, *10*, 9970–9983. [CrossRef] [PubMed]
184. Hu, Y.; Rao, S.S.; Wang, Z.X.; Cao, J.; Tan, Y.J.; Luo, J.; Li, H.M.; Zhang, W.S.; Chen, C.Y.; Xie, H. Exosomes from human umbilical cord blood accelerate cutaneous wound healing through miR-21-3p-mediated promotion of angiogenesis and fibroblast function. *Theranostics* **2018**, *8*, 169–184. [CrossRef] [PubMed]
185. Zhang, J.; Guan, J.; Niu, X.; Hu, G.; Guo, S.; Li, Q.; Xie, Z.; Zhang, C.; Wang, Y. Exosomes released from human induced pluripotent stem cells-derived MSCs facilitate cutaneous wound healing by promoting collagen synthesis and angiogenesis. *J. Transl. Med.* **2015**, *13*, 49. [CrossRef]
186. Kim, Y.J.; Yoo, S.M.; Park, H.H.; Lim, H.J.; Kim, Y.L.; Lee, S.; Seo, K.W.; Kang, K.S. Exosomes derived from human umbilical cord blood mesenchymal stem cells stimulates rejuvenation of human skin. *Biochem. Biophys. Res. Commun.* **2017**, *493*, 1102–1108. [CrossRef]
187. Wang, L.; Hu, L.; Zhou, X.; Xiong, Z.; Zhang, C.; Shehada, H.M.A.; Hu, B.; Song, J.; Chen, L. Exosomes secreted by human adipose mesenchymal stem cells promote scarless cutaneous repair by regulating extracellular matrix remodelling. *Sci. Rep.* **2017**, *7*, 13321. [CrossRef]
188. Marino, D.; Luginbuhl, J.; Scola, S.; Meuli, M.; Reichmann, E. Bioengineering dermo-epidermal skin grafts with blood and lymphatic capillaries. *Sci. Transl. Med.* **2014**, *6*, 221ra214. [CrossRef]
189. Zhang, J.; Chen, C.; Hu, B.; Niu, X.; Liu, X.; Zhang, G.; Zhang, C.; Li, Q.; Wang, Y. Exosomes Derived from Human Endothelial Progenitor Cells Accelerate Cutaneous Wound Healing by Promoting Angiogenesis Through Erk1/2 Signaling. *Int. J. Biol. Sci.* **2016**, *12*, 1472–1487. [CrossRef]
190. Chen, B.; Sun, Y.; Zhang, J.; Zhu, Q.; Yang, Y.; Niu, X.; Deng, Z.; Li, Q.; Wang, Y. Human embryonic stem cell-derived exosomes promote pressure ulcer healing in aged mice by rejuvenating senescent endothelial cells. *Stem Cell Res. Ther.* **2019**, *10*, 142. [CrossRef]
191. Zhang, B.; Wu, X.; Zhang, X.; Sun, Y.; Yan, Y.; Shi, H.; Zhu, Y.; Wu, L.; Pan, Z.; Zhu, W.; et al. Human umbilical cord mesenchymal stem cell exosomes enhance angiogenesis through the Wnt4/beta-catenin pathway. *Stem Cells Transl. Med.* **2015**, *4*, 513–522. [CrossRef] [PubMed]
192. Shi, Q.; Qian, Z.; Liu, D.; Sun, J.; Wang, X.; Liu, H.; Xu, J.; Guo, X. GMSC-Derived Exosomes Combined with a Chitosan/Silk Hydrogel Sponge Accelerates Wound Healing in a Diabetic Rat Skin Defect Model. *Front. Physiol.* **2017**, *8*, 904. [CrossRef] [PubMed]
193. Tao, S.C.; Guo, S.C.; Li, M.; Ke, Q.F.; Guo, Y.P.; Zhang, C.Q. Chitosan Wound Dressings Incorporating Exosomes Derived from MicroRNA-126-Overexpressing Synovium Mesenchymal Stem Cells Provide Sustained Release of Exosomes and Heal Full-Thickness Skin Defects in a Diabetic Rat Model. *Stem Cells Transl. Med.* **2017**, *6*, 736–747. [CrossRef] [PubMed]
194. Sze, S.K.; de Kleijn, D.P.; Lai, R.C.; Khia Way Tan, E.; Zhao, H.; Yeo, K.S.; Low, T.Y.; Lian, Q.; Lee, C.N.; Mitchell, W.; et al. Elucidating the secretion proteome of human embryonic stem cell-derived mesenchymal stem cells. *Mol. Cell Proteom. MCP* **2007**, *6*, 1680–1689. [CrossRef]
195. Timmers, L.; Lim, S.K.; Hoefer, I.E.; Arslan, F.; Lai, R.C.; van Oorschot, A.A.; Goumans, M.J.; Strijder, C.; Sze, S.K.; Choo, A.; et al. Human mesenchymal stem cell-conditioned medium improves cardiac function following myocardial infarction. *Stem Cell Res.* **2011**, *6*, 206–214. [CrossRef] [PubMed]
196. Timmers, L.; Lim, S.K.; Arslan, F.; Armstrong, J.S.; Hoefer, I.E.; Doevendans, P.A.; Piek, J.J.; El Oakley, R.M.; Choo, A.; Lee, C.N.; et al. Reduction of myocardial infarct size by human mesenchymal stem cell conditioned medium. *Stem Cell Res.* **2007**, *1*, 129–137. [CrossRef]

197. Lai, R.C.; Arslan, F.; Lee, M.M.; Sze, N.S.; Choo, A.; Chen, T.S.; Salto-Tellez, M.; Timmers, L.; Lee, C.N.; El Oakley, R.M.; et al. Exosome secreted by MSC reduces myocardial ischemia/reperfusion injury. *Stem Cell Res.* **2010**, *4*, 214–222. [CrossRef]
198. Khan, M.; Nickoloff, E.; Abramova, T.; Johnson, J.; Verma, S.K.; Krishnamurthy, P.; Mackie, A.R.; Vaughan, E.; Garikipati, V.N.; Benedict, C.; et al. Embryonic stem cell-derived exosomes promote endogenous repair mechanisms and enhance cardiac function following myocardial infarction. *Circ. Res.* **2015**, *117*, 52–64. [CrossRef] [PubMed]
199. Oskouei, B.N.; Lamirault, G.; Joseph, C.; Treuer, A.V.; Landa, S.; Da Silva, J.; Hatzistergos, K.; Dauer, M.; Balkan, W.; McNiece, I.; et al. Increased potency of cardiac stem cells compared with bone marrow mesenchymal stem cells in cardiac repair. *Stem Cells Transl. Med.* **2012**, *1*, 116–124. [CrossRef] [PubMed]
200. Li, T.S.; Cheng, K.; Malliaras, K.; Smith, R.R.; Zhang, Y.; Sun, B.; Matsushita, N.; Blusztajn, A.; Terrovitis, J.; Kusuoka, H.; et al. Direct comparison of different stem cell types and subpopulations reveals superior paracrine potency and myocardial repair efficacy with cardiosphere-derived cells. *J. Am. Coll. Cardiol.* **2012**, *59*, 942–953. [CrossRef]
201. Kervadec, A.; Bellamy, V.; El Harane, N.; Arakelian, L.; Vanneaux, V.; Cacciapuoti, I.; Nemetalla, H.; Perier, M.C.; Toeg, H.D.; Richart, A.; et al. Cardiovascular progenitor-derived extracellular vesicles recapitulate the beneficial effects of their parent cells in the treatment of chronic heart failure. *J. Heart Lung Transpl.* **2016**, *35*, 795–807. [CrossRef] [PubMed]
202. Liu, B.; Lee, B.W.; Nakanishi, K.; Villasante, A.; Williamson, R.; Metz, J.; Kim, J.; Kanai, M.; Bi, L.; Brown, K.; et al. Cardiac recovery via extended cell-free delivery of extracellular vesicles secreted by cardiomyocytes derived from induced pluripotent stem cells. *Nat. Biomed. Eng.* **2018**, *2*, 293–303. [CrossRef] [PubMed]
203. Bittel, D.C.; Jaiswal, J.K. Contribution of Extracellular Vesicles in Rebuilding Injured Muscles. *Front. Physiol.* **2019**, *10*, 828. [CrossRef] [PubMed]
204. Choi, J.S.; Yoon, H.I.; Lee, K.S.; Choi, Y.C.; Yang, S.H.; Kim, I.S.; Cho, Y.W. Exosomes from differentiating human skeletal muscle cells trigger myogenesis of stem cells and provide biochemical cues for skeletal muscle regeneration. *J. Control. Release* **2016**, *222*, 107–115. [CrossRef] [PubMed]
205. Kharraz, Y.; Guerra, J.; Pessina, P.; Serrano, A.L.; Munoz-Canoves, P. Understanding the process of fibrosis in Duchenne muscular dystrophy. *Biomed. Res. Int.* **2014**, *2014*, 965631. [CrossRef]
206. Fry, C.S.; Kirby, T.J.; Kosmac, K.; McCarthy, J.J.; Peterson, C.A. Myogenic Progenitor Cells Control Extracellular Matrix Production by Fibroblasts during Skeletal Muscle Hypertrophy. *Cell Stem Cell* **2017**, *20*, 56–69. [CrossRef]
207. Zanotti, S.; Gibertini, S.; Blasevich, F.; Bragato, C.; Ruggieri, A.; Saredi, S.; Fabbri, M.; Bernasconi, P.; Maggi, L.; Mantegazza, R.; et al. Exosomes and exosomal miRNAs from muscle-derived fibroblasts promote skeletal muscle fibrosis. *Matrix Biol.* **2018**, *74*, 77–100. [CrossRef]
208. Wang, H.; Wang, B.; Zhang, A.; Hassounah, F.; Seow, Y.; Wood, M.; Ma, F.; Klein, J.D.; Price, S.R.; Wang, X.H. Exosome-Mediated miR-29 Transfer Reduces Muscle Atrophy and Kidney Fibrosis in Mice. *Mol. Ther.* **2019**, *27*, 571–583. [CrossRef]
209. Wang, B.; Zhang, A.; Wang, H.; Klein, J.D.; Tan, L.; Wang, Z.M.; Du, J.; Naqvi, N.; Liu, B.C.; Wang, X.H. miR-26a Limits Muscle Wasting and Cardiac Fibrosis through Exosome-Mediated microRNA Transfer in Chronic Kidney Disease. *Theranostics* **2019**, *9*, 1864–1877. [CrossRef]
210. Aminzadeh, M.A.; Rogers, R.G.; Fournier, M.; Tobin, R.E.; Guan, X.; Childers, M.K.; Andres, A.M.; Taylor, D.J.; Ibrahim, A.; Ding, X.; et al. Exosome-Mediated Benefits of Cell Therapy in Mouse and Human Models of Duchenne Muscular Dystrophy. *Stem Cell Rep.* **2018**, *10*, 942–955. [CrossRef]
211. Murphy, C.; Withrow, J.; Hunter, M.; Liu, Y.; Tang, Y.L.; Fulzele, S.; Hamrick, M.W. Emerging role of extracellular vesicles in musculoskeletal diseases. *Mol. Asp. Med.* **2018**, *60*, 123–128. [CrossRef] [PubMed]
212. Gao, X.; Ran, N.; Dong, X.; Zuo, B.; Yang, R.; Zhou, Q.; Moulton, H.M.; Seow, Y.; Yin, H. Anchor peptide captures, targets, and loads exosomes of diverse origins for diagnostics and therapy. *Sci. Transl. Med.* **2018**, *10*. [CrossRef]
213. Rome, S.; Forterre, A.; Mizgier, M.L.; Bouzakri, K. Skeletal Muscle-Released Extracellular Vesicles: State of the Art. *Front. Physiol.* **2019**, *10*, 929. [CrossRef]
214. Cho, H.; Blatchley, M.R.; Duh, E.J.; Gerecht, S. Acellular and cellular approaches to improve diabetic wound healing. *Adv. Drug Deliv. Rev.* **2019**, *146*, 267–288. [CrossRef] [PubMed]
215. Paganini, C.; Capasso Palmiero, U.; Pocsfalvi, G.; Touzet, N.; Bongiovanni, A.; Arosio, P. Scalable Production and Isolation of Extracellular Vesicles: Available Sources and Lessons from Current Industrial Bioprocesses. *Biotechnol. J.* **2019**, *14*, e1800528. [CrossRef] [PubMed]
216. Ludwig, N.; Whiteside, T.L.; Reichert, T.E. Challenges in Exosome Isolation and Analysis in Health and Disease. *Int. J. Mol. Sci.* **2019**, *20*, 4684. [CrossRef]
217. Li, J.; Lee, Y.; Johansson, H.J.; Mager, I.; Vader, P.; Nordin, J.Z.; Wiklander, O.P.; Lehtio, J.; Wood, M.J.; Andaloussi, S.E. Serum-free culture alters the quantity and protein composition of neuroblastoma-derived extracellular vesicles. *J. Extracell. Vesicles* **2015**, *4*, 26883. [CrossRef] [PubMed]
218. Whitford, W.; Guterstam, P. Exosome manufacturing status. *Future Med. Chem.* **2019**, *11*, 1225–1236. [CrossRef]
219. Shelke, G.V.; Lasser, C.; Gho, Y.S.; Lotvall, J. Importance of exosome depletion protocols to eliminate functional and RNA-containing extracellular vesicles from fetal bovine serum. *J. Extracell. Vesicles* **2014**, *3*. [CrossRef]
220. Eitan, E.; Zhang, S.; Witwer, K.W.; Mattson, M.P. Extracellular vesicle-depleted fetal bovine and human sera have reduced capacity to support cell growth. *J. Extracell. Vesicles* **2015**, *4*, 26373. [CrossRef]

221. Mateescu, B.; Kowal, E.J.; van Balkom, B.W.; Bartel, S.; Bhattacharyya, S.N.; Buzas, E.I.; Buck, A.H.; de Candia, P.; Chow, F.W.; Das, S.; et al. Obstacles and opportunities in the functional analysis of extracellular vesicle RNA—An ISEV position paper. *J. Extracell. Vesicles* **2017**, *6*, 1286095. [CrossRef] [PubMed]
222. Lener, T.; Gimona, M.; Aigner, L.; Borger, V.; Buzas, E.; Camussi, G.; Chaput, N.; Chatterjee, D.; Court, F.A.; Del Portillo, H.A.; et al. Applying extracellular vesicles based therapeutics in clinical trials—An ISEV position paper. *J. Extracell. Vesicles* **2015**, *4*, 30087. [CrossRef] [PubMed]
223. Li, P.; Kaslan, M.; Lee, S.H.; Yao, J.; Gao, Z. Progress in Exosome Isolation Techniques. *Theranostics* **2017**, *7*, 789–804. [CrossRef] [PubMed]
224. Gardiner, C.; Di Vizio, D.; Sahoo, S.; Thery, C.; Witwer, K.W.; Wauben, M.; Hill, A.F. Techniques used for the isolation and characterization of extracellular vesicles: Results of a worldwide survey. *J. Extracell. Vesicles* **2016**, *5*, 32945. [CrossRef] [PubMed]
225. Brennan, K.; Martin, K.; FitzGerald, S.P.; O'Sullivan, J.; Wu, Y.; Blanco, A.; Richardson, C.; Mc Gee, M.M. A comparison of methods for the isolation and separation of extracellular vesicles from protein and lipid particles in human serum. *Sci. Rep.* **2020**, *10*, 1039. [CrossRef]
226. Lamparski, H.G.; Metha-Damani, A.; Yao, J.Y.; Patel, S.; Hsu, D.H.; Ruegg, C.; Le Pecq, J.B. Production and characterization of clinical grade exosomes derived from dendritic cells. *J. Immunol. Methods* **2002**, *270*, 211–226. [CrossRef]
227. Lobb, R.J.; Becker, M.; Wen, S.W.; Wong, C.S.; Wiegmans, A.P.; Leimgruber, A.; Moller, A. Optimized exosome isolation protocol for cell culture supernatant and human plasma. *J. Extracell. Vesicles* **2015**, *4*, 27031. [CrossRef]
228. Andriolo, G.; Provasi, E.; Lo Cicero, V.; Brambilla, A.; Soncin, S.; Torre, T.; Milano, G.; Biemmi, V.; Vassalli, G.; Turchetto, L.; et al. Exosomes From Human Cardiac Progenitor Cells for Therapeutic Applications: Development of a GMP-Grade Manufacturing Method. *Front. Physiol.* **2018**, *9*, 1169. [CrossRef]
229. Watson, D.C.; Yung, B.C.; Bergamaschi, C.; Chowdhury, B.; Bear, J.; Stellas, D.; Morales-Kastresana, A.; Jones, J.C.; Felber, B.K.; Chen, X.; et al. Scalable, cGMP-compatible purification of extracellular vesicles carrying bioactive human heterodimeric IL-15/lactadherin complexes. *J. Extracell. Vesicles* **2018**, *7*, 1442088. [CrossRef]
230. Nordin, J.Z.; Lee, Y.; Vader, P.; Mager, I.; Johansson, H.J.; Heusermann, W.; Wiklander, O.P.; Hallbrink, M.; Seow, Y.; Bultema, J.J.; et al. Ultrafiltration with size-exclusion liquid chromatography for high yield isolation of extracellular vesicles preserving intact biophysical and functional properties. *Nanomedicine* **2015**, *11*, 879–883. [CrossRef]
231. Mol, E.A.; Goumans, M.J.; Doevendans, P.A.; Sluijter, J.P.G.; Vader, P. Higher functionality of extracellular vesicles isolated using size-exclusion chromatography compared to ultracentrifugation. *Nanomedicine* **2017**, *13*, 2061–2065. [CrossRef] [PubMed]
232. Yanez-Mo, M.; Siljander, P.R.; Andreu, Z.; Zavec, A.B.; Borras, F.E.; Buzas, E.I.; Buzas, K.; Casal, E.; Cappello, F.; Carvalho, J.; et al. Biological properties of extracellular vesicles and their physiological functions. *J. Extracell. Vesicles* **2015**, *4*, 27066. [CrossRef] [PubMed]
233. Wiklander, O.P.B.; Bostancioglu, R.B.; Welsh, J.A.; Zickler, A.M.; Murke, F.; Corso, G.; Felldin, U.; Hagey, D.W.; Evertsson, B.; Liang, X.M.; et al. Systematic Methodological Evaluation of a Multiplex Bead-Based Flow Cytometry Assay for Detection of Extracellular Vesicle Surface Signatures. *Front. Immunol.* **2018**, *9*, 1326. [CrossRef]
234. Gorgens, A.; Bremer, M.; Ferrer-Tur, R.; Murke, F.; Tertel, T.; Horn, P.A.; Thalmann, S.; Welsh, J.A.; Probst, C.; Guerin, C.; et al. Optimisation of imaging flow cytometry for the analysis of single extracellular vesicles by using fluorescence-tagged vesicles as biological reference material. *J. Extracell. Vesicles* **2019**, *8*, 1587567. [CrossRef]
235. Jeyaram, A.; Jay, S.M. Preservation and Storage Stability of Extracellular Vesicles for Therapeutic Applications. *AAPS J.* **2017**, *20*, 1. [CrossRef] [PubMed]
236. Kusuma, G.D.; Barabadi, M.; Tan, J.L.; Morton, D.A.V.; Frith, J.E.; Lim, R. To Protect and to Preserve: Novel Preservation Strategies for Extracellular Vesicles. *Front. Pharmacol.* **2018**, *9*, 1199. [CrossRef] [PubMed]

cells

MDPI

Review

Stem Cells and Extrusion 3D Printing for Hyaline Cartilage Engineering

Océane Messaoudi [1], Christel Henrionnet [1], Kevin Bourge [1], Damien Loeuille [1,2], Pierre Gillet [1,3] and Astrid Pinzano [1,4,*]

1 UMR 7365 CNRS-UL, IMoPA (Ingénierie Moléculaire et Physiopathologie Articulaire), Biopôle de l'Université de Lorraine, Campus Brabois-Santé, 9, Avenue de la Forêt de Haye, BP20199, 54505 Vandœuvre-Lès-Nancy, France; oceane.mess.96@gmail.com (O.M.); christel.henrionnet@univ-lorraine.fr (C.H.); kevin.bourge@univ-lorraine.fr (K.B.); d.loeuille@chru-nancy.fr (D.L.); pierre.gillet@univ-lorraine.fr (P.G.)

2 Service de Rhumatologie, CHRU de Nancy, Hôpitaux de Brabois, Bâtiment des Spécialités Médicales, 5 rue du Morvan, F54511 Vandœuvre-Lès-Nancy, France

3 Laboratoire de Pharmacologie, Toxicologie et Pharmacovigilance, Bâtiment de Biologie Médicale et de Biopathologie, CHRU de Nancy-Brabois, 5 Rue du Morvan, F54511 Vandœuvre-Lès-Nancy, France

4 Contrat d'Interface, Service de Rhumatologie, Hôpital de Brabois, Bâtiment Spécialités Médicales, F54511 Vandœuvre Lès Nancy, France

* Correspondence: astrid.pinzano@univ-lorraine.fr; Tel.: +33-(0)372-746-565

Abstract: Hyaline cartilage is deficient in self-healing properties. The early treatment of focal cartilage lesions is a public health challenge to prevent long-term degradation and the occurrence of osteoarthritis. Cartilage tissue engineering represents a promising alternative to the current insufficient surgical solutions. 3D printing is a thriving technology and offers new possibilities for personalized regenerative medicine. Extrusion-based processes permit the deposition of cell-seeded bioinks, in a layer-by-layer manner, allowing mimicry of the native zonal organization of hyaline cartilage. Mesenchymal stem cells (MSCs) are a promising cell source for cartilage tissue engineering. Originally isolated from bone marrow, they can now be derived from many different cell sources (e.g., synovium, dental pulp, Wharton's jelly). Their proliferation and differentiation potential are well characterized, and they possess good chondrogenic potential, making them appropriate candidates for cartilage reconstruction. This review summarizes the different sources, origins, and densities of MSCs used in extrusion-based bioprinting (EBB) processes, as alternatives to chondrocytes. The different bioink constituents and their advantages for producing substitutes mimicking healthy hyaline cartilage is also discussed.

Keywords: stem cells; 3D printing; cartilage engineering; bio-ink

Citation: Messaoudi, O.; Henrionnet, C.; Bourge, K.; Loeuille, D.; Gillet, P.; Pinzano, A. Stem Cells and Extrusion 3D Printing for Hyaline Cartilage Engineering. *Cells* **2021**, *10*, 2. https://doi.org/10.3390/cells10010002

Received: 3 November 2020
Accepted: 19 December 2020
Published: 22 December 2020

1. Introduction

Articular cartilage is a specialized tissue that lines the ends of the epiphyses and allows joint movement. It is a layered tissue consisting of 4 zones—the superficial, transitional, deep, and calcified areas separated from the underlying bone. Differences in cell morphology, the constitution of the extracellular matrix (ECM), and the collagen fibers' orientation in each layer are responsible for the overall structure's physical and biomechanical properties. The primary cartilage cell is the chondrocyte, whose prominent role is to maintain the ECM's integrity. The physiology, morphology, and metabolism of the chondrocyte also vary from one area to another. Unfortunately, cartilage has limited self-repair capabilities due to its avascular nature. Focal or diffuse cartilage damages can lead to pain, joint dysfunction, or even secondary osteoarthritis. Apart from arthroplasty, the repair or replacement of hyaline cartilage is a significant challenge in orthopedic surgery. Current orthopedical methods such as microfracture, autologous chondrocyte implantation, or mosaicplasty might offer short-term solutions. Still, none of them provide lasting repair, as the quality of the scarred neocartilage remains poor. As such, tissue engineering presents itself as a promising alternative for the restoration of cartilage lesions.

Joint engineering is an interdisciplinary field that aims to recreate a neo-tissue whose physical and biochemical property are close to those of the native tissue. It combines cells, biomaterials, and environmental factors. It represents a potential tool for cartilage regeneration. The main criteria to be taken into account are:

- homogeneous distribution of cells into the biomaterial,
- adapted porosity for homing, nutrient diffusion,
- cell differentiation optimized towards chondrocytes-like cells,
- preservation of the chondrocyte phenotype in situ,
- synthesis of a peri-cellular chondral-like ECM,
- persistent cell viability despite progressive biodegradation of the bioprinted biomaterial,
- biomechanical properties progressively close to those of the native cartilage
- bio-integration of the implants into the joint.

It is worth noting that conventional cartilage engineering methods have minimal control over the shape, size, and organization of engineered products.

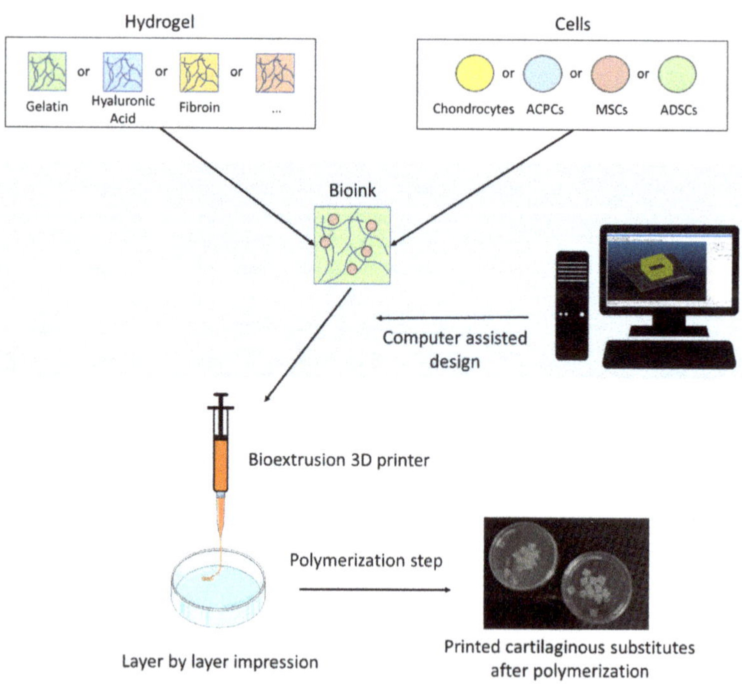

Figure 1. Schematic illustration of the extrusion-based 3D bioprinting process for articular tissue engineering using various hydrogels and cell types.

In contrast, the development of 3D bio-imaging technologies represents a recent revolution in personalized regenerative medicine. This technique makes it possible to obtain a well-defined, sometimes complex structure of a custom size, using a layer-by-layer bio-manufacturing strategy, guided by computer-aided design. Different 3D bioprinting processes are available, depending on the mechanical and biochemical properties of the native tissues. The three primary techniques currently developed are laser-assisted printing, inkjet, and bio-extrusion. The restoration of a layered structure such as a native hyaline cartilage is a complicated technological lock. The bio-extrusion process presents itself as the best alternative to recreate layered structures, such as skin and cartilage tissue.

One of the advantages is the direct encapsulation of living cells in the bio-ink, during the printing process, which allows the production of customized composite biomaterials. In this review, after a brief synopsis of normal and pathological cartilage, as well as the leading 3D printing methods, we focused on the interest of bioextrusion in cartilage engineering, according to the biomaterial composition of the bioinks, and the nature of the cell contingent, mainly mesenchymal stromal stem cells (MSCs), which are pluripotent, depending on the environment used.

To this day, many exhaustive review already exist concerning the use of 3D printing for the reconstruction of various tissues [1,2], or specifically on 3D printing for cartilage regeneration [3,4]. The originality of our work is the particular focus on extrusion bioprinting of cellularized hydrogels for articular cartilage tissue engineering (Figure 1).

2. Methodology

To perform this review, we searched for articles published up to 1 December 2020, in PubMed, with no start date restriction, with the keywords '3D printing cartilage'. About 442 references were found. Second, a manual search of reference lists from selected articles was conducted, with the keywords '3D printed AND cartilage', 'extrusion AND cartilage', 'biofabrication AND cartilage', '3D printing AND chondrogenesis' 'extrusion AND chondrogenesis' and 'biofabrication AND chondrogenesis'. We only selected the references that specifically uses a bioextrusion process of a cellularized hydrogel. The studies using a PCL scaffold for support, post-printing molding, or aiming to recreate tissues other than articular cartilage were eliminated. Only 25 were selected based on the selected criteria. We apologize for those excluded from those criteria, and therefore not cited in this review.

3. Articular Cartilage Lesions and Their Surgical Treatment

3.1. Osteoarthritis

Osteoarthritis (OA) is the most common joint affliction, and its frequency and socioeconomic impacts make it a public health challenge around the world, particularly in the context of overall population aging [5]. Its symptoms are pain, swelling, occasionally inflammation, and articular rigidity [6]. It is characterized by progressive degeneration of cartilage that can lead to subchondral bone damage. Cartilage loss causes bone remodeling, which is associated with acute pain [7]. There are four stages of pathological attack, depending on the extent of the remodeling [8,9]. OA management encompasses the prevention and treatment of pain, and includes palliative measures such as anti-inflammatory drugs and analgesics [10,11]. Pharmacological treatments are usually paired with physical therapy and weight control, to maintain or improve joint function [6,12,13]. When those measures fail to improve the patient's lifestyle, the last resort is surgery to substitute the damaged joint with a synthetic prosthetic [10]. Early OA is characterized by a lack of existing lesions. The exact causes of its appearance are still unknown [14]. Some already identified factors are age, sex, weight, and metabolic dysfunction [11,15,16]. Secondary arthrosis follows repetitive or abnormal loading; traumas can damage articular chondral surfaces [17]. These traumas can lead to mechanical dysfunction, swelling, or pain. The depth of their focal lesion classifies the different articulation injuries—(1) chondral lesion leading to mechanical malfunction and (2) osteochondral lesion with damaged cartilage, and subchondral bone causing articular disruption [18].

In chondral lesions, only the articular cartilage is injured. Chondrocytes near the lesion react by increasing the synthesis of extracellular matrix proteins, but due to their low proliferation rate, the cells are unable to restore cartilage integrity [19]. The increased synthesis is quickly stopped, leaving the articular surface with a chondral defect that can degenerate [20]. When the damage reaches the subchondral bone, undifferentiated cells from the bone marrow can infiltrate the injured zone to start the healing process. Mesenchymal stem cells (MSCs) differentiate into chondrocyte-like cells and initiate extracellular matrix synthesis (ECM), but the organization and biomechanical properties of this newly synthesized matrix differ from those of hyaline cartilage [20]. This fibrous repair tissue is unstable and leads to long-term degradation of the articular surface and function. Cartilaginous defects tend to progress over time and might lead to OA [21,22]. One of the significant challenges of cartilage regenerative medicine is treating these traumatic cartilage lesions early, to prevent long-term degradation and secondary OA.

3.2. Actual Management and Its Limitations

Numerous surgical techniques were developed to address focal cartilage defects [23]. Here, we present some of the most commonly used surgical procedures and their processes. Abrasion is a technique developed by Johnson in 1980 [24]; it is based on superficial debridement of the exposed bone to expose the vascularity underneath and obtain a viable bone surface to permit fibrin clot formation and attachment. The newly formed tissue on the exposed bone surface is a fibrocartilage-type tissue resulting from blood clot differentiation [25]. Microfracture and similar methods aim to stimulate the natural healing properties of the body. Microfracture involves the perforation of the subchondral bone, to allow MSCs and growth factors to escape the bone and fill the defect with newly synthesized tissue. Microfracture can only be applied to treat full-thickness defects with healthy subchondral bone. Furthermore, the repair is made of fibrocartilaginous tissue and is not stable in the long-term, generally leading to joint surface degradation [26]. Full-thickness osteochondral grafts are usually allografts. In this case, a unique cylindrical sample is harvested from a tissue donor and reimplanted in the defect to fill the lesion. This method allows for partial reparation of the surface through the formation of fibrocartilage between the native tissue and graft [27].

In mosaicplasty, multiple cylindrical cartilage grafts are harvested from a healthy, nonbearing zone of the patient's joint [28], and then reimplanted to fill the defect. This surgical procedure aims to permit repair of the articular surface, by producing neocartilage in the gaps separating the edges of the lesion and cartilage shreds. It causes donor-site morbidity but provides good long-term stability [29]. Brittberg's technique, or autologous chondrocyte implantation (ACI), was first performed in 1987 [30]. It consists of multiplying the patient's chondrocytes in vitro and then reinjecting them into the injured area with support, allowing them to fill the cartilaginous defect. These different techniques generally result in insufficient quality repair tissue, with low type II collagen content. This fibrocartilaginous tissue does not possess the phenotype of native hyaline cartilage [31] and might not support the necessary constraints and biomechanical loads. Hence, finding alternatives to those surgical procedures is a public health issue challenge. Tissue engineering (TE) approaches offer the potential to recreate hyaline-like cartilage in vitro, making them a promising tool for cartilage rehabilitation.

3.3. Healthy Cartilage Structure and Composition

Articular cartilage is a living, specialized connective tissue found in diarthrodial joints such as knee or hip joints. Its primary role is to provide a smooth and lubricated surface to permit load transmission during movement with a low friction coefficient [32]. Cartilaginous tissues can support movement and resist shear stress and deformation. These tissues need to store energy to prevent lasting compression [33,34]. Cartilage is a nonvascular and noninnervated tissue that possesses limited self-regeneration properties. It comprises a single cell type, chondrocytes, and a dense extracellular matrix. Thus, the sole resident cells of the cartilage are chondrocytes. Chondrocytes represent only 10% of the articular cartilage tissue volume [35]. The chondrocytes are spread across a dense matrix and have no cell-to-cell contact. They are responsible for the synthesis and degradation of the ECM component and maintain the homeostasis of the tissue; they secrete integrins as mediators to control cell differentiation, proliferation, and survival and matrix remodeling [36]. Chondrocytes specifically synthesize proteoglycans (PGs), collagens, and other noncollagenous proteins [32,37]. Chondrocytes are isolated in hypoxic niches, making hypoxia-inducible factor 1-alpha a key regulator of differentiation and homeostasis [38].

The hyaline ECM comprises water, PGs, and collagens, particularly type II collagen. To a lesser extent, noncollagenous proteins and glycoproteins are present in the ECM [39]. The cartilaginous matrix is highly hydrated (65 to 80% of the total weight) with a specific repartition of the water between the intra- and extrafibrillar compartments [33]. Less than one-third of the water content is linked with collagenous fibrils, the rest is located in the free space of the ECM [40]. PGs are highly glycosylated proteins, composed of a core protein and glycosaminoglycan chains, such as chondroitin sulfate (CS) or keratan sulfate [33,41]. The most common PG in cartilage is aggrecan, which plays a role in load-bearing. PG binds hyaluronan, forming a complex that retains a high amount of water in the ECM [37,42]. Collagen is a fibrillar protein; its primary role is to form a complex and organized network supporting the matrix structure [43,44]. Type II collagen is one of the main components of the ECM. It is distributed in a gradient with a higher density in the superficial zone of the hyaline cartilage and the lowest density in the deep zone. Type II collagen interacts with other collagen types, such as IX, XI, and III, for structural purposes [45].

Hyaline cartilage possesses a unique zonal organization in four differentiated layers [32]. The superficial zone is the thinnest layer, containing collagen fibers (primarily type II and IX) oriented parallel to the articular surface. This layer is responsible for most of the resistance properties of the articular surface [39]. In the transitional zone, collagen fibrils are thicker and less organized, with more PGs [35]. The deep zone is characterized by the highest density of PGs and the lowest water content; the collagen structures perpendicular to the tidemark separate the deep zone from the subchondral bone [46]. The chondrocyte density, morphology, and gene expression also vary depending on the depth within the hyaline cartilage [47–49]. These differences are also associated with the different biomechanical stresses exerted on the different layers [43,50]. This particular zonal organization and cellular distribution are a crucial component of cartilage tissue repair. The reconstruction of cartilaginous tissue with good repair, integration, and biomechanical properties is the main challenge of cartilage tissue engineering.

3.4. Tissue Engineering for Cartilage Repair

Cartilage tissue engineering is currently considered a promising alternative to classic treatment strategies [51–53]. It aims to recreate cartilaginous substitutes with properties similar to those of natural cartilage. The three main axes to consider for cartilage TE are biomaterials, cells, and the environment [54]. The biomaterial needs to have optimal porosity, reticulation, biointegration, cell-seeded scaffolds, cytocompatibility, and good cell adhesion properties [55,56]. Different cell types were investigated for cartilage tissue engineering, the most common being chondrocytes and MSCs. The last important factor is the environment. Chondrogenic matrix synthesis is driven by growth factors [57–61], oxygen levels [62–66], maturation time [67], and mechanical stimuli, including dynamic compression and shear stress, to mimic the natural diarthrodial environment [68–73]. Classical TE usually produces homogenous constructs. A new field of tissue engineering development is the field of 3D printing. It offers new potential to produce stratified products as well as innovations in personalized regenerative medicine.

4. 3D Printing for Cartilage Tissue Engineering

4.1. History of 3D Printing

Three-dimensional (3D) printing (or additive manufacturing) was invented in 1983 by Chuck Hull. Initially, the printing process was based on the stereolithography method. A computer-aided design (CAD) is created and sent to a 3D printer. Sequential coats of material are solidified in a layer-by-layer manner, until the full product is produced [74]. Numerous 3D printing processes were developed, based either on the solidification of materials via an energy source or by the deposition of a liquid material that is polymerized postprinting. First used exclusively in industry, 3D printing reached the medical field in the 2000s to produce synthetic surgical models. Subsequently, the concept of bioprinting emerged, raising the possibility of printing biological tissues and organs [75]. A decade later, the first printing process with live cell structures was successfully executed [76]. The concept of bioprinting is a promising perspective for modern tissue engineering. It has only begun to influence medicine and surgery and revolutionize health care [77].

4.2. 3D Printing Processes for Cartilage Reconstruction

Significant advances were made in the field of cartilage and bone tissue engineering over the past two decades [3]. 3D printing is a revolutionary process for the field of personalized medicine. It allows for the layer-by-layer deposition of a biomaterial as specified by a CAD, making it possible to adapt the constructs to specific lesions, unlike other classical TE methods [78,79]. Two main printing classes can be distinguished—acellular processes and bioprinting. In this review concerning articular cartilage engineering, we focus on bioprinting strategies.

Different bioprinting strategies are available for cartilage tissue engineering—inkjet printing, laser-assisted printing, and bioextrusion (Figure 2) [80]. Inkjet bioprinting is based on the deposition of droplets directly onto a support by thermal or piezoelectric methods [81]. Laser-assisted bioprinting consists of the deposition of droplets from a specific material onto a receiving substrate, under the influence of a laser-based energy source [82]. Bioextrusion (or microextrusion) is a process based on the direct deposition of a bioink onto a support via a printing needle (screw-based, pneumatic, or piston-driven), following a CAD [83]. Bioextrusion is used in association with bioinks composed of natural (alginate, gelatin, chitosan, hyaluronan) or synthetic (PCL, PGA, PEG) polymers and embedded cells.

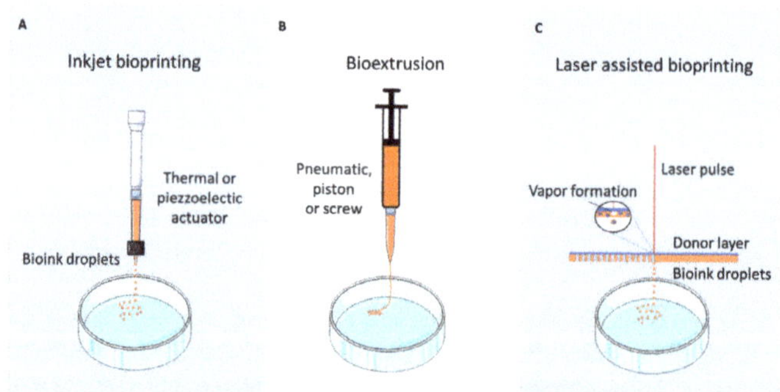

Figure 2. Different bioprinting strategies are available for cartilage tissue engineering. (**A**) Inkjet bioprinting based on the deposition of droplets formed by a thermal or piezoelectric actuator on a support. (**B**) Bioextrusion (or microextrusion) based on the extrusion of a continuous bioink filament through a printing needle driven by a screw, pneumatic or a piston. (**C**) Laser-assisted bioprinting uses a laser-based energy source to produce droplets via the production of a vapor bubble in the substrate.

The bioextrusion process allows the printing of large and bulky substitutes with high cell density, making it an excellent candidate to reproduce full-thickness cartilage tissue [84,85]. It allows the deposition of different biomaterials and cell types throughout the different printed layers of bioink to better mimic the natural osteochondral organization, and more precisely, the four different hyaline cartilage layers [86]. This innovative process could allow for personalized constructs with the zonal organization of native cartilage directly adaptable to the patient's lesion sites.

5. Bioextrusion Processes for Cartilage Tissue Engineering

5.1. Bioinks for Extrusion-Based Bioprinting

Biomaterials used for classic tissue engineering need to have specific characteristics. The main factors are biocompatibility, biodegradability, and porosity [87]. The biomaterial needs to have biomechanical properties compatible with those of the native tissue it aims to recreate. It also needs to be loose enough to permit ECM development but stable enough to maintain a three-dimensional environment for the cells [55]. The biomaterial needs to be adapted to the target tissue, the cell type, and mechanical constraints. Specifically in cartilage TE, biomaterials play different roles depending on the embedded cell types. The environments required to maintain chondrocyte differentiation and permit MSC differentiation induction are different [88–90]. The integration of the engineered substitute with the surrounding healthy tissue needs to be assessed.

The two main classes of biomaterials used for tissue engineering are synthetic and natural polymers. Synthetic polymers are human-made materials that are already widely used in cartilage tissue engineering for their well-characterized and stable chemical properties. Polymers such as poly(ethylene)-glycol (PEG), polycaprolactone (PCL), or polyglycolic acid (PGA) can be combined or coated with hydrogels or natural polymers to enhance their biocompatibility [91–94]. Another class of material developing is nanomaterials, such as carbon nanotubes (CNTs) for their physico-chemical properties [95]. Natural polymers are also considered promising for TE, with alginate, gelatin, and agarose being widely studied for their properties [96]. The different biomaterial concentrations can be tuned to optimize the final construct's biological and mechanical properties [97,98]. They can be combined to form a complex bioink, taking advantage of their different biomaterial characteristics. To improve the stability of some natural polymers, such as gelatin or hyaluronic acid (HA), modifications such as methacrylation are often used [99,100]. Hydrogels are natural polymers widely used for their excellent biocompatibility and ECM mimetism [101]. Hydrogels are usually polysaccharides (e.g., alginate, HA) [102,103] or protein-based (collagen, fibrin) [89,103].

Hydrogels can also be based on a decellularized extra-cellular matrix (dECM). First used in biological sheets or coating for bioengineered scaffold, dECM can now also be used for cartilage 3D-printing [104]. The aim while producing dECM is to eliminate the cellular component, while maintaining the structure and composition of the native ECM [105]. It can easily be made into a soft gel, making it a promising feature for bioextrusion. The main advantage of using dECM as a biomaterial is the mimicking of the structure and biological cues of the native tissue that allows for the induction of growth and differentiation of the cellular contingent. For cartilage TE, dECM is already used as a bioink to produce 3D printed cartilaginous substitutes [106–108].

One crucial characteristic for any material used in three-dimensional bioprinting is printability. Bioinks designed for EBB processes are based on biocompatible and bioprintable hydrogels. The advantages of the different biomaterials used in EBB processes are presented in Table 1. Critical criteria include viscosity and viscoelasticity of the bioink to achieve the optimal printing process [109,110]. The stability and mechanical properties of printed gels are also essential considerations for the final construct [111]. To ensure three-dimensional stability, hydrogel-based bioinks can be solidified by temperature change, photocrosslinking, or chemical crosslinking [112–114]. Another method consists of printing a heterogeneous scaffold composed of a structural PCL scaffold and a cytocompatible hydrogel containing the cells [115,116].

The advantage of bioextrusion processes is the ability to print bioinks and cells to simultaneously produce functionalized substitutes. The biocompatibility of the bioink is often assessed by evaluating the viability of the printed cells [117–119]. In comparing different cell-seeded bioinks and printing parameters, the yield stress, shear stress, and viscosity were highlighted as crucial printing factors [120]. Shear stress can be impacted by different printing parameters, such as needle geometric shape and diameter [117]. These parameters need to be finely tuned to promote cell viability and differentiation potential in the final constructs. The biomaterial used to produce cellularized hydrogels also needs to be adapted to the construct's cellular content.

5.2. Mesenchymal Stem Cells as an Alternative to Native Chondrocytes

As the sole resident of the cartilage, chondrocytes seem to be the most suitable cell type for cartilage tissue engineering. However, apart from the fact that they are hardly available within the joint, their amplification in a monolayer generates cell dedifferentiation from the first passages, with a significant decrease in type 2 collagen synthesis [121,122]. Of all adult stem cells present in the body, mesenchymal stem cells currently represent a promising candidate substitute for chondrocytes in cartilage tissue engineering. Many different factors characterize human MSCs, including their adhesion to plastic supports; their expression of stemness markers, such as CD105, CD73, CD29, and CD90; and their lack of expression of hematopoietic surface proteins CD45 or CD34 [123–126]. MSCs are self-renewable and multipotent-capable of differentiating into multiple cell lineages [127,128]. The signaling pathways that affect the differentiation of MSCs are well characterized [129]. Their multipotent potential can be influenced by their origin, depending on the cell source. Differences in chondrogenic and osteogenic properties were already highlighted [130–132]. First identified in 1970 by Friedenstein, MSCs were initially isolated from bone marrow [133], but further investigation showed that they could be easily isolated from other source tissues, the most common being adipose tissue [134], synovial membrane, synovial fluid [135,136], dental pulp [137], Wharton's jelly, and umbilical cord blood [138].

Bone marrow-derived MSCs (BM-MSCs) are now well characterized because they are the most commonly used cell type in tissue engineering. They possess good proliferation properties but can also be induced to differentiate into various cell types, including osteocytes, adipocytes, chondrocytes, and neural or muscular cells [139,140]. To be induced in the chondrogenic lineage, they require a combination of differentiation factors (mainly growth factors such as TGF-β1 and TGF-β3) and a 3D environment to promote and stabilize the chondrogenic phenotype [57]. Adipose-derived stem cells (ADSCs) are widely used in cartilage tissue engineering for their chondrogenic properties [141–143]. Their main advantage over other MSCs is that they are easily accessible via minimally invasive procedures. Their differentiation capacity differs from that of BM-MSCs, which have better osteogenic properties, while ADSCs synthesize more collagen. They can be isolated from different fat tissues. Indeed, one of the sites most commonly used in cartilage TE is the infrapatellar fat pad, already in the knee joint [144,145].

MSCs are also present in the synovial membrane. Their multipotency was investigated to prove that they can differentiate into chondrogenic, osteogenic, adipogenic, and sometimes myogenic pathways [135]. They have healing potential for articular tears [146], making them good candidates for cartilage tissue engineering. Similar MSCs can be isolated from synovial fluid [59], presenting an MSC phenotype and surface markers. They possess the same multilineage potential as synovial membrane-derived cells. Synovial fluid mesenchymal stem cells show the highest chondrogenic potential among osteoarticular cell types [131]. MSCs isolated from dental pulp possess different differentiation properties. They primarily differentiate into the odontoblast pathway but can also be induced to become adipocytes, osteoblasts, chondrocytes, and neural cells [147,148]. Already used in cartilage TE, dental pulp MSCs exhibited potential for hyaline-like cartilage formation with the synthesis of ECM components, such as aggrecan or collagen [149–151]. Umbilical cord blood and Wharton's jelly also contain mesenchymal-like cells expressing MSC markers and lacking hematopoietic markers [138,152]. Their multipotent potential for cartilage TE was already studied, and they showed good hyaline-like cartilage neosynthesis under different conditions with lower type X collagen synthesis than BM-MSCs [62,132,153].

As not all cell sources were used in EBB processes for hyaline cartilage regeneration, we focus only on the extruded cellularized constructs and their cell contingents.

5.3. Cell Types Used in Extrusion-Based Bioprinting

5.3.1. Cartilage-Derived Cells

Chondrocytes are intensively investigated for cartilage regeneration. The EBB process allows for the direct bioprinting of chondrocytes embedded in bioinks to print cellularized constructs. They are usually associated with natural polymers. Some studies use alginate [117], HA [118], gelatin [154] or dECM-based bioinks [108]. Chondrocytes can be used to assess the biocompatibility and printability properties of different polymers and test different concentrations of polymers [120] or different biomaterial blends [155]. To further recreate the cartilage's zonal organization, EBB systems were used to create layer-by-layer substitutes. The gradient can be tuned by modifying the cell density within the gridded construct [156]. Adding constituents to the deepest layers of the printed substitutes, such as calcium, can improve a calcified zone [157]. Once printed and matured, cartilaginous constructs can be used to assess chondrogenesis inside the biomaterial, by measuring gene expression and matrix synthesis [154]. Mechanical properties such as compressive stress are also key factors that need to be assessed to recreate native cartilage [158].

An alternative to mature chondrocytes for tissue engineering is a subpopulation of chondrocytes, articular chondroprogenitor cells (ACPCs) found in the surface zone of mature cartilage [159]. ACPCs maintain good chondrogenic potential after extending the monolayer culture, unlike mature chondrocytes [160]. ACPCs were already used in the EBB process developed with a gelatin-based bioink playing two key roles—the ink provides a scaffold to encapsulate the cells and acts as glue so that the extruded material directly adheres to the defect surface in situ [161]. ACPCs must be more deeply investigated to evaluate the chondrogenic potential of those cells. As seen previously, autologous chondrocytes are very limited in number, and while undergoing expansion in vitro, they might lose their phenotype, morphology, and expression of specific markers. Therefore, the limitations encountered in chondrocyte-based therapies instigated alternative cell searches as tools in cartilage regeneration.

5.3.2. Mature MSCs

MSCs represent a promising alternative for cartilage tissue engineering due to their many advantages, as discussed previously. The most common MSCs used are BM-MSCs. By using extrusion-based 3D printing, BM-MSCs can be directly seeded into the biomaterial and extruded into a compact three-dimensional construct. Due to their potential, BM-MSCs could engineer different layers of native cartilage in vitro. MSCs are usually embedded in a hydrogel to reproduce the hyaline-like cartilaginous matrix, due to their excellent hydration properties, such as alginate [162] GelMA [163] or dECM-based bioinks [106]. An important aspect of tissue mimetism is the fiber organization within the different layers; the bioextrusion process can print layers with different alignments, making it possible to reproduce the collagen fibers' natural organization within the cartilaginous ECM [103]. The addition of compounds that lead to differentiation was investigated at length. CS can induce cartilaginous matrix production, especially type II collagen, while the presence of HA in the hydrogel enhances cell viability and chondrogenesis [164]. HA also favors the hypertrophic differentiation of MSCs [165]. To further reproduce the calcified layer, calcium can be added to the bioink to increase the expression of hypertrophic cartilage markers [166]. To fine-tune the chondrogenesis of the embedded MSCs in 3D-printed constructs, the use of growth factors, especially TGF-β family members, is essential [60].

ADSCs can also be used in EBB processes to reproduce cartilaginous tissue through additive manufacturing. A specific device, the BioPen, was developed to directly print substitutes seeded with ADSCs [167]. Different studies assessed the process [168] and its potential for cartilage reparation therapy in vitro [143]. The next step was to directly print into a full-thickness defect with the handheld device, to assess in vivo printing in a large animal model, highlighting the ability of the ADSC combined with a bioextruded hydrogel, to promote the reparation of the cartilage by enhancing a more hyaline-like cartilaginous reparation [142]. Other MSC sources (synovium, Wharton's jelly) were already used in classical cartilage tissue engineering, but their chondrogenic potential in an extrusion-based 3D printing process is yet to be assessed [146].

5.3.3. Coculture of MSCs and Chondrocytes

The chondrocyte phenotypes vary from one zonal area to the next in joint cartilage [169]. To recreate this complex structural cellular organization, recent findings highlighted the fact that coculturing MSCs and chondrocytes in the same constructs could favor the induction of chondrogenesis, especially for BM-MSCs [170]. Studies showed that when combining a nanocellulose-based biomaterial with a coculture of BM-MSCs and chondrocytes at a ratio of 8:2, the presence of MSCs enhances the proliferation of chondrocytes in vivo [171]. Using the same parameters, it was also proven that MSCs could improve cartilaginous ECM synthesis in vivo, especially type II collagen synthesis [172]. To mimic the zonal organization of native cartilage, specific cell contingents can be used to reproduce different layers. Alginate and gelatin-based bioinks were used to produce a composite construct. The top layer comprises BM-MSCs and chondrocytes cocultured (7.5:2.5) with CS and is designed to reproduce hyaline-like cartilage. The bottom layer contains only MSCs embedded in a bioink to which HA was added to reproduce the calcified zone. These constructs demonstrate the potential for zone-specific cartilage tissue engineering [173]. Many factors in addition to cell origin need to be assessed to optimize neocartilage production in the constructs.

5.4. *Cell Density for Cartilage Tissue 3D Printing*

The cell seeding density to use during 3D bioprinting remains an open question. Due to chondrocytes' low proliferation rate, a higher cell density generally yields a better engineered cartilage tissue [174]. The cell density used for the extruded products generally ranges from 5 to 20×10^6 cells/mL. Very few studies aimed to compare those densities to assess induced chondrogenesis [156]. In most cases, researchers work with a standard density optimized within that range [108,117–120,155,157]. In one study working with chondroprogenitors, a high density of 20×10^6 was also selected [161]. Very few studies used low cell densities, but they achieved good chondrogenic results [154,158]. For BM-MSCs, the range of densities is comparable to that of chondrocytes, starting from 4×10^6 cells/mL [175] and increasing to as high as 20×10^6 cells/mL [163]; most of the studies were between those two values, thus achieving the best chondrogenic induction possible [106,162,165,166]. Recently, some research aimed to compare two very low cell densities, 1 and 2×10^6 cells/mL, to assess better options for obtaining good chondrogenesis, and they showed that the lowest density seemed to be optimal [60]. This could lead to a new approach in cartilage tissue engineering by better mimicking the chondrocytes' natural repartition in cartilage.

ADSCs delivered with BioPen technology are seeded at a slightly lower range of densities than the other cell types, between 2 and 10×10^6 cells/mL [142,143,168]. These conditions seemed optimal to obtain good chondrogenic properties of the final construct in vitro and in vivo. In cocultured substitutes, although the cell density used is crucial. The most important cell number is usually 10×10^6 cells/mL. Another important aspect is the ratio between the two cell types and the use of a higher density of stem cells than chondrocytes in all printed substitutes. The ratio most often used is 8:2 [171,172]. The ratio between MSCs and chondrocytes can vary between the layers, depending on the desired characteristics [173]. The chondrogenic differentiation of the embedded cells is impacted by their origin, their density, the nature of the biomaterial they are seeded in and environmental factors.

Table 1. Summary of the advantages of the various biomaterials used to date in extrusion-based 3D printing depending on the embedded cell types for cartilage tissue engineering.

Biomaterial	Advantages	Cell Types	References
Alginate	Biocompatible Good printability Ionic gelation Low cost	Chondrocytes	[117,119,155,157,158]
		BM-MSCs	[60,162,166]
		Chondrocytes + BM-MSCs	[173]
Gelatin	Biocompatible Non immunogenic Thermic gelation Biodegradable	Chondrocytes	[120,154]
		ACPCs	[161]
		BM-MSCs	[60,163,165,166,175]
		ADSC	[142,143,168]
		Chondrocytes + BM-MSCs	[173]
Hyaluronic acid	Biocompatible Promote proliferation Bio printability Chondrogenic signalling Chemical crosslinking Anti-inflammatory	Chondrocytes	[118]
		BM-MSCs	[165]
		ADSC	[142,143,168]
		Chondrocytes + BM-MSCs	[173]
Collagen	ECM component Good printability Promote cell adhesion Chondro-induction	Chondrocytes	[155,156]
Chondroitin sulphate	Component of ECM Anti-inflammatory Gelation by chemical modification	Chondrocytes	[118]
		BM-MSCs	[165]
		Chondrocytes + BM-MSCs	[173]
Nanocellulose	Biocompatibility Shear thinning properties High stiffness	Chondrocytes	[117,119,158]
		Chondrocytes + BM-MSCs	[171,172]
Agarose	Biocompatible High stability Thermic gelation Low cost	Chondrocytes	[155]
Fibrinogen	Biocompatible Strong 3D network Chemical crosslinking	BM-MSCs	[60]
dECM	Biocompatible Native ECM structure Biological cues Promote cell growth and differentiation	Chondrocytes	[108]
		BM-MSCs	[106]

NB: We presented only the studies using the 3D bioextrusion printing method, where the hydrogel-based bioink is directly seeded by the cells before printing. Therefore, we eliminated acellular printing methods, publications associating a PCL scaffold with hydrogels, and studies using extrusion coupled with post-printing molding.

5.5. Cell Viability

The cells' ability to proliferate and differentiate inside the 3D-printed cartilaginous construct is directly correlated with cell viability. The cell viability of the embedded cells depends on the cell type and density used for 3D printing of cartilaginous constructs, as presented in Table 2. Printing parameters such as shear stress and pressure applied to the cells during the extrusion process can affect cell viability, proliferation, and chondrogenic properties. To reduce shear stress during extrusion, nozzle diameter and geometry are vital factors [117]. Other printing parameters, such as cross-linking (UV light, ionic, or enzymatic gelation), also need to be tuned to permit maximum cell survival [60,103,119,165,166]. To assess the long-term effects of the overall printing process, many researchers evaluated viability at different times postprinting [162,165,166]. The biomaterials and additives in the bioink can also affect cell viability, exerting a cytotoxic effect on the embedded cells. Previous studies highlighted that adding components to the bioink, such as nanocellulose or calcium derivative, can reduce the cell viability [117,119,157,173]. This decrease can be explained either by a slightly toxic effect of the added molecule [157,173] or by a too highly concentrated hydrogel [118]. In contrast, other biomaterials, such as gelatin, silk, or collagen, can improve cell nesting and survival [154,155]. Other studies used cell aggregates or spheroids to seed into the bioink instead of single cells [103,175]. A comparison of embedded spheroids to isolated cells in bioextruded substitutes demonstrated a better survival of cells inside spheroids [175]. Globally, the bioextrusion process using biocompatible biomaterials and optimized printing parameters maintains good viability, generally as high as >90% viability, after a sufficient maturation time [108,118,142,156,163,168,173].

5.6. Environmental Factors

First, oxygenation plays a crucial role in the chondrogenic pathway. The articular cartilage is avascular; the chondrocytes are therefore exposed to low oxygen contents, which vary from 5% at the level of the superficial zone to 2% in the deep zone [176]. The surrounding medium's oxygen content impacts the proliferation, maintenance of the chondrogenic phenotype, and chondrogenic differentiation of MSCs in tissue engineering of cartilage [177]. Comparing normoxic to hypoxic conditions is of interest to determine the best culture conditions for the cells [59,178,179]. This parameter is yet to be assessed in EBB processes.

Supplementation of the culture media with growth factors can also impact cell differentiation. Members of the TGF-β superfamily are essential regulators of chondrogenic differentiation during embryonic development in chondrogenesis and osteogenesis. The TGF-β superfamily is composed of 5 members (TGF-β1 to TGF-β5). TGF-β1 remains the most widely used and is known to stimulate the synthetic activity of chondrocytes and induce the chondrogenic differentiation of bone-marrow MSC, by decreasing the expression of type I collagen and increasing the production of type II collagen and aggrecan [180]. Bone morphogenetic proteins (BMPs) are glycoproteins of the TGF-β superfamily. BMP-2 is the most widely used in vitro to induce cartilage-type ECM production, with the synthesis of PGs and type II collagen [181]. Other factors such as IGF-1 showed an interest to maintain chondrocyte anabolic activities [182].

In extrusion-based constructs, culture media supplementation with growth factors is essential for the redifferentiation of embedded chondrocytes or MSC differentiation induction. TGF-β1 is used in many different studies involving embedded ACPCs [161] or MSCs [163,175]. Extruded gels can also be cultured in medium supplemented with TGF-β3 to induce chondrocyte redifferentiation [157] or MSC chondrogenesis [143,165], or in coculture substitutes containing both cell types [173]. For cartilage TE, BMP2 is usually associated with a TGF-β factor. A study aiming to compare TGF-β1, TGF-β3, BMP-2, or the association of these factors demonstrated the benefit of BMP-2 and TGF-β1 [60]. TGF-β2 was also used in 3D printed substitutes fed with chondrocytes, to support cartilaginous matrix formation [120].

Table 2. Comparison of cell viability and chondrogenic evaluation depending on cell types, origins, and densities used to date in extrusion-based processes for cartilage tissue engineering.

Cell Type	Bioink Components	Species	Cellular Density (cells/mL)	Cell Viability	Chondrogenic Evaluation				Reference
					Biochemical Assays	Gene Expression	Matrix Synthesis	Biomechanical Testing	
Chondrocytes	Silk fibroin Gelatin	Porcine	1 M	Good Viability	DNA and GAG content	*Col2*, *Sox9*, *ACAN* and *Col10*	H&E staining	Uniaxial compressive test	[154]
	Alginate Nanocellulose	Human	2 M	71.6–97.3%	N.D	N.D	N.D	Unconfined compression test	[158]
	Type II collagen	Rabbit	5 M 10 M 20 M	93%	DNA and GAG content	*Col1A1*, *Col2A1* and *ACAN*	H&E and Alcian blue staining Type I and II collagen and PRG4 IHC	N.D	[156]
	Alginate Nanocellulose	Calves	6 M	>65%	N.D	N.D	Type I and II collagen and Proteoglycan-HA IHC	N.D	[117]
	Hyaluronic acid Chondroitin sulphate	Bovine	6 M	91%	N.D	N.D	N.D	N.D	[118]
	Alginate Methylcellulose	Human	6.5 - 7 M	45–75%	GAG and type II collagen content	*Col2*, *ACAN*, *COMP*, *Col10*, *Col1* and *Sox9*	Aggrecan IF	N.D	[157]
	Sodium Alginate Agarose Type I collagen	Rat	10 M	70–95%	DNA and GAG content	*ACAN*, *Sox9*, *Col2A1* and *Col1A1*	H&E staining	Uniaxial compressive test	[155]
	GelMA	Equine	10-20 M	N.D	DNA and GAG content	N.D	Safranin-O/Fast green staining Type II collagen IHC	N.D	[120]
	Alginate Nanocellulose	Human	15 M	72.8–93%	N.D	N.D	N.D	N.D	[119]
	dECM	Rabbit	20 M	90–98%	GAG and collagen content	N.D	H&E, Safranin-O/Fast-green and Alcian blue staining	N.D	[108]

Table 2. *Cont.*

Cell Type	Bioink Components	Species	Cellular Density (cells/mL)	Cell Viability	Chondrogenic Evaluation				Reference
					Biochemical Assays	Gene Expression	Matrix Synthesis	Biomechanical Testing	
ACPCs	GelMA	Equine	20 M	>70%	DNA and GAG content	*Col1A1*, *Col2A1* and *PRG4*	Safranin-O staining Type I and II collagen IHC	N.D	[161]
BM-MSCs	Alginate, Gelatin and Fibrinogen	Human	1 M	>90%	N.D	*Col2A1*, *Col10A1*, *ACAN*, *VCAN*, *Sox9*, *COMP*, *ALP*, *BGLAP* and *OSX*	HES, Alcian blue, Alizarin red and Sirius Red staining Type II collagen IHC	N.D	[60]
	Hyaluronic acid	Human	3 M	Good viability	DNA and GAG content	*ACAN*, *Col2A1*, *Sox9*, *Col1A1*, *Col10A1* and *RunX2*	Safranin-O/Fast green staining Type I collagen IF	N.D	[103]
	GelMA	Human	4 M	Good viability	N.D	N.D	H&E, Alcian blue and picrosirius staining Type I and II collagen IHC	N.D	[175]
	Sodium alginate	Human	6 M	73–87%	N.D	N.D	Alizarin red, Sirius red and Safranin-O staining Type I and II collagen IHC	Unconfined compression test	[162]
	GelMA and alginate	Human	10 M	>80-85%	N.D	*Col1*, *Col2*, *Col10A1*, *ACAN*, *ALPL* and *BGLAP*	Type I, II and X collagen and aggrecan ICC	N.D	[166]
	dECm, silk-fibroin and PEG	Rabbit	10 M	>80%	DNA, GAG and collagen	*Col1*, *Col2*, *ACAN* and *Sox9*	Safranin-O and Masson's trichrome staining	N.D	[106]
	Alginate, HAMA and CS-AEMA	Human	10–15 M	85–90%	N.D	*ACAN*, *Col1*, *Col2* and *Col10*	Type I, II and X collagen and aggrecan ICC	N.D	[165]
	HAMA and GelMA	Rat	20 M	>90%	DNA and GAG content	*Col2*, *Col1*, *Col10*, *Sox9* and *ACAN*	Alcian blue, H&E, Safranin-O staining	Dynamic mechanical compressive test	[163]

Table 2. *Cont.*

Cell Type	Bioink Components	Species	Cellular Density (cells/mL)	Cell Viability	Chondrogenic Evaluation				Reference
					Biochemical Assays	Gene Expression	Matrix Synthesis	Biomechanical Testing	
ADSCs	HAMA-GelMA	Human	2 M	97%	N.D	N.D	N.D	N.D	[168]
	HAMA-GelMA	Sheep	2.5 M	97%	N.D	N.D	Safranin-O/Fast green staining Type I and II collagen IHC	Indentation test	[142]
	HAMA-GelMA	Human	10 M	N.D	N.D	*Col2A1*, *Col1A2*, *ACAN*, *Sox9*, *RunX2* and *Col10A1*	Alizarin red, Safranin-O staining Type I, II and X collagen and proteoglycan IF	Unconfined compression test	[143]
BM-MSCs + Chondrocytes	NFC	Human	10 M	N.D	N.D	N.D	Alcian blue, Von Gieson and Safranin-O staining Type II collagen IHC	N.D	[171]
	NFC and alginate	Human	10 M	N.D	N.D	N.D	Alcian blue and Von Gieson staining Type II collagen IHC	Unconfined compression test	[172]
	GelMA, HAMA, CS-AEMA	Human	10 M	88–90%	N.D	*Col2A1*, *ACAN*, *Col1A1*, *Col10A1* and *ALPL*	Type I, II and X collagen and aggrecan IHC	Dynamic mechanical compressive test	[173]

N.D = No Data; Good Viability means that no precise value is available; *ACAN*: Aggrecan (gene), *BGLAP*: Osteocalcin (gene); *COMP*: Cartilage oligomeric matrix protein (gene); CS-AEMA: Chondroitin sulphate 2-aminoethyl Methacrylate; *Col1A1*: Collagen Type I Alpha 1 Chain (gene); *Col2A1*: Collagen Type II Alpha 1 Chain (gene); *Col1*: Collagen Type I (gene); *Col2*: Collagen Type II (gene), *Col10*: Collagen type X (gene); dECM: decellularized ExtraCellular Matrix; DNA: Deoxyribonucleic acid; GAG: Glycosaminoglycan; GelMA: Gelatin methacrylamide; H&E: Hematoxylin Eosin staining; HA: Hyaluronic Acid; HAMA: Methacrylated hyaluronic Acid; HES: Hematoxylin Eosin and Saffron staining; ICC: Immunocytochemistry, IF: Immunofluorescence; IHC: Immunohistochemistry; M: 10^6 (for Million); NFC: Nanofibrillated Cellulose; PEG: Poly(Ethylene)-Glycol; *PRG4*: Proteoglycan 4 (gene); *RunX2*: Runt-related transcription factor 2 (gene); *Sox9*: Sex-determining region-related HMG-box9 (gene); and *VCAN*: Versican (gene). NB: We presented only the studies using the 3D bioextrusion printing method, where the hydrogel-based bioink is directly seeded by the cells before printing. Therefore, we eliminated acellular printing methods, publications associating a PCL scaffold with hydrogels, and studies using extrusion coupled with post-printing molding.

5.7. Biomechanical Properties

Healthy articular cartilage needs to have biomechanical properties for physiological load bearing. The compressive modulus of native cartilage ranges from 240 to 1000 kPa but many hydrogels fail to meet this criterion [183]. Thus, 3D-printed cartilaginous biomechanical properties need to be investigated. Only a third of the studies presented in this review evaluated the compressive strength of the constructs either directly after printing [154], after some maturation time [173], or by comparing different time points [143,163,172]. Most commonly, the mechanical process used for evaluation of the properties are a uniaxial compressive test [154,155], unconfined compression test [143,158,162,172], dynamical mechanical compression test [163,173], or indentation [142]. The aim of those measures being to compare the tissue-engineered construct to native cartilage but also to evaluate the strength of different bioprinted biomaterials [162], to compare acellular constructs to cellularized ones [172], or even to evaluate the most efficient culture conditions for chondrogenesis [143]. The different biomechanical parameters such as compressive strength but also stiffness and elasticity need to be further investigated to better mimic native cartilage properties.

We resume all parameters discussed above, such as cell origins, densities, and viability, and chondrogenic and biomechanical properties in Table 2.

6. Conclusions and Future Directions

Currently, extrusion-based 3D printing can be used to produce cartilaginous constructs. Future applications still present challenges and limitations. Most studies aim to produce standardized structures, generally cubes or rings, but in order to treat cartilaginous defects, the 3D constructs need to be adapted to the depth and shape of a unique lesion [93]. Furthermore, the biomaterial used to produce those structures needs to integrate with the native cartilage edges to form a strong and stable joint surface [142]. New noninvasive methods are being developed to assess cartilage thickness, composition, and functional integrity [184]. The most challenging requirement of TE is that the neosynthesized matrix must have sufficient stability to bear the physiological loads of the joint. Thus, numerous biomechanical parameters need to be assessed after 3D printing (compression modulus, deformation) [120,143,163]. The gelation process usually necessary to maintain printed construct integrity remains an obstacle to direct in situ printing. New methods were developed to directly crosslink the ink during the printing process [185] or to polymerize the bioink in situ [142]. The ideal cell type to use for cartilage TE is still controversial. The sole resident cells of cartilage present many disadvantages to in vitro expansion, such as in vitro dedifferentiation. Adult MSCs are a promising alternative. With good proliferation and differentiation potential, they offer new cell types and density possibilities for engineered constructs. Other cell types were extruded with subsequent evaluation of their chondrogenic properties, such as ATDC5 cells [186,187] or IPS cells [188]. Today's research aims to create layered substitutes that reproduce the natural zonal organization of hyaline cartilage, by optimizing cell types and densities, biomaterials, and environmental factors (growth factors, oximetry), to mimic both the structural and biomechanical properties of the natural material. However, many barriers still remain for the clinical translation of those research [189].

Author Contributions: O.M. contributed to the redaction of the manuscript. O.M., C.H. and K.B. were involved in bibliographic research. C.H. and A.P. were involved in redaction supervision. D.L. was involved in intellectual contribution. P.G. and A.P. participated to the editing of the manuscript. All authors have read and agreed to the published version of the manuscript.

Funding: This research received no external funding.

Acknowledgments: This work was supported by "Direction Générale des Armées (DGA), Grant/Award Number: ANR-16-ASTR-0021"; "Fondation de l'Avenir pour la Recherche Médicale Appliquée, Grant/Award Number: AP-RM-16-042"; and "Université de Lorraine-Région Alsace-Champagne-Ardennes-Lorraine 2016, Grant/Award Number: AAP-002-037." Océane MESSAOUDI is supported by a Ph.D. scholarship granted by the Fondation pour la Recherche Médicale (FRM). Kevin BOURGE is supported by a Ph.D. scholarship granted by the Ministère de l'Education Nationale, de l'Enseignement Supérieur et de la Recherche (MENESR).

Conflicts of Interest: The authors declare no conflict of interest.

Abbreviations

3D	Three Dimension
ACI	Autologous chondrocyte implantation
ACPC	Articular Chondroprogenitor Cells
ADSC	Adipose Derived Stem Cells
BM-MSC	Bone Marrow-derived Mesenchymal Stem Cells
BMP	Bone Morphogenetic Protein
CAD	Computer-Aided Design
CNTs	Carbon nanotubes
CS	Chondroitin Sulfate
dECM	decellularized ExraCellular Matrix
EBB	Extrusion-Based Bioprinting
ECM	ExtraCellular Matrix
GelMA	Gelatin Methacryloyl
HA	Hyaluronic Acid
IGF	Insulin-like Growth Factor
IPSc	Induced Pluripotent Stem cells
MSC	Mesenchymal Stem Cells
OA	OsteoArthritis
PCL	Polycaprolactone
PEG	Poly(Ethylene)-Glycol
PG	Proteoglycan
PGA	Polyglycolic acid
TE	Tissue engineering
TGF-β	Transforming Growth Factor Beta
UV	UltraViolet

References

1. Askari, M.; Naniz, M.A.; Kouhi, M.; Saberi, A.; Zolfagharian, A.; Bodaghi, M. Recent progress in extrusion 3D bioprinting of hydrogel biomaterials for tissue regeneration: A comprehensive review with focus on advanced fabrication techniques. *Biomater. Sci.* **2020**. [CrossRef] [PubMed]
2. Dzobo, K.; Thomford, N.E.; Senthebane, D.A.; Shipanga, H.; Rowe, A.; Dandara, C.; Pillay, M.; Motaung, K.S.C.M. Advances in Regenerative Medicine and Tissue Engineering: Innovation and Transformation of Medicine. Available online: https://www.hindawi.com/journals/sci/2018/2495848/ (accessed on 14 December 2020).
3. Daly, A.C.; Freeman, F.E.; Gonzalez-Fernandez, T.; Critchley, S.E.; Nulty, J.; Kelly, D.J. 3D Bioprinting for Cartilage and Osteochondral Tissue Engineering. *Adv. Healthc. Mater.* **2017**, *6*. [CrossRef] [PubMed]
4. Guo, T.; Lembong, J.; Zhang, L.G.; Fisher, J.P. Three-Dimensional Printing Articular Cartilage: Recapitulating the Complexity of Native Tissue. *Tissue Eng. Part B Rev.* **2017**, *23*, 225–236. [CrossRef] [PubMed]
5. Vina, E.R.; Kwoh, C.K. Epidemiology of osteoarthritis: Literature update. *Curr. Opin. Rheumatol.* **2018**, *30*, 160–167. [CrossRef]
6. Glyn-Jones, S.; Palmer, A.J.R.; Agricola, R.; Price, A.J.; Vincent, T.L.; Weinans, H.; Carr, A.J. Osteoarthritis. *Lancet* **2015**, *386*, 376–387. [CrossRef]
7. Martel-Pelletier, J.; Barr, A.J.; Cicuttini, F.M.; Conaghan, P.G.; Cooper, C.; Goldring, M.B.; Goldring, S.R.; Jones, G.; Teichtahl, A.J.; Pelletier, J.-P. Osteoarthritis. *Nat. Rev. Dis. Primers* **2016**, *2*, 16072. [CrossRef]
8. Aho, O.-M.; Finnilä, M.; Thevenot, J.; Saarakkala, S.; Lehenkari, P. Subchondral bone histology and grading in osteoarthritis. *PLoS ONE* **2017**, *12*, e0173726. [CrossRef]
9. Kohn, M.D.; Sassoon, A.A.; Fernando, N.D. Classifications in Brief: Kellgren-Lawrence Classification of Osteoarthritis. *Clin. Orthop. Relat. Res.* **2016**, *474*, 1886–1893. [CrossRef]
10. Robinson, P.D.; McEwan, J.; Adukia, V.; Prabhakar, M. Osteoarthritis and arthroplasty of the hip and knee. *Br. J. Hosp. Med.* **2018**, *79*, C54–C59. [CrossRef]

11. Taruc-Uy, R.L.; Lynch, S.A. Diagnosis and treatment of osteoarthritis. *Prim. Care* **2013**, *40*, 821–836. [CrossRef]
12. Kan, H.S.; Chan, P.K.; Chiu, K.Y.; Yan, C.H.; Yeung, S.S.; Ng, Y.L.; Shiu, K.W.; Ho, T. Non-surgical treatment of knee osteoarthritis. *Hong Kong Med. J.* **2019**, *25*, 127–133. [CrossRef] [PubMed]
13. Paglia, D.N.; Kanjilal, D.; Kadkoy, Y.; Moskonas, S.; Wetterstrand, C.; Lin, A.; Galloway, J.; Tompson, J.; Culbertson, M.D.; O'Connor, J.P. Naproxen Treatment Inhibits Articular Cartilage Loss in a Rat Model of Osteoarthritis. *J. Orthop. Res.* **2020**. [CrossRef] [PubMed]
14. Siva, C.; Velazquez, C.; Mody, A.; Brasington, R. Diagnosing acute monoarthritis in adults: A practical approach for the family physician. *Am. Fam. Physician* **2003**, *68*, 83–90.
15. Wang, Y.; Wluka, A.E.; English, D.R.; Teichtahl, A.J.; Giles, G.G.; O'Sullivan, R.; Cicuttini, F.M. Body composition and knee cartilage properties in healthy, community-based adults. *Ann. Rheum. Dis.* **2007**, *66*, 1244–1248. [CrossRef]
16. Cicuttini, F.M.; Wluka, A.; Bailey, M.; O'Sullivan, R.; Poon, C.; Yeung, S.; Ebeling, P.R. Factors affecting knee cartilage volume in healthy men. *Rheumatology (Oxford)* **2003**, *42*, 258–262. [CrossRef]
17. Bader, D.L.; Salter, D.M.; Chowdhury, T.T. Biomechanical influence of cartilage homeostasis in health and disease. *Arthritis* **2011**, *2011*, 979032. [CrossRef]
18. Buckwalter, J.A.; Mow, V.C.; Ratcliffe, A. Restoration of Injured or Degenerated Articular Cartilage. *J. Am. Acad. Orthop. Surg.* **1994**, *2*, 192–201. [CrossRef]
19. van Haaften, E.E.; Ito, K.; van Donkelaar, C.C. The initial repair response of articular cartilage after mechanically induced damage. *J. Orthop. Res.* **2017**, *35*, 1265–1273. [CrossRef]
20. Buckwalter, J.A. Articular cartilage injuries. *Clin. Orthop. Relat. Res.* **2002**, 21–37. [CrossRef]
21. Wang, Y.; Ding, C.; Wluka, A.E.; Davis, S.; Ebeling, P.R.; Jones, G.; Cicuttini, F.M. Factors affecting progression of knee cartilage defects in normal subjects over 2 years. *Rheumatology (Oxford)* **2006**, *45*, 79–84. [CrossRef]
22. Roos, E.M. Joint injury causes knee osteoarthritis in young adults. *Curr. Opin. Rheumatol.* **2005**, *17*, 195–200. [CrossRef] [PubMed]
23. Richter, D.L.; Schenck, R.C.; Wascher, D.C.; Treme, G. Knee Articular Cartilage Repair and Restoration Techniques: A Review of the Literature. *Sports Health* **2016**, *8*, 153–160. [CrossRef] [PubMed]
24. Johnson, L.L. Arthroscopic abrasion arthroplasty historical and pathologic perspective: Present status. *Arthroscopy* **1986**, *2*, 54–69. [CrossRef]
25. Johnson, L.L. Arthroscopic abrasion arthroplasty: A review. *Clin. Orthop. Relat. Res.* **2001**, *391*, S306–S317. [CrossRef]
26. Steadman, J.R.; Rodkey, W.G.; Rodrigo, J.J. Microfracture: Surgical Technique and Rehabilitation to Treat Chondral Defects. *Clin. Orthop. Relat. Res.* **2001**, *391*, S362. [CrossRef]
27. Torrie, A.M.; Kesler, W.W.; Elkin, J.; Gallo, R.A. Osteochondral allograft. *Curr. Rev. Musculoskelet Med.* **2015**, *8*, 413–422. [CrossRef]
28. Hangody, L.; Kish, G.; Kárpáti, Z.; Szerb, I.; Udvarhelyi, I. Arthroscopic autogenous osteochondral mosaicplasty for the treatment of femoral condylar articular defects. A preliminary report. *Knee Surg. Sports Traumatol. Arthrosc.* **1997**, *5*, 262–267. [CrossRef]
29. Hangody, L.; Berta, Á. Mosaicplasty for Articular Cartilage Defects. In *European Surgical Orthopaedics and Traumatology: The EFORT Textbook*; Bentley, G., Ed.; Springer: Berlin/Heidelberg, Germany, 2014; pp. 2913–2924. ISBN 978-3-642-34746-7.
30. Brittberg, M.; Lindahl, A.; Nilsson, A.; Ohlsson, C.; Isaksson, O.; Peterson, L. Treatment of deep cartilage defects in the knee with autologous chondrocyte transplantation. *N. Engl. J. Med.* **1994**, *331*, 889–895. [CrossRef]
31. Makris, E.A.; Gomoll, A.H.; Malizos, K.N.; Hu, J.C.; Athanasiou, K.A. Repair and tissue engineering techniques for articular cartilage. *Nat. Rev. Rheumatol.* **2015**, *11*, 21–34. [CrossRef]
32. Bhosale, A.M.; Richardson, J.B. Articular cartilage: Structure, injuries and review of management. *Br. Med. Bull.* **2008**, *87*, 77–95. [CrossRef]
33. Flik, K.R.; Verma, N.; Cole, B.J.; Bach, B.R. Articular Cartilage. In *Cartilage Repair Strategies*; Williams, R.J., Ed.; Humana Press: Totowa, NJ, USA, 2007; pp. 1–12. ISBN 978-1-59745-343-1.
34. Silver, F.H.; Bradica, G. Mechanobiology of cartilage: How do internal and external stresses affect mechanochemical transduction and elastic energy storage? *Biomech. Model. Mechanobiol.* **2002**, *1*, 219–238. [CrossRef] [PubMed]
35. Cohen, N.P.; Foster, R.J.; Mow, V.C. Composition and dynamics of articular cartilage: Structure, function, and maintaining healthy state. *J. Orthop. Sports Phys. Ther.* **1998**, *28*, 203–215. [CrossRef]
36. Loeser, R.F. Integrins and chondrocyte–matrix interactions in articular cartilage. *Matrix Biol.* **2014**, *39*, 11–16. [CrossRef] [PubMed]
37. Roughley, P.J. Articular cartilage and changes in arthritis: Noncollagenous proteins and proteoglycans in the extracellular matrix of cartilage. *Arthritis Res.* **2001**, *3*, 342–347. [CrossRef]
38. Duval, E.; Baugé, C.; Andriamanalijaona, R.; Bénateau, H.; Leclercq, S.; Dutoit, S.; Poulain, L.; Galéra, P.; Boumédiene, K. Molecular mechanism of hypoxia-induced chondrogenesis and its application in in vivo cartilage tissue engineering. *Biomaterials* **2012**, *33*, 6042–6051. [CrossRef]
39. Fox, A.J.S.; Bedi, A.; Rodeo, S.A. The Basic Science of Articular Cartilage: Structure, Composition, and Function. *Sports Health* **2009**. [CrossRef]
40. Maroudas, A.; Wachtel, E.; Grushko, G.; Katz, E.P.; Weinberg, P. The effect of osmotic and mechanical pressures on water partitioning in articular cartilage. *Biochim. Biophys. Acta* **1991**, *1073*, 285–294. [CrossRef]
41. Yanagishita, M. Function of proteoglycans in the extracellular matrix. *Acta Pathol. Jpn.* **1993**, *43*, 283–293. [CrossRef]
42. Roughley, P.J.; Mort, J.S. The role of aggrecan in normal and osteoarthritic cartilage. *J. Exp. Orthop.* **2014**, *1*. [CrossRef]
43. Wong, M.; Carter, D.R. Articular cartilage functional histomorphology and mechanobiology: A research perspective. *Bone* **2003**, *33*, 1–13. [CrossRef]

44. Quiroga, J.M.P.; Wilson, W.; Ito, K.; van Donkelaar, C.C. Relative contribution of articular cartilage's constitutive components to load support depending on strain rate. *Biomech. Model. Mechanobiol.* **2017**, *16*, 151–158. [CrossRef] [PubMed]
45. Eyre, D.R.; Weis, M.A.; Wu, J.-J. Articular cartilage collagen: An irreplaceable framework? *Eur. Cell Mater.* **2006**, *12*, 57–63. [CrossRef] [PubMed]
46. Mansfield, J.C.; Peter Winlove, C. A multi-modal multiphoton investigation of microstructure in the deep zone and calcified cartilage. *J. Anat.* **2012**, *220*, 405–416. [CrossRef]
47. Grogan, S.P.; Duffy, S.F.; Pauli, C.; Koziol, J.A.; Su, A.I.; D'Lima, D.D.; Lotz, M.K. Zone-specific Gene Expression Patterns in Articular Cartilage. *Arthritis Rheum.* **2013**, *65*, 418–428. [CrossRef] [PubMed]
48. Schuurman, W.; Gawlitta, D.; Klein, T.J.; ten Hoope, W.; van Rijen, M.H.P.; Dhert, W.J.A.; van Weeren, P.R.; Malda, J. Zonal Chondrocyte Subpopulations Reacquire Zone-Specific Characteristics during in Vitro Redifferentiation. *Am. J. Sports Med.* **2009**. [CrossRef] [PubMed]
49. Amanatullah, D.F.; Yamane, S.; Reddi, A.H. Distinct patterns of gene expression in the superficial, middle and deep zones of bovine articular cartilage. *J. Tissue Eng. Regen. Med.* **2014**, *8*, 505–514. [CrossRef] [PubMed]
50. Quinn, T.; Häuselmann, H.; Shintani, N.; Hunziker, E. Cell and Matrix Morphology in Articular Cartilage from Adult Human Knee and Ankle Joints Suggests Depth-Associated Adaptations to Biomechanical and Anatomical Roles. Available online: https://pubmed.ncbi.nlm.nih.gov/24455780/ (accessed on 31 August 2020).
51. Kuo, C.K.; Li, W.-J.; Mauck, R.L.; Tuan, R.S. Cartilage tissue engineering: Its potential and uses. *Curr. Opin. Rheumatol.* **2006**, *18*, 64–73. [CrossRef]
52. Huang, B.J.; Hu, J.C.; Athanasiou, K.A. Cell-based tissue engineering strategies used in the clinical repair of articular cartilage. *Biomaterials* **2016**, *98*, 1–22. [CrossRef]
53. Lalan, S.; Pomerantseva, I.; Vacanti, J.P. Tissue engineering and its potential impact on surgery. *World J. Surg.* **2001**, *25*, 1458–1466. [CrossRef]
54. Gugjoo, M.B.; Amarpal; Sharma, G.T.; Aithal, H.P.; Kinjavdekar, P. Cartilage tissue engineering: Role of mesenchymal stem cells along with growth factors & scaffolds. *Indian J. Med. Res.* **2016**, *144*, 339–347. [CrossRef]
55. Mir, T.A.; Nakamura, M. Three-Dimensional Bioprinting: Toward the Era of Manufacturing Human Organs as Spare Parts for Healthcare and Medicine. *Tissue Eng. Part B Rev.* **2017**, *23*, 245–256. [CrossRef]
56. Schon, B.S.; Hooper, G.J.; Woodfield, T.B.F. Modular Tissue Assembly Strategies for Biofabrication of Engineered Cartilage. *Ann. Biomed. Eng.* **2017**, *45*, 100–114. [CrossRef] [PubMed]
57. Johnstone, B.; Hering, T.M.; Caplan, A.I.; Goldberg, V.M.; Yoo, J.U. In vitro chondrogenesis of bone marrow-derived mesenchymal progenitor cells. *Exp. Cell Res.* **1998**, *238*, 265–272. [CrossRef] [PubMed]
58. Alegre-Aguarón, E.; Sampat, S.R.; Xiong, J.C.; Colligan, R.M.; Bulinski, J.C.; Cook, J.L.; Ateshian, G.A.; Brown, L.M.; Hung, C.T. Growth Factor Priming Differentially Modulates Components of the Extracellular Matrix Proteome in Chondrocytes and Synovium-Derived Stem Cells. *PLoS ONE* **2014**, *9*, e88053. [CrossRef] [PubMed]
59. Neybecker, P.; Henrionnet, C.; Pape, E.; Mainard, D.; Galois, L.; Loeuille, D.; Gillet, P.; Pinzano, A. In vitro and in vivo potentialities for cartilage repair from human advanced knee osteoarthritis synovial fluid-derived mesenchymal stem cells. *Stem. Cell Res. Ther.* **2018**, *9*, 329. [CrossRef]
60. Henrionnet, C.; Pourchet, L.; Neybecker, P.; Messaoudi, O.; Gillet, P.; Loeuille, D.; Mainard, D.; Marquette, C.; Pinzano, A. Combining Innovative Bioink and Low Cell Density for the Production of 3D-Bioprinted Cartilage Substitutes: A Pilot Study. Available online: https://www.hindawi.com/journals/sci/2020/2487072/ (accessed on 22 June 2020).
61. Gupta, M.S.; Nicoll, S.B. Duration of TGF-β3 Exposure Impacts the Chondrogenic Maturation of Human MSCs in Photocrosslinked Carboxymethylcellulose Hydrogels. *Ann. Biomed. Eng.* **2015**, *43*, 1145–1157. [CrossRef]
62. Gómez-Leduc, T.; Desancé, M.; Hervieu, M.; Legendre, F.; Ollitrault, D.; de Vienne, C.; Herlicoviez, M.; Galéra, P.; Demoor, M. Hypoxia Is a Critical Parameter for Chondrogenic Differentiation of Human Umbilical Cord Blood Mesenchymal Stem Cells in Type I/III Collagen Sponges. *Int. J. Mol. Sci.* **2017**, *18*, 1933. [CrossRef]
63. Henrionnet, C.; Liang, G.; Roeder, E.; Dossot, M.; Wang, H.; Magdalou, J.; Gillet, P.; Pinzano, A. Hypoxia for Mesenchymal Stem Cell Expansion and Differentiation: The Best Way for Enhancing TGFß-Induced Chondrogenesis and Preventing Calcifications in Alginate Beads. *Tissue Eng. Part A* **2017**, *23*, 913–922. [CrossRef]
64. Pattappa, G.; Schewior, R.; Hofmeister, I.; Seja, J.; Zellner, J.; Johnstone, B.; Docheva, D.; Angele, P. Physioxia Has a Beneficial Effect on Cartilage Matrix Production in Interleukin-1 Beta-Inhibited Mesenchymal Stem Cell Chondrogenesis. *Cells* **2019**, *8*, 936. [CrossRef]
65. Meretoja, V.V.; Dahlin, R.L.; Wright, S.; Kasper, F.K.; Mikos, A.G. The effect of hypoxia on the chondrogenic differentiation of co-cultured articular chondrocytes and mesenchymal stem cells in scaffolds. *Biomaterials* **2013**, *34*, 4266–4273. [CrossRef]
66. Wernike, E.; Li, Z.; Alini, M.; Grad, S. Effect of reduced oxygen tension and long-term mechanical stimulation on chondrocyte-polymer constructs. *Cell Tissue Res.* **2008**, *331*, 473–483. [CrossRef] [PubMed]
67. Mohanraj, B.; Farran, A.J.; Mauck, R.L.; Dodge, G.R. Time-dependent functional maturation of scaffold-free cartilage tissue analogs. *J. Biomech.* **2014**, *47*, 2137–2142. [CrossRef] [PubMed]
68. Aisenbrey, E.A.; Bilousova, G.; Payne, K.; Bryant, S.J. Dynamic mechanical loading and growth factors influence chondrogenesis of induced pluripotent mesenchymal progenitor cells in a cartilage-mimetic hydrogel. *Biomater. Sci.* **2019**, *7*, 5388–5403. [CrossRef] [PubMed]

69. Wang, W.; Wan, Y.; Fu, T.; Zhou, T.; Tang, X.; Wu, H.; Liu, C.; Jagodzinski, M. Effect of cyclic compression on bone marrow mesenchymal stromal cells in tissue engineered cartilage scaffold. *J. Biomed. Mater. Res. A* **2019**, *107*, 1294–1302. [CrossRef]
70. Carroll, S.F.; Buckley, C.T.; Kelly, D.J. Cyclic hydrostatic pressure promotes a stable cartilage phenotype and enhances the functional development of cartilaginous grafts engineered using multipotent stromal cells isolated from bone marrow and infrapatellar fat pad. *J. Biomech.* **2014**, *47*, 2115–2121. [CrossRef]
71. Davisson, T.; Kunig, S.; Chen, A.; Sah, R.; Ratcliffe, A. Static and dynamic compression modulate matrix metabolism in tissue engineered cartilage. *J. Orthop. Res.* **2002**, *20*, 842–848. [CrossRef]
72. Vainieri, M.L.; Wahl, D.; Alini, M.; van Osch, G.J.V.M.; Grad, S. Mechanically stimulated osteochondral organ culture for evaluation of biomaterials in cartilage repair studies. *Acta Biomater.* **2018**, *81*, 256–266. [CrossRef]
73. Choi, J.R.; Yong, K.W.; Choi, J.Y. Effects of mechanical loading on human mesenchymal stem cells for cartilage tissue engineering. *J. Cell Physiol.* **2018**, *233*, 1913–1928. [CrossRef]
74. Hull, C.W.; Gabriel, S. Apparatus for Production of three-dimensional obejcts by stereolithography. U.S. Patent 4575330A, 11 March 1986.
75. Klebe, R.J. Cytoscribing: A method for micropositioning cells and the construction of two- and three-dimensional synthetic tissues. *Exp. Cell Res.* **1988**, *179*, 362–373. [CrossRef]
76. Boland, T.; Crisp Wilson, W.; Xu, T. Ink-Jet Printing of Viable Cells. U.S. Patent No. 7,051,654B2, 30 May 2006.
77. Ventola, C.L. Medical Applications for 3D Printing: Current and Projected Uses. *Pharm. Ther.* **2014**, *39*, 704–711.
78. Li, L.; Yu, F.; Shi, J.; Shen, S.; Teng, H.; Yang, J.; Wang, X.; Jiang, Q. In situ repair of bone and cartilage defects using 3D scanning and 3D printing. *Sci. Rep.* **2017**, *7*. [CrossRef] [PubMed]
79. Cohen, D.L.; Lipton, J.I.; Bonassar, L.J.; Lipson, H. Additive manufacturing for in situ repair of osteochondral defects. *Biofabrication* **2010**, *2*, 035004. [CrossRef] [PubMed]
80. Borovjagin, A.V.; Ogle, B.M.; Berry, J.L.; Zhang, J. From Microscale Devices to 3D Printing: Advances in Fabrication of 3D Cardiovascular Tissues. *Circ. Res.* **2017**, *120*, 150–165. [CrossRef] [PubMed]
81. Rider, P.; Kačarević, Ž.P.; Alkildani, S.; Retnasingh, S.; Barbeck, M. Bioprinting of tissue engineering scaffolds. *J. Tissue Eng.* **2018**, *9*. [CrossRef]
82. Guillemot, F.; Guillotin, B.; Fontaine, A.; Ali, M.; Catros, S.; Kériquel, V.; Fricain, J.-C.; Rémy, M.; Bareille, R.; Amédée-Vilamitjana, J. Laser-assisted bioprinting to deal with tissue complexity in regenerative medicine. *MRS Bull.* **2011**, *36*, 1015–1019. [CrossRef]
83. Ozbolat, I.T.; Hospodiuk, M. Current advances and future perspectives in extrusion-based bioprinting. *Biomaterials* **2016**, *76*, 321–343. [CrossRef]
84. Yilmaz, B.; Tahmasebifar, A.; Baran, E.T. Bioprinting Technologies in Tissue Engineering. *Adv. Biochem. Eng. Biotechnol.* **2019**. [CrossRef]
85. Derakhshanfar, S.; Mbeleck, R.; Xu, K.; Zhang, X.; Zhong, W.; Xing, M. 3D bioprinting for biomedical devices and tissue engineering: A review of recent trends and advances. *Bioact. Mater.* **2018**, *3*, 144–156. [CrossRef]
86. Shim, J.-H.; Jang, K.-M.; Hahn, S.K.; Park, J.Y.; Jung, H.; Oh, K.; Park, K.M.; Yeom, J.; Park, S.H.; Kim, S.W.; et al. Three-dimensional bioprinting of multilayered constructs containing human mesenchymal stromal cells for osteochondral tissue regeneration in the rabbit knee joint. *Biofabrication* **2016**, *8*, 014102. [CrossRef]
87. Armiento, A.R.; Stoddart, M.J.; Alini, M.; Eglin, D. Biomaterials for articular cartilage tissue engineering: Learning from biology. *Acta Biomater.* **2018**, *65*, 1–20. [CrossRef]
88. Salamon, A.; van Vlierberghe, S.; van Nieuwenhove, I.; Baudisch, F.; Graulus, G.-J.; Benecke, V.; Alberti, K.; Neumann, H.-G.; Rychly, J.; Martins, J.C.; et al. Gelatin-Based Hydrogels Promote Chondrogenic Differentiation of Human Adipose Tissue-Derived Mesenchymal Stem Cells In Vitro. *Materials (Basel)* **2014**, *7*, 1342–1359. [CrossRef]
89. Bosnakovski, D.; Mizuno, M.; Kim, G.; Takagi, S.; Okumura, M.; Fujinaga, T. Chondrogenic differentiation of bovine bone marrow mesenchymal stem cells (MSCs) in different hydrogels: Influence of collagen type II extracellular matrix on MSC chondrogenesis. *Biotechnol. Bioeng.* **2006**, *93*, 1152–1163. [CrossRef] [PubMed]
90. Jin, G.-Z.; Kim, H.-W. Efficacy of collagen and alginate hydrogels for the prevention of rat chondrocyte dedifferentiation. *J. Tissue Eng.* **2018**, *9*, 2041731418802438. [CrossRef] [PubMed]
91. He, Y.; Liu, W.; Guan, L.; Chen, J.; Duan, L.; Jia, Z.; Huang, J.; Li, W.; Liu, J.; Xiong, J.; et al. A 3D-Printed PLCL Scaffold Coated with Collagen Type I and Its Biocompatibility. *Biomed. Res. Int.* **2018**, *2018*. [CrossRef] [PubMed]
92. Park, J.Y.; Choi, Y.-J.; Shim, J.-H.; Park, J.H.; Cho, D.-W. Development of a 3D cell printed structure as an alternative to autologs cartilage for auricular reconstruction. *J. Biomed. Mater. Res. Part B Appl. Biomater.* **2017**, *105*, 1016–1028. [CrossRef] [PubMed]
93. Li, Z.; Wu, N.; Cheng, J.; Sun, M.; Yang, P.; Zhao, F.; Zhang, J.; Duan, X.; Fu, X.; Zhang, J.; et al. Biomechanically, structurally and functionally meticulously tailored polycaprolactone/silk fibroin scaffold for meniscus regeneration. *Theranostics* **2020**, *10*, 5090–5106. [CrossRef]
94. Sun, Y.; You, Y.; Jiang, W.; Wang, B.; Wu, Q.; Dai, K. 3D bioprinting dual-factor releasing and gradient-structured constructs ready to implant for anisotropic cartilage regeneration. *Sci. Adv.* **2020**, *6*, eaay1422. [CrossRef]
95. Szymański, T.; Mieloch, A.A.; Richter, M.; Trzeciak, T.; Florek, E.; Rybka, J.D.; Giersig, M. Utilization of Carbon Nanotubes in Manufacturing of 3D Cartilage and Bone Scaffolds. *Materials* **2020**, *13*, 4039. [CrossRef]
96. Jin, M.; Shi, J.; Zhu, W.; Yao, H.; Wang, D.-A. Polysaccharide-based Biomaterials in Tissue Engineering: A review. *Tissue Eng. Part B Rev.* **2020**. [CrossRef]

97. Ewa-Choy, Y.W.; Pingguan-Murphy, B.; Abdul-Ghani, N.A.; Jahendran, J.; Chua, K.H. Effect of alginate concentration on chondrogenesis of co-cultured human adipose-derived stem cells and nasal chondrocytes: A biological study. *Biomater. Res.* **2017**, *21*. [CrossRef]
98. Schuurman, W.; Levett, P.A.; Pot, M.W.; van Weeren, P.R.; Dhert, W.J.A.; Hutmacher, D.W.; Melchels, F.P.W.; Klein, T.J.; Malda, J. Gelatin-methacrylamide hydrogels as potential biomaterials for fabrication of tissue-engineered cartilage constructs. *Macromol. Biosci.* **2013**, *13*, 551–561. [CrossRef] [PubMed]
99. Mouser, V.H.M.; Levato, R.; Mensinga, A.; Dhert, W.J.A.; Gawlitta, D.; Malda, J. Bio-ink development for three-dimensional bioprinting of hetero-cellular cartilage constructs. *Connect. Tissue Res.* **2020**, *61*, 137–151. [CrossRef] [PubMed]
100. Ma, K.; Zhao, T.; Yang, L.; Wang, P.; Jin, J.; Teng, H.; Xia, D.; Zhu, L.; Li, L.; Jiang, Q.; et al. Application of robotic-assisted in situ 3D printing in cartilage regeneration with HAMA hydrogel: An in vivo study. *J. Adv. Res.* **2020**, *23*, 123–132. [CrossRef] [PubMed]
101. Balakrishnan, B.; Banerjee, R. Biopolymer-based hydrogels for cartilage tissue engineering. *Chem. Rev.* **2011**, *111*, 4453–4474. [CrossRef] [PubMed]
102. Antich, C.; de Vicente, J.; Jiménez, G.; Chocarro, C.; Carrillo, E.; Montañez, E.; Gálvez-Martín, P.; Marchal, J.A. Bio-inspired hydrogel composed of hyaluronic acid and alginate as a potential bioink for 3D bioprinting of articular cartilage engineering constructs. *Acta Biomater.* **2020**, *106*, 114–123. [CrossRef]
103. Schwab, A.; Hélary, C.; Richards, R.G.; Alini, M.; Eglin, D.; D'Este, M. Tissue mimetic hyaluronan bioink containing collagen fibers with controlled orientation modulating cell migration and alignment. *Mater. Today Bio.* **2020**, *7*. [CrossRef]
104. Chen, W.; Xu, Y.; Li, Y.; Jia, L.; Mo, X.; Jiang, G.; Zhou, G. 3D printing electrospinning fiber-reinforced decellularized extracellular matrix for cartilage regeneration. *Chem. Eng. J.* **2020**, *382*, 122986. [CrossRef]
105. Dzobo, K.; Motaung, K.S.C.M.; Adesida, A. Recent Trends in Decellularized Extracellular Matrix Bioinks for 3D Printing: An Updated Review. *Int. J. Mol. Sci.* **2019**, *20*, 4628. [CrossRef]
106. Zhang, X.; Liu, Y.; Luo, C.; Zhai, C.; Li, Z.; Zhang, Y.; Yuan, T.; Dong, S.; Zhang, J.; Fan, W. Crosslinker-free silk/decellularized extracellular matrix porous bioink for 3D bioprinting-based cartilage tissue engineering. *Mater. Sci. Eng. C* **2021**, *118*, 111388. [CrossRef]
107. Pati, F.; Jang, J.; Ha, D.-H.; Won Kim, S.; Rhie, J.-W.; Shim, J.-H.; Kim, D.-H.; Cho, D.-W. Printing three-dimensional tissue analogues with decellularized extracellular matrix bioink. *Nat. Commun.* **2014**, *5*, 3935. [CrossRef]
108. Visscher, D.O.; Lee, H.; van Zuijlen, P.P.M.; Helder, M.N.; Atala, A.; Yoo, J.J.; Lee, S.J. A photo-crosslinkable cartilage-derived extracellular matrix (ECM) bioink for auricular cartilage tissue engineering. *Acta Biomater.* **2020**. [CrossRef]
109. Chung, J.H.Y.; Naficy, S.; Yue, Z.; Kapsa, R.; Quigley, A.; Moulton, S.E.; Wallace, G.G. Bio-ink properties and printability for extrusion printing living cells. *Biomater. Sci.* **2013**, *1*, 763–773. [CrossRef] [PubMed]
110. Chimene, D.; Lennox, K.K.; Kaunas, R.R.; Gaharwar, A.K. Advanced Bioinks for 3D Printing: A Materials Science Perspective. *Ann. Biomed. Eng.* **2016**, *44*, 2090–2102. [CrossRef] [PubMed]
111. Hölzl, K.; Lin, S.; Tytgat, L.; Van Vlierberghe, S.; Gu, L.; Ovsianikov, A. Bioink properties before, during and after 3D bioprinting. *Biofabrication* **2016**, *8*, 032002. [CrossRef] [PubMed]
112. Trachsel, L.; Johnbosco, C.; Lang, T.; Benetti, E.M.; Zenobi-Wong, M. Double-Network Hydrogels Including Enzymatically Crosslinked Poly-(2-alkyl-2-oxazoline)s for 3D Bioprinting of Cartilage-Engineering Constructs. *Biomacromolecules* **2019**, *20*, 4502–4511. [CrossRef] [PubMed]
113. Fischetti, T.; Celikkin, N.; Contessi Negrini, N.; Farè, S.; Swieszkowski, W. Tripolyphosphate-Crosslinked Chitosan/Gelatin Biocomposite Ink for 3D Printing of Uniaxial Scaffolds. *Front. Bioeng. Biotechnol.* **2020**, *8*, 400. [CrossRef]
114. Gao, G.; Schilling, A.F.; Hubbell, K.; Yonezawa, T.; Truong, D.; Hong, Y.; Dai, G.; Cui, X. Improved properties of bone and cartilage tissue from 3D inkjet-bioprinted human mesenchymal stem cells by simultaneous deposition and photocrosslinking in PEG-GelMA. *Biotechnol. Lett.* **2015**, *37*, 2349–2355. [CrossRef]
115. Lee, J.-S.; Hong, J.M.; Jung, J.W.; Shim, J.-H.; Oh, J.-H.; Cho, D.-W. 3D printing of composite tissue with complex shape applied to ear regeneration. *Biofabrication* **2014**, *6*, 024103. [CrossRef]
116. Bae, S.-W.; Lee, K.-W.; Park, J.-H.; Lee, J.; Jung, C.-R.; Yu, J.; Kim, H.-Y.; Kim, D.-H. 3D Bioprinted Artificial Trachea with Epithelial Cells and Chondrogenic-Differentiated Bone Marrow-Derived Mesenchymal Stem Cells. *Int. J. Mol. Sci.* **2018**, *19*, 1624. [CrossRef]
117. Müller, M.; Öztürk, E.; Arlov, Ø.; Gatenholm, P.; Zenobi-Wong, M. Alginate Sulfate-Nanocellulose Bioinks for Cartilage Bioprinting Applications. *Ann. Biomed. Eng.* **2017**, *45*, 210–223. [CrossRef]
118. Kesti, M.; Müller, M.; Becher, J.; Schnabelrauch, M.; D'Este, M.; Eglin, D.; Zenobi-Wong, M. A versatile bioink for three-dimensional printing of cellular scaffolds based on thermally and photo-triggered tandem gelation. *Acta Biomater.* **2015**, *11*, 162–172. [CrossRef]
119. Markstedt, K.; Mantas, A.; Tournier, I.; Martínez Ávila, H.; Hägg, D.; Gatenholm, P. 3D Bioprinting Human Chondrocytes with Nanocellulose-Alginate Bioink for Cartilage Tissue Engineering Applications. *Biomacromolecules* **2015**, *16*, 1489–1496. [CrossRef] [PubMed]
120. Mouser, V.H.M.; Melchels, F.P.W.; Visser, J.; Dhert, W.J.A.; Gawlitta, D.; Malda, J. Yield stress determines bioprintability of hydrogels based on gelatin-methacryloyl and gellan gum for cartilage bioprinting. *Biofabrication* **2016**, *8*, 035003. [CrossRef] [PubMed]
121. Von Der Mark, K.; Gauss, V.; Von Der Mark, H.; Müller, P. Relationship between cell shape and type of collagen synthesised as chondrocytes lose their cartilage phenotype in culture. *Nature* **1977**, *267*, 531–532. [CrossRef] [PubMed]

122. Cournil-Henrionnet, C.; Huselstein, C.; Wang, Y.; Galois, L.; Mainard, D.; Decot, V.; Netter, P.; Stoltz, J.-F.; Muller, S.; Gillet, P.; et al. Phenotypic analysis of cell surface markers and gene expression of human mesenchymal stem cells and chondrocytes during monolayer expansion. *Biorheology* **2008**, *45*, 513–526. [CrossRef]
123. Lennon, D.P.; Haynesworth, S.E.; Bruder, S.P.; Jaiswal, N.; Caplan, A.I. Human and animal mesenchymal progenitor cells from bone marrow: Identification of serum for optimal selection and proliferation. *Vitr. Cell Dev. Biol. Anim.* **1996**, *32*, 602–611. [CrossRef]
124. Li, X.; Wang, M.; Jing, X.; Guo, W.; Hao, C.; Zhang, Y.; Gao, S.; Chen, M.; Zhang, Z.; Zhang, X.; et al. Bone Marrow- and Adipose Tissue-Derived Mesenchymal Stem Cells: Characterization, Differentiation, and Applications in Cartilage Tissue Engineering. *Crit. Rev. Eukaryot. Gene Expr.* **2018**, *28*, 285–310. [CrossRef]
125. Gardner, O.F.W.; Alini, M.; Stoddart, M.J. Mesenchymal Stem Cells Derived from Human Bone Marrow. *Methods Mol. Biol.* **2015**, *1340*, 41–52. [CrossRef]
126. Lee, T.-C.; Lee, T.-H.; Huang, Y.-H.; Chang, N.-K.; Lin, Y.-J.; Chien, P.-W.C.; Yang, W.-H.; Lin, M.H.-C. Comparison of surface markers between human and rabbit mesenchymal stem cells. *PLoS ONE* **2014**, *9*, e111390. [CrossRef]
127. Han, Y.; Li, X.; Zhang, Y.; Han, Y.; Chang, F.; Ding, J. Mesenchymal Stem Cells for Regenerative Medicine. *Cells* **2019**, *8*, 886. [CrossRef]
128. Tsutsumi, S.; Shimazu, A.; Miyazaki, K.; Pan, H.; Koike, C.; Yoshida, E.; Takagishi, K.; Kato, Y. Retention of multilineage differentiation potential of mesenchymal cells during proliferation in response to FGF. *Biochem. Biophys. Res. Commun.* **2001**, *288*, 413–419. [CrossRef] [PubMed]
129. Almalki, S.G.; Agrawal, D.K. Key Transcription Factors in the Differentiation of Mesenchymal Stem Cells. *Differentiation* **2016**, *92*, 41–51. [CrossRef] [PubMed]
130. Richardson, S.M.; Kalamegam, G.; Pushparaj, P.N.; Matta, C.; Memic, A.; Khademhosseini, A.; Mobasheri, R.; Poletti, F.L.; Hoyland, J.A.; Mobasheri, A. Mesenchymal stem cells in regenerative medicine: Focus on articular cartilage and intervertebral disc regeneration. *Methods* **2016**, *99*, 69–80. [CrossRef] [PubMed]
131. Sakaguchi, Y.; Sekiya, I.; Yagishita, K.; Muneta, T. Comparison of human stem cells derived from various mesenchymal tissues: Superiority of synovium as a cell source. *Arthritis Rheum.* **2005**, *52*, 2521–2529. [CrossRef]
132. Reppel, L.; Schiavi, J.; Charif, N.; Leger, L.; Yu, H.; Pinzano, A.; Henrionnet, C.; Stoltz, J.-F.; Bensoussan, D.; Huselstein, C. Chondrogenic induction of mesenchymal stromal/stem cells from Wharton's jelly embedded in alginate hydrogel and without added growth factor: An alternative stem cell source for cartilage tissue engineering. *Stem. Cell Res. Ther.* **2015**, *6*, 260. [CrossRef]
133. Friedenstein, A.J.; Chailakhjan, R.K.; Lalykina, K.S. The development of fibroblast colonies in monolayer cultures of guinea-pig bone marrow and spleen cells. *Cell Prolif.* **1970**, *3*, 393–403. [CrossRef]
134. Zuk, P.A.; Zhu, M.; Mizuno, H.; Huang, J.; Futrell, J.W.; Katz, A.J.; Benhaim, P.; Lorenz, H.P.; Hedrick, M.H. Multilineage cells from human adipose tissue: Implications for cell-based therapies. *Tissue Eng.* **2001**, *7*, 211–228. [CrossRef]
135. De Bari, C.; Dell'Accio, F.; Tylzanowski, P.; Luyten, F.P. Multipotent mesenchymal stem cells from adult human synovial membrane. *Arthritis Rheum.* **2001**, *44*, 1928–1942. [CrossRef]
136. Jones, E.A.; English, A.; Henshaw, K.; Kinsey, S.E.; Markham, A.F.; Emery, P.; McGonagle, D. Enumeration and phenotypic characterization of synovial fluid multipotential mesenchymal progenitor cells in inflammatory and degenerative arthritis. *Arthritis Rheum.* **2004**, *50*, 817–827. [CrossRef]
137. Gronthos, S.; Mankani, M.; Brahim, J.; Robey, P.G.; Shi, S. Postnatal human dental pulp stem cells (DPSCs) in vitro and in vivo. *Proc. Natl. Acad. Sci. USA* **2000**, *97*, 13625–13630. [CrossRef]
138. Erices, A.; Conget, P.; Minguel, J. Mesenchymal progenitor cells in human umbilical cord blood. *Br. J. Haematol* **2000**, *109*, 235–242. [CrossRef] [PubMed]
139. Pittenger, M.F.; Mackay, A.M.; Beck, S.C.; Jaiswal, R.K.; Douglas, R.; Mosca, J.D.; Moorman, M.A.; Simonetti, D.W.; Craig, S.; Marshak, D.R. Multilineage potential of adult human mesenchymal stem cells. *Science* **1999**, *284*, 143–147. [CrossRef]
140. Hernández, R.; Jiménez-Luna, C.; Perales-Adán, J.; Perazzoli, G.; Melguizo, C.; Prados, J. Differentiation of Human Mesenchymal Stem Cells towards Neuronal Lineage: Clinical Trials in Nervous System Disorders. *Biomol. Ther. (Seoul)* **2020**, *28*, 34–44. [CrossRef] [PubMed]
141. Ogawa, R.; Mizuno, S. Cartilage regeneration using adipose-derived stem cells. *Curr. Stem. Cell Res. Ther.* **2010**, *5*, 129–132. [CrossRef] [PubMed]
142. Di Bella, C.; Duchi, S.; O'Connell, C.D.; Blanchard, R.; Augustine, C.; Yue, Z.; Thompson, F.; Richards, C.; Beirne, S.; Onofrillo, C.; et al. In situ handheld three-dimensional bioprinting for cartilage regeneration. *J. Tissue Eng. Regen. Med.* **2018**, *12*, 611–621. [CrossRef]
143. Onofrillo, C.; Duchi, S.; O'Connell, C.D.; Blanchard, R.; O'Connor, A.J.; Scott, M.; Wallace, G.G.; Choong, P.F.M.; Di Bella, C. Biofabrication of human articular cartilage: A path towards the development of a clinical treatment. *Biofabrication* **2018**, *10*, 045006. [CrossRef]
144. Felimban, R.; Ye, K.; Traianedes, K.; Di Bella, C.; Crook, J.; Wallace, G.G.; Quigley, A.; Choong, P.F.M.; Myers, D.E. Differentiation of stem cells from human infrapatellar fat pad: Characterization of cells undergoing chondrogenesis. *Tissue Eng. Part A* **2014**, *20*, 2213–2223. [CrossRef]
145. Ye, K.; Felimban, R.; Traianedes, K.; Moulton, S.E.; Wallace, G.G.; Chung, J.; Quigley, A.; Choong, P.F.M.; Myers, D.E. Chondrogenesis of infrapatellar fat pad derived adipose stem cells in 3D printed chitosan scaffold. *PLoS ONE* **2014**, *9*, e99410. [CrossRef]

146. Nakagawa, Y.; Muneta, T.; Kondo, S.; Mizuno, M.; Takakuda, K.; Ichinose, S.; Tabuchi, T.; Koga, H.; Tsuji, K.; Sekiya, I. Synovial mesenchymal stem cells promote healing after meniscal repair in microminipigs. *Osteoarthr. Cartil.* **2015**, *23*, 1007–1017. [CrossRef]
147. Gronthos, S.; Brahim, J.; Li, W.; Fisher, L.W.; Cherman, N.; Boyde, A.; DenBesten, P.; Robey, P.G.; Shi, S. Stem cell properties of human dental pulp stem cells. *J. Dent. Res.* **2002**, *81*, 531–535. [CrossRef]
148. Hilkens, P.; Gervois, P.; Fanton, Y.; Vanormelingen, J.; Martens, W.; Struys, T.; Politis, C.; Lambrichts, I.; Bronckaers, A. Effect of isolation methodology on stem cell properties and multilineage differentiation potential of human dental pulp stem cells. *Cell Tissue Res.* **2013**, *353*, 65–78. [CrossRef] [PubMed]
149. Fernandes, T.L.; Cortez de SantAnna, J.P.; Frisene, I.; Gazarini, J.P.; Gomes Pinheiro, C.C.; Gomoll, A.H.; Lattermann, C.; Hernandez, A.J.; Franco Bueno, D. Systematic Review of Human Dental Pulp Stem Cells for Cartilage Regeneration. *Tissue Eng. Part B Rev.* **2020**, *26*, 1–12. [CrossRef] [PubMed]
150. Dai, J.; Wang, J.; Lu, J.; Zou, D.; Sun, H.; Dong, Y.; Yu, H.; Zhang, L.; Yang, T.; Zhang, X.; et al. The effect of co-culturing costal chondrocytes and dental pulp stem cells combined with exogenous FGF9 protein on chondrogenesis and ossification in engineered cartilage. *Biomaterials* **2012**, *33*, 7699–7711. [CrossRef] [PubMed]
151. Westin, C.B.; Trinca, R.B.; Zuliani, C.; Coimbra, I.B.; Moraes, Â.M. Differentiation of dental pulp stem cells into chondrocytes upon culture on porous chitosan-xanthan scaffolds in the presence of kartogenin. *Mater. Sci. Eng. C Mater. Biol. Appl.* **2017**, *80*, 594–602. [CrossRef] [PubMed]
152. Wang, H.-S.; Hung, S.-C.; Peng, S.-T.; Huang, C.-C.; Wei, H.-M.; Guo, Y.-J.; Fu, Y.-S.; Lai, M.-C.; Chen, C.-C. Mesenchymal stem cells in the Wharton's jelly of the human umbilical cord. *Stem. Cells* **2004**, *22*, 1330–1337. [CrossRef]
153. Liu, S.; Jia, Y.; Yuan, M.; Guo, W.; Huang, J.; Zhao, B.; Peng, J.; Xu, W.; Lu, S.; Guo, Q. Repair of Osteochondral Defects Using Human Umbilical Cord Wharton's Jelly-Derived Mesenchymal Stem Cells in a Rabbit Model. *Biomed. Res. Int.* **2017**, *2017*, 8760383. [CrossRef]
154. Singh, Y.P.; Bandyopadhyay, A.; Mandal, B.B. 3D Bioprinting Using Cross-Linker-Free Silk-Gelatin Bioink for Cartilage Tissue Engineering. *ACS Appl. Mater. Interfaces* **2019**, *11*, 33684–33696. [CrossRef]
155. Yang, J.; Zhang, Y.S.; Yue, K.; Khademhosseini, A. Cell-laden hydrogels for osteochondral and cartilage tissue engineering. *Acta Biomater.* **2017**, *57*, 1–25. [CrossRef]
156. Ren, X.; Wang, F.; Chen, C.; Gong, X.; Yin, L.; Yang, L. Engineering zonal cartilage through bioprinting collagen type II hydrogel constructs with biomimetic chondrocyte density gradient. *BMC Musculoskelet. Disord.* **2016**, *17*, 301. [CrossRef]
157. Kilian, D.; Ahlfeld, T.; Akkineni, A.R.; Bernhardt, A.; Gelinsky, M.; Lode, A. 3D Bioprinting of osteochondral tissue substitutes—In vitro-chondrogenesis in multi-layered mineralized constructs. *Sci. Rep.* **2020**, *10*, 8277. [CrossRef]
158. Jessop, Z.M.; Al-Sabah, A.; Gao, N.; Kyle, S.; Thomas, B.; Badiei, N.; Hawkins, K.; Whitaker, I.S. Printability of pulp derived crystal, fibril and blend nanocellulose-alginate bioinks for extrusion 3D bioprinting. *Biofabrication* **2019**, *11*, 045006. [CrossRef] [PubMed]
159. Dowthwaite, G.; Bishop, J.; Redman, S.; Khan, I.; Rooney, P.; Evans, D.; Haughton, L.; Bayram-weston, Z.; Boyer, S.; Thomson, B.; et al. The Surface of Articular Cartilage Contains a Progenitor Cell Population. *J. Cell Sci.* **2004**, *117*, 889–897. [CrossRef] [PubMed]
160. Khan, I.M.; Bishop, J.C.; Gilbert, S.; Archer, C.W. Clonal chondroprogenitors maintain telomerase activity and Sox9 expression during extended monolayer culture and retain chondrogenic potential. *Osteoarthr. Cartil.* **2009**, *17*, 518–528. [CrossRef] [PubMed]
161. Lim, K.S.; Abinzano, F.; Bernal, P.N.; Albillos Sanchez, A.; Atienza-Roca, P.; Otto, I.A.; Peiffer, Q.C.; Matsusaki, M.; Woodfield, T.B.F.; Malda, J.; et al. One-Step Photoactivation of a Dual-Functionalized Bioink as Cell Carrier and Cartilage-Binding Glue for Chondral Regeneration. *Adv. Healthc. Mater.* **2020**, e1901792. [CrossRef] [PubMed]
162. Armstrong, J.; Burke, M.; Carter, B.; Davis, S.; Perriman, A. 3D Bioprinting Using a Templated Porous Bioink. *Adv. Healthc. Mater.* **2016**, *5*, 1681. [CrossRef]
163. Luo, C.; Xie, R.; Zhang, J.; Liu, Y.; Li, Z.; Zhang, Y.; Zhang, X.; Yuan, T.; Chen, Y.; Fan, W. Low temperature 3D printing of tissue cartilage engineered with gelatin methacrylamide. *Tissue Eng. Part C Methods* **2020**. [CrossRef]
164. Stichler, S.; Böck, T.; Paxton, N.; Bertlein, S.; Levato, R.; Schill, V.; Smolan, W.; Malda, J.; Teßmar, J.; Blunk, T.; et al. Double printing of hyaluronic acid/poly(glycidol) hybrid hydrogels with poly(ε-caprolactone) for MSC chondrogenesis. *Biofabrication* **2017**, *9*, 044108. [CrossRef]
165. Costantini, M.; Idaszek, J.; Szöke, K.; Jaroszewicz, J.; Dentini, M.; Barbetta, A.; Brinchmann, J.E.; Święszkowski, W. 3D bioprinting of BM-MSCs-loaded ECM biomimetic hydrogels for in vitro neocartilage formation. *Biofabrication* **2016**, *8*, 035002. [CrossRef]
166. Kosik-Kozio\l, A.; Costantini, M.; Mróz, A.; Idaszek, J.; Heljak, M.; Jaroszewicz, J.; Kijeńska, E.; Szöke, K.; Frerker, N.; Barbetta, A.; et al. 3D bioprinted hydrogel model incorporating β-tricalcium phosphate for calcified cartilage tissue engineering. *Biofabrication* **2019**, *11*, 035016. [CrossRef]
167. Duchi, S.; Onofrillo, C.; O'Connell, C.; Wallace, G.G.; Choong, P.; Di Bella, C. Bioprinting Stem Cells in Hydrogel for In Situ Surgical Application: A Case for Articular Cartilage. *Methods Mol. Biol.* **2020**, *2140*, 145–157. [CrossRef]
168. O'Connell, C.D.; Di Bella, C.; Thompson, F.; Augustine, C.; Beirne, S.; Cornock, R.; Richards, C.J.; Chung, J.; Gambhir, S.; Yue, Z.; et al. Development of the Biopen: A handheld device for surgical printing of adipose stem cells at a chondral wound site. *Biofabrication* **2016**, *8*, 015019. [CrossRef] [PubMed]
169. Yin, L.; Wu, Y.; Yang, Z.; Denslin, V.; Ren, X.; Tee, C.A.; Lai, Z.; Lim, C.T.; Han, J.; Lee, E.H. Characterization and application of size-sorted zonal chondrocytes for articular cartilage regeneration. *Biomaterials* **2018**, *165*, 66–78. [CrossRef] [PubMed]

170. Acharya, C.; Adesida, A.; Zajac, P.; Mumme, M.; Riesle, J.; Martin, I.; Barbero, A. Enhanced chondrocyte proliferation and mesenchymal stromal cells chondrogenesis in coculture pellets mediate improved cartilage formation. *J. Cell. Physiol.* **2012**, *227*, 88–97. [CrossRef] [PubMed]
171. Apelgren, P.; Amoroso, M.; Lindahl, A.; Brantsing, C.; Rotter, N.; Gatenholm, P.; Kölby, L. Chondrocytes and stem cells in 3D-bioprinted structures create human cartilage in vivo. *PLoS ONE* **2017**, *12*. [CrossRef] [PubMed]
172. Möller, T.; Amoroso, M.; Hägg, D.; Brantsing, C.; Rotter, N.; Apelgren, P.; Lindahl, A.; Kölby, L.; Gatenholm, P. In Vivo Chondrogenesis in 3D Bioprinted Human Cell-laden Hydrogel Constructs. *Plast. Reconstr. Surg. Glob. Open* **2017**, *5*, e1227. [CrossRef]
173. Idaszek, J.; Costantini, M.; Karlsen, T.A.; Jaroszewicz, J.; Colosi, C.; Testa, S.; Fornetti, E.; Bernardini, S.; Seta, M.; Kasarełło, K.; et al. 3D bioprinting of hydrogel constructs with cell and material gradients for the regeneration of full-thickness chondral defect using a microfluidic printing head. *Biofabrication* **2019**, *11*, 044101. [CrossRef]
174. Cao, C.; Zhang, Y.; Ye, Y.; Sun, T. Effects of cell phenotype and seeding density on the chondrogenic capacity of human osteoarthritic chondrocytes in type I collagen scaffolds. *J. Orthop. Surg. Res.* **2020**, *15*. [CrossRef]
175. De Moor, L.; Fernandez, S.; Vercruysse, C.; Tytgat, L.; Asadian, M.; De Geyter, N.; Van Vlierberghe, S.; Dubruel, P.; Declercq, H. Hybrid Bioprinting of Chondrogenically Induced Human Mesenchymal Stem Cell Spheroids. *Front. Bioeng. Biotechnol.* **2020**, *8*, 484. [CrossRef]
176. Silver, I.A. Measurement of pH and ionic composition of pericellular sites. *Philos. Trans. R. Soc. Lond. Ser. B Biol. Sci.* **1975**, *271*, 261–272. [CrossRef]
177. Lafont, J.E.; Talma, S.; Hopfgarten, C.; Murphy, C.L. Hypoxia Promotes the Differentiated Human Articular Chondrocyte Phenotype through SOX9-dependent and -independent Pathways. *J. Biol. Chem.* **2008**, *283*, 4778–4786. [CrossRef]
178. Ronzière, M.C.; Perrier, E.; Mallein-Gerin, F.; Freyria, A.-M. Chondrogenic potential of bone marrow- and adipose tissue-derived adult human mesenchymal stem cells. *Bio. Med Mater. Eng.* **2010**, *20*, 145–158. [CrossRef] [PubMed]
179. Dennis, J.E.; Whitney, G.A.; Rai, J.; Fernandes, R.J.; Kean, T.J. Physioxia Stimulates Extracellular Matrix Deposition and Increases Mechanical Properties of Human Chondrocyte-Derived Tissue-Engineered Cartilage. *Front. Bioeng. Biotechnol.* **2020**, *8*. [CrossRef] [PubMed]
180. Fortier, L.A.; Barker, J.U.; Strauss, E.J.; McCarrel, T.M.; Cole, B.J. The role of growth factors in cartilage repair. *Clin. Orthop. Relat. Res.* **2011**, *469*, 2706–2715. [CrossRef] [PubMed]
181. Sekiya, I.; Larson, B.L.; Vuoristo, J.T.; Reger, R.L.; Prockop, D.J. Comparison of effect of BMP-2, -4, and -6 on in vitro cartilage formation of human adult stem cells from bone marrow stroma. *Cell Tissue Res.* **2005**, *320*, 269–276. [CrossRef]
182. Shi, S.; Wang, C.; Acton, A.; Eckert, G.; Trippel, S. Role of Sox9 in Growth Factor Regulation of Articular Chondrocytes. *J. Cell. Biochem.* **2015**, *116*. [CrossRef]
183. Beck, E.C.; Barragan, M.; Tadros, M.H.; Gehrke, S.H.; Detamore, M.S. Approaching the Compressive Modulus of Articular Cartilage With a Decellularized Cartilage-Based Hydrogel. *Acta Biomate* **2016**, *38*, 94–105. [CrossRef]
184. Kafian-Attari, I.; Nippolainen, E.; Semenov, D.; Hauta-Kasari, M.; Töyräs, J.; Töyräs, J.; Afara, I.O. Tissue optical properties combined with machine learning enables estimation of articular cartilage composition and functional integrity. *Biomed. Opt. Express* **2020**, *11*, 6480–6494. [CrossRef]
185. Galarraga, J.H.; Kwon, M.Y.; Burdick, J.A. 3D bioprinting via an in situ crosslinking technique towards engineering cartilage tissue. *Sci. Rep.* **2019**, *9*, 19987. [CrossRef]
186. Kim, S.W.; Kim, D.Y.; Roh, H.H.; Kim, H.S.; Lee, J.W.; Lee, K.Y. Three-Dimensional Bioprinting of Cell-Laden Constructs Using Polysaccharide-Based Self-Healing Hydrogels. *Biomacromolecules* **2019**, *20*, 1860–1866. [CrossRef]
187. Irmak, G.; Gümüşderelioğlu, M. Photo-activated platelet-rich plasma (PRP)-based patient-specific bio-ink for cartilage tissue engineering. *Biomed. Mater.* **2020**, *15*, 065010. [CrossRef]
188. Nguyen, D.; Hägg, D.A.; Forsman, A.; Ekholm, J.; Nimkingratana, P.; Brantsing, C.; Kalogeropoulos, T.; Zaunz, S.; Concaro, S.; Brittberg, M.; et al. Cartilage Tissue Engineering by the 3D Bioprinting of iPS Cells in a Nanocellulose/Alginate Bioink. *Sci. Rep.* **2017**, *7*, 658. [CrossRef] [PubMed]
189. Francis, S.L.; Di Bella, C.; Wallace, G.G.; Choong, P.F.M. Cartilage Tissue Engineering Using Stem Cells and Bioprinting Technology-Barriers to Clinical Translation. *Front. Surg.* **2018**, *5*, 70. [CrossRef] [PubMed]

cells

MDPI

Perspective

When the Search for Stemness Genes Meets the Skin Substitute Bioengineering Field: KLF4 Transcription Factor under the Light

Nicolas O. Fortunel [1,2,3,4,*] and Michèle T. Martin [1,2,3,4,*]

1 Laboratoire de Génomique et Radiobiologie de la Kératinopoïèse, Institut de Biologie François Jacob, CEA/DRF, Institut de Radiobiologie Cellulaire et Moléculaire, 91000 Evry, France
2 INSERM U967, 92260 Fontenay-aux-Roses, France
3 Université Paris-Saclay, 91190 Saint-Aubin, France
4 Université Paris-Diderot, 75013 Paris, France
* Correspondence: nicolas.fortunel@cea.fr (N.O.F.); michele.martin@cea.fr (M.T.M.); Tel.: 33-1-60-87-34-92 (N.O.F.); 33-1-60-87-34-91 (M.T.M.); Fax: 33-1-60-87-34-98 (N.O.F. & M.T.M.)

Received: 27 August 2020; Accepted: 25 September 2020; Published: 28 September 2020

Abstract: The transcription factor "Kruppel-like factor 4" (KLF4) is a central player in the field of pluripotent stem cell biology. In particular, it was put under the spotlight as one of the four factors of the cocktail originally described for reprogramming into induced pluripotent stem cells (iPSCs). In contrast, its possible functions in native tissue stem cells remain largely unexplored. We recently published that KLF4 is a regulator of "stemness" in human keratinocytes. We show that reducing the level of expression of this transcription factor by RNA interference or pharmacological repression promotes the ex vivo amplification and regenerative capacity of two types of cells of interest for cutaneous cell therapy: native keratinocyte stem and progenitor cells from adult epidermis, which have been used for more than three decades in skin graft bioengineering, and keratinocytes generated by the lineage-oriented differentiation of embryonic stem cells (ESCs), which have potential for the development of skin bio-bandages. At the mechanistic level, *KLF4* repression alters the expression of a large set of genes involved in TGF-β1 and WNT signaling pathways. Major regulators of TGF-β bioavailability and different TGF-β receptors were targeted, notably modulating the ALK1/Smad1/5/9 axis. At a functional level, *KLF4* repression produced an antagonist effect on TGF-β1-induced keratinocyte differentiation.

Keywords: KLF4; adult epidermal keratinocytes; ESC-derived keratinocytes; stemness; skin grafts; TGF-β1; WNT

1. Biomedical Context

This perspective provides a highlight on a recent publication of our group in Nature Biomedical Engineering [1]. Human epidermis is naturally endowed with remarkable capacities for renewal and regeneration, due to the presence of resident epithelial stem cells within its keratinocyte basal layer. These capacities have enabled the development of different research and clinical models of skin organoids, including skin substitutes that have proved efficient for more than three decades in their use as a treatment for severely burned patients by autologous grafts [2,3]. More recently, an approach combining cell and gene therapies has been successfully achieved to genetically correct the entire regenerated epidermis of an epidermolysis bullosa patient, in the context of a compassionate clinical trial [4]. The bioengineering of large surfaces of skin substitutes, up to one meter squared obtained from skin biopsies of a few centimeters squared, requires a massive ex vivo expansion of patient keratinocytes, during which the preservation of functional stem and progenitor cells is critical to ensure a successful graft take and long-term outcome. Better deciphering the molecular networks that ensure the control of stemness and self-renewal in keratinocytes is thus required for the conception of next-generation skin substitute bioengineering strategies. In addition to native keratinocytes extracted from skin biopsies, keratinocytes produced by the lineage-oriented differentiation of embryonic stem cells (ESCs) may constitute a complementary cell source, as they have the potential to generate three-dimensional (3D) epidermis organoids [5]. Induced pluripotent stem cells (iPSCs) can be obtained by the reprogramming of skin fibroblasts or keratinocytes, from the perspective of re-differentiation towards the keratinocyte lineage (Figure 1). A key input is that autologous iPSCs can thus provide a source of keratinocytes for modeling genodermatoses, such as epidermolysis bullosa, to improve their cellular and molecular characterization, and to test new therapeutic strategies [6,7]. However, the reliability of these alternative keratinocyte sources will depend on the robustness of production methods, and the possibility of obtaining the high proliferative capacity that characterizes native immature keratinocyte stem and progenitor cells, which is not demonstrated today.

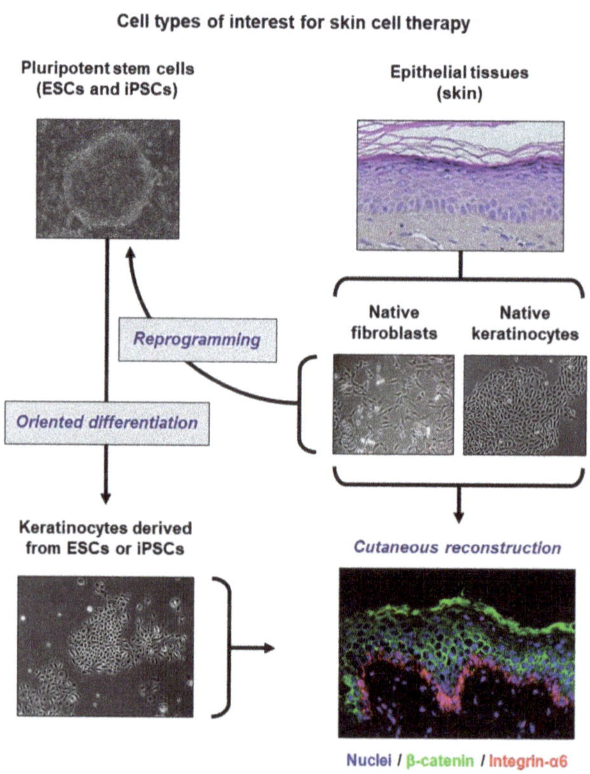

Figure 1. Different sources of keratinocytes of interest for cutaneous reconstruction (adapted from [8]). Native keratinocytes extracted from adult skin biopsies currently constitute the major source of keratinocytes for medical uses. They have been used for more than three decades for the bioengineering of grafts intended for the treatment of severe burns, and more recently for gene therapy. Keratinocytes produced by oriented differentiation of pluripotent stem cells are currently investigated as a complementary source for the development of skin bio-bandages. Embryonic stem cells (ESCs) can be differentiated into keratinocytes capable of generating an epidermis. Pluripotency-induced stem cells (iPSCs), which are obtained by reprogramming adult cells (for example, fibroblasts or skin keratinocytes), can also be differentiated into keratinocytes.

2. The KLF4 Candidate

The transcription factor "Kruppel-like factor 4" (KLF4) is far from being unknown in the field of stem cell biology. Notably, the KLF family is involved in the regulation of self-renewal and immaturity in ESCs [9], and KLF4 is one of the four factors of the reprogramming cocktail originally described for the generation of iPSCs [10]. In the epidermis, KLF4 is known to exert a regulatory role in keratinocyte terminal differentiation [11], which is essential for the establishment of the barrier function of this epithelium [12]. In contrast, KLF4 functions in native tissue stem cells have been unexplored until now. We performed transcriptomic profiling screens on subpopulations of human keratinocyte precursors, enriched in stem cells or in progenitor cells [13,14] (complete micro-array datasets available in the GEO database, accession no GSE68583), and determined that *KLF4* gene was differentially expressed according to cell immaturity or differentiation. We thus explored the working hypothesis that this transcription factor might regulate epidermal keratinocyte stem cell functions and notably their proliferation and skin regenerative capacity.

3. KLF4 Functions in Native Adult Keratinocytes

The regulatory functions of the transcription factor KLF4 have been firstly investigated in "holoclone" keratinocytes [1], which are representative of an immature population of cultured precursors containing functional stem and progenitor cells [15]. These cells correspond to the clonal progeny of single keratinocyte stem cells that were functionally characterized by their extensive growth potential exceeding 100 population doublings through successive subcultures, and the capacity for epidermis reconstruction in vitro and regeneration in vivo. A functional genomics approach was designed to study the effects of *KLF4* repression in holoclone keratinocytes. These cells were transduced with lentiviral vectors designed for "short hairpin RNA" (shRNA)-mediated *KLF4* stably repressed cells, or with a control vector to generate a comparable *KLF4* wild-type cellular context. Using clonal assays and long-term cultures, we demonstrated that maintaining *KLF4* expression at a low level preserves an undifferentiated status in keratinocyte precursors, and promotes their self-renewal. This repression led to the maintenance of high levels of characteristic markers of an immature precursor state, notably integrin $\alpha6$ and $\Delta Np63\alpha$, and to an improvement of cellular expansion in bidimensional culture. Moreover, the regenerative capacity of *KLF4* stably repressed keratinocytes was higher than that of *KLF4* wild-type keratinocytes. This gain-of-function was demonstrated in vitro by the generation of epidermis organoids, and in vivo by the augmented capacity of *KLF4* stably repressed keratinocytes to ensure iterative xenografting in immuno-deficient athymic nude $Foxn1^{nu}$ mice [16]. In a first cycle of xenografting, *KLF4* repressed and *KLF4* wild-type keratinocytes were equally potent for regeneration, indicating an absence of deleterious effects of *KLF4* repression on epidermis stratification and differentiation. In contrast, when keratinocytes extracted from primary grafts were tested for a secondary cycle of epidermis reconstruction and xenografting, the success rate obtained with *KLF4* repressed cells was three-fold higher than that obtained with *KLF4* wild-type cells, indicating a better long-term maintenance of functional keratinocyte stem cells in response to *KLF4* repression. The correct differentiation of epidermises regenerated by *KLF4* repressed and *KLF4* wild-type keratinocytes was checked by a histological examination of tissue sections stained with hematoxilin-eosin-saffron (HES), and similar expression patterns of typical epidermis markers such as involucrin, keratin 10 and keratin 5 were observed.

4. Insights into the Mechanisms

To explain the gain-of-function promoted by *KLF4* repression, a comparative transcriptome profiling of *KLF4* wild-type and *KLF4* stably repressed keratinocyte precursor cells was conducted and analyzed to search for signaling pathways that depend on *KLF4* expression level (complete RNA-seq datasets are available in the GEO database, accession no GSE111786). Analysis of the differentially expressed genes pinpointed the TGF-β1 and WNT signaling pathways (Figure 2), which are major regulatory elements in stem cell biology. A global repression of transcripts related to the TGF-β1 pathway was detected in *KLF4* repressed keratinocytes. Notably, the expression of extracellular modulators of TGF-β bioavailability and of four major membrane receptors of the factor was impaired (Figure 2A) and the ALK1/Smad1/5/9 axis was particularly repressed (Figure 2A). As a known effect of TGF-β1 on keratinocytes is commitment to differentiation [17], the functional relations between this factor and KLF4 were explored. As expected, wild-type keratinocyte precursors entered into differentiation in response to TGF-β1, which was documented by the loss of integrin α6. In contrast, the pro-differentiation effect of TGF-β1 was markedly attenuated in *KLF4* repressed keratinocytes, showing that this repression promotes a better maintenance of stemness. The WNT pathway was also a molecular target of *KLF4* repression, at all regulation levels of the network. As performed for *KLF4*, dedicated functional genomics approaches using RNA interference tools for stable and transient repression will be necessary to elucidate the specific functions of selected WNT candidates pinpointed by our screen (Figure 2B).

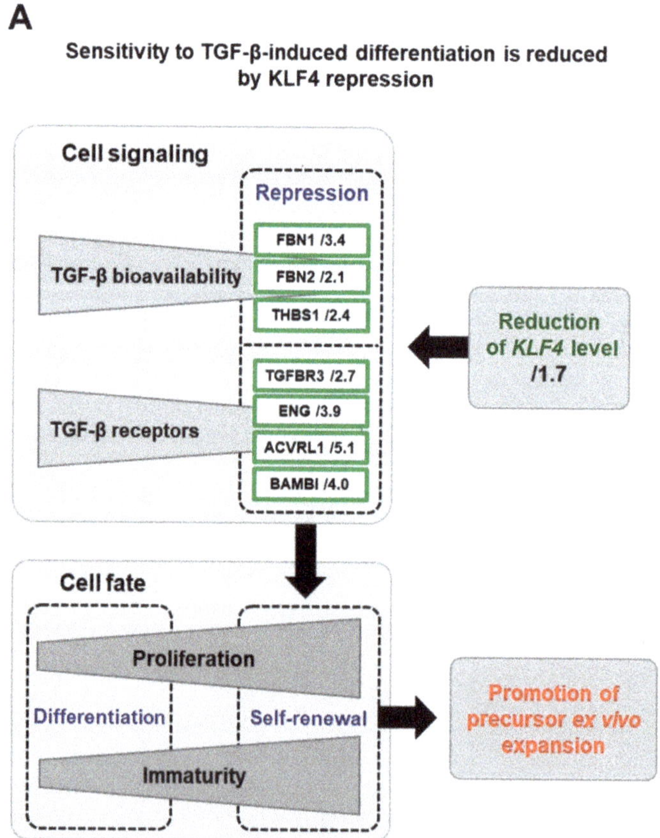

Figure 2. *Cont.*

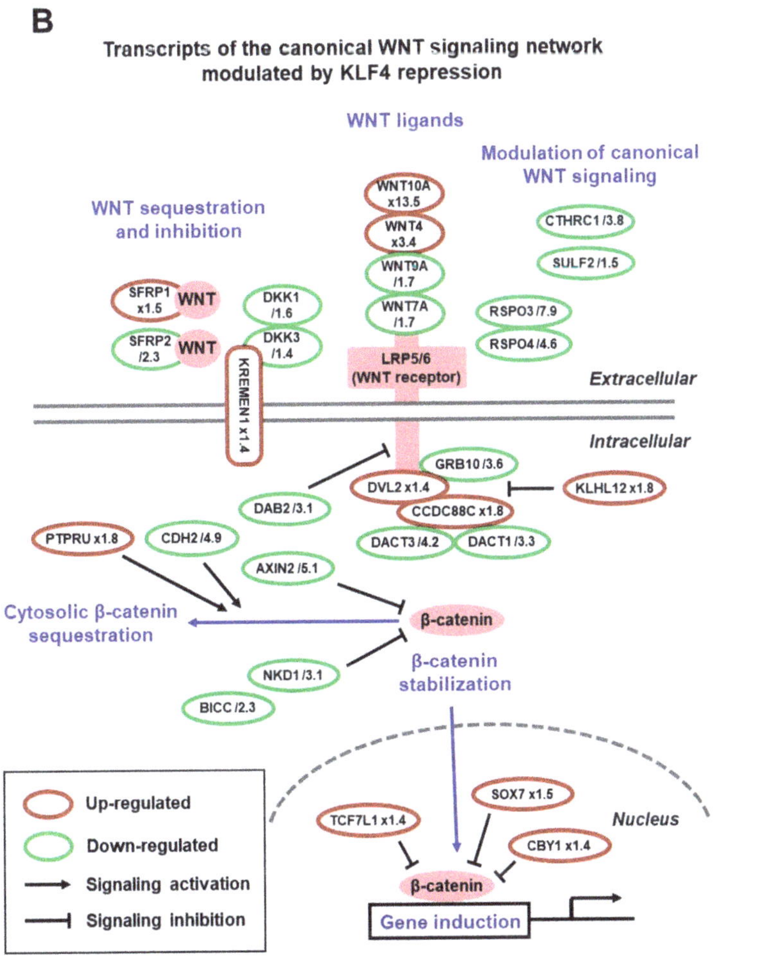

Figure 2. *KLF4* repression impacts the expression of transcripts related to TGF-β and WNT signaling. The comparison of the transcriptional profiles of *KLF4* wild-type and *KLF4* stably repressed keratinocytes, characterized by RNA sequencing (complete datasets are available in the GEO database, accession no GSE111786), showed that the repression of *KLF4* gene expression modulates the expression of numerous genes encoding effectors of the TGF-β (**A**) and WNT (**B**) signaling networks. Notably, a decreased expression was detected for extracellular modulators and membrane receptors of TGF-β signaling. The indicated values correspond to expression fold-changes in *KLF4* stably repressed versus *KLF4* wild-type keratinocytes.

5. From Basic Research to Pre-Clinical Models

To get closer to a clinically applicable situation, we have then explored the effects of a transitory repression of *KLF4*. Total basal keratinocytes were used [18] as this cellular material is more representative of the clinical samples used for the production of grafts than the research model of holoclones. Firstly, treatment of keratinocytes with "small interfering RNAs" (siRNAs) directed against *KLF4* showed that transient inhibition was efficient at promoting the expression of immaturity-associated markers. In a next step, the feasibility and efficacy of pharmacological KLF4 inhibition was demonstrated using the small molecule kenpaullone, which has been shown to decrease

KLF4 mRNA level [19]. Kenpaullone decreases the *KLF4* level in keratinocyte precursors at both mRNA and protein levels and was efficient to augment their clonogenic capacity and growth in mass cultures. Considering the biosafety issue, exome sequencing did not detect any deleterious effects of kenpaullone treatment on keratinocyte genome integrity. Importantly, reconstructed epidermises generated with kenpaullone-treated cells exhibited normal expression profiles of the classical epidermal markers, laminin 5, integrin α6, filaggrin, and involucrin, showing that kenpaullone treatment did not alter epidermis differentiation. Moreover, using the lucifer-yellow diffusion assay, we could show that epidermises generated by kenpaullone-treated keratinocytes exhibited a lower permeability, compared to those produced with untreated cells, which suggested a more efficient barrier function. The gain-of-function demonstrated by kenpaullone-induced *KLF4* repression may result from a combination of mechanisms. One of them is *KLF4*-mediated attenuation of sensitivity to TGF-β1, as demonstrated in stably transduced keratinocytes. As kenpaullone is also known to target the WNT/β-catenin signaling component GSK-3 [20] at concentrations in the μM range corresponding to those used in our study to repress *KLF4*, modulation of WNT pathway activities can also participate to the observed beneficial effects. Both mechanisms will require further investigations after transient *KLF4* repression.

6. Extrapolation to ESC-Derived Keratinocytes

To widen the perspectives that may emerge from the concept of *KLF4* repression in keratinocyte stem cells, the effects of transient *KLF4* downmodulation were also studied on bioengineered keratinocytes obtained by the lineage-oriented differentiation of human ESCs (Ker-ESCs) [1]. Although Ker-ESCs can ensure the reconstruction of correctly differentiated 3D epidermises [5], they do not reproduce all the characteristics of native keratinocytes. Notably, their capacity for proliferation is very limited, probably due to the lack of immature precursor cell status. Kenpaullone also showed efficacy to obtain *KLF4* repression in Ker-ESCs, with a beneficial effect on their proliferative capacity, as described in native keratinocytes. Importantly, kenpaullone-treated Ker-ESCs exhibited a more immature cellular phenotype, which at a functional level resulted in an improvement of growth potential and quality for 3D epidermis reconstruction. In a classical high-density culture condition, untreated and kenpaullone-treated Ker-ESCs equally gave rise to correctly differentiated epidermises, as assessed by a histological examination and visualization of marker expression (involucrin and filaggrin). In contrast, in a stressed condition obtained by low-density seeding, stratification occurred only with kenpaullone-treated Ker-ESCs.

7. Conclusions

We identified the transcription factor KLF4 as a regulator of the balance between immaturity and differentiation in the human keratinocyte lineage, which is set notably via interactions with the TGF-β1 signaling pathway. One mechanistic interpretation of our data is that keratinocyte precursor cells exhibiting *KLF4* repression are less sensitive to TGF-β1-mediated commitment to differentiation, through a reduction in TGF-β1 extracellular bioavailability and an impairment of its cell-surface receptor machinery. At the basic research level, further deciphering these molecular links will be needed to elucidate the related mechanisms. More generally, the molecular networks and cascades involved constitute an elegant model for further dissecting the molecular identity of stemness [21,22]. In addition, our study suggests that interactions between KLF4 and WNT signaling also participate in the regulation of epidermal precursor and stem cell proliferation and immaturity maintenance during in vitro amplification (Figure 2B). A major objective is the full dissection of the specific roles of KLF4 in the different epidermis layers. In the upper part of mouse epidermis, KLF4 has been shown to be essential for keratinocyte terminal differentiation and barrier function [12]. In a human model, such an induction of differentiation genes has been shown to be driven notably by interactions between KLF4 and epigenetic factors such as chromatin modifiers and long non-coding RNAs [23–25]. We propose that KLF4 might be a gate keeper of stemness in the basal layer of epidermis, thus exerting different roles through specific partners in different epidermis layers. A recent study performed on N/TERT-immortalized human keratinocytes and mouse skin has shown that regulatory loops involving KLF4 together with the YAP1/TAZ-TEAD transcriptional network participate to the control of the equilibrium between proliferation and differentiation [26]. In these models, YAP1/TAZ-TEAD promote keratinocyte proliferation, while KLF4 drives cells towards differentiation, which is in agreement with our finding that lowering KLF4 level improves the proliferative capacity of immature keratinocyte precursor cells.

At the translational research level, the key finding that emerged from the study is that maintaining KLF4 at a low level by transient pharmacological repression constitutes a promising approach to promote the ex vivo expansion of two types of cells of interest for cutaneous graft bioenginering: native epidermal keratinocytes and ESC-derived keratinocytes. Moreover, the ex vivo expansion of keratinocytes for clinical purposes is generally performed in undefined culture media containing bovine serum in the presence of growth-arrested feeder fibroblasts [27]. This situation is expected to progressively evolve towards undefined and animal-component-free culture conditions to avoid the risk of introducing unknown pathogens or any kind of deleterious biomolecules. Considering this demand of regulatory agencies, the substitution of undefined culture components by active molecules promoting stemness, such as KLF4 inhibitors, will open original perspectives for the development of next-generation models of skin substitutes. In this context, the use of PAK1-ROCK-Myosin II and TGF-β signaling inhibitors has been proposed to promote the preservation of stemness in expansion cultures of different types of epithelial cells [28]. Clues will certainly emerge from the combination of these converging studies (Figure 3).

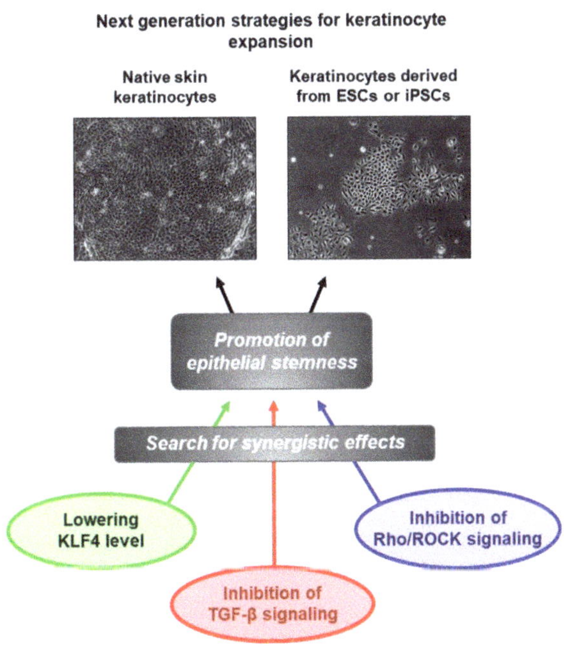

Figure 3. Improvement of keratinocyte ex vivo amplification using molecules promoting epithelial stemness is currently an active research field. The work of our laboratory has identified the transcription factor KLF4 as a relevant target, which could be used in combination with other candidate targets, in order to develop culture media of defined composition for biomedical uses.

Author Contributions: N.O.F.: conceptualization, writing—original draft preparation, and writing review and editing; funding acquisition. M.T.M.: conceptualization, writing—original draft preparation, and writing review and editing; funding acquisition. All authors have read and agreed to the published version of the manuscript.

Funding: This work was supported by grants from: CEA and INSERM (UMR967); "Délégation Générale de l'Armement" (DGA) grants; FUI-AAP13 and the "Conseil Général de l'Essonne" within the STEMSAFE grant; EURATOM (RISK-IR, FP7, grant 323267), and "Electricité de France" (EDF). Genopole® (Evry, France) also provided support for equipment and infrastructures.

Acknowledgments: We wish to thank all the co-authors and contributors of the original publication. Our thanks also go to K. Alves, S. de Bernard, and L. Buffat from AltraBio (Lyon), who contributed to the analysis of RNA-seq data, notably signaling pathway building. We also thank T. Jouault and J.P. Hardelin, co-editors-in-chief of Med/Sci (Paris) (https://www.medecinesciences.org/fr/) for their permission for using in an international publication parts of our previous French contribution referenced as [8].

Conflicts of Interest: Authors declare no conflict of interest.

References

1. Fortunel, N.O.; Chadli, L.; Coutier, J.; Lemaître, G.; Auvré, F.; Domingues, S.; Bouissou-Cadio, E.; Vaigot, P.; Cavallero, S.; Deleuze, J.-F.; et al. KLF4 inhibition promotes the expansion of keratinocyte precursors from adult human skin and of embryonic-stem-cell-derived keratinocytes. *Nat. Biomed. Eng.* **2019**, *3*, 985–997. [CrossRef]
2. Gallico, G.G., III; O'Connor, N.E.; Compton, C.C.; Kehinde, O.; Green, H. Permanent coverage of large burn wounds with autologous cultured human epithelium. *N. Engl. J. Med.* **1984**, *311*, 448–451. [CrossRef] [PubMed]
3. Ronfard, V.; Rives, J.-M.; Neveux, Y.; Carsin, H.; Barrandon, Y. Long-term regeneration of human epidermis on third degree burns transplanted with autologous cultured epithelium grown on a fibrin matrix. *Transplantation* **2000**, *70*, 1588–1598. [CrossRef] [PubMed]
4. Hirsch, T.; Rothoeft, T.; Teig, N.; Bauer, J.W.; Pellegrini, G.; De Rosa, L.; Scaglione, D.; Reichelt, J.; Klausegger, A.; Kneisz, D.; et al. Regeneration of the entire human epidermis using transgenic stem cells. *Nature* **2017**, *551*, 327–332. [CrossRef] [PubMed]
5. Guenou, H.; Nissan, X.; Larcher, F.; Feteira, J.; Lemaitre, G.; Saidani, M.; Del Rio, M.; Barrault, C.C.; Bernard, F.-X.; Peschanski, M.; et al. Human embryonic stem-cell derivatives for full reconstruction of the pluristratified epidermis: A preclinical study. *Lancet* **2009**, *374*, 1745–1753. [CrossRef]
6. Itoh, M.; Kiuru, M.; Cairo, M.S.; Christiano, A.M. Generation of keratinocytes from normal and recessive dystrophic epidermolysis bullosa-induced pluripotent stem cells. *Proc. Natl. Acad. Sci. USA* **2011**, *108*, 8797–8802. [CrossRef]
7. Kolundzic, N.; Khurana, P.; Hobbs, C.; Rogar, M.; Ropret, S.; Törmä, H.; Ilic, D.; Liovic, M. Induced pluripotent stem cell (iPSC) line from an epidermolysis bullosa simplex patient heterozygous for keratin 5 E475G mutation and with the Dowling Meara phenotype. *Stem Cell Res.* **2019**, *37*, 101424. [CrossRef]
8. Fortunel, N.O.; Martin, M.T. When stemness genes meet skin graft bioengineering: Focus on KLF4. *Med. Sci.* **2020**, *36*, 565–568. [CrossRef] [PubMed]
9. Chia, N.Y.; Chan, Y.S.; Feng, B.; Lu, X.; Orlov, Y.L.; Moreau, D.; Kumar, P.; Yang, L.; Jiang, J.; Lau, M.S.; et al. A genome-wide RNAi screen reveals determinants of human embryonic stem cell identity. *Nature* **2010**, *468*, 316–320. [CrossRef]
10. Takahashi, K.; Tanabe, K.; Ohnuki, M.; Narita, M.; Ichisaka, T.; Tomoda, K.; Yamanaka, S. Induction of pluripotent stem cells from adult human fibroblasts by defined factors. *Cell* **2007**, *131*, 861–872. [CrossRef]
11. Sen, G.L.; Boxer, L.D.; Webster, D.E.; Bussat, R.T.; Qu, K.; Zarnegar, B.J.; Johnston, D.; Siprashvili, Z.; Khavari, P.A. ZNF750 is a p63 target gene that induces KLF4 to drive terminal epidermal differentiation. *Dev. Cell* **2012**, *22*, 669–677. [CrossRef] [PubMed]
12. Segre, J.A.; Bauer, C.; Fuchs, E. Klf4 is a transcription factor required for establishing the barrier function of the skin. *Nat. Genet.* **1999**, *22*, 356–360. [CrossRef] [PubMed]
13. Li, A.; Simmons, P.J.; Kaur, P. Identification and isolation of candidate human keratinocyte stem cells based on cell surface phenotype. *Proc. Natl. Acad. Sci. USA* **1998**, *95*, 3902–3907. [CrossRef] [PubMed]
14. Rachidi, W.; Harfouche, G.; Lemaitre, G.; Amiot, F.; Vaigot, P.; Martin, M.T. Sensing radiosensitivity of human epidermal stem cells. *Radiother. Oncol.* **2007**, *83*, 267–276. [CrossRef] [PubMed]

15. Fortunel, N.O.; Cadio, E.; Vaigot, P.; Chadli, L.; Moratille, S.; Bouet, S.; Roméo, P.-H.; Martin, M.T. Exploration of the functional hierarchy of the basal layer of human epidermis at the single-cell level using parallel clonal microcultures of keratinocytes. *Exp. Dermatol.* **2010**, *19*, 387–392. [CrossRef] [PubMed]
16. Fortunel, N.O.; Bouissou-Cadio, E.; Coutier, J.; Martin, M.T. Iterative three-dimensional epidermis bioengineering and xenografting to assess long-term regenerative potential in human keratinocyte precursor cells. *Methods Mol. Biol.* **2020**, *2109*, 155–167. [PubMed]
17. Fuchs, E. Epidermal differentiation and keratin gene expression. *J. Cell Sci. Suppl.* **1993**, *17*, 197–208. [CrossRef]
18. Fortunel, N.O.; Chadli, L.; Bourreau, E.; Cadio, E.; Vaigot, P.; Marie, M.; Deshayes, N.; Rathman-Josserand, M.; Leclaire, J.; Martin, M.T. Cellular adhesion on collagen: A simple method to select human basal keratinocytes which preserves their high growth capacity. *Eur. J. Dermatol.* **2011**, *21* (Suppl. 2), 12–20. [CrossRef]
19. Yu, F.; Li, J.; Chen, H.; Fu, J.; Ray, S.; Huang, S.; Zheng, H.; Ai, W. Kruppel-like factor 4 (KLF4) is required for maintenance of breast cancer stem cells and for cell migration and invasion. *Oncogene* **2011**, *30*, 2161–2172. [CrossRef]
20. Lange, C.; Mix, E.; Frahm, J.; Glass, A.; Müller, J.; Schmitt, O.; Schmöle, A.-C.; Klemm, K.; Ortinau, S.; Hübner, R.; et al. Small molecule GSK-3 inhibitors increase neurogenesis of human neural progenitor cells. *Neurosci. Lett.* **2011**, *488*, 36–40. [CrossRef]
21. Fortunel, N.O.; Otu, H.H.; Ng, H.H.; Chen, J.; Mu, X.; Chevassut, T.; Li, X.; Joseph, M.; Bailey, C.; Hatzfeld, J.A.; et al. Comment on "'Stemness': Transcriptional profiling of embryonic and adult stem cells" and "a stem cell molecular signature". *Science* **2003**, *302*, 393. [CrossRef] [PubMed]
22. Vogel, G. 'Stemness' genes still elusive. *Science* **2003**, *302*, 371. [CrossRef] [PubMed]
23. Boxer, L.D.; Barajas, B.; Tao, S.; Zhang, J.; Khavari, P.A. ZNF750 interacts with KLF4 and RCOR1, KDM1A, and CTBP1/2 chromatin regulators to repress epidermal progenitor genes and induce differentiation genes. *Genes Dev.* **2014**, *28*, 2013–2026. [CrossRef] [PubMed]
24. Bao, X.; Tang, J.; Lopez-Pajares, V.; Tao, S.; Qu, K.; Crabtree, G.R.; Khavari, P.A. ACTL6a enforces the epidermal progenitor state by suppressing SWI/SNF-dependent induction of KLF4. *Cell Stem Cell* **2013**, *12*, 193–203. [CrossRef]
25. Lopez-Pajares, V.; Qu, K.; Zhang, J.; Webster, D.E.; Barajas, B.C.; Siprashvili, Z.; Zarnegar, B.J.; Boxer, L.D.; Rios, E.J.; Tao, S.; et al. LncRNA-MAF:MAFB transcription factor network regulates epidermal differentiation. *Dev. Cell* **2015**, *32*, 693–706. [CrossRef] [PubMed]
26. Yuan, Y.; Park, J.; Feng, A.; Awasthi, P.; Wang, Z.; Chen, Q.; Iglesias-Bartolome, R. YAP1/TAZ-TEAD transcriptional networks maintain skin homeostasis by regulating cell proliferation and limiting KLF4 activity. *Nat. Commun.* **2020**, *11*, 1472. [CrossRef]
27. Alexaline, M.M.; Trouillas, M.; Nivet, M.; Bourreau, E.; Leclerc, T.; Duhamel, P.; Martin, M.T.; Doucet, C.; Fortunel, N.O.; Lataillade, J.-J. Bioengineering a human plasma-based epidermal substitute with efficient grafting capacity and high content in clonogenic cells. *Stem Cells Transl. Med.* **2015**, *4*, 643–654. [CrossRef]
28. Zhang, C.; Lee, H.J.; Shrivastava, A.; Wang, R.; McQuiston, T.J.; Challberg, S.S.; Pollok, B.A.; Wang, T. Long-term in vitro expansion of epithelial stem cells enabled by pharmacological inhibition of PAK1-ROCK-myosin II and TGF-β signaling. *Cell Rep.* **2018**, *25*, 598–610.e5. [CrossRef]

MDPI
St. Alban-Anlage 66
4052 Basel
Switzerland
www.mdpi.com

Cells Editorial Office
E-mail: cells@mdpi.com
www.mdpi.com/journal/cells